D1764929

INTEGRATED PEST AND DISEASE MANAGEMENT IN GREENHOUSE CROPS

Developments in Plant Pathology

VOLUME 14

Integrated Pest and Disease Management in Greenhouse Crops

Edited by

R. ALBAJES
University of Lleida,
Lleida, Spain

M. LODOVICA GULLINO
University of Torino,
Torino, Italy

J. C. VAN LENTEREN
University of Wageningen,
Wageningen, The Netherlands

and

Y. ELAD
The Volcani Center,
ARO, Bet Dagen, Israel

CIHEAM

KLUWER ACADEMIC PUBLISHERS
DORDRECHT / BOSTON / LONDON

A C.I.P. Catalogue record for this book is available from the Library of Congress.

20326017

ISBN 0-7923-5631-4

Published by Kluwer Academic Publishers,
P.O. Box 17, 3300 AA Dordrecht, The Netherlands.

Sold and distributed in North, Central and South America
by Kluwer Academic Publishers,
101 Philip Drive, Norwell, MA 02061, U.S.A.

In all other countries, sold and distributed
by Kluwer Academic Publishers,
P.O. Box 322, 3300 AH Dordrecht, The Netherlands.

Covercredits:

Photographers: Y. Elad and E. Fischer (The Volcani Center,
Bet Dagan, Israel)
1. Use of yeasts in the biocontrol of plant diseases: Yeast
cells associated with a conidium of Botrytis
2. Classic parasitism among fungi: Fulvia fulva with a
micoparasite on it

Photographer: P. Sutherland (The Horticultural and Food Research New
Zealand Ltd, Auckland, New Zealand)
Reproduced by permission of New Zealand Institute for Crop & Food Research
Ltd, Lincoln, Canterbury, New Zealand
Insect parasitoids: Encarsia formosa inspecting or ovipositing in a nymph
of greenhouse whitefly (Trialeurodes vaporariorum) on a tobacco leaf

Printed on acid-free paper

All Rights Reserved
©1999 Kluwer Academic Publishers
No part of the material protected by this copyright notice may be reproduced or
utilized in any form or by any means, electronic or mechanical,
including photocopying, recording or by any information storage and
retrieval system, without written permission from the copyright owner

Printed in the Netherlands.

CONTENTS

Part IV: Biological and Microbial Control of Greenhouse Pests and Diseases

IV(A) Biological and Microbial Control of Arthropod Pests

CONTRIBUTORS

Ramon Albajes
Universitat de Lleida
Centre UdL–IRTA
Rovira Roure 177
25006 Lleida
Spain

Oscar Alomar
IRTA – Centre de Cabrils
Ctra. de Cabrils, s/n
08348 Cabrils, Barcelona
Spain

Richard R. Bélanger
University Laval
Dept. Phytopatologie – FSAA
Cité Universitaire
Québec G1K 7P4
Canada

Menachem J. Berlinger
Agricultural Reseach Organization (ARO)
Gilat Regional Experiment Station
Entomology Laboratory
Mobile Post Negev 85-280
Israel

Mohamed Besri
Institut Agronomique et Vétérinaire Hassan II
B.P. 6202
10101 Rabat-Instituts
Morocco

Sylvia Blümel
Institut für Phytomedizin (BFL)
Spargelfeldstrasse 191
P.O. Box 400
A-1226 Wien
Austria

Karel J.F. Bolckmans
Koppert Biological Systems B.V.
Veilingweg 17
P.O. Box 155
2650 AD Berkel en Rodenrijs
The Netherlands

Henrik F. Brødsgaard
Danish Institute of Agricultural Sciences
Research Centre Flakkebjerg
Dept. of Crop Protection
Research Group Entomology
DK–4200 Slagelse
Denmark

Cristina Castañé
IRTA – Centre de Cabrils
Ctra. de Cabrils, s/n
08348 Cabrils, Barcelona
Spain

Elzbieta Ceglarska
Debrecen University of Agricultural Sciences
Faculty of Agriculture
P.O. Box 79
6801 Hódmezõvasárhely
Hungary

Norman D. Clarke
AI Solutions
47 Tomlin Crescent
Richmond Hill, Ontario L4C 7T1
Canada

Jesús Cuartero
Consejo Superior de Investigaciones
Científicas (CSIC)
Estación Experimental "La Mayora"
Agarrobo-Costa s/n
29750 Algarrobo-Costa, Málaga
Spain

Aleid J. Dik
Research Station for Floriculture
and Glasshouse Vegetables (PBG)
Kruisbroekweg 5
P.O. Box 8
2670 AA Naaldwijk
The Netherlands

Yigal Elad
Agricultural Research Organization (ARO)
The Volcani Center
Institute of Plant Protection
Dept. of Plant Pathology
P.O. Box 6
Bet-Dagan 50250
Israel

Deborah R. Fravel
USDA – Agricultural Research Service
Beltsville Agricultural Research Center
Biocontrol of Plant Diseases Laboratory
Bldg. 011A, Room 275, BARC- West
Beltsville, Maryland 20705-2350
USA

Stanley Freeman
Agricultural Research Organization (ARO)
The Volcani Center
Institute of Plant Protection
Dept. of Plant Pathology
P.O. Box 6
Bet-Dagan 50250
Israel

Dan Funck Jensen
The Royal Veterinary and
Agricultural University (KVL)
Dept. of Plant Biology
Plant Pathology Section
40, Thorvaldsensvej
DK-1871 Frederiksberg C
Copenhagen
Denmark

Rosa Gabarra
IRTA – Centre de Cabrils
Departamento de Protección Vegetal
Ctra. de Cabrils, s/n
08348 Cabrils, Barcelona
Spain

Abraham Gamliel
Agricultural Research Organization (ARO)
The Volcani Center
Institute of Agricultural Engineering
Bet-Dagan 50250
Israel

Dimitris E. Goumas
Plant Protection Institute
P.O. Box 1802
71110 Heraklio, Crete
Greece

Don A. Griffiths
Novartis BCM Ltd
Aldham Business Centre
New Road, Aldham
Colchester, Essex
England CO6 3PN
United Kingdom

Avi Grinstein
Agricultural Research Organization (ARO)
The Volcani Center
Institute of Agricultural Engineering
P.O. Box 6
Bet-Dagan 50250
Israel

M. Lodovica Gullino
Università degli Studi di Torino
Dipartimento di Valorizzazione e Protezione
delle Risorse Agroforestali – Patologia Vegetale
Via Leonardo da Vinci 44
10095 Grugliasco (Torino)
Italy

Zoltan Ilovai
Ministry of Agriculture and
Regional Development
Plant Health and Soil Conservation Station
Coordination Unit
Plant Protection Department
P.O. Box 340
H-1519 Budapest
Hungary

William R. Jarvis
Agriculture and Agri-Food Canada
Greenhouse and Processing
Crops Research Centre
Harrow, Ontario N0R 1G0
Canada

Tom J. Jewett
Agriculture and Agri-Food Canada
Greenhouse and Processing
Crops Research Centre
Harrow, Ontario N0R 1G0
Canada

Sonja Sletner Klemsdal
The Norwegian Crop Research Institute
Plant Protection Centre
Fellesbygget, N-1432 Ås
Norway

Jürgen Köhl
DLO Research Institute for
Plant Protection (IPO-DLO)
Binnenhaven 5
P.O.Box 9060
NL- 6700 GW Wageningen
The Netherlands

Laurent Lapchin
INRA – Centre de Recherches d'Antibes
37, Boulevard du Cap
B.P. 2078
06606 Antibes Cedex
France

Robert P. Larkin
USDA – Agricultural Research Service
Beltsville Agricultural Research Center
Biocontrol of Plant Diseases Laboratory
Bldg. 011A, Room 275, BARC- West
Beltsville, Maryland 20705-2350
USA

Henri Laterrot
INRA – Centre d'Avignon
Unité de Génétique et d'Amélioration
des Fruits et Légumes
B.P. 94
84143 Montfavet Cedex
France

Sara Lebiush-Mordechi
Agricultural Reseach Organization (ARO)
Gilat Regional Experiment Station
Entomology Laboratory
Mobile Post Negev 85-280
Israel

Jerzy J. Lipa
Institute of Plant Protection
Dept. of Biocontrol & Quarantine
Miczurina 20
60-318 Poznan
Poland

Marisol Luis-Arteaga
Diputación General de Aragón
Servicio de Investigación Agroalimentaria
Ctra. de Montañana 177
Apdo. Correos 727
50080 Zaragoza
Spain

Robert D. Lumsden
USDA Agricultural Research Service
Beltsville Agricultural Research Center
Plant Sciences Institute
Biocontrol of Plant Diseases Laboratory
Rm 275 Bldg 011A BARC W
Beltsville, Maryland 20705-2350
USA

Nikolaos E. Malathrakis
Technological Education Institute of Heraklio
P.O. Box 140
71510 Heraklio, Crete
Greece

Giuseppe Manzaroli
Biolab. Centrale Ortofrutticola
Centro Servizi Avanzati per l'Agricultura,
Soc. Coop. A.R.L.
Via Masiera Prima 1191
47020 Martorano, Cesena, Forlí
Italy

Nicholas A. Martin
New Zealand Institute for Crop & Food Research
Mount Albert Research Centre
120 Mount Albert Road
Private Bag 92 169
Auckland
New Zealand

Alberto Matta
Università degli Studi di Torino
Dipartimento di Valorizzazione e Protezione
delle Risorse Agroforestali – Patologia Vegetale
Via Leonardo da Vinci 44
10095 Grugliasco (Torino)
Italy

Graham A. Matthews
Imperial College of Science,
Technology and Medicine
International Pesticide Application Research
Centre (IPARC)
Dept. of Biology
Silwood Park, Ascot
Berkshire SL5 7PY
United Kingdom

Quirico Migheli
Università degli Studi di Torino
Dipartimento di Valorizzazione e Protezione
delle Risorse Agroforestali – Patologia Vegetale
Via Leonardo da Vinci 44
10095 Grugliasco (Torino)
Italy

Enrique Moriones
Consejo Superior de Investigaciones
Científicas (CSIC)
Estación Experimental "La Mayora"
Algarrobo-Costa s/n
29750 Algarrobo-Costa, Málaga
Spain

Giorgio Nicoli
Università di Bologna
Istituto di Entomologia "Guido Grandi"
Via Filippo Re, 6
40126 Bologna
Italy

Timothy M. O'Neill
ADAS Arthur Rickwood
Mepal
Ely
Cambs CB6 2BA
United Kingdom

Jean-Claude Onillon
INRA – Centre de Recherches d'Antibes
Laboratoire de Biologie des Invertébrés
Unité de Recherches sur les Parasitoïdes
d'Aleurodes
37, Boulevard du Cap
B.P. 2078
06606 Antibes Cedex
France

Timothy C. Paulitz
McGill University
Faculty of Agricultural and
Environmental Sciences
Dept. of Plant Science
Macdonald Campus of McGill Univ.
21,111 Lakeshore
Ste. Anne de Bellevue
Québec H9X 3V9
Canada

Jean-Michel Rabasse
INRA – Centre de Recherches d'Antibes
Unité de Biologie pour la Santé
des Plantes et l'Environnement
37, Boulevard du Cap
B.P. 2078
06606 Antibes Cedex
France

Pierre M.J. Ramakers
Research Station for Floriculture
and Glasshouse Vegetables
Kruisbroekweg 5
Postbus 8
2670 AA Naaldwijk
The Netherlands

David J. Rhodes
Zeneca Agrochemicals
Fernhurst Haslemere
Surrey GU27 3JE
United Kingdom

Jordi Riudavets
IRTA – Centre de Cabrils
Ctra. de Cabrils, s/n
08348 Cabrils, Barcelona
Spain

J. Leslie Shipp
Agriculture and Agri-Food Canada
Greenhouse and Processing
Crops Research Centre
Harrow, Ontario N0R 1G0
Canada

Dan Shtienberg
Agricultural Research Organization (ARO)
The Volcani Center
Institute of Plant Protection
Dept. of Plant Pathology
P.O. Box 6
Bet-Dagan 50250
Israel

Peter H. Smits
Research Institute for
Plant Protection (IPO-DLO)
Binnenhaven 5
P.O. Box 9060
6700 GW Wageningen
The Netherlands

Elefterios C. Tjamos
Agricultural University of Athens
Dept. of Plant Pathology
Iera Odos 75
Votanikos 11855, Athens
Greece

Maria Grazia Tommasini
Biolab. Centrale Ortofrutticola
Centro Servizi Avanzati per l'Agricoltura,
Soc. Coop. A.R.L.
Via Masiera Prima 1191
47020 Martorano – Cesena, Forlí
Italy

Arne Tronsmo
Agricultural University of Norway
Dept. of Biotechnological Sciences
P.O. Box 5040
1432 Ås
Norway

Jan Dirk van Elsas
Research Institute for Plant Protection (IPO-DLO)
Binnenhaven 5
P.O. Box 9060
6700 GW Wageningen
The Netherlands

Joop C. van Lenteren
Wageningen Agricultural University
Laboratory of Entomology
Binnenhaven 7
P.O. Box 8031
6700 EH Wageningen
The Netherlands

Machiel J. van Steenis
Brinkman B.V.
Woutersweg 10
P.O. Box 2
NL-2690 AA 's-Gravenzande
The Netherlands

Soledad Verdejo-Lucas
IRTA – Centre de Cabrils
Ctra. de Cabrils, s/n
08348 Cabrils, Barcelona
Spain

Leslie R. Wardlow
L.R. Wardlow Ltd
Horticultural Pest Advice
Miranda, Marsh Lane, Ruckinge
Ashford, Kent TN26 2NZ
United Kingdom

Eizi Yano
National Institute of
Agro-Environmental Sciences
Division of Entomology
Kannondai 3-1-1, Tsukuba
Ibaraki 305-8604
Japan

FOREWORD

The International Centre for Advanced Mediterranean Agronomic Studies (CIHEAM), established in 1962, is an intergovernmental organization of 13 countries: Albania, Algeria, Egypt, France, Greece, Italy, Lebanon, Malta, Morocco, Portugal, Spain, Tunisia and Turkey.

Four institutes (Bari, Italy; Chania, Greece; Montpellier, France; and Zaragoza, Spain) provide postgraduate education at the Master of Science level. CIHEAM promotes research networks on Mediterranean agricultural priorities, supports the organization of specialized education in member countries, holds seminars and workshops bringing together technologists and scientists involved in Mediterranean agriculture and regularly produces diverse publications including the series *Options Méditerranéennes*. Through these activities, CIHEAM promotes North/South dialogue and international co-operation for agricultural development in the Mediterranean region.

Over the past decade, the Mediterranean Agronomic Institute of Zaragoza has developed a number of training and research-supporting activities in the field of agroecology and sustainability of agricultural production systems. Some of these activities have been concerned with the rational use of pesticides and more particularly with the implementation of integrated control systems in order to gain in efficacy and decrease both the environmental impact and the negative repercussions for the commercialization of agricultural products. Stemming from the organization of a course on "Integrated Pest and Disease Management in Protected Crops", and as a consequence of the enthusiasm of the lecturers who took part in the course and its scientific co-ordinators, we decided to publish a book based on the contents of the course to provide professionals interested in updating their knowledge with a comprehensive vision of the state of the art of IPM.

Several objective reasons convinced us of our decision. On one hand, the growing economic and social importance of protected crops in the countries of the Mediterranean area. On the other, the fragility of the ecosystems on which they are grown, very often close to areas of urban concentration and tourist development. Therefore, integrated management must be incorporated into the present production systems and appropriate research and experimentation programmes must be developed in order to generate a pest and disease control technology adapted to the ecological conditions and predominant species in each circumstance. We felt that this book could contribute in this task. The Mediterranean Agronomic Institute of Zaragoza has experience from similar publications arising from their professional-training programmes and this also encouraged us to undertake this ambitious project.

The magnitude of our ambition only became clear to us when, compiling the book, we were confronted with the large number of authors, their diverse specialities and origins (from researchers to extensionists, from both the public sector and private firms), and the multidisciplinary nature of the approach, addressing both basic and applied aspects. Accommodating such diversity into the different parts of the book has been our most difficult task. Therefore, it is with great satisfaction and gratitude that we acknowledge and thank the editors, R. Albajes, M.L. Gullino, J.C. van Lenteren and Y. Elad for their inspired and efficient work in orienting and co-ordinating the book. Likewise, we would like to express our gratitude to each and every one of the 62 authors for their contribution to this team effort.

The design and development of this book are yet another example of the results that can be achieved through co-operation, and as such, contributes to CIHEAM's objective of promoting co-operation for the development of the agro-food sector in the

Mediterranean area. We hope this example will encourage the same co-operative attitude amongst readers.

Finally we should like to express our satisfaction of the efficacious collaboration from Kluwer Academic Publishers and wish to thank them for their interest in this project.

Miguel Valls
Director
Mediterranean Agronomic Institute of Zaragoza, Spain

PREFACE

This book originated from an international course that was organized on "Integrated Pest and Disease Management in Protected Crops" at the Mediterranean Agronomic Institute of Zaragoza of the CIHEAM. Thirteen guest speakers lectured to some thirty participants, and the idea of publishing the contributions to the course arose as a result of their enthusiasm. The project soon became more ambitious with the purpose of enriching the publication's objectives and contents. Thus, the variety of ways in which protected crops are cultivated world-wide demanded the collaboration, not only of European authors, but of authors from all those regions that have developed the greenhouse crop industry. Likewise it was necessary, on this occasion, to count on the multi-disciplinarity of integrated control, therefore new entomologists and plant pathologists working in different disciplinary environments, such as ecology, molecular biology, statistics, information systems and plant breeding, were incorporated into the project. It was also considered necessary to count on the collaboration of specialists from the public and private sectors involved in the different links of the chain necessary for the technological innovation of integrated control: researchers, extensionists, natural enemy producers, consultants. This diversity of authors is probably what we are most satisfied with as editors. Nevertheless, this has also complicated the edition work as we have tried to keep a maximum of homogeneity without falling into too much uniformity. As the basic elements of integrated control need to make use of local conditions favourable to pest and disease control, one cannot expect the points of view, practices, even scientific backgrounds to be common throughout all the chapters of the book when very often the authors work in areas which are geographically very different. Whenever possible, we have entrusted each chapter to authors whose activity and perspectives could be complementary: entomologists together with pathologists, from both public and private sectors, differentiated geographical areas, etc. It is our sincere belief that no text published to date has offered such a diverse yet integrated approach to pest and disease control in greenhouse crops.

The book opens with an initial chapter describing the scenario where integrated pest and disease control operates, that is, the greenhouse and its environment. Ensuing chapters provide the basic strategies and tactics of integrated control, with special reference to greenhouse crops. Further chapters include the different facets of biological pest and disease control – its scientific bases, its development in practice, its commercialization and quality control. The pre-eminence of biological control in the book is not surprising since without a doubt it is the cornerstone of integrated insect pest control and is also becoming increasingly more important in disease control. The concluding chapters of the book show us the present situation of integrated pest and disease control in the most important greenhouse crops world-wide. This final section opens with a chapter discussing the technology transfer process from research to the consumer; this chapter is by no means superfluous, as the lack of an efficient technology transfer is often the main cause of the slow adoption of integrated control.

This book is neither a manual nor a guide. We have attempted to provide post-graduate and professional readers already familiar with the subject, with a means to acquire deeper knowledge on integrated control of pests and diseases in greenhouse crops and furthermore suggest possible roads to take in future tasks. It is evident, however, that each situation and each problem requires a particular solution. Integrated control in greenhouses first developed in England and The Netherlands in the 60s. The success reached in both countries led the research, extension and application of this type of control system to become generalized throughout northern Europe in the 70s and 80s.

This experience, so positive in the North of Europe, stimulated the adaptation of integrated systems for other areas such as the Mediterranean, North America, Oceania and Asia at various rates and degrees of success. It has been shown that a mere transposition of northern European solutions is not valid in other parts of the world. Each new situation demands further research, development, extension, training and new forms of application. Without this local effort, it will be very difficult for integrated control to progress at a faster rate. We trust that this work will contribute to stimulating and guiding this effort.

We have many people to thank. The Mediterranean Agronomic Institute of Zaragoza organized and hosted the course that gave rise to this book and subsequently undertook the co-ordination of the edition and technical editing. Had we not been able to count on their experience, professionalism and enthusiasm, we would not have been able to embark on this endeavour. The participants in the mentioned course have also permitted us to enrich the content of this work with their suggestions and constructive criticism. The authors have shown at all times a great patience and comprehension on reacting to our requests and revisions with good will and wisdom. The IOBC/WPRS, "International Organization for Biological and Integrated Control of Noxious Animals and Plants, West Palaeartic Regional Section" likewise deserves a special mention of gratitude. In two of their working groups on "Integrated Control in Greenhouse Crops", these editors and many of the authors have been collaborating and continue to do so, thus facilitating the edition of the book.

To publish a book is an arduous task. The mere conviction of the need to divulge and teach what has been learnt from others and our own sense of duty can compensate such an undertaking. Fortunately, we are convinced that the effort of the hundred people who have collaborated, in one way or another, in this book has been worthwhile. Another decisive stimulant for this endeavour was the realization of the growing need to incorporate integrated systems of protection from arthropod pests and diseases for the thousands of hectares of protected crops in the world. Both the fruit, vegetable and ornamental plant markets and the technical and economic efficiency of crop protection require these integrated control systems.

Ramon Albajes
M. Lodovica Gullino
Joop C. van Lenteren
Yigal Elad

SETTING THE STAGE: CHARACTERISTICS OF PROTECTED
CULTIVATION AND TOOLS FOR SUSTAINABLE CROP PROTECTION
M. Lodovica Gullino, Ramon Albajes and Joop C. van Lenteren

1.1. Protected Cultivation and the Role of Crop Protection

Attempts to adapt crop production to the environment with protective devices or
practices date back to ancient times. Structures for crop production were first used in
the early period of the Roman Empire, under Emperor Tiberius Caesar, 14–37 AD.
Such structures consisted of mobile beds of cucumber placed outside on favourable
days and inside during bad weather. Covers were slate-like plates or sheets of mica or
alabaster (Dalrymple, 1973). Greenhouses in the UK and The Netherlands developed
from glass structures built to protect plants imported from tropical Asia and America in
the 16th and 17th century during the winter period. However, such methods of
cultivation ceased with the decline of the Roman Empire and it was not until the 15th to
18th centuries that simple forms of greenhouses appeared, primarily in England, The
Netherlands, France, Japan and China. By the end of the 19th century, commercial
greenhouse crop production was well-established (Wittwer and Castilla, 1995).

The purpose of growing crops under greenhouse conditions is to extend their
cropping season and to protect them from adverse environmental conditions, such as
extreme temperatures and precipitation, and from diseases and pests (Hanan *et al.*,
1978). Greenhouse structures are essentially light scaffolding covered by sheet glass,
fibreglass or plastic. Such materials have a range of energy-capturing characteristics, all
designed to maximize light transmission and heat retention. Crops may be grown in
groundbed soil, usually amended with peat or farmyard manure, in benches, in pots
containing soil or soil mixtures or soil substitutes, and in hydroponic systems, such as
sand or rock wool cultures and flowing nutrient systems, without a matrix for the roots.

Modern technology has given the grower some powerful management tools for
production. Generally, added-value crops are grown under protection. Most of them are
labour-intensive and energy-demanding during cold weather. Greenhouse production
therefore normally requires a high level of technology to obtain adequate economic
returns on investments. Quality is a high priority for greenhouse crops, requiring much
care in pest and disease management, not only to secure yields but also to obtain a high
cosmetic standard. Although technological changes are ultimately intended to reduce
production costs and maximize profits, precise environmental and nutritional control
push plants to new limits of growth and productivity. This can generate chronic stress
conditions, which are difficult to measure, but apparently conducive to some pests and
diseases. Historically, not enough attention has been paid to exploiting and amending
production technology for the control of pests and diseases. This makes the control of
pests and diseases in protected crops even more challenging, with many important

1

R. Albajes et al. (eds.), Integrated Pest and Disease Management in Greenhouse Crops, 1-15.
© 1999 *Kluwer Academic Publishers. Printed in the Netherlands.*

problems being unresolved and new ones arising as the industry undergoes more changes in production systems.

Additionally, the international trade in ornamental and flower plants facilitates the spread of pests and diseases around the world and their establishment in new areas. In Europe, for example, at least 40 new pests have been recorded in protected crops in the last 25 years. The increasing complexity of pest and disease problems and the high cosmetic standards of vegetable, ornamental and flower products have led growers to apply intensive preventive chemical programmes, which result in pests and pathogens becoming resistant to the most frequently used pesticides in a few years, which, in turn, increases control costs. In southern Spain, the average cost of pesticide application in 1992 in protected vegetables was estimated as US$0.14/m^2 (16.5% of the total production cost) (Cabello and Cañero, 1994), and several whitefly, thrips, aphid and fungus species are suspected to be resistant to several active ingredients. A similar figure is valid for Italy, where the most sophisticated structures are located in the northern part of the country: pesticides are widely applied and pest and disease resistance is quite widespread (Gullino, 1992). In The Netherlands, pest and disease control costs for vegetables are still limited and are normally below 3% of the total costs to produce a crop (van Lenteren, 1995).

As control costs increase, pesticide-resistance spreads and consumers become aware of the risks of pesticide-residues in fresh vegetables, a strong demand for non-chemical control methods is emerging in many countries. Integrated systems for greenhouse pest and disease control have been developed and implemented in northern Europe and Canada, but implementation is still cumbersome in other parts of the world.

1.2. Importance of Protected Crops for Plant Production

During the late 50s and early 60s the use of greenhouses spread: initially they were mostly used for vegetable production, with an emerging cut-flower and ornamental plant industry starting, particularly in the UK and in The Netherlands. By 1960, The Netherlands had the most concentrated production of glass-house grown crops, estimated as 5000–6000 ha (75% of which grew tomatoes). At the same time, the UK had 2000 hectares of greenhouses (Wittwer, 1981). Hydroponic cultivation started in The Netherlands in the 60s and spread to many countries. In the USA, hydroponic cultivation became widespread (Jensen and Collins, 1985): in the late 60s and early 70s, there were more than 400 ha devoted to hydroponic vegetable production (tomato, followed by cucumber and lettuce), although this surface area has diminished to less than 100 ha today (Wittwer and Castilla, 1995). Moreover, there has been a strong shift from vegetables to ornamentals grown in glasshouses. Nowadays, in the USA, of the total greenhouse production (estimated as 2000 ha), 95% is represented by flowers, potted plants, ornamentals and bedding plants (Wittwer and Castilla, 1995). There has also been a shift in northern Europe, with a delay of about 15 years compared to the USA, from vegetables to added-value ornamental crops (Wittwer and Castilla, 1995). For example, more than 80% of the greenhouses in The Netherlands were used for vegetables in the 60s, whereas now 60% of the approximate 10,000 ha are used for production of ornamentals.

By 1980, there was an estimated 150,000 ha of greenhouses (glass, fibreglass, plastic) world-wide producing high-value crops (Wittwer, 1981). In 1995, the surface area had increased to about 280,000 ha (Bakker, 1995; Wittwer and Castilla, 1995) (Table 1.1). New areas, particularly in Asian and Mediterranean countries, showed a strong increase in protected areas, attracted by cultivation of high-value vegetable crops. The expansion in plasticulture in the Mediterranean area is still going on, again with a gradual transition from the production of vegetables to ornamentals. Spain and Italy have been the leading countries in the 80s and 90s. At present, the North African countries are experiencing a very rapid increase in the area covered with plastic houses, often with very simple structures. This development has been accompanied by a spread in drip irrigation (Wittwer and Castilla, 1995). At the same time, the use of plastic row tunnels, covers and plastic soil mulches has expanded world-wide. These structures will not be discussed further in this book, but it is interesting to know that, for example, in China an area of more than 2.8 million ha of crops was covered with plastic soil mulch in 1995 (Wittwer and Castilla, 1995).

TABLE 1.1. Distribution of protected cultivation world-wide (from Wittwer and Castilla, 1995)

Structure	ha (× 1000) in geographical area				
	Asia	Mediterranean[1]	North/South America	North Europe	Total
Direct cover (floating types)	5.5	10.3	1.5[2]	27.0	44.3
Low tunnels (row-covers)	143.4	90.5	20.0[2]	3.3	257.2
High tunnels	–	27.6[3]	–	–	27.6
Plastic-houses	138.2	67.7	15.6	16.7	238.2
Glass-houses	3.0	7.9	4.0	25.8	40.7

[1]Including France
[2]Figures are crude estimates
[3]High tunnels are often taken together with plastic-houses in countries other than Mediterranean

The world greenhouse area is now estimated as 307,000 ha, 41,000 ha of which is covered with glass, 266,000 ha with plastic. The global status of protected cultivation (*sensu lato*) is reported in Table 1.1. The distribution and types of crops grown in greenhouses are outlined in Table 1.2. Vegetable crops are grown in about 65% of greenhouses, and ornamentals in the remaining 35%.

1.3. Type of Structures Adopted for Protected Cultivation and their Impact on Cultivation Techniques and Crop Protection

Structures adopted for covering crops vary a lot, from the simple to the sophisticated:
(i) Low tunnels (row-covers). These are small structures that provide temporary

protection to crops. Their height is generally 1 m or less, with no aisle for walking, so that cultural practices must be performed from the outside. Their use enhances early yields and yield volume; they also protect against unfavourable weather. The thermal films of infra-red polyethylene (PE), ethylene vinylacetate (EVA), copolymer, polyvinylchloride (PVC) and conventional PE are used.

(ii) High tunnels (walk-in tunnels). Such structures use the same cover materials as low tunnels and are high enough to perform cropping practices inside. Moderately tall crops are grown. Statistics concerning high tunnels are often included in the same category as low cost plastic houses (Table 1.1) since the materials used are similar.

(iii) Greenhouses. These differ from other protection structures in that they are sufficiently high and large to permit a person to conveniently stand upright and work within (Nelson, 1985). Greenhouses appeared when glass became available for covering. Later, the introduction of plastic films permitted world-wide expansion of the greenhouse industry.

TABLE 1.2. World distribution of crops most commonly grown under protected structures (tunnels, greenhouses) (modified from Wittwer and Castilla, 1995)

Crop	Leading countries
Cucurbits	Argentina, Belgium, Canada, Chile, China, Egypt, Finland, France, Germany, Greece, Hungary, Israel, Italy, Japan, Morocco, Turkey, Korea, Poland, Portugal, Saudi Arabia, Scandinavia, Spain, Taiwan, The Netherlands, Tunisia, Ukraine, former URSS, USA
Strawberry	Argentina, Belgium, Canada, China, Finland, France, Greece, Israel, Italy, Japan, Jordan, Morocco, Portugal, Spain, Tunisia, Turkey, USA
Solanaceous + green beans	Algeria, Argentina, Belgium, Bulgaria, Canada, Chile, China, Denmark, Egypt, Finland, France, Germany, Great Britain, Greece, Hungary, Israel, Italy, Japan, Jordan, Korea, Morocco, Poland, Portugal, Saudi Arabia, Scandinavia, Spain, Taiwan, Tunisia, Ukraine, The Netherlands, Turkey, former URSS, USA
Grapes and tree fruits	Italy, Japan, Morocco, Portugal, Spain
Lettuce, cabbage, celery, radish, asparagus	Belgium, China, France, Germany, Great Britain, Hungary, Italy, Japan, Korea, Poland, Spain, The Netherlands, former URSS, USA
Flowers, ornamentals	Argentina, Belgium, Canada, Denmark, Egypt, Finland, France, Germany, Great Britain, Hungary, Israel, Italy, Japan, Poland, Scandinavia, Spain, The Netherlands, USA
Bedding and potted plants	Belgium, Canada, Denmark, Finland, France, Germany, Great Britain, Hungary, Israel, Italy, Japan, Scandinavia, Spain, The Netherlands, USA

Greenhouses protect crops against cold, rain, hail and wind, providing plants with improved environmental conditions compared to the open field. In greenhouses, crops can be produced out-of-season year-round with yields and qualities higher than those produced in the open field. Greenhouses have also allowed the introduction of new crops, normally foreign to the region (Germing, 1985).

There are two basic types of greenhouse. The first type seeks maximum control in an environment to optimize productivity. In Europe, optimal conditions for year-round production are provided in the glasshouses of The Netherlands, Belgium, the UK and Scandinavia. The other type of greenhouse, which is very common throughout the Mediterranean area, provides minimal climatic control, enabling the plants grown inside to adapt to suboptimal conditions, survive and produce an economic yield (Enoch, 1986; Tognoni and Serra, 1989; Castilla, 1994).

The choice of greenhouse depends on location, crop and financial resources. There is a strong relationship between local conditions, greenhouse design, cladding materials and insulation needs.

The structure of a greenhouse depends on the climate and the cladding used. There are various roof, space and height geometries with single-span materials such as bamboo, used in low cost structures, particularly in China and in semi-tropical and tropical areas. Cladding materials were limited to glass until the middle of the 20th century. From 1950, plastic films, because of their low cost, light weight and adaptability to different frame designs, became available, permitting world-wide development of the greenhouse industry, particularly in the semi-tropical areas (Nelson, 1985). But plastic covers are not acceptable in northern Europe because of low light transmission compared to glass.

A full range of conventional and modified plastic films is now available (Giacomelli and Roberts, 1993): all coverings can perform well, depending on the desired use and location. Single plastic films prevail in warm climates; inflated double plastic film or rigid single plastic panels are more common in cool areas. A combination of high and low technology may be seen in countries such as Korea and Israel.

Nets are used in tropical areas or during hot weather in temperate zones: they may reduce pest damage and the extremes of temperature and air humidity. Moreover, nets have a windbreak effect and reduce the damage from heavy rain and hail (Castilla, 1994) (see Chapter 18 for a further description of the use of nets for pest control).

The greenhouse design (particularly its height, shape, opening systems and cladding material) strongly influences climatic conditions inside, thus having a profound impact on pest and disease development. Plastic houses almost always have a more humid climate, large diurnal temperature variation and are more difficult to ventilate. Typically, they result in more problems with high humidity-dependent diseases, such as grey mould, downy mildews and rusts (Jarvis, 1992). Regulating the atmosphere throughout the day and night is important for disease control and for reducing the total amount of chemicals sprayed. This has been demonstrated in the case of grey mould (*Botrytis cinerea* Pers.:Fr.) in tomato (Gullino *et al.*, 1991) and cucumber (Yunis *et al.*, 1994), and of downy mildew (*Bremia lactucae* Regel) in lettuce (Morgan, 1984).

With respect to the cladding material used, in some cases a possible effect on diseases has been reported, mostly through the direct influence of radiation on sporulation (Jarvis, 1992). Certain UV-absorbing plastic coverings for greenhouses that absorb light at 340 nm have been exploited to inhibit the sporulation of *Sclerotinia sclerotiorum* (Lib.) de Bary (Honda and Yunoki, 1977), and species of *Alternaria* and *Botrytis squamosa* J.C. Walker (Sasaki *et al.*, 1985). Reuveni *et al.* (1989) observed a reduction in the number of infection sites of *B. cinerea* on tomato and cucumber when a

UV-absorbing material was added to polyethylene film to increase the ratio of blue light to transmitted UV light. Recently, blue photoselective polyethylene sheets have been suggested for their ability to reduce grey mould on tomato (Reuveni and Raviv, 1992) and downy mildew on cucumber (Reuveni and Raviv, 1997). Green-pigmented polyethylene reduced the conidial load and grey mould in commercial tomato and cucumber greenhouses by 35–75%. *Sclerotinia sclerotiorum* on cucumber, *Fulvia fulva* (Cooke) Cif. (= *Cladosporium fulvum* Cooke) on tomato and cucumber powdery mildew were also reduced (Elad, 1997).

The technologies for environmental control in the most sophisticated greenhouses have been characterized by many new developments over the past three decades. The variables of light, temperature, air and soil humidity, and CO_2 content of the atmosphere are computer-programmed 24 h a day to achieve maximum crop yield (Nederhoff, 1994). Further refinements and improvements for adjusting the greenhouse climate to optimal crop productivity can be expected. In the less sophisticated structures of the sub-tropical and tropical regions, it is much more difficult to manipulate the greenhouse climate (Gullino, 1992). In tropical and subtropical areas greenhouses often simply have an umbrella effect, using just roofs, with sides left open.

The influence of greenhouse structures and covers on greenhouse climatic regimes may have strong consequences for pests and their natural enemies, as they have for diseases. A typical case of climate influence on pests and natural enemies concerns the spider mite and its predator *Phytoseiulus persimilis* Athias-Henriot: low humidity regimes may constrain effective use of *P. persimilis* (Stenseth, 1979). In high-tech greenhouses, regulation of temperature and water pressure deficit enables the creation of conditions less favourable to pathogens and, in some cases, more favourable to biocontrol agents. The use of heating to limit development of a number of pathogens is well known (Jarvis, 1992): however, heating is not economically feasible in all greenhouse systems. Recently, with the development of soilless systems, the effect of managing the temperature of the circulating solution has been studied, and has proven to be effective against certain pathogens. The use of high root temperatures in winter-grown tomatoes in rock wool offers a non-chemical method of controlling root rot caused by *Phytophthora cryptogea* Pethybr. & Lafferty. The high temperature was shown to enhance root growth while simultaneously suppressing inoculum potential and infection, and, consequently, reducing or preventing aerial symptoms (Kennedy and Pegg, 1990). Careful control of the temperature also proved important in the case of hydroponically grown spinach and lettuce, in which it prevented or reduced attack by both *Pythium dissotocum* Drechs. and *Pythium aphanidermatum* (Edson) Fitzp. (Bates and Stanghellini, 1984). Recently, attacks of *P. aphanidermatum* on nutrient film technique (NFT) grown lettuce in Italy were related to the high temperature (>29°C) of the nutrient solution. Root rot was inhibited by reducing the temperature below 24°C (Carrai, 1993).

Much less exploited are the effects of temperature and water pressure deficit on biocontrol agents, although the first models, resulting in advice for optimal climate control for insect natural enemies, are now becoming available (van Roermund and van Lenteren, 1998). In the case of biological control of plant pathogens, most of the studies carried out are related to the effect of environmental conditions on *Trichoderma harzianum* Rifai, used as biocontrol agent of *B. cinerea* and of several hyperparasites of *Sphaerotheca fusca* (Fr.) Blumer. [= *Sphaerotheca fuliginea* (Schlechtend.:Fr.) Pollacci]. In the case of *T. harzianum*, populations of the antagonist are promoted by low vapour pressure deficit; in commercial greenhouses significant control of grey mould of cucumber has been correlated with low water pressure deficit but not with conditions of air saturation and dew deposition (Elad and Kirshner, 1993). In the case of *Ampelomyces quisqualis* Cesati:Schltdl., hyperparasite of *S. fusca*, a period of 24 h with low vapour pressure deficit is necessary (Philipp *et al.*, 1984). Low vapour pressure deficit also favours the activity of *Sporothrix flocculosa* Traquair, Shaw & Jarvis (Hajlaoui and Bélanger, 1991). More studies in this field are necessary, both in order to keep conditions close to the optimum for biocontrol agents within the greenhouse and for selecting biocontrol agents more adapted to the greenhouse environment (Elad *et al.*, 1996).

Greenhouses were initially built in areas with long, cold seasons to produce out-of-season vegetables, flowers and ornamental plants. Northern Europe is the paradigm of pioneering areas of greenhouse cultivation. The development of international exchanges of agricultural products and the availability of a variety of cheap plastic materials for covering simple structures has led to a spectacular increase in the area of protected crops in warmer regions like the Mediterranean basin and East and Southeast Asia (Wittwer and Castilla, 1995). These new regions are commonly characterized by low or irregular annual precipitation and poor vegetation development. The insertion of greenhouse patches leads to drastic changes in the structure and ecology of the landscape. In early stages of greenhouse cultivation in a new area, greenhouses are isolated spots, like oases, where some phytophagous insects find good seasonal conditions for rapid increases in density. But optimal weather and host-plant conditions rarely last throughout the year and for a few months – usually the hottest – the increase in the herbivore population is interrupted. When greenhouses become more common in the area, the mosaic pattern may evolve to a large area of protected crops, with a succession of crops throughout most of the year and with polyphagous pests. These pests are able to feed on many agricultural plants and migrate between greenhouses. Additionally, field crops may be excellent refuges for pests in hot seasons, when the temperature is too high for greenhouse cultivation. This has several consequences, as the immigration of pests into the greenhouse causes sudden and largely unpredictable pest density increases.

Exotic pests quickly become established, especially if ornamental plants are cultivated. Polyphagous pests (like whiteflies, spider mites, thrips, leafminers, several aphids species, especially *Aphis gossypii* Glover, leaf-eating caterpillars and soilworms), which may exploit several crops successively, become prevalent. As pest densities increase, crops are increasingly sprayed with insecticides, native natural enemies become very rare, and natural control loses effectiveness. Unexpected and high pest pressure from the outside makes biological control very difficult. Under such conditions, a more holistic approach would consider the fields outside the greenhouse and the crop inside the greenhouse as a single entity for applying integrated strategies against pests and diseases. Programmes for conserving native or introduced natural enemies in the area should both lower pest pressure on greenhouse crops and incorporate beneficial fauna into the outside-inside greenhouse cycle of the pest-natural enemy complex.

1.4. Cultural Techniques Used in Protected Cultivation

In most greenhouses of northern Europe continuous cropping is practised, without a fallow crop-free interval. This has profound implications for diseases and pests. In the case of plant pathogens, it leads to the build-up of soilborne pathogens and an increased importance of foliar pathogens with a broad host-spectrum (i.e. *B. cinerea*). The same can be said for insects that pupate in the soil such as leafminers and thrips.

Greenhouse crops are grown in various soils and soilless media whose physical and chemical properties are adjusted to obtain maximum productivity. These properties, such as heat conservation, water-holding capacity, fertilizer levels and pH can also be manipulated to reduce the amount of inoculum of pests and pathogens and the probability of infection (Jarvis, 1992). Systems for growing crops in the greenhouse vary widely in terms of complexity. The most common rooting media are soil and various soil mixtures, incorporating peat, vermiculite, perlite and several other materials which are added to the soil in order to modify its structure.

In the 60s, bench cultivation was adopted for high value crops (i.e. carnations), permitting better results in soil disinfestation. In the 80s and 90s, soilless substrates gained more and more importance, particularly in the northern European countries, because they eliminate or reduce the need for soil disinfestation. Among soilless substrates, rock wool has been widely used in northern Europe, while in the tropics and sub-tropics cheaper substrates have been exploited. The nutrient film technique, originally devised to improve precision in crop nutrition, reduces soilborne diseases and removes the cost of soil disinfestation. In fact, it confers relative freedom from diseases, although severe epidemics can still occur (Stanghellini and Rasmussen, 1994).

During the past two decades, various composted organic wastes and sewage sludges have partially replaced peat in container media used for production of ornamentals. Recycling of these wastes has been adopted for economic and production reasons. The cost of these composts can be lower than peat. Production costs may also be decreased because some of the compost-amended media, particularly those amended with composted bark, suppress major soilborne plant pathogens, thus reducing plant losses (Hoitink and Fahy, 1986). As discussed later, not only chemical and physical, but also biotic factors affect disease suppressiveness (see Chapter 23). The low pH of sphagnum peat, pine bark and composts could theoretically have beneficial side effects for some plants. For example, Phytophthora root rot of rhododendron (*Phytophthora cinnamomi* Rands) is suppressed at pH<4.0, because the low pH reduces sporangium formation, zoospore release and motility. This may be important during propagation of rhododendron cuttings under mist. Moreover, chemical inhibitors of *Phytophthora* spp. have been identified in composted hardwood bark. These inhibitors do not affect *Rhizoctonia solani* Kühn (Hoitink and Fahy, 1986).

Soilless cultivation can affect pests that need the soil/substrate to complete their development, as in the case of leafminers or thrips.

The thermal and gas exchange properties of rooting media affect the growth of roots as well as the activities of pathogens. Peat, a common rooting medium, used either alone or in mixture, often suppresses pathogen activity, depending on its origin (Tahvonen, 1982). However, pathogens, including species of pathogenic *Pythium* and *Fusarium* (including *Fusarium oxysporum* Schlechtend.:Fr. f. sp. *radicis-lycopersici* W.R. Jarvis & Shoemaker) have been isolated from commercial peat compost (Couteaudier *et al.*, 1985; Gullino and Garibaldi, 1994).

The design of benches is important due to the effect on the ventilation of seedling trays and potted plants.

Correct spacing prevents the establishment of a microclimate conducive to foliar diseases and the rapid spread of pathogens and pests from plant to plant in crops grown in groundbeds. Altered greenhouse and bench design can improve air movement, thus reducing the risk of diseases. Bottom heating of benches, a traditional means of avoiding Phytophthora, Pythium and Rhizoctonia root rots, is enhanced in cutting and seedling trays with upward air movement between the young plants. Through-the-bench air movement is perhaps the most neglected and simplest means of reducing seedling rots in tangled plant masses (Jarvis, 1989).

Every crop species and cultivar requires a special fertilizer regime in order to obtain maximum productivity and to prevent stress on the plant. Fertilizer requirements change as the crop ages from seeding to harvest. In general, excessive nitrogen leads to excessive foliage that is intrinsically more succulent and susceptible to damage and necrotrophic pathogens, such as *B. cinerea*, and also stimulates development of pests such as aphids and leafminers. Nitrogen generally has to be balanced with potassium; for many diseases, susceptibility decreases as the potassium-nitrogen ratio increases. Calcium generally enhances resistance, due to its role in the integrity of the cell wall.

No general practical recommendations can be made for controlling diseases by adjusting the fertilizer levels supplied to plants: each host-pathogen combination reacts differently. However, optimal, instead of maximal fertilization, results in lower pest and disease pressure. General recommendations can be given concerning irrigation. First of all, the factors that determine irrigation demand in greenhouse crops can all be closely regulated. From a general point of view, overhead irrigation must be carried out early in the day and should be limited late in the afternoon in order to avoid long periods of leaf wetness, which favour diseases such as downy mildews, rusts, grey mould, leaf spots, etc. When it is necessary to wet foliage for any reason (including pesticide spraying), it is always essential to maintain environmental conditions under which the foliage can dry out within a very short period of time. Also, it is important to avoid excess water in the soil: this creates conditions that are very favourable for the development of root rots. The effects of irrigation on pests are mainly through the relative humidity of the environment or through the water-status of the plants. For instance, plants under stress are more easily colonized by thrips and spider mites.

1.5. Factors Favourable to Pest and Disease Development

Well-grown and productive crops are generally less susceptible to diseases, but in many cases compromises have to be made between optimum conditions for economic productivity and conditions for disease and pest prevention. Well-fertilized and irrigated crops are, however, often more sensitive to pests, like aphids, whiteflies and leafminers.

Groundbed crops are rarely rotated, so soilborne pathogens and pests pupating in the soil accumulate if the soil is not disinfested. Soil disinfestation, although effective, creates a "biological vacuum" (Katan, 1984) (see Chapter 10). Major changes in cultural techniques include the use of hydroponic and soilless cultures and artificial substrates controlled by computerized systems. Although these changes are ultimately intended to reduce production costs and maximize profits, precise environmental and nutritional control that pushes plants to new limits of growth and productivity can generate chronic stress conditions, which are difficult to measure, but are apparently conducive to diseases caused by pathogens such as *Penicillium* spp. or *Pythium* spp. (Jarvis, 1989). Some soil substitutes and soilless systems do not always provide sufficient competition for pathogens, due to their limited microflora.

High host plant densities and the resulting microclimate are favourable to disease spread. Air exchange with the outside is restricted, so water vapour transpired by the plants and evaporated from warm soil tends to accumulate, creating a low vapour pressure deficit (high humidity). Therefore, the environment is generally warm, humid and wind-free inside the greenhouse.

Such an environment promotes the fast growth of most crops, but it is also ideal for the development of bacterial and fungal diseases (Baker and Linderman, 1979; Fletcher, 1984; Jarvis, 1992), of insects vectoring viruses and of herbivorous insects. For bacteria and many fungi (causal agents of rusts, downy mildews, anthracnose, grey mould, etc.) infection is usually accomplished in a film or drop of water on the plant surface. Unless temperature, humidity and ventilation are well regulated, this surface water can persist in the greenhouse until infection becomes assured.

Many of the energy saving procedures adopted during the past three decades are favourable to disease development, since they favour increases in relative humidity (Jarvis, 1992), but they may lead to pest suppression as temperatures are generally somewhat lower (see Chapter 8).

Most greenhouse crops are labour-intensive, and for long periods require daily routine operations (such as tying, pruning, harvesting). The risks of spreading pathogens through workers and machinery are increased by the risks deriving from accidental wounds and from the exposure of large areas of tissues by pruning.

Greenhouses are designed to protect crops from many adverse conditions, but most pathogens and several pests are impossible to exclude. Windblown spores and aerosols containing bacteria enter doorways and ventilators; soilborne pathogens enter in windblown dust, and adhere to footwear and machinery. Aquatic fungi can be present in irrigation water; insects that enter the greenhouse can transmit viruses and can carry bacteria and fungi as well. Once inside a greenhouse, pathogens and pests are difficult to eradicate.

1.6. Factors Stimulating Sustainable Forms of Crop Protection in Protected Cultivation

Protected cultivation is an extremely high-input procedure to obtain food and other agricultural products per unit of land, although inputs are the lowest when related to the yield per area. Crop protection activities contribute to the total input in variable proportions mainly through the application of pesticides. Several features of protected cultivation are delaying the adoption of more sustainable ways to control pests and diseases. In areas where protected cultivation is most intensive, crop protection costs rarely exceed 5% of the total production costs. In these circumstances, growers are not stimulated to make decisions based on economically founded criteria, and chemicals are frequently applied to prevent pest occurrence rather than to control real pest problems. This is particularly true in ornamental and flower crops, which can lose their value at extremely low pest densities (see Chapter 34). In addition, pesticides may be applied easily and little expertise is needed to spray or to recommend pesticides so that no specialized advisory personnel is usually employed by growers who rely on this "simple" technology.

Consequently, innovative crop protection methods become difficult to implement in practice. From a general point of view, vegetable crops, due to their limited diversity, are most suitable for IPM (see Chapters 30–33). In the case of ornamentals, the enormous crop diversity and the many cultivars per species grown make the development of IPM strategies more complicated (see Chapter 34).

Several stimuli are pushing growers to use less pesticides and to adopt more sustainable ways to protect crops from noxious organisms as world marketing becomes more global. Among the factors stimulating sustainable forms of crop protection are the following:

(i) Consumer concern about chemical residues. This is a general stimulus for growers wishing to adopt IPM systems (Wearing, 1988), but it is particularly relevant in fresh-consumed products like the majority of vegetables grown in greenhouses. Consumers not only demand high quality products, but are also concerned with how they are grown to judge them from the environmental aspects. Food marketers and European regional administrations are developing auditing procedures to sell vegetables under IPM or Integrated Production (IP) labels. In some cases, a surplus price is achieved by growers who produce vegetables under established IPM/IP technology.

(ii) Pesticide-resistance in pests and pathogens. As protected cultivation allows pest and pathogen populations to increase faster than in the open air, and as protected crops receive a great number of pesticide treatments, pesticide-resistance develops rapidly. Dozens of greenhouse pests or pathogens are suspected to have developed resistance to the most common active ingredients and this has been observed in many pests (aphids, whiteflies) and pathogens such as *B. cinerea* (see Chapter 11).

(iii) Side-effects of chemical application are increasingly observed in old and new growing areas (see Chapter 11). Because society in general and governments in particular are aware of the impact of chemicals on soil, water and air, several initiatives to restrict the use of chemicals in Europe and North America are being undertaken (van Lenteren, 1997).

(iv) Efficacy. Some pests and diseases are difficult – sometimes impossible – to control if an integrated approach is not adopted. On the other hand, natural control can prevent several pests from building-up high populations under the action of predators, parasitoids and entomopathogens that naturally establish on greenhouse crops if chemicals are not intensively applied, and several cultural practices allow enhancement of their effectiveness (see Chapters 18 and 19 for the role of parasitoids in leafminer control and polyphagous predators for a potentially broader effect on pests).

A first step towards sustainability in greenhouse crop protection is to analyse why and which phytophages and pathogens are able to increase their population densities until reaching damaging levels. Methods to improve the accuracy and speed of diagnosis are needed, particularly for diseases, and may be one of the most useful applications of biotechnology. Once the pest or disease is correctly diagnosed, environmental factors that allow or prevent such a pest or pathogen to reach economic injury levels should be identified.

Such knowledge may help us to design integrated methods to take advantage of the whole environment. If an action threshold is determined, accurate techniques for pest and disease sampling and monitoring should permit intervention at the best moment (see Chapters 6 and 7) and prevent unnecessary treatments. The identification of key factors governing pest or pathogen population dynamics may allow modification of the greenhouse and crop environment – including greenhouse-surroundings – to adversely affect a pest or pathogen or to favour the effectiveness of the natural enemies or antagonists.

Sometimes this can be achieved cheaply – in both economic and energetic terms – by means of correct crop and management practices (see Chapter 8). As mentioned before, the most damaging pests and many pathogens in greenhouses are polyphagous; although they are able to develop on many host plants, their negative effect on yield varies with host plant species and cultivar. The development of cultivars which are less susceptible to pests and diseases or that favour the activity of pest natural enemies is undoubtedly one of the most sustainable ways to control diseases in greenhouses and its potential for pests has been shown in a few but significant cases (see Chapter 9).

Many of the arthropod pests and diseases that affect greenhouse crops are exotic and became established in greenhouse growing areas from accidental importation of infested crops, mainly ornamentals. In some cases, as for Liriomyza trifolii (Burgess) and Liriomyza huidobrensis (Blanchard), native natural enemies have been able to greatly contribute to the natural control of these pests, but in other cases exotic parasitoids or predators have to be released in the environment to control them, as is done for Trialeurodes vaporariorum (Westwood) by means of Encarsia formosa Gahan. Natural and biological control is nowadays the basis of most of the integrated pest management strategies adopted in northern Europe (van Lenteren, 1995) and its practical achievements are particularly emphasized in this book (see Chapters 13–22). The history of biological control of diseases in greenhouses is more recent, but significant advances have also been achieved here in the last few years (see Chapters 23–28). Given the very high cosmetic demands and the low pest and disease thresholds applied by greenhouse growers, the progress in application of Integrated Pest and Disease Management is remarkable, as described in Chapters 30–34. Until recently, biological and integrated control was seen as a cost factor. Nowadays, however, it is considered as a beneficial marketing factor.

1.7. Concluding Remarks

The greenhouse industry faces many new crop protection problems as a consequence of modification of production procedures and crops. The major changes will include more widely adopted mechanization and automation systems for improved crop management and the use of biotechnology in plant production. These modifications will affect the severity of pests and diseases.

Strong cooperation among plant pathologists, entomologists and horticulturists is necessary in order to assure that new management practices have a beneficial effect on plant health. Methods to improve the accuracy and speed of diagnosis are needed and may be one of the best applications of biotechnology. Improved and widely used monitoring and diagnosis systems to determine the degree of infestation and economic thresholds of pathogens and pests will enable rational management decisions. A high priority should be given to the production of pathogen and pest-free propagation material, obtained through sanitation. The use of pest and pathogen-free material, and growing media disinfested with steam or naturally suppressive to soilborne pathogens will help to reduce the impact of important pests and diseases considerably.

When all such measures are integrated with the use of resistant germplasm, with modern techniques for applying pesticides and with biological control of several diseases and pests, a greatly reduced input of chemicals becomes realistic for protected cultivation.

References

Baker, K.F. and Linderman, R.G. (1979) Unique features of the pathology of ornamental plants, *Annual Review Phytopathology* **17**, 253–277.

Bakker, J.C. (1995) *Greenhouse Climate Control: An Integrated Approach*, Wageningen Press, Wageningen.

Bates, M.L. and Stanghellini, M.E. (1984) Root rot of hydroponically grown spinach caused by *Pythium aphanidermatum* and *P. dissotocum*, *Plant Disease* **68**, 989–991.

Cabello, T. and Cañero, R. (1994) Technical efficiency of plant protection in Spanish greenhouses, *Crop Protection* **13**, 153–159.

Carrai, C. (1993) Marciume radicale su lattuga allevata in impianti NFT, *Colture Protette* **22**(6), 77–81.

Castilla, N. (1994) Greenhouses in the Mediterranean area: Technological level and strategic management, *Acta Horticulturae* **361**, 44–56.

Couteaudier, Y., Alabouvette, C. and Soulas, M.L. (1985) Nécrose du collet et pourriture des racines de tomate, *Revue Horticole* **254**, 39–42.

Dalrymple, D.G. (1973) *Controlled Environment Agriculture: A Global Review of Greenhouse Food Production*, Foreign Agricultural Economic Report No. 89, Economic Research Service, USDA, Washington, DC.

Elad, Y. (1997) Effect of filtration of solar light on the production of conidia by field isolates of *Botrytis cinerea* and on several diseases of greenhouse-grown vegetables, *Crop Protection* **16**, 635–642.

Elad, Y. and Kirshner, B. (1993) Survival in the phylloplane of an introduced biocontrol agent (*Trichoderma harzianum*) and populations of the plant pathogen *Botrytis cinerea* as modified by abiotic conditions, *Phytoparasitica* 21, 303–313.

Elad, Y., Malathrakis, N.E. and Dik, A.J. (1996) Biological control of Botrytis-incited diseases and powdery mildews in greenhouse crops, *Crop Protection* 15, 229–240.

Enoch, H.Z. (1986) Climate and protected cultivation, *Acta Horticulturae* 176, 11–20.

Fletcher, J.T. (1984) *Diseases of Greenhouse Plants*, Longman, London.

Germing, G.H. (1985) Greenhouse design and cladding materials: A summarizing review, *Acta Horticulturae* 170, 253–257.

Giacomelli, G.A. and Roberts, W.J. (1993) Greenhouse covering systems, *HortTechnology* 3, 50–58.

Gullino, M.L. (1992) Integrated control of diseases in closed systems in the sub-tropics, *Pesticide Science* 36, 335–340.

Gullino, M.L., Aloi, C. and Garibaldi, A. (1991) Integrated control of grey mould of tomato, *IOBC/WPRS Bulletin* 14, 211–215.

Gullino, M.L. and Garibaldi, A. (1994) Influence of soilless cultivation on soilborne diseases, *Acta Horticulturae* 361, 341–354.

Hajlaoui, M.R. and Bélanger, R.R. (1991) Comparative effects of temperature and humidity on the activity of three potential antagonists of rose powdery mildew, *Netherlands J. of Plant Pathology* 97, 203–208.

Hanan, J.J., Holley, W.D. and Goldsberry, K.L. (1978) *Greenhouse Management*, Springer-Verlag, Berlin.

Hoitink, H.A.J. and Fahy, P.C. (1986) Basis for control of soilborne pathogens with composts, *Annual Review of Phytopathology* 24, 93–114.

Honda, Y. and Yunoki, T. (1977) Control of Sclerotinia disease of greenhouse eggplant and cucumber by inhibition of development of apothecia, *Plant Disease Reporter* 61, 1036–1040.

Jarvis, W.R. (1989) Managing diseases in greenhouse crops, *Plant Disease* 73, 190–194.

Jarvis, W.R. (1992) *Managing Diseases in Greenhouse Crops*, American Phytopathological Society Press, St Paul, Minn.

Jensen, M.H. and Collins, W.L. (1985) Hydroponic vegetable production, *Horticultural Reviews* 7, 483–558.

Katan, J. (1984) The role of soil disinfestation in achieving high production in horticulture crops, in *Proceedings Brighton Crop Protection Conference*, Vol. 3, BCPC, Farnhan, pp. 1189–1196.

Kennedy, R. and Pegg, G.F. (1990) *Phytophthora cryptogea* root rot of tomato in rock wool nutrient culture. III. Effect of root zone temperature on infection, sporulation and symptom development, *Annals of Applied Biology* 117, 537–551.

Morgan, W.M. (1984) Integration of environmental and fungicidal control of *Bremia lactucae* in a glasshouse lettuce crop, *Crop Protection* 3, 349–361.

Nederhoff, E.M. (1994) *Effects of CO$_2$ Concentration on Photosynthesis, Transpiration and Production of Greenhouse Fruit Vegetable Crops*, Glasshouse Crops Research Station Report, Naaldwijk.

Nelson, P.V. (1985) *Greenhouse Operation and Management*, Prentice–Hall, New Brunswick, NJ.

Philipp, W.D., Grauer, U. and Grossmann, F. (1984) Erganzende Untersuchugen zur biologischen und integrierten Bekampfung von Gurkenmehltau unter Glas durch *Ampelomyces quisqualis, Zeitschrift für Pflanzenkrankheiten und Pflanzenschutz* 93, 384–391.

Reuveni, R. and Raviv, M. (1992) The effect of spectrally-modified polyethylene films on the development of *Botrytis cinerea* in greenhouse grown tomato plants, *Biological Agriculture & Horticulture* 9, 77–86.

Reuveni, R. and Raviv, M. (1997) Control of downy mildew in greenhouse-grown cucumbers using blue photoselective polyethylene sheets, *Plant Disease* 81, 999–1004.

Reuveni, R., Raviv, M. and Bar, R. (1989) Sporulation of *Botrytis cinerea* as affected by photoselective sheets and filters, *Annals of Applied Biology* 115, 417–424.

Sasaki, T., Honda, Y., Umekawa, M. and Nemoto, M. (1985) Control of certain diseases of greenhouse vegetables with ultraviolet-absorbing vinyl film, *Plant Disease* 69, 530–533.

Stanghellini, M.E. and Rasmussen, S.L. (1994) Hydroponics a solution for zoosporic pathogens, *Plant Disease* 78, 1130–1137.

Stenseth, C. (1979) The effect of temperature and humidity on the development of *Phytoseioulus persimilis* and its ability to regulate populations of *Tetranychus urticae* (Acarina: Phytoseiidae, Tetranychidae), *Entomophaga* 24, 311–317.

Tahvonen, R. (1982) The suppressiveness of Finnish light coloured sphagnum peat, *J. of the Scientific Agricultural Society of Finland* **54**, 345–356.

Tognoni, F. and Serra, G. (1989) The greenhouse in horticulture: The contribution of biological research, *Acta Horticulturae* **245**, 46–52.

van Lenteren, J.C. (1995) Integrated pest management in protected crops, in D. Dent (ed.), *Integrated Pest Management*, Chapman & Hall, London, pp. 311–343.

van Lenteren, J.C. (1997) Benefits and risks of introducing exotic macrobiological control agents into Europe, *OEPP/EPPO Bulletin* **27**, 15–27.

van Roermund, H.J.W. and van Lenteren, J.C. (1998) Simulation of the whitefly-*Encarsia formosa* interaction, based on foraging behaviour of individual parasitoids, in J. Baumgaertner, P. Brandmayr and B.F.J. Manly (eds.), *Population and Community Ecology for Insect Management and Conservation*, Proceedings 20th International Congress of Entomology, Florence, Italy, 25–31 August 1996, Balkema, Rotterdam, pp. 175–182.

Wearing, C.H. (1988) Evaluating the IPM implementation process, *Annual Review of Entomology* **33**, 17–38.

Wittwer, S.H. (1981) Advances in protected environments for plant growth, in *Advances in Food Producing Systems for Arid and Semi-arid Lands*, Academic Press, New York.

Wittwer, S.H. and Castilla, N. (1995) Protected cultivation of horticultural crops worldwide, *HortTechnology* **5**, 6–23.

Yunis, H., Shtienberg, D., Elad, Y. and Mahrer, Y. (1994) Qualitative approach for modelling outbreaks of grey mould epidemics in non-heated cucumber green-houses, *Crop Protection* **13**, 99–104.

CHAPTER 2

VIRAL DISEASES

Enrique Moriones and Marisol Luis-Arteaga

2.1. Introduction

Viruses are a major problem in greenhouse crops especially in temperate regions. However, most efforts in programmes for integrated pest and disease management are focused on pest and fungal or bacterial disease control and few recommendations are given for viral diseases. In general, viruses are not considered at all or are treated in a very simplistic manner. The main reason for this is the lack of information about viral disorders to give recommendations to deal with plant virus problems. In addition, in contrast to pests, fungi, or bacteria, no direct control methods can be used against viruses. Nevertheless, in recent years a significant progress in knowledge on plant viruses has occurred and valuable information has been obtained that will facilitate the development of control strategies. Because of difficulties and costs of reducing the spread of viruses by controlling their vectors and sources of infection, the introduction of resistance to a particular virus into commercially useful cultivars is the best control method but, unfortunately, the exception. Most virus management programmes involve the integration of indirect measures to avoid or reduce the sources of infection and dispersal of the virus, or the minimization of the effect of infections on crop yield. When confronted to a virus problem, the understanding of the ecology and epidemiology of the disease will provide the information needed to make strategic decisions for virus disease control.

In many circumstances control strategies are based on the dispersal procedures used by viruses in nature and similar control measures are recommended for viruses with equivalent dispersal manners. Therefore, virus dispersal mechanisms and the deduced control methods will be briefly reviewed in the next section before major diseases caused by plant viruses in protected crops are described.

2.2. Plant Virus Dispersal Mechanisms

The ability of a virus to be disseminated and perpetuated in time and space depends upon which methods are used for dispersal. Figure 2.1 summarizes the main transmission mechanisms of plant viruses; one or several of them can be exploited by a specific virus. The knowledge about the main dispersal procedures of a virus in nature will provide a means to prevent and control viral diseases: to minimize sources of infection, to reduce dissemination during growing practices, and/or to limit spread by vectors. Some aspects of virus dispersal and their importance in virus control are analysed below.

R. Albajes et al. (eds.), Integrated Pest and Disease Management in Greenhouse Crops, 16-33.
© 1999 *Kluwer Academic Publishers. Printed in the Netherlands.*

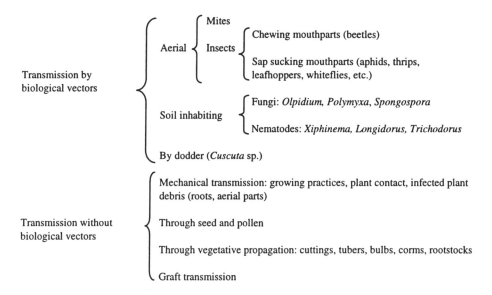

Figure 2.1. Main transmission strategies used by plant viruses.

2.2.1. SOURCES OF INFECTION

As a general rule, virus-infected plants are sources for secondary spread by mechanical or biological vector means and, therefore, should be eliminated as soon as possible. When existing, mechanical transmission is one of the most dangerous dispersal methods for viruses in protected crops due to the frequent handling of plants during the intensive cropping practices. Some viruses are extremely important in protected crops because of their efficient transmission by mechanical inoculation during cultural operations. If plants infected with some of these viruses are suspected to be present in a crop, secondary spread can be reduced by adequate treatment of hands and implements during plant handling. In these cases, plant debris in soil and greenhouse structures are important sources for primary infections in subsequent sensitive crops and, therefore, as long as possible, they should be eliminated and soil and structures disinfected.

The propagation material used for planting can be a very effective means of introducing viruses into a crop at an early stage, giving randomized foci of infection within the planting. If other transmission methods (e.g. mechanical, insects) are coupled, which may rapidly spread the virus within the crop, then infected seeds, plantlets, etc. can be of significant importance in the epidemic of the disease. In these cases, certified virus-free material should be used as the basis to control the virus.

Approximately 18% of the known plant viruses are seed-transmitted in one or more hosts (Mink, 1993; Johansen *et al.*, 1994). The rate of seed transmission is very variable depending on the virus/host combination and is not necessarily a good indicator of the epidemiological importance; low transmission rates combined with efficient secondary spread can be very important epidemically. Tolerance levels in a seed certification programme will depend, therefore, on the kind of secondary spread. For example, only very low infection levels are permitted in lettuce seed lots for an effective control of lettuce mosaic virus (LMV) because of its efficient secondary spread by aphids; good control was obtained in California if less of one seed in 30,000 was infected (Grogan, 1980; Dinant and Lot, 1992).

For many vegetatively propagated crops like ornamentals (carnation, tulip, etc.) the main virus sources are infected plants themselves and their vegetative derivatives (cuttings, tubers, bulbs, corms, rootstocks). In these cases, control may be done by using virus-free stocks and certification schemes to produce propagation material free of virus.

Soil may be another source of virus infection. Soilborne viruses can be transmitted by fungi or nematodes or can have no biological vector like tobamoviruses, that are very stable and are maintained in infected plant debris mixed with the soil. Control usually is through soil disinfection if no resistant cultivars are available.

The maintenance of virus-sensitive crops continuously throughout the year will ensure the permanent presence of significant levels of inoculum and, then, of virus infection. Therefore, crop rotations should incorporate non-sensitive species. However, although a rupture of the infection cycle is done, the presence of alternative hosts for the virus in the surroundings of the protected crop can be of special relevance to perpetuate the virus. The management of these hosts will help to the control of the virus.

2.2.2. VECTOR TRANSMISSION

Many important viruses in protected crops are transmitted from plant to plant by invertebrates. Sap-sucking insects are the main vectors, mostly Homoptera, and among them, aphids are the most important, transmitting 43% of known viruses.

Control of insect-transmitted viruses has been traditionally done by spraying insecticides to reduce the vector populations. However, the effectiveness of treatments in controlling the virus depends on virus/vector transmission relationships. Table 2.1 summarizes the main properties of the different kinds of relationships based on the feeding times needed by the vector to acquire (*acquisition time*) and inoculate (*inoculation time*) the virus, on the *latent period* from acquisition until the vector is able to transmit the virus, and on the *retention time* during which the vector remains infective following inoculative feeding without further access to the virus. This classification is mainly based on aphid-transmitted viruses. No evidence for virus in hemocoel or salivary system exists in the *noncirculative* transmission. In the *circulative transmission*, virus is acquired by feeding, enters the hemocoel via the hindgut, circulates in hemolymph, and enters the salivary gland. Inoculation results from transport of virus into the salivary duct, and introduction of saliva into the plant during feeding. If virus multiplies in the insect cells then the transmission is called *propagative*.

Insecticide treatments may be ineffective in controlling nonpersistently-transmitted viruses (short acquisition and inoculation times, no latent period, Table 2.1) because acquisition, latent, and inoculation times are so short that the virus is acquired and transmitted before the vector can be affected by most insecticides. However, especially in protected crops, chemical treatments can help to reduce the overall vector populations and therefore secondary spread of the disease. For nonpersistently-transmitted viruses, oils or tensioactive film-forming products have been reported to be effective in controlling virus acquisition and inoculation in outdoor crops. Insecticidal treatments used to control semipersistently- (long acquisition and inoculation times, no latent period, Table 2.1) or circulatively- (long acquisition and inoculation times, latent period, Table 2.1) transmitted viruses can be effective in controlling the virus because longer acquisition, inoculation and/or latent times are needed, and the vector may die before the virus can be transmitted. In any case, it should be noted that the small percentages of insects that usually survive the treatments are enough to cause important infections if virus sources are present. Accurate knowledge of disease epidemiology in a certain region will provide information about the critical periods of infection, which will facilitate decisions on when treatments should be done, or the adjustment of planting dates to avoid high vector populations in young plantings (Zitter and Simons, 1980).

TABLE 2.1. Classification of virus/vector relationships based on time needed for acquisition and inoculation of the virus, time after an acquisition feed for which the vector is unable to transmit (latent period), and time of retention of inoculativity following inoculative feeding, without further access to virus

	Noncirculative		Circulative (Persistent)	
	Nonpersistent	Semipersistent	Nonpropagative	Propagative
Acquisition time	Seconds	Minutes/hours	Minutes/hours	Minutes/hours
Latent period	0	0	Hours/days	Hours/days
Inoculation time	Seconds	Minutes/hours	Hours	Hours
Retention time	Minutes	Hours	Days	Entire life

2.3. Major Virus Diseases in Greenhouse Crops

Table 2.2 summarizes the characteristics of the main virus species that cause diseases in protected crops, for which comprehensive reviews are available (Smith *et al.*, 1988; Dinant and Lot, 1992; German *et al.*, 1992; Coffin and Coutts, 1993; Shukla *et al.*, 1994; Murphy *et al.*, 1995; Brunt *et al.*, 1996). Some of these species have been further reviewed in the text.

2.3.1. APHID-TRANSMITTED VIRUSES

Cucumber Mosaic Virus (CMV)

Description. CMV is the type species of the genus Cucumovirus of the family Bromoviridae of plant viruses. CMV virions are 29 nm icosahedral particles that

TABLE 2.2. Major virus species affecting protected crops

Natural transmission[1]	Family	Genus[2]	Species[2]	Achronym	Main infected crops[3]
Aphids (np)	Bromoviridae	Cucumovirus	**Cucumber mosaic virus**	CMV	cu, le, me, pe, to, wa, zu
Aphids (np), seed (be)	Potyviridae	**Potyvirus**	Bean common mosaic virus	BCMV	be, fa
Aphids (np)	Potyviridae	Potyvirus	Bean yellow mosaic virus	BYMV	be, fa
Aphids (np), seed (le)	Potyviridae	Potyvirus	Lettuce mosaic virus	LMV	le
Aphids (np)	Potyviridae	Potyvirus	Potato virus Y	PVY	pe, to
Aphids (np)	Potyviridae	Potyvirus	Papaya ringspot virus-W	PRSV-W	cu, me, wa, zu
Aphids (np)	Potyviridae	Potyvirus	Watermelon mosaic virus2	WMV2	cu, me, wa, zu
Aphids (np)	Potyviridae	Potyvirus	Zucchini yellow mosaic virus	ZYMV	cu, me, wa, zu
Aphids (np)	Potyviridae	Potyvirus	Zucchini yellow fleck virus	ZYFV	cu, me, wa, zu
Aphids (p)	–[4]	**Luteovirus**	Beet western yellows virus	BWYV	le, cu, pe, to, wa, zu
Aphids (p)	–	Luteovirus	Cucurbit aphid-borne yellows virus	CABYV	cu, me, zu
Thrips (p)	Bunyaviridae	Tospovirus	**Tomato spotted wilt virus**	TSWV	fa, le, pe, pea, to
Bemisia tabaci (p)	Geminiviridae	Geminivirus	**Tomato yellow leaf curl virus**	TYLCV	to
Bemisia tabaci (sp)	–	–	Cucumber vein yellowing virus	CVYV	cu, me
Bemisia tabaci	–	**Closterovirus**	Lettuce chlorosis virus	LCV	le
Bemisia tabaci (sp)	–	Closterovirus	Lettuce infectious yellows virus	LIYV	le, me, wa, zu
Bemisia tabaci (sp)	–	Closterovirus	Cucurbit yellow stunting disorder virus	CYSDV	cu, me

[1]np, non persistent; sp, semipersistent; p, persistent
[2]In bold are indicated those species or virus genus that are reviewed in this chapter
[3]be, French bean; cu, cucumber; eg, eggplant; fa, faba bean; le, lettuce; me, melon; pe, pepper; to, tomato; wa, watermelon; zu, zucchini squash
[4]Not defined

TABLE 2.2. Major virus species affecting protected crops (cont.)

Natural transmission[1]	Family	Genus[2]	Species[2]	Achronym	Main infected crops[3]
Trialeurodes vaporariorum (sp)	–[4]	Closterovirus	Beet pseudoyellows virus	BPYV	cu, me, le
Trialeurodes vaporariorum (sp)	–	Closterovirus	Tomato infectious chlorosis virus	TICV	to
Trialeurodes vaporariorum, *Trialeurodes abutiloneus*, *Bemisia tabaci*	–	Closterovirus	Tomato chlorosis virus	ToCV	to
Beetles (np)	Comoviridae	Comovirus	Squash mosaic virus	SqMV	me, wa, zu
Olpidium bornovanus	Tombusviridae	Carmovirus	**Melon necrotic spot virus**	MNSV	cu, me, wa
Olpidium brassicae	–	–	Lettuce big vein virus	LBVV	le
Olpidium brassicae	–	–	Lettuce ring necrosis virus	LRNV	le
Fungus	–	–	Pepper yellow vein disease		pe
Mechanical	–	**Tobamovirus**	Tobacco mosaic virus	TMV	pe, to
Mechanical, seed (to)	–	Tobamovirus	Tomato mosaic virus	ToMV	pe, to
Mechanical, seed (pe)	–	Tobamovirus	Pepper mild mottle virus	PMMV	pe
Mechanical, seed (cu)	–	Tobamovirus	Cucumber green mottle mosaic virus	CGMMV	cu, me, wa
Mechanical, fungus	–	Potexvirus	Potato virus X	PVX	to
Unknown (soil)	Tombusviridae	Tombusvirus	Tomato bushy stunt virus	TBSV	eg, pe, to
Unknown (soil), seed (cu)	Tombusviridae	Tombusvirus	Cucumber leaf spot virus	CLSV	cu

[1]np, non persistent; sp, semipersistent; p, persistent

[2]In bold are indicated those species or virus genus that are reviewed in this chapter

[3]be, French bean; cu, cucumber; eg, eggplant; fa, faba bean; le, lettuce; me, melon; pe, pepper; to, tomato; wa, watermelon; zu, zucchini squash

[4]Not defined

encapsidate a single-stranded RNA genome of messenger sense divided in three molecules, RNA 1, 2 and 3 (1.3×10^6, 1.1×10^6, and 0.8×10^6 daltons, respectively). Some CMV isolates encapsidate an additional small RNA called satellite RNA (0.1×10^6 daltons), that depends on virus for replication, encapsidation and movement. RNA satellites are able to modulate the symptoms induced by CMV (Palukaitis *et al.*, 1992).

A great variability among CMV isolates has been reported. According to biological properties of symptomatology, thermosensitivity *in vivo*, molecular and serological characteristics, most CMV isolates have been assigned to two main groups.

Transmission, Host Range and Diseases. In nature, CMV is transmitted in a nonpersistent manner by more than 60 aphid species including *Aphis gossypii* Glover, *Macrosiphum euphorbiae* (Thomas) and *Myzus persicae* (Sulzer). Variable rates of seed transmission have been described in 20 species including some vegetable crops like bean or spinach, or weeds like *Stellaria media* Cyrill. There is no evidence of seed transmission in cucurbits. CMV can be mechanically transmitted in experimental conditions.

CMV has an extremely wide host range that comprises more than 1000 species of dycotyledons and monocotyledons. Host range includes many important vegetable crops like melon, cucumber, zucchini squash, watermelon, tomato, pepper, eggplant, lettuce, carrot, celery, spinach, pea, etc.; ornamentals like anemone, aster, dahlia, delphinium, geranium, lily, periwinkle, primula, petunia, viola, zinnia, etc.; and woody and semiwoody plants like banana, ixora, passion fruit, etc. Symptoms are extremely variable depending on the CMV isolate, host species or cultivar, plant age at infection time, and environmental conditions. Early infected plants can show marked stunting. Symptoms in leaves are mosaic, mottle and/or distortion. Necrosis is induced by certain isolates. Flower abortion and fruit discoloration and malformations are caused.

Economic Importance and Control. CMV is distributed worldwide, predominantly in temperate regions but with increasing importance in tropical countries. It causes serious diseases in many important crops grown in the open but also in protected conditions (tomato, pepper, cucurbits, etc.) (Jordá *et al.*, 1992). Yield reductions are mainly due to decreased fruit set, and production of non-marketable fruits because reduced size, or presence of symptoms like mosaics, malformations or necrosis. Control of CMV is difficult because of the wide host range and its rapid natural transmission by aphids. Integrated control measures are recommended in protected crops to reduce CMV incidence: (i) elimination of infected plants; (ii) avoidance of aphid entrance in the greenhouse by covering entrances with aphid-proof nets; (iii) reduction of aphid populations by using insecticides; (iv) use of virus-free seeds (for example in bean and spinach); and (v) elimination of alternative spontaneous hosts present in and around the crop. Resistance to CMV is available in cucumber and programmes are in course in melon using Korean and Chinese varieties. Sources of resistance or tolerance have been found in most cultivated or related species. However, in most cases resistance or tolerance is not absolute, and is overcome by some CMV species. Aphid vector tolerance or resistance incorporated in the plant can be combined with other control methods. Transgenic melon, cucumber and squash plants expressing the coat protein gene of CMV offer a good level of resistance to several strains of the virus.

Potyvirus Genus

Description. The Potyvirus genus of the family Potyviridae is by far the largest of the plant virus groups. Many members cause important economic losses in protected crops and can be a major limiting factor for production. Virus particles are elongated and flexuous (680–900 × 11 nm) with one molecule of messenger sense single-stranded RNA (3.0–$3.5 × 10^6$ daltons) attached covalently to a protein. The genomic RNA codes for a large polyprotein that is proteolytically cleaved to yield the mature viral proteins. Virus infections are associated with characteristic cytoplasmic and nuclear inclusions, pinwheels, bundles and laminated aggregates (Shukla *et al.*, 1994).

Transmission, Host Range and Diseases. Potyviruses are transmitted in nature by aphids in a non-persistent manner. Some aphid species (especially those of the genera *Myzus, Aphis* and *Macrosiphum*) are associated with high virus incidences in crops. Seed transmission is important epidemiologically in certain potyviruses, like bean common mosaic virus (BCMV) in French bean, or LMV in lettuce. Potyviruses can be transmitted experimentally by mechanical inoculation.

In nature, most potyviruses have relatively narrow host ranges, few species within one genus or closely related genera; for example: BCMV is restricted to *Phaseolus* species; potato virus Y (PVY) to members of the Solanaceae; watermelon mosaic virus2 (WMV2), zucchini yellow mosaic virus (ZYMV) and zucchini yellow fleck virus (ZYFV) mainly to species of the Cucurbitaceae; and LMV to species mainly in Compositae.

Potyviruses can induce severe diseases in important crops. Symptoms may vary depending on host species, virus strain, environmental conditions and plant age at infection time. Potyviruses like ZYMV, WMV2 and papaya ringspot virus-W strain (PRSV-W) can cause severe diseases in zucchini, squash, melon, cucumber and watermelon, inducing stunting, chlorosis, mosaic, leaf malformation, flower abortion, and fruit and seed malformation. Vein clearing, mosaic, yellow mottling and growth reduction are often observed in LMV infections of lettuce, endive and spinach. Legume infecting potyviruses like BCMV cause abnormal formation of seeds that are smaller, discoloured and/or distorted.

Economic Importance and Control. The Potyvirus genus is the most devastating among plant viruses. Damaging members like BCMV, bean yellow mosaic virus (BYMV), ZYMV, WMV2, PRSV-W and PVY are spread worldwide and cause economically important problems where present. Several authors reported losses up to 100% in squash, cucumber and watermelon caused by ZYMV. Potyviruses are mainly a problem in outdoor crops, however, can also be a severe threat in protected crops.

Control should be done by an adequate management of the crop, integrating different control measures: if seed-transmitted, the use of certified virus-free seeds is the basis for effective control; use of virus-free plantlets will avoid primary infections; because transmitted in a nonpersistent manner, spraying insecticides is not effective for preventing virus spread, however, effective control has been obtained in some cases by spraying with light mineral oils in outdoor crops. Successful breeding programmes for resistant cultivars have been done in lettuce to LMV (resistance breaking strains have recently been described), in French bean to BCMV, and in melon to PRSV-W. Transgenic approaches have also been explored, overcoming difficulties associated with conventional breeding methods. Cross protection using an attenuate poorly aphid-transmissible strain of ZYMV (ZYMV WK) have been successfully used to control ZYMV in cucumber, melon and squash.

Luteovirus Genus

Description, Transmission, Host Range, Diseases and Economic Importance. There are a number of yellowing diseases transmitted in nature by aphids that are caused by viruses in the Luteovirus genus. This is the case of beet western yellows (BWYV) and cucurbit aphid-borne yellows (CABYV) viruses. Viral particles are 25–30 nm icosahedral, and encapsidate a monopartite, single stranded, messenger sense RNA genome. Transmission in nature is by aphids in a circulative, nonpropagative, persistent manner. BWYV infects lettuce, cucumber, watermelon, squash, sugarbeet, carrot, spinach, pepper and tomato, symptoms being mild chlorotic spotting, yellowing, thickening and brittleness of older leaves. It has been reported in North America, Europe and Asia, and is probably distributed worldwide. CABYV causes a yellowing disease of melon, cucumber and zucchini squash; symptoms are initial chlorotic patches, leaf thickening and general bright yellowing of leaves. In melon and cucumber important yield losses are reported, due to reduced number of fruits per plant caused by flower abortion but not by altering fruit shape or quality. It was first described in France in outdoor and protected crops and has been found through the Mediterranean area, Asia, Africa and California.

Control. Disease management should be by integrating measures to reduce aphid populations within the greenhouse via avoidance of insect entrance (nets in windows) and chemical spraying, with measures to reduce infection foci (virus-free planting material, elimination of infected plants). Sources of resistance have been found for CABYV in melon germplasm, and for BWYV in lettuce (Dogimont *et al.*, 1996).

2.3.2. WHITEFLY-TRANSMITTED VIRUSES

Tomato Yellow Leaf Curl Virus (TYLCV)

Description. TYLCV is a member of the geminivirus genus of plant viruses whose virions have twin isometric particle morphology that encapsidate a circular, single-stranded, monopartite DNA genome. Based on its transmission by the whitefly *Bemisia tabaci* (Gennadius) to dycotiledons, TYLCV belongs to the subgroup III of geminiviruses. Similar to other monopartite geminiviruses of this group, TYLCV genome contains six partially overlapping open reading frames (ORFs) organized bidirectionally, with two ORFs (V1 and V2) in the virion-sense, and four (C1, C2, C3, and C4) in the complementary sense. These ORFs encode proteins involved in replication, movement, transmission and encapsidation of the virus, and are separated by an intergenic region of approximately 300 nucleotides that contains signals for replication and transcription of the viral genome.

Transmission, Host Range and Diseases. TYLCV is transmitted from plant to plant by *B. tabaci* in a circulative manner (Mehta *et al.*, 1994); propagation in insect cells is still under discussion. TYLCV has a very narrow host range that covers some solanaceae species like tomato, *Datura stramonium* L. and different *Nicotiana* spp., and has also been described in French bean and *Malva parviflora* L. (Mansour and Al-Musa, 1992; Cohen and Antignus, 1994). In nature, TYLCV-caused diseases mainly affect to tomato crops. Symptoms in tomato consist in stunting, curling of leaflet margins with or without yellowing, reduction in leaf size and flower abortion.

Economic Importance and Control. TYLCV causes devastating damages in tomato crops of the Mediterranean basin, subtropical Africa and Central America. Losses are caused by reduced fruit yield and by the limitation of the economically feasible growing areas and periods. Effective control through crop management measures to avoid the vector and inoculum sources is possible in greenhouse crops. In the semiprotected crops typical of the Mediterranean regions, chemical control of vectors is ineffective to limit the spread of TYLCV. In these cases, control should be based on crop management following recommendations derived from the epidemiological knowledge of the disease and/or the use of the resistant/tolerant cultivars commercially available.

Clostero and Clostero-like Viruses

Description, Transmission, Host Range, Diseases and Economic Importance. In recent years there is an emerging threat in worldwide agriculture, particularly in temperate regions, that is caused by a number of viruses that are transmitted by whiteflies and induce yellowing symptoms in plants. This is probably related to the increasing importance of whitefly populations worldwide and to changes in the relative predominance of existing species. These viruses are not generally well characterized, however most of them seem to be members of the Closterovirus genus of plant viruses. This is the case of beet pseudo yellows virus (BPYV) and tomato infectious chlorosis virus (TICV), transmitted by *Trialeurodes vaporariorum* (Westwood), and of cucumber yellow stunting disorder virus (CYSDV), lettuce infectious yellows virus (LIYV) and lettuce chlorosis virus (LCV), transmitted by *B. tabaci* and *Bemisia argentifolii* Bellows & Perring (Célix *et al.*, 1996; Duffus, 1996a,b). A semipersistent transmission manner has been demonstrated in certain cases. Whitefly-transmitted closteroviruses have flexuous particles of variable length depending on species (900 × 12 nm). The genome is composed by two molecules of single stranded, messenger sense RNA, with a size of about 8 kilobases each. This is opposed to the monopartite genome characteristic of the aphid-transmitted closteroviruses such as beet yellows (BYV) or citrus tristeza (CTV) viruses. Most of these viruses have been first described in USA and cause important diseases in outdoor and protected crops. Symptoms usually consist in interveinal yellowing of the leaves, stunting and/or necrosis. LIYV infects lettuce, sugar beet, melon, squash, watermelon and carrot; yield losses of up to 50–75% occur in lettuce affected crops. TICV was found infecting field and greenhouse tomato crops in California. LCV infects lettuce crops and does not infect cucurbits. CYSDV is present in the Mediterranean area, causing disease in cucurbits, and has not been described in America.

Control. Integrated management of the disease in protected crops should be based on the early elimination of primary infected plants, avoidance of entrance of whiteflies, and rationale insecticide treatments to reduce overall vector populations in the greenhouse. In melon, resistance to BPYV has been described few years ago and, recently, to CYSDV (Gómez-Guillamón *et al.*, 1995).

2.3.3. THRIPS-TRANSMITTED VIRUSES

Tomato Spotted Wilt Virus (TSWV)

Description. TSWV is the type species of the Tospovirus genus of the family Bunyaviridae. TSWV has isometric, membrane-bound particles of approximately 80 nm in diameter that contain two ambisense, S (small) and M (medium), and one negative sense, L (large) linear single stranded RNA segments. The L RNA encodes the viral RNA polymerase, the M RNA encodes a non-structural (NsM) protein and a precursor to the G1 and G2 glycoproteins associated with the lipid membrane of the virus particle, and the S RNA encodes an additional non-structural (NsS) protein and the nucleocapsid (N) protein.

Transmission, Host Range and Diseases. TSWV is transmitted by several species of thrips of which *Frankliniella occidentalis* (Pergande) is the most important worldwide. Transmission is circulative and propagative and is unique in that the virus is only acquired by first stage larvae and is transmitted by second stage larvae and adults. Adults are the most important epidemiologically because are more mobile and remain viruliferous for their entire life (German *et al.*, 1992; Aramburu *et al.*, 1997).

TSWV has a wide host range, infecting more than 250 species in 70 different families of both monocotyledons and dycotyledons including important cultivated species (Edwardson and Christie, 1986). The symptomatology vary from no symptoms to chlorotic or necrotic local lesions, ring spots, line patterns, mosaic, mottling, bronzing, chlorosis, necrosis, leaf or stem malformation, and stunting. Flower abortion is observed and fruits can exhibit malformation, necrosis and abnormal coloration. Symptoms vary depending on host-virus isolate combination, plant age at infection time and environmental conditions.

Economic Importance and Control. TSWV causes serious diseases worldwide in both outdoor and protected economically important crops. Significant yield losses are caused in vegetable crops like tomato, pepper or lettuce, and in different ornamental species.

Control of TSWV is difficult because of the wide host ranges of both the virus and the vector and the efficient natural transmission by thrips. The use of insecticides to reduce virus incidence by controlling the vector is ineffective and crop management practices are difficult to implement. In this situation, the use of resistant cultivars is the best solution. Genetic resistance to TSWV has been difficult to identify, characterize and incorporate into commercial cultivars. Some important progress has been done in this field in tomato, where resistant cultivars are available, and in pepper and lettuce. However, the durability of resistance depends upon the biological variability that seems to exist among TSWV isolates (Roca *et al.*, 1997). The development of genetically-engineered virus-resistant plants is also under investigation. While efforts to produce resistant crops are going on, control in protected crops should be done integrating measures to limit the spread of the disease using certified virus-free vegetal material, roguing infected plants, and by biological or chemical control of thrips.

2.3.4. BEETLE-TRANSMITTED VIRUSES

Squash Mosaic Virus (SqMV)

Description. SqMV is a member of the Comovirus genus. Virions are 30 nm isometric particles that encapsidate two single-stranded RNA segments of 1.6×10^6 and 2.4×10^6 daltons, respectively. Comoviruses produce polyproteins from which the non-structural and structural proteins are generated by proteolitic cleavage. RNA1 carries all information for RNA replication, including the polymerase. Non-structural proteins include a putative cell-to-cell movement protein (encoded by RNA2), an NTP-binding motif-containing protein, a Vpg, a proteinase, and a polymerase. Two coat polypeptides are encoded by the RNA2. SqMV has several pathogenically different strains. Isolates could be grouped into 2 serological groups that differ in seed transmissibility and, to a certain extent, in host range and symptomatology (Campbell, 1971).

Transmission, Host Range and Diseases. SqMV is naturally transmitted by chewing insects, especially chrysomelid beetles, in a nonpersistent manner, and, like all comoviruses, is seed-borne (embryo-borne). Subgroup 1 isolates are seed-transmitted in pumpkin, squash, melon and watermelon, and subgroup 2 isolates in pumpkin and squash. Mechanical transmission easily occurs by plant contact and during cultural operations. Commercial and experimental seed lots generally yield about 1–10% infected seedlings but up to 94% transmission has been reported in melon. Natural host-range is narrow, restricted to the Cucurbitaceae, in which most species are susceptible. Experimentally, it also infects plants in other families. In cucurbits, SqMV cause symptomless infection or may induce ringspots, systemic mosaic, malformation and vein-banding, depending on virus strain, host and environmental conditions. Symptoms in fruits vary from small chlorotic areas to severe malformation with dark green areas. Isolates in subgroup 1 cause severe symptoms in melon, and mild ones in pumpkin; some strains infect watermelon. Subgroup 2 isolates do not infect watermelon and cause mild symptoms in melon and severe in pumpkin.

Economic Importance and Control. SqMV is widely distributed in the western hemisphere and also occurs in other countries throughout the world, probably introduced through seed lots. Control is achieved by testing seed lots to prevent seed transmission (Nolan and Campbell, 1984). If present, mechanical transmission should be avoided by elimination of symptomatic plants, and reducing handling and pruning transmission possibilities.

2.3.5. FUNGI-TRANSMITTED VIRUSES

Melon Necrotic Spot Virus (MNSV)

Description. MNSV belongs to the genus Carmovirus of the family Tombusviridae. Virions are 30 nm icosahedral particles that encapsidate a monopartite, single-stranded RNA genome (1.5×10^6 daltons). Two putative proteins (p29 and its read-through p89) are expressed from the genomic-length RNA, and another two (p7A and its read-through p14) from a 1.9 kilobases (kb) subgenomic RNA. Coat protein is expressed from a 1.6 kb subgenomic RNA (Riviere and Rochon, 1990).

Transmission, Host Range and Diseases. MNSV is naturally transmitted by the zoospores of the fungal vector *Olpidium bornovanus* (Sahtiyanchi) Karling (= *Olpidium radicale* Schwartz & Cook *fide* Lange & Insunza). Seed-transmission is reported: 10–40% of the seedlings from seeds of muskmelon affected plants became infected when grown in presence of *Olpidium* contaminated soil. Mechanical transmission is possible experimentally and has been reported during cultural operations. MNSV isolates have a narrow experimental host range mainly restricted to cucurbits and differ in the systemic infection of certain hosts: watermelon isolates failed to infect melon and cucumber plants systemically, melon isolates systemically infect melon plants but not watermelon and cucumber, and cucumber isolates infect melon and cucumber plants systemically and inoculated but not uninoculated leaves of watermelon plants. In melon, cucumber and watermelon, MNSV causes small chlorotic spots in young leaves that turn into necrotic spots and large necrotic lesions. In melon and watermelon, necrotic streaks appear along the stems and petioles and sometimes are the only visible symptoms. In fruits, discoloration, necrosis and malformation both externally and internally are observed.

Economic Importance and Control. MNSV has been found as a natural pathogen in melon, cucumber and watermelon protected crops in Japan, USA and Europe in which it causes significant yield losses. Apart from recommended control methods for soil-, seed- and mechanically-transmitted viruses (soil, seeds and tools disinfection, etc.), grafting on immune *Cucurbita ficifolia* Boucé rootstocks has been used in cucumber to control MNSV. Melon cultivars resistant to this virus are commercially available.

2.3.6. MECHANICALLY-TRANSMITTED VIRUSES

Tobamovirus Genus

Description. The genus Tobamovirus of plant viruses includes species that cause devastating diseases in protected crops. Virions are elongated rigid rod-shaped particles about 300×18 nm that encapsidate one molecule of single-stranded RNA of messenger sense (2×10^6 daltons). The type member is tobacco mosaic virus (TMV): the genome contains five open reading frames, four of which encode proteins (126K, 183K, 30K and 17.5K) found *in vivo* that have been associated with replication, encapsidation, movement and symptoms induction. A fifth protein (54K) is obtained by *in vitro* translation but has not been found *in vivo*. Homologous genetic organization and genome expression is found in the tobamoviruses that have been sequenced to date.

Transmission, Host Range and Diseases. In nature, tobamoviruses are the most infectious and persistent disease agents; they are transmitted and easily spread between plants by contact, and during cultural operations, through contaminated implements. The viruses can survive over years in plant debris that may be source for new infections via the roots or aerial parts if infected remains are present in the greenhouse structures. In certain cases (Table 2.2), these viruses are seed-transmitted: the virus is carried in the external seed surface, testa, and sometimes in the endosperm (Johansen *et al.*, 1994). Seed samples with endosperm infection can remain infected for years. No natural vectors are known; presence in irrigation water has been reported for tomato mosaic virus (ToMV). Tobamoviruses are easily transmitted experimentally by mechanical inoculation.

Natural host range is very narrow, usually restricted to specific hosts; however, experimentally can be transmitted to numerous species of different families. For example, pepper mild mottle virus (PMMV) naturally infects pepper, ToMV tomato and pepper (Brunt, 1986), and cucumber green mottle mosaic virus (CGMMV) some cucurbits like cucumber, watermelon and melon, and spontaneous perennial hosts like *Lagenaria siceraria* (Molina) Standl. (Okada, 1986).

Tobamoviruses cause severe diseases in susceptible species especially in protected crops because of the intensive production that implies high density of plants and frequent cultural operations which favour mechanical transmission. PMMV induces a faint mosaic in pepper leaves whereas fruits are severely malformed with distorted coloration and often exhibit depressed necrotic areas. ToMV causes a wide range of symptoms on tomato depending on virus strain, cultivar, plant age at infection time, and environmental conditions: mottle or mosaics are observed in leaves, that are malformed, plants are stunted, and fruits show external mottling and, sometimes, internal browning. In pepper, symptoms vary with cultivar and can be mosaics, systemic chlorosis, necrotic local lesions, leaf abscission, and/or systemic leaf and stem necrosis. In cucurbits, CGMMV causes more or less prominent leaf symptoms (mosaic, mottling, malformation), stunting, flower abortion, and fruit mottling, distortion, and/or internal discoloration.

Economic Importance and Control. Tobamoviruses are a first order problem in protected crops. Most tobamoviruses are easily distributed worldwide via infected seeds. PMMV is one of the most destructive pathogens of protected pepper crops; infections may reach 100% of the plants and the yield of marketable fruit be drastically reduced. ToMV has been for years a virus of great economic importance in protected tomato crops; however, the development of resistant cultivars has reduced considerably the incidence of the disease, but it is still a serious threat where resistant cultivars are not grown. In pepper, ToMV can also cause severe losses on susceptible cultivars (Brunt, 1986).

In first term, control methods are addressed to eliminate or reduce primary inoculum sources. Virus-free seeds should be used: sanitation of seeds can be done by soaking seeds in different solutions of active reagents (trisodium phosphate, hydrochloric acid, sodium hypochlorite) or by dry heat treatment (Rast and Stijger, 1987). Removal of plant debris from previous susceptible crops and steam treatment of the soil and greenhouse structures will aid to avoid primary infections. Secondary spread can be reduced by washing hands and implements with soap and water before and during plant handling, and/or frequent dipping into skim milk solutions. Cross protection has been largely used in greenhouse tomato crops to control ToMV by inoculation of tomato seedlings with an attenuated strain obtained by Rast (1972) in The Netherlands, thus avoiding ulterior infection with virulent ToMV strains. Other solanaceous crops that are susceptible to the mild strain (like pepper) must not be grown in proximity. Resistant genes have been described and incorporated in commercial tomato against ToMV, and in pepper against different tobamoviruses [TMV, ToMV, PMMV, and paprika mild mottle virus (PaMMV)]. However, resistant breaking strains can be detected (Tenllado et al., 1997).

2.4. Current Perspectives for Plant Virus Control within Integrated Management of Greenhouse Crops

Greenhouse crops represent a singular case for disease management. They are closed systems where external exchanges are reduced to the minimum, although the intermediate situation present in the protected crops grown under the simple and less hermetic structures typical of the Mediterranean area, should also be considered. The most damaging viruses in protected crops are soilborne viruses [MNSV, PMMV, ToMV, tomato bushy stunt virus (TBSV)], or those imported via contaminated seed (TMV, ToMV, PMMV, CGMMV, SqMV, MNSV, BCMV, LMV, etc.), or contaminated plantlets. The precise knowledge about which virus problems are affecting in a specific crop, the dispersal mechanisms, and the epidemiology of the disease induced will help to make strategic management decisions within an integrated control strategy.

The means to prevent and control viral diseases based on the knowledge of their dispersal mechanisms have been discussed in Section 2.2. Other strategies for virus control are focused to the minimization of the impact of the infection on crop yield; breeding for resistance and cross protection are two of these strategies. When possible, the best control method against plant viruses would be the development of resistant cultivars (Sherf and Macnab, 1986). However, experience has shown that breeding for resistance or the development of transgenic plants is unlikely to give permanent solutions for any particular virus and crop. Variable virus populations may be present (Pink et al., 1992; Luis-Arteaga et al., 1996; Tenllado et al., 1997) and/or virus can mutate (Aranda et al., 1997) in the field with respect to virulence and the range of crops and cultivars they can infect. Cross protection is based in that mild virus strains can be used to protect plants against infection by severe strain(s) of the same virus. Basic criteria for selection of cross protection as a disease control strategy are well known (Fulton, 1986; Gonsalves and Garnsey, 1989). Mildness of a strain is usually relative to a certain target crop and this should be taken into account if cross protection want to be used in greenhouses where other crops that may be sensitive to the protective virus strain are grown simultaneously. The same applies for precautions to be taken to avoid dispersal of the mild strain to sensitive crops grown in the vicinity of the protected greenhouse crop. Due to possible virus mutations, the reversion of the mild strain used in the cross protection programme to a severe one must be continuously verified. When using cross protection, the risk of coinfection with other virus(es) that may have synergistic effects with the protective strain should also be evaluated. Cross protection alone is not enough to give a high level of control of the disease because protection depends on the homology of the severe strain and on challenge pressure (Gonsalves and Garnsey, 1986). Therefore, the combination of various virus management practices compatible with an integrated management of the greenhouse is often desirable. Indirect measures for virus control have been discussed, e.g.: (i) adjustment of planting dates to avoid high vector populations in young plantings if epidemiological data of the disease are available; (ii) use of virus-free propagation material; (iii) disinfection of soil and greenhouse structures; (iv) minimization of external entrance of insects; (v) rapid elimination of virus-infected plants; (vi) adequate plant handling; and (vii) avoidance of overlapping or continuous cultivation of sensitive species in the rotation.

References

Aramburu, J., Riudavets, J., Arnó, J., Laviña, A. and Moriones, E. (1997) The proportion of viruliferous individuals in field populations of *Frankliniella occidentalis*: Implications for tomato spotted wilt virus epidemics in tomato, *European J. of Plant Pathology* 103, 623–629.

Aranda, M.A., Fraile, A., Dopazo, J., Malpica, J.M. and García-Arenal, F. (1997) Contribution of mutation and RNA recombination to the evolution of a plant pathogenic RNA, *J. of Molecular Evolution* 44, 81–88.

Brunt, A.A. (1986) Tomato mosaic virus, in M.H.V. van Regenmortel and H. Fraenkel-Conrat (eds.), *The Plant Viruses. Vol. II: The Rod-shaped Plant Viruses*, Plenum Press, New York, pp. 181–204.

Brunt, A.A., Crabtree, K., Dallwitz, M.J., Gibbs, A.J. and Watson, L. (1996) *Viruses of Plants. Descriptions and Lists from the VIDE Database*. CAB International, Wallingford.

Campbell, R.N. (1971) *Squash Mosaic Virus*, CMI/AAB Descriptions of Plant Viruses No. 43, CMI, Kew.

Célix, A., López-Sesé, A., Almarza, N., Gómez-Guillamón, M.L. and Rodríguez-Cerezo, E. (1996) Characterization of cucurbit yellow stunting disorder virus, a *Bemisia tabaci*-transmitted closterovirus, *Phytopathology* 86, 1370–1376.

Coffin, R.S. and Coutts, R.H.A. (1993) The closteroviruses, capilloviruses and other similar viruses: A short review, *J. of General Virology* 74, 1475–1483.

Cohen, S. and Antignus, Y. (1994) Tomato yellow leaf curl virus, a whitefly-borne geminivirus, *Advances in Disease Vector Research* 10, 259–288.

Dinant, S. and Lot, H. (1992) Lettuce mosaic virus, *Plant Pathology* 41, 528–542.

Dogimont, C., Slama, S., Martin, J., Lecoq, H. and Pitrat, M. (1996) Sources of resistance to cucurbit aphid-borne yellows luteovirus in melon germ plasm collection, *Plant Disease* 80, 1379–1382.

Duffus, J.E., Lin, H-Y. and Wisler, G.C. (1996a) Tomato infectious chlorosis virus – A new clostero-like virus transmitted by *Trialeurodes vaporariorum, European J. of Plant Pathology* 102, 219–226.

Duffus, J.E., Lin, H-Y. and Wisler, G.C. (1996b) Lettuce chlorosis virus – A newly whitefly-transmitted closterovirus, *European J. of Plant Pathology* **102**, 591–596.

Edwardson, J.R. and Christie, R.G. (1986) Tomato spotted wilt virus, in J.R. Edwardson and R.G. Christie (eds.), *Viruses Infecting Forage Legumes,* Vol. III, Monograph Series No. 14, Florida Agricultural Experiment Stations, Gainesville, pp. 563–580.

Fulton, R.W. (1986) Practices and precautions in the use of cross protection for plant virus control, *Annual Review of Phytopathology* **24**, 67–81.

German, T.L., Ullman, D.E. and Moyer, W.M. (1992) Tospoviruses: Diagnosis, molecular biology, phylogeny, and vector relationships, *Annual Review of Phytopaholygy* **30**, 315–348.

Gómez-Guillamón, M.L., Torés, J.A., Soria, C. and Sesé, A.I.L. (1995) Searching for resistance to *Sphaeroteca fulginae* and two yellowing diseases in *Cucumis melo* and related *Cucumis* species, in G.E. Lester and J.R. Dunlap (eds.), *Proceedings Cucurbitaceae '94: Evaluation and Enhancement of Cucurbit Germplasm,* Gateway Printing, South Padre Island, Tex., pp. 205–208.

Gonsalves, D. and Garnsey, S.M. (1989) Cross-protection techniques for control of plant virus diseases in the Tropics, *Plant Disease* **73**, 592–597.

Grogan, R.G. (1980) Control of lettuce mosaic with virus-free seed, *Plant Disease* **64**, 446–449.

Johansen, E., Edwards, M.C. and Hampton, R.O. (1994) Seed transmission of viruses: Current perspectives, *Annual Review of Phytopathology* **32**, 363–386.

Jordá, C., Alfaro, A., Aranda, M.A., Moriones, E. and García-Arenal, F. (1992) An epidemic of cucumber mosaic virus plus satellite RNA in tomatoes in Eastern Spain, *Plant Disease* **76**, 363–366.

Luis-Arteaga, M., Rodríguez-Cerezo, E., Fraile, A., Sáez, E. and García-Arenal, F. (1996) Different tomato bushy stunt virus strains cause disease outbreaks on solanaceous crops in Spain, *Phytopathology* **86**, 535–542.

Mansour, A. and Al-Musa, A. (1992) Tomato yellow leaf curl virus: Host range and virus-vector relationships, *Plant Pathology* **41**, 122–125.

Mehta, P., Wyman, J.A., Nakhla, M.K. and Maxwell, D.P. (1994). Transmission of tomato yellow leaf curl geminivirus by *Bemisia tabaci* (Homoptera: Aleyrodidae), *J. of Economical Entomology* **87**, 1291–1297.

Mink, G.I. (1993). Pollen- and seed-transmitted viruses and viroids, *Annual Review of Phytopathology* **31**, 375–402.

Murphy, F.A., Fauquet, C.M., Bishop, D.H.L., Ghabrial, S.A., Jarvis, A.W., Martelli, G. P., Mayo, M.A. and Summers, M.D. (eds.) (1995) Virus taxonomy: Classification and nomenclature of viruses, *Archives of Virology* Suppl. 10.

Nolan, P.A. and Campbell, R.N. (1984) Squash mosaic virus detection in individual seeds and seed lots of cucurbits by enzyme-linked immunosorbent assay, *Plant Disease* **68**, 971–975.

Okada, Y. (1986) Cucumber green mottle mosaic virus, in M.H.V. van Regenmortel and H. Fraenkel-Conrat (eds.), *The Plant Viruses. Vol. II: The Rod-shaped Plant Viruses,* Plenum Press, New York, pp. 267–281.

Palukaitis, P., Roossinck, M.J., Dietzgen, R.G. and Francki, R.I.B. (1992) Cucumber mosaic virus. Advances, *Disease Vector Research* **41**, 281–348.

Pink, D.A.C., Kostova, D. and Walkey, D.G.A. (1992) Differentiation of pathotypes of lettuce mosaic virus, *Plant Pathology* **41**, 5–12.

Rast, A.T.B. (1972) MH-16, an artificial symptomless mutant of tobacco mosaic virus for seedling inoculation of tomato crops, *Netherland J. of Plant Pathology* **78**, 110–112.

Rast, A.T.B. and Stijger, C.C.M.M. (1987) Disinfection of pepper seed infected with different strains of capsicum mosaic virus by trisodium phosphate and dry heat treatment, *Plant Pathology* **36**, 583–588.

Riviere, C.J. and Rochon, D.M. (1990) Nucleotide sequence and genomic organization of melon necrotic spot virus, *J. of General Virology* **71**, 1887–1896.

Roca, E., Aramburu, J. and Moriones, E. (1997) Comparative host reactions and *Frankliniella occidentalis* transmission of different isolates of tomato spotted wilt tospovirus from Spain, *Plant Pathology* **56**, 407–415.

Sherf, A.F. and Macnab, A.A. (1986) *Vegetable Diseases and their Control,* John Wiley & Sons, Inc., New York.

Shukla, D.D., Ward, C.W. and Brunt, A.A. (1994) *The Potyviridae*, CAB International, Wallingford.

Smith, I.M., Dunez, J., Lelliot, R.A., Phillips, D.H. and Archer, S.A. (1988) *European Handbook of Plant Diseases,* Blackwell Scientific Publications, Oxford.

Tenllado, F., García-Luque, I., Serra, M.T. and Díaz-Ruiz, J.R. (1997) Pepper resistance-breaking tobamoviruses: Can they co-exist in single pepper plants? *European J. of Plant Pathology* **103**, 235–243.

Zitter, T.A. and Simons, J.N. (1980) Management of viruses by alteration of vector efficiency and by cultural practices, *Annual Review of Phytopathology* **18**, 289–310

FUNGAL AND BACTERIAL DISEASES
Nikolaos E. Malathrakis and Dimitris E. Goumas

3.1. Introduction

Greenhouse cropping is the most intensive agricultural industry. It is suitable wherever land is limited or where early produce is required under adverse environmental conditions. Greenhouse cropping poses complex challenges in the field of plant protection. In such an intensive cropping system several factors, explained in other chapters of this volume, favour the development of a large number of fungal and bacterial diseases; if no proper control measures are taken in time, losses may be very high. This chapter provides information relevant to the diagnosis and biology of the pathogen, and epidemiology of several diseases, key information if control strategy is to be effective. Disease control is dealt with in other chapters and so is only briefly discussed here. Diseases are grouped arbitrarily and the main characteristics of each group are described. Due to the huge number of diseases reported in greenhouse crops, several of them, of minor or local interest, have been omitted. Additional information can be found in books dealing specifically with greenhouse diseases (Fletcher, 1984; Jarvis, 1992) or in books on vegetables or floral crop diseases (Strider, 1985; Sherf and Macnab, 1986; Blancard, 1988; Horst, 1989; Blancard *et al.*, 1991; Jones *et al.*, 1993).

3.2. Fungal Diseases

3.2.1. DAMPING OFF – CROWN AND ROOT ROTS

Plants in seedbeds may be diseased, either before or after their emergence from the soil, and the disease is called pre- or post-emergence damping off, respectively. In the first case, seedlings do not emerge in patches of the seedbeds. In the second case, plants rot quickly and drop down on the soil. Low temperatures and very wet soils, which delay the growth of the plants, favour infection. A large number of fungi may cause damping off, but *Pythium* spp., *Phytophthora* spp., *Fusarium* spp. and *Rhizoctonia solani* Kühn are the most common. Nowadays, due to the use of improved technology, damping off is no longer a severe disease in greenhouses (Sherf and Macnab, 1986; Blancard, 1988; Blancard *et al.*, 1991). However, root rots and crown rots are still destructive in soil, though not in soilless cultures (Davies, 1980). The most widespread diseases are as follows.

Pythium and Phytophthora Rots
Various *Pythium* spp. and *Phytophthora* spp. may damage the lower part of tomato, pepper, cucumber, carnation, poinsettia, gerbera, etc. both in soil and soilless cultures.
  In tomato a root and crown rot extending a considerable height above the soil level may

34

R. Albajes et al. (eds.), Integrated Pest and Disease Management in Greenhouse Crops, 34-47.
© 1999 *Kluwer Academic Publishers. Printed in the Netherlands.*

occur. The infected area has a dark discoloration and the pith is usually destroyed. *Phytophthora nicotianae* Breda de Haan var. *parasitica* (Dastur) G.M. Waterhouse is the most common pathogen. In pepper a similar disease caused by *Phytophthora capsici* Leonian is very common. Collar, stem and fruit rot as well as leaf spots may occur. In cucumber a soft rot of the young plants at the soil level may occur soon after transplanting. Infected tissues shrink and in wet weather a white mycelium develops. Infected plants wilt and die quickly. Poinsettia grown in pots also suffers from Pythium rot. Severe root rot, extending above ground in succulent plants, and quick death are the main symptoms. In cucumber and poinsettia, *Pythium ultimum* Trow, *Pythium irregulare* Buisman, *Pythium debaryanum* Auct. Non R. Hesse and *Pythium aphanidermatum* (Edson) Fitzp. are mostly involved (Tompkins and Middleton, 1950). Carnations infected by *Pythium* and *Phytophthora* species develop soft rot at the collar and in the root system, resembling Rhizoctonia stem rot.

Rhizoctonia Stem Rot (R. solani)
This infects a large number of plants, such as tomato, carnation, poinsettia, etc. causing symptoms resembling Pythium or Phytophthora rots. Rhizoctonia stem rot is mainly confined to the collar. Carnation is very susceptible. Infected plants show pale brown dry lesions, with circular rings, at soil level. Growth is stunted and leaves become dull green. Complete wilting soon follows. Strands of the pathogen develop on the lesions and stems break easily at the infected area (Parmeter, 1970).

All *Pythium* and *Phytophthora* species as well as *R. solani* are common soil inhabitants. They survive in the soil. Infection usually takes place at the time of planting and symptoms appear very soon in Pythium and Phytophthora rot or several weeks later in Rhizoctonia stem rot. *Rhizoctonia solani* may infect at moderate soil moisture levels, but *Pythium* spp. and *Phytophthora* spp. infect only in water-saturated soils (Strider, 1985).

Corky Root Rot of Tomato (Pyrenochaeta lycopersici *R. Schneider & Gerlach)*
The pathogen damages mostly tomato, but also eggplant, melon, etc. Initially tomato leaves turn dull green and growth is stunted. Later, leaves take on a bronze colour and curl downwards. Necrosis of the leaflets follows. Young roots are brown and poorly developed. Scattered lesions appear on the surface of the larger roots which become corky with cracks of different sizes. Yield may be severely reduced. The pathogen survives on the infected root debris due to the presence of minute sclerotia. It is a cool weather disease. In subtropical countries it progresses during the winter and plants start to recover by early spring (Ebben, 1974; Malathrakis *et al.*, 1983).

Crown and Root Rot of Tomato (Fusarium oxysporum *Schlechtend.:Fr. f. sp.* radicis-lycopersici *W.R. Jarvis & Shoemaker)*
In plastic greenhouses, a yellowing of the lower leaves appears in infected plants during late winter, when many fruits have already set. In severe infections the whole plant becomes chlorotic and wilts. A dry lesion up to 10 cm long appears on part of or all around the collar. There is a brown discoloration on the root system, predominantly at the end of the main root, the base of the stem and the vascular region of the central root.

A large number of microconidia, which disseminate the pathogen, appear on the infected stem. The fungus survives by chlamydospores which develop in the soil. The disease is favoured by cool weather (Jarvis *et al.*, 1975, 1983).

Black Root Rot of Cucurbits (Phomopsis sclerotioides *van Kestern*)
The disease has been recorded in several countries of northwestern Europe and elsewhere. It infects cucumber and melon, causing a brown rot in the cortical tissue of the root system. Soon a large number of sclerotia develop and the infected tissues turn black. Severely infected plants wither and die. Infection is favoured by cool weather. The pathogen survives in the soil for several years by means of sclerotia (Blancard *et al.*, 1991).

Control
Effective control of the above-mentioned diseases may be obtained selectively by the following means: (i) use of naturally or artificially suppressive substrates; (ii) early drenching by effective fungicides; (iii) soil disinfestation; (iv) use of resistant cultivars (cvs); (v) grafting on resistant rootstocks; and (vi) biological control (Ginoux *et al.*, 1978; Jarvis *et al.*, 1983; Hoitink and Fahy, 1986; Tjamos, 1992) (see Chapter 23).

3.2.2. WILTS

All major greenhouse crops suffer from one or more wilts. In several crops wilts are the main diseases due to the damage they cause and the difficulty of controlling them.

Fusarium Wilt (F. oxysporum)
The most common Fusarium wilts in greenhouses appear on: tomato [*Fusarium oxysporum* Schlechtend.:Fr. f. sp. *lycopersici* (Sacc.) W.C. Snyder & H.N. Hans.], cucumber (*Fusarium oxysporum* Schlechtend.:Fr. f. sp. *cucumerinum* J.H. Owen), melon (*Fusarium oxysporum* Schlechtend.:Fr. f. sp. *melonis* W.C. Snyder & H.N. Hans.), carnation [*Fusarium oxysporum* Schlechtend.:Fr. f. sp. *dianthi* (Prill. & Delacr.) W.C. Snyder & H.N. Hans.], gladiolus [*Fusarium oxysporum* Schlechtend.:Fr. f. sp. *gladioli* (L. Massey) W.C. Snyder & H.N. Hans.], cyclamen (*Fusarium oxysporum* Schlechtend.:Fr. f. sp. *cyclaminis* Gerlach) and chrysanthemum (*Fusarium oxysporum* Schlechtend.:Fr. f. sp. *chrysanthemi* G.M. Armstrong, J.K. Armstrong & R.H. Littrell).

Wilt, yellowing, chlorosis, drooping (mostly of the lower leaves), stunting and brown discoloration of the vascular bands up to the top of the stem are the dominant symptoms. Wilting of the lateral shoots and large lesions on the lower part of the stem are also common in Fusarium wilt of carnation and melon. All the above *F. oxysporum* formae have more than one race. Each of them infects cvs of one host but may colonize the root system of other plants as well. They survive in the soil for several years, due to the production of thick-walled chlamydospores, but inoculum is reduced over the years. Fusarium wilt in tomato, watermelon, carnation, cyclamen, chrysanthemum and gladiolus is favoured by higher temperatures than Fusarium wilt in melon (Walker, 1971; Nelson *et al.*, 1981; Strider, 1985; Sherf and Macnab, 1986).

Verticillium – Phialophora Wilt [Verticillium dahliae *Kleb.*, Verticilium albo-atrum *Reinke & Berthier*, Phialophora cinerescens *(Wollenweb.) van Beyma* (= Verticillium cinerescens *Wollenweb.*)]
This infects a huge number of plants and among them the majority of the plants grown in greenhouses. It is more severe in Solanaceae such as tomato, eggplant and pepper. Of the

floral crops, chrysanthemum seems to be more susceptible. Symptoms are very similar to those of Fusarium wilt. Verticillium wilt is favoured by moderate temperatures. *Verticillium dahliae*, which is more common, survives in the soil for many years due to the abundant production of black resistant microsclerotia, while *V. albo-atrum* survives by producing dark dormant mycelium. A similar wilt caused by *P. cinerescens* damages carnations in several areas (Strider, 1985; Sherf and Macnab, 1986).

3.2.3. POWDERY MILDEWS

Powdery mildews are very destructive of several greenhouse crops. The following are some of the powdery mildew fungi which most attack greenhouse-grown plants: (i) on cucurbits *Sphaerotheca fusca* (Fr.) Blumer. [= *Sphaerotheca fuliginea* (Schlechtend.:Fr.) Pollacci], *Erysiphe cichoracearum* DC. and *Leveillula taurica* (Lév.) G. Arnaud (only on cucumber); (ii) on solanaceous plants *L. taurica* and *Oidium lycopersicum* Cook & Massee (only on tomato); (iii) on roses *Sphaerotheca pannosa* (Wallr.:Fr.) Lév.; (iv) on begonia *Microsphaera begoniae* Sivan.; and (v) on gerbera *E. cichoracearum*.

Powdery mildew fungi, except for *L. taurica*, may attack all green tissues. Initially, white powdery spots, which enlarge and coalesce to cover large areas, are the dominant symptoms. *Leveillula taurica* infects only leaves. Light yellow or yellow-green spots on the upper leaf surface, which later become brown, and scarce white mould on the lower surface are the main characteristics. Powdery mildew-infected plant parts may be chlorotic and distorted. Premature defoliation and poor growth are common features of severely infected plants (Palti, 1971; Sitterly, 1978; Braun, 1995).

Infections take place by conidia. Under favourable conditions powdery mildew progresses rapidly. By the end of the season some powdery mildew fungi, such as *S. fuliginea*, *E. cichoracearum*, etc. may develop cleistothecia with ascospores, but these do not play an important role in the epidemiology of the disease (Braun, 1995).

Conidia are mostly discharged and transferred by wind currents. Animal pests may also disseminate conidia in greenhouse crops. Young conidia readily germinate on plant surfaces depleted of nutrients. The relative humidity (RH) favouring infection by powdery mildew fungi and development of the disease differs from species to species. For instance, high RH is more favourable for *S. fuliginea* than for *E. cichoracearum*. Therefore, the first fungus is more frequent on greenhouse cucurbits than the second. High RH may favour spore germination of powdery mildew fungi, but free water may be deleterious. RH at 97–99% is optimal for spore germination of *S. pannosa* and *S. fuliginea*. At RH below 75% spores of *S. pannosa* do not germinate, but mycelium development and sporulation may occur at RH as low as 21–22%. Powdery mildew fungi overwinter on cultivated plants or weeds, which survive in or outside the greenhouse (Coyier, 1985a,b).

Chemicals such as demethylation inhibitors (DMIs) (triadimefon, fenarimol, etc.), pyrimidines (ethirimol, bupirimate, etc.), pyrazophos and dinocap remain the main means of controlling powdery mildews in greenhouses. Biological control agents have also been effectively tested against *S. fuliginea* and *S. pannosa*. Finally, fully resistant cvs of melon and partially resistant cvs of long-type cucumber are available (Coyier, 1985b; Molot and Lecoq, 1986).

3.2.4. DOWNY MILDEWS

Downy mildews of tomato [*Phytophthora infestans* (Mont.) de Bary], cucurbit [*Pseudoperonospora cubensis* (Berk. & M.A. Curtis) Rostovzev], lettuce (*Bremia lactucae* Regel), rose (*Peronospora sparsa* Berk.) and snapdragon (*Peronospora antirrhini* J. Schröt.) are the most destructive in greenhouse-grown plants.

In tomato, leaves and young shoots are infected first. Fruit infection starts mostly near the stalk and spreads very quickly to the whole fruit. Infected tissues of fruits and shoots are firm and brown (Sherf and Macnab, 1986).

In cucurbits downy mildew appears as yellow, angular or circular spots on the upper surface of the mature leaves of the plant. Soon the tissues at the centre of the spots die and become light brown. Cucumber and melon are more susceptible than watermelon.

Downy mildew of lettuce causes scattered light-green to yellow spots on the upper leaf surface. Old spots become brown and dry up.

Downy mildew of rose damages all green plant parts, but leaves are more susceptible. Leaf infection resembles the effect of toxins. Infected leaves have purplish red to dark-brown irregular spots and shed readily (Strider, 1985). Snapdragon plants infected by *P. antirrhini* are stunted and the top internodes of the young plants are short. The borders of the lower leaves curl down and then dry. Eventually the entire plant dies (Garibaldi and Rapetti, 1981). A white fungal growth (brown for cucurbit downy mildew) on the infected tissues under moist conditions is typical of all downy mildews.

Plant infection takes place through stomata and mycelium develops intercellularly. Soon branched conidiophores are produced and protrude through the stomata. Infection progresses in the periphery of the spot which gradually enlarges. Conidiospores of downy mildews are ovoid and hyaline, except for *P. cubensis* which are brown. They are discharged by hygroscopic changes and disseminate in greenhouses by wind currents and water splashes. Initial infection may take place by spores transferred long distances on the wind. Abundant oospores of *P. antirrhini* develop on dead plant stems. Oospores of *P. sparsa* also very often develop on infected roses, whereas *P. cubensis* and *P. infestans* oospores are rare.

Phytophthora infestans survives on seed potato tubers and spreads to young potato plants after they have been planted. Inoculum is disseminated from potatoes to neighbouring tomato crops. Cucurbit downy mildew can infect all year round several species of cucurbits, grown either in greenhouses or open fields. There is evidence that *P. sparsa* survives as a dormant mycelium on the infected stems of roses. *Peronospora antirrhini* perennates as dormant thick-walled oospores in dead plant parts and soil (Garibaldi and Rapetti, 1981; Sherf and Macnab, 1986).

Free water on plant tissues is necessary for downy mildew fungi to cause infection. High RH is also required for good sporulation. *Peronospora antirrhini* is favoured by low temperature and high RH. Free water or high relative humidity is not often a factor limiting downy mildew development in plastic greenhouses. It seems that temperature is more critical. For instance, *P. cubensis*, with a high maximum temperature for development and infection, may, under certain conditions, infect all year round, whereas *P. infestans* and *P. sparsa* do not infect during the hot period of the year. Downy mildews complete a cycle within about 6-8 days. Thus, under favourable weather conditions they may have several cycles and spread rapidly (Palti and Cohen, 1980; Strider, 1985).

Chemical fungicides remain the major means of control of downy mildews. Dithiocarbamates, chlorothalonil and the systemic phenylamides (metalaxyl, etc.) are the most commonly used in greenhouses. There are some tomato cvs fully resistant to downy mildew and some partially resistant cucumber cvs suitable for greenhouses, but all rose cvs grown for cut flowers are susceptible to downy mildew. Ventilation of the greenhouses may also effectively prevent infection (Palti and Cohen, 1980; Fletcher, 1984; Strider, 1985).

3.2.5. BOTRYTIS DISEASES

Botrytis cinerea Pers.:Fr., *Botrytis tulipae* (Lib.) Lind and *Botrytis gladiolorum* Timmermans are *Botrytis* spp. that most damage greenhouse crops.

Botrytis cinerea causes grey mould on a large range of hosts, including nearly all the major greenhouse plants. All plant parts at different growth stages may be damaged. Due to the diversity of the infected plant parts, several types of symptoms appear on one or on various hosts. On young stems, leaves, flowers and fruits, initially water-soaked spots occur, which rapidly enlarge under favourable weather conditions. In tomato fruits green-white circular spots called "ghost spots" also appear. On hard plant parts, such as stems and collars, *B. cinerea* causes cankers and parts above them may die. These symptoms are very common on vegetables such as tomato, eggplant, pepper and cucumber. Infected tissues die soon and a grey mould which consists of conidiophores with clusters of spores develops on their surface. In plants, like tomato, black sclerotia develop inside the infected stems. *Botrytis cinerea* also causes very characteristic collar rot in lettuce. The infected plants usually develop large brown necrotic lesions on the stem near the soil surface and the lower leaves. The infection gradually progresses upwards. Infected plants may wither and die in a short time (Sherf and Macnab, 1986).

Botrytis tulipae causes tulip fire blight. Spots of various types on leaves and flowers, lesions on the stem, blossom blight and bulb rot are the dominant characteristics. *Botrytis gladiolorum* damages gladiolus and some other Iridaceae. Large spots on leaves and the stem, pinpoint spots on the flowers, neck rot and soft rot of corms are the most common symptoms. *Botrytis* spp. also infect all types of propagating material, which are either destroyed before planting out or become weak plants which may die before or after emergence. Finally, *Botrytis* spp. may cause severe post-harvest losses in plant products during storage or transportation (Trolinger and Stider, 1985).

Botrytis cinerea develops and sporulates profusely on any organic material. Spores are disseminated by wind over long distances or by water splashes. Healthy plants are infected through wounds, senescent tissues, directly through the epidermis and rarely through stomata. Symptoms may appear very quickly or infection may remain quiescent and symptoms appear later when tissues age or during storage. In greenhouses, initial infection depends on spores transferred from outdoors. Later, the inoculum established in the greenhouse is the main source of infection. In plants grown in non-heated greenhouses, low temperature, high RH and low light intensity, prevalent from late November till late March, create good conditions for infection by *B. cinerea* (Elad *et al.*, 1992; Jarvis, 1992).

Botrytis-incited diseases are prevented by ventilation and heating of greenhouses. Fungicides, mostly benzimidazoles and dicarboxymides, are also used extensively. Nowadays, due to the predominance of resistant strains of the pathogen, they are only marginally effective and growers are advised to combine dicarboxymides with other means of control such as biocontrol preparations. New fungicides have recently been released, but in greenhouses they are used on a limited scale. Formulations of biological control agents such as Trichodex (*Trichoderma harzianum* Rifai T39) are also available (Elad *et al.,* 1992; Gullino, 1992).

3.2.6. SCLEROTINIA ROT [*Sclerotinia sclerotiorum* (Lib.) de Bary and *Sclerotinia minor* Jagger]

This is a common greenhouse disease that damages lettuce, eggplant, tomato, cucumber, pepper, etc. Infection on lettuce begins close to the soil, where a water-soaked area appears. Infection may spread downwards to the roots or upwards to the heart of the plants. Infected leaves fall onto the soil and dry up. The other plants are infected along the stem, leaves, flowers and fruits. Infected areas become water-soaked. Stem infection is more severe. Leaves above the infection area become yellow, wither and die. In wet weather a white mass of mycelia appears on the infected areas, which gradually develops into black sclerotia. *Sclerotinia sclerotiorum*, which is the most common pathogen, produces sclerotia up to the size of bean seeds, whereas *S. minor* produces smaller sclerotia. Sclerotia fall onto the soil where they can survive for several years. When weather conditions are favourable they germinate to produce apothecia which release ascospores and cause new infection. High RH and moderate temperature is required for infection (Purdy, 1979; Fletcher, 1984).

The elimination of sclerotia and the control measures recommended against grey mould are effective against Sclerotinia rot as well.

3.2.7. ALTERNARIA DISEASES

The following diseases, caused by *Alternaria* spp., seriously affect vegetable and floral crops in greenhouses.

Tomato Early Blight (Alternaria solani *Sorauer*)
A collar rot of the young plants before or after transplanting may be the first symptom. In mature plants small irregular brown spots, with or without a yellow halo and concentric rings, appear mainly on leaves. Severely infected leaves are ragged and senescent. Similar spots without a yellow ring appear along the stem, leaf stalks, pentucles and the calyx. On fruits, brown to black spots with a leathery surface appear at the stem end. Severely infected plants may be defoliated (Sherf and Macnab, 1986).

Alternaria Branch Rot and Leaf Spot of Carnation (Alternaria dianthi *Stev. and Hall.*)
This mostly infects carnation cuttings during mist propagation and in wet parts of greenhouses. Small purple spots on the leaves are the first symptoms. Soon they enlarge, and their centre turns brown and then black due to the masses of spores which develop. Stem infection usually appears on the knots (Strider, 1978).

Alternaria diseases of minor importance for greenhouses
As well as the Alternaria diseases described above, strains of *Alternaria alternata* (Fr.:Fr.)

Keissl. have been recorded: (i) causing cankers in tomato crops; (ii) causing leaf spotting in cucumber; and (iii) causing mostly post-harvest rotting on tomato fruits. Also *Alternaria cucumerina* (Ellis & Everh.) J.A. Elliot may on occasion infect cucumber, melon, watermelon and squash (Grogan *et al.*, 1975; Fletcher, 1984; Vakalounakis and Malathrakis, 1987). At present, none of them has any economic impact on greenhouse crops.

All *Alternaria* species are facultative parasites mostly infecting weak plants. They survive in the soil on plant debris, but their black spores may also survive on several surfaces in greenhouses. *Alternaria solani* may survive on potato, which is an alternative host. Spores growing on dead material or on host plants are easily disseminated by wind or by splashed water. Plant infection takes place through stomata or directly through leaf surface. Spore germination and subsequent infection take place under a wide range of temperature. RH needs to be higher than 97% for rapid germination, but germination may take place in some cases at RH >75%. Senescent tissues are preferentially infected. The optimal temperature reported for *A. solani* is 18–25°C and for *A. cucumerina* 20–32°C. However, temperatures prevailing during the growing period of the respective hosts are not a factor limiting infection.

Control
Alternaria diseases can be prevented by dithiocarbamates, chlorotholonil, iprodione, etc. Hygienic measures and use of healthy propagating material are very important, especially when crops are grown in the soil. Inoculum surviving on plant debris in the soil and spores remaining on the greenhouse frames should be eradicated.

3.2.8. DIDYMELLA DISEASES

Two very severe diseases of greenhouse crops are caused by *Didymella* spp.: Didymella stem rot or canker in tomato and eggplant {*Didymella lycopersici* Kleb [teleomorph of *Phoma lycopersici* Cooke (= *Diplodina lycopersici* Hollós)]} and gummy stem blight in cucurbits {*Didymella bryoniae* (Auersw.) Rehm [anamorph *Phoma cucurbitacearum* (Fr.:Fr.) Sacc.]}.

Both diseases damage all aerial plant parts of their hosts in greenhouses when weather is cool and RH high. They may infect the collar and root system causing yellowing and withering of the plants, which may later die. Cankers along the stem and the petioles are also very common. Plant parts above cankers may die. Both diseases cause large spots on the leaves which may cover the entire leaf surface. Tomato fruits are infected at the stem end. Initially, the infected area is light brown but it soon turns pink due to the large amount of pycnidio-spores released. Infected parts may cover one third of the fruit surface. Infection of cucumber and melon fruits by *D. bryoniae* appears mostly at the blossom end. Infection may occur only inside the fruit without being visible on the surface. Soon after infection, a lot of pycnidia appear on the infected areas and their colour turns dark brown. Dark perithecia also appear a little later than pycnidia produced by *D. bryoniae*, while those of *D. lycopersici* are rare (Anonymous, 1971; Blancard *et al.*, 1991).

The inoculum remains in plant residues inside and outside greenhouses. In the first case infection starts through the collar. There is good evidence that infection of the aerial parts by *D. bryoniae* is initiated by ascospores released from infected plant material left outside greenhouses. In greenhouses the two diseases are rapidly spread by water splashes and cultural practices. Soil disinfestation, destruction of plant residue and strict hygienic conditions delay the outbreak of the diseases. However, disinfested soil is readily reinfested. The fungicides commonly used in greenhouse against other fungal diseases are also effective. Moreover, the reduction of the RH and of free water on the leaf surfaces is very effective (Anonymous, 1971; Sherf and Macnab, 1986) .

3.2.9. RUST DISEASES

These are a very important group, with many common characteristics. The following are the main rusts affecting greenhouse crops.

Carnation Rust [Uromyces dianthi *(Pers.:Pers.) Niessl* (= Uromyces caryophyllinus *G. Wint.*)]
The disease is more severe on leaves, but other green plant parts are infected as well. Initially, small light green spots appear. They gradually turn to powdery brown blisters due to the urediospores developed. Severely infected plant parts are twisted.

Healthy crops are infected by urediospores transferred from neighbouring crops. They are wind or water-splash disseminated and germinate readily on free water. The cycle of the pathogen lasts about two weeks. In greenhouses, where leaves may remain wet for several hours, there may be many disease cycles per crop season (Strider, 1985).

Rose Rust [Phragmidium mucronatum *(Pers.:Pers.) Schlechtend.*]
The disease is easily identified by the yellow orange rust pustules which develop profusely on the lower surface of older leaves. In greenhouses it is not very destructive. Several species of *Phragmidium* have been reported to infect rose, but *P. mucronatum* is the most common. It is an autoecious, macrocyclic fungus producing telia by the end of the crop season in the same place as uredospores. They serve as overwintering structures and initiate infection during spring. Free water and temperature 9–27°C are necessary for the uredospores to germinate (Horst, 1989).

Chrysanthemum Rust [Puccinia tanaceti *DC.* (Puccinia chrysanthemi *Roze*)] *and White Rust of Chrysanthemum* (Puccinia horiana *Henn.*)
Pale yellow flecks on the leaves followed by dark brown pustules with urediospores are the dominant symptom. Leaves with several pustules may wither and die. No stem infection has been reported. It is a low to moderate temperature disease requiring free water for infection. It survives on infected leaves and is disseminated by wind. Chrysanthemum white rust is a new and destructive disease of chrysanthemum in Europe and the Mediterranean. Initially, circular white or yellow cushions develop on the lower leaf surface and then soon turn brown. The disease is favoured by high RH and moderate temperatures (Strider, 1985). Snapdragon rust (*Puccinia antirrhini* Dietel & Holw.), geranium rust (*Puccinia pelargonii-zonalis* Doidge), etc. are also destructive diseases, but the respective crops are not grown in large acreage (Strider, 1985).

Regular applications of protective fungicides, such as dithiocarbamates and chlorothalonil, or systemic fungicides, such as oxycarboxin and members of the DMIs, are mostly recommended for rust control. Prevention of water condensation is also very effective (Strider, 1985; Horst, 1989).

3.2.10. CLADOSPORIUM DISEASES

Tomato Leaf Mould [Fulvia fulva *(Cooke) Cif.* (= Cladosporium fulvum *Cooke*)]
This causes light green to yellow spots on the upper surface of mature leaves. Soon the sporulating fungus growth appears as an olive-green velvety growth on the underside of the yellow spots. The pathogen survives for several months on the greenhouse frame, on the materials used for cropping and in plant debris. It is disseminated by wind or splashed by water drops. The optimal temperature for infection is 20 to 25°C. If weather conditions are favourable, leaf mould has several cycles in a season and can destroy the crop completely. There are several races of the pathogen (Blancard, 1988; Jones *et al.*, 1993).

Cucurbit Scab (Cladosporium cucumerinum *Ellis & Arth.*)
This mostly attacks cucumber, but also squash, melon, etc. It causes nearly circular or angular leaf spots on the leaves, which look water-soaked. Fruit infection is more serious. Initially, water-soaked lesions about 1 cm long, with gummy exudations, develop. A corky tissue usually develops around the lesions, which finally develop a scabby appearance. The pathogen survives on plant debris and spores are air-disseminated. Temperatures of about 15 to 25°C and RH over 86% favour the disease (Sherf and Macnab, 1986; Blancard *et al.*, 1991).

Control
For both diseases, greenhouse ventilation is the best control measure. The disinfestation of greenhouse soil and frames is also very important. Regular application of dithiocarbamates, iprodione, benzimidazoles, etc. are recommended as well. There are several resistant cvs against some races of the pathogens.

3.3. Bacterial Diseases

Several bacterial diseases damage all types of greenhouse crops. The most common are the following.

3.3.1. WILTS

Tomato Bacterial Canker {Clavibacter michiganensis *(Smith) Davis* et al. *ssp.* michiganensis *(Smith) Davis* et al. [= Corynebacterium michiganense *(Smith) Jensen ssp.* michiganense *(Smith) Jensen*]}
Initially, infected plants show a sudden unilateral wilting of leaflets, entire leaves or shoots. Young plants are more susceptible to wilting. Stem vessels at the side of the wilted leaves develop a yellow-brown discoloration. In the more severely infected places, the cortex splits and cankers several centimetres long may develop. Such plants usually die prematurely. Systemic fruit infection leads to yellow or brown discoloration of vascular strands and infected seeds are often shrivelled and black. Birds-eye spots, up to 6 mm in diameter, often appear on fruits.
    The pathogen is a typical seed-borne organism. It can also survive for several months on

cultivation equipment, on plant debris and in the soil. It can also maintain large populations on leaves of tomato and other plant species. It may infect at 16–36°C, with optimum at about 24–28°C. It is disseminated by seed or transplants, which remain symptomless until transplanted. In greenhouses, it spreads mostly during cultural practices (Strider, 1969; Gleason *et al.*, 1993).

Slow Wilt, Bacterial Stunt of Carnation [Erwinia chrysanthemi *Burkholder, McFadden & Dimock pv.* dianthicola *(Hellmers) Dickey*]
Infected plants become grey-green and may be stunted without any obvious wilting. Plants eventually wilt and in a period of 6–8 months may die. Vascular tissues, and pith mainly at the base of the stem, may show a yellow discoloration. Occasionally, stem cracks and root rot may occur (Fletcher, 1984).

Control
Soil disinfestation, use of resistant cvs, grafting on resistant root stocks, use of clean propagating material and application of strict hygienic conditions are recommended against wilts (Walker, 1971; Ginoux *et al.*, 1978; Sherf and Macnab, 1986).

3.3.2. ROTS

Tomato Soft Rots [Erwinia carotovora *(Jones) Bergey* et al. *ssp.* carotovora *(Jones) Bergey* et al., Erwinia carotovora *(Jones) Bergey* et al. *ssp.* atroseptica *(van Hall) Dye,* Pseudomonas viridiflava *(Burkholder) Dowson*] *and Tomato Pith Necrosis* [Pseudomonas corrugata *(ex Scarlett* et al.*) Roberts & Scarlett,* P. viridiflava, Pseudomonas cichorii *(Swingle) Stapp*]
Infected plants are stunted, their lower leaves show yellowing at the edges and on the veins and become flaccid. Initially the pith turns yellow to light brown, but later it disintegrates. The stem becomes hollow, splits and may exude bacterial slime. Brown to black blotches may also appear along the stem and the leaf stalks. A yellow to light-brown discoloration usually appears along the vascular system. Plants with severe stem rot may wilt and die, but very often even plants with split stems survive and yield normally. Several reports indicate that the above bacteria can cause similar symptoms under similar conditions in tomato plants. Plants with lush growth, grown under conditions of high RH, are more susceptible. Infection starts from leaf scars on the lower part of the stem, but may also appear in plants which have never been pruned (Scarlett *et al.*, 1978; Malathrakis and Goumas, 1987).

Bacterial Blight of Floral Crops (Pathovars of E. chrysanthemi)
This causes various rotting, necrotic and systemic diseases of several floral crops, such as chrysanthemum, cyclamen and saintpaulia, in greenhouses. The pathogen comes from affected stock plants and is disseminated by cultural practices. Infected plants should be discarded and knives disinfected (Fletcher, 1984).

3.3.3. LEAF AND STEM SPOTS

Tomato speck [Pseudomonas syringae *van Hall pv.* tomato *(Okabe) Young* et al. *and*

Bacterial Spot of Tomato and Pepper [Xanthomonas vesicatoria *(ex Doidge) Vauterin* et al.]
Bacterial speck causes small dark brown spots with bright yellow halo on tomato leaves. Necrotic tissues tear off and leaves appear ragged. Small dark brown spots develop on stem and petioles. Spots may coalesce to cause dark brown-black blotches on the surface of the infected plant parts. Small (up to 1 mm) black spots also appear on the fruits. Severely infected leaves turn yellow and finally dry out. The symptoms of bacterial spot are similar to those of bacterial speck. The spots on the fruits are initially raised and at the end look scabby. Both pathogens survive on plant debris in the greenhouse or outdoors, as well as on seeds. They are splashed from plant to plant by water drops from condensation and infect plants through stomata and injuries. Infection requires free water on plant surfaces (Schneid and Grogan, 1977; Goode and Sasser, 1980; Gitaitis *et al.*, 1992).

Angular Leaf Spot of Cucurbits [Pseudomonas syringae *van Hall pv.* lachrimans *(Smith & Bryan) Young* et al.]
This mostly damages cucumber, zucchini and melon causing small, angular, light-grey leaf spots. They may coalesce to cover large areas. Severely infected leaves become chlorotic; infected areas tear off and appear ragged. Water-soaked spots also appear on the stem and fruits. In humid conditions, tear drops form on leaves, stem and fruit spots. The causal organism survives on the infected plant debris and in the seed coat. The bacterium is splashed from the soil by water and infects plants. It spreads from plant to plant during the cultural practices (van Gundy and Walker, 1957; Fletcher, 1984).

Control
Strict hygiene, soil disinfestation, use of healthy seeds and reduction of the wetness period are recommended measures against bacterial diseases in greenhouses. Reduction of nitrogen fertilizers is also important for tomato soft rot. Copper fungicides are the most effective chemicals. Some resistant cvs have also been released for tomato speck and bacterial spot of tomato, but none of them is suitable for greenhouses (van Gundy and Walker, 1957; Fletcher, 1984).

3.4. Future Prospects

Bacterial and fungal diseases will remain serious problems in protected crops in the future, particularly in the case of plastic-houses. The severity and even the relative importance of diseases may vary as a consequence of the introduction of new crops/cultivars and/or cropping systems.

The shift to control strategies which rely less on chemicals and the application of the most recent fungicides with a specific mode of action favoured, in some cases, the development of some foliar pathogens, formerly of secondary importance.

Disease control is complex and necessarily relies on the integration of several measures. While fungicides played a major role in the past, recently, for technical, economical and environmental reasons, a big effort has been made to integrate disease management. Such an approach is also necessary because fewer and fewer chemicals are now registered for use on crops such as most of those grown under protection, which are considered "minor".

Better diagnostic tools, for early and quick disease detection, a wider use of resistant cultivars, a more considered adoption of cultural practices, coupled with the use, whenever possible, of biocontrol agents, will enable our dependence on chemicals to be reduced in the near future.

References

Anonymous (1971) *Didymella Stem and Fruit Rot of Tomato*, Ministry of Agriculture Fisheries and Food Advisory Leaflet 560, Her Majesty's Stationery Office, Edinburgh.

Blancard, D. (1988) *Maladies de la tomate*, Revue Horticole, INRA, Paris.

Blancard, D., Lecoq, H. and Pitrat, M. (1991) *Maladies des cucurbitacées*, Revue Horticole, INRA, Paris.

Braun, W. (1995) *The Powdery Mildews (Erysiphales) of Europe*, Gustav Fischer, New York.

Coyier, D.L. (1985a) Powdery mildews, in D.L. Strider (ed.), *Diseases of Floral Crops*, Vol. 1, Praeger, New York, pp. 103–140.

Coyier, D.L. (1985b) Roses, in D.L. Strider (ed.), *Diseases of Floral Crops*, Vol. 2, Praeger, New York, pp. 405–417.

Davies, J.M.L. (1980) Diseases in NFT, *Acta Horticulturae* **98**, 299–305.

Ebben, M.H. (1974) Brown root rot of tomato, in *1973 Annual Report of the Glasshouse Crops Research Institute*, Glasshouse Crops Research Institute, Littlehampton, pp. 127–135.

Elad, Y., Shtienberg, D., Yunis, H. and Mahrer, Y. (1992) Epidemiology of grey mould, caused by *Botrytis cinerea* in vegetable greenhouses, in K. Verhoeff, N.E. Malathrakis and B. Williamson (eds.), *Recent Advances in Botrytis Research*, Pudoc Scientific Publishers, Wagenigen, pp. 147–158.

Fletcher, J.T. (1984) *Diseases of Greenhouse Plants*, Longman, Inc., New York.

Garibaldi, A. and Rapetti, S. (1981) Grave epidemia di peronospora su antirrino, *Colture Protette* **9**, 35–38.

Gitaitis, R.D., McCarter, S. and Jones, J. (1992). Disease control in tomato transplants produced in Georgia and Florida, *Plant Disease* **76**, 651–656.

Gleason, M.L., Gitaitis, R.D. and Ricker, M.D. (1993) Recent progress in understanding and controlling bacterial canker of tomato in eastern north America, *Plant Disease* **77**, 1069–1076.

Ginoux, G., Dauple, P. and Lefebvre, J.M. (1978) Greffage de la tomate, *PHM-Revue Horticole* **192**, 33–44.

Goode, M.J. and Sasser, M. (1980) Prevention – The key to controlling bacterial spot and bacterial speck of tomato, *Plant Disease* **64**, 831–834.

Grogan, R.G., Kimble, K.A. and Misaghi, I.J. (1975) A stem canker of tomato caused by *Alternaria alternata* f. sp. *lycopersici*, *Phytopathology* **65**, 880–886.

Gullino, M.L. (1992) Chemical control of *Botrytis* spp., in K. Verhoeff, N.E. Malathrakis and B. Williamson (eds.), *Recent Advances in Botrytis Research*, Pudoc Scientific Publishers, Wagenigen, pp. 217–222.

Hoitink, H.A.J and Fahy, P.C. (1986) Basis for the control of soilborne plant pathogen with composts, *Annual Review of Phytopathology* **24**, 93–114.

Horst, R.K. (ed.) (1989) *Compendium of Rose Diseases*, APS Press, St Paul, Minn.

Jarvis, W.N., Thorpe, H.J. and MacNeill, B.H. (1975) A foot and root rot disease of tomato caused by *Fusarium oxysporum*, *Canadian Plant Disease Survey* **55**, 25–26.

Jarvis, W.N., Thorpe, H.J. and Meloche, R.B. (1983) Survey of greenhouse management practices in Essex County, Ontario, in relation to Fusarium foot and root rot of tomato, *Plant Disease* **67**, 38–40.

Jarvis, W.N. (1992) *Managing Diseases in Greenhouse Crops*, APS Press, St Paul, Minn.

Jones, J.P., Stall, R.E. and Zitter, T.A. (eds.) (1993) *Compendium of Tomato Diseases*, APS Press, St Paul, Minn.

Malathrakis, N.E. and Goumas, D. (1987) Bacterial soft rot of greenhouses in Crete, *Annals of Applied Biology* **111**, 115–123.

Malathrakis, N.E., Kapetanakis, G.E. and Linardakis, D.C. (1983) Brown root rot of tomato, and its control, in Crete, *Annals of Applied Biology* **102**, 251–256.

Molot, P.M. and Lecoq, H. (1986) Les oidiums des cucurbitacées. I. Données bibliographiques. Travaux préliminaires, *Agronomie* **6**(4), 355–362.

Nelson, P.E., Toussoun, T.A. and Cook, R.J. (eds.) (1981) *Fusarium Diseases, Biology and Taxonomy,* The Pennsylvania State University Press, University Park, Pa.

Palti, J. (1971) Biological characteristics, distribution and control of *L. taurica, Phytopathologia Mediterranea* **10**, 139–153.

Palti, J. and Cohen, Y. (1980) Downy mildew of cucurbits (*Pseudoperonospora cubensis*): The fungus and its hosts, distribution, epidemiology and control, *Phytoparasitica* **82**(2), 109–147.

Parmeter, J.R., Jr (1970) *Rhizoctonia solani: Biology and Pathology,* University of California Press, Berkeley, Calif.

Purdy, L.H. (1979) *Sclerotinia sclerotiorum*: History diseases and symptomatology, host range, geographic distribution, and impact, *Phytopathology* **69**, 875–880.

Scarlett, C.M., Fletcher, J.T., Roberts, P. and Lelliott, R.A. (1978) Tomato pith necrosis caused by *Pseudomonas corrugata* n. sp., *Annals of Applied Biology* **88**, 105–114.

Schneid, R.W. and Grogan, R.G. (1977) Bacterial speck of tomato: Sources of inoculum and establishment of a resident population, *Phytopathology* **67**, 388–394.

Sherf, A.F. and McNab, A.A. (1986) *Vegetable Diseases and their Control,* John Wiley and Sons, New York.

Sitterly, W.R. (1978). Powdery mildew of cucurbits, in D.M. Spencer (ed.), *The Powdery Mildews,* Academic Press, New York.

Strider, D.L. (1969) *Bacterial Canker of Tomato Caused by* Corynebacterium michiganense: *A Literature Review and Bibliography,* Technical Bulletin No. 193, North Carolina Agricultural Experimental Station, Raleigh, NC.

Strider, D.L. (1978) Alternaria blight of carnation in the greenhouse and its control, *Plant Disease Reporter* **62**, 24–28.

Strider, D.L. (ed.) (1985) *Diseases of Floral Crops,* Vols 1 and 2, Praeger Special Studies, New York.

Tjamos, E.C. (1992) Selective elimination of soilborne plant pathogens and enhancement of antagonists by steaming, sublethal fumigation and solarization, in E.C. Tjamos, G.C. Papavizas and R.J. Cook (eds.), *Biological Control of Plant Diseases*, Plenum Press, London.

Tompkins, C.M. and Middleton, J.T. (1950) Etiology and control of poinsettia root and stem rot caused by *Pythium* spp. and *Rhizoctonia solani, Hilgardia* **20**, 171–182.

Trolinger, J.C. and Strider, D.L. (1985) Botrytis diseases, in D.L. Strider (ed.), *Diseases of Floral Crops,* Vol. 1, Praeger Special Studies, New York, pp. 17–101.

Vakalounakis, D.J. and Malathrakis, N.E. (1987) A cucumber disease caused by the fungus *Alternaria alternata, J. of Phytopathology* **121**, 325–336.

van Gundy, S.D. and Walker, D.W. (1957) Seed transmission, overwintering and host range of cucurbit angular leaf spot pathogen, *Plant Disease Reporter* **41**, 137–140.

Walker, J.C. (1971) *Fusarium Wilt of Tomato*, Monograph No. 6, Amer. Phytopath. Soc., St Paul, Minn.

CHAPTER 4

INSECT AND MITE PESTS

Henrik F. Brødsgaard and Ramon Albajes

4.1. Introduction

The greenhouse environment is characterized by conditions that optimize plant growth. However, it is not only plants that benefit from the stable greenhouse environment, but also herbivorous insects and mites. In addition, greenhouses often provide herbivores with an unlimited amount of food plants in monoculture and lack of natural regulating factors such as predators, parasitoids and diseases. On top of this, crop cultivars have for generations been selected for quick growth and maximum yield, often resulting in reduction or even loss of resistance mechanisms against herbivores. Hence, greenhouse crops are very vulnerable to herbivore attack.

Herbivores that are accidentally introduced into greenhouses or migrate into greenhouses through open vents will most likely find almost all biotic and abiotic conditions in favour of rapid population increase. Hence, herbivores from quite a number of insect and mite orders have obtained pest status in greenhouse crops. Although belonging to various taxonomic groups, the major greenhouse arthropod pests share several traits in their biology – they are mostly polyphagous, are able to develop continuously with no diapause, and have high rates of increase – that allow them to quickly exploit ephemeral but extremely favourable habitats.

Traditionally, i.e. up to the 50s and 60s, these pests were easily controlled with pesticides, but in the 70s problems with pesticide resistance in greenhouse pests rapidly developed. Most of the major pests are today characterized by very high pesticide resistance levels to various types of active ingredient. Pesticide resistance combined with the intensive international trade in plant material have given a number of herbivorous insects and mites global pest status. Pesticide resistance first, and consumer and environmental concerns second, are leading growers to replace simple pesticide-based control programmes by other, more sophisticated and tactically broader integrated control systems. In this chapter a brief review of the major insect and mite pests and the current status of their control is presented, followed by some ideas about developments in the coming decades.

4.2. Major Insect and Mite Pests

The characteristics and control of major greenhouse insect and mite pests are described in specific chapters later in this book. Here only the most significant features of groups of pests are included. Pests, to which no specific chapters are devoted, like scale insects, Lepidoptera and fungus gnats, are discussed in more detail. In Table 4.1 the reader is

48

R. Albajes et al. (eds.), Integrated Pest and Disease Management in Greenhouse Crops, 48-60.
© 1999 *Kluwer Academic Publishers. Printed in the Netherlands.*

referred to those chapters in which the biology and control of each pest group are discussed in more detail.

TABLE 4.1. Chapters in this book where the reader can find more details on the biology and biological and integrated control of each group of pest

Group of pests	Detailed discussion in Chapter	Biological and integrated control discussed in Chapters
Whiteflies	14	14, 19, 21, 29, 30, 31, 33, 34
Spider mites	15	15, 19, 30, 31, 32, 33, 34
Tarsonemid mites	15	15, 32, 34
Rust mites	15	15, 30
Aphids	16	16, 21, 30, 31, 32, 33, 34
Thrips	17	17, 19, 21, 31, 32, 33, 34
Leafminers	18	18, 30, 34
Lepidoptera	4	4, 21, 30, 32, 33, 34
Scale insects	4	4, 21, 34
Fungus gnats	4	4, 34

4.2.1. WHITEFLIES

The suborder Hemiptera of the insect order Homoptera includes a range of families that have pest species on greenhouse crops. The economically important families are: Aleyrodidae (whiteflies), Aphididae (aphids), Pseudococcidae (mealybugs) and Coccidae (scale insects).

Whiteflies belong to the insect family Aleyrodidae. Of the 1200 described species in the family only three are major pests in greenhouse crops. The vast majority of whitefly species (>85%) are oligophagous, but all three greenhouse pests are highly polyphagous, each with approximately 300 recorded host plants. Common to the three greenhouse pest species is that they also have very high levels of insecticide resistance. *Trialeurodes vaporariorum* (Westwood), the greenhouse whitefly is presently one of the most widespread greenhouse pests. It attacks a large number of ornamental crops and most vegetables grown in greenhouses, although it shows clear preferences. Although observed causing problems on greenhouse crops in several regions of the world for many years, *Bemisia tabaci* (Gennadius) has become a serious world-wide pest since the early 80s. *Bemisia argentifolii* Bellows & Perring, the silverleaf whitefly, has apparently evolved from a race of *B. tabaci*, perhaps on poinsettia. It has recently expanded its distribution area enormously and is now the number one pest in many crops, such as cucurbits, tomatoes and cotton in many warm regions and in ornamental crops in temperate climates (for a review, see Gerling, 1990).

Both whitefly adults and nymphs feed by inserting their rostrum into phloem cells and sucking the sap of the host plant. Direct damage such as reduced growth and leaf fall occur due to removal of starch and chlorophyll by nymphs on heavily infested plants. However, indirect damage may occur at lower infestation levels mainly because of the excreted honeydew from both nymphs and adults. The honeydew, like in aphids, acts as a growth medium for sooty moulds with severe economic consequences. Furthermore, all three pest species – particularly *Bemisia* species – are vectors for a number of very serious plant viruses. In addition, the silverleaf whitefly is phytotoxic to some cucurbit crops.

Control of *T. vaporariorum* by the parasitoid *Encarsia formosa* Gahan provides a good example of successful and popular biological control in greenhouses, and it is at present applied in Europe on more than 4000 ha (see Chapter 14). Unfortunately, this parasitoid is not as effective for the control of *Bemisia* species. Several parasitoids and predators are currently being tested, and some of them are commercially available for the biological control of *Bemisia* pests. New pesticides, which are initially active against whiteflies, appear regularly on the market, but whiteflies have shown a high capacity to develop resistance quickly.

4.2.2. SPIDER MITES

Spider mites is the common name of mites belonging to the family Tetranychidae. This includes thousands of species, many of which are economically very important. Spider mites are phytophagous and feed on several parts of the plant, mainly on the under part of leaves, by inserting stylets and sucking epidermal and mesophyl cell contents. Damaged cells have a reduced number of chloroplasts and they can be seen from the upper part of the leaf as yellowing punctures that later coalesce and form bigger yellowing areas; the leaf falls off prematurely. This cytological damage is accompanied by chemical and physiological alterations in the plant that lead to growth retardation and yield loss (for a review, see Helle and Sabelis, 1985).

Several *Tetranychus* species damage a large number of greenhouse vegetables and ornamentals world-wide. Cucurbits, French beans, and a variety of foliage and flowering ornamental plants are among the most affected crops in greenhouses. Under intensive chemical control, many other crops, such as tomatoes, may be severely infested by spider mites. Economic injury levels (EIL) that relate spider mite densities (number of mites, mite-days) or index of leaf area damaged (Scopes, 1985) to yield loss have been determined. Values of EIL found in the literature, mostly obtained in *Tetranychus urticae* Koch, are quite variable, as they depend on many factors, such as crop species, variety and growth conditions.

Many organic acaricides are nowadays available for the chemical control of spider mites, but they commonly have to be repeatedly applied to achieve good control. Most of the acaricides interfere with naturally occurring or released natural enemies of mites and other pests. As noted for other greenhouse pests, correct cultural practices – fertilization, irrigation, greenhouse and crop hygiene, greenhouse humidity and daylength – may prevent or, at least, reduce spider mite problems. But biological control is the only permanent and durable method with which to control spider mites in greenhouses. The phytoseiid predator *Phytoseiulus persimilis* Athias-Henriot has become, together with *E. formosa* for the control of the greenhouse whitefly, the paradigm of the success of biological control in greenhouses in many parts of the world. Several thousands of hectares of greenhouse crops in the world are today protected with different strains of this predator, and more phytoseiid species are being tested for biological control of *Tetranychus* spp. (see Chapter 15). General predators that naturally occur in greenhouses, such as anthocorid and mirid bugs, or that are released, may have an important role in preventing spider mite population outbreaks.

4.2.3. TARSONEMID MITES

Tarsonemid mites belong to the superfamily Tarsonemoidea. Two main species attack protected vegetable and ornamental crops: the broad mite [*Polyphagotarsonemus latus* (Banks)] and the cyclamen mite [*Phytonemus pallidus* (Banks)]. They feed preferably on young or succulent plant tissue. *Polyphagotarsonemus latus* is largely polyphagous, but causes particularly severe damage to sweet peppers and, to a lesser extent, to tomatoes and cucumbers in warm greenhouse areas. Terminal shoots and the underside of leaves turn bronzy and shiny a few days after their colonization by the broad mite, and later the higher part of the plant dries and appears burnt-like. *Phytonemus pallidus* has a narrower host range, and affects strawberries and some ornamental plants in many parts of the world. Its feeding causes stunting and distortion of newly emerging strawberry leaves, and causes flowers to wither and die. In heavy infestations, strawberry plants become stunted with a compact mass of crinkled leaves in the centre, and fruit is dwarfed and appears seedy. Within and between-field movement of the cyclamen mite is facilitated by wind, flying insects, machines and other cultivation tools.

Control of tarsonemid mites by acaricides is hard and largely ineffective. Mite-free seedlings and propagating nursery stock are essential to prevent introducing the initial population into greenhouses or strawberry tunnels. Biological programmes based on phytoseiid mites have not been developed for broad mite control, but have shown some encouraging results for the cyclamen mite (see Chapter 32).

4.2.4. RUST MITES

Rust, gall or eriophyoid mites belong to the superfamily of Eriophyoidea, with nearly 3000 described species. They are phytophagous and most are quite host specific. They feed on epidermal cells causing morphological and physiological alterations. Common morphological alterations include gall formation and other tissue distortions, toxaemias, and several non-distortive effects such as rusting, browning or silvering of leaves and other green plant parts. Physiological alterations may interfere with photosynthetic function and alter plant nutrient and hormone contents. Secondarily, rust mites may also be vectors of plant pathogens. These morphological and physiological effects of rust mite feeding may lead to a reduction in crop yield or to diminished aesthetic value.

Among eriophyoid mites attacking greenhouse-grown vegetables, the tomato russet mite *Aculops lycopersici* (Massee) is probably the most widespread and harmful. It affects various solanaceous greenhouse crops, but particularly tomato, in warm and temperate areas. In warm areas it is also a serious pest in outdoor tomatoes. It feeds on epidermal cells, mainly on the upper surface of the leaf, which turns bronze after a few days following the first mite infestation. Heavy infestation may lead to fruit russeting and plant dessication and eventually to death as a result of water loss through the destroyed epidermis. In susceptible tomato cultivars, the attack may also occur on flower peddles, causing flower bud death.

Damage begins near the ground and spreads upwards. Rust mite pests in greenhouse tomato usually appear at a few foci and then spread quickly to nearby plants, spreading throughout the greenhouse in a few weeks in low humidity environments. In addition, a variety of eriophyoid mites, particularly *Aceria* species, may cause severe losses in ornamental plants grown in greenhouses.

The tomato russet mite can be managed by cultural practices that include a correct irrigation regime, prevention of mite introduction into the greenhouse through infested plants, and destruction of solanaceous weeds (also bindweed, *Convolvulus arvensis* L.). As low mite numbers are difficult to observe in routine inspections, particularly intensive sampling is needed to detect initial infestation foci early. Biological control of *A. lycopersici* is poorly developed at the moment but current trials with phytoseiid mites have shown encouraging results (Chapter 15).

4.2.5. APHIDS

Aphids are among the most important insect pest groups both in greenhouses and in field crops, especially those belonging to the family Aphididae. Most damaging aphids in greenhouses are more or less polyphagous, with the major exception of the lettuce aphid, *Nasonovia ribisnigri* (Mosley). *Aphis gossypii* Glover is a serious pest of many ornamentals and especially plants in the Cucurbitaceae. *Macrosiphum euphorbiae* (Thomas) and *Macrosiphum rosae* (L.) are rather polyphagous although they cause major problems on solanaceous crops and on roses respectively. In addition, they are vectors for a range of serious plant viruses. *Myzus persicae* (Sulzer) is characterized by an ability to develop very high levels of insecticide resistance which, in combination with a broad host range and a high reproductive capacity, makes this species a serious pest, especially on sweet peppers and various ornamentals. *Nasonovia ribisnigri* causes increasingly serious problems in Mediterranean lettuce production (for a review, see Minks and Harrewijn, 1987).

The life history of many species is very complex involving shifts between winter and summer hosts, and different reproduction strategies in relation to time of year. In greenhouses, however, many aphids may reproduce parthenogenetically throughout the year and build up high density populations on successive overlapping annual crops or on year-round ornamentals. Most aphids of economic importance suck the phloem sap of leaves or young shoots, causing distortion, stunting and premature leaf fall. Some species also feed on flowers and flower buds, causing flower discoloration or abortion of the buds. A few pest species feed on roots, causing the plants to wilt. Apart from the direct damage to plants caused by feeding, aphids also damage the host plants indirectly by vectoring a vast number of serious plant viruses. Furthermore, their feeding punctures provide means of entry for plant pathogens such as bacteria or fungi. Like many sap-sucking insects, aphids excrete copious amounts of honeydew that is deposited on the foliage making it shiny and sticky. This provides an ideal medium for the growth of sooty moulds, which in turn reduce photosynthesis and downgrade the aesthetic value of fruits and ornamentals.

The extremely high reproductive capacity of aphids, particularly in greenhouse environments, makes their control very difficult. An integration of control methods (host resistance, cultural, biological and microbial methods, and selective chemicals) is needed for sufficient aphid control (see Chapter 16).

4.2.6. THRIPS

Thrips are insects belonging to the order Thysanoptera. This order includes more than 5000 species among which half are phytophagous, but only a handful may cause serious economic damage in greenhouse crops. Despite the limited number of pest species, thrips have for a number of years been ranked as the number one pest in greenhouse crops in Europe, the USA and Southeast Asia, mainly due to the spread of insecticide resistant tropical or subtropical polyphagous species into greenhouse crops in colder regions. The most damaging three species in greenhouse crops world-wide are: *Frankliniella occidentalis* (Pergande), *Thrips tabaci* Lindeman and *Thrips palmi* Karny. Other species are pests of ornamentals of more or less local importance but are rarely widely distributed (for a review, see Lewis, 1997).

Frankliniella occidentalis is the most important thrips pest world-wide. Its original distribution area was in the USA, but it spread during the 70s and 80s and has obtained an almost cosmopolitan distribution in greenhouse crops. *Thrips tabaci* probably originates from the Middle East although it is now a cosmopolitan pest both in greenhouse and outdoor crops. *Thrips palmi* originates from Sumatra, but has spread rapidly throughout the Pacific and Orient. In 1978 it was introduced into greenhouse crops in Japan, and it is now also widespread in Hawaii, the West Indies and Florida. In Europe, this species is considered as a quarantine organism.

All thrips species have piercing-sucking mouthparts. The phytophagous species feed by puncturing epidermal and parenchymal cells, and sucking out the cell contents leaving typical grey or silvery chlorotic spots on infested plant parts. The most serious thrips pests of greenhouse crops are characterized by being highly polyphagous on both flower and leaf tissue. Besides the direct feeding damage that causes reduction of photosynthetic tissue and especially ornamental growth disorders or other cosmetic damage, several of the most severe thrips pest species are vectors of plant viruses of the family Tospovirus, some of which are devastating and polyphagous (see Chapter 2).

As a result of rapid insecticide resistance build-up in several thrips species and reduced numbers of insecticides registered for use in greenhouse crops, the interest in biological pest control among growers has recently increased in many countries (see Chapter 17). Furthermore, increasing numbers of more or less effective beneficial species as well as a constant supply of these natural enemies have increased the prospects of successful biological control of thrips.

4.2.7. LEAFMINERS

The dipterous leafminers belong to the family Agromyzidae, which includes 1800 described species of which 156 species are reported as pests. However, under greenhouse conditions only a handful of polyphagous species are of economic importance (for a review, see Spencer, 1973). Leafminers are parasitized by a wide range of parasitoid species that normally control the leafminer populations effectively. Leafminers only tend to become pests of economic importance if their natural parasitoids are hindered. That is the case when pesticides are applied, if crops are grown too early in the season for the natural populations of parasitoids to build up, or if the parasitoids are excluded from the crop by the greenhouse construction.

Four main leafminer species will be mentioned which sometimes cause severe problems in greenhouse crops. *Liriomyza bryoniae* (Kaltenbach) is a European species that is a pest of tomatoes and occasionally of cucurbits and lettuce. *Liriomyza trifolii* (Burgess) is a polyphagous American species on a wide range of greenhouse crops. It was introduced into Europe by the early 80s and is now widely distributed around the world as a result of plant trade. *Liriomyza huidobrensis* (Blanchard) is a polyphagous South American pest on many greenhouse crops. *Phytomyza syngenesiae* (Hardy), the chrysanthemum leafminer, is a widely distributed Western European species that was introduced to North America. It has a broad host-plant range among ornamental crops but is a serious pest mainly on chrysanthemum.

Larvae of Agromyzidae may feed on all plant parts, but the greenhouse pests are all leaf feeders causing apparent mines on the leaf blade. The mining causes reduction in leaf photosynthesis. The actual impact on the plant is of course related to the position of the leaf in question (e.g. for tomato crops the top one metre of the canopy is by far the most important metabolizing part of the entire canopy). Larval feeding in young plants and especially cotyledons can seriously weaken and even destroy the plants, but if leafminers occur in the lower canopy, damage is limited. Adult females injure plants by puncturing leaf tissue, and damage mesophyll cells by means of the ovipositor. Besides the direct damage caused by larval mining and adult host feeding, leafminers may cause severe indirect damage by providing infection sites for damaging fungi such as *Verticillium* spp. and *Fusarium* spp.

Chemical control of leafminers is complicated by their endophytic habits and high reproductive capacity. Resistance to insecticides in leafminers has been repeatedly reported, particularly in *L. trifolii*. Furthermore, most insecticides are toxic for the complex of parasitoids that hold leafminers in check and that have shown an excellent capacity to naturally parasitize exotic leafminers that have been successively introduced into Europe (see Chapter 18). When natural parasitism is not sufficient to keep leafminer population densities under economic thresholds, several parasitoid species are commercially available for seasonal inoculative releases in greenhouses.

4.2.8. LEPIDOPTERA

This order contains many serious pests. On greenhouse ornamentals, a large variety of Lepidoptera from several families may cause severe damage. However, in greenhouse vegetables only a few of them are regularly harmful, most belonging to the family Noctuidae. Below, we only refer to the most common species.

Economically Important Greenhouse Lepidoptera Species
Three main groups of Noctuidae that damage greenhouse vegetables may be

distinguished by their damaging potential on three parts of the host plant: (i) leaves (occasionally also flowers); (ii) fruits; and (iii) roots and stem at ground level. The first group of leaf feeders is the most important world-wide in greenhouses: *Spodoptera littoralis* (Boisduval) extends through Africa and the Mediterranean basin; *Spodoptera exigua* (Hübner) has a world-wide distribution in greenhouse areas, except for South America and Japan; *Autographa gamma* (L.) affects greenhouse crops across the Palaearctic region from Europe to North Africa and Asia; two widespread *Chrysodeixis* species, the Afrotropical and European *Chrysodeixis chalcites* (Esper) and its Indo-Australian relative *Chrysodeixis eriosoma* (Doubleday), may also injury vegetable fruit; and finally, *Trichoplusia ni* (Hübner) is a common vegetable pest in North America. Most of these leaf-feeding species are migrant and largely polyphagous on vegetables and herbaceous ornamentals. *Lacanobia oleracea* (L.) also feeds on foliage of several plants, but in greenhouses the major damage is caused on tomatoes. Larvae of the second group feed on vegetables, fruit and ornamental flowers, and include *Helicoverpa* (= *Heliothis*) *armigera* (Hübner) which causes severe damage on greenhouse tomatoes and carnations. The range of this partially migrant species extends through the Mediterranean basin, to Central Europe, Africa, South and eastern Asia, and Australasia. Larvae of the third group feed on roots and the stem at ground level and are commonly called cutworms. Several cutworms – mainly *Agrotis* spp. – may cause important damage on young plants of greenhouse crops cultivated on non disinfested soils. Many Lepidoptera other than noctuids affect greenhouse ornamental crops. Among these, the tortricid moth *Cacoecimorpha pronubana* (Hübner) is probably one of the most polyphagous. The reader is referred to specialized textbooks on ornamental pests for a complete list of Lepidoptera species involved (i.e. Alford, 1991; van de Vrie, 1991) and also to Chapter 34.

Main Features of the Biology of Leaf Feeders

Light attracts adults, which may be stimulated to come into lightened greenhouses where susceptible ornamental crops are grown. The diapausing and migration capacities of most of the Lepidoptera species mentioned condition their biology, phenology and, consequently, monitoring and population forecasting. Eggs are laid on leaves at different plant heights according to the species. Young larvae chew the epidermis and parenchyma of the lower leaf surface usually during the night. As they develop, the caterpillars cause larger holes on leaves, and excrement becomes more apparent on chewed leaves. They feed throughout the day – except in the case of *Spodoptera* spp. – and night. As a result, older larvae and their damage are easily detected by the grower. Older larvae may also feed on immature green fruit, creating a series of shallow wounds as they move from one fruit to another. On cut flower crops, one larva may damage a large number of buds and flowers. Pupation takes place on foliage or, in the case of *Spodoptera* spp., in the soil a few centimetres from the ground. The number of generations varies with pest species, latitude and availability of host crops outside the greenhouse. Adults can mate inside or outside the greenhouse, a feature that may influence the applicability of monitoring and control methods based on pheromones.

Damage
Leaf injury caused by foliage feeders is particularly damaging in young plants. Mature plants may tolerate several larvae with no yield loss. On flowers and fruit, cosmetic damage may reduce their value. Fruit-eating caterpillars have a high damaging potential because one larva may consume several fruits and cause injury, which may result in pathogen infection later. Cutworms are mainly damaging in seedlings and soon after transplanting.

Control
Insecticides applied to young larvae are effective for control of leaf- and fruit-eating caterpillars. However, most insecticides cannot be integrated in IPM systems. Several insect growth regulators can be used for caterpillar control, but these insecticides are also toxic for many beneficials. Selective control of caterpillars in the framework of IPM systems may be achieved by sprays of *Bacillus thuringiensis* Berliner ssp. *kurstaki* Dulmage, *Bacillus thuringiensis* Berliner ssp. *aizawai* de Barjac & Bonnefoi, *Bacillus thuringiensis* Berliner ssp. *thuringiensis* Heimpel & Angus and some entomopathogenic viruses (see Chapter 21 for further details on *Bacillus thuringiensis* Berliner and virus uses). To be fully effective, bacterial and viral insecticides need to be applied on young larvae, which are the most susceptible stage. Later, larvae become more tolerant and treatment loses efficacy. However, since pheromone traps have not been shown to be reliable enough to monitor and forecast most noctuid populations in greenhouses, tedious on-plant sampling is needed to optimally time *B. thuringiensis* sprayings. Another selective method to control Lepidopteran pests in greenhouses is based on sex pheromones. These have been successfully tested for mating disruption purposes to control *C. chalcites* and *S. exigua*, but they are not used on a commercial scale yet. Biological control is commercially applied to several dozens of hectares: the heteropteran predator *Podisus maculiventris* (Say) is released mainly for the control of *S. exigua* and *T. ni*. In the last few years, the egg parasitoid *Trichogramma* sp. has been released to control *L. oleracea* and, to a lesser extent, *C. chalcites*. The current research on biological control of lepidopteran pests in greenhouses deals with larval parasitoids that attack several Lepidopteran pest species. Cutworm problems are unlikely to occur in soilless cultures or correctly managed soils. Insecticidal baits against cutworms must be used with caution because active ingredients may sublimate at high greenhouse temperatures and create a toxic environment for natural enemies.

4.2.9. SCALE INSECTS

Greenhouse environments, which are usually humid and warm, are very suitable for scale insect population development. Active international trade of ornamentals means that a large variety of insect scales poses serious.pest problems in greenhouse-grown ornamental crops. Harmful scale insect (superfamily Coccoidea) pests in greenhouses belong to three main families: soft scales (Coccidae), mealybugs (Pseudococcidae) and armoured scales (Diaspididae) (for a review, see Rosen, 1990).

Economically Important Greenhouse Scale Insects
The soft scale, *Coccus hesperidum* L., is a highly polyphagous species recorded on hosts belonging to 37 plant families, ranging from ferns to orchids. This soft scale is the most common species in interior plantings. The adults measure 4 mm in length and the size, shape and markings of the scale vary depending on the host plant. A female will produce 80–250 offspring and the life cycle at 20°C is approximately 60 days. It produces considerably more honeydew than the other greenhouse soft scale species. The hemispherical scale, *Saissetia coffeae* (Walker), is also a very polyphagous species that attacks a range of greenhouse ornamentals. It varies in size from 2 to 4.5 mm depending on the host plant. In spite of only three nymphal instars, the life cycle is long: approximately 90 days at 20°C. However, the fertility is high. Each female may produce 500 to 2500 eggs depending on the host plant and temperature. The citrus mealybug, *Planococcus citri* (Risso), is the most common and the most damaging species in greenhouses. It infests plants from more than 25 plant families including a range of important ornamentals. The citrus mealybug produces both sexes in equal numbers and is the only pest scale that has sexual reproduction. More than 400 eggs are produced per female at 18°C, and development at this temperature is completed in approximately 90 days. The long tailed mealybug, *Pseudococcus longispinus* (Targioni-Tozzetti), which is a pest of many ornamentals, differs in biology from the rest of the mealybugs by being viviparous. Many armoured scales affect ornamental plants, shrubs and trees that are grown in greenhouses. Two cosmopolitan species of armoured scales that occur outdoors in the tropics and subtropics and in indoor greenhouses, both in warm and cold areas, are serious pests: *Aspidiotus nerii* Bouché, which is largely polyphagous, and *Diaspis boisduvalii* Signoret, which is also polyphagous but mainly affects glasshouse-grown orchids, bromeliads and palms.

Main Features of their Biology
Morphologically they differ so much from the rest of the Hemiptera that the resemblance may be difficult to see. The adult males are winged and quite minute and do not feed. Their sole function is to mate with the females. However, they are absent or rare in most species, and so reproduction is by oviparous or viviparous parthenogenesis. The females are covered by a protective secretion. In most species, all female stages but the first instar nymphs (the crawlers) have no legs or eyes except among mealybugs, which possess legs during all stages of development. In the species where males are present, the male nymphs resemble the females but are smaller and more elongate. Scale insects are relatively slow-growing but they produce large numbers of offspring. Depending on the species, a female may produce from 80 to 2500 eggs. In some species, the eggs are protected by a covering of white woolly wax, but more generally the eggs are retained under the female scale. The female dies shortly after the eggs have been laid, but her scale remains adhering to the surface of the plant for a long time afterwards. After hatching, the crawlers leave the scale and move over the plant for a few hours until they find a suitable site and become sedentary. Within a short time, the nymphs become covered by waxy secretions or scales of thickened integuments, and gradually become larger and thicker as they grow. Wind and trade of plant material are the most common dispersal vehicles of scale insects.

Feeding and Damage

Scale insects are serious pests on woody ornamentals especially in interior plantings in warm climates and in greenhouses in cooler climates. Scale insects feed by piercing the tissues of leaves and stems and withdraw the phloem sap. This causes loss of vigour of the host plant, stunting of young growth and yellowing of leaves. The ornamental value of the plants is also reduced by deposits of honeydew and consequent growth of sooty moulds, or by the mere presence of scales or white wax secretions on the plant. Some armoured scales may inject toxic substances into the plant tissue during the feeding process and this may also affect plant vigour and cause cosmetic damage.

Control

The economic thresholds of scale insects on ornamental crops are practically zero due to cosmetic demands and quarantine regulations. This has led growers to spray chemicals as the main control tactic for scale insects. Insecticide resistance, risks of phytotoxicity on ornamental plants, and a more restricting legislation for pesticide use are increasingly pushing growers to adopt safer methods for controlling scale insects. The biological control of scale insects has a long history for crops grown in the open air. However, in greenhouse-grown ornamentals, which include hundreds of plant species and varieties with extremely low economic thresholds of dozens of scale species, the potential market for each crop and scale species is rather narrow and does not justify high investments in research, development, production and marketing of specific natural enemies. Only a few natural enemies are commercially available for scale insect control, mostly against scales that are important pests in crops other than greenhouse ornamentals. This is the case of: (i) *Cryptolaemus montrouzieri* Mulsant, a predator used against *P. citri* and other mealybugs; (ii) *Leptomastix dactylopii* Howard and *Leptomastix epona* (Walker), a parasitoid of citrus mealybug; and (iii) *Metaphycus helvolus* (Compère), a parasitoid of *Saissetia oleae* (Olivier). Note that the three examples deal with natural enemies principally used for biological control of citrus pests.

4.2.10. FUNGUS GNATS

Most economically important fungus gnats or sciarids belong to the dipterous families Lycoriidae and Sciaridae. Under natural conditions, the larvae of fungus gnats inhabit wild fungi, leaf mould, manure piles, rotting wood and other decaying material where they have a predominantly saprophagous habit. In cool areas, fungus gnats are serious pests of mushroom production and are increasingly recognized as pests of ornamentals. The major pest species are *Lycoriella solani* (Winnertz) and some species of the genus *Bradysia*, mainly *Bradysia paupera* Tuomikoski, *Bradysia impatiens* (Johannsen), *Bradysia coprophila* (Lintner) and *Bradysia tritici* (Coquillet) (for a review, see Wilkinson and Daugherty, 1970; Freeman, 1983).

Direct plant damage may occur through larval feeding on root hairs and roots. The damage is most severe on seedlings, which may die due to the attack, and on unrooted cuttings where the larvae may tunnel up through the stem and, thus, also be lethal to the plant. Older well rooted plants will tolerate very heavy infestations of fungus gnats but may be damaged indirectly by the larvae. Indirect damage is caused by infection by pathogens through the damaged root parts or by pathogenic fungi (e.g. *Fusarium* spp. or *Pythium* spp.) that use fungus gnats as vectors. Furthermore, the number of fungus gnats in greenhouses may build up to such large numbers that the flying adults are a major nuisance to the greenhouse workers.

Chemical control of fungus gnats has been increasingly difficult due to the development of insecticide resistance and statutory reductions of available insecticides. Several biological control agents are available against fungus gnats: toxins from *B. thuringiensis*, entomophagous nematodes [*Steinernema carpocapsae* (Weiser) and *Steinernema feltiae* (Filipjev)] and soil dwelling predatory mites [*Hypoaspis miles* (Berlese) and *Hypoaspis aculeifer* (Canestrini)]. Biological control based on a combination of these agents and a high hygienic standard in the greenhouses are normally effective.

4.3. Prospects for the Future

The increasing globalization of international trade in vegetable and particularly ornamental crops will probably result in an acceleration of the establishment of exotic insect and mite pests in old and new greenhouse areas. *Frankliniella occidentalis*, *T. palmi*, *B. tabaci/argentifolii*, *Liriomyza* spp. and *S. exigua* are among the many examples of such a globalization process in recent years. Despite efforts devoted to preventing or reducing the entrance of exotic phytophagous species, the catalogue of new pests in greenhouses becomes longer every year. A global distribution of all important greenhouse pests in the future seems difficult to avoid.

Globalization of pest occurrence is a common phenomenon in all crops, but particularly prevalent in protected cultivation. The international trade in an increasing variety of ornamentals, the favourable conditions of greenhouse environment for a rapid increase in the pest population, and the capacity of some insects and mites to establish in open field explain the special relevance of the problem in protected cultivation.

Several actions to mitigate the problem may be undertaken. First of all, on the basis of pest characteristics and potential vehicles, a list of potential pests that are likely to arrive and establish in new areas should be compiled. Then, biological data on the potential pests should be collected in order to identify ways to prevent their importation and establishment. A complete dossier on the potential pests with accurate data on their life history, plant hosts, climatic preferences, natural enemies and other control tools would allow the areas and crops with the major risks to be defined. Effective quarantine programmes do not completely prevent pest entry but may delay it and provide time to prepare the actions to be undertaken in the case of pest occurrence. Quick and reliable tools for detection and identification would help in adopting the right ways to avoid the spread of the pest from the infestation origin and eventually its eradication. If finally the pest becomes established, the potential control methods that have been previously defined may be more rapidly tested *in situ*.

As a final remark we could say that greenhouse pests – indeed any pest – do not know anything about political and administrative frontiers and are less and less national

– or continental – problems but are becoming global pests. Restrictions to international trade in ornamental and food products will be restricted under the pressure of economic and social needs. Only international co-operation among scientists and scientific institutions may help to mitigate the globalization of pest problems.

Acknowledgements

W. Ravensberg, from Koppert Biological Systems, is thanked for providing us with valuable information on the current state of biological control of lepidopteran pests in greenhouses.

References

Alford, D.V. (1991) *Pests of Ornamental Trees, Shrubs and Flowers*, Wolfe Publishing Ltd, London.
Freeman, P. (1983) *Sciarid Flies. Diptera, Sciaridae*, Handbooks of the Identification of British Insects, Vol. 9, No. 6, Royal Entomological Society, London.
Gerling, D. (ed.) (1990) *Whiteflies: Their Bionomics, Pest Status and Management*, Intercept Ltd, Andover.
Helle, W. and Sabelis, M.W. (eds.) (1985) *Spider Mites: Their Biology, Natural Enemies and Control*, Vols. A and B, Elsevier, Amsterdam.
Lewis, T. (ed.) (1997) *Thrips as Crop Pests*, CAB International, Wallingford.
Minks, A.K. and Harrewijn, P. (1987) *Aphids: Their Biology, Natural Enemies and Control*, Vols. A, B and C, Elsevier, Amsterdam.
Rosen, D. (ed.) (1990) *Armored Scale Insects: Their Biology, Natural Enemies and Control*, Vol. B, Elsevier, Amsterdam.
Scopes, N.E.A. (1985) Red spider mite and the predator *Phytoseiulus persimilis*, in N.W. Hussey and N.E.A. Scopes (eds.), *Biological Pest Control, the Glasshouse Experience*, Blandford Press, Poole, Dorset, pp. 43–52.
Spencer, K.A. (1973) *Agromizidae (Diptera) of Economic Importance*, Dr. W. Junk B.V. Publishers, The Hague.
van de Vrie, M. (1991) Tortricids in ornamental crops in greenhouses, in L.P.S. van der Geest and H.H. Evenhuis (eds.), *Tortricid Pests: Their Biology, Natural Enemies and Control*, Elsevier, Amsterdam, pp. 515–539.
Wilkinson, J.D. and Daugherty, D.M. (1970) The biology and immature stages of *Bradysia impatiens* (Diptera: Sciaridae), *Annals of the Entomological Society of America* **63**, 656–660.

CHAPTER 5

NEMATODES

Soledad Verdejo-Lucas

5.1. Introduction

Plant-parasitic nematodes are microscopic obligate parasites of an aquatic nature that feed on plant roots. They cause severe damage to major food and fibre crops, whose estimated annual yield loss is about 12%. Nematode impact is probably underestimated because symptoms of nematode damage are usually non-specific and the damage is only measurable as yield loss. The nematodes that harm horticultural crops most are the root-knot nematodes, *Meloidogyne* spp., of which *Meloidogyne javanica* (Treub) Chitwood, *Meloidogyne incognita* (Kofoid & White) Chitwood and *Meloidogyne arenaria* Neal are the most common species. These nematodes cause yield losses of between 15 and 60% in the Mediterranean region (Lamberti, 1981; Ibrahim, 1985; Verdejo-Lucas *et al.*, 1997). Other nematode genera associated with horticultural crops in greenhouses are *Tylenchorhynchus, Pratylenchus, Helicotylenchus, Paratylenchus, Ditylenchus, Trichodorus, Paratrichodorus* and *Heterodera*. They may significantly affect plant growth in specific environmental niches, but cause little overall economic loss. This chapter only refers to *Meloidogyne* spp.

To alleviate nematode problems, procedures based on integrated pest management (IPM) and principles of prevention, population reduction and tolerance can be implemented. IPM seeks to stabilize nematode populations at acceptable levels resulting in favourable long-term socio-economic and environmental consequences (Bird, 1981). IPM, however, has so far received little attention for nematode control due to the availability of reliable broad-spectrum soil fumigants such as methyl bromide (Roberts, 1993). This situation may change if broad-spectrum pesticides are further restricted or phased out. According to the Montreal Protocol, methyl bromide will be phased out by 2001 in the USA and 2010 in the European Union due to its role in stratospheric ozone depletion.

5.2. Description and Biology

Meloidogyne spp. are sedentary endoparasitic nematodes. The infective second-stage juvenile hatches from the egg in moist soil, moves freely and penetrates into the root just behind the root tip. With *M. incognita, M. arenaria* and *M. javanica*, invasion does not occur below about 15°C (McKenry and Roberts, 1985). Once inside the cortical tissue, the juvenile establishes a feeding site. Several root cells around the nematode's head region enlarge in response to feeding to form giant cells that have several nuclei and constitute a nutrient sink from which the feeding nematode draws nutrient.

61

R. Albajes et al. (eds.), Integrated Pest and Disease Management in Greenhouse Crops, 61-68.
© 1999 *Kluwer Academic Publishers. Printed in the Netherlands.*

Juveniles enlarge and become saccate in shape as they develop. Most host plants react to *Meloidogyne* feeding by rapid cell division and expansion in the cortical area surrounding the point of infection; this results in characteristic knots or galls. A pear-shaped female lays eggs into a gelatinous matrix, known as the egg mass, which holds the eggs together at the posterior end of the female body. Under normal conditions, males are rare and mating is not necessary for *M. incognita*, *M. arenaria* and *M. javanica*, all of which reproduce by parthenogenesis. The nematode is active throughout the year in warm, moist soils that support growth of host plants.

The distribution of the nematodes forms an aggregation pattern within a field. The spread of nematodes in the soil by their own movement is slow and dissemination mainly occurs with movement of soil, plant material, machinery, containers, water and wind. Development of *Meloidogyne* spp. is influenced by several factors such as temperature, the suitability of the host plant and soil texture. Thus, the nematode needs to accumulate 600–700 degree-days of soil temperature above 10°C to complete a generation, which corresponds to 3–4 weeks when the soil is warm and moist. *Meloidogyne* spp. has a broad host range that includes many cultivated and non-cultivated plants including weeds. The nematode is favoured by sandy soils, but may cause damage in almost any kind of soil.

5.3. Symptoms and Damage

As in other root diseases and some nutrient deficiencies, above-ground symptoms include stunting, slow growth, yellowing, early senescence and abnormal wilting even when the soil is wet. These symptoms result from damage to the root system and disruption reducing the plant's ability to take up the water, minerals and nutrients necessary for normal growth. Symptoms are most pronounced when fruit development stresses the plant. Uneven plant growth is an early symptom of nematode attack and is caused by simultaneous root invasion by many juveniles which causes a retardation in plant growth due to great tissue injury. Nematode-damaged plants are usually located in patches or along the planting row, reflecting their aggregation pattern.

As for below-ground symptoms, root galls are the most distinctive symptom of root-knot nematode attack. Galls are produced as a specific response of the plant to the nematode. Their size and number vary, depending on host susceptibility and nematode population densities at planting. However, high root-knot nematode populations may be present in some plants even if no galls appear. In contrast, individual plants with large galls may be found scattered and adjacent to unaffected plants. Infected plants have poor root systems with an abnormal number of shallow lateral roots. The predisposition of nematode-infected plants to secondary infections by fungal and bacterial organisms may have additional adverse effects on plant growth.

5.4. Sampling and Monitoring

The objectives of sampling for IPM programmes are to relate numbers and kinds of nematodes to crop performance, and to aid evaluation and selection of management strategies. Due to the typical aggregation pattern of nematode populations, collection of composite samples with varying numbers of soil cores is necessary. Samples are collected with a sampling tube, and the populations of the taxa present are extracted and counted. Important considerations in soil sampling are: (i) the number, diameter and depth of cores needed to provide an adequate sample; (ii) a representative pattern of sampling to obtain reliable data on population densities; (iii) sampling at a time that reflects population density at critical stages of the growing season; (iv) soil conditions; and (v) proper handling and storage of samples. Nematodes should be extracted and identified in properly equipped laboratories by trained personnel. For extraction procedures and identification of major nematode genera, see Hooper and Evans (1993). Identification of *Meloidogyne* is based on morphological characteristics of the adult males and females and juveniles (Eisenback, 1985), on enzyme phenotypes (Esbenshade and Triantaphyllou, 1985) and on DNA-based techniques such as RAPD-PCR (Cenis, 1993).

Examination of roots for root-knot nematode galls is a means of detecting the presence of *Meloidogyne* spp. in a field. The degree of galling also provides information on severity of damage. Root-galling indices are based on the proportion of roots with galls in the entire root system. Root gall indices and yield losses have a linear relationship.

Ferris and Noling (1987) and McSorley and Phillips (1993) have discussed the use of modelling population dynamics and yield losses for nematode management. For annual crops, the critical population density is that at the time of planting, which enables predictive relationships between preplanting populations and crop yield to be calculated. Such relationships can be easily utilized because most management strategies, including varietal selection and soil fumigation, are preplanting decisions. Increasing nematode population densities can progressively affect crop performance, though there is a minimal density threshold below which no measurable loss yield occurs. This is referred to as the tolerance limit. Root-knot nematode population densities of one juvenile or less per cm^3 of soil usually do not cause measurable yield losses. Nematode densities higher than 3.5 juveniles per cm^3 of soil at planting cause yield losses of about 50% of the crop. Temperature, moisture, soil type, age of the plant at infection and other stress factors seriously affect the amount of damage, when nematode population densities at planting are between 2–3 juveniles per cm^3 of soil.

Because of the relative immobility of nematodes, immigration and emigration do not play an important role in population growth within a field and so the final population (Pf) at harvest can be predicted from the initial population (Pi) before planting. The maximum multiplication rate (Pf/Pi) is seen at low initial densities when resources are unlimited. As the initial population density increases, the multiplication rate decreases, owing to increasing competition between individuals and a decreasing supply of food. At higher population densities, an equilibrium may be reached, at which the final and initial nematode populations are equal. Initial nematode densities may decrease at harvest if plants are severely damaged.

5.5. Control Strategies

Crop losses caused by nematodes can be avoided by preventing the introduction of specific nematodes or nematode problems into areas where they have not existed before. Exclusion procedures include sanitation, the use of certified plant material, nematode-free soil and planting media, and regulatory activities (quarantine). Successful management of established populations begins by reducing nematode population densities in the field before planting, since there is no consistently effective way to rescue a crop once it has been infected by nematodes (Noling and Becker, 1994). The strategies available for nematode control can be applied sequentially or simultaneously and their effect should be considered over multiple seasons because actions taken in one crop may affect subsequent crops.

5.5.1. CULTURAL PRACTICES

Soilless Cultivation
Inorganic and organic media are used for soilless cultivation. Important factors that have contributed to the development and commercial application of soilless techniques include the difficulty and cost of controlling soilborne pests and diseases, soil salinity, lack of fertile soils and water shortage (Olympios, 1993). The technique is increasingly applied to protected vegetable crops and ornamentals. It is widely used in greenhouses in northern Europe and its use is expanding to the Mediterranean region.

Crop Rotation
Crop rotation with poor or non-host crops can provide adequate suppression of nematode build-up in annual crops. In an optimal rotation sequence, the preceding crop reduces the population density of the nematode and prevents damage to the subsequent crop (Ornat *et al.*, 1997). Crop rotation for nematode control in high-value cash crops, such as vegetable production in greenhouses, is often not a realistic option due to the polyphagous nature of *Meloidogyne* spp. However, it can be a useful strategy in subsistence agriculture (Bridge, 1987) and field crops (Rodriguez-Kabana, 1992).

Weed Control
Weeds can act as reservoirs of infection and as hosts that maintain or even build up nematode populations. Weed control is especially important in rotation programmes, because the success of a programme depends on the absence of host roots to prevent nematode feeding and reproduction. Therefore, the presence of weed hosts will negate the benefits of growing a resistant or non-host crop plant.

Fallowing
Fallowing is a simple method which controls nematode populations by starvation. This strategy aims to reduce population densities of all kinds of plant-parasitic nematodes. The fallow period is limited to a few weeks in intensive agriculture. However, when coupled with root destruction, even short-term fallowing has a significant and immediate impact on total nematode population densities in soil. Unfortunately,

fallowing has unfavourable effects on soil organic matter and soil structure. Because of the wide range of root-knot nematode hosts, weed growth should be controlled during the fallow period; otherwise fallowing will be less effective. Frequent tillage is usually required to maintain clean fallow soil conditions.

Crop Destruction
Post-harvest nematode reproduction can be prevented by root systems being mechanically removed from soil and being exposed to the drying effects of sun and wind. Nematode densities will continue increasing if roots are not destroyed. Crop destruction is more effective than fallowing for population reduction.

Organic Soil Amendments
Organic soil amendments play an important role in restricting populations of plant-parasitic nematodes. They may enhance the activity of natural enemies and improve soil fertility and structure. Soil amendments tested include oil cakes, plant crop residues, green manure, and agro-industrial, animal and urban wastes. Some of these amendments have been shown to be beneficial by increasing yields with or without decreasing nematode populations. Research is needed to identify and characterize amendments that are locally available in the large quantities needed for nematode control.

5.5.2. RESISTANT CULTIVARS

Host plant resistance can be an effective, economic and environmentally acceptable method of controlling diseases and pests in greenhouses. This topic is discussed in Chapter 9. Plants resistant to *Meloidogyne* spp. can be grown on infested land without significant yield reduction because they considerably reduce nematode reproduction. Genetic resistance to *Meloidogyne* spp. has been developed commercially only in tomato, cowpea, lima bean and sweet potato. Genetic engineering will be a useful tool for the development of new nematode-resistant crop cultivars. Several classes of potential anti-nematode genes, encoding lectins, enzymes and enzyme inhibitors, are being evaluated for their ability to confer broad-spectrum nematode resistance. Plantibodies (antibodies produced in plants) are also being investigated as means of conferring pest and disease resistance (Stiekema *et al.*, 1997). The frequent cultivation of resistant cultivars may encourage the development of new pathotypes able to reproduce on such cultivars. The occurrence of these pathotypes in the Mediterranean region has already been reported (Tzortzakakis and Gowen, 1996; Eddaoudi *et al.*, 1997). They may also occur without previous exposure to resistant cultivars. Hence, resistant cultivars will be more effective if used in combination with other management practices. Further restrictions on the use of resistant cultivars should also be considered. For instance, in tomato, a single dominant gene is responsible for the plant's resistance, but this resistance is overcome at soil temperatures above 30°C. Therefore, in parts of the Mediterranean region, the use of these resistant cultivars will have to be restricted to autumn-winter planting when cooler temperatures prevail.

5.5.3. BIOLOGICAL CONTROL

The natural antagonists of nematodes most widely studied include: (i) the bacterial obligate parasite *Pasteuria penetrans* (Thorne) Sayre & Starr; (ii) the nematode trapping-fungi *Arthrobotrys oligospora* Fresen., *Arthrobotrys dactiloides* Drechsler and *Monacrosporium ellipsosporum* (Grove) Cooke & Dickinson; and (iii) the fungal egg parasites *Verticillium chlamydosporium* Goddard and *Paecilomyces lilacinus* (Thom) R.A. Samson (Stirling, 1991). Greenhouse and microplot experiments have shown that these organisms can reduce population densities of *Meloidogyne* spp. and, in some instances, match chemical nematicides in their results. Research continues on ways of exploiting these agents on a field scale and some progress has been made recently. Thus, *P. penetrans* has been commercialized as a soil improvement agent in Japan. DiTera®, produced by the submerged fermentation of the fungus *Myrothecium* spp. Tode, has been approved as a biological nematicide in some states of the USA and has been registered in Chile, Mexico and Panama.

5.5.4. STEAM

Steam at 80–100°C effectively controls most soilborne pathogens, weeds and insects (see Chapter 10), as discussed by Maas (1987). After the banning of methyl bromide in northern Europe, new and more effective steam application methods were developed for greenhouse soil disinfection (Runia, 1984). Steaming is considered by many growers as effective and economical as chemical disinfestation, but this strategy, used in the more sophisticated heated glasshouses of northern Europe, may not be so attractive for soil disinfestation in the unheated plastic structures predominant in the Mediterranean region.

5.5.5. SOIL SOLARIZATION

Soil solarization by covering moistened soil with a clear plastic sheet is an attractive control strategy for warm areas of the Mediterranean region (see Chapter 10). Its major advantages are the simultaneous control of insects, soilborne pathogens, weeds and nematodes, and the improvement of crop performance through modification of the physicochemical properties of the soil. The use of this technique for nematode control has been reviewed by Gaur and Perry (1991). The main limitations of soil solarization are its dependence on climatic conditions, the duration of the treatment and the cost. Research to determine the applicability, timing and economics of soil solarization is needed in specific geographical areas.

5.5.6. NEMATICIDES

Nematicides belong to two groups of pesticides: fumigants and non-fumigants. Fumigant compounds are volatile chemicals and broad-spectrum pesticides, with the exception of 1.3-dichloropropene which is only nematicidal. The success of fumigation depends on the application method, dosage and timing relative to temperature and moisture content of the soil. Fumigants are applied before planting because of their phytotoxicity. Non-fumigant nematicides are non-volatile compounds that include organophosphate and carbamate compounds. They are known as nematostats since they do not kill nematodes directly at field concentrations but do affect their behaviour by delaying hatching, interfering with movement and root invasion, impairing feeding behaviour and neuromuscular activity, and disorienting the behaviour of males towards females. Non-fumigant nematicides can be used at or just before planting. They have not always provided consistent results either for controlling the nematode or for increasing yield. A decrease in effectiveness due to microbial degradation has been reported for carbofuran, fenamiphos and oxamyl.

5.6. Integrated Management

Control of *Meloidogyne* spp. in greenhouses is difficult, mainly because of high soil temperatures that favour nematode development and the presence of susceptible host plants all year round. The characterization and correct identification of populations of *Meloidogyne* that occur locally is essential if management strategies that do not rely on the use of chemicals are to be developed successfully. Also, the relationship between numbers of nematodes, and crop performance and yield needs to be worked out. In the near future, growers will have to adopt IPM systems due to the restrictions or loss of broad-spectrum soil fumigants. The combination of two or more strategies will be needed to alleviate nematode problems, since there is seldom one single effective control method. Thus, soil solarization may be combined with soil fumigants, resistant cultivars or soil amendments to improve its efficacy. Resistant plants reduce nematode reproduction considerably, but the epidemiological implications of the residual populations left after their cultivation need to be determined in the subsequent crops. Regulation of nematode populations will be possible in areas with a distinctive cold season because low temperatures delay nematode development and prevent juveniles of *M. incognita*, *M. arenaria* and *M. javanica* from entering the roots. Cropping systems that leave the land uncultivated in the summer months will accelerate the rate of nematode mortality because of high soil temperature and the absence of host plants.

Acknowledgement

Thanks are given to the Instituto Nacional de Investigación y Tecnología Agraria y Alimentaria of Spain for economic support through grant No. SC95–049.

References

Bird, G.W. (1981) Integrated nematode management for plant protection, in B.M. Zuckerman, W.F. Mai and R.A. Rohde (eds.), *Plant Parasitic Nematodes*, Vol. III, Academic Press, New York, pp. 355–375.
Bridge, J. (1987) Control strategies in subsistence agriculture, in R.H. Brown and B.R. Kerry (eds.), *Principles and Practice of Nematode Control in Crops*, Academic Press, Sidney, pp. 389–420.

Cenis, J.L. (1993) Identification of four major *Meloidogyne* spp. by random amplified polymorphic DNA (RAPD-PCR), *Phytopathology* **83**, 76–78.

Eddaoudi, M., Ammati, M. and Rammah, A. (1997) Identification of the resistance breaking populations of *Meloidogyne* on tomatoes in Morocco and their effect on new sources of resistance, *Fundamental and Applied Nematology* **20**, 285–289.

Eisenback, J.D. (1985) Diagnostic characters useful in the identification of the four most common species of root-knot nematodes (*Meloidogyne* spp.), in J.N. Sasser and C.C. Carter (eds.), *An Advanced Treatise on Meloidogyne*, Vol. I, *Biology and Control*, North Carolina State University, Raleigh, NC, pp. 95–112.

Esbenshade, P.R. and Triantaphyllou, A.C. (1985) Use of enzyme phenotypes for identification of *Meloidogyne* species, *J. of Nematology* **17**, 6–20.

Ferris, H. and Noling, J.W. (1987) Analysis and prediction as a basis for management decisions, in R.H. Brown and B.R. Kerry (eds.), *Principles and Practice of Nematode Control in Crops*, Academic Press, Sidney, pp. 49–85.

Gaur, H.S. and Perry, R.N. (1991) The use of soil solarization for control of plant-parasitic nematodes, *Nematological Abstracts* **60**, 153–167.

Hooper, D.J. and Evans, K. (1993) Extraction, identification and control of plant parasitic nematodes, in K. Evans, D.L. Trudgill and J.M. Webster (eds.), *Plant Parasitic Nematodes in Temperate Agriculture*, CAB International, Wallingford, pp. 1–59.

Ibrahim, Y.K.A. (1985) The status of root-knot nematodes in the Middle East, region VII of the international *Meloidogyne* project, in J.N. Sasser and C.C. Carter (eds.), *An Advanced Treatise on Meloidogyne*, Vol. I, *Biology and Control*, North Carolina State University, Raleigh, NC, pp. 373–378.

Lamberti, F. (1981) Plant nematode problems in the Mediterranean region, *Helminthological Abstracts, Series B, Plant Nematology* **50**, 145–166.

Maas, P.W.T. (1987) Physical methods and quarantine, in R.H. Brown and B.R. Kerry (eds.), *Principles and Practice of Nematode Control in Crops*, Academic Press, Sydney, pp. 265–291.

McKenry, M.V. and Roberts, P.A. (1985) *Phytonematology Study Guide*, Publication 4045, Cooperative Extension University of California, Division of Agriculture and Natural Resources, Riverside, Cal.

McSorley, R. and Phillips, M.S. (1993) Modelling population dynamics and yield losses and their use in nematode management, in K. Evans, D.L. Trudgill and J.M. Webster (eds.), *Plant Parasitic Nematodes in Temperate Agriculture*, CAB International, Wallingford, pp. 61–85.

Noling, J.W. and Becker, J.O. (1994) The challenge of research and extension to define and implement alternatives to methyl bromide, *J. of Nematology* **26**, 573–586.

Olympios, C.M. (1993) Soilless media under protected cultivation: Rockwool, peat, perlite, and other substrates, *Acta Horticulturae* **323**, 215–234.

Ornat, C., Verdejo-Lucas, S. and Sorribas, F.J. (1997) Effect of the previous crop on population densities of *Meloidogyne javanica* and yield of cucumber, *Nematropica* **27**, 83–88.

Roberts, P.A. (1993) The future of Nematology: Integration of new and improved management strategies, *J. of Nematology* **25**, 383–394.

Rodriguez-Kabana, R. (1992) Cropping systems for the management of phytonematodes, in F.J. Gommers and P.W.T. Maas (eds.), *Nematology from Molecule to Ecosystem*, European Society of Nematologists, Wageningen, pp. 219–233.

Runia, W.T. (1984) A recent development in steam sterilization, *Acta Horticulturae* **152**, 195–200.

Stiekema, W.J., Bosch, D., Wilmink, A., de Boer, J.M., Schouten, A., Roosien, J., Goverse, A., Smant, G., Stokkermans, J., Gommers, F.J., Schots, A. and Bakker, J. (1997) Towards plantibody-mediated resistance against nematodes, in C. Fenoll, F.M.W. Grundler and S.A. Ohl (eds.), *Cellular and Molecular Aspects of Plant-Nematode Interactions*, Kluwer Academic Publishers, Dordrecht, pp. 262–271.

Stirling, G.R. (1991) *Biological Control of Plant Parasitic Nematodes*, CAB International, Wallingford.

Tzortzakakis, E.A. and Gowen, S.R. (1996) Occurrence of a resistant breaking pathotype of *Meloidogyne javanica* on tomatoes in Crete, *Fundamental and Applied Nematology* **19**, 283–288.

Verdejo-Lucas, S., Ornat, C. and Sorribas, F.J. (1997) Management of root-knot nematodes in protected crops of northeast Spain, *Bulletin OILB /SROP* **20**, 94–98.

CHAPTER 6

PRINCIPLES OF EPIDEMIOLOGY, POPULATION BIOLOGY, DAMAGE RELATIONSHIPS AND INTEGRATED CONTROL OF DISEASES AND PESTS

Aleid J. Dik and Ramon Albajes

6.1. Introduction

Epidemiology and population biology study the development and spread of plant diseases and arthropod pests and the factors affecting these processes. The level of disease or pest infestation is the result of many interacting factors and this level determines the yield loss that the grower suffers from the pathogen or pest. In many respects, the methodology of research and the underlying concepts are very similar for bacterial and fungal plant pathogens, insects, mites, viruses and nematodes. In this chapter, we will: (i) introduce the reader to these concepts; (ii) explain how they can be used in integrated control; and (iii) illustrate how damage relationships can be established.

6.2. The Disease/Pest Tetrahedron

The disease/pest tetrahedron is used to envisage the interaction of diseases and pests with their environment. The tetrahedron consists of four components, which can all influence each other and together determine the level of the disease or pest. The four components are the plant pathogen/pest, the host plant, the environment, and human activity. Generally, it can be said that chemical control is only aimed at influencing the pest or pathogen directly, whereas integrated control may reduce the level of disease or pest by influencing any or several of the four components of the tetrahedron. A thorough knowledge on the influence of different factors on pests and diseases offers the basis for integrated control. The four components of the tetrahedron will be discussed below.

6.2.1. THE PATHOGEN/PEST

Infection Cycle of Plant Pathogens
The infection cycle of fungal plant pathogens consists of the following phases (Zadoks and Schein, 1979): (i) infection (germination of spores, penetration of plant tissue and colonization of plant tissue); (ii) sporulation (production of spores and maturation of spores); and (iii) dissemination (spore liberation, spore dispersal and spore deposition). Bacterial pathogens and viruses generally have a similar but simpler infection cycle. Some pathogens only complete one infection cycle per season, but most pathogens complete several or sometimes many cycles per season. The amount of disease that

R. Albajes et al. (eds.), Integrated Pest and Disease Management in Greenhouse Crops, 69-81.
© 1999 *Kluwer Academic Publishers. Printed in the Netherlands.*

develops is the sum of the successes of the different phases in the cycle. All phases are subject to influences from the environment, the host plant and human activity.

Life Cycle of Pest Organisms
Although pest organisms belong to various taxonomic groups and consequently their life cycle varies accordingly, a general pattern for phytophagous arthropods may be described. Dispersal usually occurs at adult stage; adults look for an exploitable habitat and then a plant to feed and oviposit on. Once a plant is recognized as suitable for feeding and ovipositing, an adult female lays eggs (ovipary) or deposits nymphs or larvae (vivipary). Progeny commonly feed and develop on the plant where oviposition took place or on neighbouring host plants until reaching the mature stage. The adults may feed and oviposit on the same plant or move to younger and more suitable host plants. Many arthropods spend all their life on a plant but others, particularly among holometabolous insects, either have soil-inhabiting stages or the adults feed on a different host plant than the immatures did. Most greenhouse pests are multivoltine (several generations within a year), and univoltinism (one generation per year) is rather rare.

Research on disease epidemiology and pest population biology, can study the effect of different factors on the phases in the infection/pest cycle or can determine the effect on the resulting amount of disease or pest density directly. Studying the effect on each of the processes described – for instance germination and spore production in the infection cycle, or host plant selection and exploitation in a pest cycle – often yields a good understanding of the effect of a factor on a disease or amount of damage caused by a pest. This kind of research is best done under controlled conditions, which allows the fluctuation of only one or two factors. Research on epidemiology of a disease or pest population biology in a whole crop is usually based on monitoring many factors and analysing their respective effects through regression analysis. It is important that in such research all the relevant factors are monitored.

6.2.2. THE HOST PLANT

Traditionally, the host plant has been considered as a more important factor for the development of a disease than for determining the damage caused by an arthropod pest. This may be one of the reasons why plant resistance has been used more frequently to control diseases than to control pests. In fact, the tetrahedron scheme is common in plant pathology texts but rare in applied entomology books. However, this situation has changed recently as entomologists have intensified studies on insect-plant relationships.

Different cultivars may vary in their susceptibility to certain diseases or pests, even if none of them is completely resistant. This partial resistance may influence disease development by decreasing the number of successful infections by the pathogen, by increasing the latency period, by reducing the rate of lesion expansion or sporulation, or by any combination of these processes. The result will be a slower development of the epidemic. Similarly, different cultivars may influence the host plant selection by a pest and/or its oviposition, development and survival, and thus the rate of population increase. Many partially resistant cultivars express the same resistance during their entire life, but some resistance may depend on the physiological age of the plant. Age-related resistance can be either adult plant resistance or young plant resistance. On another scale, certain parts of the plant may be more or less susceptible to disease because of their age (see Chapter 9 for the use of plant resistance in IPM).

The physiological status of the plant is affected by temperature, humidity and nutrition. Nutrient deficiencies or certain climatic conditions may predispose plants to the development of diseases and pests, but unbalanced fertilization may also increase this kind of problems. For example, excess of nitrogen amendments renders the plant more susceptible to *Botrytis cinerea* Pers.:Fr. and enhances the population increase of homopteran pests like aphids and whiteflies. A thorough knowledge of the factors affecting the incidence of diseases and pests through their host plant may help to prevent outbreaks by applying correct crop management practices (see Chapter 8 for crop management practices and pest and disease control in greenhouses).

Because the host plant is such an important factor in disease epidemics and pest population dynamics, it is important to carefully monitor the host plant in any epidemiological research. This means not only that cultivar and planting date should be recorded, but also plant spacing, nutrition, development stage of the plant and plant growth. Recording plant growth is also important for another reason: whether the disease is assessed as a percentage of total plant area – for example leaf area covered – or as a pest density – for example, the number of insects per leaf area – it is important to know the size of the host plant. Also, a count of the number of lesions for example may yield the same number for different crops, but if the number of leaves or other aspects of the size of the host plant are different, the impact of this number of lesions will be different and results difficult to compare between crops.

6.2.3. THE ENVIRONMENT

Many components of the environment can indirectly influence the severity of disease and pest injury through the host plant or by a direct effect on the pathogen and pest. Here, only the direct impact of the environment on the pathogen and pest population is dealt with.

The main influences on pathogens stem from temperature, relative humidity (RH), radiation and wind. Pathogens are often affected by climatic conditions in most of the phases of the infection cycle. Germination of spores and superficial germ tube growth often show an optimum curve for temperature. For most pathogens, germination only occurs above a certain (high) level of RH or in the presence of wetness on the plant surface. Lesion expansion is often influenced by air temperature, since this also affects the temperature of the plant tissue, but less by RH. Sporulation is affected by temperature and RH, whereas spore dispersal is often mostly influenced by air humidity and movement. It is important to realize that all these environmental factors will be different at different crop heights. It is therefore necessary to measure microclimatic conditions at a height where the pathogen is expected to attack the host plant. Manipulation of the environment by, for example, changing radiation with different covers, and/or changing temperature, RH and ventilation by using different heating and/or ventilation regimes, may influence diseases and pests. This will be discussed in more detail in Chapter 8.

Concerning pests, their density is directly affected by many biotic and abiotic factors of the environment other than the host plant, such as competing herbivores, natural enemies and climatic factors. Among the latter, temperature, RH, light (quantity, quality and periodicity) and air movement are usually the most decisive in determining behaviour, phenology, survival, fecundity and dispersal of pests. All these abiotic factors may also affect natural enemy populations and, therefore, pests in an indirect, but important, way. Understanding all these complex interactions is crucial for detecting the most decisive factors that govern pest population dynamics and managing the environment for pest control accordingly.

6.2.4. HUMAN ACTIVITY

Humans can affect both the host plants, the greenhouse environment and the pathogen or pest organism by cultural practices and by application of chemical, biological and other types of control. Cultural practices may be aimed at rendering the plant less susceptible or acceptable to diseases and pests, or eliminating (or at least reducing) infection sources. Furthermore, any cultural practice changing plant growth, such as leaf area, will influence the microclimatic conditions for the pathogens and pest organisms in the crop. Greenhouse crops tend to be labour-intensive and this can lead to a more frequent spread of pathogens and pests over the crop by workers than in field crops.

6.3. Disease Epidemics and Pest Population Dynamics: Bases for Intervening in Agroecosystems to Reduce Losses

The amount of disease or the pest density resulting from the interactions between environment, plant and pathogen/pest, and the influence of humans on these factors, is the subject of plant disease epidemiology and pest population dynamics research. Plant disease epidemiologists and agricultural entomologists have approached these studies in seemingly different ways. However, given that both types of scientist deal with populations of living organisms that are subjected to common phenomena such as birth, death, development and migration, approaches of plant disease epidemiology and pest population dynamics are basically similar. Whereas the plant pathologist deals mainly with the effects of the harmful agent (the pathogen), that is the disease, the entomologist is more concerned with the agent itself (the pest). Consequently, plant pathologists usually measure the disease incidence or severity, whereas entomologists generally estimate the number of pest individuals.

6.3.1. DISEASE EPIDEMICS

The amount of disease changes over time. The curve of the amount of disease against time is called the disease progress curve (DPC). It is typically of a sigmoid shape, with a slow increase in the beginning, followed by a logistic increase and a levelling off at the maximum level of disease. Vanderplank (1963) showed that DPCs can be described by logit:

$$(y_t) = \text{logit}(y_0) + r * t$$

where $\text{logit}(y) = \ln [y/(1 - y)]$, y_t = disease on time t, y_0 = initial disease on time 0, r = rate of increase, and t = time. The logit transformation will turn a sigmoid curve into a straight line, which enables an easier comparison of DPCs than the original sigmoid curves. For some diseases, transformation by gompits is better than logits:

$$\text{gompit}(y) = -\ln [-\ln(y)]$$

Disease can be reduced by reducing initial inoculum (y_0), lowering the rate of increase of the disease (r) or limiting the duration of the epidemic (t) by delaying its start through preventive measures. Reduction of initial inoculum is the purpose of sanitation measures before planting the new crop, but inoculum can also be reduced during the epidemic by removing diseased plants or plant parts. Reduction of inoculum will delay the epidemic. It depends on the kind of pathogen whether this delay provides sufficient control. In the case of a pathogen whose spores are abundantly present in the environment, for example powdery mildew fungi, the reduction of initial inoculum will only give a small delay in disease progress. In the case of rare inocula, sanitation may provide almost complete control. Sanitation takes place both in the nurseries and in the greenhouses.

Most manipulations of either the pathogen, the environment or host-plant susceptibility are aimed at a reduction of the rate of disease progress. This is achieved by slowing down any of the processes in the infection cycle.

6.3.2. PEST POPULATION DYNAMICS

Malthus' equation, initially developed to describe human population growth, was soon adopted by entomologists to study insect demography. The equation predicts that a population will grow exponentially according to:

$$N_t = N_0 e^{rt}$$

where N_t is the number of pest organisms at a specified time, N_0 is that number at an initial time (0), e is the base of Naperian logarithms, r is the rate of population increase, and t is the elapsed time. If r is assumed to be constant and independent of conditions that affect pest development, survival and reproduction, population growth is unlimited. This rate of increase r is also called intrinsic or maximal rate of increase and as such is referred to as r_m and depicts the rate at which the population would increase under permanently favourable conditions. In nature, however, favourable conditions are never indefinitely maintained and several – usually many – factors limit or retard population growth. To reflect this, the so-called Verhulst' model predicts that populations will grow until reaching a maximum following a logistic or sigmoid curve that can be mathematically expressed by:

$$N_t = K/\{1 + [(K - N_0)/N_0] e^{-rt}\}$$

where K, called *carrying capacity* (the maximum population size that the environment's resources can sustain), is the asymptote of the sigmoid curve and the rest of elements as in the formula of exponential growth. The parameter K is a measure of the global effect of all environmental factors that limit the growth of a population, the so-called environmental resistance. The shape of population growth in Verhulst' model is represented in Fig. 6.1. Note that it can also represent the logistic model of Vanderplank and, although biologically unrealistic, it allows to show how pest control procedures may prevent pest populations to reach damaging densities. Like in disease control, pest population growth may be reduced by decreasing or delaying immigration of first pest individuals into crop plants (lowering N_0 or t), or by decreasing the rate of population increase via integrated enhancement of environmental resistance, for example by release of natural enemies (lowering K). In the chapter on plant resistance (Chapter 9), the reader will find a discussion on how different effects on pest development and reproduction influences r_m.

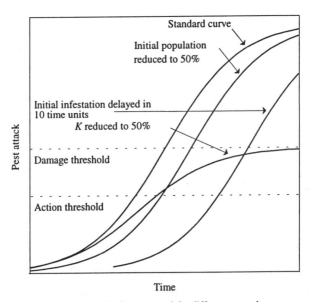

Figure 6.1. The logistic curve of population increase and the different procedures to prevent pest population to reach damage threshold. Initial population can be reduced (e.g. by planting uninfested seedlings) or K can be lowered (e.g. host-plant resistance or biocontrol) or initial infestation delayed (e.g. sanitation).

6.4. Damage Relationships

Knowledge of the impact of diseases and pests on crop yield is needed for decision-making in disease/pest control. Decision-making is based on the damage relationship, that is, the level of yield loss associated with different amounts of disease or levels of pest attack. Yield loss can be either loss of quality or loss of quantity or both. Usually, damage relationships for a certain pathosystem or pest/plant interaction are expressed in quantity of yield, either in relative or absolute units. Most frequently, damage relationships are assessed empirically and analysed with regression analysis to estimate yield or yield loss from observations on the amount of infestation. The main drawback of this approach is that the resulting models are descriptive and do not take into account the physiological processes underlying the yield. Despite this, the descriptive models can be very useful in integrated control. Following Campbell and Madden (1990), five descriptive models to calculate the amount of disease and yield loss relationship may be distinguished:

(i) The single-point models or critical-point models. These models estimate yield loss by determining the amount of disease on one given moment, usually determined by the physiological status of the crop, for example the onset of flowering. Less frequently, these models have been developed with time variables, for example the number of disease-free days or the time until a certain level of disease is reached. Single-point models have been developed mainly for diseases in cereals. Their use is limited in crops in which yield accumulates over a longer period of time or harvesting takes place more than once, as for example in greenhouse vegetables.

(ii) Multiple-point models. These models use several disease assessments to estimate yield loss. This type of model is most useful in situations where disease progress can be highly variable, depending on the host plant or the environment.

(iii) Integral models. These models use the summed disease pressure over a specific period of crop growth which is relevant to yield. This is determined by calculating the area under the disease progress curve (AUDPC). These models can not distinguish between an early moderate epidemic and a more severe epidemic which starts later with the same AUDPC. This can be overcome by assigning weighting factors to the disease assessments made on different times or by incorporating another factor, for example the number of disease-free days, into the model.

(iv) Response surface models. These models predict yield loss by using two different types of variables, for example disease severity and crop growth stage.

(v) Synoptic models. They are multivariate models that estimate yield loss by incorporating all independent variables in one equation.

Much of the conceptual framework to estimate the relationship between amount of disease and yield loss may be applied as well to damage relationship concerning arthropod pests. For decision-making purposes, a linear function of the yield response to insect infestation is generally assumed. In case the crop is able to compensate limited injury, there is a level of tolerance associated with low pest density. Crop tolerance to pest attack may be relatively high when pests injure the leaves of fruit vegetables like tomato, pepper, cucurbits or egg plants; often even 30–40% of leaf injury does not result in yield reduction. Sigmoid yield responses to pest infestation are more difficult to fit, but logarithmic or power transformations may linearize the damage relationship. The consideration of more than one pest or disease and crop variables render complex polynomial relationships (the synoptic models mentioned above) which are difficult to interpret and to use in decision-making. If a linear yield response may be assumed or derived, damage relationship can be "easily" found with field data as it has been mentioned above in single- and multiple-point models. When pests are multivoltine and their numbers are quite variable along the season, the use of insect*days instead of seasonal mean insect densities may be more meaningful as noted in the above-mentioned integral models. Methods and techniques for this kind of studies may be found in Teng (1987).

A different approach to determining crop loss is the use of dynamic simulation models. Generally, a model for the development of the pest or disease is combined with a model for crop growth and production. This approach generates explanatory models, which are expected to have a greater predictive value than descriptive models. However, the development of simulation models requires a lot more basic information on the physiological processes and on the effect of environmental parameters on epidemics, and are therefore more difficult to develop than models based on regression analysis.

6.5. Damage and Action Thresholds

In combination with the damage relationship for a given pathosystem or pest/crop interaction, it is important to assess the damage and action thresholds. The damage threshold is the maximum level of disease or pest attack below which yield losses do not occur. The action threshold is the level of disease or pest attack at which control action should be taken to prevent the epidemic or pest to reach the damage threshold. Because there are often no fully effective techniques to control a disease or a pest immediately, the action threshold is lower than the damage threshold. Generally speaking, action thresholds for diseases lie below the logistic increase in the DPC.

Damage and action thresholds are an important tool in integrated control when several control alternatives are available. Whereas the damage threshold essentially depends on the disease/pest level and yield loss relationship, the action threshold may greatly vary according to the efficacy of each of the control alternatives and how long they take to be effective. Action threshold to control a disease – or a pest – will be probably higher if we choose a quick acting pesticide than with biological control, where natural enemies need time to react. For example, the action threshold for greenhouse whitefly control may be up several adults per leaf if we rely on insecticides, whereas it is around one adult per leaf when the parasitoid *Encarsia formosa* Gahan must be used in seasonal inoculative releases for the biological control of the whitefly. The knowledge and application of action thresholds generally reduce the amount of control inputs compared to general practice.

Determination of the thresholds is not always easy. Yield loss can be defined as loss in weight of the harvested product or the loss in economic value. This potential economic loss can be compared to the cost of a control measure. The translation of yield loss in weight to economic yield loss depends on the expected price of the harvested product and is therefore difficult to perform. In greenhouse crops that are harvested continuously, this is further complicated by the fluctuating prices within one growing season. For example, one kilo yield loss in cucumber or tomato in The Netherlands will be much more costly for the grower at the beginning (early spring) and end (late fall) of the season when prices are higher than in summer. Further complexity in the determination of the action thresholds leads to some other considerations: (i) long term consequences of the current decisions for the disease or pest levels (instead of considering just one generation or infection cycle); (ii) influence of control actions on crop revenue (it may be different if the decision is made at farmer or regional level); and (iii) the risk attitude of the grower. Regarding the latter, stochastic models (in which an occurrence probability is associated to each decision) are more useful thresholds for quantifying risks than the deterministic damage and action model. Readers specially interested in the subject may consult the book by Norton and Mumford (1993).

6.6. Damage Relationships and Thresholds in Greenhouse Crops

Despite the importance of knowing damage relationships and damage and action thresholds for integrated control, very limited specific information is available for most greenhouse diseases and pests. This can be explained in flower and pot plant crops for the extremely low tolerance of their aesthetic value to the most common diseases and pests. When known, damage thresholds in ornamental crops are near zero, as in the case of powdery mildew in roses, where the damage threshold is only 5 pustules/m^2 (Pieters et al., 1994). The same pest may have quite different damage thresholds if vegetable or ornamental crops are considered. For example, tomato may tolerate relatively high leafminer infestation (e.g. several dozens of mines per plant) with no yield loss, whereas 1–2 mines on chrysanthemum leads to cosmetic damage.

Some thresholds are available for greenhouse vegetable diseases and pests. Currently, the damage relationship for powdery mildew fungi in greenhouse vegetables is determined in The Netherlands. For cucumber, the best fitting model to describe the damage relationship for powdery mildew caused by *Sphaerotheca fusca* (Fr.) Blumer. [= *Sphaerotheca fuliginea* (Schlechtend.:Fr.) Pollacci] was an integral model using AUDPC. The slope of the regression line between yield and AUDPC was similar in several seasons and for different cultivars. In this case, early disease was also compared to late, more severe disease. Similar AUDPC values and similar yield losses were found for these situations (Dik, 1995), so the inability of integral models to distinguish between early and late epidemics does not seem to be very important here.

An additional problem for the determination of damage relationships is the fact that some pests can inflict more than one type of damage concurrently. This is the case of greenhouse whitefly, that feeds on phloem sap, with the consequent debilitating effect on the plant, but it also damages leaves and fruits by producing honeydew on which sooty mould develops, hampering photosynthesis and respiration and rendering fruits unmarketable. Damage thresholds may be quite different depending on which type of damage is considered to first occur as whitefly populations increase and this is decisively influenced by RH. In very humid greenhouse environments – particularly common in northern Europe – damage by sooty mould development occur at whitefly densities lower than those needed for damage derived from whitefly feeding activity. A density of 2500 greenhouse whitefly nymphs per leaf has been reported to cause yield reduction on tomatoes as direct consequence of phloem extraction, whereas a much lower density of 60 nymphs/leaf has been observed to produce sufficient honeydew to induce sooty mould development on tomatoes if RH reaches at least 80% for eight hours (Hussey and Scopes, 1977). For *Bemisia* species, a third type of damage is known which is that related to their ability to transmit tomato yellow leaf curl virus (TYLCV). This adds even more complexity to the pest density and yield loss relationship. The damage threshold is highly dependent on the amount of virus inoculum present in or near the greenhouse. Examples of damage thresholds for thrips pests in vegetable crops may illustrate the variability of values that can be found in the literature. For *Thrips palmi* Karny on aubergine the damage threshold is 0.08 individuals/leaf, whereas the same author (Kawai, 1990) gives a value of 4.4 thrips/leaf for cucumber. For sweet pepper, Kawai (1986) gives, for the same pest, a threshold of 0.11 thrips/flower (all details in Parrella and Lewis, 1997). There is unlikely to be a single damage threshold for a given pest on a given crop, but many, depending on the market and climatic conditions (Pedigo *et al.*, 1986). Even more variable are the action thresholds of greenhouse pests in various IPM programmes implemented in the world. If natural control must be considered in biological control of greenhouse pests, abundance of naturally occurring predators and parasitoids is an additional element to monitor and consider in decision-making. This is the case of action thresholds to release *Diglyphus isaea* (Walker) in Mediterranean greenhouses for the control of leafminers; the standard action threshold can be lowered if the parasitoid, that occurs naturally in the area, comes into greenhouses and establishes early in the season (Albajes *et al.*, 1994).

6.7. Research on Damage Relationships

In order to establish damage relationships, several issues should be considered.

6.7.1. MONITORING AND SAMPLING: WHAT, HOW AND WHEN?

A decision has to be made on what and how to monitor and sample. Accurate assessment of disease severity or pest density is essential. Decisions have to be made on the size, method and frequency of sampling. This topic will be discussed in more detail in Chapter 7. Besides monitoring the pest or disease, it is also important to monitor the host plants and the environmental parameters. Plant growth may vary from season to season and thus potential yield in a crop free of pests and diseases will vary. Often, disease or pest infestation are assessed as number of lesions, pustules or insects, for example, without monitoring the size of the host plant. From a physiological viewpoint, the amount of healthy plant tissue is more important than the amount of affected plant tissue, since the healthy tissue produces the yield.

6.7.2. REGRESSION ANALYSIS: WHICH VARIABLES TO USE?

When yield loss is expressed as a percentage of the yield in the uninfested control, a problem may arise when comparing different growing seasons or predicting future losses. When the yield in the uninfested control varies, the slopes of the regression lines describing percentage yield loss will be different from each other, since both lines are forced through the origin. When the damage relationship is described as yield compared to the control, the level of the lines will vary from season to season, but the slopes of the regression lines will be similar and give a more accurate description of the actual situation (Pace and MacKenzie, 1987).

Modern computer programs enable fairly easy stepwise multiple regression analyses. However, the choice of parameters should be restricted to those that can logically be expected to play a role in the damage relationship in order to provide a more predictive relationship.

6.7.3. HOW TO CREATE DIFFERENT EPIDEMICS/PEST DENSITIES FOR DETERMINING THRESHOLDS?

Various methods can be used to create different epidemics or pest densities. The time of inoculation/infestation can be modified, which will result in different levels of disease and pest attack at any given time point. However, climatic conditions will also vary and may interfere with an adequate analysis of the impact of severity on yield. Furthermore, it is possible to use different levels of inoculum/initial pest density or create a gradient of disease/pest attack by putting infested plants on one side of the greenhouse. Disease or pest level can also be modified by a variation in environmental parameters, but this method is not preferred because the environmental parameters may influence yield regardless of disease or pest. The most frequently used method is the utilization of selective pesticides. A disease or pest is allowed to reach a certain predetermined level at which time it is stopped with a chemical pesticide.

6.7.4. SIZE OF THE EXPERIMENTS

The design of the experiments largely depends on the type of model to be developed. For descriptive models based on regression analysis, the experiment should resemble commercial practices as much as possible. To assess yield, the plots should be large enough to rule out significant edge-effects. Sometimes, damage relationships are derived from a comparison of different greenhouses. However, this is not recommended, since factors other than the level of disease or pest may also vary.

For the development of simulation models, the experiments are usually of smaller scale and can partly be done under controlled conditions. The effect of one or two factors on plant physiology and on pests or pathogens can thus be determined and this effect is then quantified and incorporated into the model. Thus, prediction of yield is done by using information from an immediately lower level.

6.8. Integrated Control

Knowledge of the epidemiology of plant pathogens and population biology of pest organisms in greenhouse crops enables the development of integrated control measures. More than in field crops, cultural practices and the environment can be manipulated to prevent epidemics. Until now, growers were mostly interested in high yields and therefore, cultural practices and greenhouse climate regimes would not primarily be chosen for disease/pest damage prevention. However, the increasing awareness of the need to limit the input of energy and chemical pesticides, as well as the increasing problems with pesticide resistance, have made growers more willing to consider adaptations to limit diseases and pests. It is important that this knowledge is available.

Integrated control can consist of any combination of control measures, including chemical control. Usually, chemical control is limited to an absolute minimum in IPM systems and it is considered as the last defensive barrier. Integration of cultural practices, such as cultivar choice, the composition of the nutrient solution and climate control, together with biological control measures offer good perspectives for the future. It depends on the crop and on the greenhouse facilities to determine which measures can be incorporated into an IPM programme. In general, all components of the tetrahedron may be modified. As long as the control measures have no negative influence on each other, generally speaking, the amount of control will be greater when more than one component of the tetrahedron is modified. Biological control will generally be enhanced by cultural practices that prevent a too explosive disease epidemic or pest outbreak, or by practices that favour the activity of natural enemies. For example, biological control of powdery mildew fungi is more suitable in partially resistant cucumber cultivars than in very susceptible cultivars (Dik *et al.*, 1998), and the biological control of greenhouse whitefly works better in cultivars that are less prone to pest development. In heated greenhouses, biological control can be combined with a climate regime that promotes the development of the biocontrol agents (see Chapter 14). This combination of climate control and biological control is also a form of IPM.

As mentioned before, integrated control is more complicated than chemical control because more components of the tetrahedron are usually involved and more detailed knowledge on interactions is needed. However, several successful integrated control programmes have been developed (see Chapters 30 to 34).

6.9. Concluding Remarks

In this text, several aspects of plant disease epidemiology and population biology of pest organisms have been discussed, mainly to show how many factors play a role in the occurrence and management of an epidemic or pest. The factors which should be studied largely depend on the disease or pest concerned. In general, a combination of small-scale experiments under controlled conditions and larger-scale experiments under semi-commercial conditions gives a good insight into the relationships between the host plant, the environment, the pathogen or pest and human activity. The knowledge of such relationships is fundamental to understand why and when a pathogen or pest population may grow and cause yield loss. The identification of the factors and relationships that permit a species to achieve pest status can help the researcher to design and evaluate methods to manipulate such factors in an integrated way to prevent diseases and pests from reaching the damage threshold.

Damage thresholds are a basic tool for decision-making in integrated pest management. Reliable damage thresholds are derived from a full understanding of damage relationships. However, the complexity of damage relationships means that relatively few damage thresholds are nowadays available, even in greenhouse crops in which several IPM programmes have been successfully implemented. Further knowledge on damage relationships would permit pest control decisions to be based on a cost/benefit analysis. This is particularly relevant for diseases and pests that leave visible injuries on the plant and force growers to spray chemicals – or to adopt any other control measure – unnecessarily. Furthermore, awareness of the plant tolerance to certain levels of diseases and pests would help to apply control methods – like plant resistance or biological control – that do not seek to eradicate the disease or the pest, but to optimize their control in an economical, social and environmental context.

References

Albajes, R., Gabarra, R., Castañé, C., Alomar, O., Arnó, J., Riudavets, J., Ariño, J., Bellavista, J., Martí, M., Moliner, J. and Ramírez, M. (1994) Implementation of an IPM program for spring tomatoes in Mediterranean greenhouses, *IOBC/WPRS Bulletin* **17**(5), 14–21.

Campbell, C.L. and Madden, L.V. (1990) *Introduction to Plant Disease Epidemiology*, John Wiley & Sons, New York.

Dik, A.J. (1995) Integrated control of cucumber powdery mildew, in *6th International Symposium on the Microbiology of Aerial Plant Surfaces*, Bandol, France, September 1995, p. 40 (abstract).

Dik, A.J., Verhaar, M.A. and Bélanger, R.R. (1998) Comparison of three biological control agents against cucumber powdery mildew (*Sphaeroteca fuliginea*) in semi-commercial scale greenhouse trials, *European J. of Plant Pathology* **104**, 413–423.

Hussey, N.W. and Scopes, N. (1977) The introduction of natural enemies for pest control in glasshouse: Ecological considerations, in R.L. Ridgway and S.B. Vinson (eds.), *Biological Control by Augmentation of Natural Enemies*, Plenum Press, New York, pp. 349–377.

Kawai, A. (1986) Studies on population ecology of *Thrips palmi* Karny. X. Differences in population growth on various crops, *Japanese J. of Applied Entomology and Zoology* **30**, 7–11 (cited in Parrella and Lewis, 1997).

Kawai, A. (1990) Life cycle and population dynamics of *Thrips palmi*, *Japan Agriculture Research Quarterly* **23**, 282–283 (cited in Parrella and Lewis, 1997).

Norton, G.A. and Mumford, J.D. (eds.) (1993) *Decision Tools for Pest Management*, CAB International, Wallingford.

Pace, M.E. and MacKenzie, D.R. (1987) Modeling of crop growth and yield for loss assessment, in P.S. Teng (ed.), *Crop Loss Assessment and Pest Management*, APS Press, St Paul, Minn.

Parrella, M.P. and Lewis, T. (1997) Integrated Pest Management (IPM) in field crops, in T. Lewis (ed.), *Thrips as Crop Pests*, CAB International, Wallingford, pp. 595–614.

Pedigo, L.P., Hutchins, S.H. and Highey, L.G. (1986) Economic injury levels in theory and practice, *Annual Review of Entomology* **31**, 342–368 (cited in Parrella and Lewis, 1997).

Pieters, M.M.J., Kerssies, A. and van der Mey, G.J. (1994) *Epidemiologisch Onderzoek Naar Echte Meeldauw* (Sphaerotheca pannosa*) Bij De Kasroos*, Research Station for Floriculture and Glasshouse Vegetables Report 194, Research Station for Floriculture and Glasshouse Vegetables, Aalsmeer.

Teng, P.S. (ed.) (1987) *Crop Loss Assessment and Pest Management*, APS Press, St Paul, Minn.

Vanderplank, J. (1963) *Plant Diseases: Epidemics and Control*, Academic Press, New York.

Zadoks, J.C. and Schein, R.D. (1979) *Epidemiology and Plant Disease Management*, Oxford University Press, New York.

CHAPTER 7

SAMPLING AND MONITORING PESTS AND DISEASES
Laurent Lapchin and Dan Shtienberg

Integrated Pest Management strategies require detailed studies which can be broken down into three steps: (i) a precise description of pest population dynamics in space and time in order to assess damage thresholds, to determine key points for control (possibly by modelling), and to evaluate control efficiency; (ii) a general survey to estimate the variability in the first step between seasons or across a region; and (iii) the control strategy, including a survey by the grower of population dynamics. Each of these steps requires particular sampling methods that differ in accuracy: precise measurements for detailed studies, less precise measurements but which can be used on a larger scale for variability evaluation in the second step, and quick and simple methods for final use by the growers.

Pest and disease intensity may be quantified using two different measurements: (i) estimation of the population size, e.g. number of aphids per leaf, number of fungal spores in a cubic meter of air, etc.; and (ii) quantification of the injury caused to the host plant, e.g. the proportion of leaf tissue infested by larvae, the relative leaf area covered with disease symptoms, etc. The methods should be easy to use, allow rapid estimation, be applicable over a wide range of conditions, and most of all, be accurate and reproducible. This chapter presents some of these methods which may be used for greenhouse crops.

7.1. Insect Pests

7.1.1. ESTIMATING INSECT NUMBERS IN SAMPLES

We will examine different ways of reducing pest assessment time. Methods based on visual abundance indices will be developed in particular, and examples of their application to insect pests will be given. Since many authors have developed methods which may be used with species other than those that they have studied, the references do not always concern greenhouse species.

At each step of a study, the spatial distribution of most pests will be very patchy. Sampling plans will thus require numerous data to reach the required level of accuracy. Particular attention should be given to the evaluation of pest densities at each sampling point. This is a bottleneck which will define the "cost" of the sampling. Methods are often available to reduce the cost of counts, but, except when automatic counting (i.e. picture analysis) is possible, these methods lead to a drastic decrease in accuracy. This loss of accuracy is cumulated with the error induced by the sampling itself, to define the final value of the density estimates. Moreover, several methods that are used to

82

R. Albajes et al. (eds.), Integrated Pest and Disease Management in Greenhouse Crops, 82-96.
© 1999 *Kluwer Academic Publishers. Printed in the Netherlands.*

accelerate pest counts (e.g. field visual observation) underestimate pest densities per area or volume. Such systematic bias, as well as the accuracy of estimates, must be evaluated before using this kind of method.

The most accurate way of obtaining quantitative data on a pest population is to collect the substrate (e.g. host plants) and to take these samples to the laboratory where individuals may be isolated and counted under a stereoscopic microscope (see, for instance, Baumgärtner et al., 1983). As this method is time-consuming, numerous authors have attempted to reduce the time required. The first step is the mechanical extraction of the individuals from the substrate. They can be extracted by washing (Banks, 1954; Taylor, 1970; Halfhill et al., 1983; Raworth et al., 1984; McLeod and Gonzalez, 1988), by flotation in high-density medium such as saccharose (Lapchin and Ingouf-Le Thiec, 1977), or by the use of Berlesse-Tullgren funnels (Wright and Cone, 1983, 1986). Once the insects are separated from mud, sand or plant fragments, the clean extract can be fractionated into sub-samples (Banks, 1954; Waters, 1969; Taylor, 1970; Raworth et al., 1984). Both steps give results of varying precision, depending on the medium surrounding the insects and the species involved. The time required for counting the insects is reduced two to five times, but is often still too great (e.g. up to two hours per leaf, including washing, sub-sampling and counting, for aphids on cucumber).

Collecting insect substrate is not efficient for species such as thrips, which are very mobile. In this case, the extraction has to be made directly in the field, by using sweep-nets (Cuperus et al., 1982; Senanayake and Holliday, 1988), direct picking, mouth vacuum devices (Lapchin et al., 1987) or vacuum nets like the Dietrick vacuum (D-vac) (Rohitha and Penman, 1981; Cuperus et al., 1982; Dewar et al., 1982; Hand, 1986). The numbers are then calculated from the part of the population which can be recovered. This part is often highly variable and the precision of the method is difficult to evaluate. However, successive sampling at the same sites may enable insect density to be estimated with accuracy. The method of Suber and Le Cren (1967), frequently used to evaluate fish densities in rivers (Laurent and Lamarque, 1974), was adapted to benthic insect counts by Lapchin and Ingouf-Le Thiec (1977), and then by Lapchin et al. (1987) to estimate larval and adult coccinellid densities in wheat fields. It has recently been used in cucumber greenhouses by Boll et al. (1997a) to estimate the number of thrips on leaves after several successive strikes.

7.1.2. ESTIMATING INSECT POPULATION DENSITIES

Insect densities may be estimated in situ, without collecting samples (see, for instance, Dewar et al., 1982). The characteristics of the species distribution on the host plants should be taken into account and the most representative leaves or stems should be observed (Addicott, 1978; Hull and Grimm, 1983; Bues et al., 1988; Steiner, 1990). This method eliminates the laboratory stage of density estimation, but does not significantly reduce counting time.

Observation time may sometimes be drastically decreased if rough counting is performed. This method is useful if the strong systematic under-estimation of the numbers it produces is constant or depends on known parameters. In the previous example of *Coccinella septempunctata* L. in wheat fields, Lapchin *et al.* (1987) developed a "quick visual method", i.e. the observer walked within the 25 m² sub-plot for 2 min and counted all the adult coccinellids he saw. The numbers obtained by this method correlated well with the density estimated with the Seber and Le Cren method, and allowed the development of a sequential sampling plan for adult coccinellids in wheat fields (Iperti *et al.*, 1988). However, such a quick method is not automatically appropriate, even for closely related species, as Frazer and Raworth (1985) did not find that "walking counts" of adult coccinellids in strawberry fields were reliable.

Variance of insect numbers is closely related to the mean (Taylor, 1961). This property implies that variations in numbers may be more easily perceived by observers who follow a geometric rather than arithmetic scale of density. In the field, such orders of magnitude can be easily translated into abundance classes. This was probably the reason why the use of categorical data was soon considered to be a good way of drastically reducing sampling costs. In the case of aphid populations, precise density estimates of *Myzus persicae* (Sulzer) were related to the proportion of infested leaves in different parts of potato plants (Broadbent, 1948). This method was used on other host plants and statistically developed by several authors (Tamaki and Weiss, 1979; Hull and Grimm, 1983; Ward *et al.*, 1985a,b, 1986; Bues *et al.*, 1988). As far back as 1954, presence-absence methods were improved by use of a set of abundance classes (Banks, 1954). Such classes can be purely arbitrary and, for instance, the sampling units distributed into poor, medium or heavy infestation classes (Srikanth and Lakkundi, 1988). Several authors used more precisely defined classes, according to the number of colonies, their size and their localization (Banks, 1954; Baggiolini, 1965; Leclant and Remaudière, 1970; Anderson, 1981; Lapchin, 1985; Lapchin *et al.*, 1994). Different kinds of abundance class systems may be developed, according to the insect studied and its environment.

Building a System of Abundance Classes
Three main types of class can be considered. Firstly, there are classes whose limits are defined by the number of individuals that are seen during one sample unit of observation. A logarithmic scale of these limits was first used by Leclant and Remaudière (1970) to estimate *M. persicae* densities on peach trees. Another scale which is based on the approximate powers of √10 has successive classes such as: no insect seen, 1 to 3, 4 to 10, 11 to 30, etc. This scale was used by Ferran *et al.* (1996) to evaluate the density of the rose aphid [*Macrosiphum rosae* (L.)] on rose bushes, by Boll and Lapchin (1997) for *Macrosiphum euphorbiae* (Thomas) in tomato greenhouses, and by Lapchin *et al.* (1997) to estimate mummified *Aphis gossypii* Glover on cucumber plants. Secondly, purely qualitative classes, based on size and number of insect patches (Lapchin, 1985) or on the percentage of contaminated shoots (Lapchin *et al.*, 1994), may be used. Such classes are generally used for large sampling units such as trees. Finally, there are intermediate systems which are based on the number of sub-units (e.g. leaves of a plant) in each class of a set of qualitative classes. This system was used by Lapchin *et al.* (1997) to evaluate non-mummified *A. gossypii* on cucumber plants and by Boll *et al.* (1997b) for *A. gossypii* in open-field melon crops.

A visual class system must be both simple and complete. Ease of use depends on the number of classes and therefore there should be a sufficient number to describe accurately the trends of variability in insect density, but not so high as to be difficult to remember. An optimal class number is generally between five and eight. Another condition required for the use of visual classes is that there must be a biological basis for their definition. To be representative, a qualitative class set must cover the different kinds of patchiness of the species which may be encountered in the field. For example, on cucumber plants isolated colonies of *A. gossypii* a few centimetres in diameter will first develop around winged immigrants (slightly infested leaves). After several days of development, the colonies suddenly spread all over the leaf area (heavily infested leaves). These simple characteristics, which are associated with the size of the leaves that are heavily infested, define the classes of abundance. Simplicity of the class system determines both the robustness of the results and the time required for field observations. In the example cited above, the observation of one sampling unit takes approximately 30 sec for each cucumber plant.

The class system must cover the whole range of insect densities per sampling unit which may be encountered within the observation period and under different conditions. Thus, this range must be evaluated either from previous knowledge or from trials prior to defining the classes.

Calibration of Visual Abundance Classes
The results of visual observations may often be used without any reference to the number of individuals that they represent and, as such, these ranked qualitative data may be analysed using a large set of non-parametric statistic tools. This method has been used, for example, to evaluate the efficiency of biological control of the rose aphid on rose bushes in public gardens (Ferran *et al.*, 1996). When more precise data are needed, each visual class must be calibrated by computing the mean and variability of the number of individuals actually present in the sampling units. This step is very time-consuming because a large set of precise counts must be gathered so as to represent accurately the variability of the situations in which a given class may be chosen.

Improving the Calibration of the Classes Using Environmental Descriptive Variables
Calibration of the classes may be viewed as a statistical model having as a response variable the density of aphids, and as a categorical explanatory variable the visual abundance classes. A complex multivariate regression method, "projection pursuit regression", was adapted to this statistical data (Lapchin *et al.*, 1997). Predictions of these models may be further improved by complementary explanatory variables. This work, including calibration with complementary variables, was performed with four different class systems which were used to evaluate the density of the aphid *A. gossypii* and its parasitoid *Lysiphlebus testaceipes* (Cresson) on cucumber plants in greenhouses. Two visual methods were used to estimate densities: "the detailed visual method" (DVM) for a leaf, and the "quick visual method" (QVM) for the whole plant. The class sets of DVM and QVM were built according to the apparent numbers of individuals in the observed sampling units. When the QVM was applied to healthy aphids, the four classes were based on the proportion of the area of leaf infested and on the size of the leaves.

Precise counts were also made on the same sampling units and used as the response variable, and the data were divided into reference and validation sets. The reference sets were used to develop the regression models, and the validation sets to test their robustness.

The choice of complementary explanatory variables was crucial to the development of these regression models. These variables were selected for their influence on the goodness of fit of the models as well as for the time required for collecting. For example, when using the detailed visual method and when the target population of the model was either healthy or mummified aphids, seven explanatory variables were used: (i) the visual class of the leaf; (ii) the visual class of the non-target population on the leaf; (iii) and (iv) the visual classes of the target population on the upper and lower neighbouring leaves on the same plant; (v) the vertical rank of the leaf on the plant; (vi) the number of leaves on the plant; and (vii) the number of leaves infested by the target aphid on the plant. Such data can be easily gathered during sampling without significant additional cost. QVM sampling of whole plants yielded a mean error of approximately one class per plant (the limits of each class are in the ratio of √10). The DVM had a mean error of less than one class (Fig. 7.1). The range of the residuals was generally the same for both the reference and the validation data sets, confirming the robustness of these models.

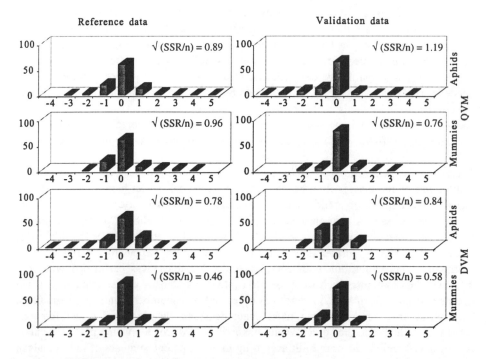

Figure 7.1. Robustness of the projection pursuit regression models used to fit precise counts of healthy or mummified aphids on cucumber plants (QVM) or leaves (DVM) against visual abundance classes and environmental complementary variables. Differences between observed and fitted values are expressed in number of classes, according to the scale based on the powers of √10. SSR: sum of squared residuals.

The same method has now been used to calibrate visual class systems for different pest species on vegetable crops in greenhouses (thrips on cucumber plants, aphids on tomato, melon, eggplant and sprout). Each time that a new regression model is tested, particular attention must be given to the development and sampling of the reference data sets, i.e. they must include the same combinations of variables which will be used in further field sampling.

7.1.3. REDUCING THE TIME OF SAMPLING

Reducing the time spent on evaluating insect densities in sampling units has a cost, which is a decrease in the precision of the estimation. However, the time saved allows the observer to increase the number of units taken into account in a sampling plan, and thus to increase the precision of the mean and variance estimates of the density.

The gain in time is greatly increased when visual methods are used (a ratio of 1:10, when compared with precise counts). Mechanical methods, such as washing (ratio of 1:2), are much slower. However, the evaluation of the precision of visual methods requires a time-consuming calibration of the classes, which must be repeated for each study that deals with a different species, a different environment of the insects or a different scale of observation (i.e. plant or leaf). The decision to undertake such work will depend on the chance of building a "good" visual system. We can summarize the four criteria of this evaluation as follows: (i) insect density must be highly variable from one sampling unit (often a host plant) to another, and from one sampling date to another (this condition is easily met for numerous phytophagous insects whose densities vary on logarithmic scales from one host plant to another); (ii) most insects must be visible (for instance, such methods cannot be used for certain aphid species mainly located inside rolled leaves); (iii) the visual classes must be simple and distinct, i.e. their boundaries have to be easily recognizable in the field; and (iv) these boundaries must be stable in time and space (for example, independent of the host plant growth stage).

The benefit of building such calibrated visual scales depends on which species is being observed and its environment. The scales are particularly useful for most aphid species, as they can reach very high densities and have strongly aggregated distribution patterns. Since the sampling units are not destroyed, crops can be easily monitored and the population dynamics studied separately in different fields. Such a method permits large-scale surveys. An example is given in Fig. 7.2 (Boll *et al.*, 1994): a set of cucumber greenhouses in Provence (France) was sampled weekly by using visual abundance classes and number modelling (see Section 7.1.2). A regular sampling grid was used in every greenhouse and required less than one hour of observation by two people on each sampling occasion.

7.1.4. PEST MONITORING

Most phytophagous insects are highly aggregative. Thus, the number of elementary units that are needed in a sampling plan to reach a given reliability of density estimates is drastically increased. This problem is particularly serious at the beginning of the crop season when insects are clumped around early immigrants and for species with a very high rate of increase. This is the case for most insect pests that exist under greenhouse conditions. Moreover, the efficiency of a biological control will often depend on the detection of such initial foci.

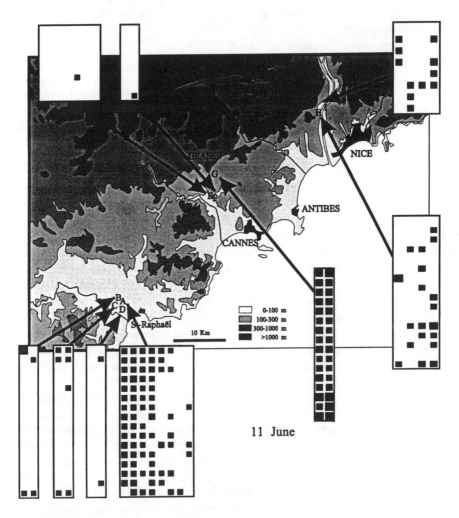

Figure 7.2. Example of regional cartography of *A. gossypii* infestation in cucumber greenhouses in Provence (France) on one sampling date. The squares are proportional to the number of aphids per sampling plant, estimated by projection pursuit regression modelling.

Early trials have been performed in tomato greenhouses to develop whitefly sampling schemes that are compatible with the time constraints imposed by the grower. Eggenkamp-Rotteveel Mansveld *et al.* (1978) used stratified random sampling in which absolute counts were performed on 0.6% of the plants, spread evenly through the greenhouse. These data were compared with an enormous set of absolute counts obtained from the 18,000 plants in the greenhouse. The results demonstrated that the random sampling did not accurately reflect the actual numbers and distribution of the whitefly and also that, in practice, absolute counts were not useful. The same conclusions were drawn by Ekbom (1980). She suggested that some device should be used to detect at an early stage the presence of whiteflies. This was tried out by Guldemont and den Belder (1993) in chrysanthemum greenhouses. They simultaneously used yellow sticky traps and incidence counts (percentage of infested plants) to detect the moment and the level of the attacks by the major pests of the crop: leafminers, thrips, aphids, whiteflies, spider mites and caterpillars. They concluded that traps were still useful for monitoring the number of leafminers and thrips during the entire season and for aphids in the winter season, but that less emphasis should be placed on the use of traps and more on crop sampling. The low density of pests and their aggregated distribution, however, makes the use of fixed sampling sites less suitable.

Different approaches to aphid sampling have been tested, but unfortunately most of these experiments were performed in open-field cereal crops, and not in greenhouses. The spatial heterogeneity of populations was incorporated into the sequential sampling plans, based on the relationships of variance and the mean of density (Ba-Angood and Stewart, 1980; Ekbom, 1985; Elliott and Kieckhefer, 1986). The sample size may be adapted according to the reliability desired. More recently, incidence counts (percentage of infested/noninfested tillers) have replaced precise counts (Ekbom, 1987). Incidence counts will improve aphid sampling efficiency if there is a strong relationship between the incidence and the precise counts at the scale of the sampling unit, and if the loss of precision for each unit is more than compensated by the increase in the number of units observed in a given time. It is useful to combine the errors that are induced by the representativeness of the sampling scheme and by the use of incidence counts, as has been done for aphid predators in cereal crops (Iperti et al., 1988).

The monitoring of insect pests in greenhouses thus remains a complex problem. The most accurate and least expensive methods need to be developed in each situation and then adjusted to give the necessary precision for each particular biological question to be answered.

7.2. Plant Pathogens

7.2.1. MEASURING DISEASE INTENSITY

The intensity of disease may be estimated by two distinct measurements: disease incidence and disease severity. Disease incidence is defined as the number of units infected, expressed as a proportion of the total number of units assessed, e.g. the percentage of infected plants, leaves, fruits, tubers, twigs, etc. This is a quintal measurement (i.e. the unit is infected or it is not infected). Disease severity expresses the intensity of the symptoms, e.g. the area of plant tissue affected by disease expressed as a proportion of the total leaf area, number of lesions per plant unit, etc. (Horsfall and Cowling, 1978).

Measurement of disease intensity in a crop is fundamental for IPM. Disease incidence is generally easy to assess with considerable accuracy, but accurate estimates of the severity of many diseases are much more difficult to obtain. Moreover, a farmer concerned about his crop readily overestimates severity. Thus for decision-making, disease incidence rather than disease severity is the preferred measurement. However, disease severity generally correlates better with yield and crop loss. Because of the relative ease of obtaining most incidence values with accuracy, many attempts have been made to correlate severity to incidence. At low disease levels, good correlation between disease severity and incidence has been found (Seem, 1984). At high disease levels, the relationship between incidence and severity becomes insufficient. When the correlation is significant, the similarity of the two measurements is confirmed and more easily measured incidence values for disease assessment may be used. When this relationship is not linear, an appropriate transformation may be employed. A square root transformation of the severity values is often used to create regression equations that predict severity from incidence (Seem, 1984). Thus, many schemes that warn against pests and diseases depend on enumeration rather than estimation procedures.

Estimating Disease Severity in Field Situations
Visual estimation of disease severity is almost exclusively used for estimating disease severity in the field. Methods for visual assessment of disease generally fall into two categories (Lindow, 1983). The first category contains descriptive keys that utilize arbitrary scales, indices, ratings, grades or percentages to quantify disease (James and Teng, 1979). Such keys have been successfully used to estimate disease severity of host plants with differing disease resistance, or of host plants subjected to different environmental conditions or cultivation procedures. For example, disease can be described using categories of 1–5 to denote incidence (none to extreme) or severity (none to heavy). It is not appropriate to perform mathematical manipulations such as averaging on these records because values between two adjacent categories have no meaning (Berger, 1980).

The second category for visual assessment of disease involves the use of standard area diagrams. Pictorial representations of the host plant with known and graded amounts of disease are compared with diseased leaves to allow estimation of disease severity. Estimates of disease severity are proportional to the absolute area of the leaf that is diseased, and are not expressed as a percentage of an arbitrary maximum severity value. In contrast to descriptive keys, standard area diagrams allow estimation of intermediate levels of disease severity by comparing a diseased plant with diagrams that show both more and less disease (Lindow, 1983).

Horsfall and Barratt (1945), while noting the Weber-Feckner law, emphasized the limitation of the eye in the assessment of plant disease. The Weber-Feckner law states that the visual acuity of the eye is proportional to the logarithm of the intensity of the stimulus. These authors also noted that in visually estimating disease severity, the observer actually assesses the diseased proportion of leaves having <50% injury and the healthy portion of leaves having >50% injury (Horsfall and Cowling, 1978). Horsfall and Barratt (1945) developed a disease-rating scale that contained 12 equal divisions of disease severity on a logarithmic scale with a median value of 50%. Thus, divisions of this scale included decreasing ranges of disease severity when either increasing or decreasing from 50% disease severity (Horsfall and Cowling, 1978). This scale and many standard diagrams constructed thereafter account for the logarithmic decrease in acuity of the eye in estimating severities approaching 50% by their selection of representative keys. Estimations of disease severity intermediate between two keys are made by careful interpolation.

Accuracy, Repeatability and Reliability of Disease Assessments
Visual estimation of disease severity can differ significantly from the actual amount of disease. If the observer is not aware of the limitation in visual acuity at the midrange of disease severity, estimated disease severity and actual disease severity will be linearly related, and the variance of estimates will be independent of disease severity. However, the Weber-Feckner law indicates that the true confidence interval of estimates of disease severity will approach the expected linear relationship at both low and high disease levels, but will increasingly depart from this line with increasing disease severity, with a maximum variance at 50% disease (Lindow, 1983).

Inter-rater reliability has been operationally defined as the ratio of true variance to total variance, which includes a variance component for the error among raters (Shokes *et al.*, 1987). Although improved sampling designs and increased sample size can lower the actual and total variance, limited resources often restrict sample size. In addition, when more than one rater is involved, it is difficult to quantify the bias attributable to any one individual. Shokes *et al.* (1987) proposed measuring intra-rater repeatability with the test-retest correlation procedure. The correlation coefficient (r) provides a statistical measure of the relationship between repeated assessments of the same sampling units by the same individual or instrument. However, correlation analysis between two variables cannot be used to infer a cause-and-effect relationship, nor can one variable (repeated assessments) be used to predict the value of another variable (first-time assessments).

Least-squares regression can be used to determine if there is a significant linear relationship between disease assessment performed by different raters and whether there is a statistical relationship between related assessments performed by the same individual (Nutter *et al.*, 1993). Regression-equation parameters, such as the slope and the intercept, could be used to evaluate and compare the accuracy and precision of disease assessment raters and methods. Slopes that are significantly different from one indicate the presence of systematic bias among rates, whereas intercepts significantly different from zero indicate the presence of a constant source of error among raters.

7.2.2. DISTRIBUTION OF DISEASE

Spatial distribution of diseased units in a pathosystem is the most important factor affecting the field estimation of disease intensity. Spatial distribution includes the way in which disease lesions are distributed among healthy units and the way in which diseased host units are distributed among healthy units. Distribution of diseased units may be random, aggregated or regular (Teng, 1983). With randomly distributed disease, the variance is theoretically equal to the mean. In aggregated patterns, the variance in the number of lesions per leaf is greater than the mean number of lesions per leaf, but when there is a regular pattern the variance is smaller than the mean.

When a large number of host units are sampled for disease, a frequency distribution showing the number of diseased units in each severity category may be determined. The sample frequency distribution can be compared with theoretical distributions using the goodness-of-fit test, and the parameters of the empirical distribution may be defined. Theoretical distributions applied to biological systems include the normal, log normal, Poisson, Weibull, Gamma and negative binomial ones. Knowledge of the frequency distribution is essential for the design of sound sampling procedures.

When estimating disease intensity per field, the sampling unit, sample size, sampling point, sampling fraction and sampling method must be considered. In most disease assessments, the sampling unit is a plant. Often, only selected parts (such as individual leaves) may be assessed for disease intensity. For each field, a predetermined number of sampling units is selected to give a mean value representative of that field; this is the sample size. Sample size is determined by the cost of sampling, the precision required and the available time, as well as by the spatial distribution of the disease; that is, the sample size should be empirically defined.

Many sampling methods have been reported for plant disease assessments. Samples may be taken at intervals along predetermined lines in the field or greenhouse and these may be either one diagonal, both diagonals (forming a big letter X), or (if a more representative sample is required) a large W or Z pattern. With a disease that is randomly distributed, all the above methods will give comparable results and reduced variance in the sample mean may be better achieved by increasing sample size. If the diseased units are aggregated, the sampling method will be more important than sample size, and the large X or W sampling pattern is preferable to the single diagonal (Teng, 1983).

7.2.3. MONITORING PATHOGEN POPULATIONS

Monitoring the pathogen, primarily by trapping air-borne spores, has been used as a measurement of disease intensity and development and could serve either as an alternative, or a complement, to disease assessment. Given the current technology, the use of spore counts of pathogen populations for field measurement of disease is unlikely to replace the main conventional methods of measuring disease severity (disease symptoms), unless its accuracy can be shown to override the ease and low cost of symptom assessment (Teng, 1983).

The monitoring of pathogen populations may serve another purpose. As fungicides still remain an important tool for control of plant pathogens in the greenhouse, it is important to monitor populations of the pathogen for their resistance to potential fungicides. The term "monitoring resistance" is used to denote testing for sensitivity of target organisms in field populations. This can range in scope from continuous surveillance programmes over several years and involving many locations to short-term investigations into individual cases of suspected resistance. Good monitoring is the cornerstone of fungicide resistance research. Without such work, we would know virtually nothing about the occurrence of resistance in crop pathogens. Moreover, resistance monitoring, together with monitoring for changes in practical performance, is a vital component of integrated resistance management (Gullino and Garibaldi, 1986; Brent, 1988). Several tools have been developed for such a purpose. For example, a tool for estimating the resistance of populations of *Botrytis cinerea* Pers.:Fr. to common fungicides has recently been developed (Elad and Shtienberg, 1995). Tested fungicides are added to a selective medium in Petri dishes. The plates are exposed in the greenhouse at approximately midday, when *B. cinerea* conidia are released into the air. Plates are exposed for 30–60 min, according to the intensity of the disease in the greenhouse and then incubated for 4–7 days. Counts of typical *B. cinerea* colonies in the media supplemented with the fungicides are compared with those from fungicide-free plates. The data may then be used to make a recommendation on fungicide use.

7.3. Concluding Remarks

Studies on population dynamics of insect pests or beneficials and plant pathogens, which have been performed to improve the efficiency of IPM, have followed parallel paths and run into similar obstacles. Because of the speed of the dynamics and the strong spatial heterogeneities of these populations, control strategies have had to be designed to include the large amounts of data that may be generated over different temporal and spatial scales. In both disciplines, methods have been designed to evaluate quickly insect densities or levels of disease injury in large and frequent samples. Moreover, the need to sample commercial crops to take into account large-scale variations implies the use of non-destructive methods. Pathologists and entomologists have independently concluded that visual indices could be practical and efficient. Initially, both of these groups have tried the two-class (presence/absence) indices, and then later the several-class indices. After it was discovered that the logarithmic scale is a natural tool of the human eye discriminating among different kinds of intensities, statistical approaches were developed to evaluate the precision of such evaluations.

In the future, many new methods will need to be constructed to advance IPM strategies. A good idea would be to synchronize some of these developments in the two disciplines (i.e. for all the major insects and pathogens of a given protected crop), and to pool the statistical approaches which, as a matter of fact, deal with very similar problems. Such an integration would allow the pathologists and entomologists to propose standardized "toolboxes" to professional and technical partners.

References

Addicott, J.F. (1978) The population dynamics of aphids on fireweed: A comparison of local populations and metapopulations, *Canadian J. of Zoology* **56**, 2554–2564.

Anderson, J.M.E. (1981) Estimating abundance of the aphid *Aphis eugeniae* van der Goot on *Glochidion ferdinandi* (Muell. Arg.) (Euphorbiaceae), using a subjective and a biomass method, *J. of the Australian Entomological Society* **20**, 221–222.

Ba-Angood, S.A. and Stewart, R.K. (1980) Sequential sampling for cereal aphids on barley in Southwestern Quebec, *J. of Economic Entomology* **73**, 679–681.

Baggiolini, M. (1965) Méthode de contrôle visuel des infestations d'arthropodes ravageurs du pommier, *Entomophaga* **10**, 221–229.

Banks, C.J. (1954) A method for estimating populations and counting large numbers of *Aphis fabae* Scop., *Bulletin of Entomological Research* **45**, 751–756.

Baumgärtner, J., Bieri, M. and Delucchi, V. (1983) Sampling *Acyrthosiphon pisum* Harris in pea fields, *Mitteilungen der Schweizerischen Entomologischen Gesellschaft* **56**, 173–181.

Berger, R.D. (1980) Measuring disease intensity, in P.S. Teng and S.V.C. Krupa (coords.), *Crop Loss Assessment*, Proceedings of E.C. Stakman Commemorative Symposium, Miscellaneous Publication 7, Agricultural Experiment Station, University of Minnesota, St Paul, Minn., pp. 28–31

Boll, R. and Lapchin, L. (1997) Dénombrement visuel du puceron *Macrosiphum euphorbiae* Thomas en serres de tomate, International Symposium on Integrated Production and Protection in Horticultural Crops, Agadir, 6–9 May 1997 (in press).

Boll, R., Lapchin, L., Franco, E. and Géria, A.-M. (1997a) Dénombrement visuel des thrips en serre de concombre, Journées d'Informations Mutuelles en Cultures Légumières, Service Régional de la Protection des Végétaux, Montpellier, 30–31 January 1997.

Boll, R., Lapchin, L., Géria, A.-M. and Franco, E. (1997b) Contribution à la mise au point d'un système de dénombrement visuel du puceron *Aphis gossypii* Glover en culture de melon en plein champ, Journées d'Informations Mutuelles en Cultures Légumières, Service Régional de la Protection des Végétaux, Montpellier, 30–31 January 1997.

Boll, R., Rochat, J., Franco, E. and Lapchin, L. (1994) Variabilité inter-parcellaire de la dynamique des populations du puceron *Aphis gossypii* Glover en serres de concombre, *IOBC/WPRS Bulletin* **17**(5), 184–191.

Brent, K.J. (1988) Monitoring for fungicide resistance, in C.J. Delp (ed.), *Fungicide Resistance in North America*, APS Press, St Paul, Minn., pp. 9–11.

Broadbent, L. (1948) Methods of recording aphid populations for use in research on potato virus diseases, *Annals of Applied Biology* **35**, 551–566.

Bues, R., Toubon, J.F. and Poitout, H.-S. (1988) Development of a control programme against the aphids (*Macrosiphum euphorbiae* Thomas and *Myzus persicae* Sulzer) in processing tomato cultures in the south-east of France, in R. Cavalloro and C. Pelerents, *Progress on Pest Management in Field Vegetables*, Balkema, Rotterdam, pp. 209–219.

Cuperus, G.W., Radcliffe, E.B., Barnes, D.K. and Marten, G.C. (1982) Economic injury levels and economic thresholds for pea aphid, *Acyrthosiphon pisum* (Harris), on alfalfa, *Crop Protection* **1**, 453–463.

Dewar, A.M., Dean, G.J. and Cannon, R. (1982) Assessment of methods for estimating the number of aphids (Hemiptera: Aphididae) in cereals, *Bulletin of Entomological Research* **72**, 675–685.

Eggenkamp-Rotteveel Mansveld, M.H., Ellenbroek, F.J.M., van Lenteren, J.C. and Woets, J. (1978) The parasite-host relationship between *Encarsia formosa* Gah. (Hym., Aphelinidae) and *Trialeurodes vaporariorum* (Westw.) (Homoptera, Aleyrodidae). VIII. Comparison and evaluation of an absolute count and a stratified random sampling programme, *J. of Applied Entomology* **85**, 133–140.

Ekbom, B.S. (1980) Some aspects of the population dynamics of *Trialeurodes vaporariorum* and *Encarsia formosa* and their importance for biological control, *IOBC/WPRS Bulletin* **3**(3), 25–34.

Ekbom, B.S. (1985) Spatial distribution of *Rhopalosiphum padi* (L.) (Homoptera: Aphididae) in spring cereals in Sweden and its importance for sampling, *Environmental Entomology* **14**, 312–316.

Ekbom, B.S. (1987) Incidence counts for estimating densities of *Rhopalosiphum padi* (Homoptera: Aphididae), *J. of Economic Entomology* **80**, 933–935.

Elad, Y. and Shtienberg, D. (1995) *Botrytis cinerea* in greenhouse vegetables: Chemical, cultural, physiological and biological controls and their interaction, *International Pest Management Reviews* **1**, 15–29.

Elliott, N.C. and Kieckhefer, R.W. (1986) Cereal aphid populations in winter wheat: Spatial distributions and sampling with fixed levels of precision, *Environmental Entomology* **15**, 954–958.

Ferran, A., Nikmam, H., Kabiri, F., Picart, J.-L., De Herge, C., Brun, J., Iperti, G. and Lapchin, L. (1996) The use of *Harmonia axyridis* larvae (Coleoptera: Coccinellidae) against *Macrosiphum rosae* (Hemiptera: Sternorrhyncha: Aphididae) on rose bushes, *European J. of Entomology* **93**, 59–67.

Frazer, B.D. and Raworth, D.A. (1985) Sampling for adult coccinellids and their numerical response to strawberry aphids (Coleoptera: Coccinellidae; Homoptera: Aphididae), *Canadian Entomologist* **117**, 153–161.

Guldemont, J.A. and den Belder, E. (1993) Supervised control in chrysanthemums: One year's experience, *IOBC/WPRS Bulletin* **16**(2), 51–54.

Gullino, M.L. and Garibaldi, A. (1986) Fungicide resistance monitoring as an aid to tomato grey mould management, in *1986 British Crop Protection Conference – Pests and Diseases*, BCPC, Thornton Heath, pp. 499–505.

Halfhill, J.E., Gefre, J.A., Tamaki, G. and Fye, R.E. (1983) Portable device to remove aphids and eggs from asparagus ferns, *J. of Economic Entomology* **76**, 1193–1194.

Hand, S.C. (1986) The capture efficiency of the Dietrick vacuum insect net for aphids on grasses and cereals, *Annals of Applied Biology* **108**, 233–241.

Horsfall, J.G. and Barratt, R.W. (1945) An improved grading system for measuring plant diseases, *Phytopathology* **35**, 655.

Horsfall, J.G. and Cowling, E.B. (1978) Pathometry: The measurement of plant disease, in J.G. Horsfall and E.B. Cowling (eds.), *Plant Disease: An Advanced Treatise*, Academic Press, New York, pp. 119–136.

Hull, L.A. and Grimm, J.W. (1983) Sampling schemes for estimating populations of the apple aphid, *Aphis pomi* (Homoptera: Aphididae), on apple, *Environmental Entomology* **12**, 1581–1586.

Iperti, G., Lapchin, L., Ferran, A., Rabasse, J.M. and Lyon, J.P. (1988) Sequential sampling of adult *Coccinella septempunctata* L. in wheat fields, *Canadian Entomologist* **120**, 773–778.

James, W.C. and Teng, P.S. (1979) The quantification of production constraints associated with plant Diseases, in T.H. Coaker (ed.), *Applied Biology*, Vol. 4, Academic Press, Inc., London, pp. 210–267.

Lapchin, L. (1985) Echantillonnage des populations aphidiennes en vergers de pêchers. Comparaison de la précision de deux méthodes à l'aide de modèles de simulation d'échantillonnage, *Agronomie* **5**, 217–226.

Lapchin, L., Boll, R., Rochat, J., Géria, A.M. and Franco, E. (1997) Using projection pursuit nonparametric regression for predicting insect densities from visual abundance classes, *Environmental Entomology* **26**, 736–744.

Lapchin, L., Ferran, A., Iperti, G., Rabasse, J.M. and Lyon, J.P. (1987) Coccinellids (Coleoptera: Coccinellidae) and syrphids (Diptera: Syrphidae) as predators of aphids in cereal crops: A comparison of sampling methods, *Canadian Entomologist* **119**, 815–822.

Lapchin, L., Guyot, H. and Brun, P. (1994) Spatial and temporal heterogeneity in population dynamics of citrus aphids at a regional scale, *Ecological Research* **9**, 57–66.

Lapchin, L. and Ingouf-Le Thiec, M. (1977) Le dépouillement des échantillons d'invertébrés benthiques: Etude comparée de différentes méthodes de tri, *Annales d'Hydrobiologie* **8**, 231–245.

Laurent, M. and Lamarque, P. (1974) Utilisation de la méthode des captures successives (De Lury) pour l'évaluation des peuplements piscicoles, *Annales d'Hydrobiologie* **5**, 121–132.

Leclant, F. and Remaudière, G. (1970) Eléments pour la prise en considération des aphides dans la lutte intégrée en vergers de pêchers, *Entomophaga* **15**, 53–81.

Lindow, S.E. (1983) Estimating disease severity of single plants, *Phytopathology* **73**, 1576–1581.

McLeod, P. and Gonzalez, A. (1988) A rotary leaf washer for sampling *Myzus persicae* (Homoptera: Aphididae) on spinach, *J. of Economic Entomology* **81**, 741–742.

Nutter, F.W. Jr., Gleason, M.L., Jenco, J.H. and Christians, N.C. (1993) Assessing the accuracy, intra-rater repeatability, and inter-rater reliability of disease assessment systems, *Phytopathology* **83**, 806–812.

Raworth, D.A., Frazer, B.D., Gilbert, N. and Wellington, W.G. (1984) Population dynamics of the cabbage aphid, *Brevicoryne brassicae* (Homoptera: Aphididae) at Vancouver, British Columbia. I. Sampling methods and population trends, *Canadian Entomologist* **116**, 861–870.

Rohitha, B.H. and Penman, D.R. (1981) A comparison of sampling methods for bluegreen lucerne aphid (*Acyrthosiphon kondoi*) in Canterbury, New Zealand, *New Zealand J. of Zoology* **8**, 539–542.

Seem, R.C. (1984) Disease incidence and severity relationships, *Annual Review of Phytopathology* **22**, 133–150.

Senanayake, D.G. and Holliday, N.J. (1988) Comparison of visual, sweep-net, and whole plant bag sampling methods for estimating insect pest populations on potato, *J. of Economic Entomology* **81**, 1113–1119.

Shokes, F.M., Berger, R.D., Smith, D.H. and Rasp, J.M. (1987) Reliability of disease assessment procedures: A case study with late leafspot of peanuts, *Oleagineux* **42**, 245–251.

Srikanth, J. and Lakkundi, N.H. (1988) A method for estimating populations of *Aphis craccivora* Koch on cowpea, *Tropical Pest Management* **34**, 335–337.

Steiner, M.Y. (1990) Determining population characteristics and sampling procedures for the western flower thrips (Thysanoptera: Thripidae) and the predatory mite *Amblyseius cucumeris* (Acari: Phytoseiidae) on greenhouse cucumber, *Environmental Entomology* **19**, 1605–1613.

Suber, G.A.F. and Le Cren, E.D. (1967) Estimating population parameters from catches large relative to the population, *J. of Animal Ecology* **36**, 631–643.

Tamaki, G. and Weiss, M. (1979) The development of a sampling plan on the spatial distribution of the green peach aphid on sugarbeets, *Environmental Entomology* **8**, 598–605.

Taylor, L.R. (1961) Aggregation, variance and the mean, *Nature* **189**, 732–735.

Taylor, L.R. (1970) Aggregation and the transformation of counts of *Aphis fabae* Scop. on beans, *Annals of Applied Biology* **65**, 181–189.

Teng, P.S. (1983) Estimating and interpreting disease intensity and loss incommercial fields, *Phytopathology* **73**, 1576–1581.

Ward, S.A., Rabbinge, R. and Mantel, W.P. (1985a) The use of incidence counts for estimation of aphid populations. 2. Confidence intervals from fixed sample sizes, *Netherlands J. of Plant Pathology* **91**, 100–104.

Ward, S.A., Rabbinge, R. and Mantel, W.P. (1985b) The use of incidence counts for estimation of aphid populations. 1. Minimum sample size for required accuracy, *Netherlands J. of Plant Pathology* **91**, 93–99.

Ward, S.A., Sunderland, K.D., Chambers, R.J. and Dixon, A.F.G. (1986) The use of incidence counts for estimation of cereal aphid populations. 3. Population development and the incidence-density relation, *Netherlands J. of Plant Pathology* **92**, 175–183.

Waters, T.F. (1969) Subsampler for dividing large samples of stream invertebrate drift, *Limnology and Oceanography* **14**, 813–815.

Wright, L.C. and Cone, W.W. (1983) Extraction from foliage and within-plant distribution for sampling of *Brachycolus asparagi* (Homoptera: Aphididae) on asparagus, *J. of Economic Entomology* **76**, 801–805.

Wright, L.C. and Cone, W.W. (1986) Sampling plan for *Brachycorynella asparagi* (Homoptera: Aphididae) in mature asparagus fields, *J. of Economic Entomology* **79**, 817–821.

CHAPTER 8

MANAGING THE GREENHOUSE, CROP AND CROP ENVIRONMENT

Menachem J. Berlinger, William R. Jarvis, Tom J. Jewett and Sara Lebiush-Mordechi

8.1. Introduction

Greenhouses vary in structural complexity from simple plastic film-covered tunnels, with no assisted ventilation, to tall, multispan, glass or plastic-covered structures covering several hectares and having sophisticated, computer-controlled environments. Essentially, however, all have climates inside that are rain-free, warm, humid and windless, ideal for raising crops but at the same time also ideal for many diseases and arthropod pests (Hussey *et al.*, 1967; Jarvis, 1992).

Though it is restricted, the climate within the greenhouse forms a continuum with the climate outside the greenhouse, and there are gradients in temperature, humidity, light and carbon dioxide. Depending on the needs of the crop, the need to exclude pests and pathogens, and the need to implement biological control programmes, these gradients can be manipulated to certain extents by such devices as screening, shading, cooling, heating and ventilation. At the other end of the scale, the climate at the immediate plant surface, the so-called boundary layer (Burrage, 1971), whether of shoots or roots, is of paramount importance in the avoidance of pests and diseases. It extends 1–2 mm for arthropod pests, about 30 μm for fungi and even less for bacteria. Its climate, the true microclimate, forms a continuum with the climate within the intercellular spaces of leaves on the one hand, and with the macroclimate of the greenhouse and its environs on the other hand. While most stages of most arthropod pests and beneficial insects are free to enter and leave the boundary layer if it is inimical to their activity, most micro-organisms enter passively and leave as wind-dispersed or water-splashed secondary propagules. In order to escape arthropod pests and pathogens, the microclimates of phyllosphere and rhizosphere must be made inimical to their activity but at the same time biological control organisms have to be encouraged with appropriate microclimates. It is often overlooked that biological control organisms have their own hyperparasite and predator chains extending theoretically indefinitely and acting alternately counter to effective biological control on the crop or beneficially with it (Jarvis, 1989, 1992). They also have their own adverse environments. It is apparently an insoluble task to manage boundary layer microclimates without detriment to the crop or to biological control, at the same time not permitting primary pests and diseases to become established.

8.2. Managing the Greenhouse

The local climate, the external disease and insect pressures, the greenhouse structural design, the climate-control equipment available, and the skill level of greenhouse workers have a major bearing on how a greenhouse is managed to control insects and diseases.

R. Albajes et al. (eds.), Integrated Pest and Disease Management in Greenhouse Crops, 97-123.
© 1999 *Kluwer Academic Publishers. Printed in the Netherlands.*

From the outset, it is important to have the input of a greenhouse manager to ensure that the physical facilities are properly designed for IPM when building a new greenhouse operation. Once a greenhouse is in operation, greenhouse managers have to be forever mindful of how activities in and around a greenhouse will affect IPM.

8.2.1. SITING AND ORIENTATION

On a world-wide basis, commercial greenhouse production is concentrated in regions between 25° and 65° latitude where the climate is moderate and local weather patterns are favourable. At high latitudes solar irradiance is low, day length is short and temperatures are low during the winter months resulting in poor growth and increased susceptibility to disease. Under such conditions, diapause of predatory insects may make biological control difficult. Large inputs of energy are required to maintain greenhouse temperatures, and humidification is often necessary to overcome the drying effect of continual heating. At low latitudes, high solar irradiance stresses crops making them more susceptible to disease. More outside ventilation air is required which brings with it more pathogen propagules and insect pests.

Within the most favourable latitudes, greenhouse production is concentrated in maritime areas where large bodies of water moderate the local climate. In continental areas, large swings in outdoor temperature and maximum solar-irradiance levels (Short and Bauerle, 1989) on a day-to-day basis create crop stresses that make greenhouse management more difficult. In summer, cooling of greenhouses is difficult if ambient air temperatures are above the desired greenhouse temperature, and if the relative humidity is so high that evaporative cooling is not effective.

Within any given region, the siting of a particular greenhouse operation makes a significant difference in the management of disease and insect problems. Field crops and natural vegetation growing in close proximity to a greenhouse create disease and insect pressure, especially if those crops and the vegetation are susceptible to the same disease and insect pests as the greenhouse crop. This pressure is intensified when pathogen propagules are stirred up by field operations, or when the outdoor crop is harvested or senesces and insects are forced to find a new host. Low temperatures force insects to seek out warmer climates indoors. On the other hand, freezing outdoor temperatures reduce pest pressures by inactivating pathogens and arthropod pests. Insects and pathogen propagules are carried into greenhouses through vents and doors by wind. By locating a greenhouse away from and/or upwind of outdoor crops, many pest problems can be reduced to manageable levels.

Out of concern for maximizing productivity and crop uniformity, greenhouses are oriented for maximum light penetration. This usually means an east-west orientation for free standing greenhouses and gutter-connected complexes (Harnett and Sims, 1979). Achieving good lighting uniformity over the course of a day is also important for IPM because insects and diseases proliferate in shaded areas and on stunted plants. In addition to orientation for optimal lighting, greenhouses should be oriented to take advantage of the prevailing winds. High wind speeds, if not reduced by windbreaks, increase heat loss and increase static pressures against which ventilation fans must operate. Moderate wind velocity at right angles to ridge, gutter and side vents is optimal for natural ventilation air movement through vents.

As said before, the environs of the greenhouse may be reservoirs of pathogens and pests. Greenhouses are often in an arable area, with trash piles, weeds and crops botanically related to the crop being grown in the greenhouse to provide ample inoculum and infestations of pathogen vectors (Harris and Maramorosch, 1980; Jarvis, 1992). Entry into the greenhouse can be rapid and on a massive scale: wind-blown dust carries spores and bacteria, air currents with or without forces ventilation carry spores and viruliferous insects from trash piles and weeds, water run-off into the greenhouse can carry soilborne pathogens such as *Pythium* and *Phytophthora* species and chytrid vectors of viruses, and dirt on feet and machinery carries pathogens. A foot bath containing a disinfectant reduces this latter risk when placed at the doorway. To surround greenhouses by a 10-m band of weed-free lawn and to eliminate trash piles may prevent or delay pest and pathogen inoculum entrance into greenhouses. Though whitefly-proof screens can keep out most insects (and keep in pollinator insects) fungal spores and bacteria cannot be excluded. Diseases of tomato such as Verticillium wilt, Fusarium crown and root rot, and bacterial canker are often first noticed directly beneath root vents or just inside doorways, as is the *Diabrotica*-borne bacterial wilt of cucumber [*Erwinia tracheiphila* (Smith) Bergey *et al.*].

Overlapping of cropping, i.e. raising seedlings and transplants alongside production crops, is unsound hygiene, inviting infection and infestation of the new crop from large reservoirs in the old crop.

8.2.2. STRUCTURES AND EQUIPMENT

The structural complexity of successful greenhouse operations tends to increase with time as older structures are replaced with more advanced designs, as the operations increase in size, as profits are reinvested, and as the need for improved climate-control becomes apparent. The low cost, low height, plastic film-covered structures that are often first built by growers provide some protection from outdoor weather and pests, but without any means for climate-control, conditions inside are often more favourable for diseases and pests than outside. Higher structures with more substantial framing members are required to accommodate climate-control equipment.

The trend in greenhouse structural design in recent years has been towards large gutter-connected complexes with high (4–5 m) gutter heights. As the size of operations under one roof has increased, increased gutter heights have become necessary to create the chimney effect needed to ventilate these structures naturally. With increased air space between the crop and the greenhouse cover, the uniformity of horizontal and vertical air movement has improved, temperature gradients in the crop canopy have been reduced and the uniformity of lighting of the crop has improved because shadows cast by higher overhead structural members move around more throughout the day. Increased gutter heights have also been beneficial for IPM because they increase the height that insects and pathogen propagules must be transported by wind to find their way into greenhouses through vents.

With larger complexes and the economies of scale they provide, it is feasible to incorporate features in a greenhouse design that favour IPM. With large-scale operations, it is practical to build header-house facilities that restrict access to the greenhouse. Separate shower and lunch room facilities, foot baths, refuse handling facilities, concrete floors, etc., that reduce the transport of insects and pathogen propagules into the growing areas can be justified. The costs of pressure washing equipment and specialized potting and growing medium sterilizing equipment are easier to justify. Also, for large scale operations, it is feasible to have separate propagation facilities (Section 8.3.2) specially designed for the production of disease-free transplants. On the other hand, because of the increased number of nooks and crannies, it is more difficult to eradicate insects and disease propagules from large complexes once they have gained a foothold.

Covers

The radiation transmission characteristics and the air tightness of greenhouse cover materials have a major effect on the climate for IPM inside a greenhouse. Ideally a cover material should have a high photosynthetically active radiation (PAR) transmission to maximize productivity and solar gain, low infra-red (IR) transmission to minimize radiation heat loss, and low ultraviolet (UV) transmission to inhibit sporulation of fungi (see Section 8.4.4). Unfortunately, no material has all these radiation transmission characteristics. Depending on latitude and local climate, some cover materials have been found better than others for IPM.

Glass is the preferred greenhouse cover material at high latitudes, where winter light levels are limiting and outdoor temperatures are low, because of its high PAR and low IR transmission characteristics. Glass, however, does transmit the UV radiation necessary for the sporulation of fungi and has relatively high air leakage which can lead to very low humidity during cold periods with high heat demand. During these periods it is necessary to humidify glass greenhouses to ensure the continued activity of biological control agents.

Polyethylene is the preferred greenhouse cover material at lower latitudes where high PAR transmission is not as critical and where retention of humidity for IPM is important. Some manufacturers include admixtures in their polyethylene films to block the UV wavelengths necessary for sporulation of fungi. The effectiveness of these blockers decreases as the films age. Polyethylene-film covered greenhouses are tighter than glass houses and therefore are better at retaining humidity during hot dry periods. During cool wet periods, high humidity and condensation on the underside of polyethylene films is a problem that can lead to indiscriminate dripping and spread of diseases in the crop. Surfactant sprays have been developed for polyethylene films that cause a film-wise condensation and runoff at the gutter. In recent years, roof arches used for polyethylene greenhouses have been modified from a semi-circular shape to a gothic shape to enhance film-wise condensation and runoff at the gutter.

Heating Systems

A carefully designed heating system to maintain air and root zone temperatures close to recommended levels is essential for an effective IPM programme in greenhouses. In the northern hemisphere greenhouse heating systems should be designed to maintain the desired indoor temperature when the outdoor temperature is at the 2.5% January design temperature (i.e. the temperature below which 2.5% of the hours in January occur on average) for a given location. If it is expected the greenhouse will be heated from a cold start in January, then it is common practice to add another 25% of pick-up capacity to the calculated 2.5% January design heating load so that the greenhouse can be fully warmed up before plants are transplanted.

Centralized hot-water or steam pipe heating systems are the most practical for commercial greenhouses. Fan-forced unit heaters are practical for small greenhouses or in greenhouses where it is only desirable to maintain temperatures above freezing, but heat delivery from fan-forced units is too costly and very non-uniform on a large scale. With hot-water or steam heating systems, heat is delivered to the base of the plants via radiation pipes running between the crop rows approximately 15 cm above floor level. Low-level positioning of heat pipes is important to provide heat to the root zone and to induce vertical air movement via natural convection. The temperature of water circulating in hot-water heating pipes is adjusted from 40 to 90°C depending on heating demand, thus heat is always applied at the base of the plants for a uniform temperature distribution. The flow of steam at 100°C through steam pipes is cycled on and off as required to maintain air temperature. This cycling leads to a non-uniform heating of the base of the plants and more temperature variability in steam-heated greenhouses. During very cold weather, operation of additional heating pipes around the perimeter and under gutters in hot-water and steam heated greenhouses is required to prevent cold spots where diseases are prone to develop. In hot-water heated greenhouses, especially those with tomato crops, an additional small-bore heating pipe is often used to apply heat at the growing tip of the plants to enhance growth and to prevent condensation on developing fruit.

Misting Systems

A common reason for failure of biological disease and insect controls early in the greenhouse growing season, and later on when outdoor conditions become hot and dry, is very low humidity levels in the greenhouse air. Under these conditions, transpiration of the crop is not adequate to maintain humidity levels in the optimum range for biological controls and it is necessary to add humidity to the air. Under hot and dry conditions, addition of humidity to the greenhouse has the added benefit of evaporatively cooling the greenhouse air. The theoretical and practical management of greenhouse humidity has been discussed by Stanghellini (1987) and Stanghellini and de Jong (1995).

The best humidification systems for greenhouses are those that create small water droplets that evaporate before they have a chance to settle out on leaves where they could provide the moisture necessary for germination of fungal spores. High-pressure (4–7 MPa) misting systems with 10 μm diameter nozzles and sonic misting systems that require a compressed air supply have been developed to create 10 μm diameter water droplets for greenhouse humidification. When properly maintained, these systems create a fog that gradually disperses as the water droplets evaporate in the air.

Ventilation Systems

Intake of outdoor air and exhaust of indoor air is necessary to prevent excessive solar-heat gain or humidity build-up inside greenhouses. Most large scale greenhouse operations are passively naturally ventilated through vents in the roofs and side walls. Small greenhouses, and polyethylene covered structures that are not equipped with roof vents, are actively ventilated with fans. Gutter vent systems have recently been developed for polyethylene covered greenhouses that allow them to be ventilated passively. Ventilation rates required for summertime temperature control are 0.75–1.0 air changes per hour (ASAE, 1989). Winter ventilation rate requirements are typically 10–15% of summer requirements. The relationships between greenhouse geometry, vent geometry, wind speed, wind direction, temperature and natural ventilation rates have been established by Kittas *et al.* (1997).

When greenhouse vents are closed, natural convection air movement inside greenhouses is often not sufficient for good air mixing and mass transport in the crop canopy. At low wind speeds leaf boundary layer resistance increases, resulting in decreased transpiration (Stanghellini, 1987) and increased relative humidity at the leaf surface. In large greenhouse complexes overhead fans strategically placed above the crop are required to bring horizontal air velocities up to approximately 0.5 m/s for good air mixing and to minimize boundary layer effects.

Air pressure differentials between inside and outside are necessary to move air actively through greenhouses. In actively and passively ventilated greenhouses, the pressure differential between inside and outside is usually negative, and it is easy for airborne pathogens and insects to enter the greenhouse, particularly if doors and ventilators are left open in hot weather. In special circumstances where it is essential to exclude pests and disease propagules, it may be necessary to maintain a positive pressure differential. With such a ventilation system, air can be filtered as it is drawn into the greenhouse to remove insects (Section 8.2.3) but removing airborne fungal spores and bacteria is impracticable. With a positive pressure differential, there is less tendency for infiltration of insects and disease propagules from outside through cracks in the greenhouse cover.

Regardless of type of ventilation system, any obstructions that reduce the vent openings increase the pressure differential and/or reduce the air flow through vents. If screens are placed over vent openings (Section 8.2.3) then the area of the vent openings must be increased by a factor equal to the reciprocal of the percent free area of the screen material to maintain the same pressure differential. If screens are used in established greenhouses, it would be necessary to build boxes over vents, add screened-in bays or screen the entire head space of a greenhouse to provide adequate intake air for good ventilation.

Thermal/Shade Curtains

Thermal curtains and shade curtains are generally beneficial for IPM because they reduce the extremes in climate that stress the crop and biological controls. Thermal curtains, aside from saving energy in the winter, reduce the net radiation from leaves through a greenhouse cover to a clear sky. For this reason leaf temperatures are higher and condensation on leaves is less under thermal curtains.

Shading of greenhouses is necessary in hot climates to reduce solar radiation and heat stress on crops. Paints can be applied on the exterior surface of the greenhouse cover (Grafiadellis and Kyritsis, 1978) or shade curtains can be deployed inside or outside (Willits *et al.*, 1989) to attenuate the radiation reaching the crop. Moveable shading systems (Jewett and Short, 1992) are also useful for acclimatizing crops and biological controls to rapidly changing solar radiation conditions.

Control Systems

The climate inside a greenhouse at any given time is determined by a complex interaction between outside climate variables, status of the crop and operating state of the climate-control equipment. Because of highly variable solar energy fluxes, the climate can change rapidly and climate-control equipment has to be manipulated quickly and frequently to maintain optimum conditions. The complex climate-control requirements of modern greenhouses can realistically only be met with computer-control systems.

Climate-control computers have been specially developed to meet the demanding requirements of greenhouse operations. The hardware used in greenhouses has been specially designed to withstand the high humidity and high levels of electrical noise. Special temperature and humidity sensing systems have been designed to monitor the inside and outside climate for control purposes. These sensors are shielded from the sun and are aspirated so that control is based on measurements of true ambient air temperature and relative humidity.

The software in commercial greenhouse computers has been specially developed to be fault-tolerant and to integrate the operation of climate-control equipment. In most cases the software has to be configured and control loops for each piece of climate-control equipment have to be tuned by the installer to give satisfactory performance. Currently available greenhouse control software enables greenhouse operators to schedule climate setpoints for the conditions that they believe are best for production and IPM. The actual climate-control achieved is limited by the capabilities of the climate-control equipment and the operator's skill and knowledge.

8.2.3. INSECT SCREENING

In the Mediterranean basin, protecting crops from arthropods is regarded as more important than protecting them from the weather, so the physical exclusion of insects from the greenhouse should help in reducing the incidence of direct crop damage and also of insect-transmitted virus diseases, theoretically this exclusion can be done by fitting fabric screens of mesh aperture smaller than the insects' body width over ventilators and doorways, or by insect-repellent fabrics, but in practice there still can be significant insect penetration. Moreover, screens impede ventilation and reduce light transmission, so compromises in the management of light, temperature and humidity are necessary to avoid adverse effects on crops and their susceptibility to diseases.

Screens do not suppress or eradicate pests, they merely exclude most of them; therefore, they must be installed prior to their appearance, and supplementary pest control measures, such as biocontrol, are still required (Berlinger et al., 1988). Insect parasitoids and predators that are smaller than their prey can still immigrate through pest screens into the greenhouse but larger ones have to be introduced. Since they offer an economical method of biological control of pests, they must be preserved, and destructive insecticides should be avoided. Screens impede ventilation (Robb, 1991; Price and Evans, 1992; Baker and Shearin, 1994), resulting in overheating and increased humidity. Increased humidity necessitates more frequent fungicide sprays than were required previously in an unscreened greenhouse. In Israel, 5–6 sprays per season (as opposed to 2–3 previously) are required in screened greenhouses (Y. Sachs, pers. com.). To minimize these harmful effects, growers add forced ventilation but this only helps to pull whiteflies through the screen, while exhausting air from the screenhouse increases the intake of small insects. Application of positive air pressure, pushing air into the structure through an insect-proof filter, reduces whitefly influx (Berlinger and Lebiush-Mordechi, 1995).

Thus, while screens can reduce immigrant populations of pests, they also reduce the immigration of beneficial arthropods. In neither case is exclusion total. Screens are disadvantageous in that temperatures and humidities tend to rise, promoting plant stress and susceptibility to diseases, and they also reduce light. Access to the greenhouse by workers and machinery is more difficult.

Types of Screens
Various types of screens and plastic covers have been developed to protect crops from insects; the challenge for the grower is to match the proper type of screen to local insect populations.

Woven Screens. The conventional woven screens are made from plain woven plastic yarns. Weaving leaves gaps (slots) between the yarns both in the warp and in the weft. In commercial screens the slot is rectangular whose width must be smaller than the whitefly's body size, about 0.2 mm, but it must allow maximum air and light transmission. Elongating the slot to improve ventilation is not feasible, since the threads slide apart, allowing insect penetration.

Bethke and Pain (1991) found that screens designed to exclude *Bemisia tabaci* (Gennadius) still permitted some to penetrate, and they failed to exclude *Frankliniella occidentalis* (Pergande). They did, however, exclude most larger insects such as moths, beetles, leafminers, aphids and leafhoppers, and they retained bumble bee pollinators.

Unwoven Sheets. These are made of porous, unwoven polyester and polypropylene or of clear, microperforated, polyethylene fabric. All are very light materials which can be applied loosely and directly over transplants or seeded soil, without the need of mechanical support. They have been used primarily in the open field, in early spring, as spun-bonded row covers, to enhance plant growth and to increase yield. At the same time they also proved to protect plants from insects. A polypropylene perforated sheet protected tomatoes from Tomato Yellow Leaf Curl Virus (TYLCV) transmission by *B. tabaci* (Berlinger *et al.*, 1988).

Knitted Screens. Because of irregularity in the shape of the holes, whiteflies are not excluded (Berlinger, unpublished). Reducing slot size to block whiteflies reduced ventilation to an impractical level. However, knitted screens can exclude larger insects.

Knitted-Woven Screen. This plastic screen is produced by a technique that combines knitting and weaving. The slot is almost 3 times longer than in the commercial woven screen, while the width is smaller than the whitefly body size. The insect cannot pass, but ventilation is improved. A laboratory test confirmed the screen's high blockage capacity for whiteflies, which was similar to that of a conventional screen (0.1% vs. 0.5% penetration, respectively; Berlinger, unpublished).

UV-Absorbing Plastic Sheets. These are claimed to protect crops from insect pests and from virus diseases vectored by insects, by modifying insect behaviour (Antignus *et al.*, 1996) but Berlinger (unpublished) was unable to confirm those claims. Nevertheless, these UV-absorbing plastic sheets have become available for commercial use. Their role in controlling diseases is discussed in Section 8.4.4.

Whitefly Exclusion
The sweetpotato whitefly (*B. tabaci*) is a small insect, about 0.2 mm wide, which transmits TYLCV, and has become the limiting factor in vegetable and flower production in Israel (Cohen and Berlinger, 1986; Zipori *et al.*, 1988). Its physical exclusion from greenhouses

is crucial, and accordingly whitefly-proof screens were developed (Berlinger *et al.*, 1991). While the rate of whitefly exclusion is generally proportional to the screen's mesh ($R^2 =$ 0.85) (Berlinger and Lebiush-Mordechi, 1995), the insect's ability to pass through any barrier could not be predicted solely from thoracic width and mesh size (Bethke and Pain, 1991). There is an unexpectedly high rate of whitefly penetration resulting from a great variability among the samples of the same screen resulting from uneven and slipping weave (Berlinger, unpublished).

Thrips Exclusion
Whitefly-proof (50 mesh) woven screens are by far the most widely used covers for the exclusion of whiteflies and bigger insects. In laboratory tests, thrips, with a body width of only 245 μm, moved freely through this screen. However, in the field, a high proportion (50%) are excluded, possibly because of the optical features of the plastic (Berlinger *et al.*, 1993).

Western flower thrips are strongly affected by colour. A loose shading net of aluminium colour, through which even whiteflies penetrated freely in the laboratory test, was tested in the field and in a walk-in tunnel. The aluminium screen reduced thrips penetration by 55% over an identically shading net but white in colour (Berlinger *et al.*, 1993). The closer aluminium fabric is placed around the entrance the more effectively it works (Mcintyre *et al.*, 1996).

8.2.4. OPERATION AND MAINTENANCE OF EQUIPMENT

Proper operation and maintenance of climate-control equipment is essential for healthy crops and avoidance of disease and insect problems in greenhouses. Mistakes in climate-control settings or failures of key pieces of equipment can lead to devastating losses in a matter of minutes. Even if such events do not cause immediate crop losses, physiological, disease and insect problems often show up some time later. The key to avoiding such problems is skilled operators and preventive maintenance programmes. Regardless of the level of equipment sophistication and maintenance, alarm systems together with backup power units and fuel supplies are essential to guard against losses during equipment break-downs or service interruptions.

Computer-control systems have taken much of the manual labour out of operating greenhouse climate-control equipment. A greenhouse manager should review climate data collected by the computer on a daily basis and make adjustments to setpoints to keep the climate conditions within desired ranges. It is critical that the temperature and humidity sensors used as the basis for control in each greenhouse compartment be cleaned and checked on at least a monthly basis. Greenhouse boiler systems need to be kept on line and in peak operating condition, not only during the winter heating months, but also in the summer months when it may be necessary to provide heat in the morning hours to avoid condensation on the crops. Vents and vent drives have to be kept in good working order to ensure they open when needed or close under high wind conditions when they could be damaged. Misting systems require stringent water treatment programmes to prevent nozzle blockages. The mechanisms for thermal and/or shade curtains have to be kept in alignment so that the curtains can be deployed quickly without snags or tears of the material. Insect screens have to be repaired if damaged. Also, insect screens have to be cleaned periodically to prevent blockages of light and air flow.

8.2.5. WORKER EDUCATION

For an effective IPM programme, greenhouse workers have to be trained to recognize nutrient deficiencies and disease and insect problems, and to take appropriate action. Personal protective gear, disinfectants, disposal bins, markers, etc. have to be made available to workers so that they can play their part in an IPM programme. In large operations, it is necessary to have a large site map of the greenhouses and a good record-keeping system so that disease and pest outbreaks as well as control actions that have been taken can be noted for the information of all greenhouse staff. New decision-support software programs (Clarke *et al.*, 1994) (Chapter 12) offer great potential for education of workers and record-keeping of all greenhouse activities, including IPM.

8.3. Managing the Crop

8.3.1. SANITATION

After genetic resistance, prophylaxis is by far the most effective and cheapest way of escaping major disease epidemics and pest infestations. It reduces the need for multiple applications of pesticides (which stress the crop), the risks of pesticide resistance, and pesticide contamination of the produce, the operator and the environment. Physical screening against immigrant pests has already been discussed (Section 8.2.3), which, coupled with aggressive control of insects in the environs of the greenhouse and in adjacent weeds and field crops, is very effective prophylaxis against both direct damage and insect-transmitted diseases. Some growers rely on old crop prunings to perpetuate populations of biocontrol insects. This is not a good practice because they constitute a reservoir of pathogens and non-parasitized pests. New introductions of biocontrol insects are a better practice.

Reducing inoculum is also important in early crop management (Baker and Chandler, 1957; Jarvis, 1992), with such tactics as quarantine, seed disinfestation, the use of healthy mother plants for cuttings, micropropagation, removing and properly disposing of all previous crop debris, pasteurizing or solarizing soil and soilless media, and disinfesting the greenhouse structure, benches, trays, stakes and other materials.

Disinfestants include formaldehyde (as formalin) and hypochlorites but both materials are hazardous to humans and residues are phytotoxic. A persulphate oxidising agent (Virkon®; Antec International), however, destroys viruses and micro-organisms without such side effects (Anonymous, 1992; Avikainen *et al.*, 1993; Jarvis and Barrie, unpublished results).

8.3.2. CROP SCHEDULING

Seeding, pricking-out and sticking cuttings should all be done in a greenhouse separate from the main production areas, and on mesh or slatted benches allowing through-the-bench ventilation (Section 8.4.6). The benches should be well above the level of soil-splash and there should be no overhead pots from which contaminated soil and drainage-water fall.

Where there is risk of diseases more destructive in cool soils, for example, Fusarium crown and root rot and corky root rot of tomatoes (Section 8.4.1), transplanting should be delayed until the root zone has warmed up, and insulating mulch materials put down later.

Where two or more monocrops are grown each year, overlapping of transplant production and marketable crop production means that pest and pathogen populations are perpetuated unless special care is taken to keep the young and cropping plants entirely separate. There is further risk if adjacent field crops constitute a reservoir of pathogens and pests.

8.3.3. SPACING

Close horizontal and vertical spacing of plants both on the bench and in the ground bed invites rapid plant-to-plant spread of walking insects, and of pathogens as diverse as *Pythium* spp., tomato mosaic virus, *Clavibacter michiganensis* (Smith) Davis *et al.* ssp. *michiganensis* (Smith) Davis *et al.* [= *Corynebacterium michiganense* (Smith) Jensen ssp. *michiganense* (Smith) Jensen], the downy mildews and *Botrytis cinerea* Pers.: Fr. (Burdon and Chilvers, 1982; Trolinger and Strider, 1984; Burdon *et al.*, 1989). The agents of virus spread are mainly water and soil splash, insects, and workers handling plants with contaminated tools and fingers (Thresh, 1982). Since air movement is restricted in dense plantings, the movement of airborne propagules is restricted, giving patchy distribution of diseases (Burdon *et al.*, 1989) and insects. Moreover, close spacing results in undue interplant competition for water, nutrients, light and CO_2, and undue damage by workers.

8.3.4. THE GROWING MEDIUM

Growing media cover a wide spectrum of substrates: soil and soil-mix composts, organic materials such as sawdust and coconut fibre, inorganic materials such as rockwool and synthetic foams and aggregates, and the nutrient film technique (NFT). Soilborne diseases are no less prevalent in soilless substrates than in soil (Zinnen, 1988; Jarvis, 1992). All substrates must be substantially free of insects and pathogens at planting and must be kept so throughout the life of the crop, thus demanding a high standard of hygiene.

Soils are usually heavily amended with peat, farmyard manure, straw or crop residues. Ploughing or rotovating the soil should be done in order to comminute plant root debris and other organic matter, and so expose pathogen propagules to natural biological control. Getting the soil into good tilth with optimum temperature, water content and aeration promotes this microbial activity. Soils also harbour several insects, such as pupae of leaf miners and thrips, as well as fungus gnat and shore fly larvae, both of which vector *Pythium* and *Fusarium* spp. Their populations, as well as populations of predatory microarthropods, are determined by soil organic matter, soil type and pore size (Vreeken-Buis *et al.*, 1998). Populations of omnivorous collembola and non-cryptostigmatic mites, for example, are enhanced by the organic matter usually plentifully added to greenhouse soils. Fungal parasites of insects and nematodes are also encouraged in soils of good tilth. The root-knot nematode *Meloidogyne incognita* (Kofoid & White) Chitwood, however, survives at 1–2 m, well below soil disturbance levels (Johnson and McKeen, 1973).

Most substrates can be fumigated or heat-sterilized but pasteurization to about 70°C (Baker, 1957) or solarization to about 40–55°C (Katan, 1981) is preferred over total steam sterilization to 100°C because it preserves thermophilic biocontrol organisms. The whole greenhouse can be closed in sunny conditions for solarization of both substrate and superstructure (Shlevin et al., 1995; Jarvis and Slingsby, unpublished). High temperature and vapour pressure deficit in closed greenhouses can kill the western flower thrips (F. occidentalis but unfortunately also its predator Neoseiulus (= Amblyseius) cucumeris (Oudemans) (Shipp and Gillespie, 1993; Shipp and van Houten, 1996).

As with the original ideas that soilless cultivation would eliminate soilborne pathogens, crops in rockwool or other inert substrate, or in NFT are no less free of soilborne arthropods. Fungus gnats, leafminers and thrips are numerous in rockwool and shore flies are always present in pools of water on plastic sheets. Even if soil is covered with plastic sheet, there are always gaps around stems, and tears and displacement of the cover readily permit insect access.

8.3.5. NUTRITION

Deficiencies and excesses of macro- and micronutrients, and imbalances in relative amounts of fertilizers can predispose plants to most diseases (Schoenweiss, 1975; Jarvis, 1977, 1992; Engelhard, 1989). In addition, fertilizers that increase foliage density at the expense of flowers and fruit not only reduce yield but tend to lower the vapour pressure deficit (VPD) in the boundary layer by restricting transpiration and wind-assisted evaporation, and consequently increase the risks of infection.

High nitrogen rates in fertilizers generally increase foliage density and softness, with increasing susceptibility to leaf and flower pathogens. For example, Hobbs and Waters (1964) found a quadratic increase in grey mould (B. cinerea) in chrysanthemum flowers (Dendranthema grandiflora Tzvelev) with nitrogen supplied with 1.5, 3.8 and 6.0 g/m^2. Nitrate nitrogen combined with liming gives excellent control of Fusarium wilt of several crops (Jones et al., 1989). Because of its role in the integrity of cell walls, calcium imparts resistance if balanced with potassium in a high ratio. A low Ca:K ratio permits susceptibility to B. cinerea in tomato (Stall et al., 1965). The K:N ratio is important in the susceptibility of tomato stems to the soft rot bacterium Erwinia carotovora (Jones) Bergey et al. ssp. carotovora (Jones) Bergey et al. (Dhanvantari and Papadopoulos, 1995). The incidence of soft rot was low at a K:N ratio of 4:1, increasing at 2:1 and 1:1. Verhoeff (1968) noted similar trends in tomato stems infected by B. cinerea. Paradoxically Verhoeff noted that high soil nitrogen can delay the development of latent lesions of B. cinerea in tomato, possibly because stem senescence is delayed.

Over-luxuriant foliage is conducive to greater damage by sap-sucking insects such as aphids (Scriber, 1984).

8.3.6. PRUNING AND TRAINING

Pruning and training tall staked and wire-supported crops like peppers, tomatoes and cucumbers not only modify the microclimate by altering spacing (Section 8.3.3) but pruning alters the fruit:foliage ratio and hence source-sink relationships in photosynthates (Section 8.3.7) and the disease-susceptibility of various tissues.

Removal of leaves bearing prepupal and pupal stages of pests can reduce their populations, but premature removal of leaves bearing parasitized stages can result in loss of biocontrol.

8.3.7. FRUIT LOAD

Closely related to the management of pruning is the distribution of photosynthates in heavily cropping plants (Gifford and Evans, 1981) in relation to the susceptibility of tissues to fungal and bacterial pathogens (Grainger, 1962, 1968). As Jarvis (1989) pointed out, modern technology has increased yields of greenhouse vegetables several-fold in the last two decades, with accompanying source-sink stresses on cultivars that have not changed very much. Thus, diseases such as Fusarium crown and root rot (*Fusarium oxysporum* Schlechtend.:Fr. f. sp. *radicis-lycopersici* W.R. Jarvis & Shoemaker) of tomatoes and Penicillium stem and fruit rot (*Penicillium oxalicum* Currie & Thom) of cucumbers have become serious in that same period. Both have been shown to be stress-related (Jarvis, 1988; Barrie, unpublished; Jarvis, unpublished) and there has been a resurgence in the incidence of corky root rot (*Pyrenochaeta lycopersici* R. Schneider & Gerlach) of tomatoes that might be related to a diminished flow of photosynthates to roots (Jarvis, unpublished observations). Grainger (1962, 1968) referred the "plunderable" carbohydrates available to certain pathogens – the so-called high-sugar pathogens (Horsfall and Dimond, 1957) – which include *B. cinerea,* whereas other pathogens, notably *Fusarium* spp., are classed as low-sugar pathogens principally attacking tissues starved of photosynthates. It is therefore incumbent on the grower to manage the nutrition, light and pruning of fruit and foliage so that a balanced partition of assimilates is attained without unduly compromising yield.

8.3.8. MANAGING PESTICIDES

Pesticides are a component of integrated pest management systems but are used too freely as insurance applications rather than judiciously as almost agents of last resort. Pesticides are significant agents of stress (Schoenweiss, 1975) whose over-use leads to problems of resistance (Regev, 1984; van Lenteren and Woets, 1988), to interference with microbial, insect biocontrol organisms (see Chapter 11) and bee pollinators, and so to an increase in iatrogenic diseases, diseases normally held in check by indigenous biological controls (Griffiths, 1981).

Unlike the pesticides on crops outdoors, pesticides in the greenhouse remain unweathered and persist longer, thus putting edible produce at risk of exceeding legally-tolerated residues, and exposing workers to higher concentrations for longer. There are no well-established economic threshold populations of insect pests and pathogens and the grower must thus rely largely on his own experience and on the experience of his advisors. It is at present difficult, if not impossible, to predict the course of disease epidemics in the greenhouse because the complex sequence of events in the life cycles of pathogens is dependent on a succession of different microclimates occurring in the correct order (Fig. 8.1). At best, therefore, fungicides can be used only in expensive and often unnecessary insurance programmes or within a very few hours of the requisite microclimate for spore germination occurring. On foliage this can usually simply mean leaf wetness (Section 8.4.2).

Pesticides are discussed at length in Chapter 11.

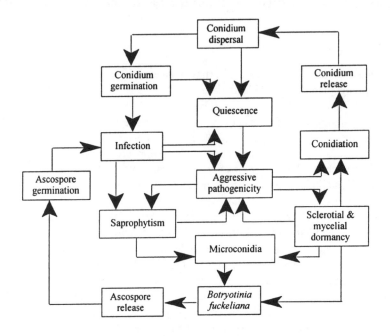

Figure 8.1. The life cycle of *B. cinerea*. Each stage is differently affected by microclimate factors, and control of grey mould is achieved by interrupting as many pathways as possible with environmental and cultural manipulations. Reprinted with permission from Jarvis (1992).

8.4. Managing the Crop Environment

8.4.1. TEMPERATURE

In very general terms, diseases as well as arthropods can be said to have optimum temperatures for their dispersal and development (Avidov and Harpaz, 1969; Jarvis, 1989, 1992; Chase, 1991) but these cardinal points are the integral of the optima of several growth phases of the pathogen as well as of different defence reactions of the host. Jarvis (1992) cited different temperature optima for different growth processes in the grey mould pathogen *B. cinerea*: mycelium growth, sporulation, conidium germination, germ tube growth, appressorium formation, sclerotium formation and sclerotium germination. All have different temperature optima, most of which lie above the general optimum range for grey mould development, 15–20°C. In most of its hundreds of hosts, resistance to *B. cinerea* is probably least within that range.

The temperatures of leaves and fruit can vary markedly from ambient air temperatures as determined by conventional greenhouse instruments, and so the temperature within the boundary layer can be assumed also to be different. At night, energy lost by radiation from leaves can result in temperatures 1–3°C cooler than ambient air and temperatures frequently reach the dew point. In crops transpiring well, evaporative cooling can also reduce leaf temperature but insolated leaves not transpiring can become considerably warmer, by as much as 2–8°C, than ambient air (Curtis, 1936; Shull, 1936).

Similarly, Schroeder (1965) found that the temperature of red tomato fruits rose from about 20 to over 50°C in air that rose from 26 to 37°C in the same period. On the other hand, green fruits exposed to the same conditions remained 4–8°C cooler than the red ones.

Temperatures of leaves, flowers and fruit can be considerably decreased by shading from direct sun and by increasing evaporative cooling by adequate ventilation and forced air flow (Carpenter and Nautiyal, 1969; von Zabeltitz, 1976). Eden et al. (1996) discussed the possibilities of raising flower truss temperatures in tomato crops to avoid grey mould. Whereas higher temperatures resulted in increased numbers of flowers infected by B. cinerea, the fungus was less likely to grow proximally to the main stem where the damage would be far more severe than one infected flower. On the other hand, higher temperatures (20–25°C) resulted in fewer infections of stem wounds than at 15°C. Eden et al. (1996) interpreted these results in terms of changing balances between fungal aggression and host defence reactions.

Just as with diseases of shoots, temperatures can be to some extent selected to minimize diseases of roots; for example corky root rot (P. lycopersisci) of tomato can be largely avoided by transplanting into warm media at 20°C (Last and Ebben, 1966), as can Fusarium crown and root rot (F. oxysporum f. sp. radicis-lycopersici) (Jarvis, 1988). By contrast, the optimum temperature for the expression of Fusarium wilt [Fusarium oxysporum Schlechtend.:Fr. f. sp. lycopersici (Sacc.) W.C. Snyder & H.N. Hans.] is 27°C. Similarly, Pythium aphanidermatum (Edson) Fitzp. is most pathogenic to spinach in hydroponic culture at 27°C, whereas Pythium dissotocum Drechs. is most pathogenic at 17–22°C (Bates and Stanghellini, 1984). It is therefore important to know exactly which of closely related pathogens is present.

Insects and mites, like diseases, have also an optimum temperature for their activity, dispersal and development. Generally, greenhouse pests are thermophilic and perform best within 20–30°C night-day ambient temperatures. The preferred temperature for aphids and the greenhouse whitefly is somewhat lower, 15–25°C. The interaction between temperature and VPD on the survival of western flower thrips was determined by Shipp and Gillespie (1993).

Of course, temperature affects not only arthropod pests but also their natural enemies. Natural enemies may perform poorly if temperatures are too high or too low which may occur during summer and winter respectively in the Mediterranean area. Then, the more temperature-tolerant Diglyphus isaea (Walker) or Dacnusa sibirica Telenga can be used according to thermal regimes expected in greenhouses. Excessive heat, combined with high VPD is a serious constraint for Phytoseiulus persimilis Athias-Henriot in warmer Mediterranean areas. Shipp and van Houten (1996) determined optimum temperatures and VPD for the use of N. cucumeris in Canadian cucumber houses, and these types of studies serve as guides to more intelligent biological control.

8.4.2. HUMIDITY

The effects of humidity on greenhouse crops have been reviewed by Grange and Hand

(1987), and their direct and indirect effects on diseases by Jarvis (1992). Uncertainty about VPD and temperatures in the boundary layer raises considerable suspicion about the validity of countless experiments on the infective abilities of fungal spores and disease prediction systems at low VPDs and inadequately measured or inadequately controlled temperatures (Schein, 1964). Fungal spores and bacteria require a wet substrate in which to initiate infection, and the water on leaves and fruits is provided by dew, guttation or overhead irrigation. This last can be discounted in well-managed greenhouses as an invitation to pathogens. Fogging systems cooling the air by evaporation are permissible if all the droplets evaporate before they land on plants (Section 8.2.2).

Measuring the onset and disappearance of dew is very difficult without the sensors themselves altering the boundary layer microclimate by heat conduction, shading, etc. (Wei et al., 1995a). Wei et al. (1995a), however, developed a copper-coated polyamide film sensor that could be wrapped around a tomato fruit and which had a response time of only a few seconds from dry to wet, and a response of less than 2 minutes to Peltier cooling of the surface to dewpoint. Connected to microclimate modifiers (heating, ventilation), this device could obviate much of the risk of infection.

Predicting the onset of condensation and its evaporation is even more difficult using atmospheric variables such as relative humidity, temperature, airspeed and radiation. Most predictions have errors in excess of 0.8 h and as much as ±3h (Wei et al., 1995b). Clearly this is unacceptable in a cucumber house where infection of flowers by *Didymella bryoniae* (Auersw.) Rehm can occur in 1–2 h (van Steekelenburg, 1985). Modelling the duration of dew in situations other than greenhouses has been done but with wide differences between predicted and observed durations of wetness (Wei et al., 1995b). When the dewpoint temperature of the air falls below the temperature of the plants in a greenhouse, they become covered with water droplets and films, perhaps with hydrophilic fungal spores as nuclei, especially in still air at low VPD. Wei et al. (1995b) developed a model from heat transfer theory that accurately simulated condensation and evaporation from tomato fruits still attached to the plant:

$$E_L = \frac{\rho_a C_P \left[e_a - e_s(\mathrm{T}) \right]}{r_d \gamma}$$

where E_L is the latent heat flux, ρ_a is the density of air, C_P is the specific heat of air, e_a and e_s (T) are vapour pressures of air and saturated vapour pressure of air, respectively at T°, r_d is the boundary layer resistance to vapour transfer between the wet surface and the air, and γ is a psychometric constant. Using the wetness sensor of Wei et al. (1995a), Wei et al. (1995b) obtained excellent agreement between simulated and measured fruit surface temperatures during condensation and evaporation, within 0.3–0.5°C (standard deviation 0.4°C). The model predicted wetness within 5 minutes of its detection, and dryness came as predicted. Clearly, this precision gives ample time for preventive action to be taken against most fungal infections.

While free water and low VPD are to be avoided if pathogens are present, those very conditions are needed to establish epidemics of fungal pathogens of insects, such as *Verticillium lecanii* (A. Zimmerm.) Viégas, *Beauveria bassiana* (Balsamo) Vuillemin and *Paecilomyces fumosoroseus* (Wize) Brown & Smith (Quinlan, 1988) (see Chapter 21).

Similar contrary indications have been obtained for arthropod pests and their predators. While spider mites are most active at relatively high temperatures and low VPDs, their predator *P. persimilis* is inhibited in those same conditions. Optimum humidity conditions for the predatory activity of *N. cucumeris* has been established by Shipp and van Houten (1996).

8.4.3. WATER STRESS

Guttation results when the rate of water supply osmotically pumped by the roots exceeds the rate of water lost by transpiration and used in growth (Hughes and Brimblecombe, 1994). To prevent guttation, the osmotic potential of the root xylem must be more negative than that of the nutrient solution (Kaufmann and Eckard, 1971; Bradfield and Guttridge, 1984). In poorly-managed greenhouses, guttation frequently happens at night when VPDs are low and root temperatures maintain high metabolic activity and root pressure. Tissues become waterlogged (oedema) and water guttates from stomata and from hydathodes at leaf margins with profound effects on the phylloplane micro-organisms (Frossard, 1981). Water continuous with the surface and substomatal vesicles facilitates the entry of bacteria into leaves of for example *Pelargonium* spp. (Lelliott 1988), particularly when resumed transpiration leads to resorption of the water. Wilson (1963) described how reversal of transpiration flow permits conidia of *B. cinerea* to enter tomato stem xylem, there to remain a latent inoculum.

Water alternately accumulating and evaporating from hydrothodes leaves toxic deposits of salts (Curtis, 1943; Ivanoff, 1963), a ready entry point for necrotrophic pathogens (Yarwood, 1959a,b). Lesions of gummy stem blight (*D. bryoniae*) are frequently seen originating from such points on cucumber leaves.

Guttation damage can easily be eliminated by regulating atmospheric humidity, ventilating effectively, reducing evening watering and adjusting the osmoticum of nutrient solutions (Slatyer, 1961).

8.4.4. LIGHT

Setting aside the effects of daylength on flowering in florists' crops, photosynthetically active radiation (PAR) (400–700 nm) is the part of the spectrum with the greatest effect on crop growth and productivity (Cockshull, 1985). Low and high light intensities are important agents of stress in crops (Schoenweiss, 1975) that induce physiologic strains predisposing the crops to disease. Particularly important is the effect of light combined with crop management procedures, such as plant spacing, row orientation, training and pruning systems, irrigation and nutrition, on the partition of assimilates, and the relative susceptibility of different tissues and organs to disease (Yarwood, 1959b; Grainger, 1968; Jarvis, 1989, 1992).

Daylength, however, is important in determining diapause in both arthropod pests and their predators. Early diapause may be a major constraint in their use. Non-diapausing strains can, to some extent, overcome this problem.

Light also has direct effects on fungal sporulation, germination and sclerotium formation. In *B. cinerea*, most isolates are stimulated to form conidia by light in the near-UV band (320–380 nm), an effect temporarily reversed by blue light (Epton and Richmond, 1980). Some isolates, however, form conidia in the dark (Hite, 1973;

Stewart and Long, 1987). All fungi grow mycelium in the dark, and *B. cinerea* forms its sclerotia in darkness, or in yellow or red light, or when irradiated for less than 30 min with near-UV light (Tan and Epton, 1973).

The requirement of *B. cinerea* and some other fungi for near-UV light for sporulation has led to the development of greenhouse covering materials that screen out that band as a means of disease control. Tuller and Peterson (1988) found fibreglass to transmit much less light of 315–400 nm than did polyethylene but in a comparative assessment of grey mould in Douglas fir seedlings [*Pseudotsuga menziesii* (Mirb.) Franco] it was concluded that the principal effect of low irradiance transmitted by fibreglass was in inducing needle senescence in dense canopies and thus susceptibility to grey mould, rather than on any direct effect on fungal sporulation. In both types of greenhouse, the mean intensity of light that inhibited sporulation (430–490 nm) exceeded that that promoted sporulation (300–420 nm). In those greenhouses, too, predisposing conditions of temperature (15–20°C) and humidity (>90% RH) persisted 14.5 times longer in fibreglass than in polyethylene-covered houses.

Humidity effects also seem to have outweighed effects of light wavelength in a series of trials with coloured cloches covering strawberries (Jordan and Hunter, 1972). Grey mould was most severe under pink and blue plastic covers, where VPDs were lower (0.41 and 0.64 kPa, respectively) than under clear plastic (1.14 kPa), or under glass (1.74 kPa). The effects of light are evidently not simple. Nevertheless, attempts have been made to filter out the near-UV light that induces sporulation in some fungi. Reuveni *et al.* (1989) incorporated hydroxybenzophenone into polyethylene, which increased the ratio of inhibitory blue light (480 nm) to UV (310 nm), and reduced the sporulation of *B. cinerea* in polystyrene petri dishes. Under the treated plastic, grey mould lesions were fewer in tomato and cucumber (17 and 15, respectively) than under untreated plastic (41 and 29, respectively) (Reuveni *et al.*, 1988). Similarly, plastic coverings absorbing light at 340 nm inhibited the sporulation and reduced the incidence of grey mould lesions on cucumber and tomato (Honda *et al.*, 1977) as well as white mould lesions caused by *Sclerotinia sclerotiorum* (Lib.) de Bary (Honda and Yunoki, 1977). Many isolates of *Alternaria solani* Sorauer also depend on near-UV light for sporulation, and Vakalounakis (1991) used vinyl films filtering out light of <385 nm to reduce the incidence of early blight in tomato greenhouses to less than 50% of that under unamended vinyl film.

Except as an agent of stress on the host, light has little direct effect on the rhizosphere microflora.

8.4.5. CARBON DIOXIDE AND OXYGEN

Carbon dioxide enrichment is a standard procedure in many commercial greenhouses (Porter and Grodzinski, 1985) but because it necessarily involves some restriction in ventilation to achieve the concentrations of CO_2 required, of the order of 1000 vpm, there is often increased danger of unmanageable low VPD (Watkinson, 1975; Ferare and Goldsberry, 1984). The concentrations of CO_2 that impair the growth of *B. cinerea* are 2–3 orders of magnitude greater than those found even in CO_2-enriched greenhouses (Brown, 1922; Svircev *et al.*, 1984) and so reports, for example, of Winspear *et al.* (1970), of increased incidences of grey mould in CO_2-enriched greenhouses, can be interpreted in terms of enhanced levels of assimilates (Grainger, 1962, 1968), or a denser canopy, with its increased risks of disease-susceptible wet plants (Grange and Hand, 1987).

While CO_2 is a prominent component of the rhizosphere atmosphere as a product of root and microbial respiration, it has little direct effect on pathogens.

Oxygen deficiency stress readily occurs in compacted and waterlogged soils and in over-warm hydroponic solutions in which both increasing temperature and increasing solute concentration decrease oxygen solubility. Further, increased temperatures lead to higher root and microbial respiration rates which further deplete oxygen tensions (Stolzy et al., 1975). Low oxygen tension has been advanced as an explanation for physiological root death (Daughtrey and Schippers, 1980; van der Vlugt, 1989) as well as decreased host resistance to root pathogens.

8.4.6. AIR MOVEMENT

The primary purposes of directing and regulating air movement in the greenhouse are: (i) to reduce the steepness of gradients in temperature, vapour pressure deficits and CO_2; (ii) to assist in the evaporation of infection droplets; and (iii) to induce thigmomorphogenesis in bench-grown crops. This last results in sturdier plants (Biro and Jaffe, 1984) and resistance to Fusarium wilts (Shawish and Baker, 1982).

Through-the-bench air movement and plant spacing on the bench are important factors in escape of forest seedlings (Peterson et al., 1988) and Exacum affine I.B. Balf. ex Regel (Trolinger and Strider, 1984) from grey mould.

Counter to the generally beneficial effects of air movement are its effects on pathogen spore dispersal. Most fungi sporulate best in still air at VPD of 1.2–0.6 kPa but fungi of the Peronosporales, like Bremia lactucae Regel and Pseudoperonospora cubensis (Berk. & M.A. Curtis) Rostovzev sporulate on wet surfaces (Rotem et al., 1978; Crute and Dixon, 1981). Airborne conidia are often liberated from conidiophores by hygroscopic mechanisms (Ingold, 1971) and are dispersed by air currents. Both mechanisms rely on disturbance of the microenvironment such as is readily provided by worker activity (Peterson et al., 1988; Hausbeck and Pennypacker, 1991).

The same mechanisms that control the liberation and dispersal of pathogen spores also apply to spores of biocontrol fungi when control is by enhancement of indigenous populations (Jarvis, 1992).

Air movement also effects the passive transport of spider mites on webs floating through the air and being trapped on neighbouring plants (Avidov and Harpaz, 1969). Forced air flows can transport larger insects into the greenhouse, even through some screens (Section 8.2.3). Aggregation of insects is controlled by airborne semiochemicals, while the dispersal of pheromones on excessive air currents can interfere with mating disruption as a means of biological control, or attraction into sticky traps.

8.4.7. INTEGRATION OF ENVIRONMENTAL FACTORS

Epidemics of diseases are the result of a complex sequence of biological events each with a different set of permissive environments that have to occur in sequence, and coupled with hosts in a receptive state. Jarvis (1977, 1992) outlined the complexity of those events in the case of grey mould epidemics (Fig. 8.1). Beginning with sporulation, conidia are formed at temperatures around 15°C and in moderate VPD; they are liberated by hygroscopic movements of the conidiophore in rapidly changing conditions of humidity, and are dispersed on air currents or by water-splash; infection occurs on wet surfaces at 15–20°C; and colonization of the host is fastest at 25–30°C. Marois *et al.* (1988) found that epidemics of grey mould on rose depend as well on inoculum concentration, a relationship that was different in winter and summer, and affected by temperature, relative humidity and VPD, the latter the far more meaningful parameter for describing epidemiology of *B. cinerea* in roses.

It has been possible to construct working models of grey mould epidemics in cucumber (Yunis *et al.*, 1990, 1994; Elad *et al.*, 1992; Elad and Shtienberg, 1995; Shtienberg and Elad, 1997); tomatoes (Eden *et al.*, 1996; Shtienberg and Elad, 1997); gerbera and rose (Salinas *et al.*, 1989; Kerssies, 1992); and conifer seedlings (Zhang and Sutton, 1994a,b). The value of epidemic models such as BOTMAN (Shtienberg and Elad, 1997), an integrated chemical and biological control program, in predicting the onset and course of epidemics, however, is severely compromised by the rapidity with which infection occurs – 9–10 h for grey mould (Yunis *et al.*, 1994) and only 1 h for gummy stem blight in cucumber flowers (van Steekelenburg, 1985; Arny and Rowe, 1991) – and by the wide variability of the greenhouse climate typically served by only one psychrometer in several hundred cubic metres of space (Jarvis, 1992). Shtienberg and Elad (1997) found that over three years, a rain forecasting system did not enable BOTMAN to perform significantly better than a weekly fungicide insurance program in unheated tomato and cucumber crops. However, a 4-day weather forecast proved more useful than immediate past records of weekly averages of surface wetness (calculated from dewpoint) of 7 h/d and 9.5 h/d at night temperatures between 9 and 21°C. By the time the requisite data have been collected and analysed, infection has already begun, and is an irreversible action even with the use of fungicides, which act mostly on germinating spores and thus too late to stop infection. Surface wetness is the key factor in all infections, and so its prediction from rates of change in surface and ambient air temperatures combined, by data processor, with simultaneous rates of change in VPD would be more timely in the immediate application of environmental control measures (Section 8.4.2).

Powdery mildew epidemics have a somewhat less complicated sequence of events prior to infection than grey mould epidemics but they, too, are ultimately dependent on the deposition of dew (Cobb *et al.*, 1978; Quinn and Powell, 1982; Powell, 1990; Jewett and Cerkauskas, unpublished results).

Control of any fungus-incited disease is achieved by breaking any of the pathways in life cycles similar to those of Fig. 8.1 (Jarvis, 1992) but the denial of water to germinable spores is the most important.

Computer models can be used to optimize greenhouse climate for both crop production and pest and disease control. For example, in The Netherlands a climate management program was developed for optimal production of tomatoes and is linked to a model for biological control of greenhouse whitefly by *Encarsia formosa* Gahan (van Roermund *et al.*, 1997). Further, the model can be extended with a humidity management module which prevents the development of fungal diseases.

Integration of pest and disease control primarily by manipulating the environment is a highly complex problem (Shipp *et al.*, 1991). Clarke *et al.* (1994), in describing a computer-managed system, considered the holistic production system as a six-hierarchy complex of factors in which any change at one level affected the other five levels. Thus, any change in greenhouse climate, whether engineered or not, effects changes in pesticide efficacy, biological control agents, pests and disease vectors, diseases, and ultimately productivity and profit.

There are a number of electronic decision support systems for various facets of greenhouse pest and disease control and production strategy (Papadopoulos et al., 1997). Jones et al. (1986, 1988) described an expert system with grower selection of climate set points based on his experience; Jacobson (1987) further developed an expert system with pre-set points for tomato production; and Dayan et al. (1983) developed TOMGRO that modelled physiological processes in tomato. Only Martin-Clouaire et al. (1993) considered disease escape in their model for tomato. Van Roermund et al. (1997), however, described the apposition of a whitefly control model to a production model, to which can be added a disease-avoidance model. Clarke et al. (1994) and Jewett et al. (1996) described a holistic Harrow Greenhouse Crop Management System (HGCMS) for both greenhouse tomato and cucumber. In addition to providing blueprints for production in which the grower has his own input, HGCMS provides user-friendly diagnoses for diseases, pests, biological controls and physiological disorders. It accepts climate monitoring. In addition, HGCMS allows the grower to enter economic data, and will analyse it for him. Conflict resolution, as far as can be agreed among experts, is a feature of HGCMS but ultimately the grower can accept or reject the advice of HGCMS.

The use and analysis of computer models and controls depends, of course, on a reasonable degree of computer literacy among growers, together with a basic understanding of plant growth and pest and disease biology. Otherwise reliance on expert advisory services is obligatory.

8.4.8. ENVIRONMENTS FOR MICROBIAL CONTROLS

In general, the microclimates for the successful deployment of fungal antagonists and parasites are close to those that promote plant infection by pathogens. Ideally, then, pre-emptive colonization of the phylloplane, as it is for rhizosphere, is the preferred strategy (Andrews, 1992). Adaptation to that microenvironment is a prerequisite (Dickinson, 1986). This colonization can also be achieved by enhancing indigenous populations of phylloplane antagonists (Jarvis and Atkey, unpublished results, in Jarvis, 1992). Similarly, the use of green manures and composts can achieve control in the rhizosphere without the necessity of isolating, registering and redeploying specific antagonists (Jarvis and Thorpe, 1981; Hoitink and Fahy, 1986; Ebben, 1987). McPherson and Harriman (1994) have suggested that in recirculating hydroponic systems, antagonist populations build up naturally in a disease-suppressive system that is reminiscent of take-all decline in wheat.

8.4.9. CONCLUSIONS

The primary objective of the commercial greenhouse grower is to obtain maximum yield per unit area of space with the least financial input. However, in order to achieve this, certain minimum standards in environment management have to be maintained in such matters as crop spacing, pruning and training, irrigation, fertilization, CO_2 supplies, and temperature and humidity regimes. While much is known about disease epidemiology and insect behaviour, scant attention, however, has been paid to the manipulation of greenhouse environments expressly to avoid disease epidemics and insect infestations, which together can easily account for 30% crop losses (Pimentel, 1991). This is a significant factor in a grower's balance sheet which is often overlooked, and usually dealt with simplistically by indiscriminate pesticide applications (Regev, 1984).

Careful analyses of epidemiological and epizootic data can indicate environments to be avoided or encouraged in greenhouse operations but integrating the desired environments into those wanted by the grower solely to maximize yields by physiological means is extremely difficult. The solution of these problems requires the consensus of several specialized experts, experienced crop advisors and, not least, good growers, whose experience and intuition are not to be ignored. The construction of predictive models can provide valuable insight into how environments affect diseases and insects, but experts can differ widely on which environment is best to escape, for example, lettuce downy mildew, or grey mould, or whiteflies or thrips. Resolution of these apparent conflicts can now be attained, or at least reasonable compromises achieved, by the inference engine in a computer expert system (see Chapter 12). One developed by Clarke *et al.* (1994) and Jewett *et al.* (1996) is a decision support system for greenhouse tomatoes and cucumbers that collates expert opinions on all aspects of crop production, including disease and pest management, the grower's own input, and internal and external environmental parameters. It can also provide the financial consequences of various actions, as well as of no action. Ultimately, the grower, whose brain no-one can replace, has the final decision.

References

ASAE, American Society of Agricultural Engineers (1989) Heating, ventilating and cooling greenhouses, in *Standards,* 36th edn, American Society of Agricultural Engineers, St Joseph, Mich., pp. 452–455.

Andrews, J.H. (1992) Biological control in the phyllosphere, *Annual Review of Phytopathology* **30**, 603–635.

Anonymous (1992) *Virkon Manual,* Antec International, Sudbury.

Antignus, Y., More, O., Ben Yoseph, R., Lapidot, M. and Cohen, S. (1996) UV-absorbing plastic sheets protect crops from insect pests and from virus diseases vectored by insects, *Environmental Entomology* **25**, 919–924.

Arny, C.J. and Rowe, R.C. (1991) Effects of temperature and duration of surface wetness on spore production and infection of cucumbers by *Didymella bryoniae, Phytopathology* **81**, 206–209.

Avidov, Z. and Harpaz, I. (1969) *Plant Pests of Israel,* Israel Universities Press, Jerusalem.

Avikainen, H., Kopenen, H. and Tahvonen, R. (1993) The effect of disinfectants on fungal diseases of cucumber, *Agricultural Science in Finland* **2**, 179–188.

Baker, J.R. and Shearin, B.A. (1994) *Insect Screening: Ornamentals and Turf,* Information Note, North Carolina Cooperative Extension Service, NC State University, Raleigh, NC.

Baker, K.F. (ed.) (1957) *The U.C. System for Producing Healthy Container-Grown Plants,* Manual 23, California Agricultural Experiment Station, Berkeley, Cal.

Baker, K.F. and Chandler, P.A. (1957) Development and maintenance of healthy planting stock, in K.F. Baker (ed.), *The U.C. System for Producing Healthy Container-Grown Plants,* Manual 23, California Agricultural Experiment Station, Berkeley, Cal., pp. 217–236.

Bates, M.L. and Stanghellini, M.E. (1984) Root rot of hydroponically grown spinach caused by *Pythium aphanidermatum* and *P. dissotocum, Plant Disease* **68**, 989–991.

Berlinger, M.J., Dahan, R. and Mordechi, S. (1988) Integrated pest management of organically grown tomato in Israeli greenhouses, *Applied Agricultural Research* **3**, 233–238.

Berlinger, M.J. and Lebiush-Mordechi, S. (1995) Physical methods for the control of *Bemisia,* in D. Gerling and R.T. Mayer (eds.), *Bemisia 1995: Taxonomy, Biology, Damage, Control and Management,* Intercept Ltd, Andover, Hants, pp. 617–634.

Berlinger, M.J., Lebiush-Mordechi, S., Fridja, D. and Mor, Neta (1993) The effect of types of greenhouse screens on the presence of western flower thrips: A preliminary study, *IOBC/WPRS Bulletin* **16**, 13–16.

Berlinger, M.J., Mordechi, Sara and Leeper, A. (1991) Application of screens to prevent whitefly penetration into greenhouses in the Mediterranean Basin, *IOBC/WPRS Bulletin* **14**(5), 105–110.

Bethke, J.A. and Pain, T.D. (1991) Screen hole size and barriers for exclusion of insect pests of glasshouse crops, *J. of Entomological Science* **26**, 169–177.

Biro, R.L. and Jaffe, J.J. (1984) Thigmomorphogenesis: Ethylene evolution and its role in the changes observed in mechanically perturbed bean plants, *Physiologia Plantarum* **62**, 284–296.

Bradfield, E.G. and Guttridge, C.G. (1984) Effects of night-time humidity and nutrient solution concentration on the calcium content of tomato fruit, *Scientia Horticulturae* (Amsterdam) **22**, 207–217.

Brown, W. (1922) On the germination and growth of fungi at various temperatures and in various concentrations of oxygen and carbon dioxide, *Annals of Botany* **36**, 257–283.

Burdon, J.J. and Chilvers, G.A. (1982) Host density as a factor in plant disease ecology, *Annual Review of Phytopathology* **20**, 143–166.

Burdon, J.J., Jarosz, A.M. and Kirby, G.C. (1989) Pattern and patchiness in plant-pathogen interactions – Causes and consequences, *Annual Review of Ecology and Systematics* **20**, 119–136.

Burrage, S.W. (1971) The microclimate at the plant surface, in T.F. Preece and C.H. Dickinson (eds.), *Ecology of Leaf Surface Micro-Organisms,* Academic Press, London, pp. 92–101.

Carpenter, W.J. and Nautiyal, J.P. (1969) Light intensity and air movement effects on leaf temperatures and growth of shade-requiring greenhouse crops, *J. of the American Society of Horticultural Science* **70**, 490–500.

Chase, A.R. (1991) Greenhouse ornamental crops – Pest management systems for disease, in D. Pimental (ed.), *Handbook of Pest Management in Agriculture,* CRC Press, Boca Raton, Fla., pp. 709–728.

Clarke, N.D., Shipp, J.L., Jarvis, W.R., Papadopoulos, A.P. and Jewett, T.J. (1994) Integrated management of greenhouse crops – A conceptual and potentially practical model, *Horticultural Science* **29**, 846–849.

Cobb, G.S., Hanan, J.J. and Baker, R. (1978) Environmental factors affecting rose powdery mildew in greenhouses, *HortScience* **13**, 464–466.

Cockshull, K. E. (1985) Greenhouse climate and crop response, *Acta Horticulturae* **174**, 285–292.

Cohen, S. and Berlinger, M.J. (1986) Transmission and cultural control of whitefly-borne viruses, *Agriculture, Ecosystems & Environment* **17**, 89–97.

Crute, J.R. and Dixon, G.R. (1981) Downy mildew diseases caused by the genus *Bremia,* in D.M. Spencer (ed.), *The Downy Mildews,* Academic Press, London, pp. 421–460

Curtis, L.C. (1943) Deleterious effects of guttated fluid on foliage, *American J. of Botany* **30**, 778–781.

Curtis, O.F. (1936) Leaf temperature and the cooling of leaves by radiation, *Plant Physiology* **11**, 343–364.

Daughtrey, M.L. and Schippers, P.A. (1980) Root death and associated problems, *Acta Horticulturae* **98**, 283–291.

Dayan, E., van Keulen, H., Jones, J. W., Zipori, I., Shmuel, D. and Challa, H. (1993) Development, calibration and validation of a greenhouse tomato growth model: I. Description of the model, *Agricultural Systems* **43**, 145–163.

Dhanvantari, B.N. and Papadopoulos, A.P. (1995) Suppression of bacterial stem rot (*Erwinia carotovora* subsp. *carotovora*) by high potassium-to-nitrogen ratio in the nutrient solution of hydroponically grown tomato, *Plant Disease* **79**, 83.

Dickinson, C.H. (1986) Adaptations of micro-organisms to climatic conditions affecting aerial plant surfaces, in N.J. Fokkema and J. van den Heuvel (eds.), *Microbiology of the Phyllosphere,* Cambridge University Press, Cambridge, pp. 77–100.

Ebben, M.H. (1987) Observations on the role of biological control methods with integrated system, with reference to three contrasting diseases of protected crops, in R. Cavalloro (ed.), *Integrated and Biological Control in Protected Crops,* A.A. Balkema, Rotterdam, pp. 197–208.

Eden, M.A., Hill, R.A., Beresford, R. and Stewart, A. (1996) The influence of inoculum concentration, relative humidity, and temperature on infection of greenhouse tomatoes by *Botrytis cinerea, Plant Pathology* **45**, 795–806.

Elad, Y. and Shtienberg, D. (1995) *Botrytis cinerea* in greenhouse vegetables: Chemical, cultural, physiological and biological controls and their integration, *Integrated Pest Management Reviews* **1**, 15–29.

Elad, Y., Shtienberg, D., Yunis, Y. and Mahrer, Y. (1992) Epidemiology of grey mould, caused by *Botrytis cinerea* in vegetable greenhouses, in K. Verhoeff, N.E. Malathrakis and B. Williamson (eds.), *Recent Advances in Botrytis Research*, Pudoc Scientific Publishers, Wageningen, pp. 147–158.

Engelhard, A.W. (ed.) (1989) *Soilborne Plant Pathogens: Management of Diseases with Macro- and Micronutrients*, American Phytopathological Society, St Paul, Minn.

Epton, H.A.S. and Richmond, D.V. (1980) Formulation, structure and germination of conidia, in J.R. Coley-Smith, K. Verhoeff and W.R. Jarvis (eds.), *The Biology of Botrytis*, Academic Press, London, pp. 41–83.

Ferare, J. and Goldsberry, K.L. (1984) Environmental conditions created by plastic greenhouse covers, *Acta Horticulturae* 148, 675–682.

Frossard, R. (1981) Effect of guttation fluids on growth of microorganisms on leaves, in J.P. Blakeman (ed.), *Microbial Ecology of the Phyllosphere*, Academic Press, London, pp. 213–226.

Gifford, R.M. and Evans, L.T. (1981) Photosynthesis, carbon partitioning and yield, *Annual Review of Plant Physiology* 32, 485–509.

Grafiadellis, M. and Kyritsis, S. (1978) New developments in shading plastic greenhouses, *Acta Horticulturae* 76, 365–368.

Grainger, J. (1962) The host plant as a habitat for fungal and bacterial pathogens, *Phytopathology* 52, 140–152.

Grainger, J. (1968) C_p/R_s and the disease potential of plants, *Horticultural Research* 8, 1–40.

Grange, R.K. and Hand, D.W. (1987) A review of the effects of atmospheric humidity on growth of horticultural crops, *J. of Horticultural Science* 62, 125–134.

Griffiths, E. (1981) Iatrogenic plant diseases, *Annual Review of Phytopathology* 19, 69–82.

Harnett, R.F. and Sims, T.V. (1979) Comparison of glasshouse types and their orientation, *Experimental Horticulture* 31, 59–66.

Harris, K. F. and Maramorosch, K. (1980) *Vectors of Plant Pathogens*, Academic Press, New York.

Hausbeck, M.K. and Pennypacker, S.P. (1991) Influence of grower activity and disease incidence on concentrations of airborne conidia of *Botrytis cinerea* among geranium stock plants, *Plant Disease* 75, 798–803.

Hite, R.E. (1973) The effect of irradiation on the growth and asexual reproduction of *Botrytis cinerea*, *Plant Disease Reporter* 57, 131–135.

Hobbs, E. L. and Waters, W. E. (1964) Influence of nitrogen and potassium on susceptibility of *Chrysanthemum morifolium* to *Botrytis cinerea*, *Phytopathology* 54, 674–676.

Hoitink, H.A.J. and Fahy, P.C. (1986) Basis for the control of soilborne plant pathogens with composts, *Annual Review of Phytopathology* 24, 93–114.

Honda, Y., Toki, T. and Yunoki, T. (1977) Control of gray mold of greenhouse cucumber and tomato by inhibiting sporulation, *Plant Disease Reporter* 61, 1041–1048.

Honda, Y. and Yunoki, T. (1977) Control of Sclerotinia disease of greenhouse eggplant and cucumber by inhibition of development of apothecia, *Plant Disease Reporter* 61, 1036–1040.

Horsfall, J.G. and Dimond, A.E. (1957) Interaction of tissue sugar, growth substances, and disease susceptibility, *Zeitschrift für Pflanzenkrankheiten, Pflanzenpathol, Pflanzenschutz* 64, 415–421.

Hughes, R.N. and Brimblecombe, P. (1994) Dew and guttation: Formation and environmental significance, *Agricultural and Forest Meteorology* 67, 173–190.

Hussey, N.W., Read, H.W. and Hesling, J.J. (1967) *The Pests of Protected Cultivation*, Elsevier, New York.

Ingold, C.T. (1971) *Fungal Spores: Their Liberation and Dispersal*, Clarendon Press, Oxford.

Ivanoff, S.S. (1963) Guttation injuries of plants, *Botanical Review* 29, 202–209.

Jacobson, B.K., Jones, J.W. and Jones, P. (1987) Tomato greenhouse environment controller: Real-time expert system supervisor, Paper No. 87-5022, American Society of Agricultural Engineers, Univ. Florida, Gainesville, Fla.

Jarvis, W.R. (1977) *Botryotinia and Botrytis species: Taxonomy, Physiology and Pathogenicity*, Monograph No. 15, Research Branch Canada Department of Agriculture.

Jarvis, W.R. (1988) Fusarium crown and root rot of tomatoes, *Phytoprotection* 69, 49–64.

Jarvis, W.R. (1989) Managing diseases in greenhouse crops, *Plant Disease* 73, 190–194.

Jarvis, W.R. (1992) *Managing Diseases in Greenhouse Crops*, APS Press, St Paul, Minn.

Jarvis, W.R. and Thorpe, H.J. (1981) Control of Fusarium foot and root rot of tomatoes by soil amendments with lettuce residues, *Canadian J. of Plant Pathology* 3, 159–162.

Jewett, T.J., Clarke, N.D., Shipp, J.L., Papadopoulos, A.P. and Jarvis, W.R. (1996) Greenhouse climate control: Past, present and future (Abs.), *Canadian Agricultural Engineering* **38**: 239.

Jewett, T.J. and Short, T.H. (1992) Computer control of a five stage greenhouse shading system, *Transactions of the ASAE* **35**, 651–658.

Johnson, P.W. and McKeen, C.D. (1973) Vertical movement and *Meloidogyne incognita* (Nematodea) under tomato in a sandy loam greenhouse soil, *Canadian J. of Plant Science* **53**, 837–841.

Jones, J.P., Engelhard, A.W. and Woltz, S.S. (1989) Management of Fusarium wilt of vegetables and ornamentals by macro- and microelement nutrition, in A.W. Engelhard (ed.), *Soilborne Pathogens: Management of Diseases with Macro- and Microelements*, American Phytopathological Society, St Paul, Minn., pp. 18–32.

Jones, P., Jacobson, B.K. and Jones, J.W. (1988) Applying expert systems concepts to real-time greenhouse controls, *Acta Horticulturae* **230**, 201–208.

Jones, P., Jacobson, B.K., Jones, J.W. and Paramore, J.A. (1986) Real-time greenhouse monitoring and control with an expert system, Paper No. 86–4515, American Society of Agricultural Engineers, Univ. Florida, Gainesville, Fla.

Jordan, V.W.L. and Hunter, T. (1972) The effects of glass cloche and coloured polyethylene tunnels on microclimate growth and disease severity of strawberry plants, *J. of Horticultural Science* **47**, 419–426.

Katan, J. (1981) Solar heating (solarization) of soil for control of soilborne pests, *Annual Review of Phytopathology* **19**, 211–236.

Kaufmann, M.R. and Eckard, A.N. (1971) Evaluation of water stress control with polyethylene glycols by analysis of guttation, *Plant Physiology* **47**, 453–456.

Kerssies, A. (1992) Epidemiology of *Botrytis cinerea* in gerbera and rose grown in glasshouses, in K. Verhoeff, N.E. Malathrakis and B. Williamson (eds.), *Recent Advances in* Botrytis *Research*, Pudoc Scientific Publishing, Wageningen, pp. 159–166.

Kittas, C., Boulard, T. and Papadakis, G. (1997) Natural ventilation of a greenhouse with ridge and side openings: Sensitivity to temperature and wind effects, *Transactions of the ASAE* **40**, 415–425.

Last, F.T. and Ebben, M.H. (1966) The epidemiology of tomato brown root rot, *Annals of Applied Biology* **57**, 95–112.

Lelliott, R.A. (1988) *Xanthomonas campestris* pv. *pelargonii* (Brown) Dye, in I.M. Smith, J. Dunez, R.A. Lelliott, D.H. Phillips and S.A. Archer (eds.), *European Handbook of Plant Diseases*, Blackwell, Oxford, pp. 164–165.

Marois, J.J., Redmond, J.C. and MacDonald, J.D. (1988) Quantification of the impact of the environment on the susceptibility of *Rosa hybrida* flowers to *Botrytis cinerea*, *J. of the American Society for Horticultural Science* **113**, 842–845.

Martin-Clouaire, R., Kovats, K. and Cros, M. (1993) Determinations of greenhouse set points by SERRISTE: The approach and its object-oriented implementation, *AI Applications* **7**, 1–15.

McIntyre, J.A., Hopper, D.A. and Cranshaw, W.S. (1996) Aluminized fabrics deter thrips from entering greenhouses, *Southwestern Entomologist* **21**, 135–140.

McPherson, M. and Harriman, M. (1994) Root diseases in hydroponic systems and their control, in *Annual Report of the Horticulture Research International*, Horticulture Research International, Wallesbourne, p. 22.

Papadopoulos, A.P., Pararajasingham, S., Shipp, J.L., Jarvis, W.R., Jewett, T.J. and Clarke, N.D. (1997) Integrated management of vegetable crops, *Horticultural Reviews* **21**, 1–39.

Peterson, M.J., Sutherland, J.R. and Tuller, S.E. (1988) Greenhouse environment and epidemiology of grey mould of container-grown Douglas fir seedlings, *Canadian J. of Forest Research* **18**, 974–980.

Pimentel, D. (ed.) (1991) *Handbook of Pest Management in Agriculture*, CRC Press, Boca Raton, Fla.

Porter, M.A. and Grodzinski, B. (1985) CO_2 enrichment of protected crops, *Horticultural Reviews* **7**, 345–398.

Powell, C.C. (1990) Studies on the chemical and environmental control of powdery mildew on greenhouse roses, *Roses Inc. Bulletin* September, 57–65.

Price, J.E. and Evans, M.R. (1992) Proper screen use can exclude insect and mite pests from your production area, *Foliage Digest* January, 3–4.

Quinlan, R.J. (1988) Use of fungi to control insects in glasshouses, in M.N. Burge (ed.), *Fungi in Biological Control Systems*, Manchester University Press, Manchester, pp. 19–36.

Quinn, J.A. and Powell, C.C. (1982) Effects of temperature, light and relative humidity on powdery mildew of begonia, *Phytopathology* **72**, 480–484.

Regev, U. (1984) An economic analysis of man's addiction to pesticides, in G.R. Conway (ed.), *Pest and Pathogen Control: Strategic, Tactical and Policy Models,* John Wiley & Sons, Chichester, pp. 441–453.

Reuveni, R., Raviv, M., Allingham, Y. and Bar, R. (1988) Development of a solar radiation-selective greenhouse cover for foliar pathogen control (Abs.), *Phytoparasitica* **16**, 74.

Reuveni, R., Raviv, M. and Bar, R. (1989) Sporulation of *Botrytis cinerea* as affected by photoselective sheets and filters, *Annals of Applied Biology* **115**, 417–424.

Robb, K. (1991) Incorporating physical barriers can limit pesticide use in IPM programs, *Greenhouse Manager* February, 87.

Rotem, J., Cohen, Y. and Bashi, E. (1978) Host and environmental influences on sporulation *in vivo, Annual Review of Phytopathology* **16**, 83–101.

Salinas, J., Glandoff, D.C.M., Picavet, E.D. and Verhoeff, K. (1989) Effects of temperature, relative humidity, and age of conidia on the incidence of spotting on gerbera flowers caused by *Botrytis cinerea, Netherlands J. of Plant Pathology* **95**, 51–64.

Schein, R.D. (1964) Comments on the moisture requirements of fungus germination, *Phytopathology* **54**, 1427.

Schoenweiss, D.F. (1975) Predisposition, stress, and plant disease, *Annual Review of Plant Pathology* **13**, 193–211.

Schroeder, C.A. (1965) Temperature relationships of fruit tissues under extreme conditions, *Proceedings of the American Society for Horticultural Science* **87**, 199–203.

Scriber, J.M. (1984) Nitrogen nutrition of plants and insect invasion, in R.D. Hauck (ed.), *Nitrogen in Crop Production,* American Society of Agronomy, Madison, Wis., pp. 441–460.

Shawish, O. and Baker, R. (1982) Thigmomorphogenesis and predisposition of hosts to Fusarium wilt, *Phytopathology* **72**, 63–68.

Shipp, J.L., Boland, G.J. and Shaw, L.A. (1991) Integrated pest management of disease and arthropod pests of greenhouse vegetable crops in Ontario: Current status and future possibilities, *Canadian J. of Plant Science* **71**, 887–914.

Shipp, J.L. and Gillespie, T.J. (1993) Influence of temperature and vapor pressure deficit on survival of *Frankliniella occidentalis* (Thysanoptera: Thripidae), *Environmental Entomology* **22**, 726–732.

Shipp, J.L. and van Houten, Y.M. (1996) Effects of temperature and vapor pressure deficit on the rate of predation by the predatory mite, *Amblyseius cucumeris, Entomologia Experimentalis et Applicata* **78**, 31–38.

Shlevin, E., Katan, J. and Mahrer, Y. (1995) Space solarization for sanitation of inocula of plant pathogens in the greenhouse structure (Abs.), *Phytopathology* **85**, 1209.

Short, T.H. and Bauerle, W.L. (1989) Climatic advantages for greenhouses located in Ohio (USA), The Netherlands, and eastern Australia, *Acta Horticulturae* **257**, 87–92.

Shtienberg, D. and Elad, Y. (1997) Incorporation of weather forecasting in integrated, biological-chemical management of *Botrytis cinerea, Phytopathology* **87**, 332–340.

Shull, C.A. (1936) Rate of adjustment of leaf temperature to incident energy, *Plant Physiology* **11**, 181–188.

Slatyer, R.O. (1961) Effects of several osmotic substrates on the water relationships of tomato, *Australian J. of Biological Sciences* **14**, 519–540.

Stall, R.E., Hortenstine, C.C. and Iley, J.R. (1965) Incidence of Botrytis gray mold of tomato in relation to a calcium-phosphorus balance, *Phytopathology* **55**, 447–449.

Stanghellini, C. (1987) *Transpiration of Greenhouse Crops. An Aid to Climate Management,* PhD Dissertation, IMAG (Instituut voor Mechanisatie, Arbeid en Gebouwen), Wageningen.

Stanghellini, C. and de Jong, T. (1995) A model of humidity and its applications in a greenhouse, *Agricultural and Forest Meteorology* **76**, 129–148.

Stewart, T.M. and Long, P.G. (1987) Sporulation of *Botrytis cinerea* in the dark, *New Zealand J. of Experimental Agriculture* **5**, 389–392.

Stolzy, L.H., Zentmeyer, G.A. and Roulier, M.H. (1975) Dynamics and measurement of oxygen diffusion and concentration in the root zone and other microsites, in G.W. Bruehl (ed.), *Biology and Control of Soil-Borne Plant Pathogens,* American Phytopathological Society, St Paul, Minn., pp. 50–54.

Svircev, A.M., McKeen, W.E. and Berry, J.W. (1984) Sensitivity of *Peronospora hyoscyami* f. sp. *tabacina* to carbon dioxide, compared to that of *Botrytis cinerea* and *Aspergillus niger, Phytopathology* **74**, 445–447.

Tan, K.K. and Epton, H.A.S. (1973) Effect of light on the growth and sporulation of *Botrytis cinerea, Transactions of the British Mycological Society* **61**, 145–147.

Thresh, J.M. (1982) Cropping practices and virus spread, *Annual Review of Plant Pathology* **20**, 193–218.

Trolinger, J.C. and Strider, D.L. (1984) Botrytis blight of *Exacum affine* and its control, *Phytopathology* **74**, 1181–1188.

Tuller, S.E. and Peterson, M.J. (1988) The solar radiation environment of greenhouse-grown Douglas-fir seedlings, *Agricultural and Forest Meteorology* **44**, 49–65.

Vakalounakis, D.J. (1991) Control of early blight of greenhouse tomato, caused by *Alternaria solani*, by inhibiting sporulation with ultraviolet-absorbing vinyl film, *Plant Disease* **75**, 795–797.

van Roermund, H.J.W., van Lenteren, J.C. and Rabbinge, R. (1997) Biological control of greenhouse whitefly with the parasitoid *Encarsia formosa* on tomato: An individual-based simulation approach, *Biological Control* **9**, 25–47.

van der Vlugt, J.L.F. (1989) A literature review concerning root death in cucumber and other crops, *Norwegian J. of Agricultural Sciences* **3**, 265–274.

van Lenteren, J.C. and Woets, J. (1988) Biological and integrated pest control in greenhouses, *Annual Review of Entomology* **33**, 239–269.

van Steekelenberg, N.A.M. (1985) Influence of humidity on incidence of *Didymella applanata* on cucumber leaves and growing tips under controlled conditions, *Netherlands J. of Plant Pathology* **91**, 277–283.

Verhoeff, K. (1968) Effect of soil nitrogen level and methods of deleafing upon the occurrence of *Botrytis cinerea* under commercial conditions, *Netherlands J. of Plant Pathology* **74**, 184–194.

von Zabeltitz, C. (1976) Temperaturverteilung bei einer Luft-Vegetationsheizung aus PE-Schläuchen, *Deutscher Gartenbau* **30**, 7–8.

Vreeken-Buis, M.J., Hassink, J. and Brussaard, L. (1998) Relationships and soil microarthropod biomass with organic matter and pore size distribution in soils under different land use, *Soil Biology & Biochemistry* **30**, 97–106.

Watkinson, A. (1975) Insulating fully creates other problems, *Grower* (London) **84**, 611–613.

Wei, Y.Q., Bailey, B.J. and Stenning, B.C. (1995a) A wetness sensor for detecting condensation on tomato plants in greenhouses, *J. of Agricultural Engineering Research* **61**, 197–204.

Wei, Y.Q., Bailey, B.J., Thompson, A.K. and Stenning, B.C. (1995b) Prediction of condensation on tomatoes, in *Proc. 1ˢᵗ IFAC/CIGR/EURAGENG/ISHS Workshop: Control Applications in Post-Harvest and Processing Technology*, Ostend, Belgium, 1995, pp. 51–56.

Willits, D.H., Bottcher, R.W., Marshall, J.L. and Overcash, M.R. (1989) Factors affecting the performance of external shade cloths, in *Summer Meeting Paper No. 89–4034*, ASAE/Canadian Society of Agricultural Engineers, Winnipeg, pp. 1–30.

Wilson, A.R. (1963) Some observations on the infection of tomato stems by *Botrytis cinerea* (Abs.), *Annals of Applied Biology* **51**, 171.

Winspear, K.W., Postlethwaite, J.D. and Cotton, R.E. (1970) The restriction of *Cladosporium fulvum* and *Botrytis cinerea*, attacking glasshouse tomatoes, by automatic humidity control, *Annals of Applied Biology* **65**, 75–83.

Yarwood, C.E. (1959a) Predisposition, in J.G. Horsfall and E.B. Cowling (eds.), *Plant Pathology. An Advanced Treatise*, Vol. 4, Academic Press, New York, pp. 521–562.

Yarwood, C.E. (1959b) Microclimate and infection, in C.S. Halton, G.W. Fischer, R.H. Fulton, H. Hart and S.E.A. McCallan (eds.), *Plant Pathology: Problems and Progress 1908–1958*, University of Wisconsin Press, Madison, Wis., pp. 548–556.

Yunis, H., Elad, Y. and Mahrer, Y. (1990) The effect of air temperature, relative humidity, and canopy wetness on gray mold of cucumbers in unheated greenhouses, *Phytoparasitica* **18**, 203–215.

Yunis, H., Shtienberg, D., Elad, Y. and Mahrer, Y. (1994) Qualitative approach for modelling outbreaks of gray mold of cucumbers in non-heated cucumber greenhouses, *Crop Protection* **13**, 99–104.

Zhang, P.G. and Sutton, J.C. (1994a) Effects of wetness duration, temperature and light on infection of black spruce seedlings by *Botrytis cinerea*, *Canadian J. of Forest Research* **24**, 707–713.

Zhang, P.G. and Sutton, J.C. (1994b) High temperature, darkness, and drought predispose black spruce seedlings to gray mold, *Canadian J. of Botany* **72**, 135–142.

Zinnen, T.M. (1988) Assessment of plant diseases in hydroponic culture, *Plant Disease* **72**, 96–99.

Zipori, I., Berlinger, M.J., Dayan, E., Dahan, R., Shmuel, D., Mordechi, Sara and Aharon, Y. (1988) Integrated control of *Bemisia tabaci* in greenhouse tomatoes from early planting, *Hassadeh* **68**, 1710–1713 (in Hebrew with English summary).

CHAPTER 9

HOST-PLANT RESISTANCE TO PATHOGENS AND ARTHROPOD PESTS

Jesús Cuartero, Henri Laterrot and Joop C. van Lenteren

9.1. Introduction

The aim of searching for host-plant resistance or tolerance is to develop cultivars that show little or no reduction in their normal yields when they are exposed to pests and diseases. Growers profit from better yields from resistant crops that need much less use of expensive pesticides and consumers benefit from vegetables with smaller amounts of chemical residues.

The capacity of plants to adapt to abiotic and biotic factors was known even to growers in ancient times. When they selected those plants that gave the highest yields and lowest levels of pests and diseases they were unknowingly exploiting genetic resistance. Scientific plant selection started around 1900, and in the following thirty years new varieties with more and more genes of resistance were released. However, subsequent experience revealed that genetic resistance has limits and that sometimes it only serves to combat low pest populations or to delay pest infection; sometimes, resistant cultivars stimulate the selection of pest populations able to live and reproduce on previously resistant cultivars. Consequently, host-plant resistance is best exploited in combination with other techniques like crop rotation, control of weeds within the crops and surrounding areas, biological control of animal pests, etc. Host-plant resistance is then one but important link in the chain of Integrated Pest Management.

9.2. Terminology

A host plant is a species in which or on which another organism lives. An organism that obtains some advantage from a host plant without benefiting the plant is usually termed a parasite. However, because parasite is used in other chapters of this book for the arthropod species used in biological control, we shall employ the term pest from FAO terminology to denote those weeds, animal species and microorganisms that damage crops. The term pathogen applies to specific microorganisms like bacteria, fungi, mycoplasmas and viruses, that parasitize plants. Plant disorders caused by pathogens are diseases. An animal pest is any animal that usually damages crops (nematodes, insects, mites, etc.). Aggressive strains of a pest are those strains that cause severe symptoms of disease in the plant genotypes attacked. A physiological race of a fungus, bacteria or virus with genes that enable it to attack a specific host-plant genotype is a virulent race; conversely, an avirulent race cannot attack this specific host-plant genotype.

Painter (1951) defines host-plant resistance as the relative amounts of heritable

R. Albajes et al. (eds.), Integrated Pest and Disease Management in Greenhouse Crops, 124-138.
© 1999 Kluwer Academic Publishers. Printed in the Netherlands.

characteristics of a plant that influence the degree of damage produced by a pest. Host-plant resistance is then: (i) heritable and controlled by one or more genes; (ii) measurable because its magnitude can be determined; (iii) relative because measurements are comparative with those of a susceptible plant of the same species that is damaged severely by the pest; and (iv) variable because it may be modified by biotic or abiotic factors. If the particularly sensitive phases of plant development do not coincide with the optimum conditions for pest development one speaks about escape.

Against the enormous numbers of pests and plant species in the world, host-plant resistance is common and host-plant susceptibility is exceptional. The combinations of the many types of barriers to infection (resistance characteristics) in a plant species and their collective effectiveness give rise to a series of genotypes that range from highly susceptible to highly resistant. When a pest cannot establish a compatible relationship under any condition with a certain plant genotype, then the genotype is said to be immune or absolute resistant to the pest. Resistance shown by non-host plants is termed non-host resistance, basic resistance, or basic incompatibility. Non-host resistant plants can exhibit resistance to their specific pests. If a plant expresses some resistance to all isolates or races of a pest it has non-race-specific resistance. If it expresses resistance to only one isolate or pest race it has race-specific resistance.

A tolerant plant may be colonized by a pest to the same extent as susceptible plants, but there is no reduction in yield quantity and quality. The converse of tolerance is sensitivity. Tomato yellow leaf curl virus (TYLCV), for example, produces very mild or no symptoms in both *Lycopersicon chilense* Dun. LA-1969 and *Lycopersicon pimpinellifolium* (Jusl.) Mill. LA-1478, but the concentration of virus antigen in the resistant cultivar LA-1969 is less than 1% of that in the susceptible 'Moneymaker' cultivar, while the concentration in the tolerant cultivar LA-1478 is similar to that in 'Moneymaker' (Fargette *et al.*, 1996). Rapid recovery of the plant after animal-pest attack is also considered as tolerance.

9.3. Resistance Mechanisms

Defence mechanisms present in the plant before pest attack are constitutive mechanisms and those induced by the infection process are induced mechanisms. Plants do induce responses instead of only constitutive and permanently present resistance because of: (i) chemicals produced by the plant as a result of interactions with pests may be toxic not only for the pest but also for the plant itself leading to a lower plant fitness when no pests attack the plant; and (ii) to produce defence chemicals may be costly, so that plants should allocate resources to defence only when and where interaction with pest occurs.

Constitutive and induced mechanisms may be either morphological or chemical. Examples of morphological constitutive defence mechanisms are the waxes of the cuticule that form a hydrophobic surface preventing water retention and pathogen deposition and germination. Thicker cuticles impede or make difficult penetration of insects, mites and pathogens, particularly when the latter penetrate by appresorium pressure. Thick and tough epidermal cell walls make difficult or impossible direct insect and fungal penetration; lignification or suberization give additional effective protection. The size and distribution of stomata and lenticels are associated with resistance to those insects, bacteria and fungi that make their entries through these structures. Internal barriers to movement through plant tissues like leaf-vein bundle sheaths and sclerenchyma cells may limit the spread of some pathogens and may prevent penetration of the phloem by aphids and whiteflies.

Chemical constitutive defence compounds interfere with the growth and reproduction of pests. The germination of some conidia is inhibited by compounds excreted by the plant. There are also internal secretions of inhibitors like phenolic acids in coloured onions and tomatine in tomato (Isaac, 1992). Plant tissues may contain antifungal agents produced by normal plant metabolism and, because the concentration of these compounds do not increase in response to infection, they are termed phytoanticipins to distinguish them from the phytoalexins, other chemical defence compounds produced only as a response to infection and that rapidly reach effective antimicrobial levels around the site of infection (van Etten *et al.*, 1994). Different plant families produce their characteristically different types of phytoalexins. For example, Fabaceae produce isoflavonoids, and Solanaceae, sesquiterpenes. Furthermore, pest damage can also induce an indirect defence, i.e. a defence that improves the effectiveness of natural enemies of the pest. Plants respond to damage by herbivorous mites or insects with the production of volatile chemicals that attract enemies of the herbivore, such as predators or parasitoids. This plant response occurs both locally and systemically (Dicke, 1994).

The morphological and chemical induced defence mechanisms of plants to pests are sometimes associated with the hypersensitive response, a process that leads to the rapid necrosis of infected cells. The pathogen can survive for some time in the necrosed cells around the site of original infection (Milne, 1966), but the rest of the plant remains healthy. The hypersensitive response is induced by specific elicitors of the pest that interact with specific receptors of the plant (elicitor-receptor model) and, in a number of plant species, it is commonly activated by viruses, bacteria, fungi, insects or mites. The elicitor-receptor model is confirmed in the pathosystem tomato *Cf-9–Fulvia fulva* (Cooke) Cif. (= *Cladosporium fulvum* Cooke) race 9 (De Wit, 1992). However, in the pathosystem tomato–*Pseudomonas syringae* van Hall pv. *tomato* (Okabe) Young *et al.*, the hypersensitive response is initiated when the serine-threonine kinase encoded by the resistance gene of the plant interacts physically with the avirulence gene of *Pseudomonas* (Tang *et al.*, 1996).

When a virus triggers a hypersensitivity response in a resistant plant, the tissues that surround the necrotic patches develop some localized acquired resistance to further attack by the same or other viruses (Ross, 1961a). The acquired resistance can be shown also by leaves not directly infected by the inductor virus (leaves without hypersensitive necrotic patches) and Ross (1961b) called this phenomenon systemic acquired resistance. Systemic acquired resistance is not common and even, if present, it does not always protect against a second systemic virus (Roggero and Pennazio, 1988). Pathogen-related proteins and salicylic acid appear to be involved in the mechanism of systemic acquired resistance.

Changes in plants after damage by pests or stresses can either decrease or increase plant resistance. The increase in resistance is called induced resistance that is usually systemic and increases with the degree of injury to the plant and reflects complex cytological, histological and physiological changes in the plant. For example, animal pest feeding activities produce short-term responses that affect animal pest feeding behaviour (Karban and Myers, 1989), but also long-term responses that can vary from premature leaf abscission to altered morphology, like increased hair density. Induced resistance elicited by pathogens is also termed cross protection and usually occurs when a plant has been inoculated by a mild strain of the infecting pathogen sometime before the attack of an aggressive strain. Concurrent protection is a special case of virus cross protection in which the protector virus does not replicate to detectable levels (the plant seems to be immune to that virus), however, the protector virus can induce protection against the second virus (Ponz and Bruening, 1986).

In plants, the two major resistance mechanisms against herbivorous insects are antixenosis (interference with insect behaviour) and antibiosis (interference with insect physiology). The usual patterns of insect approach, landing, probing, feeding and egglaying on a susceptible plant can be disturbed by resistance and induce non-preference or non-acceptance. These disturbances modify the behaviour of the insect and so protect a plant in the initial phase of an attack. Many examples of plant substances with repellent, deterrent or antifeedant properties are known. Several groups of toxic, secondary plant compounds like alkaloids, flavonoids and terpenoids may adversely affect the growth, development, generation-time and fertility of the insects. Some plant morphological characteristics that can interfere with or modify the behaviour of the insect are colour, shape, type of cuticle wax and the hairiness of plant stalks and leaves.

9.4. Genetics of Host-Plant Resistance

The fact that in nature host plants and their pests coexist, even though the pests may sometimes severely damage the plants indicates that they have evolved together and have established a dynamic equilibrium of resistance-virulence. Should either pest virulence or host-plant resistance increase without opposition, then the particular plant or the pest will be eliminated. Consequently, genetic studies of host-plant resistance should include studies of pest virulence genetics.

9.4.1. INHERITANCE OF RESISTANCE

In a segregating plant population, variations of expressed resistance to a particular pest may be either continuous or discontinuous depending on the number of resistance genes involved. Continuous variation from susceptible to resistant plants indicates that the resistance is polygenic which means that it is the sum of the small, individual expressions of many genes. Discontinuous variation indicates that the resistance is monogenic or oligogenic (controlled by one or a few genes) that may be either dominant or recessive major genes: individual plants fall into relatively well-defined classes of resistance or susceptibility. Genes of resistance are frequently clustered in linkage groups or complex loci and sometimes comprise genes involved in the recognition of more than one taxonomically unrelated pest (Crute, 1994). The first reported genetic study of resistance was published in 1905. Since then, the enormous amount of work in this field shows that resistance in many cases is inherited in a simple way. Dominance is very common, especially in hypersensitive responses, and recessive resistance occurs much less frequently. Inter-allelic gene interaction (epistasis) is only reported in a few cases (Niks *et al.*, 1993).

9.4.2. THE GENE-FOR-GENE CONCEPT

In gene-for-gene relationships, the host-plant resistance expression to a particular pest depends on the pest genotype, and the observed virulence of the pest depends on the host genotype. Flor (1956) demonstrates that for each gene that governs resistance in the plant there is a specific gene that governs virulence in the pest. This relationship became known as the gene-for-gene concept and was first shown for a number of fungal diseases and later for viruses, bacteria, nematodes, insects and parasitic plants (e.g. *Orobanche*). Today, it is generally accepted that the interaction takes place between the, usually dominant, alleles of the resistance and the, usually dominant, alleles of the avirulence. The gene-for-gene concept might then be reworded as: any resistance gene can act only if a locus in the pest carries a matching gene for avirulence (Niks *et al.*, 1993).

Table 9.1 displays the 16 possible combinations when two genes of resistance in a homozygous diploid plant are matched by two genes of avirulence in the haploid pest. Susceptible plants without no genes of resistance, $(r_1r_1r_2r_2)$ are attacked by all races of the pest, even those without genes of virulence (A_1A_2). Pests that carry two genes of virulence (a_1a_2) attack all plants independently of their combinations of genes of resistance. The two pest/plant combinations A_1/R_1 or A_2/R_2 trigger the often hypersensitive resistance response (plant and pest are incompatible). The combination $a_1A_2/R_1R_1r_1r_2$ is compatible because the avirulence gene A_2 is not matched by the corresponding R_2 host allele. The four possible combinations given by, A_1a_2 and a_1A_1, with R_1r_1 and r_1R_2, illustrate the differential interaction that reveals the occurrence of a gene-for-gene relationship. The differential interaction is used to classify pathotypes and to differentiate genes of resistance.

TABLE 9.1. Interaction of two plant genes of resistance in a plant with two pest genes of virulence according to the gene-for-gene concept

		Resistance (R) or susceptibility (r) alleles in the plant			
		R_1R_1 R_2R_2	R_1R_1 r_2r_2	r_1r_1 R_2R_2	r_1r_1 r_2r_2
Avirulence (A)	A_1A_2	−	−	−	+
or virulence (a) genes in the pest	A_1a_2	−	−	+	+
	a_1A_2	−	+	−	+
	a_1a_2	+	+	+	+

−: Incompatibility, +: Compatibility (infection)

The gene-for-gene interaction produces absolute resistance, or absolute susceptibility, of the host plant against a race of the pest. This race-specific response is termed vertical resistance and is very effective, but only against certain genotypes of a particular pest species. If the resistance is effective against all genotypes of the pest species without differential interaction, the resistance would be race-non-specific or horizontal resistance. The gene-for-gene concept presumably also applies to horizontal (usually polygenic) resistance, although this lacks proof until now.

9.5. Durability of Resistance

Johnson and Law (1975) proposed the term durable to describe long-lasting resistance. Durability does not imply that resistance is effective against all variants of a pest, but that the resistance has merely given effective control for many years in environmental conditions favourable to the pest (Russell, 1978).

Where susceptible cultivars are grown, the pest population comprises a set of races in dynamic equilibrium, but one or two of the races will tend to predominate. If a resistant cultivar is introduced, the predominant races either will not propagate, or their propagation rate will be substantially less than normal. In both cases, if one or some races can propagate effectively in the resistant cultivar, their proportions in the pest population will increase because they no longer have competition from the other races. A new outbreak of the pest will occur because the resistance will have been effectively "broken". It is difficult to determine whether a pest population is composed of a mixture of races, some present in very small proportions, or whether the pest produces virulent mutants that disappear from the pest population unless there is a compatible resistant host plant in which they can propagate.

In theory, when the introduced resistance is complete, the predominant races will disappear and more virulent races will spread. The spread will be faster than when the introduced resistance is only partial because the virulent and dominant races will compete. Before the introduction of the first resistant tomato cultivars, the predominant if not the only tobacco mosaic virus (TMV) race was race 0. When *Tm-1* resistant cultivars were introduced, the pathogen population changed and very soon TMV race 1 progressively predominated. *Tm-2* cultivars resistant to TMV races 0 and 1 were not much better because TMV race 2 quickly spread. *Tm-1* proved to be resistant to race 2 and cultivars with *Tm-1* and *Tm-2* were released. Again, the resistance of these new cultivars was quickly broken down because TMV race 1-2 predominated. These case histories of *Tm-1*, *Tm-2* and *Tm-1-Tm-2* cultivars support the "lack of durability hypothesis" of complete resistance. However, the subsequent release of cultivars with the $Tm-2^2$ allele resistant to TMV races 0, 1, 2 and 1-2 effectively controlled TMV for 20 years. Why the *Tm-1* and *Tm-2* resistances were so ephemeral, and that of $Tm-2^2$ has lasted more than 20 years, we do not know. Other examples of durable resistances governed by major genes are resistance to *Stemphylium* in tomato and to *Cladosporium* in cucumber. Examples of low durability resistances are those to *F. fulva* in tomato and to *Bremia* in lettuce.

Resistance to insects tends often to be partial and polygenic. It appears then unlikely that more virulent populations (biotypes) adapted to partial resistant cultivars might be selected. However, transgenic cultivars that carry the *Bt* gene from *Bacillus thuringiensis* Berliner rely on a monogenic factor that has a very high expression and, as McGaughey (1988) reports, several species of insects like *Helicoverpa* (= *Heliothis*) *zea* (Boddie), quickly adapt to tolerate the *Bt*-gene toxin. The use of partially resistant cultivars reduces the selection pressures on insect populations and this effectively delays the development of virulent biotypes.

The type of reproduction of a pest greatly influences the durability of host-plant resistance. Aphids, for example, exploit their capacity to reproduce parthenogenetically to colonize resistant cultivars and large populations quickly develop from a few individuals able to overcome host-plant resistance. Soilborne pests, on the other hand, spread more slowly than airborne pests and thus virulent biotypes or races of pathogens take long times to colonize the area in which a resistant cultivar is grown.

Some kinds of host-plant resistance are more durable than others. For example, those which involve changes in plant morphology (growth of hairs or trichomes that interfere with insect movements or feeding activities, or water repellent, waxy surfaces and thickened epidermis of leaves that prevent fungal spores from sticking to the leaf or resist the penetration of some fungi, etc.) require complex changes in the pest to successfully adapt to and overcome the defensive strategy of modified plant structure, and complex changes take very long time.

9.6. Breeding to Improve Host-Plant Resistance

Resistant plant varieties are produced by breeding programs that involve: (i) search for sources of resistance; (ii) evaluation of the resistance found; and (iii) selection in segregating generations. To growers, the pest resistance of a new variety is only one characteristic out of many, and it is not the most important. Therefore, plant breeders have to bear in mind that the agronomic characteristics of a new resistant variety must be as good as, or better than, previous non-resistant varieties when the pest to which it is resistant is not present. If not, no matter how good its resistance to a particular pest is, the variety is most unlikely to be grown on a large scale.

9.6.1. SOURCES OF RESISTANCE

If the resistance to a particular pest is already present in commercial cultivars (either hybrids or open pollinated cultivars) the source of resistance for our breeding programme would be the resistant commercial cultivar most similar to our ideotype. Commercial cultivars have genes for high yield and quality, for resistances to some pests, for adaptation to specific environments like greenhouses, etc., that must be exploited. Should resistance to the target pest not be present in a commercial cultivar, the first step the plant breeder must take is to search the literature for plants described as sources of resistance, obtain seeds of those source plants, and then evaluate the level of their resistance to help decide whether the source plants might serve as the starting point of the breeding programme. If the desired resistance is not yet described, it can be searched for in accessions from germplasm banks. The usual search sequence is: landraces, wild forms, related species and related genera.

Should it be impossible to find a source of high-level resistance in germplasm collections, the breeding material might still be manipulated by mutation, tissue culture and molecular genetic techniques to produce new variability. Artificially induced mutations have produced a small number of commercial cultivars and, except in those resistances that involve recessive characters in vegetatively propagated ornamental crops, the method is not to be recommended. When cell or tissue cultures are grown for extended periods, genetic variation, termed somaclonal variation, usually takes place. Examples of useful variation from tissue culture are resistance to *Bipolaris oryzae* (Breda de Haan) Shoemaker (= *Helminthosporium oryzae* Breda de Haan) and resistance to the herbicide glyphosate. However, in spite of these examples, there are serious doubts about using somaclonal variation as a source of variability, mainly because of the unstability of the variation. To increase the variability of a species by genetic manipulation is limited principally because it is difficult to identify and clone genes. As the number of cloned genes increases, more variability will be generated by plant transformation. The expression of viral DNA sequences in transgenic plants may produce virus-resistant plants that introduce new variability into the gene pool of the plant species.

9.6.2. EVALUATION OF RESISTANCE

Plant populations must be exposed to the pest in such a way that resistant and susceptible plants can be differentiated as quickly and clearly as possible. Field screening has the advantage that the cost per plant tested is low and, more importantly, that the test conditions simulate those under which commercial crops grow. However, field screening has disadvantages, it is dependent on the weather, whether or not the pest will develop is always uncertain, and other pests may interfere with the tests. Screening under controlled conditions like glasshouses or climatized rooms gives standardized environmental conditions, and the amount of pest present and its distribution can be controlled, but the conditions of growth are not representative of those under which commercial crops grow.

The expression of resistance in a host-parasite system is not constant but it depends largely on the composition and amount of the inoculum, on the stage of development of the plant and on the conditions under which the resistance is evaluated. Small amounts of inoculum produce little or no symptoms in susceptible plants and so resistance may be overestimated. For example, in the pepper–*Phytophthora capsici* Leonian system, concentrations of 10^2 zoospores/cm^3 of some isolates produce no mortality on 'Morron' cultivar, but concentrations of 10^4 produce 100% mortality (Gil Ortega *et al.*, 1995). Breeders prefer to test plants as early as possible because seedlings need less space and time to develop and, in general, are less resistant than mature plants. The expression of resistance is greatly influenced by environmental variables (like light, temperature, soil fertility) and the distribution pattern of plant genotypes in the field. To measure resistance properly, the values of those environmental variables should all be within the range of values of the conditions under which commercial crops grow. The expression of resistance shows no constant relationship with light parameters. Host-plant resistance to *Manduca sexta* (Johannsen) in the wild tomato *Lycopersicon hirsutum* Humb. & Bonpl. f.

glabratum Mull. increases when plants grow under long-day-light conditions (Kennedy *et al.*, 1981), but low-intensity light like that of cloudy days tends to reduce the expression of resistance to insects (Smith, 1989). Temperatures outside the range of conditions under which commercial crops are grown reduce the expression of resistance in a number of host-pest systems (Smith, 1989). However, Gómez-Guillamón and Torés (1992) report that three lines of melon, when grown at normal temperatures for commercial crops, show resistance to *Sphaerotheca fusca* (Fr.) Blumer. [= *Sphaerotheca fuliginea* (Schechtend.:Fr.) Pollacci] at 26°C but are susceptible below 21°C. High doses of nitrogenous fertilizers generally increase the susceptibility while additional applications of potassium and phosphorus fertilizers increase resistance. When resistance of different genotypes is assessed in small plots, resistant genotypes will export small levels of inoculum, but will receive high levels of inoculum from susceptible genotypes that, in turn, export more inoculum than they receive and, consequently, the resistance of resistant genotypes will tend to be underestimated in comparison with that of the same genotypes measured in trials carried out in large plots or in separate plots. This phenomenon is termed interplot interference and can be mitigated by including control cultivars with different levels of resistance as references. In any case, small differences found in the level of infection in small plots should be most carefully noted (Parlevliet and van Ommeren, 1984). Pests that generally display a vertical dispersion show smaller interplot interference than pests that display a horizontal dispersion.

The principal application of *in vitro* resistance screening is to select those cells, calli, or somatic embryos that show resistance to the toxin of a pathogen. Advantages of this technique are: (i) large numbers of individuals can be processed; (ii) haploid cells reveal concealed recessive traits; (iii) it can exploit somaclonal variation; and (iv) the uniformity of the experimental environmental conditions helps discriminate slight quantitative differences of plant resistances. Disadvantages are: (i) it is limited to tests for pathogens that produce toxins; (ii) cells that survive infection may be physiologically adapted and not genetic variants; (iii) resistance at cellular level is not necessarily expressed in the whole plant; and (iv) *in-vitro* resistance screening does not detect defence mechanisms that are based on differentiated tissues.

9.6.3. SELECTION METHODS

After a source of resistance has been identified and an appropriate evaluation procedure has been set up, the next step is to integrate the resistance into the set of agronomic characters that a cultivar needs for success on the market. The donor of resistance should be selected taken into account that the closer the genotype of the donor to that of the cultivar to be improved, the shorter will be the process of introduction of resistance. Complete resistance is frequently easier to manage than partial resistances. Complete resistance is essential for pests damaging the end-product of the crop because the greenhouse-grown produce must be of prime quality without cosmetic damage like spots or scars that reduce consumer acceptability.

Most cultivars of the greenhouse-grown species are hybrids. To produce a resistant hybrid the resistance has to be introduced into one of the parents (dominant resistance), or in both parents (recessive resistance). The appropriate selection procedure for monogenic resistances is backcross and for polygenic resistances is recurrent selection.

Marker-assisted selection recovers genes linked to markers. The markers are more easily scored than the genes of resistance. To ensure that only a minor fraction of the individuals selected are recombinants, the linkage between the marker and the target gene in coupling phase should be <5 cM. A repulsion-phase marker linked at <10 cM provides higher efficiency than that of a 1 cM coupling-phase linkage (Kelly, 1995). Marker-assisted selection do not need inoculation of pests, so that it avoids the errors caused by failed infection, incomplete penetrance of the resistance and variability of aggressiveness. In addition, breeding for resistance can be carried out where inoculations of healthy plants in the field are not allowed for safety reasons. The susceptibility to Fenthion insecticide shown by tomato seedlings on detached leaves that carry the *Pto* gene of resistance to *P. syringae* pv. *tomato* is used as an indirect indicator to select for resistance to this bacteria (Laterrot, 1985). The isozyme marker *Aps-1* has been used commercially for many years as a substitute for screening with nematodes to select for the *Mi* resistance gene in tomato. *Mi* genotypes can now be selected by a PCR-based marker that is more tightly linked to *Mi* than *Aps-1* (Williamson *et al.*, 1994).

Screening tests for resistance to multiple pests are sometimes of doubtful validity because infection by one pest may interfere with the infections by other pests. Marker-assisted techniques avoid infection and can help to introduce several genes each resistant to a different pest. Marker-assisted selection also offers considerable potential to transfer polygenic (quantitative) resistance because markers have high heritability (h=1 for molecular markers) and direct selection of resistance genes is masked by environmental effects. In tomato, molecular markers have been discovered for oligogenic (Danesh *et al.*, 1994) and for polygenic resistances (Neinhuis *et al.*, 1987).

A solution to control pathogens that infect roots is to use resistant rootstocks. They are used for several greenhouse crops such as tomatoes, eggplants, melons, water-melons, cucumbers, carnations and roses. However, for roses, rootstock grafting is done to improve disease resistance and to change the vigour and longevity of the crop.

9.7. Strategies to Improve Durability

The vast majority of the resistant cultivars rely on the use of single, major genes and these have proved remarkably successful, even though severe breakdown of resistance occurs from time to time. Several strategies are proposed to reduce the risk of resistance breakdown when major genes of resistance are used.

Multilines or cultivar mixtures are formed either by phenotypically similar lines, or cultivars that each contain a different single, race-specific gene of resistance. No examples of multilines or cultivar mixtures occur among the species usually grown in greenhouses.

Gene deployment uses several cultivars each with a different gene of resistance and grown within a clearly defined area. If the pest produces a virulent race on the cultivar grown, another cultivar that carries another gene of resistance will be grown in the area from next year until a new virulent race breaks its resistance. The next cultivar grown will either be the first cultivar or a new one with resistance to the last virulent race.

Gene deployment, as multilines, exploits the diversity of the host-plant population to stabilize the pest population and avoid the appearance of virulent races. To effectively use gene deployment all the growers of the area must use cultivars with the same resistance gene.

Pyramiding resistance genes involves the introduction into the same cultivar, of all, or as many as possible, of the genes of resistance for a pest. The rationale behind pyramiding is that the pest will need several mutations from avirulence to virulence to overcome the resistance and that the probability of two or more successive mutations is extremely low because it is the product of the probability of each mutation. The gene *Pto* protects tomato against *P. syringae* pv. *tomato* race 0 and some resistant cultivars *Pto/+* have been released. Stockinger and Walling (1994) found the novel genes of resistance *Pto-3* and *Pto-4* that can withstand races 0 and 1. According to Buonaurio *et al.* (1996), pyramiding *Pto, Pto-3*, and *Pto-4,* in one cultivar may provide the optimum solution for this disease control.

Integrated pest management aims to keep the pest population continuously at a low level. Because the probability that new races of the pest will emerge is proportional to the population level of the pest, integrated pest management will reduce the possibility that a new virulent race will develop, and, consequently, the durability of race-specific resistance may increase.

9.8. Advantages and Disadvantages of Host-Plant Resistance

Some of the many advantages of pest control by resistant cultivars over control by pesticides are: (i) the technique is easy to apply because the grower only has to buy resistant cultivars; (ii) it is relatively inexpensive, seed of resistant cultivars is no more expensive than seed of non-resistant cultivars; (iii) completely resistant cultivars need no chemicals for pest control and even partially resistant cultivars need much less to control pests; (iv) resistant cultivars can be incorporated into integrated pest management programmes and when combined with biological control give a cumulative effect; (v) adverse environmental effects are minimal or nil, pesticide pollution is much reduced; and (vi) resistant cultivars, except transgenic cultivars, are acceptable to the public. Some of the disadvantages of resistant cultivars are: (i) it takes a long time to develop a resistant cultivar; (ii) resistant cultivars control only one pest, while pesticides are often effective against several pests; (iii) resistance must be introduced in each new cultivar; and (iv) the pest may adapt to the resistance and this limits the durability of resistant cultivars.

9.9. Present Situation of Host-Plant Resistance in Commercial Cultivars Adapted for Greenhouse Cultivation

Control of pests by resistant cultivars has been a generally successful approach and new resistant cultivars appear regularly on the seed market. Greenhouse crops are particularly suitable candidates for the introduction of resistance because the high income of greenhouse crops permits the cost.

Tomato is the most important vegetable world-wide and is the focus of attention of many seed companies. Commercial tomato cultivars can be crossed with wild species that offer the main source for genes of resistance. Resistance for almost any tomato pest is now known, but only some of them have been introduced into tomato cultivars (Table 9.2). Commercially available cultivars contain multiple resistances to several diseases, but almost all their resistances are monogenic and complete.

TABLE 9.2. Host-plant resistance in commercial greenhouse cultivars and their level of utilization (*low, **medium, ***high)

Tomato		Cucumber	
Fusarium oxysporum	***	*Cladosporium cucumerinum*	***
f. sp. *lycopersici*		*Corynespora cassiicola*	***
Fusarium oxysporum f. sp.	**	*Erysiphe cichoracearum*	*
radicis-lycopersici		*Sphaerotheca fusca*	*
Verticillium albo-atrum and	***	*Pseudoperonospora cubensis*	*
Verticillium dahliae		Cucumber mosaic virus (CMV)	*
Pyrenochaeta lycopersici	*		
Fulvia fulva	***	Melon	
Phytophthora infestans	*	*Fusarium oxysporum* f. sp. *melonis*	***
Stemphylium spp.	***	*Erysiphe cichoracearum*	**
Leveillula taurica	*	*Sphaeroteca fusca*	**
Pseudomonas syringae pv. *tomato*	**	*Pseudoperonospora cubensis*	*
Ralstonia solanacearum	*	Zucchini yellow mosaic virus (ZYMV)	*
Tobacco mosaic virus (TMV)	***	Papaya ring spot virus (PRSV)	*
Tomato yellow leaf curl virus (TYLCV)	*	Melon necrotic spot virus (MNSV)	*
Tomato spotted wilt virus (TSWV)	*	*Aphis gossypii*	**
Meloidogyne spp.	***		
		Bean	
Pepper		*Colletotrichum lindemuthianum*	***
Phytophthora capsici	*	*Pseudomonas syringae* pv. *phaseolicola*	*
Xanthomonas vesicatoria	*	Bean common mosaic virus (BCMV)	***
Tobacco mosaic virus (TMV)	***		
Cucumber mosaic virus (CMV)	*	Lettuce	
Potato virus Y (PVY)	**	*Bremia lactucae*	**
		Lettuce mosaic virus (LMV)	**

In sweet pepper, resistant sources are widely available in wild relatives. Currently, the resistance in cultivars is principally for viruses (Table 9.2). Most insect pests are under good biological control and so breeding for resistance is not pursued.

Cucumber, in contrast with tomato and sweet pepper, has a narrow genetic base. No wild relatives are available to provide genes of resistance. Nevertheless, some important successes have been achieved against cucumber mosaic virus (CMV), *Corynespora cassiicola* (Berk. & M.A. Curtis) C.T. Wei and *Cladosporium cucumerinum* Ellis & Arth. (Table 9.2). Downy-mildew [*Pseudoperonospora cubensis* (Berk. & M.A. Curtis) Rostovzev] is a serious problem in cucumber. Although genes of resistance are present, commercial cultivars only have partial resistance. A combination of partial resistance, biological control and other acceptable control measures of this disease seems to offer the best solution.

Melon has a number of botanical varieties that have provided the resistances introduced in commercial cultivars (Table 9.2). Powdery-mildew is the main fungal disease in greenhouse cultivation and almost completely resistant cultivars for the races 1 and 2 are available in the market. Resistance to papaya ring spot virus (PRSV) has been bred in melons for tropical and subtropical conditions where the virus assumes more importance. Resistance to zucchini yellow mosaic virus (ZYMV) is race specific and not effective against a second pathotype of the virus. Partial resistance to *Aphis gossypii* Glover prevents colony formation and may reduce the incidence of aphid-borne viruses.

Lettuce shows wide genetic variation, and wild species are available to carry out crosses with commercial material. Biological control is more difficult in leaf vegetables than in fruit vegetables because very short cropping cycles and, therefore, resistance breeding is more needed here. Complete monogenic resistance is present against *Bremia lactucae* Regel, based on a gene-for-gene system, but resistance is not durable (Table 9.2).

In floriculture, resistance breeding is a recent development. There are less incentives to breed resistant cultivars due to zero-tolerance, high cosmetic demands, fashion products with a short commercial life-span (a few years), many species and cultivars mostly grown on a small acreage and fewer restrictions on use of pesticides in floriculture than for food crops. In chrysanthemum, complete monogenic resistance against *Puccinia horiana* Henn. is known and commercially exploited; in addition, partial resistances against leafminers and thrips have been found. *Fusarium oxysporum* Schlechtend.:Fr. f. sp. *dianthi* (Prill. & Delacr.) W.C. Snyder & H.N. Hans. severely affects carnations mainly during the hot season and two races are known. Host-plant resistance to race 1 due to a single gene is now introduced into most commercial cultivars. Host-plant resistance to race 2 is polygenic and it is expressed when all the resistance loci are heterozygous or homozygous for the dominant alleles that confer the resistance; susceptibility would occur when there are one or more homozygous recessive alleles (Arús *et al.*, 1992). However, in spite of the complexity of the genetic basis of this resistance, resistant cultivars with good field resistance have been released.

9.10. Perspectives

The durability of a resistance increases when as many as possible genes of resistance are introduced into a cultivar. However, most of the resistances introduced in commercial cultivars to date are only monogenic, mainly because to pyramid several resistance genes for one pest in the same cultivar is difficult and costly. Appropriate molecular markers would make this task easier. Future improvement of screening techniques and indirect selection will make it easier to breed host plants with polygenic resistances.

Partial resistance is controlled by many genes with small individual effects and, although it is potentially more durable than monogenic complete resistance, it is rarely used because it is difficult: (i) to distinguish and to select the individual effect of each gene in segregating generations; (ii) to evaluate commercially the advantage of the partial resistance; (iii) to convince the growers about the benefits of resistant cultivars that show some disease symptoms. Partial resistance, in combination with biological control, can lead to sufficient control.

Public concern about the effects of pesticides have resulted in governments to make laws to reduce the use of pesticides. The best way to avoid or reduce the use of pesticides in greenhouse crops is to introduce integrated pest management techniques that include the use of resistant cultivars. The disadvantages of resistant cultivars are much less than their advantages (as explained in Section 9.8), consequently the prospects for the future development of many more resistant cultivars appear excellent.

References

Arús, P., Llauradó, M. and Pera, J. (1992) Progeny analysis of crosses between genotypes resistant and susceptible to *Fusarium oxysporum* f. sp. *dianthi* race 2, *Acta Horticulturae* **307**, 57–64.

Buonaurio, R., Stravato, V.M. and Cappelli, C. (1996) Occurrence of *Pseudomonas syringae* p.v. *tomato* race 1 in Italy on *Pto* gene-bearing tomato plants, *J. of Phytopathology* **144**, 437–440.

Crute, I.R. (1994) Gene-for-gene recognition in plant-pathogen interactions, *Philosophical Transactions of the Royal Society of London. Series B* **346**, 345–349.

Danesh, D., Aarons, S., McGill, G.E. and Young N.D. (1994) Genetic dissection of oligogenic resistance to bacterial wilt in tomato, *Molecular Plant-Microbe Interactions* **7**, 464–471.

De Wit, P.J.G.M. (1992) Molecular characterization of gene-for-gene systems in plant-fungus interactions and the application of avirulence genes in control of plant pathogens, *Annual Review of Phytopatholgy* **30**, 391–418.

Dicke, M. (1994) Local and systemic production of volatile herbivore-induced terpenoids: Their role in plant-carnivore mutualism, *J. of Plant Physiology* **143**, 465–472.

Fargette, D., Leslie, M. and Harrison, B.D. (1996) Serological studies on the accumulation and localization of three tomato leaf curl geminiviruses in resistant and susceptible *Lycopersicon* species and tomato cultivars, *Annals of Applied Biology* **128**, 317–328.

Flor, H.H. (1956) The complementary genic systems in flax and rust, *Advances in Genetics* **8**, 29–54.

Gil Ortega, R., Palazon Español, C. and Cuartero Zueco, J. (1995) Interactions in the pepper – *Phytophthora capsici* system, *Plant Breeding* **114**, 74–77.

Gómez-Guillamón, M.L. and Torés, J.A. (1992) Influence of temperature on the resistance to *Sphaerotheca fuliginea* in Spanish melon cultivars, in R.W. Doruchowski, E. Kozik and K. Niemirowicz-Szczytt (eds.), *Proceedings of the Fifth Eucarpia Cucurbitacea Symposium*, Warsaw, Poland, 27–31 July 1992, Research Institute of Vegetable Crops, Skiernewice, pp. 155–159.

Isaac, S. (1992) *Fungal-Plant Interactions*, Chapman & Hall, London.

Johnson, R. and Law, C.N. (1975) Genetic control of durable resistance to yellow rust (*Puccinia striiformis*) in the wheat cultivar Hybride de Bersée, *Annals of Applied Biology* **81**, 385–392.

Karban, R. and Myers, J.H. (1989) Induced plant responses to herbivory, *Annual Review of Ecology and Systematics* **20**, 331–348.

Kelly, J.D. (1995) Use of random amplified polymorfic DNA markers in breeding for major gene resistance to plant pathogens, *HortScience* **30**, 461–465.

Kennedy, G.G., Yamamoto, R.T., Dimock, M.B., Williams, W.G. and Bordner, J. (1981) Effect of day length and light intensity on 2-tridecanone levels and resistance in *Lycopersicon hirsutum* f. *glabratum* to *Manduca sexta*, *J. of Chemical Ecology* **7**, 707–716.

Laterrot, H. (1985) Susceptibility of the Pto plants to Lebaycid insecticide: A tool for breeders? *Tomato Genetics Cooperative Report* **35**, 6.

McGaughey, W.H. (1988) Resistance of storage pests to *Bacillus thuringiensis*, in *Proceedings of the XVIII International Congress of Entomology*, Vancouver, Canada, 3–9 July, 1988, University of British Columbia, Vancouver, p. 449.

Milne, R.G. (1966) Electron microscopy of tobacco mosaic virus in leaves of *Nicotiana glutinosa, Virology* **28**, 527–532.

Neinhuis, J., Helentjaris, T., Slocum, M., Ruggero, B. and Schaeffer, A. (1987) Restriction fragment length polymorphism analysis of loci associated with insect resistance in tomato, *Crop Science* **27**, 797–803.

Niks, R.E., Ellis, P.R. and Parlevliet, J.E. (1993) Resistance to parasites, in M.D. Hayward, N.O. Bosemark and I. Romagosa (eds.), *Plant Breeding: Principles and Prospects*, Chapman & Hall, London, pp. 422–447.

Painter, R.H. (1951) *Insect Resistance in Crop Plants*, Macmillan, New York.

Parlevliet, J.E. and van Ommeren, A. (1984) Interplot interference and the assessment of barley cultivars for partial resistance to leaf rust, *Puccinia hordei, Euphytica* **33**, 685–697.

Ponz, F. and Bruening, G. (1986) Mechanisms of resistance to plant viruses, *Annual Review of Phytopathology* **24**, 355–381.

Roggero, P. and Pennazio, S. (1988) Effects of salicylate on systemic invasion of tobacco plants by various viruses, *J. of Phytopathology* **123**, 207–216.

Ross, A.F. (1961a) Localized acquired resistance to plant virus infection in hypersensitive hosts, *Virology* **14**, 329–339.

Ross, A.F. (1961b) Systemic acquired resistance induced by localized virus infections in plants, *Virology* **14**, 340–358.

Russell, G.E. (1978) *Plant Breeding for Pest and Disease Resistance*, Butterworths, London-Boston.

Smith, C.M. (1989) *Plant Resistance to Insects: A Fundamental Approach*, John Wiley and Sons, New York.

Stockinger, E.J. and Walling, L.L. (1994) *Pto-3* and *Pto-4*: Novel genes from *Lycopersicon hirsutum* var. *glabratum* that confer resistance to *Pseudomonas syringae* p.v. *tomato, Theoretical and Applied Genetics* **89**, 879–884.

Tang, X. Frederick, R.D., Zhou, J., Halterman, D.A., Jia, Y. and Martin, G.B. (1996) Initiation of plant disease resistance by physical interaction of AvrPto and Pto kinase, *Science* **274**, 2060–2063.

van Etten, H.D., Mansfield, J.V., Bailey, J.A. and Farmer, E.E. (1994) Two classes of plant antibiotics: Phytoalexins versus "phytoanticipins", *Plant Cell* **6**, 1191–1192.

Williamson, V.M., Ho, J.Y., Wu, F.F., Miller, N. and Kaloshian, Y. (1994) A PCR-based marker tightly linked to the nematode resistance gene, *Mi*, in tomato, *Theoretical and Applied Genetics* **87**, 757–763.

CHAPTER 10

DISINFESTATION OF SOIL AND GROWTH MEDIA
Elefterios C. Tjamos, Avi Grinstein and Abraham Gamliel

10.1. Introduction

Soilborne plant pathogens constitute a major problem of plant protection in greenhouses. This is basically due to the pathogens' ability to survive for several years in the soil (or in used container media) as dormant resting structures (sclerotia or microsclerotia, chlamydospores and resting mycelia) until a susceptible crop is introduced again into the same plot. These structures are able to withstand adverse environmental conditions and chemical applications, thus creating major control problems in the world agriculture. The same holds true for other soilborne pests such as arthropods, nematodes, parasitic plants and weeds, although different mechanisms of persistence are involved. To date, fumigation (or steaming) is the most effective approach to control soilborne pests. Soil solarization (SSOL), applied to soil or growth media alone, or in combination with reduced doses of soil fumigants or other amendments, can also control most soilborne plant pathogens effectively.

This chapter reviews the management of soilborne pathogens in glass or plastic greenhouses through a wide range of chemical and physical treatments as well as SSOL, taking into consideration the forthcoming ban (scheduled now to 2005 for most of the world) on the use of methyl bromide (MBr), and the current lack of alternatives for some of its current uses. Specific chemicals such as herbicides and other pesticides are beyond the scope of this review, although some combinations of those chemicals [e.g. Ethyl dipropil thiolcarbamate (EPTC)] with SSOL have been found to be highly effective.

10.2. Steaming

Steaming, aerated steam (Dawson and Johnson, 1965), overheated and hot water treatments are used in greenhouses, especially when container (growth) media are used. Steam has been applied for soil disinfestation for almost a century. Plant pathogens (as well as other pests) are eliminated by steaming due to heating to lethal levels or to physical damages incurred to their resting structures, even in cases of heavy soil contamination. Moreover, steaming usually shows a growth stimulation effect on the following crop.

The "classic" steaming by Hoddesdon pipes, dug into the soil, is no longer used. This holds true also for heating the soil to 80–100°C. As this treatment results, in many cases, in a biological vacuum in the treated soil, heating the soil or growth substrate to 70°C – mainly by aerated steam – is now favoured; this treatment leaves part of the saprophytic population uncontrolled (Bollen, 1985).

R. Albajes et al. (eds.), Integrated Pest and Disease Management in Greenhouse Crops, 139-149.
© 1999 *Kluwer Academic Publishers. Printed in the Netherlands.*

Careful soil preparation is essential for good steam penetration. The soil should be tilled as deep as possible, preferably by a shovel-plough, and then left for complete drying before steaming. It is important to reduce amount of plant debris, especially when steaming growth medium. Good preparation permits good steam penetration and enables pest control in heavy soils, but might still result in only partial control in very light sandy soils. Steaming of aerated growth substrates, such as tuff stones, vermiculite, etc., is usually good, but peat soils pose difficulties due to their high water content.

Soil steaming is done either by "passive" or "active" techniques. In passive steaming, steam is blown to the surface, under a covering sheet, and left to heat the upper layer. Lower layers are then heated by heat transmission. This process continues until 100°C is reached at a depth of 10 cm (Runia, 1983). Disinfestation of deep layers, especially in sandy soil, might be only partial.

Active steaming can be done by either positive or negative pressure. Both techniques employ drainage systems, based on pipes laid at a 50–70 cm depth, and approximately 80 cm apart. With the "positive pressure" technique the steam is blown through holes located along the pipes. The "negative pressure" involves an improved technique, utilizing the advantages of the two above-mentioned application methods. The steam is released over the treated area under plastic sheeting, as for passive steaming, assuring rapid and even distribution throughout the plot surface, followed by active suction to the deeper layers of the soil, achieved by negative pressure applied through the drainage system. This technique, widely used in The Netherlands, is much cheaper than the two others, due to energy saving caused by the faster heat transfer (Runia, 1983). Despite this, steaming treatments are expensive, and are feasible mainly in places where there are heating systems (used mainly for heating the greenhouse during the cold season) or if applied by contractors (Anonymous, 1994). Steaming, however, can be useful and economical for disinfestation of shallow layers of growth media placed on tables, as is usually done in nurseries.

10.3. Soil Fumigation

Soil fumigation is done by applying toxic pesticides to the soil by various means, and these fumigants move down and across the soil profile and reach the target organisms directly, or by a very efficient secondary distribution due to their relatively high vapour pressure. MBr is by far the most effective fumigant (Klein, 1996). However, current concerns regarding the possible role of MBr in ozone depletion and its forthcoming phase out have triggered research efforts to develop optional methods for soil disinfestation. Other soil fumigants used for greenhouses include methyl isothiocyanate (MIT), CS_2-releasing compounds, formaldehyde, dichloropropene, etc. (Anonymous, 1994; Ristaino and Thomas, 1997).

10.3.1. FUMIGATION WITH MBr

MBr is the most powerful soil fumigant with a very broad spectrum of activity. Many soilborne fungi (e.g. *Rhizoctonia* spp., *Pythium* spp., *Phytophthora* spp., *Sclerotinia sclerotiorum* (Lib.) de Bary, *Sclerotinia minor* Jagger, *Sclerotium rolfsii* Sacc., *Verticillium* spp. and many *Fusarium* spp.) are sensitive to MBr. In contrast, some soilborne bacteria, such as *Clavibacter michiganensis* (Smith) Davis *et al.* ssp. *michiganensis* (Smith) Davis *et al.* [= *Corynebacterium michiganensis* (Smith) Jensen ssp. *michiganensis* (Smith) Jensen], are not satisfactorily controlled at regular (commercial) rates of application (Antoniou *et al.*, 1995a). The effectiveness of MBr fumigation also depends on proper soil preparation, irrigation reaching approximately 60% of "field capacity" and a tight covering of the fumigated soil with plastic (mostly polyethylene) sheeting. MBr is applied to the soil at a rate of 50 to 110 g/m^2, either by injection as a cold liquid just before covering, or by distribution as a cold or hot gas under the mulch released through a manifold of perforated pipes or from 0.3-l disposable containers which are opened under the mulch. The duration of the application depends on soil temperature (1–2 days at 15°C, 3 days at 10–15°C at the 0–20 cm-deep soil layer, but more than 4 days at 8–10°C at the same depth) (Klein, 1996). Possible problems due to the toxicological hazards of MBr are related mainly to the health danger for applicators and to the increase in inorganic bromine residues in edible plant products. MBr was found in a few cases in water near greenhouses in The Netherlands, where PVC water pipes were improperly placed only 10 cm deep in the ground.

    In 1992, MBr was listed by the Montreal Protocol as an ozone depleting material, and a procedure for banning its use was initiated (Gamliel *et al.*, 1997b). According to this decision, MBr will not be available in developed countries after 2005, and its consumption will be gradually reduced during the period remaining until the ban goes into effect (Anonymous, 1997).

    There are some MBr uses without any known substitute yet (Anonymous, 1994). Continuous efforts are now underway, to reduce MBr dosages and minimize its emission and negative side-effects on the environment. Most solutions are based on using improved, virtually impermeable mulching films. Common low- and high-density polyethylene films are poor barriers, and allow the escape of MBr at very high rates, especially where the film temperature is higher than 40°C (as is the case in most greenhouses when the film is exposed to solar irradiation). The permeability of MBr through impermeable film (normally co-extruded with a barrier layer protected by polyethylene coating from both sides), is only 0.001–0.0001 $g/m^2/h$, depending on the barrier formula, compared with emission of 5 $g/m^2/h$ for regular low density polyethylene. Control of a pest is a factor of pesticide concentration (C) and exposure time (T). Thus, extending MBr retention in soil under impermeable films for a longer period allows the use of reduced MBr dosages with the same CT values, without reducing control efficacy. Fungal pathogens such as *Fusarium oxysporum* Schlechtend.:Fr. f. sp. *dianthi* (Prill. & Delacr.) W.C. Snyder & H.N. Hans., *Fusarium oxysporum* Schlechtend.:Fr. f. sp. *radicis-lycopersici* W.R. Jarvis & Shoemaker, *Fusarium oxysporum* Schlechtend.:Fr. f. sp. *cucumerinum* J.H. Owen, etc., were controlled by reduced dosage of MBr at 25–50% of the recommended dose under impermeable films (Antoniou *et al.*, 1997; Gamliel *et al.*, 1997b,c). Further reduction is possible by deeper burying of the film edges into the soil and by continuous mulching, or by combination with SSOL (Grinstein *et al.*, 1995; Antoniou *et al.*, 1996; Gamliel *et al.*, 1997b).

10.3.2. FUMIGANTS WITH MIT

Dazomet (3,5, dimethyl-tetrahydro-1,3,5,(2H) thiodiazino-thione)
Dazomet is a product formulated either as a powder (85% a.i.) or as granules (98% a.i.). The chemical is gradually hydrolyzed to at least four subproducts, MIT being the main one. Dazomet is effective against *Verticillium dahliae* Kleb., *Verticilium albo-atrum* Reinke & Berthier, *Rhizoctonia solani* Kühn, *S. sclerotiorum*, *Phytophthora* spp. and *Pythium* spp. at a rate of 400–600 kg a.i./ha. The fumigant can be used for the control of several diseases in seed beds, greenhouses, or in field grown vegetables, cotton, tobacco and ornamentals. It is applied to the soil by spreading or irrigating followed by mechanical mixing (such as rotovator cultivation or shovel plough) into the soil. The chemical, which is not applicable at temperatures lower than 8°C, is also partially effective against insects, various nematodes and weed seeds. One of the disadvantages of dazomet is the long period (three weeks) needed after application of the chemical before planting or sowing is permissible (Anonymous, 1994; Middleton and Lawrence, 1995).

Metham sodium (sodium methyldithiocarbamate) (MES)
MES is effective against several soilborne pathogens in both in covered and open outdoor cultivation. In water solutions MES rapidly changes to methyl isothiocyanate (MIT). The broad spectrum of controlled pathogens includes Pythiaceous fungi, races of *Fusarium oxysporum* Schlechtend.:Fr., *S. sclerotiorum*, *S. rolfsii*, *V. dahliae* and species of *Phialophora, Phoma, Botrytis*, etc. Since resting structures are present mainly in the upper 40 cm of the soil profile, and since MES is 100% water soluble, it is most effective when applied via the sprinkler irrigation system. The chemical is used at various doses according to the target pathogen and/or the soil type to be disinfected. The recommended dosages for sandy, heavy, and very heavy soils are 490–650, 800 and 1000 l/ha, respectively. Soil temperature is also a critical factor in the effective application of the chemical: fluctuating between 10 and 30°C at a soil depth of 10 cm is best. Use of MES for chemigation is an effective procedure against soilborne pathogens. However, side effects may arise under certain conditions, such as when the irrigation water is contaminated with urban sewage. *Fusarium oxysporum* Schlechtend.:Fr. f. sp. *cepae* (H.N. Hans.) W.C. Snyder & H.N. Hans. on onion has been controlled by MES application, but fumigation resulted in the eradication of endomycorrhizal fungi, reduced onion growth and increased the population of another bacterial pathogen of onions, *Pseudomonas gladioli* Severini pv. *allicola* Young *et al.*, which replaced Fusaria and caused very serious damage (Kritzman and Ben-Yephet, 1990).

10.3.3. SOIL FUMIGATION AND PROBLEMS OF ALTERNATIVES TO MBr

Fumigants other than MBr, having a much narrower range, are registered and used in various cropping systems. These include nematicides (dichloropropene), fungicides (SO_2 releasing pesticides) and other. These are used on relatively small scale and will not be dealt in this paragraph (Anonymous, 1994). It is clear that with the currently available fumigants, there is no satisfactory replacement to MBr. The use of other fumigants involves identification of the casual agent, and in many cases the use of a mixture of two or more chemicals, to control a wider range of disease agents, pest and weeds in the treated plot (Anonymous, 1994). Di-Trapex (methyl isothiocyanate 20 + dichloropropane-dichloropropene 80), may serve as an example to this tendency, as this pesticide was formulated to control both pest controlled by MES and the root-rot nematode. Furthermore, data regarding residual effect of the above mentioned fumigants before planting is needed while their environmental impact is not yet fully clear.

10.4. Soil solarization (SSOL)

SSOL represents one of the very few cases where a new non-chemical control procedure has been adopted by greenhouse growers in several parts of the world, within a relatively short period of time (Katan *et al.*, 1976, 1987). SSOL is based on trapping solar irradiation by tightly covering the wet soil, usually with transparent polyethylene or other plastic sheets (Grinstein and Hetzroni, 1991). This results in a significant elevation (10–15°C above normal, depending on the soil depth) of soil temperatures up to the point where most pathogens are vulnerable to heat when applied for 4–6 weeks and controlled either directly by the heat, or by chemical and biological processes generated in the heated soil (DeVay and Katan, 1991).

10.4.1. EFFECT OF SSOL ON FUNGAL DISEASES

Ecological observations and quantitative measurements carried out after the application of the technique have differentiated the pathogens into two main categories. It should be pointed out that a pathogen could be effectively controlled by solarization in one region but less effectively in another depending on environmental, and cultural parameters. A partial list of soilborne pathogens and pests which are controlled by solarization as reported for greenhouses and open fields is listed in Table 10.1. It is important to mention that application of SSOL in a close greenhouse, or by employing two layers mulch further improves its effects (Kodama and Fukui, 1982; Garibaldi and Tamietti, 1984; Garibaldi and Gullino, 1991).

10.4.2. BACTERIAL DISEASES CONTROLLED BY SSOL

Relatively, only few reports about SSOL and bacterial diseases were published. Application of SSOL (1–2 months soil mulching with transparent polyethylene films) in tomato plastic houses drastically reduced symptoms caused by *C. michiganensis* ssp. *michiganensis* (Antoniou *et al.*, 1995b) while MBr (68 g/m^2) was ineffective in controlling the disease. Populations of Gram-positive bacteria were reduced by 64–99%

by SSOL (Stapleton and Garza-Lopez, 1988). Bacterial populations of cultures of *C. michiganensis* ssp. *michiganensis* infiltrated into tomato stem segments were embedded at various soil depths prior to the application of SSOL. A sharp decrease or elimination of the pathogen in solarized compared to MBr-treated plots was observed. *Streptomyces* spp., causing deep pitted scab of potatoes and pod-wart disease of peanut, were successfully controlled (Grinstein *et al.*, 1995). Negative effects, due to control of beneficial Rhizobia were also reported (Abdel-Rahim, 1987).

TABLE 10.1. Fungal soilborne pathogens controlled by SSOL (a partial list)

Pest	Crop	Selected references
Fusarium spp.	Basil, cucumber, melon, strawberry, tomato, watermelon	Kodama *et al.*, 1980; Kodama and Fukui, 1982; Martyn and Hartz, 1986; Tjamos and Makrynakis, 1990; Grinstein and Ausher, 1991; Oliveira, 1992; Antoniou *et al.*, 1997; Garibaldi *et al.*, 1997
Phoma lycopersici (= *Diplodina lycopersici*)	Tomato	Cartia, 1989
Phytophthora spp.	Avocado, tomato	Cartia, 1989; Grinstein and Ausher, 1991
Pyrenochaeta spp.	Onion, tomato	Malathrakis *et al.*, 1983; Garibaldi and Tamietti, 1984; Tjamos, 1984; Cartia, 1989; Grinstein and Ausher, 1991
Pythium spp.	Various	Hilderbrand, 1985; Stapleton and Garza-Lopez, 1988
Rhizoctonia solani	Various	Tamietti and Garibaldi, 1989; Grinstein and Ausher, 1991
Sclerotinia spp.	Lettuce	Porter and Merriman, 1985; Materrazzi *et al.*, 1987; Vannacci *et al.*, 1988; Phillips, 1990
Sclerotium cepivorum	Onion	Abdel-Rahim *et al.*, 1983, Porter and Merriman, 1985
Verticillium spp.	Eggplant, tomato, strawberry	Katan *et al.*, 1976; Cartia, 1989; Tjamos *et al.*, 1989; Grinstein and Ausher, 1991

10.4.3 PARTIAL CONTROL OF FUNGAL DISEASES BY SSOL

The heat tolerant *Monosporascus* sp. and *Macrophomina phaseolina* (Tassi) Goidanich, root-knot nematode *Meloidogyne* spp. and some weeds, e.g. *Cyperus rotundus* L. and the annual weed *Melilotus sulcatus* Desf. are only partially controlled by SSOL. *Fusarium oxysporum* f. sp. *dianthi* is also considered as one of the wilt pathogens not easily controlled by SSOL (Rubin and Benjamin, 1983; Gamliel and Stapleton, 1997).

10.4.4. BIOLOGICAL CONTROL ASPECTS OF SSOL

Disturbances in the biological equilibrium of the soil microflora, following soil fumigation or steaming, are known to be drastic and undesirable. Application of SSOL, however, favours the survival and increase of several heat-tolerant micro-organisms able to act as antagonists against soilborne pathogens, such as *Talaromyces flavus* (Klöcker) A.C. Stolk & R.A. Samson, *Aspergillus terreus* Thom in Thom & Church, fluorecset pseudomonades and others (Greenberger *et al.*, 1987; Tjamos and Paplomatas, 1987; Tjamos *et al.*, 1991). Solarization favours establishment of added antagonists such as *Trichoderma* spp. and *A. terreus*, saprophytic Fusaria and other (Martyn and Hartz, 1986; Triolo *et al.*, 1988).

The survival of thermophilic genera of *Bacillus, Actinomyces*, as well as the build-up of fluorescent pseudomonads and other populations of rhizosphere bacteria were reported (Stapleton and DeVay, 1982, 1984; Kaewruang *et al.*, 1989; Gamliel and Katan, 1991; Antoniou *et al.*, 1995a). The effect of SSOL can be improved also by combination with no-pesticide organic amendments incorporated into the soil before mulching. This can be related both to the release of toxic materials by combination of heating and biological activity, and to positive changes in soil microflora. Gamliel and Stapleton (1997) reported control of root rot nematodes by mixing chicken manure or dry cabbage leaves in the plot before mulching (see also Chapter 23).

10.5. Combining Disinfestation Methods

One of the major limitations of SSOL is its climate dependence. Another problem diverts from the need to keep the treated area for 35–60 days without any crop. Partial control of some pests, as well as reduced efficacy in marginal seasons limit solarization use in many places. These constrains can be reduced, or solved, by combining solarization with other control measures at reduced dosages. The control efficacy may be increased due to additive effect. More likely it is due to synergistic effect caused by the hotter environment which increases vapour pressure and chemical activity of the added pesticide. Another reason for the improved activity of the pesticide is the weakening of the resting structure by the heat (Freeman and Katan, 1988).

Reduced doses of MBr, impermeable plastics and solarization were applied against a variety of diseases, e.g. *F. oxysporum* f. sp. *cucumerinum* of cucumbers, *C. michiganensis* ssp. *michiganensis* of tomatoes (Antoniou *et al.*, 1996, 1997), the melon sudden wilt (Gamliel *et al.*, 1997b), Verticillium of potatoes (Grinstein *et al.*, 1979), deep pitted scab of potatoes, Fusarium crown rot in tomatoes, soil sickness of *Gypsophila* sp. Reduced rates of MBr (34 g/m^2) combined with simultaneous solarization effectively controlled corky root rot disease of tomatoes (Tjamos, 1984) and Verticillium wilt of globe artichoke (Tjamos and Paplomatas, 1987).

Reduced doses of chemicals are recommended as an alternative approach to the acute toxicity of full fumigation. However, their effectiveness is dependent on combinations with other pesticides or with non-chemical procedures. Sublethal fumigation is considered here in combination with SSOL (Gamliel *et al.*, 1997b). Combining sublethal fumigation with solarization could be focused on the following: (i) MBr fumigation followed immediately by solarization; (ii) simultaneous application of solarization with reduced doses of various fumigants; and (iii) solarization followed by fumigant for pathogens that are heat tolerant.

Recent studies show that the control efficacy of reduced dose of MBr combined with solarization was highly increased when applied after a short heating period, 2–3 days after the mulching (Gamliel *et al.*, 1997a). Application of MBr after the termination of the SSOL, however, can control some of the beneficial micro-organism populations which remain in the solarized plot, and has to be considered carefully.

Current reports mainly referring to field crops with applicability to covered crops deal with combinations of chemicals with SSOL. They include MES for the control of *V. dahliae* and *Fusarium oxysporum* Schlechtend.:Fr. f. sp. *vasinfectum* (Atk.) W.C. Snyder & H.N. Hans. (Ben-Yephet., 1988), dazomet either alone or in combination with SSOL to control *Phoma terrestris* E.M. Hans. on onions (Porter and Merriman, 1985), and MBr and SSOL for the control of *Pyrenochaeta lycopersici* R. Scheneider & Gerlach on tomatoes (Tjamos, 1984). Reduced doses of MES (12.5 or 25 ml/m^2) applied singly or in combination with SSOL have destroyed propagules of *V. dahliae* and *F. oxysporum* f. sp. *vasinfectum* in a naturally infested cotton field (Ben-Yephet, 1988). The combination also reduced the time needed to kill sclerotia of *V. dahliae* by one week (Ben-Yephet, 1988). Dazomet (750 kg/ha) either alone or in combination with solarization has reduced disease incidence and severity of pink root rot (caused by *P. terrestris*) and of white rot (caused by *Sclerotium cepivorum* Berk.) of onions and increased yield by at least 100% (Abdel-Rahim *et al.*, 1983). Reduced rates of MBr (34 g/m^2) combined with simultaneous solarization effectively controlled corky root rot disease of tomatoes (Tjamos, 1984) and Verticillium wilt of globe artichokes (Tjamos and Paplomatas, 1987).

Synergism in reducing disease incidence can be observed between fumigants and fungal antagonists of soilborne pathogens. Solarization in combination with *Gliocladium virens* J.H. Miller, J.E. Giddens & A.A. Foster proved to be a potential control strategy against southern blight of tomatoes (Ristaino *et al.*, 1991).

10.6. Prospects and Difficulties of Soil Disinfection

Soil fumigation with chemicals may have negative effects on the environment, could be extremely dangerous to humans, and may leave toxic residues in plant products. Thus, innovative approaches are desperately needed by the farmers and are under great demand by the consumers. Research towards exploiting SSOL by combining reduced doses of allowed fumigants, or various antagonists, could be one of the most promising approaches. This could also result in reducing duration of solarization thus making the method more acceptable by the farmers. Furthermore, sublethal fumigation in combination with solarization could solve many problems, since the combination is suitable for areas marginal for the application of solarization, and is able to reduce the duration of solarization to one half. SSOL in combination with biocontrol agents could exploit the weakening effect imposed by solar heating and could prolong its effectiveness.

References

Abdel-Rahim, M.F., Satour, M.M., Mickail, K.Y., El-Eraki, S.A., Grinstein, A., Chen, Y. and Katan, J. (1987) Effectiveness of soil solarization in furrow-irrigated Egyptian soils, *Plant Disease* 72, 143–146.

Anonymous (1994) *Montreal Protocol on Substances which Deplete the Ozone Layer,* Report of the Methyl Bromide Technical Options Committee, Nairobi.

Anonymous (1997) *Ninth Meeting of the Parties to the Montreal Protocol on Substances That Deplete the Ozone Layer,* UN Environmental program, UNEP/OzL. Pro 9/12.

Antoniou, P.P., Tjamos, E.C., Andreou, M.T. and Panagopoulos, C.G. (1995a) Effectiveness, modes of action and commercial application of soil solarization for control of *Clavibacter michiganensis* ssp. *michiganensis* of tomatoes, *Acta Horticulturae* 382, 119–124.

Antoniou, P.P., Tjamos, E.C. and Panagopoulos, C.G. (1995b) Use of soil solarization for controlling bacterial canker of tomato in plastic houses in Greece, *Plant Pathology* 44, 438–447.

Antoniou, P.P., Tjamos, E.C. and Panagopoulos, C.G. (1996) Sensitivity of propagules of *Fusarium oxysporum* f.sp. *cucumerinum, Verticillium dahliae, Clavibacter michiganensis* ssp. *michiganensis* to reduced doses methyl bromide fumigation in combination with impermeable plastics, *Phytopathologia Mediterranea* (abs.).

Antoniou, P.P., Tjamos, E.C. and Panagopoulos, C.G. (1997) Reduced doses of methyl bromide, impermeable plastics and solarization against *Fusarium oxysporum* f. sp. *cucumerinum* of cucumbers and *Clavibacter michiganensis* ssp. *michiganensis* of tomatoes, *Proceedings of the 10th Congress of the Mediterranean Phytopathological Union,* Montpellier, France, 1–5 June 1997, pp. 653–655.

Ben-Yephet, Y. (1988) Control of sclerotia and apothecia of *Sclerotinia sclerotiorum* by metham-sodium, methyl bromide and soil solarization, *Crop Protection* 7, 25–27.

Bollen, G.J. (1985) Lethal temperatures of soil fungi, in C.A. Parker, A.D. Rovira, K.J. Moore and P.T.W. Wong (eds.), *Ecology and Management of Soilborne Plant Pathogens,* APS Press, St Paul, Fla., pp. 191–193.

Cartia, G. (1989) La solarizzazione del terreno: Esperienze maturate in Sicilia. [Soil solarization: Experiments in Sicily], *Informatore Fitopatologico* 39, 49–52.

Dawson, J.R. and Johnson, R.A.H. (1965) Influence of steam air-mixtures, when used for heating soil, on biological and chemical properties that effect seedling growth, *Annals of Applied Biology* 56, 243–247.

DeVay, J.E. and Katan, J. (1991) Mechanisms of pathogen control in solarized soils, in J. Katan and J.E. DeVay (eds.), *Soil Solarization,* CRC Publications, Boca Raton, Fla., pp. 87–102.

Freeman, S. and Katan, J. (1988) Weakening effect on propagules of Fusarium by sublethal heating, *Phytopathology* 78, 1656–1661.

Gamliel, A., Grinstein, A., Eshel, D., Di Primo, P. and Katan, J. (1997a) Combining solarization and fumigants at reduced dosages for effective control of soilborne pathogens: Controlled environment study I, *Phytoparasitica* 25, 252–253.

Gamliel, A., Grinstein, A. and Katan, J. (1997b) Improved technologies to reduce emissions of methyl bromide from soil fumigation, in A. Grinstein, K.R.S. Ascher, G. Mathews, J. Katan, and A. Gamliel (eds.), *Improved Application Technology for Reduction of Pesticide Dosage and Environmental Pollution, Phytoparasitica* 25S, 21–30

Gamliel, A., Grinstein, A., Peretz, Y., Klein, L., Nachmias, A., Tzror, L. and Katan, J. (1997c) Reduced dosage of methyl bromide for controlling Verticillium wilt of potato in experimental and commercial plots, *Plant Disease* 81, 469–474

Gamliel, A. and Katan, J. (1991) Involvement of fluorescent Pseudomonas and other micro-organisms in increased growth response of plants in solarized soils, *Phytopathology* 81, 494–502.

Gamliel, A. and Stapleton, J.J. (1997) Improved soil disinfestation by biotoxic volatile compounds generated from solarized, organic amended soil, in A. Grinstein, K.R.S. Ascher, G. Mathews, J. Katan and A. Gamliel (eds.), *Improved Application Technology for Reduction of Pesticide Dosage and Environmental Pollution, Phytoparasitica* 25S, 31–38.

Garibaldi, A. and Gullino, M.L. (1991) Soil solarization in southern European countries, with emphasis on soilborne disease control of protected crops, in J. Katan and J.E. DeVay (eds.), *Soil Solarization,* CRC Publications, Boca Raton, Fla., pp. 228–235.

Garibaldi, A., Gullino, M.L. and Minuto, G. (1997) Diseases of basil and their control, *Plant Disease* **81**, 124–131.

Garibaldi, A. and Tamietti, G. (1984) Attempts to use soil solarization in closed glass houses in northern Italy for controlling corky root of tomato, *Acta Horticulturae* **152**, 237–243.

Greenberger, A., Yogev, A. and Katan, J. (1987) Induced suppressiveness in solarized soils, *Phytopathology* **77**, 1663–1667.

Grinstein, A. and Ausher, R. (1991) The utilization of soil solarization in Israel, in J. Katan and J.E. DeVay (eds.), *Soil Solarization*, CRC Publications, Boca Raton, Fla., pp. 193–204.

Grinstein, A. and Hetzroni, A. (1991) The technology of soil solarization, in J. Katan and J.E. DeVay (eds.) *Soil Solarization*, CRC Publications, Boca Raton, Fla., pp. 159–170.

Grinstein, A., Katan, J., Abdul-Razik, A., Zeidan, O. and Elad, Y. (1979) Control of Sclerotium rolfsii and weeds in peanuts by solar heating of the soil, *Plant Disease Reporter* **63**, 991–994.

Grinstein, A., Kritzman, G., Hetzroni, A., Gamliel, A., Mor, M. and Katan, J. (1995) The border effect of soil solarization, *Crop Protection* **14**, 315–320.

Hilderbrand, D.M. (1985) Soil solar heating for control of damping-off fungi and weeds at the Colorado State Forest Service Nursery, *Tree Planters' Notes* **36**, 28–34.

Kaewruang, W., Sivasithamparam, K. and Hardy, G.E. (1989) Use of soil solarization to control root rots in gerberas (*Gerbera jamesonii*), *Biology and Fertility of Soils* **8**, 38–47.

Katan, J., Greenberger, A., Alon, H. and Grinstein, A. (1976) Solar heating by polyethylene mulching for control of diseases caused by soilborne pathogens, *Phytopathology* **66**, 683–688

Katan, J., Grinstein, A., Greenberger, A., Yarden, O. and DeVay, J.E. (1987) First decade (1976–1986) of soil solarization (solar heating) – A chronological bibliography, *Phytoparasitica* **15**, 229–255.

Klein, L. (1996) Methyl bromide as a soil fumigant, in C.H. Bell, N. Price and B. Chakrabarti (eds.), *The Methyl Bromide Issue*, John Wiley and Sons Ltd, Chichester, pp. 191–235.

Kodama, T. and Fukui, T. (1982) Solar heating in closed plastic house for control of soilborne diseases. V. Application for control of Fusarium wilt of strawberry, *Annals of the Phytopathological Society of Japan* **48**, 570–577.

Kodama, T., Fukui, T. and Matsumoto, Y. (1980) Soil sterilization by solar heating against soil-borne diseases in a closed vinyl house. III. Influence of the treatment on the population level of soil microflora and the behaviour of strawberry yellows pathogen, *Fusarium oxysporum* f.sp. *fragariae*, *Bulletin of the Nara Agricultural Experiment Station* **11**, 41–52.

Kritzman, G. and Ben-Yephet, Y. (1990) Disease exchange in onions as influenced by metham-sodium, *Hassadeh* **1**, 26–29.

Malathrakis, N.E., Kapetanakis, G.E. and Linardakis, D.C. (1983) Brown root rot of tomato and its control in Crete, *Annals of Applied Biology* **102**, 251–256.

Martyn, R.D. and Hartz, T.K. (1986) Use of soil solarization to control Fusarium wilt of watermelon, *Plant Disease* **70**, 762–766.

Materrazzi, A., Triolo, E., Vannacci, G. and Scaramuzzi, G. (1987) The use of soil solar heating for controlling neck rot of greenhouse lettuce, *Colture Protette* **16**, 51–54.

Middleton, L.A. and Lawrence, N.J. (1995) The use of dazomet via the "planting through" technique in horticultural crops, *Acta Horticulturae* **382**, 86–103.

Oliveira, H. (1992) Evaluation of soil solarization for the control of Fusarium wilt of tomato, in E.C. Tjamos, G.C. Papavizas. and R.J. Cook (eds.), *Biological Control of Plant Diseases. Progress and Challenges for the Future*, NATO ASI Series A, Life Sciences, 230, Plenum Press, New York, pp. 69–73.

Phillips, A.J.L. (1990) The effects of soil solarization on sclerotial populations of *Sclerotinia sclerotiorum*, *Plant Pathology* **39**, 38–43.

Porter, I.J. and Merriman, P.R. (1985) Evaluation of soil solarization for control of root diseases of row crop in Victoria, *Plant Pathology* **34**, 108–118.

Ristaino, J.B., Perry, K.B. and Lumsden, R.D. (1991) Effect of solarization and *Gliocladium virens* on sclerotia of *Sclerotium rolfsii*, soil microbiota and incidence of southern blight of tomato, *Phytopathology* **81**, 1117–1124.

Ristaino, J.B. and Thomas, W. (1997) Agricultural MBr, and the ozone hole: Can we fill the gaps? *Plant Disease* **81**, 964–977.

Rubin, B. and Benjamin, A. (1983) Solar heating of the soil: Effect on soil incorporated herbicides and on weed control, *Weed Science* **31**, 819–825.

Runia, W.Th. (1983) A recent developments in steam sterilization, *Acta Horticulturae* **152**, 195–200.

Stapleton, J.J. and DeVay, J.E. (1982) Effect of soil solarization on populations of selected soilborne microorganisms and growth of deciduous fruit tree seedlings, *Phytopathology* **72**, 323–326.

Stapleton, J.J. and DeVay, J.E. (1984) Thermal components of soil solarization as related to changes in soil and root microflora and increased plant growth response, *Phytopathology* **74**, 255–259.

Stapleton, J.J. and Garza-Lopez, J.G. (1988) Mulching of soil with transparent (solarization) and black polyethylene films to increase growth of annual and perennial crops in Southern Mexico, *Tropical Agriculture* **65**, 29–35.

Tamietti, G. and Garibaldi, A. (1989) The use of solarization against Rhizoctonia under greenhouse conditions in Liguria, *Informatore Fitopatologico* **39**, 43–45

Tjamos, E.C. (1984) Control of Pyrenochaeta lycopersici by combined soil solarization and low dose of methyl bromide in Greece, *Acta Horticulturae* **52**, 253–258.

Tjamos, E.C., Biris, D.A. and Paplomatas, E.J. (1991) Recovery of Verticillium wilted olive trees after individual application of soil solarization in established olive orchards, *Plant Disease* **75**, 557–562.

Tjamos, E.C., Karapappas, V. and Bardas, D. (1989) Low cost application of soil solarization in covered plastic houses for the control of Verticillium wilt of tomatoes in Greece, *Acta Horticulturae* **255**, 139–149.

Tjamos, E.C. and Makrynakis, N. (1990) Control of fungal wilt diseases of melon by application of soil solarization in the field, *Proceedings of the 8th Congress of the Mediterranean Phytopathological Union*, Agadir, Morocco, November 1990, pp. 423–425.

Tjamos, E.C. and Paplomatas, E.J. (1987) Effect of soil solarization on the survival of fungal antagonists of *Verticillium dahliae, European Mediterranean Plant Protection Organization Bulletin* **17**, 645–653.

Triolo, E., Vannacci, G. and Materazzi, A. (1988) Solar heating of the soil in vegetable production. Part 2. Studies of possible mechanisms of the effect, *Colture Prottete* **17**, 59–62.

van Steekelenburg, N. (1996) Physical control, in J.C. van Lenteren (ed.), *Integrated Pest Management in Protected Cultivation*, Wageningen Agricultural University, Wageningen, pp. 17–20.

Vannacci, G., Triolo E. and Materazzi, A. (1988) Survival of *Sclerotinia minor* sclerotia in solarized soil, *Plant Soil* **109**, 49–55.

PESTICIDES IN IPM: SELECTIVITY, SIDE-EFFECTS, APPLICATION AND RESISTANCE PROBLEMS

Sylvia Blümel, Graham A. Matthews, Avi Grinstein and Yigal Elad

11.1. Importance of Selective Pesticides in IPM Programmes

The success of released or naturally occurring biological control agents in preventing pest outbreaks in protected crops has led the greenhouse industry to be particularly conscious of the necessity of applying selective pesticides. The activity of a selective pesticide is confined to a narrow range of specific pests (Heitefuß, 1975). In IPM, the process of developing the selectivity of a pesticide aims to maximize its specific effect against pests and diseases and minimize its effect on non-target organisms (Hull and Beers, 1985). Thus the selectivity of a pesticide is often used to express its harmlessness for beneficial organisms. The selectivity of the action and of the toxicity of a pesticide is dependent on its physiological selectivity and/or on the application procedures (Poehling, 1989). Physiological selectivity is expressed by reduced sensitivity of an organism to the pesticide due to pesticide metabolism and to the availability of the appropriate enzymes in the target organisms (Hassall, 1982). Application procedures comprise the dose rate, mode of action, method and timing.

The use of chemical pesticides that cause undesired side effects on non-target beneficial organisms may lead to pest outbreaks. In tomatoes, multiple application of the broad-spectrum carbamate methomyl for the control of leafminer infestation (*Liriomyza sativae* Blanchard) eliminated the naturally occurring beneficial parasitoid complex, which, without chemical treatment, reduced the pest population to 50% of the level found in pesticide-treated plots (Oatman and Kennedy, 1976). To avoid these consequences the harmful effects of pesticides on the natural enemies of target pests must be avoided or minimized for successful implementation of biological control agents within IPM strategies. Some pests and pathogens have developed resistance towards certain chemical pesticides, and this must also be considered in order to prevent misuse of pesticides.

In this chapter we will deal with the selectivity of pesticides in relation to effects on beneficial organisms that can be used in greenhouses, the potential for improving applications for better performance and selectivity, and the problems of resistance of the pests or diseases to the chemicals used in greenhouses.

11.2. Types of Side-Effects on Beneficial Organisms

Pesticides can exhibit primary or secondary effects on predators, parasitoids and pathogens of target pests. Primary effects are direct or indirect, depending on their

R. Albajes et al. (eds.), Integrated Pest and Disease Management in Greenhouse Crops, 150-167.
© 1999 *Kluwer Academic Publishers. Printed in the Netherlands.*

exposure and on the biological parameter influenced. Direct mortality of beneficial organisms may be caused by direct contact during application, pesticide residues, taking up contaminated prey, intoxication by fumigants, and contact or contamination with soil disinfectants.

Indirect or sublethal effects on beneficial arthropods include decreases in reproduction, oviposition, parasitization, predation, longevity and egg viability, and a delay in development and shifting of the sex-ratio. Morphological and behavioural changes may also occur (Elzen, 1989).

Secondary effects due to pesticides include killing the prey/host of a beneficial organism or of species which produce alternative food like honeydew (Huffaker, 1990), taking up contaminated food (Sell, 1984; Celli et al., 1997), and directly stimulating the pest; for example, some pyrethroids enhance reproduction in Tetranychus urticae Koch.

Pesticides directly affect entomopathogenic fungal biocontrol agents by inhibition of spore germination and vegetative development (mycelial growth), and they also reduce the viability of conidia (McCoy et al., 1988) and their survival and activity on plant surfaces. Viability and infectivity of the infective juveniles (J3) of entomopathogenic nematodes are also adversely affected (Rovesti et al., 1988).

Side-effects of pesticides on natural enemies may vary between and within taxonomic groups. From their comprehensive data on the side-effects of pesticides, Theiling and Croft (1988) concluded that predators were more tolerant to pesticide treatment than parasitoids, except for fungicides, towards which susceptibility was not greatly affected. The tolerance of aphid natural enemies decreases from Coccinellids > Chrysopids > Syrphids > Hemiptera > Hymenoptera (Hodek, 1973). Evaluation of effects within taxonomic groups revealed that the classification of the effects of 74 compounds tested against the parasitoids Encarsia formosa Gahan, Aphidius matricariae Haliday and Leptomastix dactylopii Howard corresponded by more than 78% (Hassan et al., 1983, 1987, 1988, 1991, 1994). In a comparison of trial results with 81 test compounds for predatory mite species occurring in orchards and vineyards with Phytoseiulus persimilis Athias-Henriot, the same level was reached in 64% of the test compounds.

Differences in susceptibility have been recorded between taxonomically close species, and even between strains within the same species. Eretmocerus mundus Mercet adults were less susceptible to residues of amitraz, thiodicarb and cypermethrin than E. formosa or Encarsia pergandiella Howard (Jones et al., 1995). Among Aphidius species, A. matricariae was more tolerant to dimethoate than Aphidius rhopalosiphi de Stefani Perez or Aphidius colemani Viereck (Maise et al., 1997). The response of several species of entomopathogenic fungi to copper incorporated in agar differed. Paecilomyces farinosus (Holmsk.) A.H.S. Brown & G. Sm. was more tolerant than Verticilium lecanii (A. Zimmerm.) Viégas, Beauveria bassiana (Balsamo) Vuillemin and Metarhizium anisopliae (Metschnikoff) Sorokin (Baath, 1991). The entomopathogenic nematodes Steinernema carpocapsae (Weiser), Steinernema feltiae (Filipjev) and Heterorhabditis HP88 exhibited different tolerance levels to 9 tested pesticides (Zimmerman and Cranshaw, 1990). Repeated exposure of local strains to chemicals may cause natural enemies to develop tolerance to pesticides. This is the case of P. persimilis and organophosphorous compounds (OPs) (Goodwin and Welham, 1992) and of Aphidoletes aphidimyza (Rondani) and azinphos-methyl (Warner and Croft, 1982). Developmental stage may greatly influence the response of natural enemies to pesticides. The susceptibility of *A. aphidimyza* and *Chrysoperla carnea* (Stephens) to pesticides with contact mode of action increased from the egg stage to the adults (Bartlett, 1964). In contrast, pesticide susceptibility was lowest in treated adults of *Coccinella septempunctata* L. (Zeleny *et al.*, 1988) and in eggs of *P. persimilis*, while in the coccinellid the egg stage and in the predatory mite the larvae or protonymph stage were the least tolerant (van Zon and Wysoki, 1978; Blümel and Stolz, 1993). However, compounds with modes of action that regulate or inhibit insect growth resulted in high mortality of *C. carnea* larvae, but not of the adults, whose fertility was only slightly affected (Vogt, 1992).

The host may offer parasitoids different degrees of protection against pesticides; unprotected stages of parasitoids (e.g. adult hymenoptera) and protected stages (e.g. different developmental stages in aphid mummies) show different levels of mortality after the same pesticide treatment. Avermectin B killed 50% of *E. formosa* protected in the whitefly scales in a direct contact test, but 79% of the adult wasps after contact with the dried residue (Zchori-Fein *et al.*, 1994). *Leptomastix dactylopii* protected in *Planococcus citri* (Risso) were barely affected by topical treatment of endosulfan, while the adults were severely damaged in residual tests (Reddy and Bhat, 1993). Even sexes of the same species may present different susceptibility against pesticides. In 5 different populations of *Diglyphus begini* (Ashmead) (Rathman *et al.*, 1992) and in predatory mites, males are less tolerant than females.

11.3. Tests and Approaches to Detect Side-Effects of Pesticides

One of the most comprehensive programmes to test side-effects of pesticides on beneficial organisms was set up by the IOBC/WPRS working group "Pesticides and Beneficial Organisms" (Hassan, 1989). In the first step, arthropod species and microorganisms that were regarded as the most important natural enemies in the different crops were identified. For these species test methods at different levels were developed. Pesticide screening is based on a sequence of three steps in laboratory, semi-field and field conditions, as shown in Fig. 11.1. The sequential programme assumes that pesticides that are harmless in the laboratory will also be safe in semi-field and field conditions, and do not need to be evaluated in further steps. When a chemical, however, is categorized as harmful in one step, its effect at the next step cannot be inferred, and the sequence must be continued until it finishes at field conditions or displays no negative effects.

The pesticides are usually tested at the highest recommended field rate as commercial formulations. The laboratory methods aim to evaluate the direct, initial toxicity of pesticide residues on susceptible and protected developmental stages of the test arthropods and are thus classified as lab-a- and lab-b-tests. The aim of the first test is the detection of pesticides which are harmless to the test organism after worst case exposure to dried pesticide residue on a defined test surface (glass or sand) after a single application of the test compound. The results of the tests should include the mortality (direct effect) and the reproduction (sublethal effect) of the test organism. Information about the duration of the effect of a pesticide is provided by the persistence test. Plant material (e.g. leaves) is sprayed with the test pesticide and left on the plant under greenhouse conditions for residue aging. Leaf samples undergo a further test, similar to the lab-a-test. The next test is the semi-field test which is carried out on pesticide residues or as a direct application on the plants with the test arthropods, and is kept under more natural conditions. Sublethal effects, behavioural changes, and the effect of more than one application of the test product are thus evaluated. The range of tests developed for a selection of organisms important in greenhouse crops is presented in Table 11.1. Most of the information that follows in this section may be found in the IOBC/WPRS Bulletin 1988,11(4); 1992,15(3); 1994,17(3).

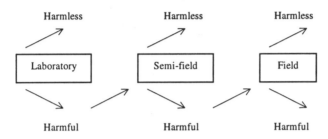

Figure 11.1. Sequential IOBC/WPRS procedure for testing effects of pesticides on natural enemies (after Hassan, 1989).

The lab-a-test for parasitoids (*E. formosa*, *A. matricariae*) and for *A. aphidimyza* adults (on leaf material) is carried out as a residual contact test with adult wasps or gall midges. Mortality and reproduction (parasitization of the host or number of eggs deposited) are evaluated. In the lab-b-test, the protected stages of the parasitoids in their hosts (aphid mummies; whitefly scales) are directly sprayed with the pesticide solution and the emergence rate from the hosts is assessed. The lab-a-test for predatory mites is a residual contact test starting with predatory larvae or protonymphs. During the test the mortality rate, escaping rate and reproduction per female are evaluated.

The same testing procedure is used as in the lab-a-test for *Orius niger* (Wolff), and the emergence from the deposited eggs is also assessed. The lab-b-test for *O. niger* is the same as the lab-a-test, but uses predatory bug adults. The lab-b-test for *A. aphidimyza* is carried out with larvae as a residual contact test on leaves and is also appropriate for a persistence test. Laboratory tests for *C. carnea* and *Syrphus corollae* Fabricius follow the same principles. Larvae are tested in a residual contact test to assess the mortality and reproduction of the test organisms. A laboratory test for Coccinellids has also been described in detail for *Hippodamia oculata* (Thunberg). A residual contact test with larval stages is carried out to evaluate mortality and duration of development. The adults deriving from this first testing phase are used to check reproduction, duration of sexual maturation of females, and emergence from the deposited eggs.

Persistence tests or tests to detect the duration of harmful effects of the pesticide residue are very similar for nearly all test organisms. Suitable plants are sprayed and kept under greenhouse conditions for different periods. Leaf samples are collected at regular intervals and are used as test surfaces, as in the lab-a-test or the lab-b-test. Mortality and reproduction are again assessed. In this case persistence must be considered as an extended laboratory test. For *C. carnea* and *Episyrphus balteatus* (DeGeer) the test is carried out on treated plants and, in addition to the above mentioned parameters, changes in the behaviour of *E. balteatus* can be examined.

TABLE 11.1. Test methods on different test levels of the sequential testing scheme developed within the IOBC/WPRS working group "Pesticides and Beneficial Organisms"

Test organism	Lab-a-test	Lab-b-test	Extended laboratory test/ persistence	Semi-field test initial tox.	Semi-field test persistence/ greenhouse
Parasitoids					
Encarsia formosa	X[1]	X	X	X	
Aphidius matricariae	X	X	X		
Leptomastix dactylopii	X				
Aphytis melinus	X				
Predatory mites					
Phytoseiulus persimilis	X		X	X	
Others	X		X		
Predators					
Chrysoperla carnea	X		X	X	X
Syrphus corollae	X				
Episyrphus balteatus				X	X
Coccinellids (*Semiadalia, Harmonia, Coccinella*)	X				
Orius niger	X	X	X	X	
Aphidoletes aphidimyza	X	X	X		
Entomopathogenic fungi					
Beauveria bassiana, Beauveria brongiartii	X	X	X	X	
Metarhizium anisopliae	X	X	X	X	
Verticillium lecanii	X				
Entomopathogenic nematodes					
Steinernema sp.	X				

[1]X: existing test methods

The sequential IOBC testing scheme for *B. bassiana* and *M. anisopliae* comprises all three test levels. In the lab-tests the mycelial growth on agar containing pesticides is measured. The production and viability of conidia is assessed with a bioassay to check virulence. It has been proposed to switch from tests on solid medium to a worst case test for growth inhibition in liquid médium, where the mycelium, as the most sensitive stage of the fungus, is immersed into the pesticide solution. In the semi-field test conidia are mixed with standard soil and treated with the test pesticide. The soil is then incubated and the number of spores per unit of soil is determined. To check the virulence of the tested fungus at each step of the sequential scheme the *Galleria*-bait-method may be used. The results of an *in vivo* assay, in which leaf discs are sprayed with the conidial suspension of the beneficial fungus on a dried residue of the test pesticide have been described. Side-effect testing at the infective juvenile J3 stage of entomopathogenic nematodes is carried out in a 2-step scheme. First, the viability and the behaviour *in vitro* in pesticide solutions is checked. In the next step, mobility and infectivity are examined in a bioassay in soil.

Compatibility of pesticides with bumble-bees used as natural pollinators in greenhouses is classified in four categories, which allow or exclude the use of bumble-bees or recommend a certain period after pesticide application during which the hives should be removed from the greenhouses.

Comprehensive data collections about side-effects of pesticides on natural enemies are available from commercial suppliers of beneficial organisms (Biobest, 1998) and also in the tables published by the IOBC/WPRS working group.

11.4. Effects of Chemical Pesticides on Beneficial Organisms Used in Greenhouses

Information on specific pesticide effects on natural enemies and pathogens may be found in the published results of the Joint Testing Programmes by the IOBC/WPRS Working Group "Pesticides and Beneficial Organisms" (Hassan *et al.*, 1983, 1987, 1988, 1991, 1994; Croft, 1990; Sterk *et al.*, 1998) and in many other references. Some examples selected from the literature are included in Table 11.2.

Generally herbicides, acaricides and fungicides have less effect than insecticides, although mycopesticides are highly susceptible to fungicides.

(i) Effect on beneficial predators. For predatory mites most pyrethroids and carbamates were harmful, both in initial toxicity and in reproduction and persistence trials with the susceptible juvenile predators. *Aphidoletes aphidimyza* showed a similar susceptibility to insecticide/acaricide treatments, and was also affected by OPs. OPs caused varying levels of mortality in predatory mites (see Section 11.3). In coccinellids, high mortality rates were caused by nearly all tested compound groups, except the microorganisms and soap. Chrysopids were not harmed by acaricides, most pyrethroids, soap or microorganisms, but were affected by most of the insect growth regulators (IGRs) and most of the OPs. For predatory bugs, pyrethroids, carbamates, most OPs and few of the IGRs proved to be harmful. Fungicides and herbicides were relatively harmless for coccinellids, chrysopids and predatory bugs, but partly harmful to predatory mites.

(ii) Effects on beneficial parasitoids. Synthetic pyrethroids and pyrethrin were very harmful to adults, regardless of the test species. In tests with the protected stages, several pyrethroids were only slightly harmful, but in combination with a persistence of more than one week this advantage was neutralized. OPs were very harmful to the unprotected stages and with few exceptions also to the protected life stages, and showed high persistence as residues. Carbamates were harmful in both types of laboratory tests, but some had a persistence shorter than three days. IGRs and most of the acaricides were harmless to both the susceptible and the protected developmental stage of the parasitoids. Plant extracts (except pyrethrin), soap and microorganisms were harmless. Fungicides belonging mainly to the group with a broad-spectrum and protective mode of action were harmful to adult parasitoids and revealed detrimental effects which persisted over one week. In tests with the protected life stage, however, all fungicides

CHAPTER 11

were considered harmless. Very few herbicides were harmful to adult wasps, but not for other developmental stages.

TABLE 11.2. Effects of pesticides on natural enemies. The type of tests are indicated in column headings: 1, lab-a-test; 2, lab-b-test; 3, persistence test; 4, semi-field or greenhouse test (see Section 11.4 in this chapter for further explanations on these types of tests). Effects have been categorized according to the following criteria. In laboratory and semi-field tests: −, <50% total effect; o, total effect between 51 and 99%; +, >99% total effect. In persistence tests: −, effect on <50% for lesser than one week; o, effect on 51–99% for lesser than one week; +, effect on >99% for more than one week. More than one effect classification indicates different test results. Compiled from published results of the Joint Testing Programmes by the IOBC/WPRS Working Group "Pesticides and Beneficial Organisms" (Hassan et al., 1983, 1987, 1988, 1991, 1994; Croft, 1990; Sterk et al., 1998) and also from many other references

Pesticide common name	Predatory mites				Coccinellids/ Chrysopids				Predatory bugs				Parasitoids				Nemat.		Fungi		
	1	2	3	4	1	2	3	4	1	2	3	4	1	2	3	4	1	2	1	2	4
Organophosphorus																					
Chlorpyrifos	o+				+	+		+	o+	+		+	+	o	+		−		o	−	
Diazinon	−+				+		−+	o+	−o	−	o		+	o	+		o	o	−		
Heptenophos	−+		−		o+	o		−	−o	−o	o		+o	−	−		o		−	+	−+
Phosmet	o+				o+	+		o+	−	o			+	o			−		o		
Triazophos	−+				+	+		+	o+	+		+	+	o+	+				o		
Organochlorines																					
Endosulfan	o	o			o+	−o							+o	−	o		−				
Lindane	−				o								+	−	+		−				
Carbamates																					
Methomyl	+				o+	+	o+	+		+		+	+	o			o	o	o		
Oxamyl	+	o			+	+		o	o	+			+	o			+	−	o−		
Pirimicarb	−o		−		o	−o	−			o			+	−	−	−					
Propoxur	o+				o	+				+			+						−		
Pyrethroids																					
Deltamethrine	+					+		o+	o	+			+	+	+	+	−		−		
Fenpropathrin	+	+		+	+	o			o	+			+	+o	+	−	o		−		
IGR's																					
Diflubenzuron	−					−	−+	o	o+	o−		−+	−				−		−		
Fenoxycarb	−					−	−+	−	−o	+			−	−						o	
Teflubenzuron	−					−	o+	+	+	o+			−	−	−		−		−	−	
Natural origin																					
Azaridachtin	−	−											−	−							
Bacillus thuringiensis ssp. *kurstaki*, Bacillus thuringiensis ssp. *tenebrionis*	−				−	−		−	−	−			−				−		−	−	
Nicotine					−+	o		−													
Pyrethrum + rotenone	+	o			o																
Verticillium lecanii	−				−	−			−				−				−		−		

TABLE 11.2. Effects of pesticides on natural enemies. The type of tests are indicated in column headings: 1, lab-a-test; 2, lab-b-test; 3, persistence test; 4, semi-field or greenhouse test (see Section 11.4 in this chapter for further explanations on these types of tests). Effects have been categorized according to the following criteria. In laboratory and semi-field tests: –, <50% total effect; o, total effect between 51 and 99%; +, >99% total effect. In persistence tests: –, effect on <50% for lesser than one week; o, effect on 51–99% for lesser than one week; +, effect on >99% for more than one week. More than one effect classification indicates different test results. Compiled from published results of the Joint Testing Programmes by the IOBC/WPRS Working Group "Pesticides and Beneficial Organisms" (Hassan *et al.*, 1983, 1987, 1988, 1991, 1994; Croft, 1990; Sterk *et al.*, 1998) and also from many other references (cont.)

Pesticide common name	Predatory mites				Coccinellids/ Chrysopids				Predatory bugs				Parasitoids				Entomopathogens Nemat.		Fungi		
	1	2	3	4	1	2	3	4	1	2	3	4	1	2	3	4	1	2	1	2	4
Other insecticides																					
Abamectin	o	+		o							–	o	o	–o			–				
Amitraz	+					–o			o	–	–		+	o	o	+	–		o		
Buprofezin	–o					–o	–o		o–	–	–		–	–			–	–	–		
Cyromazine				–+	–+	+o	o		o–	o	–		o	–			–	–	–		
Imidacloprid	+				–				+										–		
Pyriproxifen	–	–			–				–	–			–	o							
Acaricides																					
Dicofol	o+		o	+	o	–				o			+o		+		–	–			
Fenbutatin oxide	–				–	–							–						–		
Fungicides																					
Bitertanol	–				–	–		–	–	–			–			–			+–		o
Chlorothalonil	–o				–	–	+		–	–			–			–			o+	+	+
Copper oxychloride	–				–	–o	–		–	–			+	–	–				–+		
Iprodione	–				–	–			–	–			–	–		–			o–		
Quinomethionate	o				o	o							+	–	–				–		
Sulphur	–+	o			o+	–+	o		–	–o			+–	–	+–		–o		–		
Triforine	–				–	o–	–		–				–						o		o
Herbicides																					
Glyphosate	–o				o–				–				–				–		–		
2,4 D	–				–	–			–	–			–						–		

(iii) Entomopathogens. Only a small number of carbamates out of the tested insecticides/acaricides affected entomopathogenic nematodes, while fungicides proved to be mainly harmless. Insecticides, acaricides and herbicides in most cases did not adversely influence the mycelial growth or the sporulation of the fungal species *V. lecanii*, *B. bassiana*, and *M. anisopliae* in laboratory tests or during infectivity tests in the greenhouse. Half of the fungicides examined in all types of tests affected at least one of the three test fungi, whereas one fourth of the fungicides were harmless for all of them. Effects could not generally be attributed to the mode of action of the fungicides. *Verticilium lecanii* was slightly more affected than *B. bassiana*.

(iv) Sublethal effects on natural enemies. Besides direct toxicity caused by a number of "classical" insecticides, sublethal effects were also demonstrated in several investigations. Among sublethal effects of pesticide application on natural enemies which are reported in the literature are: development prolongation, reduced egg production or its total inhibition, decrease in prey consumption, changes in searching or foraging behaviour, alteration of pathogenicity in entomopathogens, and increased tendency to escape from treated surfaces. The importance of repellence of pesticide compounds for beneficials is difficult to classify. On the one hand, repellence may negatively influence natural enemies by expelling them from their host or prey which they need for further population development; on the other, beneficials can be protected from possibly hazardous contact with contaminated plant surfaces or prey/hosts. Both effects are undesirable, especially in greenhouses, where mass reared arthropods are intentionally introduced as biological control agents, and because the natural enemies would cease to be effective as control agents, particularly when untreated refuges are scarce.

Insect growth regulators, like diflubenzuron, chlorfluazuron, fenoxycarb, flufenoxuron and teflubenzuron, which are incorrectly considered as harmless to many beneficials, in fact interfere with the viability of eggs, the moulting process, and the reproduction of several predators.

The influence of different formulations of pesticides on their effects on natural enemies was shown for endosulfan, which as an emulsifiable concentrate (EC) formulation resulted in up to 17% less mortality of *P. persimilis* than the wettable powder (WP) formulation in a residual laboratory test (Blümel et al., 1993). For *E. formosa* the EC formulation of tebufenpyrad was more toxic than the WP formulation (van de Veire, 1995).

11.5. Influence of Pesticide Application on the Selectivity of a Pesticide

The relatively small areas in greenhouses – compared to arable agriculture – and high plant density dictate in many cases the use of manually operated spraying equipment. In an enclosed structure, good ambiental conditions can exist for applying very small particles and using artificial air movement to improve pesticide distribution and pest control. Conversely, improved chemical control can adversely affect bio-agents such as bumble-bees, antagonistic fungi and beneficial arthropods, factor which has to be considered when choosing a pesticide. Pesticide application in enclosed areas also imposes the risk of breathing air that contains small particles of pesticides. Personal protective clothing is often hot and uncomfortable, and farmers tend to spray unprotected.

Unfortunately, many growers continue to use high volume (HV) spraying (>1000 l/ha of spraying solution). HV spraying to run off leads to wastage to the order of 70–90% of the chemical dripping to the ground (Matthews, 1992). The low concentration of a.i. with HV applications reduces the hazard to the operator, who is often heavily contaminated by the pesticide, but may not give adequate control, and growers are thus forced to repeat sprays at frequent intervals. The whole area becomes contaminated with pesticides, making it impossible to integrate biological control with chemicals. The volume of spray and wastage due to runoff can be reduced significantly by changing nozzles to produce small droplets which do not coalesce on the target (Matthews, 1992). A widely used piece of equipment is the knapsack mistblower.

As an alternative to HV spraying, the use of thermal or cold foggers gives the grower clear savings in time and labour, although they are only suitable in totally enclosed greenhouses. Deposition is improved with cold fogging, but persistence is less. The shorter persistence obtained with cold foggers allows the introduction of natural enemies quicker after treatment than when a thermal fogger is used, and a greenhouse can be treated when parasitoids are protected inside the infested host stages (Lingappa et al., 1972). Additionally, cold fogging allows the use of a wider range of pesticides, e.g. insecticides perhaps with higher selectivity, such as *Bacillus thuringiensis* Berliner which has been used successfully by cold fogging.

Another technique, vaporization, is suitable for small areas (approximately 100 m^2). The pesticide (e.g. sulfur) is placed on a small heater installed inside a wide pipe. After evaporation or sublimation, the pesticide condenses to small particles (e.g. 2–8 μm) and is carried up by the heated air directed by the pipe. The dispersion and settling of particles of this size is influenced by the inside air circulation systems and they fall mainly on the upper side of the leaves, rendering minimal residual effect.

Alternatives to spray treatments include application of granules or drenches and chemigation by drip irrigation to the soil, when systemic pesticides can be used. Specific treatments can be combined with a pesticide or other types of lure, e.g. yellow cards in a "lure and kill" method. Thrips have been controlled with a polybutene sticky surface combined with an insecticide (Thripstick). Specific baits cause only minimal damage to non-target organisms, as their chance of exposure is very low.

The timing of the pesticide treatment is crucial in order to avoid the susceptible life stage of the non-target organism. Where chemical pesticides adversely affect the entomopathogenic fungus *V. lecanii*, they should not be applied at the same time, but after a delay (Schuler, 1991). Similarly, the alternation of chemical fungicides with the fungal biocontrol agent *Trichoderma harzianum* Rifai T39 is preferred to the use of a tank mix of this biocontrol agent with chemicals for the control of foliar pathogens (Shtienberg and Elad, 1997). Selective application can also be carried out by considering spatial factors and using the systemic pesticides as granules or seed treatment to preserve plant-inhabiting beneficials. Limited areas can be treated with hand-held air-assisted spinning disc sprayers. Multiple applications of a pesticide may cause a severe reduction in the number of natural enemies, without achieving a satisfactory control of the target pest. In contrast, a single, better timed application of the same pesticide can control the pest to the same extent, without seriously damaging the natural enemies, thus improving overall control. Keeping the pest below the economic threshold has been achieved with different use of oxamyl and methamidophos against *L. sativae* and its parasitoid complex in tomatoes (Schuster *et al.*, 1979).

Systemic fungicides, which were harmful to *V. lecanii* when applied as sprays, did not affect the fungus pathogenecity against *Aphis gossypii* Glover on cucumber when applied as a soil drench (Wilding, 1972). Another possibility for the partial preservation of natural enemies is the treatment of selected strata of the plants, e.g. flowers, and leaving the lower part of the canopy untreated, thus maintaining a significant population of natural enemies (Scopes and Biggerstaff, 1973). These localized treatments are gaining acceptance where insects are used to pollinate crops and growers release natural enemies such as *E. formosa*. In one study, the application of pyriproxifen to the upper parts of tomato plants infested with greenhouse whitefly effectively reduced the pest, but did not damage the parasitoid *E. formosa*, which, though susceptible to this compound, was situated in the whitefly pupae on the lower parts of the plants (van de Veire, 1995).

11.6. Pesticide Resistance and Anti-Resistance Strategies in IPM

Pests and pathogens may overcome the toxic effect of pesticides by metabolizing the active ingredient into less toxic components, developing a change in the target site, reducing the absorption of the chemical or by avoiding exposure to the compound. Resistance development is the most severe challenge to pesticide. In greenhouses, pesticide-resistant strains of fungi and pests have appeared frequently. This phenomenon occurs because the greenhouse is a closed system in which the population of selected strains is not diluted by the outdoor wild population. Usually, the existence of epidemic conditions in greenhouses is a prerequisite for the development of resistant populations of pathogens and pests. Moreover, the optimal conditions for their development in greenhouses prevail for long periods. The number of life cycles is increased due to the optimal conditions or the extended time they prevail, and control necessitates frequent pesticide applications. The latter result in high selection pressure towards resistance to pesticides. The main pathogens which are known to develop resistance to fungicides in greenhouses are *Botrytis cinerea* Pers.:Fr., *Pseudoperonospora cubensis* (Berk. & M.A. Curtis) Rostovzev (downy mildew of cucurbits), *Didymella bryoniae* (Auersw.) Rehm (gummy stem blight of cucurbits), *Sphaerotheca fusca* (Fr.) Blumer. [= *Sphaerotheca fuliginea* (Schlechtend.:Fr.) Pollacci] (powdery mildew of cucurbits), *Puccinia horiana* Henn., *Uromyces dianthi* (Pers.:Pers.) Niessl (= *Uromyces caryophyllinus* G. Wint.) and *Fusarium oxysporum* Schlechtend.:Fr. f. sp. *gladioli* (L. Massey) W.C. Snyder & H.N. Hans.

The benzimidazole fungicides (benomyl, carbendazim, thiophanates) have a high resistance potential against pathogens because they have a specific mode of action. The resistance is usually not associated with a significant loss of fitness of the pathogen. It occurs in populations of *B. cinerea*, *D. bryoniae*, *Fusarium* and powdery mildews. Mixtures and alternations with multi-site contact fungicides may delay this selection, before resistance becomes apparent.

Acute problems of resistance to dicarboximide fungicides (e.g. iprodione, procymidone, vinclozolin) have arisen when fungicides are used intensively and exclusively over many seasons (Gullino *et al.*, 1989). Isolates are moderately resistant and tend to be almost as fit as sensitive strains in the absence of fungicides. It is recommended to restrict the number of dicarboximide treatments to no more than three per crop in greenhouses where resistance is found, and even in the absence of detectable resistant strains. When infection pressure is high, it is usually recommended to alternate or mix these fungicides with protectants such as chlorothalonil, captan, TMTD, or with biocontrol which do not usually select for resistance. However, TMTD may interfere with natural enemies (Section 11.4).

Ergosterol biosynthesis inhibitors (EBIs) are a group of fungicides which include triazole, imidazole and pyrimidine fungicides which inhibit C14 demethylation and morpholines. Unlike the sharp, significant nature of resistance towards benzimidazoles and dicarboximides mentioned above, the resistance towards EBIs develops in the form of slow shifts in the pathogen population. For instance, powdery mildews in greenhouses were controlled for several years by benzimidazoles, hydroxypyrimidines, pyrazophos, and EBIs. Resistance is known in populations of *S. fusca* but the alternation of fungicides, which is practised in many countries, is helping to deal with the problem. It is generally recommended to rotate or mix EBI fungicides with fungicides from other groups as well as with biocontrol.

The failure of disease control in greenhouses is exemplified by the history of gray mold epidemics. Multiple resistant isolates occur in greenhouses that bear the resistance towards benzimidazole, diethofencarb, dicarboximides and ergosterol biosynthesis inhibitors (Pommer and Lorenz, 1982; Elad *et al.*, 1992). The extreme summer conditions do not interfere with the survival of fungicide-resistant isolates (Yunis and Elad, 1989). Table 11.3 illustrates the situation for Israeli vegetable greenhouses sampled in 1997 by exposing plates of *Botrytis* selective medium containing test fungicides from various groups (for method, see Elad and Shtienberg, 1995).

TABLE 11.3. Resistance towards fungicides in greenhouse populations of *B. cinerea* (number of colonies grown on exposed plates containing Botrytis selective medium indicating comparable levels)

Crop	Place	Group of fungicides, test fungicide in plate and concentration (μg/ml)			
		None	Benzimidazoles	Dicarboximides	EBIs
			Benomyl (5)	Iprodione (5)	Fenbuconazole (1)
Cucumber	Ahituv	100	98	29	21
	Yama A	23	18	3	22
	Yama B	59	7	4	72
	Tzafit	6	7	1	6
Tomato	Bet Shikma	77	100	10	100
	Sde Moshe	18	16	1	14
	Yama	46	48	1	47
	Ibtan	10	3	2	1

Phenylamide fungicides that inhibit RNA synthesis were introduced in the late 70s for Phycomycetes control. During the 70s *P. cubensis* was controlled mainly with protective applications of dithiocarbamates and chlorothalonil. In the early 80s the phenylamide metalaxyl was released and soon afterwards resistant strains were selected. Metalaxyl-resistant strains seem to be more competitive than wild-type strains (Cohen *et al.*, 1983). Resistance was found also in *Phytophthora infestans* (Mont.) de Bary on tomato and *Bremia lactucae* Regel on lettuce. Anti-resistance mixtures of metalaxyl with protectant fungicides were developed to cope with phenilamide resistance.

In order to reduce the pressure towards development of resistance in pathogen populations, it is usually better to limit the exposure of the pathogen to a group of fungicides. The number of applications of fungicides of the same mode of action has to be limited, especially against fungi with many cycles during the growing season. Moreover, the application of non-chemical methods is also recommended.

Insecticide and acaricide resistance of nearly all important arthropod greenhouse pests is well documented (Georghiou and Mellon, 1983). Besides genetic and operational factors that influence the selection of resistant individuals, biotic reasons such as generation turn-over, number of offspring per generation and type of reproduction have a major impact on resistance development. Most of the pest species on greenhouse crops favour resistance selection with regard to these biological parameters.

Recently *Bemisia tabaci* (Gennadius) and *Bemisia argentifolii* Bellows & Perring have developed resistance against a range of conventional insecticides as well as against IGRs and juvenile hormone analogs (Cahill *et al.*, 1994; Horowitz *et al.*, 1994), and *Frankliniella occidentalis* (Pergande) developed resistance against most pesticide groups (Anonymous, 1988), resulting in severe economic losses in the affected crops. Pesticide resistance can also develop in natural enemies and has been found in all taxonomic groups (Croft and Strickler, 1983). The differences in the occurrence and the level of pesticide resistance in predators and parasitoids can be explained by the influence of the factors such as food limitation and differential susceptibility to the chemical.

Chemical resistance management strategies for pests comprise different approaches classified as management by moderation (low dosages, reduced number of applications), management by saturation (suppressing detoxification) and management by multiple attack (application of mixtures) (Georghiou, 1983). For IPM programmes additionally non-target effects on natural enemies have to be considered, which might not always correspond with the aforementioned strategies.

11.7. Future Aspects

Modern techniques used in greenhouses for pesticide application allow a low input of chemicals while achieving good coverage of the right part of the plant. Selective application can also direct the active ingredient to the right target, with lowered effect on beneficial organisms. However, it is important to know the undesired side effects of chemical use in greenhouses. The use of side effect data by advisory services or growers may lead to problems due to contradictory information about the effects of the same pesticides resulting from differences in test methods, different test laboratories carrying out the tests and the formulation of the pesticide used in different countries. Therefore, uniform labelling of the non-target effects of plant protection products already during the process of authorization as proposed in the European Plant Protection Legislation (EU-Directive 414/91, including all annexes) is desirable. The basic requirements to fulfil the legislative demands were formulated during the "Workshop of European Standard Characteristics of Beneficial Arthropod Testing" (Barrett *et al.*, 1994). Resulting from this workshop 11 different ring test groups for the standardization and harmonization of existing test methods and for the development of new test methods were formed. As an outcome of this joint initiative by governmental research centres, industry, commercial test laboratories and contributions from the European and Mediterranean Plant Protection Organization (EPPO), a harmonized labelling of plant protection products concerning the non-target effects is expected.

Other topics for the implementation of side-effect data into IPM practice still need to be addressed. Most of the data about side-effects of pesticides on beneficials is derived from laboratory tests or even higher test levels with only one application of the product. However, in practice, even when natural enemies are used against arthropod pests, chemical treatment can be necessary against fungal diseases. Often these fungicides have to be applied not once, but several times at certain intervals. These applications can lead to an accumulation of the product on the plants, affecting the beneficial organisms. This situation becomes more complicated when mixtures of different active ingredients are used.

Very few chemical pesticides are selective for natural enemies. Improvements in the compatibility of beneficial organisms with pesticide application by selecting beneficials with some resistance towards chemical pesticides have been attempted, but this is often a cumbersome procedure as the pesticides used may change quickly. Besides the degree of resistance, its stability and its possible influence on the fitness of the tolerant organisms are features that must be assessed before the selected organisms can be used in pest or disease control. For phytoseiids development of pesticide resistance against several insecticide groups, acaricides and fungicides, and even against sulphur has been extensively described (Fournier et al., 1985; Croft and van den Baan, 1988). Alternatively, pesticides are applied spatially to selected areas or in frequencies which reduce the target pest to a sufficient extent, but minimize harm to natural enemies and thus allow a combination or synergized effect of both the chemical and the biological controls (Theiling and Croft, 1988; Zhang and Sanderson, 1990).

Another important topic in the assessment of side-effects is examining whether natural pesticides or natural enemies themselves affect beneficial organisms, as reported in studies of the impact of entomopathogenic nematodes on non-target organisms (Bathon, 1996). Fransen and van Lenteren (1993) could not find detrimental effects of the entomopathogenic fungi *Aschersonia aleyrodis* Webber on the parasitoid *E. formosa*, while Sterk et al. (1995) observed no effect of a commercial strain of *Paecilomyces fumosoroseus* (Wize) Brown & Smith on *P. persimilis*, *E. formosa* and *Orius insidiosus* (Say). However, Pavlyushin (1996) detected direct and sublethal effects of entomopathogenic fungi on Chrysopids in the laboratory.

The present status of resistance of pests or pathogens in greenhouses is often unknown; growers tend to apply excess amounts of chemical, and control is not achieved. The development of tools for monitoring resistance should facilitate the assessment of different management options.

References

Anonymous (1988) Meeting-Report ad hoc Panel-Meeting on *Frankliniella occidentalis* – Its biology and control, EPPO Panel Information, Paris, 25–26 January 1988.

Baath, E. (1991) Tolerance of copper by entomogenous fungi and the use of copper-amended media for isolation of entomogenous fungi from soil, *Mycological Research* **95**(9), 1140–1142.

Barrett, K.L., Grandy, N., Harrison, E.G., Hassan, S.A. and Oomen, P. (eds.) (1994) *SETAC Guidance Document on Regulatory Testing Procedures for Pesticides with Non-Target Arthropods,* ESCORT (European Standard Characteristics of Beneficials Regulatory Testing) Workshop, Wageningen, The Netherlands, 28–30 March 1994, Society of Environmental Toxicology and Chemistry, Saffron Walden.

Bartlett, B.R. (1964) Toxicity of some pesticides to eggs, larvae and adults of the green lacewing, *Chrysopa carnea, J. of Economic Entomology* **57**, 366–369.

Bathon, H. (1996) Impact of entomopathogenic nematodes on non-target hosts, *Biocontrol Science and Technology* **6**, 421–434.

Biobest (1998) *Side Effects of Pesticides on Bumblebees and Beneficials,* Biobest Biological Systems Information Booklet, Biobest, Westerloo, Belgium.

Blümel, S., Bakker, F. and Grove, A. (1993) Evaluation of different methods to assess the side effects of of pesticides on *Phytoseiulus persimilis* A.H., *Experimental and Applied Acarology* **17**(3), 161–169.

Blümel, S. and Stolz, M. (1993) Investigations on the effect of insect growth regulators and inhibitors on the predatory mite *Phytoseiulus persimilis* A.H. with particular emphasis on cyromazine, *J. of Plant Diseases and Protection* **100**(2), 150–154.

Cahill, M., Byrne, F.J., Denholm, I., Devonshire, A.L and Gorman, K.J. (1994) Insecticide resistance in *Bemisia tabaci,* in Extended Summaries SCI Pesticides Group Symposium Management of *Bemisia tabaci, Pesticide Science* **42**, 135–142.

Celli, S., Bortolotti, L., Nanni, C., Porrini, C. and Sbrenna, G. (1997) Effects of the IGR fenoxycarb on eggs and larvae of *Chrysoperla carnea* (Neuroptera, Chrysopidae). Laboratory test, in P.T. Haskell and P.K. McEwen (eds.), *New studies in Ecotoxicology,* The Welsh Pest Management Forum, Cardiff, pp. 45–49.

Cohen, Y., Reuveni, M. and Samoucha, Y. (1983) Competition between metalaxyl-resistant and -sensitive strains of *Pseudoperonospora cubensis* on cucumber plants, *Phytopathology* **73**, 1516–1520.

Croft, B.A. (1990) *Arthropod Biological Control Agents and Pesticides,* John Wiley & Sons, New York.

Croft, B.A. and Strickler, K. (1983) Natural enemy resistance to pesticides: Documentation, characterization, theory and application, in G.P. Georghiou and T. Saito (eds.), *Pest resistance to pesticides,* Plenum Press, New York, pp. 669–702.

Croft, B.A. and van de Baan, H.E. (1988) Ecological and genetic factors influencing evolution of pesticide resistance in tetranychid and phytoseiid mites, *Experimental and Applied Acarology* **4**, 277–300.

Elad, Y. and Shtienberg, D. (1995) *Botrytis cinerea* in greenhouse vegetables: Chemical, cultural, physiological and biological controls and their integration, *Integrated Pest Management Reviews* **1**, 15–29.

Elad, Y., Yunis, H. and Katan, T. (1992) Multiple fungicide resistance to benzimidazoles, dicarboximides and diethofencarb in field isolates of *Botrytis cinerea* in Israel, *Plant Pathology* **41**, 41–46.

Elzen, G.W. (1989) Sublethal effects of pesticides on beneficial parasitoids, in P. Jepson (ed.), *Pesticides and non-target invertebrates,* Incept, Wimborne Dorset, pp. 129–150.

Fournier, D., Pralavorio, M., Berge, J.B. and Cuany, A. (1985) Pesticide resistance in Phytoseiidae, in W. Helle and M.W. Sabelis (eds.), *World Crops Pests, Spider Mites, Their Biology, Natural Enemies and Control,* Vol. 1B, Elsevier Publishers, Amsterdam, pp. 423–432

Fransen, J.J. and van Lenteren, J.C. (1993) Host selection and survival of the parasitoid *Encarsia formosa* on greenhouse whitefly *Trialeurodes vaporariorum* in the presence of hosts infected with the fungus *Aschersonia aleyrodis, Entomologia Experimentalis et Applicata* **69**, 239–249.

Georghiou, G.P. (1983) Management of resistance in arthropods, in G.P. Georghiou and T. Saito (eds.), *Pest resistance to pesticides,* Plenum Press, New York, pp. 769–792.

Georghiou, G.P. and Mellon, R.B. (1983) Pesticide resistance in time and space, in G.P. Georghiou and T. Saito (eds.), *Pest Resistance to Pesticides,* Plenum Press, New York, pp. 1–47.

Goodwin, S. and Wellham, T.M. (1992) Comparison of dimethoate and methidathion tolerance in four strains of *Phytoseiulus persimilis* (Athias-Henriot) (Acarina: Phytoseiidae) in Australia, *Experimental and Applied Acarology* **16** (3), 255–261.

Gullino, M.L., Aloi, C. and Garibaldi, A. (1989) Influence of spray schedules on fungicide resistant populations of *Botrytis cinerea* Pers. on grapevine, *Netherlands J. of Plant Pathology* **95**(Suppl.1), 87–94.

Hassall, K.A. (1982) *The chemistry of pesticides,* Verlag Chemie, Weinheim, Basel.

Hassan, S.A. (1989) Testing methodology and the concept of the IOBC/WPRS working group, in P.C. Jepson (ed.), *Pesticides and Non-Target Invertebrates,* Incept, Wimborne Dorset, pp. 1–18.

Hassan, S.A., Albert, R., Bigler, F., Blaisinger, P., Bogenschütz, H., Boller, E., Brun, J., Chiverton, P., Edwards, P., Englert, W.D., Huang, P., Inglesfield, C., Naton, E., Oomen, P.A., Overmeer, W.P.J., Rieckmann, W., Samsoe-Petersen, L., Stäubli, A., Tuset, J.J., Viggiani, G. and Vanwetswinkel, G. (1987) Results of the third Joint Pesticide Testing Programme by the IOBC Working Group "Pesticides and Beneficial Organisms", *J. of Applied Entomology* 103, 92–107.

Hassan, S.A., Bigler, F., Blaisinger, P., Bogenschütz, H., Boller, E., Brun, J., Chiverton, P., Edwards, P., Mansour, F., Naton, E., Oomen, P.A., Overmeer, W.P.J., Polgar, L., Rieckmann, W., Samsoe-Petersen, L., Stäubli, A., Sterk, G., Tavares, K., Tuset, J.J., Viggiani, G. and Vivas, A.G. (1988) Results of the fourth Joint Pesticide Testing Programme carried out by the IOBC Working Group "Pesticides and Beneficial Organisms", *J. of Applied Entomology* 105, 321–329.

Hassan, S.A., Bigler, F., Bogenschütz, H., Boller, E., Brun, J., Calis, J.N.M., Chiverton, P., Coremans-Pelseneer, J., Duso, C., Grove, A., Helyer, N., Heimbach, U., Hokkanen, G.B., Lewis, H., Mansour, F., Moreth, L., Polgar, L., Samsoe-Petersen, L., Sauphanor, B., Stäubli, A., Sterk, G., van de Veire, M., Viggiani, G. and Vogt, H. (1994) Results of the sixth Joint Pesticide Testing Programme carried out by the IOBC Working Group "Pesticides and Beneficial Organisms", *Entomophaga* 39, 107–119.

Hassan, S.A., Bigler, F., Bogenschütz, H., Boller, E., Brun, J., Calis, J.N.M, Chiverton, P., Coremans-Pelseneer, J., Duso, C., Lewis, G.B., Mansour, F., Moreth, L., Oomen, P.A., Overmeer, W.P.J., Polgar, L., Rieckmann, W., Samsoe-Petersen, L., Stäubli, A., Sterk, G., Tavares, K., Tuset, J.J. and Viggiani, G. (1991) Results of the fifth Joint Pesticide Testing Programme carried out by the IOBC Working Group "Pesticides and Beneficial Organisms", *Entomophaga* 36, 55–67.

Hassan, S,A., Bigler, F., Bogenschütz, H., Brown, J.U., Firth, S.I., Huang, P., Ledieu, M.S., Naton, E., Oomen, P.A., Overmeer, W.P.J., Rieckmann, W., Samsoe-Petersen, L., Viggiani, G. and van Zon, A.Q. (1983) Results of the second Joint Pesticide Testing Programme by the IOBC/WPRS-Working Group "Pesticides and Beneficial Arthropods", *Zeitschrift fur Angewandte Entomologie* 95, 151–158.

Heitefuß, R. (1975) *Pflanzenschutz,* Georg Thieme Verlag, Stuttgart.

Hodek, I. (1973) Biology of Coccinellidae, Dr. W. Junk, The Hague, cited in D.J. Horn and R.W. Wadleigh (1993) Resistance of aphid natural enemies to insecticides, in A.K. Minks and P. Harrewijn (eds.) *World Crops Pests, Aphids, Their Biology, Natural Enemies and Control,* Vol. 2B, Elsevier Publishers, Amsterdam, pp. 337–347.

Horowitz, A.R., Forer, G. and Ishaaya, I. (1994) Managing resistance in *Bemisia tabaci* in Israel with emphasis on cotton, *Pesticide Science* 42, 113–122.

Huffaker, C.B. (1990) Effects of environmental factors on natural enemies of armoured scale insects, in D. Rosen (ed.), *World Crop Pests, Armoured Scale Insects, Their Biology, Natural Enemies and Control,* Vol. 4B, Elseviers Publishers, Amsterdam, pp. 205–220.

Hull, L.A. and Beers, E.H. (1985) Ecological selectivity: Modifying chemical control practices to preserve natural enemies, in M.A. Hoy and D.C. Herzog (eds.), *Biological Control in Agricultural IPM Systems,* Academic Press, Orlando, Fla., pp. 103–121.

Jones, W.A., Wolfenbarger, D.A. and Kirk, A.A. (1995) Response of adult parasitoids of *Bemisia tabaci* (Hom.: Aleyrodidae) to leaf residues of selected cotton insecticides, *Entomophaga* 40(2), 153–162.

Lingappa, S.S., Starks, K.J. and Eikenbary, R.D. (1972) Insecticidal effect on *Lysiphlebus testaceipes,* a parasite of greenbug at all three developmental stages, *Environmental Entomology* 1, 520–521, cited in D.J. Horn and R.W. Wadleigh (1993), in A.K. Minks and P. Harrewijn (eds.), *World Crops Pests, Aphids, Their Biology, Natural Enemies and Control,* Elsevier Publishers, Amsterdam, pp. 337–347.

Maise, S., Candolfi, M.P., Neumann, C. Vickus, P. and Mäder, P. (1997) A species comparative study: Sensitivity of *Aphidius rhopalosiphi, A. matricariae* and *A. colemani* (Hymenoptera: Aphidiidae) to Dimethoate 40EC under worst case laboratory conditions, in P.T. Haskell and P.K. McEwen (eds.), *New studies in Ecotoxicology,* The Welsh Pest Management Forum, Cardiff, pp. 45–49.

Matthews, G.A. (1992) *Pesticide Application Methods,* 2nd edn, Longman, London.

McCoy, C.W., Samson, R.A. and Boucias, D.C. (1988) Entomogenous fungi, in C.M. Ignoffo (ed.), *CRC Handbook of Natural Pesticides,* Vol. V, *Microbial Insecticides,* Part A, CRC Press, Boca Raton, Fla., pp. 151–237.

Oatman, E.R. and Kennedy, G.G. (1976) Methomyl induced outbreak of *Liriomyza sativae* on tomato, *J. of Economic Entmology* **69**(5), 667–668.

Pavlyushin, V.A. (1996) Effect of entomopathogenic fungi on entomophagous arthropods, *IOBC/WPRS-Bulletin* **19**(9), 247–249.

Poehling, H.M. (1989) Selective application strategies for insecticides in agricultural crops, in P.C. Jepson (ed.), *Pesticides and Non-Target Invertebrates*, Incept, Wimborne Dorset, pp. 151–176.

Pommer, E.H. and Lorenz, G. (1982) Resistance of *Botrytis cinerea* Pers. to dicarboximide fungicides – A literature review, *Crop Protection* **2**, 221–223.

Rathman, R.J., Johnson, M.W., Rosenheim, J.A., Tabashnik, B.E. and Purcell, M. (1992) Sexual differences in insecticide susceptibility and synergism with piperonyl butoxide in the leafminer parasitoid *Diglyphus begini* (Hymenoptera: Eulophidae), *J. of Economic Entomology* **85**(1), 15–20.

Reddy, K.B. and Bhat, P.K. (1993) Effect of endosulfan on the mealybug parasitoid *Leptomastix dactylopii* How, *J. of Coffee Research* **23**(1), 19–23.

Rovesti, L., Heinzpeter, E.W., Tagliente, F. and Deseo, K.V. (1988) Compatibility of pesticides with the entomopathogenic nematode *Heterorhabditis bacteriophora* Poinar (Nematoda: Hetereorhabditidae), *Nematologica* **34**, 462–476.

Schuler, T. (1991) *Verticillium lecanii* (Zimmermann) Viegas (Hyphomycetales: Moniliaceae): Geschichte, Systematik, Verbreitung, in T. Schuler, M. Hommes, H.P. Plate and G. Zimmermann (eds.), *Biologie und Anwendung im Pflanzenschutz, Mitteilungen aus der Biologischen Bundesanstalt für Land- und Forstwirtschaft Berlin-Dahlem* **269**, 118–125.

Schuster, D.J., Musgrave, C.A. and Jones, J.P. (1979) Vegetable leafminer and parasite emergence from tomato foliage sprayed with oxamyl, *J. of Economic Entomology* **72**, 208–210.

Scopes, N.E.A. and Biggerstaff, S.M. (1974) Progress towards integrated pest control on year round chrysanthemums, in *Proceedings of the 7th British Insecticide and Fungicide Conference*, Brighton, 19–22 November 1973, BCPC, London, pp. 227–234.

Sell, P. (1984) Wirkungen von Pflanzenschutzmitteln auf Leistungen der aphidophagen Larven von *Aphidoletes aphidimyza* (Rond.) (Diptera: Cecidomyiidae), *Zeitschrift für Angewandte Entomologie* **98**(2), 174–184.

Shtienberg, D. and Elad, Y. (1997) Incorporation of weather forecasting in integrated biological-chemical management of *Botrytis cinerea*, *Phytopathology* **87**, 332–340.

Sterk, G., Bolckmans, K., van de Veire, M., Sels, B. and Stepman, W. (1995) Side-effects of the microbial insecticide PreFeRal (*Paecilomyces fumosoroseus*, strain Apopka 97) on different species of beneficial arthropods, *Mededelingen van de Faculteit Landbouwkundige en Toegepaste Biologische Wetenschappen, Universiteit Gent* **60**(3a), 719–724.

Sterk, G., Hassan, S.A., Bakker, F., Bigler, F., Blümel, S., Bogenschütz, H., Boller, E., Bromand, B., Brun, J., Calis, J.M.N., Coremans-Pelseneer, J., Duspo, C., Grove, A., Heimbach, U., Hokkanen, H., Jacas, J., Lewis, G., Mansour, F., Moreth, L., Plogar, L., Rovesti, L., Samsoe-Petersen, L., Sauphanor, B., Stäubli, A., Tuset, J.J., Vainio, A., van de Veire, M., Viggiani, G., Vinuela, E., Vivaus, A.G. and Vogt, H. (1998) Results of the seventh Joint Pesticide Testing Programme carried out by the IOBC/WPRS Working Group "Pesticides and Beneficial Organisms", *Biocontrol* (in press).

Theiling, K.M. and Croft, B.A. (1988) Pesticide side-effects on arthropod natural enemies: A data-base summary, *Agriculture, Ecosystems and Environment* **21**, 191–218.

van de Veire, M. (1995) *Integrated Pest Management in Glasshouse Tomatoes, Sweet Peppers and Cucumbers in Belgium*, PhD Thesis, University of Gent.

van Zon, A.Q. and Wysoki, M. (1978) The effect of some fungicides on *Phytoseiulus persimilis* (Acarina: Phytoseiidae), *Entomophaga* **23**(4), 371–378.

Vogt, H. (1992) Investigations on the side effects of insecticides and acaricides on *Chrysoperla carnea* Stph. (Neuroptera, Chrysopidae), *Mededelingen van de Faculteit Landbouwwetenschappen, Rijksuniversiteit Gent* **57**(2b), 559–567.

Warner, L.A. and Croft, B.A. (1982), Toxicities of azinphosmethyl and selected orchard pesticides to an aphid predator, *Aphidoletes aphidimyza*, *J. of Economic Entomology* **75**(3), 410–415.

Wilding, N. (1972) The effect of systemic fungicides on the aphid pathogen *Cephalosporium aphidicola*, *Plant Pathology* **21**, 137–139.

Yunis, H. and Elad, Y. (1989) Survival of dicarboximide-resistant strains of *Botrytis cinerea* in plant debris during summer in Israel, *Phytoparasitica* **17**, 13–21.

Zchori-Fein, E., Roush, R.T. and Sanderson, J.P. (1994) Potential for integration of biological and chemical control of greenhouse whitefly using *Encarsia formosa* and abamectin, *Environmental Entomology* **23**(5), 1277–1282.

Zeleny, J., Vostrel, J., Ruzicka, Z. and Kalushkov, P.K. (1988) Impact of various pesticides on aphidophagous coccinellidae, in E. Niemczyk and A.F.G. Dixon (eds.), *Ecology and Effectiveness of Aphidophaga*, SPB Academic Publishing, The Hague, pp. 327–332.

Zhang, Z.Q. and Sanderson, J.P. (1990) Relative toxicity of abamectin to the predatory mite *Phytoseiulus persimilis* (Acari: Phytoseiidae) and twospotted spider mite (Acari: Tetranychidae), *J. of Economic Entomology* **83**(5), 1783–1790.

Zimmerman, R.J. and Cranshaw, W.S. (1990) Compatibility of three entomogenous nematodes (Rhabditia) in aequous solutions of pesticides used in trufgrass maintenance, *J. of Economic Entomology* **83**(1), 97–100.

DECISION TOOLS FOR INTEGRATED PEST MANAGEMENT
J. Leslie Shipp and Norman D. Clarke

12.1. Introduction

Greenhouse pest and disease problems are often the result of complex interactions among many variables such as greenhouse environment, nutrition, production practices, growing media, other pest and disease outbreaks, economics and environmental and social concerns. As a result, managing or preventing pest and disease outbreaks requires an interdisciplinary approach, which will vary according to the problem. Greenhouse industry is a very technologically-advanced agro-food industry with computerized climate control and fertigation systems in widespread commercial use. These systems offer precise and versatile tools for controlling and manipulating the greenhouse and plant environment, but also affect pest and disease outbreak dynamics. Biological control agents are commercially-available for most of the major insect and mite pests and cultural control measures are also viable management strategies to chemical control, especially for disease prevention (Clarke *et al.*, 1994a). With all these management strategies and other variables that can impact upon IPM, the grower can use as much assistance as possible to collect, collate, understand and integrate, where necessary, the information needed to choose the most viable solution for the problem at that point in time. The purpose of this chapter is to provide an overview of the decision-making process and decision tools as they apply to IPM of greenhouse crops.

12.2. Decision-Making Process

Decision-making is the process of selecting and implementing an action with the intention of producing a favourable outcome. The quality of decisions can be enhanced by using a structured, analytic methodology to decision-making. Analytic decision-making is based on logic and considers all available data and alternatives. The structured decision-making process consists of five basic steps: problem recognition and definition, alternative generation, alternative evaluation, alternative selection and decision implementation (Souder, 1980; Tregoe and Kepner, 1981). These steps do not necessarily follow one another sequentially without deviation, but often decision-makers must backtrack and repeat some steps.

Problem recognition and definition begins with recognition of a deviation between actual conditions and established standards or desired conditions. A clear, concise problem statement, defining what the variance is and is not, when and where the variance occurs, etc., should be developed. The problem statement must go beyond the symptoms and identify the true cause of the problem. For example, if *Botrytis cinerea*

R. Albajes et al. (eds.), Integrated Pest and Disease Management in Greenhouse Crops, 168-182.
© 1999 *Kluwer Academic Publishers. Printed in the Netherlands.*

Pers.:Fr. infects your crop and you only apply fungicides, you are treating the symptom (*Botrytis*) and are doing nothing to alleviate the cause of the infection. The cause of the infection may be poor sanitation, inadequate climate control or excessive plant stress.

Alternative generation is a creative process whereby alternative solutions are identified. Brainstorming at this time can result in some very novel ideas and also some non-feasible suggestions. Not all suggestions may be used, but discussion of them may help improve upon the more feasible solutions.

Alternative evaluation involves setting goals to be achieved by solving the problem and quantifying each alternative in terms of its value, cost, risk and other decision criteria. Establishing specific and measurable goals assists the decision-maker in quantifying a problem. Most greenhouse growers have many goals, including maximizing profit. Other goals may include increased productivity, increased product quality and employee safety. Decision criteria are attributes of a solution that can be measured or estimated. These attributes are used to evaluate the different alternatives that are generated in step two. Decision criteria for selecting a pesticide may include cost, efficacy, compatibility with biological control agents, safety and days to harvest. Decision aids or tools, such as decision matrix, decision tree, linear programming, simulation models, expert systems and decision support systems, can be used to more fully understand the scope of the problem, the differences among alternatives, and the relative worth of each. [This is only a partial list of the many tools that are available for decision-making. For more information on other decision tools, such as game theory, linear regression, forecasting and network models, the reader is referred to an operations management book by Heizer and Render (1991).]

Based upon the evaluation, the alternative that best satisfies the goal(s) is selected. Numerous methods or decision rules have been suggested for selecting among alternatives (Souder, 1980; Montgomery, 1983) such as the dominance rule (choose A1 over A2 if A1 is better than A2 on at least one attribute and not worse than A2 on all other attributes), lexicographic rule (choose A1 over A2 if it is better than A2 on the most important attribute; if this requirement is not met, base the choice on the next important attribute) and addition of utilities rule (choose the alternative with the greatest sum of weighted values across all attributes). Further analysis of the selected alternative may be conducted to verify the decision and identify possible adverse consequences.

Sometimes the most challenging phase of decision-making is trying to implement the selected alternative. An implementation plan that specifies the barriers and obstacles to acceptance of the decision, and ways that these can be overcome, is as important as the decision itself.

12.3. Sources of Information for Decision-Making in IPM

When making IPM decisions, it is vital that the decision-maker search for information that will help solve the problem. The search for information can help in all steps of the decision process. It may reveal facts about the situation that will result in redefinition of the problem. Valuable insight into the different alternatives and data by which they can be evaluated can be provided. The information search can also reveal how the selected alternative may be implemented.

For greenhouse growers, we only found one survey (van Lenteren, 1990) that was related to sources of information for decision-making. This survey listed growers' journals and study groups as the most important sources for Dutch growers. Surveys of other types of agricultural producers found that significant sources of information include a grower's own experience and records, extension publications and bulletins, extension specialists, grower magazines, universities, colleges and research institutions, other growers, private industry salesmen (chemical, equipment, etc.) and independent consultants (Blackburn et al., 1983; Carlson and Guenthner, 1989; Ortmann et al., 1993; Buchner et al., 1996). Greenhouse growers can and do also obtain their decision-making information from similar sources (van Lenteren, 1990).

A grower's own experience and records can be one of the most important sources of decision-making information. If a pest problem reoccurs, a grower can use their records to see how well previously implemented alternatives performed. Records can also be used to obtain evaluation data such as cost and effectiveness of chemicals and biologicals.

Extension publications can provide general recommendations for IPM in greenhouse crops (Anonymous, 1996), while detailed information on specific pests and diseases can be obtained from books (Gerling, 1990; Jarvis, 1992) or other publications (Jarvis and McKeen, 1991; Malais and Ravensburg, 1992). In addition, every grower is advised to own a good pest and disease identification reference (Hussey and Scopes, 1985; Powell and Lindquist, 1992; Howard et al., 1994) and a nutritional disorder identification reference (Winsor and Adams, 1987). These references can assist growers in quickly identifying crop disorders. Commercially produced grower magazines are widely used by growers and often report on new ideas and techniques for IPM.

Government extension advisors have traditionally been the main source of pest and disease management information for growers in many countries. Recently, government cutbacks in several countries have severely reduced the number and availability of the extension advisors. As a result, there has been an increase in the number of private consultants in the greenhouse industry. Sales representatives can also be a valuable source of information, providing advice on the use of their products. Other greenhouse growers, especially study groups, are also an important source of information. Association with other growers allows one the opportunity to obtain, discuss and compare information on new IPM practices and innovations.

12.3.1. THE INTERNET

A new source of IPM information is the Internet. The Internet has many features that can be used to assist in the management of greenhouse crops. One of the most widely used features is electronic mail (e-mail). Provided one knows the address, one can send and receive messages from anyone connected to the Internet including other growers, extension advisors and researchers. Another useful tool is the browser. A browser is an application that knows how to interpret and display documents that it finds on the Internet. Most browsers can access other Internet services including Anonymous FTP (File Transfer Protocol for downloading files), e-mail and news groups.

One way for growers to use the Internet is to find information relating to pest and disease management. Many useful sites can be found on the Internet that are related to horticulture and greenhouse management. Extension/research sites provide many extension documents and information on current research projects for growers. In addition to product information and pricing, commercial sites also provide lots of related information. All sites contain good links to other related sites.

When using information from the Internet, however, one should exercise caution. Anyone can put up a web site and publish anything on the site. Therefore, be aware of the source and quality of the information. Unlike books and journal articles, web documents are not peer reviewed so there is no guarantee that the information is accurate. As well, the main purpose of commercial sites is to advertise their products.

Another consideration when using the Internet for decision support is finding the relevant information that one requires. Searching the Internet using a search engine such as Yahoo <http://www.yahoo.com/> can generate thousands of matches. For example, a search for IPM generated 3714 matching sites. Determining which sites are truly helpful can take a considerable amount of time. This problem can be alleviated somewhat by carefully choosing keywords to search. Searching for IPM and greenhouse reduced the number of matching sites to 81. Another option is to find and search topic specific databases. The web site <http://ag.arizona.edu/Ext/MASTER-GARDENER/> is a searchable database comprising over 1000 horticultural and agricultural web sites.

12.4. Application of Decision Tools for IPM

Decision tools are techniques for modelling actual systems and are thus simplifications of actual conditions. They have become widely accepted for several reasons. Decision tools or models are less expensive and disruptive than experimenting with the actual systems and can force the decision-maker to analyse the problem in a logical and systematic manner. Decision tools allow managers to ask "what if" questions and evaluate different scenarios. They can also reduce the time needed to make a decision. On the other hand, models can be expensive and time consuming to develop. The results may be misused and misunderstood because of the complexity of models and because models may use assumptions that oversimplify actual systems.

As stated earlier, many tools are available to assist growers in making IPM decisions in the greenhouse. Practical applications of many decision tools in IPM are reviewed in Norton and Mumford (1993). Although none of the many examples presented are specific to greenhouse IPM, the techniques presented can be applied to many greenhouse IPM problems. The following sections discuss the application of decision-making tools to IPM in greenhouses.

12.4.1. DECISION TABLES AND TREES

Decision tables and trees are simple yet powerful tools to assist in the decision-making process. These tools can be used to logically and systematically select among alternatives and the structure provided by these tools can give a valuable framework for further investigations.

Decision Matrix

The decision matrix is used to select among alternatives using the addition of utilities rule. Consider a hypothetical situation where one must select a fungicide from three different alternatives (Fig. 12.1). In this example, four evaluation attributes or criteria are established and weights of importance are assigned to each. These weights reflect the beliefs, concerns and experiences of the decision-maker. Each alternative is evaluated and graded on a 0 to 10 scale on how well it satisfies the criterion. The grade is multiplied by the weight and recorded. The alternative with the greatest sum of weighted values across all criteria (chemical 2) is selected. If the lexicographic rule had been used, then chemical 1 would be selected.

Criteria	Weight	Chemical 1	Chemical 2	Chemical 3
Cost	40	10 / 400	10 / 400	7 / 280
Effectiveness	25	10 / 250	9 / 225	10 / 250
Compatibility with biologicals	20	5 / 100	9 / 180	8 / 160
Days to harvest	15	5 / 75	8 / 120	5 / 75
Total	100	825	925	765

Rating scale, R
Excellent   = 9–10
Good   = 7–8
Fair   = 5–6
Poor   = 3–4
Unsatisfactory = 0–2

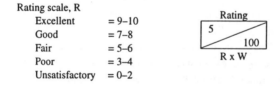

Figure 12.1. Decision matrix to select among fungicides.

Pay-off Matrix

A pay-off matrix helps the decision-maker economically evaluate alternatives. Pay-off matrices can be used both for decision-making under risk (where the decision-maker knows the probability of occurrence of the outcomes for each alternative) and decision-making under uncertainty (whether probabilities are unknown).

A possible pay-off matrix is presented for thrips control on sweet pepper in Ontario greenhouses under uncertainty (Table 12.1). The pay-off for each combination of alternative and state of nature (an occurrence or situation over which the decision-maker has little or no control) is included in the matrix. In this example, the states of nature are low, medium and high levels of thrips attack. The do nothing alternative shows the cost of damage caused by the three levels of thrips attack. The other alternatives include both the cost of the control strategy and its ability to reduce thrips levels and damage. The chemical alternative also includes an estimate of yield reduction resulting from crop injury.

TABLE 12.1. Pay-off matrix for decision-making under uncertainty

Strategy	Level of thrips infestation			Maximum outcome in row	Minimum outcome in row	Row average
	Low	Medium	High			
Do nothing	-6,000[1]	-18,000	-60,000	**-6,000[2]**	-60,000	-28,000
Chemical control	-32,500	-36,200	-49,500	-32,500	-49,500	-39,400
Biological control	-7,000	-16,000	-32,000	-7,000	**-32,000**	**-18,333**
				Maximax↑	Maximin↑	Equally likely↑

[1]Expected loss (US$/ha) for the different control strategies (control cost + damage loss)
[2]Numbers in bold indicate the preferred strategies

With decision-making under uncertainty, the decision-maker can use three different rules for selecting among the strategies. The maximax (optimistic) rule selects the alternative (do nothing) that maximizes the maximum outcome for every alternative. The maximin (pessimistic) rule selects the alternative (biological control) that maximizes the minimum outcome for every alternative. The equally likely rule finds the alternative (biological control) with the highest average outcome and assumes that each state of nature is equally likely to occur.

With a situation where a grower has kept detailed records of thrips levels in the greenhouse, the probabilities of thrips attacks can be calculated. A pay-off matrix (Table 12.2) can be developed for this situation where the decision is being made under risk. The expected monetary value (EMV) for alternative i is:

$$EMV(i) = \sum_{j=1}^{n} \$_{if} * p_j$$

where n is the total number of outcomes, $\$_{ij}$ is the payoff of alternative i for outcome j, and p_j is the probability of outcome j.

A risk-neutral grower, who is unconcerned with year to year variations in outcomes, would choose biological control, which has the highest EMV (in our example, the lowest crop loss). Most growers are more likely to be risk-adverse and choose a strategy that gives acceptable outcomes at high pest levels. In this case, an extremely risk-adverse grower would also choose biological control, which has the best outcome under the worst conditions. Similarly, a risk taker may choose to do nothing, which has the best outcome under the best conditions. Although pay-off matrices help to economically select among alternatives, they do not allow for non-economical criteria to be considered (such as compatibility with *Bombus* spp. pollinators). If a cost can be determined for these criteria, then they should be included in the analysis.

TABLE 12.2. Pay-off matrix for decision-making under risk

Strategy	Probability of thrips levels			EMV
	Low (0.2)	Medium (0.5)	High (0.3)	
Do nothing	−6,000[1]	−18,000	−60,000	−28,200
Chemical control	−32,500	−36,200	−49,500	−39,450
Biological control	−7,000	−16,000	−32,000	**−19,000**[2]

[1]Expected loss (US$/ha) for the different control strategies
[2]Number in bold indicates the preferred strategy

Decision Trees

Many problems consist of sequential decisions over time. When more than one set of decisions are necessary, a decision tree is appropriate. A decision tree is a graphic display of the decision process which indicates decision alternatives, states of nature and their respective probabilities, and pay-offs for each combination of alternative and state of nature (Heizer and Render, 1991).

A decision tree is shown for powdery mildew management in Fig. 12.2 (state of nature probabilities and pay-offs are not included). Note that as the crop season progresses, the number of options decreases. If the probabilities of powdery mildew occurrence and the pay-offs are added to the tree, the EMV can be calculated for each branch and the best decision determined (Heizer and Render, 1991). Even if probabilities and pay-offs are not known, decision trees are still very useful by laying out all possible options and providing a framework for deciding which options and strategies need further investigation.

12.4.2. DATABASE SYSTEMS

Database systems (DBS) consist of a collection of interrelated data and a set of application programs to access and manipulate the data. The different data items are stored in related files or tables. The application programs usually provide functions to enter, edit, browse, query and analysis the data.

DBS can help solve pest and disease management problems in several ways. First, the development of a DBS can help to better organize and understand the problem. One of the first steps in developing a DBS is to develop a data model. The data model identifies the paths of information flow, specific data items and relationships among data items. DBS can also assist decision-making by storing detailed records of past pest and disease management strategies, along with the outcomes and costs of these strategies. These records can help the decision-maker select among different strategies based upon past results.

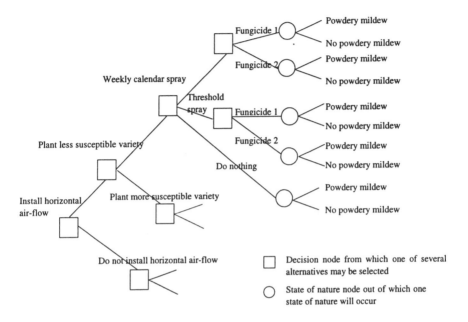

Figure 12.2. Decision tree for powdery mildew management.

There are many commercial greenhouse cost accounting and financial management DBS available. These packages usually provide facilities to record costs (including pest control) and sales throughout the cropping season (Brumfield, 1992). These DBS can assist in tracking pest management costs in the greenhouse.

A DBS for greenhouse pest surveillance, Emerald ICM, is commercially available (Van Vliet Automation Ltd, 1996). Pest survey data is collected with a hand-held computer in the greenhouse and uploaded to a personal computer. The data is used to generate colour maps of pests and their severity over time. Applications of pesticides, fungicides and biologicals are also recorded. Emerald ICM allows the grower to monitor the progress of pest and predator movement and analyse the effectiveness of different control strategies.

12.4.3. SIMULATION MODELS

A model is a description of a system. Models may be scaled physical objects, mathematical equations and relations, or graphical representations of actual systems. For purposes of this discussion, a simulation model is a mathematical-logical representation of a system which can be exercised in an experimental fashion on a digital computer (Pritsker, 1984). In terms of decision-making, simulation models allow us to examine different alternatives and how these alternatives perform under different conditions.

Simulation models have been used to study various field pests (Rabbinge *et al.*, 1989; Goodenough and McKinion, 1992). Pest population growth, fungal passive dispersal, insect active dispersal and predator-prey interactions have been simulated (Rabbinge *et al.*, 1989). Pest systems have also been modelled in the greenhouse environment.

Nachman (1991) simulated the dispersal of two-spotted spider mite (*Tetranychus urticae* Koch) and its predator *Phytoseiulus persimilis* Athias-Henriot in a greenhouse cucumber crop. Spider mite oviposition rate, death rate and emigration rate to other plants are dependent on the health of plants. The birth, death and emigration rates of the predator are linked to the predation rate. The simulation model was used to study fluctuations in overall population densities within the greenhouse.

Biological control of the leafminer species *Liriomyza trifolii* (Burgess) in greenhouse-grown chrysanthemums (Heinz *et al.*, 1993) and *Liriomyza bryoniae* (Kaltenbach) in greenhouse tomato (Boot *et al.*, 1992) by the parasitoid *Diglyphus isaea* (Walker) have been simulated. Heinz *et al.* (1993) assumed a constant greenhouse temperature of 27°C and that the population dynamics of the leafminer are independent of the quality of host plant. Simulation results indicated that successful biological control was unlikely when parasitoid releases are initiated later than 14 d after planting regardless of the release rate. Using a different approach, Boot *et al.* (1992) used ambient temperature and tomato leaf nitrogen content to determine the population dynamics of *L. bryoniae*. The timing and growth of leafminer generations were simulated and the results validated with greenhouse experiments, although no practical strategies for parasitoid were given. These two models could be used to explore different strategies for biological control of leafminers.

Disease infection and progression (Jarvis, 1992) and arthropod pest populations (Minkenberg and Ottenheim, 1990) are dependant upon plant nutrition. Crop growth simulation models, such as those developed for greenhouse tomato (Dayan *et al.*, 1993) and cucumber (Marcelis, 1994) could supply input parameters to pest population models. A feedback loop could potentially predict crop yield reductions due to the pest. Greenhouse microclimate also affects disease development (Jarvis, 1992) and insect population dynamics (Minkenberg and Helderman, 1990; Shipp and Gillespie, 1993). Microclimate models that simulate the climate within the crop canopy (Goudriaan, 1989; Yang, 1995) could be combined with disease and pest models to investigate climate control strategies for pest and disease management.

Currently, pest management simulation models are being used at the research level to better understand interactions between pests and their control agents. An example is the simulation model developed by van Roermund *et al.* (1997) to evaluate release strategies for the parasitoid, *Encarsia formosa* Gahan, for control of greenhouse whitefly, *Trialeurodes vaporariorum* (Westwood) on greenhouse crops. As personal computers become more powerful, we envision simulation models being used by growers to evaluate pest management strategies. However, before this happens more research needs to be done to develop models for other pests and diseases. As well, these models must be validated and in a format that non-modellers can understand how to operate the models and interpret the outputs. Model validation can be difficult due to inadequate experimental data, inappropriate assumptions and lack of knowledge regarding some of the physical processes being modelled. Despite these problems, simulation models certainly have a lot of potential for analysing various pest management strategies and understanding interactions between the pest, its biological control agents, the crop and the crop microclimate.

12.4.4. EXPERT SYSTEMS

Expert systems (ES) are computer programs that emulate the decision-making ability of a human expert. ES contain knowledge in one specific problem area or domain as opposed to knowledge about general problem-solving techniques. ES usually consist of a set of rules that were obtained from an expert to solve a particular problem, and an inference engine that decides which rules to execute. In terms of decision-making, ES can be used as tools for summarizing information and knowledge, selecting among alternative solutions, exploring and evaluating alternative scenarios, assessing risks, diagnosing problems, outlining approaches to problem solving and teaching non-experts the problem-solving approaches of experts (Holt, 1989).

ES technology has been applied to greenhouse pest management. The most common type of application is the diagnosis of crop diseases and pests. Several ES have been developed to identify greenhouse tomato disorders and recommend possible control actions for the identified disorder (Blancard et al., 1985; Gauy and Gauthier, 1991). With these ES, the user is prompted to enter information about the symptoms displayed by the tomato plant. The ES then uses expert rules to match the observed symptoms with a disorder.

Boulard et al. (1991) developed an ES to determine the climatic setpoints to control the climate for greenhouse tomato. In addition to information on the outside weather conditions and the current greenhouse climate, the ES also incorporated expert rules on climate and tomato diseases and physiological aspects. This system attempted to optimize conditions for energy use, crop growth and disease prevention and control. In a separate research effort and using a different approach, Manera et al. (1991) developed an ES with similar objectives for greenhouse production and pest management under Mediterranean conditions.

Although not specific to greenhouses, other pest management-related ES have been published. Logan (1988) developed an expert system to automatically assemble a model describing insect population phenology. The program offers time savings and compares well with a human expert. Messing et al. (1989) describe NERISK, an expert system that assesses the impact of pesticides on beneficial arthropod predators and parasitoids in agricultural systems. Similar ES could be developed and be useful for greenhouse pest management.

Although ES are very good at assisting in the decision-making process, certain issues should be considered before undertaking the development of one. First, it takes considerable time and resources to complete an ES. The time and commitment of an expert(s) is required as well as a knowledge engineer (an individual who extracts, organizes and programs the knowledge of an expert). The intended user of the ES must be consulted and involved during the development. Ongoing maintenance is required (McClure, 1993) as the knowledge contained in the ES may become outdated. These and other issues relating to ES development are reviewed by Clarke et al. (1994b).

12.4.5. DECISION SUPPORT SYSTEMS

Technically, any aid that assists a decision-maker could be defined as a decision support system (DSS). However, for this discussion we will consider DSS as computer programs that help decision-makers solve problems through direct interaction with data and models (Sprague and Watson, 1989). DSS usually use a combination of decision tools including ES, database systems, simulation and other computer models.

DSS have been used to solve pest management problems for field crops such as cotton (Goodell et al., 1990) and apples (Travis et al., 1992). The cotton DSS uses a cotton crop simulation model and expert advice on pests, diseases and weeds to provide recommendations on the timing of irrigation, fertilizer and pesticides. In the greenhouse, Jones et al. (1989) described a DSS where crop models were combined with an ES to choose optimal environmental setpoints for greenhouse tomato. The ES contained a knowledge base for variables that have not been well modelled, such as the length of time that humidity may remain high without a disease outbreak.

BOTMAN (Shtienberg and Elad, 1997) makes decisions concerning whether to spray the biological control agent Trichodex (developed from an isolate of *Trichoderma harzianum* Rifai T39) or fungicides for integrated biological and chemical control of *B. cinerea* in non-heated greenhouse vegetable production. BOTMAN uses weather forecasts, past weather and a *B. cinerea* risk index to predict the severity of outbreaks of grey mould. Based upon the expected severity, application of the biological control agent or a fungicide is recommended. Results show that BOTMAN controlled *B. cinerea* as well as weekly fungicide applications while significantly reducing the number of fungicide applications. Compared to a strategy of weekly Trichodex applications, BOTMAN was also significantly better. Another system, GREENMAN, was developed to deal with other greenhouse diseases. It is based on criteria that are similar to BOTMAN and, likewise, controls diseases such as leaf mould [*Fulvia fulva* (Cooke) Cif.] and white mould [*Sclerotinia sclerotiorum* (Lib.) de Bary] (Elad and Shtienberg, 1997).

A DSS for integrated management of greenhouse vegetables [Harrow Greenhouse Manager (HGM)] has been developed at the Greenhouse and Processing Crops Research Centre, Harrow, Ontario, Canada. The HGM contains modules for the following: (i) expert diagnosis of insect and mite pests of the crop; (ii) expert diagnosis of crop diseases and physiological disorders; (iii) IPM recommendations for pest, disease and physiological disorder control, and identification of conflicting recommendations; (iv) cost allocation, including pest control expenses, to crops; (v) record-keeping capabilities including crop production, labour, insect counts, disease occurrence and control measures that were implemented; (vi) tools to determine tank mixes for fertigation systems; (vii) analysis section to analyse relationships between any recorded entity (such as insect counts and crop yield); and (viii) climate data retrieval from climate control systems that are BACnet compatible. Currently, HGM contains the knowledge for greenhouse cucumber and tomato crops.

The approach in the development of the HGM was to provide a framework for integrated crop management (ICM) (Clarke *et al.*, 1994a). ICM is a multidisciplinary approach that integrates pest and disease protection strategies with routine cultural practices and environmental and fertigation regimes into a common decision-making process. It is not acceptable to manage one component of the greenhouse in isolation since the component can potentially affect all other aspects of the greenhouse crop.

DSS have a lot of potential in greenhouse pest management, particularly if greenhouse climate is integrated with control strategies. To utilize climate in controlling diseases and insects, DSS will need to control the microclimate at the leaf surface. DSS will need to integrate models that predict the microclimate at the plant surface from spatially averaged climate data with crop and pest simulation models. Computerized DSS will be required at the grower level to enable growers to interpret all the information necessary for ICM.

12.5. Conclusions

Decision-making is a very important part of greenhouse pest management. However, it is becoming more complicated and demanding in an industry that rapidly changes yearly. The grower can no longer rely on the old tools for information gathering and decision-making. Greenhouse operations are often 2–10 ha in size and must be operated more like a business corporation rather than a family-owned operation.

The Internet can provide a readily accessible and up-to-date source of information that can link a grower to current technology that is being used throughout the world. In the future, databases, DSS and other decision tools are going to become the main method for assisting the grower in making decisions as these tools can handle large data sets in an organized fashion and quickly form conclusions or solve problems.

Interpreting and managing technical information for decision-making without computers will be beyond the means of individual growers. First, the quantity of data required is large and the management of this data is impossible without computer systems. Computerized climate-control systems can quickly generate megabytes of data. Add crop production data, fertigation records, pest count records, etc. and the data quickly becomes unmanageable without computer technology. Second, the relationships between the various crop production factors are complex. For example, greenhouse climate can affect the effectiveness of entomopathogens, but, at the same time, provide conditions that are conducive to a disease outbreak, such as *B. cinerea*. In this case, climate directly affects the crop, a biological control agent and the epidemiology of a plant pathogen. Expert knowledge or simulation models can help improve our understanding of how all these factors interact.

Although the systems developed to date are useful for providing solutions to greenhouse crop management problems, the technology is still a long way from a DSS that provides true ICM strategies. To meet this goal, an improved understanding of the response of crops, pests, pathogens and biological controls to climate is required. Also those responses need to be mathematically or heuristically modelled and incorporated into DSS. Finally, DSS have to be validated in commercial greenhouses under actual production conditions.

All of the issues relating to database systems, simulation models and expert systems apply to DSS. In addition, we have found that the user interface and data entry procedures play a much bigger role with DSS. With pest control, future control actions are dependent upon previously implemented controls. For example, using *Encarsia* for whitefly control may limit the chemicals available for control of grey mould. However, the HGM does not know *Encarsia* is in the greenhouse unless the grower has recorded it in the database. Therefore, user interface and data entry procedures must be structured to streamline the time that it takes for the grower to enter the required data. Growers are showing a strong interest in databases and DSS and are beginning to incorporate them into their daily operational decision-making processes.

References

Anonymous (1996) *Pest Management Recommendations for Ontario Greenhouse Crops*, Publication 365, Ontario Ministry of Agriculture, Food and Rural Affairs, Toronto.

Blackburn, D.J., Young, W.S., Sanderson, L. and Pletsch, D.H. (1983) Farm information sources important to Ontario farmers, Research Report AEEE/83/1, School of Agricultural Economics and Extension Education, University of Guelph.

Blancard, D., Bonnet, A. and Coléno, A. (1985) TOM, un système expert en maladies des tomates, *Revue Horticole* **261**, 7–14.

Boot, W.J., Minkenberg, O.P.J.M., Rabbinge, R. and De Moed, G.H. (1992) Biological control of the leafminer *Liriomyza bryoniae* by seasonal inoculative releases of *Diglyphus isaea*: Simulation of a parasitoid-host system, *Netherlands J. of Plant Pathology* **98**, 203–212.

Boulard, T., Jeannequin, B. and Martin-Clouaire, R. (1991) Analysis of knowledge involved in greenhouse climate management – Application to the determination of daily setpoints for a tomato crop, in *Mathematical and Control Applications in Agriculture and Horticulture*, Proceedings of the IFAC/ISHS workshop, Pergamon Press, Oxford, pp. 265–270.

Brumfield, R.G. (1992) Greenhouse cost accounting: A computer program for making management decisions, *HortTechnology* **2**, 420–424.

Buchner, R.P., Grieshop, J.I., Connell, J.H., Krueger, W.H., Olson, W.H., Hasey, J.K., Pickel, C., Edstrom, J. and Yoshikawa, F.T. (1996) Growers prefer personal delivery of UC information, *California Agriculture* **50**, 20–25.

Carlson, J.E. and Guenthner, J.F. (1989) The information patterns of Idaho potato growers, *American Potato J.* **66**, 471–487.

Clarke, N.D., Shipp, J.L., Jarvis, W.R., Papadopoulos, A.P. and Jewett, T.J. (1994a) Integrated management of greenhouse crops – A conceptual and potentially practical model, *HortScience* **29**, 846–849.

Clarke, N.D., Shipp, J.L., Papadopoulos, A.P., Jarvis, W.R. and Jewett, T.J. (1994b) A knowledge-based system for integrated management of greenhouse cucumber, ASAE Paper No. 943525, American Society of Agricultural Engineers, St Joseph, Mich.

Dayan, E., van Keulen, H., Jones, J.W., Zipori, I., Shmuel, D. and Challa, H. (1993) Development, calibration and validation of a greenhouse tomato growth model: I Description of the model, *Agricultural Systems* **43**, 145–163.

Elad, Y. and Shtienberg, D. (1997) Integrated management of foliar diseases in greenhouse vegetables according to principles of a decision support system – GREENMAN, *IOBC/WPRS Bulletin* **20**(4), 71–76.

Gauy, R. and Gauthier, L. (1991) Knowledge representation in a tomato disorders diagnosis system, *Computers and Electronics in Agriculture* **6**, 21–32.

Gerling, D. (1990) *Whiteflies: Their Bionomics, Pest Status and Management*, Intercept Ltd, Hants.

Goodell, P.B., Plant, R.E., Kerby, T.A., Strand, J.F., Wilson, L.T., Zelinski, L., Young, J.A., Corgett, A., Horrocks, R.D. and Vargas, R.N. (1990) CALEX/Cotton: An integrated expert system for cotton production and management, *California Agriculture* **44**, 18–21.

Goodenough, J.L. and McKinion, J.M. (eds.) (1992) *Basics of Insect Modeling*, ASAE Monograph 10, American Society of Agricultural Engineers, St Joseph, Mich.

Goudriaan, J. (1989) Simulation of micrometeorology of crops, some methods and their problems, and a few results, *Agricultural and Forest Meteorology* **47**, 239–258.

Heinz, K.M., Nunney, L. and Parrella, M.P. (1993) Toward predictable biological control of *Liriomyza trifolii* (Diptera: Agromyzidae) infesting greenhouse cut chrysanthemums, *Environmental Entomology* **22**, 1217–1233.

Heizer, J. and Render, B. (1991) *Production and Operations Management*, Allyn and Bacon, Needham, Mass.

Holt, D.A. (1989) The growing potential of expert systems in agriculture, in J.R. Barrett and D.D. Jones (eds.), *Knowledge Engineering in Agriculture*, ASAE Monograph 8, American Society of Agricultural Engineers, St Joseph, Mich., pp. 1–11.

Howard, R.J., Garland, J.A. and Seaman, W.L. (1994) *Diseases and Pests of Vegetable Crops in Canada*, Canadian Phytopathological Society and Entomological Society of Canada, Ottawa.

Hussey, N.W. and Scopes, N. (1985) *Biological Pest Control: The Glasshouse Experience*, Cornell University Press, Ithaca, New York.

Jarvis, W.R. (1992) *Managing Disease in Greenhouse Crops*, APS Press, St Paul, Minn.

Jarvis, W.R. and McKeen, C.D. (1991) *Tomato Diseases*, Publication 1479/E, Agriculture Canada, Harrow, Ontario.

Jones, P., Roy, B.L. and Jones, J.W. (1989) Coupling expert systems and models for the real-time control of plant environments, *Acta Horticulturae* **248**, 445–452.

Logan, J.A. (1988) Towards an expert system for development of pest simulation models, *Environmental Entomology* **17**, 359–376.

Malais, M. and Ravensberg, W.J. (1992) *Knowing and Recognizing: The Biology of Glasshouse Pests and their Natural Enemies*, Koppert B.V., Berkel en Rodenrijs.

Manera, C., Di Renzo, G.C. and Damiani, P. (1991) I sistemi esperti nella gestione delle serre, *Colture Protette* **3**, 154–160.

Marcelis, L.F.M. (1994) A simulation model for dry matter partitioning in cucumber, *Annals of Botany* **74**, 43–52.

McClure, J. (1993) Costs involved in the support and maintenance of the Penn State Apple Orchard Consultant, *AI Applications in Natural Resource Management* **7**, 54–55.

Messing, R.H., Croft, B.A. and Currans, K. (1989) Assessing pesticide risk to arthropod natural enemies using expert system technology, *AI Applications in Natural Resource Management* **3**, 1–11.

Minkenberg, O.P.J.M. and Helderman, C.A.J. (1990) Effect of temperature on the life history of *Liriomyza bryoniae* (Diptera: Agromizidae) on tomato, *J. of Economic Entomology* **83**, 117–125.

Minkenberg, O.P.J.M. and Ottenheim, J.J.G.W. (1990) Effect of leaf nitrogen content of tomato plants on preference and performance of a leafmining fly, *Oecologia* **83**, 291–298.

Montgomery, H. (1983) Decision rules and the search for a dominance structure: Towards a progress model of decision making, in P. Humphreys, O. Svenson and A. Vari (eds.), *Analysing and Aiding Decision Processes*, North-Holland Publ. Co., Amsterdam.

Nachman, G. (1991) An acarine predator-prey metapopulation system inhabiting greenhouse cucumbers, *Biological J. of the Linnean Society* **42**, 285–303.

Norton, G.A. and Mumford, J.D. (eds.) (1993) *Decision Tools for Pest Management*, CAB International, Wallingford.

Ortman, G.F., Patrick, G.F., Musser, W.N. and Doster, D.H. (1993) User of private consultants and other sources of information by large cornbelt farmers, *Agribusiness* **9**, 391–402.

Powell, C.C. and Lindquist, R.K. (1992) *Ball Pest and Disease Manual*, Ball Publishing, Geneva, Ill.

Pritsker, A.A.B. (1984) *Introduction to Simulation and SLAM II*, Systems Publishing Corporation, West Lafayette, Ind.

Rabbinge, R., Ward, S.A. and van Larr, H.H. (eds.) (1989) *Simulation and Systems Management in Crop Protection*, Simulation Monograph 32, Pudoc, Wageningen.

Shipp, J.L. and Gillespie, T.J. (1993) Influence of temperature and water vapor pressure deficit on survival of *Frankliniella occidentalis* (Thysanoptera: Thripidae), *Environmental Entomology* 22, 726–732.

Shtienberg, D. and Elad, Y. (1997) Incorporation of weather forecasting in integrated, biological-chemical management of *Botrytis cinerea*, *Phytopathology* 87, 332–340.

Souder, W.E. (1980) *Management Decision Methods for Managers of Engineering and Research*, Van Nostrand Reinhold Ltd, New York.

Sprague, R.H. and Watson, H.J. (1989) *Decision Support Systems: Putting Theory into Practice*, Prentice Hall, Englewood Cliffs, NJ.

Travis, J.W., Rajotte, E., Bankert, R., Hickey, K.D., Hull, L.A., Eby, V., Heinemann, P.H., Crassweller, R., McClure, J., Browser, T. and Laughland, D. (1992) A working description of the Penn State Apple Orchard Consultant, an expert system, *Plant Disease* 76, 545–554.

Tregoe, B.B. and Kepner, C.H. (1981) *The New Rational Manager*, Princeton Research Press, Princeton, NJ.

van Lenteren, J.C. (1990) Implementation and commercialization of biological control in West Europe, International Symposium on Biological Control Implementation, 4–6 April 1989, McAllen, Texas, *NAPPO Bulletin* 6, 50–70.

van Roermund, H.J.W., van Lenteren, J.C. and Rabbinge, R. (1997) Biological control of greenhouse whitefly with the parasitoid *Encarsia formosa* on tomato: An individual-based simulation approach, *Biological Control* 9, 25–47.

Van Vliet Automation Ltd (1996) Emerald ICM and Emerald ENV100, Advertising Brochure, Van Vliet Automation Ltd, 42 Stockbridge Rd., Elloughton, Brough, North Humberside.

Winsor, G.W. and Adams, P. (1987) *Diagnosis of Mineral Disorders in Plants*, Vol. 3 *Glasshouse Crops*, Her Majesty's Stationery Office, London.

Yang, X. (1995) Greenhouse micrometeorology and estimation of heat and water vapour fluxes, *J. of Agricultural Engineering Research* 61, 227–238.

CHAPTER 13

EVALUATION AND USE OF PREDATORS AND PARASITOIDS FOR
BIOLOGICAL CONTROL OF PESTS IN GREENHOUSES
Joop C. van Lenteren and Giuseppe Manzaroli

13.1. Introduction

Until recently, the main reasons behind searching for non-chemical methods of pest control were concerns about the risks of chemicals for the environment and human health (e.g. Metcalf, 1980). Now with increasing pesticide resistance, increasing costs of pesticides and the present difficulties in developing new effective pesticides, there are also strong signals from the field of agriculture itself that the time has come to change to biologically based pest control (e.g. Lumsden and Vaughn, 1993). A powerful alternative to chemical control is biological control, which is defined as "The use of natural enemies for the control of pests, diseases and weeds". For biological control of insects and mites in greenhouses three categories of natural enemies are commercially used nowadays: parasitoids, predators and pathogens (Table 13.1).

TABLE 13.1. Examples of different categories of natural enemies used in commercial biological control programmes in greenhouses

Category of natural enemy	Pest targeted
Parasitoids	
Encarsia formosa	Whiteflies
Dacnusa sibirica/Diglyphus isaea	Leafminers
Aphidius colemani	Aphids
Predators	
Phytoseiulus persimilis	Spider mites
Macrolophus caliginosus	Whiteflies
Orius leavigatus	Thrips
Pathogen	
Spodoptera NPV (virus)	*Spodoptera exigua*
Verticillium lecanii (fungus)	Whiteflies
Steinernema spp. (nematodes)	Vine weevils/sciarids
Bacillus thuringiensis ssp. *tenebrionis* (bacterium)	*Leptinotarsa decemlineata*

Prerequisites for the development of biological control were the general acceptance that insects do not arise by spontaneous generation (documented by F. Redi in 1668), understanding of the process of predation (documented in Chinese literature 2500 years ago), the correct interpretation of behaviour of parasitic insects (documented by van Leeuwenhoek in 1700), recognition of the infection process of organisms by pathogens

R. Albajes et al. (eds.), Integrated Pest and Disease Management in Greenhouse Crops, 183-201.
© 1999 *Kluwer Academic Publishers. Printed in the Netherlands.*

(documented by Kirby in 1826) and evolution of the idea to use natural enemies for the control of pests in the 18th century. In Europe, Réaumur was the first to propose the tactic of biological control: he advised the release of lacewings in greenhouses for the control of aphids already in 1734. In 1800, Charles Darwin's grandfather, Erasmus Darwin wrote about the useful role parasitoids and predators can play in keeping down the numbers of insect pests. Shortly afterwards there were numerous publications in Europe expressing the same idea. The first practical demonstration of biological control in Europe was carried out in France in 1840: M. Boisgiraud released the carabid predator *Calosoma sycophanta* (L.) against the gypsy moth *Lymantria dispar* (L.) on poplars. At the same time in Germany, J.R.C. Ratzeburg moved heavily parasitized *Dendrolimus pini* (L.) into an outbreak area and recommended the use of ants *Formica rufa* L. against forest defoliators. Also efforts to increase insectivorous birds by providing nesting facilities were popular. The ant and bird work has typical elements of the European pattern of biological control.

The earliest – unsuccessful – attempt to establish a new natural enemy species in Europe was the importation of the predatory mite *Tyroglyphus phylloxerae* Riley & Plancon in 1873 for control of the grape phylloxera *Viteus vitifoliae* (Fitch). The first successful importation of exotic organisms into Europe dates from 1897 when the Portuguese imported and established the vedalia beetle *Rodolia cardinalis* (Mulsant) against the cottony cushion scale *Icerya purchasi* Maskell. This ladybird beetle was earlier imported into California from Australia to control the accidentally imported cottony cushion scale. The world-wide successes with biological control at the end of the previous century stimulated strong interest in this pest control method. Interest in biological control lessened with the appearance of synthetic pesticides after 1940, but insecticide resistance and the recognition of unwanted side-effects from pesticides revived interest in biological control. In Europe, several natural enemies have been imported and are still active in keeping pest populations under control on vast areas of citrus and apple orchards (Greathead, 1976). But this type of biological control, whereby natural enemies are imported and released in low numbers ("inoculated") is less often used in Europe than in countries like the USA where many pests were first accidentally introduced and where later natural enemies were deliberately imported (DeBach, 1964). Europe has served as an important source of natural enemies for export principally to North America. Currently, biological control is commercially applied to several different crop types in Europe. Natural enemies are mass produced and released to control pests in apple and olive orchards, vineyards and corn, but the greatest diversity of natural enemies is employed in greenhouses. During the past 25 years, about 80 species of natural enemies have been evaluated for use in protected cultivation (van Lenteren, 1997).

13.2. Different Strategies of Biological Control

Natural enemies can be used in the following release strategies (Fig. 13.1):

(i) The *inoculative release* method is also known as "classical" biological control and is synonymous with importation. The beneficial organisms are collected from one part of the world and introduced into the area where the pest occurs (Fig. 13.1 top). Only a relatively small number of beneficial organisms is released; the aim is long-term control. The method is usually applied in forest and orchard ecosystems where continuous existence of natural enemies can be guaranteed. An example of a successful European programme is the introduction of the parasitoid *Aphelinus mali* (Haldeman), against the apple woolly aphis, *Eriosoma lanigerum* (Hausmann) into France in 1920, and later into other European countries. This method is not used in protected crops.

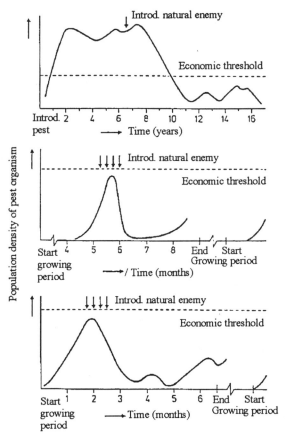

Figure 13.1. Different biological control strategies: inoculative releases (top), inundative releases (middle) and seasonal inoculative releases (bottom).

(ii) The *inundative release* method is where beneficial organisms are collected, mass reared and periodically released in large numbers to obtain immediate control of a pest (i.e. use as a biotic insecticide; Fig. 13.1 middle). Pest control is mainly obtained from the released natural enemies and not from their offspring. Inundative releases are applied to crops where viable breeding populations of the natural enemy are not possible or in crops where the damage threshold is very low and rapid control is required at very early stages of infestation. Examples are the use of *Diglyphus begini* (Ashmead) to control *Liriomyza trifolii* (Burgess) on marigolds and *Encarsia formosa* Gahan to control *Bemisia tabaci*

(Gennadius) on Poinsettia (Parrella, 1990). Inundative releases of *Chrysoperla carnea* (Stephens) larvae are applied against aphids on strawberry in northern Italy (Celli *et al.*, 1991) to obtain good control within a few days. This is achieved by releasing *Chrysoperla* at the 3rd larval stage which has the greatest predation capacity. Predation stops completely when the 3rd stage *Chrysoperla* larvae are close to pupation (see also Chapter 32). The application of the entomopathogenic fungus *Verticillium lecanii* (A. Zimmerm.) Viégas for the control of whitefly and sprays with the *Spodoptera* NPV virus can also be considered inundative releases.

(iii) The *seasonal inoculative release* method is where natural enemies are collected, mass reared and periodically released into short-term crops (6–12 months) and where many pest generations occur (Fig. 13.1 bottom). A relatively large number of natural enemies is released to obtain both immediate control and a build-up of the natural enemy population for control throughout the same growing season. This method can be applied when the growing method of a crop prevents control extending over many years, for example in greenhouses where the crop together with the pests and natural enemies are removed at the end of the growing season. The method is distinctly different from the inundative method, and more closely resembles the inoculative method because control is obtained for a number of generations of the pest and control would be permanent if the crop were grown for a much longer period. The seasonal inoculative release method has been developed in Europe during the last two decades and is applied with commercial success in greenhouses. Two well-known natural enemies used for this approach are the spider mite predator *Phytoseiulus persimilis* Athias-Henriot and the whitefly parasitoid *E. formosa*.

Another important aspect of biological control can be *conservation of natural enemies* whereby the environment is manipulated or modified to improve the effectiveness of already established natural enemies through: (i) provision of missing or inadequate requisites such as alternative hosts, supplementary food or shelter; and (ii) by elimination or mitigation of hazards or adverse environmental factors such as poor cultural practices, indiscriminate use of insecticides and other adverse physical or biotic factors. An example of (i) is the placement of alternative food (eggs of *Ephestia kuehniella* Zeller) for the nymphs and adults of the predatory bug *Macrolophus caliginosus* Wagner at times when its preferred whitefly prey is absent. The current very careful use of (selective) pesticides in greenhouses to prevent mortality of natural enemies illustrates tactic (ii).

An often neglected aspect of biological control is the phenomenon of *natural control*: many potential pest organisms are kept at densities well below the damage threshold by natural enemies that occur in the field. In natural ecosystems, a myriad of natural enemy species maintain plant-eating insects at low population densities. Even in agro-ecosystems, many potential pests are held at non-damaging levels by natural enemies which occur naturally. DeBach and Rosen (1991) estimate that more than 90% of all agricultural pest species are under natural control. Even in greenhouses natural control can play an important role: in northern Europe, parasitoids of leafminers, and predators and parasitoids of aphids invade greenhouses in April or May and result in pest control free of charge. In Mediterranean Europe, greenhouses are more open than in northern Europe, and natural control can be very important because natural enemies can easily move into the greenhouses from the field. Overlapping plantings of the same crops and abundant wild plants on which both the pest and the natural enemies can breed, creates good conditions for natural control, very often without any special intervention. High numbers of predators and parasitoids may survive and remain active during the Mediterranean winter. For example the parasitoid *Diglyphus isaea* (Walker) can develop on several leafminer species and it often migrates into early-season, newly-transplanted crops in greenhouses, and keeps the leafminers *Liriomyza bryoniae* (Kaltenbach), *L. trifolii* and *Liriomyza huidobrensis* (Blanchard) below the damage threshold (Calabretta *et al.*, 1995). Another example of natural enemies providing natural control is the whitefly predator, *M. caliginosus*. This predator is very common throughout the Mediterranean basin and can survive on wild plant species like *Inula viscosa* (L.) Ait. (Arzone *et al.*, 1990). If it is not killed by insecticides, it can be a key factor in reducing whitefly populations (Alomar *et al.*, 1994).

13.3. How to Develop a Biological Control Programme?

The planning of a biological control project and a procedure to evaluate natural enemies prior to introduction will be presented in this section.

13.3.1. PLANNING OF A PROJECT

The typical way to tackle a biological control project is as follows (Fig. 13.2):

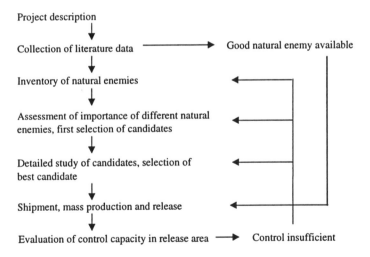

Figure 13.2. Phases in the development of a biological control project.

    (i) A project description is prepared. This includes the taxonomic and pest status of the target organism.

    (ii) Information on the biology of the pest and its natural enemies is collected through literature research and correspondence. If a good natural enemy is identified and available one may proceed to step (vi).

(iii) If an appropriate natural enemy is not available, then an area has to be selected for exploration. This is usually the area of origin of the pest organism. Inventory research can now be started. It is important to collect sufficient animals and to ensure the genetic diversity of natural enemies.

(iv) The importance of the different natural enemies in the exploration area should be estimated. The host range must be studied and negative characteristics (e.g. hyperparasitic habits) noted. These data are used to make a first selection of species for future studies. Although studies in the exploration area cannot be used to predict whether a new natural enemy species will become established or be effective in a new environment, they can show if an agent is clearly unsuitable.

(v) After the first selection a more detailed study can be made with the chosen species. Depending on the type of programme in which natural enemies will be used, a number of the characteristics mentioned in Section 13.3.2 may be studied.

(vi) The selected species of natural enemies are mass produced and released in the area where the pest has to be controlled.

(vii) After release and establishment of the natural enemy, determine its effectiveness (both biological and economic effectiveness) in the target area.

The most critical phases in any biological control programme are the steps where selection of natural enemies takes place [(iii), (iv) and (v)]. Many of the greenhouse pests in countries with a temperate climate have been introduced on the imported infested plant material. This is quite different from the pest situation in greenhouses in (sub)tropical (e.g. Mediterranean) countries, where pest organisms may migrate into greenhouses from surrounding fields. In northwestern Europe, 75% of the species of greenhouse pests, i.e. some 40 species, have been accidentally introduced into the region (for examples, see van Lenteren, 1997). The natural enemies used for biological control of these pests originate from a great variety of sources. Handbooks on biological control generally recommend that natural enemies be collected in the area where the pest is native (e.g. Huffaker and Messenger, 1976). In greenhouse biological control research we have found that it is worth trying introduced natural enemies against native pests, and endemic natural enemies against introduced pests; any dogmatism in selection of natural enemies seems to be counter productive. A good illustration of this is the discovery that several European parasitoids (*Opius*, *Dacnusa* and *Diglyphus* species) can give good control of exotic leafminer (*Liriomyza*) species which were accidentally imported. Chapter 20 gives an overview of all kinds of combinations of exotic/endemic pests and exotic/endemic natural enemies that resulted in good biological control.

An important consideration when selecting natural enemies and setting up mass production, is the quality of the starting population of the natural enemy (see Chapter 20). The initial stock for a laboratory colony should preferably be large and should contain genetically diverse material (e.g. Huffaker and Messenger, 1976). Such statements are easily formulated, but often not easy to achieve. Many of the colonies of natural enemies used for biological control in greenhouses were started from very small populations (for details see Table 10 in van Lenteren and Woets, 1988). An interesting example is the history of *P. persimilis*. This predatory mite accidentally reached Germany in 1959 on plant material imported from Chile. Less than 10 individuals reached The Netherlands in that same year, and these were the basis of research that was started to find out if the predator could be used for control of pest mites. Many commercial colonies of *P.*

persimilis in the 70s and 80s originated from this very small population, and pest control with this predator was generally very good. We do not give this example to suggest that there is no need to collect large founder populations, but rather to show that if it is difficult to obtain large numbers it might still be useful to do experiments with a new natural enemy. On the other hand, if control results are poor with a natural enemy species that was started from a very small colony, it might be worth trying it again after collecting a larger number of individuals. There are important examples in the literature showing the existence of large differences between populations of the same natural enemy species, which can result in either failure or success of biocontrol (e.g. Huffaker and Messenger, 1976).

The above outline for planning of a project needs to be adapted for each specific case of biological control and often ad hoc problems make it necessary to deviate from the general procedure.

13.3.2. PRE-INTRODUCTORY EVALUATION OF NATURAL ENEMIES

How it Was

Until now the selection of natural enemies for biological control programmes has been an empirical procedure, like the selection of the majority of the chemical pesticides. Most natural enemies have been found through trial-and-error. During the 100 or more years in which biological control has been practised, some 5500 introductions of natural enemies into new areas (168 countries) were made, and about 1500 of these introductions resulted in establishment of the species. Long lasting control was obtained in 420 cases resulting in a considerable reduction of pest problems. The success ratio of 1 out 14 in biological control is good when compared with chemical control, where it is 1 out of 20,000. Still, some biological control workers are of the opinion that the selection process should be much improved for two main reasons: first to prevent a lot of time being spent on ineffective natural enemies, and secondly, to be able to work fast and reliably during the coming decades when many new natural enemies need to be identified for use in biological control.

Many researchers have thought about ways of optimizing the pre-introductory selection from the large array of natural enemies, so as to increase the predictability of success before introductions are made (for a more detailed discussion, see van Lenteren, 1993). A biological control project can be characterized as a process whereby a diverse natural enemy complex is reduced to a few candidates for introduction. The selection process is still often highly arbitrary and not related to any aspect of an agent which might indicate its potential value. However, it is a fact that programmes usually end before all promising agents have been introduced. Hence prioritizing agents on the basis of their likely efficiency would ensure that the best species are released. It would be much better for our profession if deliberate choices between possible candidates are made, particularly if this leads to a halt in importation of useless candidates. Further, if we intend to change biological control from an art into science, we should develop a basic understanding of how biological control works and be able to make predictions about the outcome of introduction programmes.

Three approaches for the pre-release selection of natural enemies emerge from the literature: (i) evaluation based on individual attributes of natural enemies; (ii) evaluation based on integration of individual attributes; and (iii) evaluation based on ecosystem studies (Mackauer *et al.*, 1990). In the evaluation based on individual attributes of natural enemies, agents are selected on the basis of particular biological attributes or life-history characteristics (e.g. duration of development, fecundity, searching efficiency). Theory dissects natural enemies into simple sets of characters, which can be viewed and compared independently. This approach is no longer popular, although it is still used. In the evaluation based on integration of individual attributes, a composite picture is developed of the pest reduction potential of the natural enemy. When carefully applied, this method has proved to be valuable. The evaluation based on ecosystem studies proceeds from the theoretical notion of how natural enemies fit into the broad ecology of the pest and its other mortality factors. Here, community concepts predominate, expressed in arguments for density-specific agent complexes, multiple introductions and filling "empty" natural enemy niches. This approach is not often applied, but strongly supported by some biocontrol workers. Although it is scientifically attractive, it is not usable yet and it will take many more years before it is workable.

Currently, there are good evaluation criteria available to allow for a choice between useless and potentially promising natural enemies (see below). Such a choice prevents useless research on and introduction of inefficient natural enemies. With a gradual improvement of evaluation criteria and a further integration of criteria, ranking among the promising natural enemies will be possible. A pre-introductory evaluation procedure takes some 18 months per natural enemy or considerably shorter when the natural enemy shows very obvious inherent weaknesses. In that case no further money is spent on rearing, release and follow up studies of unsuitable natural enemies. The data from this research are not only useful for selection, but also provide essential information for designing a mass production method, the type of releases (inundative, seasonal inoculative), the release programme (timing, spacing and numbers to be released) and an extension programme.

Criteria for Evaluation of Natural Enemies
A compilation of the criteria which are mentioned in the biological control literature leads to the following list (Table 13.2; van Lenteren, 1986a):

TABLE 13.2. Criteria for pre-introductory evaluation of natural enemies (after van Lenteren, 1986b)

Criterion	Release programme		
	Seasonal inoculative	Inoculative	Inundative
(i) Seasonal synchronization with host	+	–	–
(ii) Internal synchronization with host	+	+	–
(iii) Climatic adaptation	+	+	+
(iv) No negative effects	+	+	+
(v) Good culture method	–	+	+
(vi) Host specificity	+	–	–
(vii) Great reproductive potential	+	+	–
(viii) Good density responsiveness	+	+	±

+ = Important; – = Not important; ± = Less important

(i) Seasonal synchronization of the natural enemy with its host/prey is important in inoculative releases ("the natural enemy has to be around when the pest occurs"). When using seasonal inoculative and inundative releases, as in greenhouses, this synchronization can be obtained by the grower through releasing natural enemies when most pest insects are in the developmental stage for optimal attack. Adjustments can be made throughout the growing season.

(ii) The natural enemy must develop to the adult stage on the pest insect in order to obtain ongoing control. If the natural enemy kills the host but cannot develop on it, the natural enemy will have to be re-introduced in each subsequent pest generation. This requires an inundative programme which is more expensive. Further, natural enemy development should be synchronous with that of the pest species so that, for example, adult parasitoids are available when suitable pest stages are present for parasitization (internal synchronization). This is especially important at the start of the growing season in greenhouses when pest generations are often still discrete. Poor synchronization can be corrected in part through repeated introductions. Later in the growing season, when generations of the pest organism overlap, this problem ceases to be important.

(iii) At an early stage of pre-introductory research, tests should be performed to determine whether the natural enemies are able to develop, reproduce and disperse in the climate conditions under which they will be used in the greenhouse.

(iv) Also at an early stage of the evaluation process, potential negative effects should be considered. The natural enemies should not attack other beneficial organisms in the same environment or non-target species in the area where they are to be introduced.

(v) Mass production of natural enemies is usually unnecessary for inoculative release programmes, but good culture methods are the basis for the successful inundative and seasonal inoculative biological control programmes used in greenhouses. Culture methods largely determine the eventual cost of the natural enemy and the probability of its commercial application.

(vi) In crops where different insect species (both non-pest and pest species) may occur it is important to introduce natural enemies that preferentially attack pest species in order to obtain adequate pest reduction. A narrow host/prey range is desirable. In greenhouses with relatively few phytophagous species this is less important than in outdoor fields.

(vii) Several biocontrol workers have stated that an efficient parasitoid should have a potential maximum rate of population increase (r_m) equal to or larger than that of its host. If the parasitoid oviposits in the host and also causes additional substantial mortality (e.g. through host feeding or host mutilation), we should reformulate the previous sentence to: "an efficient parasitoid should cause an overall host kill rate larger than the rate of population increase of the host in the absence of the natural enemy". For efficient predators this would mean that they should have a prey kill rate which is larger than the r_m of the prey. However, an r_m or host kill rate larger than the r_m of the host/prey is not by itself sufficient for natural enemy efficiency, because at low host densities the full potential may not be realized. Then searching efficiency is also of great importance.

(viii) Good density responsiveness (one aspect of searching efficiency) is often said to be an invaluable characteristic of an efficient natural enemy. The natural enemy should be able to locate and reduce pest populations before they have crossed economic threshold densities. Density responsiveness seems to be the most difficult attribute to determine. Firstly, it is not an absolute characteristic, but estimates of this response can only be compared in relation to the estimates for other natural enemies. Secondly, many methods for determining density responsiveness have been proposed but most of them are difficult to apply and do not lead to conclusive answers (van Lenteren, 1986b).

Several of the above criteria are not absolute but have relative values which enable comparison with other natural enemies [criteria (v) to (viii)]. Also, it is very important to consider in what situation the natural enemy will have to function, e.g. will it be applied in usually closed greenhouses in temperate climates, or in generally more "open" protected structures in semi-tropical conditions. In the Mediterranean basin, for example, polyphagous predators like *M. caliginosus* and *Orius* spp. can survive relatively easily even in the absence of the target pest, because alternative prey are present. This allows for early introduction without the risk of extinction of the natural enemy, and for a quick attack of the pest as soon as it occurs.

13.3.3. A PROCEDURE FOR SELECTION OF NATURAL ENEMIES FOR GREENHOUSE BIOLOGICAL CONTROL

The most relevant studies for pre-introductory evaluation criteria of natural enemies to be used in seasonal inoculative releases and inundative releases in greenhouses, are points (ii) to (v) and (vii) of Table 13.2. In Fig. 13.3, a flow diagram is presented outlining an evaluation programme. By using such a flow diagram, it is possible to separate useless from potentially useful biological control candidates at an early phase of research. In greenhouse biological control we are not interested in long-term stability *per se*, but merely aim at suppression of pest numbers below the economic threshold. It may suffice to estimate the power of a natural enemy to suppress its host by using system-specific models (van Roermund *et al.*, 1997). First, one would estimate host suppression by natural enemies searching at random. Then conduct simultaneous greenhouses experiments to determine if a natural enemy possesses any characteristics that make it perform better than random searching. Simulation models can indicate whether random searching is sufficient for pest suppression over the growing season. If so, searching efficiency does not have to be measured in more detail, and natural-enemy selection based on determination of r_m or host kill rate will suffice. If random searching is not sufficient, the selection criteria will need to be more rigorous and should include searching efficiency within and between pest patches. Behavioural ecological studies will then be needed to determine which species searches most efficiently.

The evaluation programme as described here has been used, for example, to select *Trichogramma* species/strains (Pak, 1988), to identify effective parasitoids of leafminers (Minkenberg, 1990), to evaluate natural enemies of aphids (van Steenis, 1995) and whitefly parasitoids (van Roermund, 1995; Drost *et al.*, 1996).

13.4. Improving the Evaluation and Selection of Natural Enemies

Ecological, genetic and behavioural theory might help to move the more effective biological control agents to the front of the queue of species to be introduced. In particular an understanding of variability in natural enemy behaviour may enhance selection of natural enemies and the targeting of releases. Recently, several papers have discussed how to interpret and deal with variability in natural enemy behaviour (Lewis *et al.*, 1990; Vet *et al.*, 1990; Vet and Dicke, 1992). Most ecologists are aware that variability in natural enemy behaviour occurs abundantly, often to their despair. It is important to know how natural enemies function in agro-ecosystems, because such understanding may help with the design of systems where natural enemies can play an even more important role in inundative and seasonal inoculative releases. In this section the sources of variability in behaviour are presented and we will discuss the potential for exploiting this variability to improve biological control.

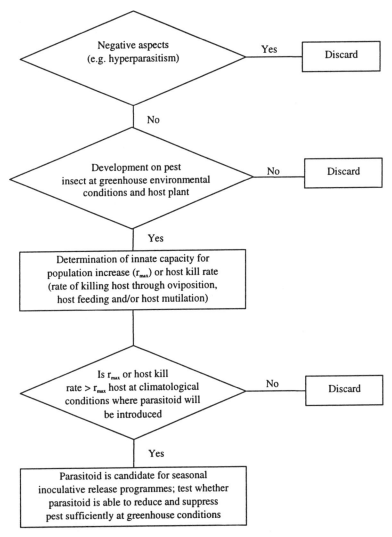

Figure 13.3. Flow diagram depicting an evaluation programme for natural enemies to be used in seasonal inoculative and inundative releases.

The very core of natural enemy behaviour, host-habitat and host location behaviour shows great variability, and is repeatedly leading to inconsistent results in biological control. Most studies aimed at understanding variability have focused on extrinsic factors as causes for inconsistencies in foraging behaviour. Typically, however, foraging behaviour remained irregular when using precisely the same set of external stimuli. These irregularities are caused by intraspecific, interindividual variation in behaviour. In order to understand erratic behaviour and to manipulate such variation, biological control researchers need to know the origins and breadth of variation. Two types of adaptive variation are distinguished in the foraging behaviour of natural enemies (Lewis *et al.*, 1990):

(i) *Genetically fixed differences* among individuals (fixed-behaviour; innate responses), e.g. natural enemy strains with different capabilities for searching in different habitats, strains with different host acceptance patterns. Such variation is now used in selection of natural enemies. Genetically different strains of the same natural enemy species may react in very different way to the same set of chemical stimuli being emitted by the host/plant complex. Knowledge of such inherited preferences for environments and matching of inherited preferences with stimuli in the environment is of vital importance when choosing correct natural enemy strains. If we want a population of natural enemies to be predictable and consistent in biological control, it must first of all have a proper blend of genetic traits appropriate to the target environment, and traits must occur sufficiently uniformly in the population. This statement has been recognized generally, but has been dealt with only on a gross level in applied programmes (e.g. climate, habitat and host matching).

(ii) *Phenotypic plasticity* (unfixed, learned, plastic behaviour), behavioural adaptation may result from the experience of foraging more effectively in one of the various circumstances that the organism may encounter. Preference develops for foraging in a habitat where suitable hosts were previously encountered. The response of a foraging natural enemy can be quite plastic, can be modified within the bounds of its genetic potential, and is dependent on the individuals experience history. Behavioural modifications can be initiated during pre-imaginal stages and at eclosion, so the response of a "naive" adult will necessarily be routinely altered as a consequence of rearing systems. Such alterations have seldom or never been quantified, although changes in preference have been observed to result from different hosts or host diets. For inundative and seasonal inoculative types of biological control, it is essential to quantify this variability due to learned behaviour. An individual can often change its inherited response range, so it can develop an increased response for particular foraging environments as a result of experience with stimuli of these environments. Absence of reinforcement (i.e. absence of contact with host-related stimuli) will result in a waning of the level of that response and a reversion to the naive preference. Natural enemies are plastic in their behaviour, but operate within genetically defined boundaries.

Only recently have we begun to appreciate the extent to which natural enemies can learn. Many parasitoid species are able to acquire by experience an increased preference for and ability to forage in a particular environmental situation (Vet *et al.*, 1990; Vet and Dicke, 1992). There is some indication for immature learning and abundant evidence for adult learning in natural enemies. Learning is mostly by association. Usually, close range, reliable, unconditional genetically fixed stimuli serve as associators and reinforcers for the longer range, more variable conditional stimuli. Foraging behaviour can continuously be modified according to the foraging circumstances encountered (Vet and Dicke, 1992).

Additionally, foraging behaviour can be strongly influenced by (iii) the *physiological condition* of the natural enemy. Natural enemies face varying situations when meeting their food, mating, reproductive and safety requirements. Presence of strong chemical, visual or auditory cues, cues related to enemy presence, and (temporary) egg depletion can all reduce or disrupt the response to host-foraging cues. For example, hunger may result in increased foraging for food and decreased attention to hosts. In that case, the reaction to food and host cues will be different than when the natural enemy is well fed.

The sources of intrinsic variation in foraging behaviour (genetic, phenotypic and those related to the physiological state) are not mutually exclusive but overlap extensively, even within a singular individual: "The resulting foraging effectiveness of a natural enemy is determined by how well the natural enemy's net intrinsic condition is matched with the foraging environment in which it operates" (Lewis *et al.*, 1990).

How can we manage variability in behaviour of natural enemies? In order to be efficient as biological control agents, natural enemies must be able to: (i) effectively locate and attack a host; and (ii) stay in a host infested area until most/all hosts are attacked. (An "efficient" biological control agent from an anthropocentric point of view, does not necessarily mean efficiency from a natural selection perspective.) Prediction of performance in efficiency is a product of proper matching of intrinsic conditions of the searching natural enemy with the target environments.

Management of the natural enemy component is particularly important in a mass production system especially when they are reared on alternative hosts (see Chapter 20). In laboratory colonies the natural enemies are removed from the context of natural selection and are exposed to artificial selection for traits not valued in the field (van Lenteren, 1986a). In addition to effects of the genetic component, associative learning may lead to many more changes in behavioural reactions. This, then, results in the need for quality control procedures in the establishment, maintenance and use of natural enemies. Quality control will have to manage both genotypic and phenotypic aspects of behavioural traits. Currently, quality control is applied on a limited scale by mass production units in Europe:

(i) *Genetic qualities.* Successful predation or parasitism of a target host in a confined situation does not guarantee that released individuals will be suitable for controlling the host under field conditions. When selecting among strains of natural enemies, we need to ensure that the traits of the natural enemies are appropriately matched with the targeted use in the field.

(ii) *Phenotypic qualities.* Without care, insectary environments lead to weak or distorted responses. When we understand the sources and mechanism of learning, we can provide the appropriate level of experience before releasing the natural enemies. Also, pre-release exposure to important stimuli can help improve the responses of natural enemies through associated learning, leading to reduction in escape response and increased arrestment in target areas.

(iii) *Physical and physiological qualities.* Natural enemies should be released in a physiological state in which they are most responsive to herbivore or plant stimuli and not be hindered in their response by e.g. food deprivation interfering with searching.

13.5. From the Laboratory to the Greenhouse: Development of Practical Biological Control

During the past 30 years 60 species of natural enemies have been commercialized for the control of 50 pest species occurring in greenhouses. Many new species of natural enemies are in the process of being evaluated for future use. Presently, biological control of the two key pests in greenhouses, whitefly [*Trialeurodes vaporariorum* (Westwood)] and spider mite (*Tetranychus urticae* Koch), is used in more than 20 countries out of 35 countries having a greenhouse industry.

If a candidate natural enemy has been identified in the laboratory, greenhouse performance testing will have to be done, a mass rearing method will have to be developed that results in reliable production of large quantities of agents which are in excellent condition for killing pest organisms, and efficient storage, shipment and release methods must be designed (see Chapter 20). After the selection process described in Section 13.3, the candidate natural enemy is considered as a "product under development". It is often difficult to determine at this phase how much time will be needed to be able to come up with a commercial product. The next stage is to evaluate the natural enemy under crop production conditions in the greenhouse and the first stage is to perform experimental releases at a range of greenhouse conditions and crop production techniques. The release programme has to be integrated with other crop management practices and evaluation is required of all operations which might interfere with the release and performance of the biological control agent.

The entire process of laboratory and greenhouse evaluation is not always performed in the sequential order as described in Section 13.3. We will present two examples of product development which included a pragmatic element. One programme failed (a parasitoid of Colorado Potato Beetle) and the other one was successful (a predator of thrips).

The egg parasitoid, *Edovum puttleri* Grissell, was evaluated for biological control of Colorado Potato Beetle in Italian greenhouses. Release experiments were carried out in commercial greenhouses before a mass rearing method was developed (Maini *et al.*, 1990). Although greenhouse performance of the parasitoid was satisfactory, it was not commercialized for reasons which included the high costs of mass rearing.

After the accidental introduction of *Frankliniella occidentalis* (Pergande) into Europe, many efforts were made to find natural enemies of this thrips species. Pirate bugs (*Orius* spp.) seemed to be the most widespread and active predators of this species of thrips in Europe. Many researchers and biocontrol companies started to investigate *Orius*, both at laboratory and field level, to determine which species would give acceptable control under the specific conditions found in greenhouses in different areas of Europe. European *Orius* species from different geographic regions were considered [*Orius niger* (Wolff), *Orius laevigatus* (Fieber), *Orius majusculus* (Reuter) and *Orius albidipennis* (Reuter)] (Nicoli and Tommasini, 1996) along with an exotic, nearctic species [*Orius insidiosus* (Say)] (Dissevelt *et al.*, 1995). Control experiments under practical cropping conditions gave variable results. Therefore, an intensive research programme was started to compare the ability of different species of *Orius* to control thrips in the laboratory and in greenhouses in several greenhouse production areas in Europe (Tommasini and Nicoli, 1993, 1995; Dissevelt *et al.*, 1995; Tommasini *et al.*, 1997). The result of this study was that researchers and biocontrol practitioners concluded that the endemic *O. laevigatus* was the best predator, providing good control under various conditions. *Orius laevigatus* is now the main natural enemy for thrips control in Europe, because it is the best for mass production and performance in greenhouses.

Testing the efficiency of natural enemies under practical growing conditions in greenhouses is complicated and expensive. It is seldom possible to realize the same environmental conditions in several greenhouses, and to obtain the same host plant quality and pest infestation levels. Often, empirical observations lead to the formulation of a practical release programme. Evidence of successful control can, for example, be deduced from situations where, after release of natural enemies, both the pest insect and its natural enemy operate at very low densities and below the economic threshold level (Stehr, 1982). Biocontrol companies usually start greenhouse tests by releasing very large numbers of natural enemies to be sure that the control will be satisfactory. The next step is to test different release rates and to determine the lowest release rate resulting in reliable control. This type of testing should be done in a situation where other species of natural enemies do not interfere. The release rates will have to be adapted to the type of greenhouse, the crop and region. In situations with low pest immigration from outside (for example in winter in northern Europe when greenhouses are closed) one or a few releases of the natural enemy may suffice. In Mediterranean areas, with open greenhouses, releases may have to continue throughout the growing season. Practical release schemes are continuously modified based on greenhouse experiences, and it normally takes several years before a standard release programme is available. Scientifically designed and statistically reliable experimentation to determine the efficiency of different natural enemies of the same pest organisms has seldom been performed because of prohibitive costs.

Candidate natural enemies may be tried out on a small scale even if the laboratory development process has not yet been completed. Trials under practical conditions will provide information about how the natural enemy can be integrated with other components of pest and disease control and information for development of practical release programmes. Such trials are conducted on properties of "pioneer" growers who like to try out new developments.

During the whole process of product development, the biocontrol company will keep an eye on the cost-effectiveness of the new product. When mass production, shipment and release of a specific natural enemy are expensive, it might be realistic to advise it for release only in ornamentals or the more expensive vegetables, where higher investments for pest control are normally made. Sometimes the high costs of mass rearing has resulted in release of low (even too low) numbers of natural enemies, with the risk of unreliable results and negative advertisement for biological control. The costs of a natural enemy may even determine the type of release programmes. A very cheap natural enemy can be used in blind, regular, inundative releases without monitoring of the pest. A more expensive natural enemy is better used in well planned, seasonal inoculative releases after the pest has been detected. When two natural enemies are available for control of the same pest, the ultimate choice may be based on very practical considerations, not just the level of performance and costs of each species. An example is the use of the parasitoid *D. isaea* for the control of leafminers instead of the parasitoid *Dacnusa sibirica* Telenga, the control effect of the ectoparasitoid *Diglyphus* is much easier to detect in the field by the grower or advisor than that of the endoparasitoid *Dacnusa*.

Once a release programme has been developed, it will have to be modified regularly, because new plant cultivars may be used, growing conditions may change, other pests and pesticides may be used in the system, etc. It is important to realize that development of biological and integrated control is knowledge intensive and that these systems need continuous modification.

13.6. Importation and Release of Exotic Natural Enemies

Quite a number of the natural enemies used for biological control of pests in Europe are exotic organisms (for an overview, see van Lenteren, 1997). Because each organism may become established, extreme care should be exerted during the evaluation phase to prevent escapes. This is always important, whether the organism is being introduced into a new region or developed for inundative or inoculative releases. Until now, introductions of several hundreds of species of insect natural enemies have not led to environmental problems. Any future problems can to a large extent be avoided by following the procedures of selection, importation and release as described above.

The use of biological control of insect pests has considerably increased during the past decades as it provides an environmentally attractive alternative to chemical pest control. Surprisingly, however, biological control practitioners are nowadays confronted with criticism from environmentalists because of the fear that the biocontrol agents may attack: (i) beneficial non-target organism like pollinators or other natural enemies; (ii) rare or endangered insects like butterflies; or (iii) other non-target organisms. Such undesirable influences on ecosystems have not, in fact, been observed, but it should be realized that the effect of biocontrol introductions on the native fauna has rarely been studied in great detail. The types of risks resulting from biological control introductions have been classified as: (i) direct effects leading to extinction or reduction in numbers of native non-target organisms; and (ii) indirect effects such as preying on or parasitizing indigenous natural enemies or competition for hosts or prey with indigenous natural enemies.

The literature of the past 100 years on introductions of natural enemies for insect control has provided no evidence of extinction of species as a consequence of such introductions, and the generally strong preference for the introduction of highly specific natural enemies may explain this. Reduction in populations of native non-target organisms is difficult to demonstrate. It is very important to realize that ecologists have long recognized the role of predators, parasitoids and pathogens in regulating populations of plant-eating organisms (in agro-ecosystems often pest insects), thereby keeping the world green. In natural and agricultural ecosystems, many herbivores occur at extremely low densities because of the action of natural enemies. Also the natural enemies themselves are normally rare when herbivores occur at low numbers. However, for each herbivore species, many different species of natural enemies may occur at such low densities without eradication of either the herbivore or its natural enemies.

Because of the demands from conservationists, there is now a tendency in some European countries to avoid all possible risks and to refuse permission for importation and release of biological control agents, or to overregulate importations. Both measures seriously hamper further development of biological control. Long before governmental demands, biological control workers themselves have developed risk assessment procedures which are based on taxonomic status and biology of natural enemy, safety screening on other organisms, and evaluation of host specificity. Such data, combined with an environmental risk analysis of other control methods, can be made to make informed decisions to choose between biological control or other control methods (for a more extensive discussion of this topic, see van Lenteren, 1997).

13.7. Conclusions

Several current trends will stimulate the application of biological control. Firstly, fewer new insecticides are becoming available because of skyrocketing costs for development and registration. Secondly, pests continue to develop resistance to any type of pesticides (conventional and high-tech modern ones), a problem particularly prevalent in greenhouses, where intensive management and repeated pesticide applications exert strong selective pressure on pest organisms. Thirdly, there is a strong demand from the general public (and in an increasing number of countries also from parliament) to reduce the use of pesticides.

Because of the desire to reduce pesticide use, the future role of biological control is expected to increase strongly. This is aided by the extensive demonstration of its positive role and because many new natural enemy species still await discovery. Cost/benefit analyses show that biological control is the most cost effective control method (van Lenteren, 1993). With the improved methods of evaluation and an increased insight into the functioning of natural enemies, the cost effectiveness may even be increased.

We should not expect that biological control will completely replace chemical control. Biological control is a powerful option and can be applied to a much larger area than at present. Biological control should be used in IPM programmes where it is combined with other pest control methods, including very careful use of certain types of chemical control. A benefit for pesticides from increased use of biological control is that this may result in extended use of chemical products because of slower development of resistance. In order to serve agriculture as well as the environment and human health, we should harvest the best from all control methods to develop effective IPM programmes.

Acknowledgement

Dr. N.A. Martin (New Zealand Institute for Crop and Food Research, Auckland, New Zealand) is gratefully acknowledged for his constructive comments on an earlier version of this text and for his language corrections.

References

Alomar, O., Goula, M. and Albajes, R. (1994) Mirid bugs for biological control: Identification, survey in non-cultivated winter plants, and colonization of tomato fields, *IOBC/WPRS Bulletin* 17(5), 217–223.

Arzone, A., Alma, A. and Tavella, L. (1990) Ruolo dei Miridi (Rhynchota, Heteroptera) nella limitazione di *Trialeurodes vaporariorum* Westw. (Homoptera, Aleyrodidae), *Bollettino di Zoologia Agraria e di Bachicoltura, Ser. II* 22(1), 43–51.

Calabretta, C., Calabro, M., Colombo, A. and Campo, G. (1995) Diffusione di *Liriomyza huidobrensis* (Blanchard) (Diptera, Agromyzidae) in colture protette della Sicilia, *Informatore Fitopatologico* **45**(6), 24–30.

Celli, G., Benuzzi, M., Maini, S., Manzaroli, G., Antoniacci, L. and Nicoli, G. (1991) Biological and integrated control in protected crops of northern Italy's Po Valley: Overview and outlook, *IOBC/WPRS Bulletin* **14**(5), 2–12.

DeBach, P. (ed.) (1964) *Biological Control of Insect Pests and Weeds*, Chapman and Hall, London.

DeBach, P. and Rosen, D. (1991) *Biological Control by Natural Enemies*, Cambridge University Press, Cambridge.

Dissevelt, M., Altena, K. and Ravensberg, W.J. (1995) Comparison of different *Orius* species for control of *Frankliniella occidentalis* in glasshouse vegetable crops in the Netherlands, *Medelingen – Faculteit Landbouwkundige en Toegepaste Biologische Wetenschappen, Universiteit Gent* **60**(3a), 839–845.

Drost, Y.C., Fadl Elmula, A., Posthuma-Doodeman, C.J.A.M. and van Lenteren, J.C. (1996) Development of selection criteria for natural enemies in biological control: Parasitoids of *Bemisia argentifolii*, *Proceedings of the Section Experimental and Applied Entomology of the Netherlands Entomological Society* **7**, 165–170.

Greathead, D.J. (ed.) (1976) A review of biological control in Western and Southern Europe, Techn. Comm. 7, Commonwealth Institute for Biological Control/ Commonwealth Agricultural Bureau, Slough, pp. 52–64.

Huffaker, C.B. and Messenger, P.S. (eds.) (1976) *Theory and Practice of Biological Control*, Academic Press, New York.

Lewis, W.J., Vet, L.E.M., Tumlinson, J.H., van Lenteren, J.C. and Papaj, D.R. (1990) Variations in parasitoid foraging behaviour: Essential elements of a sound biological control theory, *Environmental Entomology* **19**, 1183–1193.

Lumsden, R.D. and Vaughn, J.L. (eds.) (1993) *Pest Management: Biologically Based Technologies*, American Chemical Society, Washington, DC.

Mackauer, M., Ehler, L.E. and Roland, J. (eds.) (1990) *Critical Issues in Biological Control*, Intercept, Andover.

Maini, S., Nicoli, G. and Manzaroli, G. (1990) Evaluation of the egg parasitoid *Edovum putleri* Grissel (Hym.: Eulophidae) for biological control of *Leptinotarsa decemlineata* (Say) (Col.: Chrysomelidae) on eggplant, *Bollettino dell'Istituto di Entomologia "Guido Grandi" della Universita degli Studi di Bologna* **44**, 161–168.

Metcalf, R.L. (1980) Changing role of insecticides in crop protection, *Annual Review of Entomology* **25**, 219–256.

Minkenberg, O.P.J.M. (1990) On seasonal inoculative biological control, PhD Thesis, Wageningen Agricultural University.

Nicoli, G. and Tommasini, M.G. (1996) *Orius laevigatus*, *Informatore Fitopatologico* **46**(7), 21–26.

Pak, G.A. (1988) Selection of *Trichogramma* for inundative biological control, PhD Thesis, Wageningen Agricultural University.

Parrella, M.P. (1990) Biological control in ornamentals: Status and perspectives, *IOBC/WPRS Bulletin* **13**(5), 161–168.

Stehr, F.W. (1982) Parasitoids and predators in pest management, in R.L. Metcalf and W.H. Luckman (eds.), *Introduction to Insect Pest Management*, Wiley-Interscience, New York, pp. 135–173.

Tommasini, M.G., Maini, S. and Nicoli, G. (1997) Advances in the integrated pest management in protected eggplant crops by seasonal inoculative releases of *Orius laevigatus*, *Advances in Horticultural Science* **11**, 182–188.

Tommasini, M.G. and Nicoli, G. (1993) Adult activity of four Orius species reared on two preys, *IOBC/WPRS Bulletin* **16**(2), 181–184.

Tommasini, M.G. and Nicoli, G. (1996) Evaluation of *Orius* spp. as biological control agents of thrips pests: Further experiments on the existence of diapause in *Orius laevigatus*, *IOBC/WPRS Bulletin* **19**(1), 183–186.

van Lenteren, J.C. (1986a) Evaluation, mass production, quality control and release of entomophagous insects, in J.M. Franz (ed.), *Biological Plant and Health Protection*, Fischer, Stuttgart, pp. 31–56.

van Lenteren, J.C. (1986b) Parasitoids in the greenhouse: Successes with seasonal inoculative release systems, in J.K. Waage and D.J. Greathead (eds.), *Insect Parasitoids*, Academic Press, London, pp. 341–374.

van Lenteren, J.C. (1993) Parasites and predators play a paramount role in pest management, in R.D. Lumsden and J.L. Vaughn (eds.), *Pest Management: Biologically Based Technologies*, American Chemical Society, Washington, DC, pp. 66–81.

van Lenteren, J.C. (1997) Benefits and risks of introducing exotic macro-biological control agents into Europe, *Bulletin OEPP/EPPO* **27**, 15–27.

van Lenteren, J.C. and Woets, J. (1988) Biological and integrated pest control in greenhouses, *Annual Review of Entomology* **33**, 239–269.

van Roermund, H.J.W. (1995) Understanding biological control of greenhouse whitefly with the parasitoid *Encarsia formosa*: From individual behaviour to population dynamics, PhD Thesis, Wageningen Agricultural University.

van Roermund, H.J.W., van Lenteren, J.C. and Rabbinge, R. (1997) Biological control of greenhouse whitefly with the parasitoid *Encarsia formosa* on tomato: An individual-based simulation approach, *Biological Control* **9**, 25–47.

van Steenis, M. (1995) Evaluation and application of parasitoids for biological control of *Aphis gossypii* in glasshouse cucumber crops, PhD Thesis, Wageningen Agricultural University.

Vet, L.E.M. and Dicke, M. (1992) Ecology of infochemical use by natural enemies in a tritrophic context, *Annual Review of Entomology* **37**, 141–172.

Vet, L.E.M., Lewis, W.J., Papaj, D.R. and van Lenteren, J.C. (1990) A variable-response model for parasitoid foraging behaviour, *J. of Insect Behavior* **3**, 471–490.

CHAPTER 14

BIOLOGICAL CONTROL OF WHITEFLIES
Joop C. van Lenteren and Nicholas A. Martin

14.1. Introduction

Out of the 1200 described whitefly (Homoptera: Aleyrodidae) species, only some 20 species are considered potential pests. Until recently, most research on control of whiteflies was directed towards the greenhouse whitefly [*Trialeurodes vaporariorum* (Westwood)]. However, since the mid 80s sweet potato/silver leaf whitefly [*Bemisia tabaci* (Gennadius), *Bemisia argentifolii* Bellows & Perring] has created problems of such a large scale both in the field and in large, commercial greenhouses that concentration of research shifted to this pest (Byrne and Bellows, 1991; Perring, 1996).

In natural ecosystems and agroecosystems where pesticides are not (or are selectively) used, an array of natural enemies usually keeps the number of whiteflies at very low numbers: predators, parasitoids and pathogens all take their toll. Examples from pesticide free crops show that whitefly species can be kept at densities well below the economic threshold (van Lenteren *et al.*, 1996). If natural control is insufficient, inoculative or inundative releases of natural enemies can be made. Commercial biological control of greenhouse whitefly through releases of the parasitoid *Encarsia formosa* Gahan (Hymenoptera: Aphelinidae) is currently used on about 5000 ha of greenhouse crops and in most countries with an important greenhouse industry (van Lenteren, 1995). Although some predators and entomophagous pathogens show promise for control of greenhouse whitefly (Gerling, 1990), economically feasible control is based on introductions of parasitoids.

Biological control of *Bemisia* species is not as easy as that of greenhouse whitefly. *Bemisia* can be controlled by introducing a mix of *E. formosa* and *Eretmocerus eremicus* Rose & Zolnerowich (= *Eretmocerus californicus* Howard). This procedure is followed both in Europe and in the USA (W.J. Ravensberg, pers. com.). The predator *Macrolophus caliginosus* Wagner is often added to the parasitoids, as this mix of two parasitoids and one predator results in better control over a long period. In some crops *M. caliginosus* sometimes causes damage to the plant (e.g. in cherry tomatoes). In the Mediterranean area, *Eretmocerus mundus* Mercet can be released instead of *E. eremicus*. In the Mediterranean, biological control of *Bemisia* in tomato is very difficult during the hottest period of the season because of a high incidence of virus transfer (tomato yellow leaf curl virus, TYLCV) by this whitefly (M.G. Tommasini, pers. com.).

A world-wide search for new natural enemies of *Bemisia* is in progress (Gerling and Mayer, 1996). Much of this research is opportunistic and evaluation of the control potential of natural enemies is often a purely empirical process. Neither the farmer nor the scientist is much helped by such an approach because many projects are terminated prematurely if success is not obtained quickly and, thus, scientific insight does not evolve. Our biological control approach is based on understanding the functioning of natural enemies in agroecosystems and on developing criteria for a scientifically sound selection

R. Albajes et al. (eds.), Integrated Pest and Disease Management in Greenhouse Crops, 202-216.
© 1999 *Kluwer Academic Publishers. Printed in the Netherlands.*

of natural enemies (see Chapter 13; Vet and Dicke, 1992; van Lenteren, 1993). Such an approach does not have to slow down the development of practical biological control programmes, as is often claimed by biological control practitioners. This is illustrated by the successes we have had in greenhouses during the past 25 years when some 70 natural enemy species were evaluated of which 25 are commercially used for pest control today (van Lenteren *et al.*, 1997b).

In this chapter the success story of biological control of whitefly is reviewed. Background information on the biology of whiteflies and their natural enemies is provided, different strategies for whitefly control are explained and factors that may result in failure of biological control of whitefly are discussed.

14.2. Understanding Whitefly Ecology

Trialeurodes vaporariorum was first found in Europe in greenhouses in the UK during 1856 and described that year by Westwood. Westwood supposed the species to have been imported with living plants or in the packaging of Orchidaceae from Mexico. This polyphagous whitefly species is known to attack 249 genera of 84 angiosperm plant families (Russell, 1977). Nowadays the species is cosmopolitan. A survey of its pest status and biology can be found in van Lenteren and Noldus (1990).

In 1926, a tomato grower drew the attention of the English entomologist Speyer to black pupae among the normally white scales of the greenhouse whitefly. From the black pupae, parasitoids emerged that were identified as *E. formosa* (Speyer, 1927). Within a few years, a research station in England was supplying 1.5 million of these parasitoids annually to about 800 nurseries in Britain. During the 30s *E. formosa* was shipped to other European countries, Canada, Australia and New Zealand. After World War II, distribution of *E. formosa* stopped because newly introduced insecticides provided control on most greenhouse crops. A few years later, however, the first signs of resistance to pesticides were observed. Interest in whitefly parasitoids increased at the start of the 70s when enormous outbreaks of the pest took place. The commercial availability of the parasitoid *E. formosa* paved the way for the development of biological and integrated control programmes in greenhouses, but basic studies on the biology of *E. formosa* and its relationship with host and host plants were needed before this parasitoid could be exploited effectively. Presently, biological control of whitefly with *E. formosa* is achieved in more than 20 out of 35 countries having a greenhouse industry (for a review, see van Lenteren *et al.*, 1992).

For comprehensive reviews on the biology and life history related to host plant and temperature of *T. vaporariorum* and *B. tabaci* see Gerling (1990), van Lenteren and Noldus (1990), van Roermund and van Lenteren (1992) and Gerling and Mayer (1996). The selection of host plants by greenhouse whitefly before landing seems to be largely a random process. Although whiteflies exhibit colour preferences – they are attracted to yellow-green colours – this does not necessarily bring them to the most suitable hosts plants as they land on any yellow-green material. Whiteflies do not appear to use olfactory cues in host-plant selection (van Lenteren and Woets, 1977). Whiteflies can distinguish between species and cultivars of host plants only after landing on them, and primarily by probing the plant tissue (Lei *et al.*, 1996). The proximate factors mediating host-plant selection are still unknown. Although *T. vaporariorum* and *B. tabaci/argentifolii* are very polyphagous, clear oviposition preferences for certain species and cultivars exist (van Lenteren and Noldus, 1990). Within a plant, whiteflies prefer young leaves for feeding and oviposition (Noldus *et al.*, 1986c).

Emergence of whiteflies takes place on the older leaves and a dispersal phase of a few days occurs, after which the distribution patterns remain stable (Noldus *et al.*, 1986b). About 10% of the population moves up in the same host plant and starts feeding and ovipositing on the younger leaves. Whiteflies reach the younger leaves as a result of a sequential process of alighting, probing, taking off and moving upward. Most emerged whiteflies (90%) first show horizontal movement which results in dispersal of a few meters only. Consequently, patches of infestation are not static but gradually increase in area. Longevity of adult whiteflies can be considerable (maximally up to several months) and oviposition may occur over a period longer than the development time from egg to adult. Thus, generations become completely overlapping under greenhouse conditions.

As a result of the between and within host-plant selection processes, whiteflies are distinctly and very strongly aggregated at various spatial levels in the crop (Noldus *et al.*, 1986a; Martin *et al.*, 1991). This strong aggregation necessitates very large numbers of samples for reliable population estimates. In most studies on population dynamics of whitefly in either glasshouse or field situations, sampling was highly inadequate: the number of samples taken was usually decided on feasibility rather than on a calculation of the sample size to be taken for a given statistically reliable estimate and a known sampling error (Butler *et al.*, 1986).

To obtain information on the degree of infestation with whiteflies, one study of the aggregation of greenhouse whitefly puparia in commercial greenhouse tomato crops showed that evenly and widely spaced plants (every 15–20th plant) in every row should be sampled (Martin *et al.*, 1991). For a particular leaf layer with puparia in a crop, there is a correlation between the proportion of infested leaves and the population of puparia which means that only the presence or absence of puparia on a leaf needs to be recorded (Martin and Dale, 1989). A minimum sample of about 200 leaves is required but is higher if lower population densities need to be detected. In addition, presence/absence sampling can determine if the proportion of parasitized puparia is above or below a pre-set level for satisfactory control, e.g. 70%.

Greenhouse whitefly shows persistent preferences for certain plant species (van Lenteren and Woets, 1977). In experiments where plants of different quality were offered, the number of landings on various plants was random, but a redistribution of whiteflies took place after probing the plant tissue. On poor host plants (sweet pepper, for example) frequent probing occurred and whiteflies soon took off. In contrast, on a good host plant (eggplant, for example) no take offs were observed and feeding started almost immediately. This behaviour resulted in the percentage times spent on the host plants as given in Table 14.1. Characteristics such as development time, immature mortality and fecundity of whitefly differ greatly between plant species (Table 14.1). The following relationship was found for greenhouse whitefly: the more a host plant is preferred, the shorter the development time, the lower the immature mortality and the greater the fecundity. Differences in host-plant preference are thus accompanied by different rates of population growth of whiteflies (van Lenteren and Noldus, 1990). In addition to the "persistent" host-plant preferences greenhouse whitefly appears to develop local populations with special host-plant preferences, e.g. weeds like *Rumex* spp., crops like tamarilo [*Cyphomandra betacea* (Cavanilles) Sendtner], hungarian peppers (van Lenteren *et al.*, 1989) and gerbera (van Lenteren and Noldus, 1990).

TABLE 14.1. Host-plant preference (% time on plant during 7 to 8 h observation period, and average no. of eggs laid per female during life span) and host-plant suitability (% immature mortality, immature development time, and longevity; all at 22°C) of *T. vaporariorum* for three vegetables

Host plant	Sweet pepper	Tomato	Eggplant
% Time on plant	37	82	100
Average no. of eggs laid per female during life span	12	153	535
% Immature mortality	69.7	16.7	12.9
Immature development time (d)	31.9	28.6	27.6
Longevity (d)	5	36	63

The life history parameters which have been measured to determine host-plant suitability were also used to estimate population development of whiteflies on several crops using a state variable, temperature-driven simulation model (van Roermund *et al.*, 1997). After verification, the model was validated by independent greenhouse experiments. In these greenhouses whitefly infestations were created on tomato plants. Empty whitefly pupae were counted at three intervals until 83 days after infestation. The greenhouse temperature fluctuated between 18 and 35°C during the experiment. The results of one greenhouse test and the simulation are given in Fig. 14.1. The model gave reliable predictions of population growth. Population growth of whitefly is exponential as long as the physiological condition of the host plant remains good and temperature conditions are favourable. An additional feature is that the whitefly population growth model can be used to evaluate effects of intended changes in the cropping system on whitefly development, e.g. of changes in climate and crop species or cultivar.

14.3. Natural Enemies of Whitefly

Natural enemies used or tested for biological control of whitefly are listed in Table 14.2. The main groups of natural enemies are discussed below:

(i) *Predators.* About 75 species of whitefly predators have been described. But certainly many more species prey upon whiteflies, especially general predators such as spiders, beetles, etc. Individual predator species in the families Anthocoridae, Coccinellidae, Chrysopidae, Hemerobiidae and most of the Miridae are unable to maintain greenhouse whitefly numbers below damaging levels, although inundative releases of a complex of predators may do so (Heinz, 1996). Some predatory bugs in the genera *Macrolophus* or *Dicyphus* can sufficiently reduce whitefly populations (Onillon, 1990), although some also damage plants. In warm climates, where greenhouses often have large ventilation openings, generalist predators move in naturally and may cause considerable mortality of whiteflies.

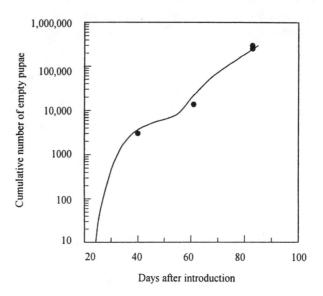

Figure 14.1. Number of empty whitefly pupae counted at three intervals on a tomato crop produced under normal cropping conditions; 100 females were introduced at the start of the experiment (black dots = numbers counted in greenhouse, line = numbers simulated).

(ii) *Pathogens*. In general, pathogens of insects belong to very different taxonomic groups like viruses, bacteria, protozoa, rickettsiae, fungi and entomophagous nematodes. The spectrum of whitefly pathogens is narrow. There are no records of nematodes parasitizing whiteflies. While it is possible that whiteflies are killed by viruses or bacteria, this is mainly due to secondary infections by entrance through existing wounds. So far, the pathogens reported from Aleyrodidae have been exclusively fungi, because only they are able to infect these plant-sucking insects by penetrating the cuticle. Three genera of fungi attacking whitefly are regularly mentioned in the literature: *Aschersonia*, *Verticillium* and *Paecilomyces*. These whitefly pathogenic fungi all germinate on the insect cuticle, penetrate the cuticle and colonize the interior of the host. The high humidity required by these fungi for germination makes it difficult to integrate them into commercial greenhouse practice, although good control has been obtained with *Aschersonia aleyrodis* Webber and *Verticillium lecanii* (A. Zimmerm.) Viégas (Fransen, 1990). *Paecilomyces fumosoroseus* (Wize) Brown & Smith is now extensively tested for control of *Bemisia* and *Trialeurodes* (Lacey *et al.*, 1996).

(iii) *Parasitoids*. Circa 100 species of whitefly parasitoids are known and more species are expected to be found. Most of the parasitoids are very host specific, but some species are hyperparasitoids and their importation might reduce the efficiency of primary parasitoids, e.g. *Encarsia pergandiella* Howard in New Zealand. Many important whitefly parasitoids belong to the genus *Encarsia*, family Aphelinidae (van Lenteren *et al.*, 1997a). They show a wide variety of reproductive behaviour (Gerling, 1990). Some species, like *E. formosa*, are primary thelytokous parasitoids, i.e. females are produced on a phytophagous host insect parthenogenetically. Other *Encarsia* species are also primary parasitoids, but produce haploid males from unfertilized eggs and diploid females from fertilized eggs, i.e. they are arrhenotokous. Still other species are arrhenotokous hyperparasitoids and they produce males and females by laying eggs in other, immature parasitoids of a different species. Further, species are known where one sex, usually the female, develops as a primary parasitoid, and the other sex, the male, develops hyperparasitically on their own or another species of parasitoid, i.e. heteronomous hyperparasitoids or facultative autoparasitoids.

TABLE 14.2.   Natural enemies commercially applied or under study for biological control of *T. vaporariorum* and *B. tabaci*

Natural enemy species	Pest species and crop	Use
Commercially applied:		
Delphastus pusillus	*Bemisia* spp. and *Trialeurodes vaporariorum*, vegetables	Limited
Encarsia formosa	*Trialeurodes vaporariorum*, vegetables, ornamentals	Frequent
Encarsia formosa	*Bemisia* spp., ornamentals	Moderate
Eretmocerus eremicus	*Bemisia* spp. and *Trialeurodes vaporariorum*, vegetables, ornamentals	Moderate
Eretmocerus mundus	*Bemisia* spp., vegetables	Limited
Macrolophus caliginosus	*Bemisia* spp. and *Trialeurodes vaporariorum*, vegetables	Moderate
Verticillium lecanii	*Bemisia* spp. and *Trialeurodes vaporariorum*, vegetables, ornamentals	Limited

Natural enemy species	Pest species	Type of natural enemy
Species under study, in addition to those mentioned above:		
Amblyseius spp.	*Bemisia argentifolii/tabaci*, *Trialeurodes vaporariorum*	Predator
Aschersonia aleyrodis	*Bemisia argentifolii/tabaci*, *Trialeurodes vaporariorum*	Pathogen
Aschersonia spp.	*Bemisia argentifolii/tabaci*	Pathogen
Amitus bennetti	*Bemisia argentifolii/tabaci*, *Trialeurodes vaporariorum*	Parasitoid
Amitus fuscipennis	*Trialeurodes vaporariorum*	Parasitoid
Beauveria bassiana	*Bemisia argentifolii/tabaci*	Pathogen
Chrysoperla carnea	*Bemisia argentifolii/tabaci*, *Trialeurodes vaporariorum*	Predator
Chrysoperla rufilabris	*Bemisia argentifolii/tabaci*	Predator
Dicyphus spp.	*Trialeurodes vaporariorum*	Predator
Encarsia inaron (= *Encarsia partenopea*)	*Bemisia argentifolii/tabaci*, *Trialeurodes vaporariorum*	Parasitoid
Encarsia luteola	*Bemisia argentifolii/tabaci*	Parasitoid
Encarsia nr. *meritoria*	*Trialeurodes vaporariorum*	Parasitoid
Encarsia pergandiella	*Bemisia argentifolii/tabaci*, *Trialeurodes vaporariorum*	Parasitoid
Encarsia transvena	*Bemisia argentifolii/tabaci*, *Trialeurodes vaporariorum*	Parasitoid
Encarsia tricolor	*Trialeurodes vaporariorum*	Parasitoid
Eretmocerus spp. (6 species)	*Bemisia argentifolii/tabaci*	Parasitoid
Euseius spp.	*Bemisia argentifolii/tabaci*, *Trialeurodes vaporariorum*	Predator
Metarhizium anisopliae	*Bemisia argentifolii/tabaci*, *Trialeurodes vaporariorum*	Pathogen
Paecilomyces fumosoroseus	*Bemisia argentifolii/tabaci*, *Trialeurodes vaporariorum*	Pathogen
Paecilomyces spp.	*Bemisia argentifolii/tabaci*	Pathogen

It has been extensively demonstrated in inoculative and seasonal inoculative biological control that introductions with individuals of one parasitoid species – and particularly with *E. formosa* – are sufficient for economically feasible whitefly control (Gerling, 1990; Onillon, 1990; Heinz, 1996). In warm climates, like the Mediterranean area, parasitoids of whitefly may immigrate into greenhouses and provide natural pest control. During the past 5 years other parasitoid species have also been tested and used to control whiteflies, like species from the genera *Eretmocerus* (Hymenoptera: Aphelinidae) and *Amitus* (Hymenoptera: Platygasteridae) (see Table 14.2). The following part of this chapter on the role of natural enemies is restricted mainly to the parasitoid *E. formosa* as this is the most important species commercially applied in augmentative releases.

14.4. Strategies Followed for Control of Whiteflies

Where pesticides are not used in agroecosystems and in natural ecosystems, an array of natural enemies usually keeps the numbers of whiteflies very low: predators, parasitoids and pathogens all contribute to whitefly mortality. Work in two cropping systems – tomatoes in the 60s in California (E.R. Oatman, pers. com.) and cotton during the period 1925–92 in Sudan (B. Munir, pers. com.) has shown that whiteflies can be kept under perfect *natural control*. When pesticides are applied, natural enemies are exterminated and whitefly pests – in the above cases *T. vaporariorum* and *B. tabaci,* respectively – are the result. Furthermore, changes in crop rotation, shortening of fallow periods, and concurrent or overlapping growth of whitefly susceptible crops may result in such a high and continuous whitefly pressure that natural enemies are not capable of adequately reducing whitefly numbers.

If natural control is insufficient one of the following biological control strategies can be pursued (van Lenteren, 1986):

(i) The *inoculative release* method, where only a relatively low number of beneficial organisms are released. This method has been successfully applied in field crops where a continuous existence of natural enemies can be guaranteed (Onillon, 1990). Because of the temporary production system in greenhouses, this method cannot be used here.

(ii) The *inundative release* method, where beneficial organisms are (periodically) released in large numbers to obtain an immediate control effect (i.e. use as biotic insecticide). An example of this method is the application of *E. formosa* against *B. tabaci* in poinsettia (Albert, 1990). Here, *E. formosa* is released weekly. The aim is not to build up a parasitoid population, because *B. tabaci* is a poor quality host for *E. formosa.* Another example of the inundative approach is the application of the entomopathogenic fungi *V. lecanii* and *A. aleyrodis* against greenhouse whitefly (Fransen, 1990).

(iii) The *seasonal inoculative release* method, where moderate numbers of natural enemies are released to obtain both an immediate control effect and also a build-up of a natural enemy population for control later during the same season. An example of this approach is the control of *T. vaporariorum* in greenhouses by *E. formosa* (van Lenteren and Woets, 1988). The parasitoid *E. eremicus* and the predator *M. caliginosus* are also used in this way.

Different methods for releasing natural enemies are used. The most common one for *E. formosa* is to hang cards with parasitized whitefly pupae in the crop. The parasitized pupae are glued onto cards (Natskova, 1987) or onto sticky tape over a hole in the card. The number of pupae is determined by the diameter of the sticky area. Alternatively, portions of plant leaf with parasitized pupae may be glued onto card or whole leaves given to growers to cut up. Unparasitized whitefly on the leaves may be killed by coating the leaf with a glue based on egg white (Nastkova, 1987) or separated from parasitized pupae by differential flotation. Parasitized pupae may also be sold to the grower loose in a bottle and measured volumes tipped into pots hanging amongst the plants (P. Walker, pers. com.). Some cards have a strip of honey agar to provide food for the wasps.

The moment of release, the number of introductions and the number of parasitoids released per introduction vary with crop, type of greenhouse and climate, and differs between biocontrol companies. In the 70s the release strategy for whitefly control in tomato in northwest Europe was as follows: four release were made of, on average, 2 wasps per plant per release, each with an interval of two weeks. The first release was planned shortly after the first whiteflies were observed in the greenhouse. This release strategy resulted in reliable control during the whole growing season. Based on long-term experience the strategy was adapted so that it would work under most conditions in tomato. Nowadays, about two months after planting the crop, five releases of, on average, 1.5 wasp per plant are made, each with an interval of one week when no or very few whiteflies are observed. When whiteflies are found, five releases of, on average, 3 wasps per plant are advised. For host plants like cucumber or eggplant, on which whiteflies develop faster than on tomato, weekly releases are made for a longer period, e.g. up to 20 weeks.

14.5. How does *Encarsia* Control Whitefly?

Intensive fundamental research on the relationship between *E. formosa*, greenhouse whitefly and host plants has provided information on how the parasitoid locates and attacks its hosts (Fig. 14.2), how temperature and host-plant architecture influence host-finding and parasitization efficiency, and how host-plant quality influences whitefly and parasitoid population dynamics (van Lenteren *et al.*, 1976, 1980). The parasitoid is not able to locate infested plants from a distance: searching is random on all levels, and after a host has been found the search pattern does not alter (Noldus and van Lenteren, 1990). The only change we detected is that, in comparison with search times on an uninfected leaf, a parasitoid keeps searching much longer on a leaf once a whitefly larva has been found or when other indicators of whitefly presence were discovered (e.g. honeydew, exuviae, dead hosts) (van Lenteren *et al.*, 1996). The efficiency of *E. formosa* in killing whiteflies once on infected plants can be explained as follows. Based on demographic data of host and parasitoid we can conclude that under greenhouse conditions the intrinsic rate of increase (r_m) of greenhouse whitefly on most plants is lower than the maximum host kill rate – the number of hosts killed by parasitism and host feeding – of the parasitoid (van Lenteren *et al.*, 1996). So that on many host plants an individual parasitoid can kill more hosts per unit of time than a whitefly female can produce offspring.

The limiting factor in parasitizing sufficient whiteflies is, thus, not the host kill rate, but the ability to find whiteflies. In order to evaluate the quantitative effects of all sorts of interrelated processes associated with host finding and parasitization, we have developed a model which is unique in that it is individual-based. The model simulates the local searching and parasitization behaviour of individual parasitoids in a whitefly-infested crop (van Roermund *et al.*, 1997). This model includes stochasticity and spatial structure. The model comprises submodels for: (i) the parasitoid's foraging behaviour; (ii) the whitefly and parasitoid population development; (iii) the spatial distribution of whitefly and parasitoid within and between plants in the crop; and (iv) leaf production. It links the population dynamics of whitefly and the parasitoid through simulation of the foraging behaviour of individual parasitoids in a crop. With the model we can simulate temporal and spatial dynamics of pest and natural enemy. Simulations with the model and verification/validation of model simulations in greenhouses have shown that with the release strategy mentioned at the end of Section 14.4, season-long biological control of whitefly is successful (van Roermund *et al.*, 1997). The model can be used: (i) to explain why the parasitoid can control whiteflies on some crops and not on others in large commercial greenhouses; (ii) to improve introduction schemes of parasitoids for crops where control was difficult; (iii) to predict effects of changes in cropping practices (e.g. greenhouse climate, choice of cultivars) on the reliability of biological control; and (iv) to develop criteria for the selection of natural enemies (van Lenteren and van Roermund, 1997).

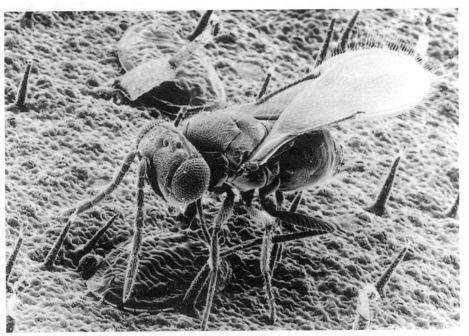

Figure 14.2. Scanning electron micrograph of an *E. formosa* female examining a greenhouse whitefly nymph.

14.6. When and Why does Biological Control of Whiteflies not Work?

A number of different factors interfered with biological control of greenhouse whitefly.

They are discussed here to illustrate the kinds of problems that may occur with biological control of *Bemisia.*

14.6.1. HOST PLANT TOO GOOD FOR WHITEFLY

On some host plants whitefly develops so fast that *E. formosa* is not able to sufficiently reduce whitefly numbers after initial inoculative releases (r_m whitefly too high). On such plants, like cucumber and eggplant, frequent inundative releases have to be applied to guarantee sufficient control. Another solution is to develop host plants that are partially resistant to whitefly.

14.6.2. HOST PLANT BAD FOR NATURAL ENEMIES

For many years, biological control of whitefly on cucumber has been difficult. Failure of good control by *E. formosa* is caused by three factors: (i) cucumber is a very good host plant for greenhouse whitefly resulting in very fast population growth of whitefly (r_m whitefly is high); (ii) cucumber has many large leaf hairs, which reduce the walking speed of *E. formosa* when searching for hosts; and (iii) the many hairs retain honeydew droplets which "catch" searching parasitoid females (van Lenteren, 1990). Intensive research involving basic studies of searching behaviour of the parasitoid and host-plant breeding resulted in the selection of a hybrid with half the number of hairs found on commercial cucumber cultivars. Laboratory and greenhouse experiments showed improved searching and higher percentages of hosts killed (host feeding + parasitization) on these "half-haired" hybrids than on the "haired" commercial cucumber cultivars (for details, see van Lenteren, 1990; van Lenteren *et al.*, 1995). As a result, whitefly were reduced to lower numbers on the "half-haired" hybrids.

14.6.3. POOR QUALITY NATURAL ENEMIES

A multitude of causes, including disease and genetic change, may result in production or delivery of natural enemies of poor quality. Due to the development of some large producers, parasitoids of good quality are currently available throughout the year. Recently, quality control guidelines have been developed which are now being implemented by the natural enemy producers (van Lenteren, 1999).

14.6.4. CROP MANAGEMENT PRACTICES DELETERIOUS FOR BIOLOGICAL CONTROL

Climate Conditions Unsuitable
When winter day time greenhouse temperatures stay below the flight threshold of *E. formosa*, poor parasitism allows whitefly populations to increase to levels which are too high for the parasitoid to control when warmer weather arrives in spring. A solution is to increase greenhouse temperatures to above the flight threshold for 2–4 hours per day. When fungi are used for whitefly control, climate conditions conflict with plant pathogen control. Entomopathogenic fungi demand a high humidity for many hours during spore germination, but such long humid periods also stimulate development of plant pathogenic fungi and can, therefore, not always be created.

Interference with Pesticides
Many examples show that application of chemical pesticides reduces the activity, or completely exterminates natural enemies of whiteflies. Some pesticides adhere to plastic in the greenhouse and are released over many weeks. Some pesticides have been identified as less harmful for *E. formosa* and can, therefore, be used with great care to control other pests and diseases (see Chapter 11).

Removal of Leaves with Immature Parasitoids
Before biological control of whitefly was first introduced, growers routinely removed the lower leaves. These leaves usually carried many parasitized pupae (Fig. 14.3). Now growers still remove the leaves but leave them in the greenhouse until parasitoids have emerged. They can also thin the leaves instead of removing them all. Some old greenhouses are too low to allow the leaves to remain on the plant long enough during winter. In addition to enabling better plant growth, taller, modern greenhouse allow leaves to be left on the plants longer.

Introduction of Parasitoids too Late
When many whitefly invade a greenhouse or the whitefly density is very high at the moment of parasitoid introduction, the build up of a natural enemy population is too slow to keep the population below the "damage" threshold. In this case an initial treatment with entomopathogenic fungi may help to greatly reduce the whitefly population, thereafter *E. formosa* can be introduced. Proper timing of the fungal spray and parasitoid introduction is essential (Fransen, 1990).

Hygiene
Too many whitefly at the start of the crop or invasions from outside can overwhelm the biological control agent. Between crops, it is important to prevent whitefly from persisting in the empty greenhouse by removing all weeds and hanging up yellow sticky traps to catch any adults. Weeds around the greenhouse, as potential hosts for whyteflies, should be killed before the old crop is pulled out so that escaping whitefly have nowhere to live. The old crop should be removed in sealed containers from the property or, if left on the property, it should be buried or covered immediately. The crop can be fumigated the night before removal to kill adult whiteflies.

14.7. Conclusions

During the past 30 years excellent results have been obtained with seasonal inoculative biological control of whitefly in greenhouses. In West Europe alone biological control of greenhouse whitefly is applied on about 4000 ha and growers consider it a more reliable method than chemical control. Biological control of another whitefly species, *B. tabaci/argentifolii*, is also possible in greenhouses, but demands more supervision for two reasons. First, control of *B. tabaci/argentifolii* is more difficult with the presently available parasitoids or predators than control of *T. vaporariorum*. Secondly, an additional difficulty with *B. tabaci/argentifolii* is its virus transmission, which is particularly problematic in the tropics and subtropics where greenhouses are prone to invasion with whiteflies from the field.

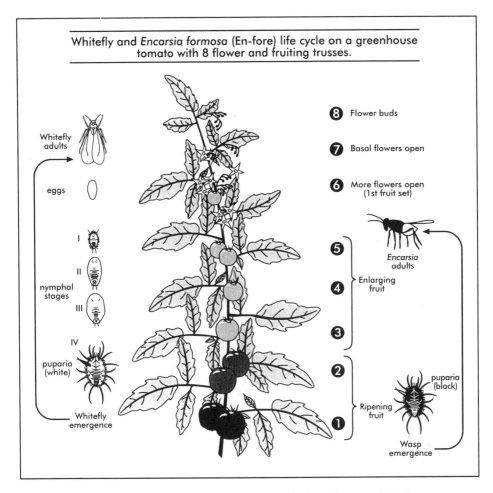

Whitefly and *Encarsia formosa* (En-fore) life cycle on a greenhouse tomato with 8 flower and fruiting trusses.

Whitefly adults

eggs

nymphal stages
I
II
III
IV

puparia (white)

Whitefly emergence

❽ Flower buds

❼ Basal flowers open

❻ More flowers open (1st fruit set)

❺

Encarsia adults

❹ Enlarging fruit

❸

❷

Ripening fruit

❶

puparia (black)

Wasp emergence

Figure 14.3. The vertical distribution of greenhouse whitefly and its parasitoid, *E. formosa*, on greenhouse tomatoes. Each life stage of whitefly and *E. formosa* is found in a particular layer of leaves. The later insect stages are found on older leaves, which are further from the top of the plant (drawing New Zealand Inst. for Crop and Food Research).

The benefits of biological control of whitefly are reduced environmental pollution, a healthier work environment for growers, healthier, more robust plants and access to markets where consumers require low or nil pesticide residue products.

Development of biological control programmes demanded interdisciplinary research. Plant anatomists, geneticists, entomologists, ecologists and modellers co-operated in acquiring a basic understanding of whitefly and *Encarsia* biology. Based on this knowledge, crop protection specialists, entomologists, plant breeders, agronomists, producers of natural enemies, extensionist specialists and growers together designed a practical, economically attractive whitefly biological control programme as part of integrated pest management for the most important vegetable crops and some ornamental crops.

References

Albert, R. (1990) Experiences with biological control measures in glasshouses in Southwest Germany, *IOBC/WPRS Bulletin* **13**(5), 1–5.

Butler, G.D., Henneberry, T.J. and Hutchison, W.D. (1986) Biology, sampling and population dynamics of *Bemisia tabaci*, *Agricultural Zoology Reviews* **1**, 167–195.

Byrne, D.N. and Bellows, T.S. (1991) Whitefly biology, *Annual Review of Entomology* **36**, 431–457.

Fransen, J.J. (1990) Natural enemies of whitefly: Fungi, in D. Gerling (ed.), *Whiteflies: Their Bionomics, Pest Status and Management*, Intercept Ltd, Andover, Hants, pp. 187–210.

Gerling, D. (ed.) (1990) *Whiteflies: Their Bionomics, Pest Status and Management*, Intercept Ltd, Andover, Hants.

Gerling, D., and Mayer, R.T. (eds.) (1996) *Bemisia 1995: Taxonomy, Biology, Damage, Control and Management*, Intercept, Andover, Hants.

Heinz, K.M. (1996) Predators and parasitoids as biological control agents of *Bemisia* in greenhouses, in D. Gerling and R.T. Mayer (eds.), *Bemisia 1995: Taxonomy, Biology, Damage, Control and Management*, Intercept, Andover, Hants, pp. 435–449.

Lacey, L.A., Fransen, J.J. and Carruthers, R. (1996) Global distribution of naturally occurring fungi of *Bemisia*, their biology and use as biological control agents, in D. Gerling and R.T. Mayer (eds.), *Bemisia 1995: Taxonomy, Biology, Damage, Control and Management*, Intercept, Andover, Hants, pp. 401–433.

Lei, H., Tjallingii, W.F. and van Lenteren, J.C. (1996) Recording of EPG's and honeydew excretion by the greenhouse whitefly, in D. Gerling and R.T. Mayer (eds.), *Bemisia 1995: Taxonomy, Biology, Damage, Control and Management*, Intercept, Andover, Hants, pp. 53–68.

Martin, N.A., Ball, R.D., Noldus, L.P.J.J. and van Lenteren, J.C. (1991) Distribution of greenhouse whitefly *Trialeurodes vaporariorum* (Homoptera, Aleyrodidae) and *Encarsia formosa* (Hymenoptera, Aphelinidae) in a greenhouse tomato crop: Implications for sampling, *New Zealand J. of Crop and Horticultural Science* **19**, 283–290.

Martin, N.A. and Dale, J.R. (1989) Monitoring greenhouse whitefly puparia and parasitism: A decision approach, *New Zealand J. of Crop and Horticultural Science* **17**, 115–123.

Nastkova, V. (1987) Economical method for mass rearing of *Encarsia formosa*, *IOBC/WPRS Bulletin* **10**(2), 121–124.

Noldus, L.P.J.J. and van Lenteren, J.C. (1990) Host aggregation and parasitoid behaviour: Biological control in a closed system, in M. Mackauer, L.E. Ehler and J. Roland (eds.), *Critical Issues in Biological Control*, Intercept Ltd, Andover, Hants, pp. 229–262.

Noldus, L.P.J.J., Xu Rumei, Eggenkamp-Rotteveel Mansveld, M.H. and van Lenteren, J.C. (1986a) The parasite-host relationship between *Encarsia formosa* (Hymenoptera: Aphelinidae) and *Trialeurodes vaporariorum* (Homoptera: Aleyrodidae). XX. Analysis of the spatial distribution of greenhouse whiteflies in a large greenhouse, *J. of Applied Entomology* **102**, 484–498.

Noldus, L.P.J.J., Xu Rumei and van Lenteren, J.C. (1986b) The parasite-host relationship between *Encarsia formosa* (Hymenoptera: Aphelinidae) and *Trialeurodes vaporariorum* (Homoptera: Aleyrodidae). XVIII. Between-plant movement of adult greenhouse whiteflies, *J. of Applied Entomology* **101**, 159–176.

Noldus, L.P.J.J., Xu Rumei and van Lenteren, J.C. (1986c) The parasite-host relationship between *Encarsia formosa* (Hymenoptera: Aphelinidae) and *Trialeurodes vaporariorum* (Homoptera: Aleyrodidae). XIX. Feeding-site selection by the greenhouse whitefly, *J. of Applied Entomology* **101**, 492–507.

Onillon, J.C. (1990) The use of natural enemies for the biological control of whiteflies, in D. Gerling (ed.), *Whiteflies: Their Bionomics, Pest Status and Management*, Intercept Ltd, Andover, Hants, pp. 287–313.

Perring, T.M. (1996) Biological differences of two species of *Bemisia* that contribute to adaptive advantage, in D. Gerling and R.T. Mayer (eds.), *Bemisia 1995: Taxonomy, Biology, Damage, Control and Management*, Intercept, Andover, Hants, pp. 3–16.

Russell, L.M. (1977) Hosts and distribution of the greenhouse whitefly, *Trialeurodes vaporariorum* (Westwood) (Hemiptera: Homoptera: Aleyrodidae), *USDA Cooperative Plant Pest Report* **2**, 449–458.

van Lenteren, J.C. (1986) Parasitoids in the greenhouse: Successes with seasonal inoculative release systems, in J.K. Waage and D.J. Greathead (eds.), *Insect Parasitoids*, Academic Press, London, pp. 341–374.

van Lenteren, J.C. (1990) Biological control in a tritrophic system approach, in D.C. Peters, J.A. Webster and C.S. Chlouber (eds.), *Aphid-Plant Interactions: Populations to Molecules*, Miscellaneous Publication No. 112, USDA/ARS – Oklahoma State University, Stillwater, Oka., pp. 3–28.

van Lenteren, J.C. (1993) Parasites and predators play a paramount role in pest management, in R.D. Lumsden and J.L. Vaughn (eds.), *Pest Management: Biologically Based Technologies*, American Chemical Society, Washington, DC, pp. 68–181.

van Lenteren, J.C. (1995) Integrated pest management in protected crops, in D. Dent (ed.), *Integrated Pest Management*, Chapman and Hall, London, pp. 311–343.

van Lenteren, J.C. (1999) Quality control of mass produced beneficial insects, in K.M. Heinz, R. van Driesche and M.P. Parrella (eds.), *Biological Control of Arthropods Pests in Protected Culture*, Ball, Batavia (accepted).

van Lenteren, J.C., Benuzzi, M., Nicoli, G. and Maini, S. (1992) Biological control in protected crops in Europe, in J.C. van Lenteren, A.K. Minks and O.M.B. de Ponti (eds.), *Biological Control and Integrated Crop Protection: Towards Environmentally Safer Agriculture*, Pudoc, Wageningen, pp. 77–89.

van Lenteren, J.C., Drost, Y.C., van Roermund, H.J.W. and Posthuma-Doodeman, C.J.A.M. (1997a) Aphelinid parasitoids as sustainable biological control agents in greenhouses. *J. of Applied Entomology* **121**, 473–458.

van Lenteren, J.C., Li Zhao hua and Kamerman, J.W. (1995) The parasite-host relationship between *Encarsia formosa* (Hymenoptera: Aphelinidae) and *Trialeurodes vaporariorum* (Homoptera: Aleyrodidae). XXVI. Leaf hairs reduce the capacity of *Encarsia* to control greenhouse whitefly, *J. of Applied Entomology* **119**, 553–559.

van Lenteren, J.C., Nell, H.W. and Sevenster-van der Lelie, L.A. (1980) The parasite-host relationship between *Encarsia formosa* (Hymenoptera: Aphelinidae) and *Trialeurodes vaporariorum* (Homoptera: Aleyrodidae). IV. Oviposition behaviour of the parasite, with aspects of host selection, host discrimination and host feeding, *J. of Applied Entomology* **89**, 442–454.

van Lenteren, J.C., Nell, H.W., Sevenster-van der Lelie, L.A. and Woets, J. (1976) The parasite-host relationship between *Encarsia formosa* (Hymenoptera: Aphelinidae) and *Trialeurodes vaporariorum* (Homoptera: Aleyrodidae). I. Host finding by the parasite, *Entomologia Experimentalis et Applicata* **20**, 123–130.

van Lenteren, J.C. and Noldus, L.P.P.J. (1990) Whitefly-plant relationships: Behavioural and ecological aspects, in D. Gerling (ed.), *Whiteflies: Their Bionomics, Pest Status and Management*, Intercept Ltd, Andover, Hants, pp. 47–89.

van Lenteren, J.C., Roskam, M.M. and Timmer, R. (1997b) Commercial mass production and pricing of organisms for biological control of pests in Europe, *Biological Control* **10**, 143–149.

van Lenteren, J.C. and van Roermund, H.J.W. (1997) Why is the parasitoid *Encarsia formosa* so successful in controlling whiteflies? in B.A. Hawkins and H.V. Cornell (eds.), *Theoretical Approaches to Biological Control*, Cambridge University Press, Cambridge, pp. 116–130.

van Lenteren, J.C., van Roermund, H.J.W. and Suetterlin, S. (1996) Biological control of greenhouse whitefly (*Trialeurodes vaporariorum*): How does it work? *Biological Control* **6**, 1–10.

van Lenteren, J.C., van Vianen, A., Hatala-Zseller, I. and Budai, Cs. (1989) The parasite-host relationship between *Encarsia formosa* (Hymenoptera: Aphelinidae) and *Trialeurodes vaporariorum* (Homoptera: Aleyrodidae). XXIX. Suitability of two cultivars of sweet pepper, *Capsicum annuum* L. for tow different strains of whiteflies, *J. of Applied Entomology* **108**, 113–130.

van Lenteren, J.C and Woets, J. (1977) Development and establishment of biological control of some glasshouse pests in the Netherlands, in F.F. Smith and R.E. Webb (eds.), *Pest Management in Protected Culture Crops*, ARS-NE-85, USDA, AS, Washington, DC, pp. 81–87.

van Lenteren, J.C. and Woets, J. (1988) Biological and integrated pest control in greenhouses, *Annual Review of Entomology* **33**, 239–269.

van Roermund, H.J.W. and van Lenteren, J.C. (1992) The parasite-host relationship between *Encarsia formosa* (Hymenoptera: Aphelinidae) and *Trialeurodes vaporariorum* (Homoptera: Aleyrodidae). XXXIV. Life-history parameters of the greenhouse whitefly, *Trialeurodes vaporariorum* as a function of host plant and temperature, *Wageningen Agricultural University Papers* **92**(3), 1–102.

van Roermund, H.J.W., van Lenteren, J.C. and Rabbinge, R. (1997) Biological control of greenhouse whitefly
    with the parasitoid *Encarsia formosa* on tomato: An individual-based simulation approach, *Biological
    Control* **9**, 25–47.
Speyer, E.R. (1927) An important parasite of the greenhouse whitefly (*Trialeurodes vaporariorum*, Westwood),
    Bulletin of Entomological Research **17**, 301–308.
Vet, L.E.M. and Dicke, M. (1992) Ecology of infochemical use by natural enemies in a tritrophic context,
    Annual Review of Entomology **37**, 141–172.

CHAPTER 15

BIOLOGICAL CONTROL OF MITES

Don A. Griffiths

15.1. Introduction

Commercial biological programmes aimed at controlling mite pests primarily use predatory mites as controlling agents. All of the available predators used in commercial crops belong to the Phytoseiidae. This group is composed of large mites ranging in size from 400 to 700 microns in length and in colour from pale grey, to shades of olive, to brown and bronze red. As predators, they move faster than their prey and have well developed searching abilities.

The pest species fall into three distinct groups. The Tetranychidae (350–500 microns long) are known as the spider mites because of their ability to produce silk webbing. The Tarsonemoidea are smaller (100–300 microns in length) and oval with colourless, white, yellow or olive brown shining cuticles. The third group, the Eriophyoidea, the russet, blister or gall mites, are the smallest of all mites (80–250 microns long) with characteristic worm or wedge shaped bodies and only two pairs of legs, positioned anteriorly.

15.2. Pest Species Taxonomy

Each pest species represents a genetic entity, with its own set of ecological, physiological and biological characteristics, which together influences its potential as a pest (such as its tolerance to pesticides or its interaction with predators). Correct pest identification at species level is thus an essential ingredient of any biological programme. Because mites are very small, however, they require specialized microscope techniques. They generally lack morphological distinguishing characters at species level within a genus, thus needing the services of experienced taxonomists. As a result, most pest identifications are conveniently limited to field examinations using a hand lens. The number of eriophyid and tarsonemid species attacking protected and semi-protected crops is small. Because they are host specific and their damage symptoms are typical of the species, microscopic examination to determine their identity is often judged to be unnecessary and is rarely undertaken.

Spider mite identification is more complex, not only for the reasons given above, but because more than fifty species have been recognized as pests, often with a wide range of hosts and a cosmopolitan distribution. Further, populations found in protected and semi-protected crops and many field situations are usually identified as one or other of the two most common spider mite pests. The first, *Tetranychus urticae* Koch, the two-spotted mite, has green summer females, while the second, *Tetranychus*

R. Albajes et al. (eds.), Integrated Pest and Disease Management in Greenhouse Crops, 217-234.
© 1999 *Kluwer Academic Publishers. Printed in the Netherlands.*

cinnabarinus (Boisduval), the carmine mite, has females which are dark red. Both mites are recognized as major pests on a global scale. A long standing argument prevails, however, as to whether they represent one genotype which is phenotypically very variable, and whether the complex of species associated with each of them should be synonymized with one or the other, or both, of the principal taxa.

It is not surprising, therefore, that the literature of the past 50 years contains some 17 major publications deliberating over the question as to whether *T. urticae and T. cinnabarinus* are discrete species. Eleven authors consider that they are, whilst six argue that they are conspecific and should be synonymized. A further eight works are concerned with the specific status of their associated sibling species namely, *Tetranychus arabicus* Attiah, *Tetranychus cucurbitacearum* (Sayed) and *Tetranychus ricinus* Sayed. Whether they are junior synonyms of one or other, or both, of the two major pests has yet to be resolved conclusively but, after careful examination of the published arguments, I believe that *T. urticae, T. cinnabarinus* and *T. ricinus* are all valid species, and the evidence put forward for synonymizing *T. arabicus* with *T. urticae*, and *T. cucurbitacearum* with *T. cinnabarinus* is insufficient to be convincing. Further work is required to prove the true standing of these two taxa. Until such studies are carried out, the *T. urticae* species complex should be considered to contain five distinct species, and this chapter is based on this premise. The argument to support these views will appear elsewhere (Griffiths and Starzewski, in preparation).

15.3. The Spider Mites

The two-spotted spider mite, *T. urticae*, is truly cosmopolitan. Its original natural range was probably within the temperate regions of the Northern Hemisphere, but the intercontinental trade in plants has made it a world species. It has been recorded from over 200 plant species (Jeppson *et al.*, 1975). In contrast, *T. cinnabarinus* occurs naturally throughout the Mediterranean region and eastwards into Asia, but is now found at an increasing frequency infesting protected crops in most temperate areas of the world (Table 15.1). Whilst in temperate zones *T. cinnabarinus* does not seem to be as prolific as *T. urticae*, it seems to be far more phytotoxic to certain plants, particularly tomato. Records for the three other members of the complex show them to be confined for the most part to the hotter areas of the Mediterranean region, where they are each considered to be a major pest (Table 15.1). In the hotter areas of the world, including the Mediterranean region, other *Tetranychus* species occur, as well as in temperate areas to a lesser degree. These are much less likely to be recorded, either because they are rare, or because they are rarely identified correctly. Examples are *Tetranychus turkestani* Ugarov & Nikolskii, *Tetranychus ludeni* Zacher, *Tetranychus canadensis* (McGregor), *Tetranychus desertorum* Banks and *Tetranychus viennensis* Zacher (see Jeppson *et al.*, 1975, for details).

15.3.1. THE IDENTIFICATION OF SPIDER MITES

Field identification of spider mites is based on the physical appearance of the summer female. In particular, the background body colour is significant, as well as the presence or absence of black spots on the antero-lateral margins of the body. Damage symptoms and the identity of the host may also be important. Thus, the body colour of the female and that of its egg, although far from definitive taxonomically, should be used as a preliminary sift. Saba (1975) provides a good guide to colours exhibited by members of the *T. urticae* complex. Jeppson *et al.* (1975) gives data for the field and microscopic identification of the other species mentioned above. Where micromorphological characters are concerned, the shape of the lobes on the dorsal opisthosomal striae can be effective in separating *T. urticae* (rounded lobes) from *T. cinnabarinus* (triangular lobes) (see Boudreaux and Dosse, 1963; Griffiths and Sheals, 1971). There is one correct area within which lobes should be viewed. This is the laterally sited "diamond" of striae which is located within the rhomboid formed by the bases of the d2 and d3 pairs of dorso-central setae. The lobes must then be viewed at right angles to their flat plane.

TABLE 15.1. Distribution/pest status records for *T. cinnabarinus* and other Mediterranean species

Pest	Region	Crop	Status	Authority
Tetranychus urticae	Temperate areas (e.g. Europe, USA, Argentina, Japan)	200 host plants Protected vegetables, ornamentals	Serious	Boudreaux (1956), Jeppson *et al.* (1975), Guppta (1993)
	India, eastern Uttar Pradesh	50 host plants Vegetables especially cucurbits	Major	Guppta (1993), Arbarbi and Singh (1996b)
Tetranychus turkestani	Europe, Middle East, USA, Japan	At least 35 hosts Vegetables, strawberry	Serious	Jeppson *et al.* (1975)
Tetranychus cinnabarinus	Mediterranean	Field and protected crops	Major	
	Turkey	Aubergines, artichokes	Major	Soysal and Yayla (1988)
	Greece (Crete)	Tomato, carnation, aubergine, beans	Major	Hatzinikolis (1969)
	Jordan	Field aubergines	Serious	Bitton and Nakash (1986)
	Israel	Field tomato	Serious	Berlinger *et al.* (1983)
	Egypt	Cotton, peppers, groundnuts, figs, lucerne, apple, pear, cucurbits	Widespread, major	Osman and El Keie (1975), Osman and Rasmy (1976)
Tetranychus cucurbitacearum	Egypt	Vegetables, cotton, groundnuts	Often	Osman and Fattah (1975)
	Morocco	Vegetables, fruit	Rarely	Saba (1979)
Tetranychus arabicus	Egypt	Vegetables, cotton	Frequent	Zaher *et al.* (1982)
Tetranychus ricinus	Morocco	Vegetables	Frequent	Saba (1979)

A major micro-morphological character for the separation of species in the Tetranychidae is the minute differences in the shape of the male aedeagus (Pritchard and Baker, 1952; Jeppson *et al.*, 1975), although some authors, notably Saba (1975), disagree as to its value. Careful specimen preparation is required if the small differences in shape are to be identified successfully. For light microscope studies, a few male specimens should be first placed on a slide. Whilst the medium is still liquid, press the top of the coverslip gently, using a blunt cocktail stick. Do this whilst the slide is under the lowest power of the microscope, examining the results frequently until an aedeagus is seen to have assumed the correct lateral position. For scanning electron microscope studies it is best to critical point dry the specimens. This sometimes causes the aedeagal apparatus to be extruded. If not, plunge a few live specimens into a small amount of very hot water beforehand. This technique also works well for phytoseiid mites. In all cases it is a matter of trial and error until the technique is perfected.

15.3.2. SPIDER MITE BIOLOGY, LIFE HISTORY, DEVELOPMENT AND CONTROL

Spider-mite colonies live on the underside of the leaf, protected by a screen of silk threads laid down by the females. All stages (larvae, nymphs and adults) feed from the plant. Feeding leads to the disappearance of chloroplasts that show up as tiny white spots on the upper surface of the leaf. Careful study of laboratory infested plants is probably the best way to learn to recognize typical damage symptoms. Particular attention should be paid to learning to identify early symptoms of attack, since predator introductions must start as soon as there is prey for them to eat. The speed of development is directly related to the temperature and humidity in the immediate vicinity of the colony and to the amount and quality of the available food. An example, comparing *T. urticae* development with that of its natural predator *Phytoseiulus persimilis* Athias-Henriot, taken from Sabelis (1981) and Osborne *et al.* (1985), is given in Table 15.2.

TABLE 15.2. Development time in days of *T. urticae* and its predator *P. persimilis*

Temperature (°C)	Development time in days of life-cycle stages					
	Egg	Larva	1st Nymph	2nd Nymph	Oviposition	Total
Tetranychus urticae						
15	14.3	6.7	5.3	6.6	3.5	36.3
20	6.7	2.8	2.3	3.1	1.7	16.6
25	2.8	1.3	1.2	1.4	0.6	7.3
Phytoseiulus persimilis						
15	8.6	3.0	3.9	4.1	5.6	25.2
20	3.1	1.1	1.4	1.6	1.9	9.1
25	1.7	0.6	0.8	0.8	1.1	5.0

Tetranychus urticae will not develop below temperatures of 14°C, nor above 40°C. However, colonies can well survive low temperatures, even under frost conditions, becoming active as soon as radiant heat is provided by the sun. Relative humidity requirements for egg development are 70% and above. Active stages appear to withstand lower humidities since they can be found on exposed foliage in hot dry conditions. The life history parameters for *T. cinnabarinus* have been well documented by Hessein (1975) and Witul (1992). The latter author showed that tomato plants reacted most severely to relatively small populations of this mite, whereas damage to the other was less, although the density of mites was higher. At 24°C, the net reproductive rate was lowest on peppers (×4), slightly higher on tomato (×8) and highest (×50) on cucumber.

Regarding diapause, again, the data refers mainly to *T. urticae*, since little is known about other species. At the end of the summer, under certain conditions of shortening day length, gravid females will seek a protected niche to go into hibernation. Their body colour darkens: reddish-orange for *T. urticae* and purple for *T. cinnabarinus*. During this time *T. urticae* can withstand severe cold conditions down to at least 10°C (Parr and Hussey, 1966). *Tetranychus urticae* requires a chilling period, of varying length depending on geographic location, before termination can occur, but according to Vaz Nunnes (1986) *T. cinnabarinus* in Greece does not need a "cold rest". The temperature prevailing at reactivation influences the length of the cold rest period. The higher the temperature, the shorter the cold rest time. Indeed Helle (1962) and Parr and Hussey (1966) showed that for *T. urticae* at a reactivation temperature of 25°C no rest period was needed. Vaz Nunnes (1986) showed this was also true for the Greek population of *T. cinnabarinus*, and further that a short day condition was sufficient to hold mites in diapause during hot Greek autumn days. The newly emerged female will start to lay eggs within a few days, gradually reverting to its summer colour. In practice, it is usual for females to emerge over a period of weeks, even in the same glasshouse. Thus, it is essential to start to monitor for emerging females very early in the spring if predator releases are to have the maximum effect.

The three predators used against spider mites are *P. persimilis*, *Neoseiulus* (= *Amblyseius*) *californicus* (McGregor) and *Feltiella acarisuga* (Vallot), which are discussed later (Sections 15.6 and 15.9). Predators can be used on a hot spot basis or preventatively. In the absence of experienced monitoring staff, the better option is to introduce several low dose applications before the pest is expected to occur. If it is thought that monitoring will be a problem, this option should always be taken.

15.4. Eriophyid Pest Species

There is at least one species of eriophyid for every plant species on earth. It is not surprising, therefore, that the state of eriophyid taxonomy, like that of the tetranychids, is unsatisfactory. However, there are compensations in that host specificity is said to be so restricted (usually one mite species to a particular plant species) that host identification automatically identifies the pest. The number of recognized pests from this group associated with protected vegetables and ornamentals and field vegetables is extremely small. The one major pest is dealt with below.

15.4.1. *Aculops lycopersici* (Massee)

The tomato russet mite is a serious pest throughout the Mediterranean region and in California, but is less common in temperate countries. It is closely associated with plants belonging to the Solanaceae family. Its common natural hosts are black nightshade (*Solanum nodiflorum* Jacq.) and bindweed (*Convolvulus arvensis* L.), from which it is able to spread to vegetable crops. Consequently, it is a particularly serious pest of tomatoes and eventually kills the infected plant if left untreated. In North African countries bordering the Mediterranean Sea, serious outbreaks occur in potato crops. It will also attack peppers, aubergines and tobacco. On tomato plants, early symptoms appear as a slight thickening of the main stem just above ground level, with some signs of pink-brown vertical striations. Later, leaves turn brown and wither to a parchment-like consistency, with fruit becoming russetted and hard. This damage is very well illustrated in Keifer *et al.* (1982).

Aculops lycopersici is a very small mite. The female in its largest stage is 150 to 180 microns long, which makes detection of the pest very difficult. Usually it is not discovered until the infestation is well established. In Mediterranean plastic houses, early attack is confined to a scattering of plants throughout the crop and, by the time it is found, workers will have distributed the mites to adjacent plants. It is recommended that monitoring begin very early, soon after the crop is planted, paying particular attention to the main stems for the appearance of the symptoms described above.

As integrated pest programmes come into more frequent use in Mediterranean protected crops, with a consequential reduction in the use of hard chemicals, the incidence of tomato russet mite may well increase. Thus, there is a real need for trial work to begin now. At present, there are no recognized biological programmes for this pest, but based on personal experience, it seems likely that *Neoseiulus* (= *Amblyseius*) *cucumeris* (Oudemans) would prove an adequate predator. It is recommended that trials include a general application of the predator at a rate of 50–100 per plant, applied around the base of the stem soon after planting. Later, whenever an infested plant is discovered, a higher dose of 500 to 1000 predators should be applied to the infected plant with a lower dose for adjacent plants. However, because there is some question as to the mobility of *N. cucumeris* on tomato plants, preliminary tests to determine dispersal and establishment may be necessary. These predators need to be placed on young leaves above the damage level, where most of the mite activity will be concentrated.

15.5. Tarsonemid Pest Species

Since there are only two recognized major pest species in this group, and one is confined to strawberries, identification relies upon field examination. Under a hand lens, the females appear as tiny hard shiny oval shapes. Often, they are pale olive in colour. The very large eggs can also be seen. Males are much smaller than the females and a different shape. They are more elongate with modifications to their posterior margin and legs, designed to capture and carry a female resting nymph, thus assuring them of a mating partner. Tarsonemids like warm temperatures, high humidities and low light hence their habit of living in the folds of unopened leaves.

15.5.1. *Polyphagotarsonemus latus* (Banks)

This pest, known as the "broad mite", the "yellow tea mite" and the "tropical mite", has a host range of over 50 plants, including tea, cotton, rubber, tobacco, potatoes and beans. However, this disparate range of hosts may well indicate the presence of a complex of species rather than an impressive distribution. Crops most affected are tomatoes, peppers, potatoes and gerbera. In tomatoes and peppers, the stems of terminal shoots and the underside of leaves become shiny and bronzed. As the infestation develops, the top of the plant appears scorched as if by a flame. The leaves dry up, parchment-like and the upper stems become swollen. During the summer period, providing moisture levels are high, a generation can be completed in as little as five days. This species will also keep breeding slowly throughout the winter period. Again, biological programmes have not been formulated, but it is strongly recommended that trials along the same lines as those suggested for the tomato russet mite be tested.

15.5.2. *Phytonemus fragariae* Zimmermann

For many years this taxon was identified as *Tarsonemus pallidus* Banks, a mite that was considered a serious pest of cyclamen as well as strawberry. It was removed into the genus *Stenotarsonemus* and then separated into two taxa, sometimes to sub-species level and sometimes recognized as discrete species, under the specific names *pallidus* – the cyclamen mite, and *fragariae* – the strawberry mite. I prescribe to the discrete species school. Lindquist (1987) then selected *T. pallidus* to be the type of a new genus *Phytonemus*, hence *P. fragariae*. Because of its tarsonemid characteristics, infestations are rarely found until symptoms are so advanced that in field strawberries the first signs are discovered about mid summer. In protected strawberries, the pest develops much earlier and serious problems can arise early in spring. Infected plants can be recognized by the absence of new, upright leaves, with curled edges to the older leaves. In extreme cases there are reflowering stems, so that the plant has the appearance of a halo of brown leaves, the older of which will be curled at the edges. Usually, infested plants are scattered throughout the crop.

Protected strawberries should be monitored from as early as January, whilst field crops will not show detectable symptoms until June. Early detection can only succeed if scouters open young closed leaves and, using a strong hand lens, examine the main vein for the presence of clusters of tiny white eggs and the slightly larger pale olive females.

Commercial control programmes employing *N. cucumeris* are now available in the UK and it has been used in California to counteract sudden outbreaks. The predator is applied in a vermiculite carrier from a shaker bottle. Infested plants must be dosed with 100 to 500 predators per plant. Depending on the severity of the attack, heavier doses should be applied to flowerless plants. Since this pest is easily spread, adjacent plants in the row for a distance of 25 m should each receive 25–50 predators. If prophylactic doses are to be applied early in the season, to succeed they must be at the rate of at least

10 predators per plant. Early monitoring of unopened leaves is the preferred recommendation for efficiency.

15.6. Commercially Available Predaceous Mites

Sadly, the range of predator species sold and actively used to control mite pests is very small. The main commercial lines are listed in Table 15.3.

TABLE 15.3. Commercially available mite predators and their target species

Predator	Main target	Commercial status
Phytoseiulus persimilis	Tetranychid spp.	World-wide
Neoseiulus (= Amblyseius) californicus	Tetranychid spp.	USA, Europe
Neoseiulus (= Amblyseius) cucumeris	Tarsonemid spp. Trials on eriophyids	World-wide
Phytoseiulus longipes	*Tetranychus* spp.	Very limited
Neoseiulus (= Amblyseius) fallacis	Orchard mites	Very limited

15.6.1. *Phytoseiulus persimilis*

This mite, the natural predator of spider mites, is the most widely used of all the commercial predators directed at controlling mites. It is an amber-red shiny pear-shaped mite that is slightly larger than its prey, compared with which it moves much faster. It is an indigenous species of the Mediterranean region, although an accidental introduction of a population from Chile is said to be the source of the original commercial material (Dosse, 1958). It is host specific, preferring the eggs and young stages of its prey. Its development rate under different temperature regimes is compared with that of *T. urticae* in Table 15.2. It should be noted that development from egg to egg is considerably shorter than that of the spider mite. Furthermore, its intrinsic rate of increase (the number of daughters produced per female per day), as well as the total number of daughters per mother, is much higher than it is for *T. urticae*. For example, at 20°C, the figures are 0.219 and 44 daughters for *P. persimilis* compared to 0.143 and 30 daughters for *T. urticae*. However, whilst this faster development is an important factor in its success as a predator, it is the initial ratio of pest to predator which decides the outcome of a control programme. Thus, in commercial programmes, the initial dose of predators must be sufficient to achieve control economically but before the damage threshold is reached. This ratio will vary between crops and with different geo-climatic conditions, so that the advice of the commercial company providing the predators should always be sought. Even so, failures do occur: these are discussed later in the text.

15.6.2. *Neoseiulus californicus*

Neoseiulus californicus is indigenous to California, where it is commercially produced.

Production of this mite has recently begun by many European producers of beneficial agents. Their objective is to use it in the expanding market of the Mediterranean region against indigenous spider mite pests. It has been recorded as occurring naturally in Italy, and is said to be widespread in Spain. The taxonomic provenance of these records requires verification, however. In some northern European countries, release licenses have been issued for its restricted commercial use in a few specified protected crops and in field strawberries.

Laboratory studies by Gilstrap and Friese (1985) indicate that at temperatures below 30°C and relative humidities between 50–70%, *P. persimilis* outperforms *N. californicus*, but commercial producers are looking to use it against spider mites in conditions of high temperatures and low humidities. Castagnoli and Simoni (1995) calculated an intrinsic rate of increase of 0.337 at a fixed temperature of 33°C. This rate of increase at such a high temperature is ideal for use in Mediterranean crops, since the maximum temperature for reproduction in *P. persimilis* is only 28°C (Sabelis, 1981). However, there is little firm evidence regarding its ability to survive low humidities. Nevertheless, in combination with *P. persimilis* it should provide a new approach to spider mite control in hot countries. At present, introduction rates for *N. californicus* equate with rates for *P. persimilis*.

15.6.3. *Neoseiulus cucumeris*

Neoseiulus cucumeris has been produced world-wide in very large quantities since the Controlled Release System (CRS) was developed to combat the western flower thrips (Sampson *et al.*, in preparation). This predator is now recommended for routine use against the strawberry tarsonemid (*P. fragariae*), both in protected and field crop situations (see Section 15.5.2 for control procedures). It is also recommended here that it be tested against eriophyid pests, in particular the tomato russet mite, and against the tarsonemid pest, the broad mite (see Sections 15.4.1 and 15.5.1). Two reasons why it should perform as a good predator of these particular pests is that it can be produced and used, in large quantities, relatively cheaply and, secondly, it has the ability to penetrate into very small crevices. Tarsonemid and eriophyid mites are secretive and small so that large numbers build up before the pest is detected, requiring large initial doses of the predator, which must then be capable of reaching the hidden populations.

15.7. Factors Influencing the Efficacy of Biological Programmes Used to Control Mite Pests

Where predatory mites have or are being used to control mites at commercial, laboratory or field trial level, failures will be encountered from time to time. There is an extensive list of factors which, acting alone or in unison, can contribute to such failures. The major contributors are considered below.

15.7.1. PREDATOR ATTRIBUTES

The predator should preferably be prey specific for the particular pest against which it is targeted. *Phytoseiulus persimilis* represents such an example, in which case success is much more certain if introductions can be made when the pest population is at a low level. If made too early, the predators will move out of the crop very quickly in search of food. Therefore, in certain cases, a predator which, in the absence of the pest, can survive on an alternative food source is equally useful in that introductions can be made before the appearance of the pest. *Neoseiulus cucumeris* and *N. californicus* are examples of mite predators that can use pollen as an alternative food source. Within the parameters of its physical environment, the predator's intrinsic rate of development must be greater than that of its prey. Another major factor in achieving success will be its ability to operate at low humidities. Thus, the parameters embracing the life style of a predator must be seriously considered before it is selected for a control programme.

15.7.2. INTRODUCTION PROCEDURES

Once the beneficial agent or agents have been selected on the basis of their suitability, certain procedures must be followed. If it is a newly planted crop, then predators must be introduced as soon as the first pests are seen, or earlier if they can exist on an alternative food source. Delaying the introduction is a major factor in programme failure since ratios of predator to pest must be realistic for cost purposes as well as control success. It is particularly important with mite predators that in an initial prophylactic introduction they are introduced evenly throughout the crop. However, if monitoring can be carried out efficiently, it is sometimes preferable to wait until the first pests are seen, and then introduce high numbers of predators into the early "hot spot". An example would be the introduction of *P. persimilis* in early spring to control populations developing from females emerging from diapause. Regular monitoring must then continue since mite pest populations can increase very rapidly.

The introduction ratio of mite predator to pest will depend on a number of factors such as the crop itself, the pest population and the prevailing physical parameters. The basic data, according to Scopes (1985), for the time *P. persimilis* is expected to take to reach control over *T. urticae* is given in Table 15.4. It clearly demonstrates the point that early application is more efficient and cost effective. It also suggests that an undetected high population or a sudden invasion from a neighbouring crop is best treated with a beneficial-compatible pesticide, after which a realistic rate of predators can be applied.

15.7.3. APPLICATION METHODS

In the early days of biological control, predators such as *P. persimilis* (reared on the pest species on bean leaves) were distributed on leaves collected from the rearing bed when the number of *T. urticae* became minimal. This technique is still preferred by some growers who believe a mixed stage population establishes better than one made up almost entirely of adults. The method can only apply to local situations since most nations ban the importation of such plant material.

TABLE 15.4. The influence of predator/prey ratios of *P. persimilis* (P.p.) and *T. urticae* (T.u.) and temperature, on the time taken (in days) to obtain control

Ratio of P.p./T.u.	Temperature (°C)		
	24	18	13
1:200	16	28	47
1:100	13	25	47
1:50	11	25	35
1:20	9	20	29

The modern alternative is to use moistened bran or vermiculite, or mixtures of both, or, for predators of insect pests, peat or buckwheat husks. A wide range of concentrations of predators is available to suit different requirements. For example, for a general application of the predator *P. persimilis* to strawberry beds, a low concentration of predators in a high volume of carrier is required. When the same predator is being applied to hot spots in, say tomatoes, it is better to have a high concentration in a small container; for example 2000 predators per 25 ml of carrier in a 30 ml vial. If a small hole is made in the lid, the predators can be accurately dispensed in small numbers onto horizontal leaves individually selected. On average, some five predators are tipped out for each tap on the side of the vial. *Neoseiulus californicus* can be distributed in the same way as *P. persimilis*, as can *N. cucumeris* when used on strawberry beds, but the need to employ large doses of *N. cucumeris* for most treatments means it is never supplied in small vials. However, a unique distribution method has been developed for distributing this predator, namely, the CRS. A breeding colony of the predator is supplied with a breeding colony of a non-pest mite species as food, all contained within a moisture retaining breathable paper sachet, supplied with or without hooks. Just before the sachets are placed in the crop, each sachet is pierced to make a small hole from which predators can emerge. The population parameters within the sachet are geared to provide an exodus of predators over five or more weeks. The main use of CRS sachets is in the control of thrips, but it should provide an excellent method of introducing the large numbers of *N. cucumeris* needed to control some of the eriophyid pests.

Since beneficial agents are living creatures, it is essential to follow the storage and application instructions provided on the label. In addition, when using these products for the first time or under new regimes, consult the supplier or failure can ensue.

15.7.4. DIAPAUSE IN PHYTOSEIID MITES

Winter diapause in phytoseiid mites is sometimes quoted as the factor causing biological programmes to fail in Mediterranean autumn and winter crops. However, such failures should not apply to the three principle agents discussed in the above paragraph. *Phytoseiulus persimilis* and *N. californicus* are not known to enter diapause, although there does not appear to have been any scientific examination of these claims. European strains of *N. cucumeris* do diapause (Morewood and Gilkeson, 1991), but the major commercial producers now use a non-diapausing strain that is said to have originated in New Zealand.

So far, scientific investigations have been limited to about ten temperate zone species (Overmeer, 1985), which show that, in each case, the inductive mechanisms involved are virtually the same as those described earlier for tetranychid mites. However, the critical inductive night temperature varies from species to species. Rock et al. (1971) showed that for *Neoseiulus* (= *Amblyseius*) *fallacis* (Garman) it is a high 27°C, compared with a low of 19°C for *Metaseiulus occidentalis* (Nesbitt) (Croft, 1971). Thus, temperate zone predators may perform inadequately when used commercially during the autumn and winter periods of southern latitudes. Therefore, selection criteria for new predators should always include an investigation of their diapause characteristics.

15.7.5. PERSISTENT PESTICIDE RESIDUES

The undetected presence of pesticide residues probably accounts for a good percentage of reported failures. Compatibility charts, giving persistence times for chemicals applied to vegetable crops, are readily available from commercial suppliers. However, for ornamental and flower crops, with woody or long term foliage, persistence is now known to be considerably longer than the published data indicate. Roses grown on a long term cropping programme of five or more years are an example of a crop where persistence periods become much extended. When introducing predators, particularly *P. persimilis*, into crops of this type for the first time, simple survival tests employing a few representative mature plants are recommended. Further information on pesticide selectivity is provided in Chapter 11.

15.8. Performance Profiles of Some Potential Candidates, Proposed for Future Use in Programmes to Control Mite Pests

McMurtry et al. (1970) and McMurtry (1982) listed some 40 species which were considered to be actually or potentially useful in the control of tetranychid mite pests. Below, I have selected species, some of which if developed to commercial realization may resolve many of the current problems caused by relying on too few products for a market that is expanding into new crops and new geographic regions.

15.8.1. *Phytoseiulus longipes* Evans

Indigenous country: Zimbabwe and Cape Province, South Africa. Introductions: Californian orchards, but did not establish (McMurtry, 1977); Egypt, cucumber plots (El Laithy et al., 1996).

Characteristics: the life history and development has been comprehensively studied by Badii and McMurtry (1984). At temperatures between 20 and 35°C its performance closely parallels that of *P. persimilis*, but importantly it performs better at lower humidities, giving egg hatch percentages shown in Table 15.5 (data from Badii and McMurtry, 1984).

TABLE 15.5. *Phytoseiulus longipes*, percentage egg hatch at various temperatures and humidities

RH (%)	Temperature (°C)			
	20	25	30	35
50	95	82	44	–
60	100	100	85	65
70	100	100	85	65

According to these authors, *P. longipes* has a higher percentage eclosion at 50% RH than most phytoseiids. Under dry, hot spring conditions in an Egyptian cucumber crop, at min./max. temperatures and RH of 17–35°C and 37.5–60%, *P. longipes* successfully controlled *T. urticae*, and in so doing out competed *P. persimilis* (El Laithy *et al.*, 1996).

15.8.2. *Neoseiulus (= Amblyseius) longispinosus* (Evans)

Indigenous countries: well distributed in India, also recorded from Japan and Taiwan. Introductions: recorded once by Zaher (1986) from cotton in Egypt, presumed to have been an introduction.

Characteristics: indigenous in the tea gardens of Japan where it has developed resistance to a range of pesticides. Arbarbi and Singh (1996a) in laboratory trials, using *T. cinnabarinus* as prey, compared its performance with seven other phytoseiid species indigenous to Uttar Pradesh region, India. It came equal first with *Amblyseius indicus* Naryan & Gear. Lo (1990) considered that of the 28 species of phytoseiids recorded in Taiwan, this species was the most promising as a control against spider mites. It gave good control against *Tetranychus kanzawai* Kishida, indigenous to Taiwan, but was less successful against *T. urticae*, an exotic species of only a few years residence in the country.

15.8.3. *Amblyseius pseudolongispinosus* Xin

Indigenous country: China. Introductions: none.

Characteristics: development times from egg to adult at 20, 30 and 35°C was 7.5, 3.4 and 3.5 days, respectively (Xin *et al.*, 1984). Eggs laid at 27°C and 75% RH, then placed at 37.5°C for up to eight hours, and returned, showed a 98% hatch rate. Larval mortality amongst these eggs was 17%, indicating that this species would have a high survival rate during hot daytime conditions. However, at 26°C and an RH of 50%, the hatching rate dropped to 14%, which suggest that compared with *N. longispinosus*, this species needs a higher humidity at the lower end of its temperature range. It has been successfully trialled in China against spider mites (species not defined) on watermelon, aubergine and cotton.

15.8.4. *Amblyseius andersoni* (Chant)

Indigenous country: north and south Europe. Recorded by various authors on vines in

France and Germany, where it has been successfully used to control phytophagous mites. Ragusa *et al.* (1996) recorded it from nine species of plants in Sicily.

Characteristics: McMurtry (1977) reported an oviposition rate of about two eggs per day, feeding on *T. urticae*. Further, Ragusa *et al.* (1995) showed it is capable of full development on a range of pollens. He also showed its performance was better when reared on *Panonychus citri* (McGregor). The above data suggests this species may be better adapted for living on arboreal crops rather than protected vegetables, even so, because it is a European species, it may be selected for use on the grounds that it could be sold into a large market.

15.8.5. *Euseius gossypi* (El Badry)

Indigenous country: widespread and common on plants in Egypt (Zaher, 1986). Introductions: none known.

Characteristics: according to Zaher (1986), it feeds on a range of spider mites including *T. urticae* and *T. cucurbitacearum*. He further reports that, when reared on *T. urticae* at 30°C, a female lives for 30 days, on average, during which time she consumes over 1000 eggs, or 900 immatures, or 300 adult spider mites. Its performance at low humidities is unknown, but its geographic location indicates it to be a favourable candidate for trials in other Mediterranean countries.

15.8.6. *Neoseiulus* (= *Amblyseius*) *idaeus* Denmark & Muma

Indigenous countries: recorded from Brazil, where it inhabits hot, dry areas (De Moraes and McMurtry, 1983). Also found in Columbia, where it was reported as surviving a period of three months drought (Herrera *et al.*, 1994).

Characteristics: Dinh *et al.* (1988a,b) studied its development at 26°C and a low RH of 55%. Under these conditions it produced an intrinsic rate of increase of 0.279, making it a potential predator candidate for use in low humidity conditions.

15.8.7. *Neoseiulus fallacis*

Indigenous countries: probably North America (USA and Canada) where it is frequently recorded in apple orchards and hop yards (Burrell and McCormick, 1964).

Introductions: this is an unusually well travelled predator, having been mass released with *P. persimilis* on strawberries in Taiwan (Lee and Lo, 1990), and into apple orchards in Ganzu Province, China (Wu *et al.*, 1991), and in 1986–1987, into apple orchards in Japan from material imported ex New Zealand (Sekita and Kinota, 1990). It has also been found on strawberries in Switzerland and Portugal. These last records probably reflect introductions on imported plants from the USA.

Characteristics: Smith and Newsome (1970), in their biological study of this species, considered it to possess the attributes required by a successful predator of tetranychids in that it achieved varying, but reasonable, degrees of control over a number of tetranychid and oligonychid species. One characteristic reported by Croft *et al.* (1993), which may restrict its use to temperate regions, is that at 20°C the lethal humidity response (LH50) was about 70%.

15.9. The Predaceous Midge *F. acarisuga*

The predaceous midge, *F. acarisuga*, in some cases sold under its junior synonym name of *Therodiplosis persicae* Kieffer, is the only insect predator of mites to receive commercial recognition, with its use being confined to spider mite control in protected crops. Like many insect predators, it is a density dependent species requiring a large well established colony of its prey to be present before it can substantially increase in numbers, after which it will then disperse to attack smaller colonies of the pest. Accordingly, it is best used within a *P. persimilis* programme in crops such as cucumber or tomato, where economic thresholds are generally able to cope with moderate populations of spider mites. It is therefore unsuitable for use in ornamental or flower crops.

Feltiella acarisuga is sold in the pupal form and should, for the reasons given above, be applied into a proportion of hot spots that must be flagged and monitored very carefully. Some commercial packs may contain the pupae plus a quantity of *P. persimilis* together with a small food source of spider mite, however there is no firm evidence that such a system improves the efficacy of *F. acarisuga*. Its role within biological programmes aimed at controlling spider mites needs further investigation. Its best function may be as a "biological pesticide" used if a chemical control suitably safe to beneficial agents is not available.

15.10. Future Requirements in Research and Commercial Development

The pool of phytoseiid species from which new beneficial agents can be selected is enormous, but its potential will never be realized unless the need for more faunal surveys is recognized, and the lack of sufficient supporting specialist taxonomists is understood and acted upon. Despite these drawbacks, which need to be addressed elsewhere, the most important task of institutional and commercial research should be to try to enhance the performance of established agents, particularly the selection of new strains, which can operate at extremes of temperature and humidity.

At the strategic level of research, new candidates, such as those discussed in Section 15.8, must be selected incorporating the above criteria into the commercial specification. There is also a need for life history and development studies on selected candidates to be increased. These should be patterned upon the methodologies such as can be found in Gilstrap and Friese (1985) and in Dinh *et al.* (1988a,b).

Since the commercial development of new agents will undoubtedly involve the use of exotic species, an important ingredient of this research must be the investigation of the physical and ecological parameters of the candidate, including their potential to enter diapause. Castagnoli and Simoni (1995) provide a good example of the required approach. Such data will be needed on an increasing scale as governments tighten their quarantine laws (see Chapter 11).

References

Arbarbi, M. and Singh, J. (1996a) The efficiency of eight phytoseiid mites (Phytoseiidae) as predators of *Tetranychus cinnabarinus* (Boisd.) (Tetranychidae), in R.M. Mitchel, D.J. Hom, G.R. Needham and W.C. Welbourn (eds.), *Proceedings of the IX International Congress of Acarology*, Ohio Biological Services, Columbus, Ohio, pp. 195–200.

Arbarbi, M. and Singh, J. (1996b) *Tetranychus cinnabarinus* (Boisd.) (Tetranychidae): A serious mite pest of vegetables in India, in R.M. Mitchel, D.J. Hom, G.R. Needham and W.C. Welbourn (eds.), *Proceedings of the IX International Congress of Acarology*, Ohio Biological Services, Columbus, Ohio, pp. 201–202.

Badii, M.H. and McMurtry, J.A. (1984) Life history of and life table for *Phytoseiulus* longipes with comparative studies on *P. persimilis* and *Typhlodromus occidentalis* (Acari: Phytoseiidae), *Acarologia* **25**, 112–123.

Berlinger, M.J., Dahan, R. and Cohen, S. (1983) Greenhouse tomato pests and their control in Israel, *IOBC/WPRS Bulletin* **6**, 7–11

Bitton, S. and Nakash, J. (1986) Control of red spider mites by the predaceous mite *Phytoseiulus persimilis* in open fields of eggplants and artichokes (globe), *Hassadeh* **66**, 682–684.

Boudreaux, H.B. (1956) Revision of the two-spotted spider mite (Acarina, Tetranychidae) complex, (*Tetranychus telarius* Linnaeus), *Annals of the Entomological Society of America* **49**, 43–48.

Boudreaux, H.B. and Dosse, G. (1963) The usefulness of new taxonomic characters in females of the genus *Tetranychus* Dufour (Acari, Tetranychidae), *Acarologia* **5**, 13–33.

Burrell, R.W. and McCormick, W.J. (1964) *Typhlodromus* and *Amblyseius* (Acarina, Phytoseiidae) as predators on orchard mites, *Annals of the Entomological Society of America* **57**, 483–487.

Castagnoli, M. and Simoni, S. (1995) The effect of high temperature on development and survival rate of *Amblyseius californicus* (McGregor) eggs (Acari: Phytoseiidae), in O. Kropcynska, J. Boczek and A. Tomczyk (eds.), *Physiological and Ecological Aspects of Acari – Host Relationships*, Oficyna DABOR, Warsaw, pp. 357–363.

Croft, B.A. (1971) Comparative studies on four strains of *Typhlodromus occidentalis* (Acarina: Phytoseiidae). V. Photoperiodic induction of diapause, *Annals of the Entomological Society of America* **64**, 962–964.

Croft, B.A., Messing, R.H., Dunley, J.E. and Strong, B.W. (1993) Effects of humidity on eggs and immatures of *Neoseiulus fallacis*, *Amblyseius andersoni*, *Metaseiulus occidentalis* and *Typhlodromus pyri* (Phytoseiidae): Implications for biological control on apple, cranberry, strawberry and hop, *Experimental & Applied Acarology* **17**, 451–459.

De Moraes, G.J. and McMurtry, J.A. (1983) Phytoseiid mites (Acari) of northwestern Brazil with descriptions of four new species, *International J. of Acarology* **9**, 131–148.

Dinh, N.V., Janssen, A. and Sabelis, M.W. (1988a) Reproductive success of *Amblyseius idaeus* and *A. anonymus* on a diet of two-spotted spider mites, *Experimental & Applied Acarology* **4**, 41–51.

Dinh, N.V., Sabelis, M.W. and Janssen, A. (1988b) Influence of humidity and water availability on the survival of *Amblyseius idaeus* and *A. anonymus* (Phytoseiidae), *Experimental & Applied Acarology* **4**, 27–40.

Dosse, G. (1958) Uber einige neue Raumilbenarten (Acari, Phytoseiidae), *Pflanzenschutzberichte* **21b**, 44–61.

El Laithy, A.Y.M., El Halawany, M.E. and Abo Donia, M.G. (1996) Effectiveness of three phytoseiid predatory mites (Phytoseiidae) for biological control of the two-spotted spider mite, *Tetranychus urticae* Koch (Tetranychidae), on cucumber grown on arid soils in Egypt, in R.M. Mitchel, D.J. Horn, G.R. Needham and W.C. Welbourn (eds.), *Proceedings of the IX International Congress of Acarology*, Ohio Biological Services, Columbus, Ohio, pp. 191–194.

Gilstrap, F.E. and Friese, D.D. (1985) The predatory potential of *Phytoseiulus persimilis*, *Amblyseius californicus* and *Metaseiulus occidentalis* (Acarina: Phytoseiidae), *International J. of Acarology* **11**, 163–168.

Griffiths, D.A. and Sheals, J.G. (1971) The scanning electron microscope in Acarine Systematics, in V.H. Heywood (ed.), *Scanning Electron Microscopy; Systematic and Evolutionary Applications*, Systematics Association Special Volume No. 4, London, pp. 67–94.

Griffiths, D.A. and Starzewski, J. (in preparation) A review of the species complex surrounding *Tetranychus urticae* Koch and *Tetranychus cinnabarinus* (Boisduval).

Guppta, S.K. (1993) *Guide to the Agriculturally Important Mites of India with Illustrated Keys and Field Keys for their Easy Identification*, Technical Bulletin No. 3, Univ. Agri. Sci., Bangalore.

Hatzinikolis, E. (1969) Preliminary notes on tetranychid and eriophyid mites infesting cultivated plants in Greece, in G.O. Evans (ed.), *Proceedings of the II International Congress of Acarology*, Hungarian Academy of Sciences, Budapest, pp. 161–167.

Helle, W. (1962) Genetics of resistance to organphosphorus compounds and its relation to diapause in *Tetranychus urticae* Koch (Acari), Tijdschrift Plantenziekten **68**, 155–195.

Herrera, F.C.J., Guerrero, J.M. and Braun, A.R. (1994) Impact of predatory mites (Acari; Phytoseiidae) associated with cassava on *Monychellus* spp. on the Atlantic coast of Columbia, *Revista Colombiana de Entomología* **20**, 137–142.

Hessein, N.A. (1975) Morphology and biology of the carmine mite *Tetranychus cinnabarinus* (Boisduval) (Acarina: Tetranychidae), *Libyan J. of Agriculture* **4**, 117–122.

Jeppson, L.R., Keifer, H.H. and Baker, E.W. (1975) *Mites Injurious to Economic Plants*, University of California Press, Berkley, Cal.

Keifer, H.H., Baker, E.W., Kono, T., Delfinado, M. and Steyer, E. (1982) *An Illustrated Guide to Plant Abnormalities Caused by Eriophyid Mites in North America*, USDA Agriculture Handbook No. 573.

Lee, W.T. and Lo, K.C. (1990) Integrated control of two-spotted mite on strawberry in Taiwan, *Chinese J. of Entomology*, Special publication No. 3, 125–137.

Lindquist, E. (1987) The world genera of the Tarsonemidae (Acari: Heterostigmata): A morphological, phylogentic and systematic revision, with a re-classification of family-group taxa in the Heterostigmata, *Memoirs of the Entomological Society of Canada* **136**, 517.

Lo, K.C. (1990) The role of a native phytoseiid mite *Amblyseius longispinosus* (Evans) in the biological control of spider mites, in D.A. Griffiths and C.E. Bowman (eds.), *Acarology VI*, Ellis and Horwood Ltd, Chichester, pp. 703–709.

McMurtry, J.A. (1977) Some predaceous mites (Phytoseiidae) on citrus in the Mediterranean region, *Entomophaga* **22**, 19–30.

McMurtry, J.A. (1982) The use of phytoseiids for biological control: Progress and future prospects, in M.A. Hoy (ed.), *Biological Control of Mite Pests*, Agr. Sci. Pubs., University of California, Berkley, Cal., pp. 23–39.

McMurtry, J.A., Huffaker, C.B. and van de Vrie, M. (1970) Ecology of tetranychid mites and their natural enemies: A Review. I. Tetranychid enemies: Their biological characters and the impact of spray practices, *Hilgardia* **40**, 331–390.

Morewood, W.D. and Gilkeson, L.A. (1991) Diapause induction in the thrips predator *Amblyseius cucumeris* (Acarina: Phytoseiidae) under greenhouse conditions, *Entomophaga* **36**, 253–263.

Osborne, L.S., Ehler, L.E. and Nechols, J.R. (1985) *Biological Control of the Two-spotted Spider Mite in Greenhouses*, Technical Bulletin No. 853, University of Florida, Gainsville, Flo.

Osman, A.A. and El Keie, I.A. (1975) On the populations and control of mites infesting pepper in Egypt (Acarina-Tetranychidae), *Bulletin of the Entomological Society of Egypt*, Economic Series **9**, 133–136.

Osman, A.A. and Fattah, M.I. (1975) Ecological studies on tetranychid mites infesting peanut in Egypt with some reference to their control (Acarina: Tetranychidae), *Bulletin of the Entomological Society of Egypt*, Economic Series **9**, 127–132.

Osman, A.A. and Rasmy, A.H. (1976) The role of alfalfa in dispersing tetranychid mites to other crops, *Bulletin of the Entomological Society of Egypt*, Economic Series **60**, 279–283.

Overmeer, W.P.J. (1985) Diapause, in W. Helle and M.W. Sabelis (eds.), *Spider Mites: Their Biology, Natural Enemies and Control*, Vol. 1B, Elsevier, Amsterdam, pp. 95–102.

Parr, W.J. and Hussey, N.W. (1966) Diapause in the glasshouse red spider mite (*Tetranychus urticae* Koch): A synthesis of present knowledge, *Horticultural Research* **6**, 1–21.

Pritchard, A.E. and Baker, E.W. (1952) A guide to the spider mites of deciduous fruit trees, *Hilgardia* **21**, 253–287.

Ragusa, S., Chiari, D. and Tsolakis, H. (1995) Influences of different kinds of food substances on the post-embryonic development and oviposition rate of *Amblyseius andersoni* (Chant) (Parasitiformes, Phytoseiidae), in D. Kropcynska, J. Boczek and A. Tomczyk (eds.), *Physiological and Ecological Aspects of Acari – Host Relationships*, Oficyna DABOR, Warsaw, pp. 411–419.

Ragusa, S., Chiari, D. and Tsolakis, H. (1996) A survey of phytoseiid mites (Phytoseiidae) associated with various plants in Sicily (Italy), in R.M. Mitchel, D.J. Hom, G.R. Needham and W.C. Welbourn (eds.), *Proceedings of the IX International Congress of Acarology*, Ohio Biological Services, Columbus, Ohio, pp. 253–256.

Rock, C., Yeargan, D.R. and Rabb, R.L. (1971) Diapause in the phytoseiid mite *Neoseiulus* (T) *fallacis, J. of Insect Physiology* 17, 1651–1659.

Saba, F. (1975) An analysis of the Tetranychid complexes in the Mediterranean area, *Zeitschrift fur Angewandte Entomologie* 79, 384–389.

Saba, F. (1979) The two Tetranychid complexes of Morocco, in E. Piffl (ed.), *Proceedings of the IV International Congress of Acarology*, Akademiai Kiado, Budapest, pp. 223–225.

Sabelis, M.W. (1981) *Biological Control of Two-spotted Spider Mites Using Phytoseiid Predators. Part 1. Modelling the Predator Prey Interaction at the Individual Level*, Agricultural Research Report No. 910, Pudoc, Wageningen.

Sampson, C., Griffiths, D.A. and Wall, C. (in preparation) The development and application of the Controlled Release System (CRS), employing *Amblyseius cucumeris*, for use against Western Flower Thrips.

Scopes, N.E.A. (1985) Biological control of spider mites, in N.W. Hussey and N.E.A. Scopes (eds.), *Biological Pest Control: The Glasshouse Experience*, Blandford Press, Poole, pp. 43–52.

Sekita, N. and Kinota, M. (1990) Use of predatory mites for controlling spider mites (Acarina, Tetranychidae) on apple in Aomori prefecture, Japan, in *FFTC-NARC International Seminar on The Use of Parasitoides and Predators to Control Agricultural Pests*, Tubaku Science City, Ibaraki-ken, 2–7 October 1990, NARC, Tukuba-gun, p. 20.

Smith, J.C. and Newsome, L.D. (1970) The biology of *Amblyseius fallacis* (Acarina: Phytoseiidae) at various temperature and photoperiod regimes, *Annals of the Entomological Society of America* 63, 460–462.

Soysal, A. and Yayla, A. (1988) Preliminary studies on the population density of *Tetranychus* spp. (Acarina: Tetranychidae), harmful on vegetable crops and their natural enemies in Antalya, *Bitki Koruma Bülteni* 28, 29–41.

Vaz Nunnes, M. (1986) Some aspects of induction and termination of diapause in a Greek strain of the mite *Tetranychus cinnabarinus* (Boisduval) Boudreaux, 1956 (Acari: Tetranychidae), *Experimental & Applied Acarology* 2, 315–321.

Witul, A. (1992) Life history parameters of *Tetranychus cinnabarinus* on glasshouse plants, *OEPP/EPPO Bulletin* 22, 521–528.

Wu, Y.S., Liu, Y.L. and Zhang, L.Y. (1991) A preliminary report on the use of *Amblyseius fallacis* (Acari, Phytoseiidae) to control *Panonychus ulmi* (Acari, Tetranychidae) in Tianshui, Ganzu Province, *Chinese J. of Biological Control* 7, 160–162.

Xin, J.-l., Liang, L.-r. and Ke, L.-s. (1984) Biology and utilisation of *Amblysieus longispinosus* (Acarina Phytoseiidae) in China, in D.A. Griffiths and C.E. Bowman (eds.), *Proceedings of the VI International Congress of Acarology*, Ellis Horwood Ltd, Chichester, pp. 693–698.

Zaher, M.A. (1986) Survey and ecological studies on phytophagus, predaceous and soil mites in Egypt, PL 480 Programme USA, Project IVo. E G-ARS-30, Grant No. FG-EG.

Zaher, M.A., Gomaa, E.A. and El-Enany, M.A. (1982) Spider mites of Egypt (Acari: Tetranychidae), *International J. of Acarology* 8, 91–114.

CHAPTER 16

BIOLOGICAL CONTROL OF APHIDS

Jean-Michel Rabasse and Machiel J. van Steenis

16.1. Introduction

Aphids rank among the most serious pests of greenhouse crops. They confront very diverse and efficient natural enemies, whose characteristics will be explained. There is increasing demand for biological control of aphids: the current solutions will be examined below.

Compared to other groups of pests, like thrips or whiteflies, there are many more pest species among aphids. The two main reasons for the demand for biological control are that insecticide pressure has led to a high level of resistance in several species [e.g. *Aphis gossypii* Glover (Furk and Hines, 1993)] and that the increasing use of biological control against other pests increases the need for compatible measures against aphids.

Numerous aphid species occurring in the fields can become greenhouse pests, as the climatic factors and plant condition are often optimal for their development and reproduction.

Based on their host-plant preference, two groups of aphids can be distinguished:

(i) Polyphagous species, attacking a wide range of plants. The most common are *Myzus persicae* (Sulzer), *Myzus nicotianae* Blackman, *Macrosiphum euphorbiae* (Thomas) and *Aulacorthum solani* (Kaltenbach), which infest mainly Solanaceae but may also attack many other plants. *Aphis gossypii* attacks Cucurbitaceae and is also widely polyphagous. *Aulacorthum circumflexum* (Buckton) lives on a wide range of ornamentals, particularly in sheltered conditions.

(ii) Oligophagous species, such as *Nasonovia ribisnigri* (Mosley), *Hyperomyzus lactucae* (Linnaeus) and *Acyrthosiphon lactucae* (Passerini) on lettuce, *Rhodobium porosum* (Sanderson) on roses, *Chaetosiphon fragaefolii* (Cockerell) on strawberries, *Myzus ascalonicus* Doncaster and *Dysaphis tulipae* (Boyer de Fonscolombe) on bulbs, *Brachycaudus helichrysi* (Kaltenbach) and *Macrosiphoniella sanborni* (Gillette) on chrysanthemums.

Aphids are known as "r strategists", i.e. they are very well adapted to exploiting a new temporary habitat by rapid population increase. Their structure is simplified to enable them to perform best in feeding and reproduction, with most of their nutrients directed to reproduction. However, they have retained their ability to walk and fly. Winged morphs are specially produced when the population needs to escape towards a new food source.

The invasion of a greenhouse in spring is often due to alate migrants entering through the vents. Vents size and time of opening influence immigration. The invading aphids may come from their winter hosts or, later in the season or further south in Europe, they may be flying between summer hosts. In both cases, flight will not occur until outdoor temperatures are high enough. Thus, in 1982, the first *M. persicae* was observed in the

R. Albajes et al. (eds.), Integrated Pest and Disease Management in Greenhouse Crops, 235-243.
© 1999 *Kluwer Academic Publishers. Printed in the Netherlands.*

suction trap network, from the south to the north of France, in Pau on 11 April, in Orleans on 9 May, in Colmar on 6 June and in Arras on 4 July. Like most pests, aphids can also be distributed to numerous crops by nurseries propagating material, e.g. cuttings. Finally, contamination may also originate in the presence of weeds, old crops or year-round crops.

The initial infestation of a crop generally happens at a small number of isolated foci. The aphids reproduce quickly in these places, build up dense populations with overlapping generations and begin to colonize neighbouring plants. As colonies become more dense, alates are formed and disseminate throughout the crop. Under greenhouse conditions, i.e. with a more or less constant temperature and in the absence of natural enemies, aphid populations are able to grow exponentially for a considerable period (Rabasse, 1980b). This means that the number of aphids increases by a fixed proportion each day: commonly 0.2 or 0.3 females per female per day and up to 0.5 in *A. gossypii* (van Steenis and El-Khawass, 1995). This value is known as the r_m (maximal rate of increase) of the population. An easier way of putting it is that they increase by 4, 8 or 33 times a week, respectively. The intrinsic rate of increase depends on the aphid species and is greatly affected by the host plant and temperature. It is roughly proportional to temperatures between 10°C or less and 25°C; it stabilizes at around 25°C and decreases sharply to an upper lethal limit around 30°C. *Aphis gossypii* is more tolerant to high temperatures than the other species. Aphids are able to survive beyond this limit, which is often exceeded in southern countries, if night temperatures are low enough.

Many aphids are efficient vectors of virus diseases. The following conditions are necessary for a transmissible virus problem to arise: availability of a virus source (infected crop, weeds, etc.), presence of dispersing aphids, availability of plants in a susceptible state, sufficient time for expression of damage by the crop. Actually, these conditions do not often occur at the same time in greenhouses and the fact that aphids are virus vectors rarely hinders biological control of them.

16.2. Characteristics of the Potential Biological Control Agents of Aphids

The biological characteristics of the different groups of natural enemies – microbials, predators and parasitoids – are very diverse. An overview is presented in Table 16.1, and more detailed information is given in the following paragraphs.

TABLE 16.1. Short overview of biological characteristics of the most important natural enemies of aphids

Group	Development time	Reproductive capacity	Specificity	Dispersal capacity
Fungi	–	–	Moderate	High
Chrysopidae	Long	High	Low	Moderate
Coccinellidae	Long	High	Low	Moderate
Cecidomyiidae	Long	Moderate	Low	High
Aphidiidae	Short	High	High	High
Aphelinidae	Moderate	High	High	Moderate

16.2.1. MICROBIALS

Fungi are considered the principal group of aphid pathogens, although some research has examined viruses as well. Fungi are good candidates for biological control as they seem very effective in both natural and experimental conditions and many species are highly specific to aphids and harmless to other beneficial and non-target organisms. The spores of these fungi, ejected from sporulating cadavers or dispersed by air, germinate on a new host. The germ tube penetrates the aphid and the fungus invades the haemocoel and the host tissues. The host dies after a few days. Under high humidity conditions, sporophores are produced on the surface of the cadaver. The fungi most usually encountered on aphids belong to the Entomophthorales (Zygomycetes); they are characterized by forcibly discharged spores. Some species also form, in the host body, resting spores that lie dormant. They often produce epizootics in humid conditions. The ubiquitous Deuteromycete *Verticillium lecanii* (A. Zimmerm.) Viégas can also be found in particular environments like greenhouses. It produces only one kind of spore on verticilliate whorls.

Although Entomophthorales are specific microbials and, as such, difficult to cultivate, great progress has been made in their production *in vitro*. However, mycelium and ballistospores are too fragile and non-dormant resting spores germinating synchronously are still not available. Like other Deuteromycetes, *V. lecanii* can be cultivated in conventional mycological media; blastospores are produced in fermenters, formulated and commercialized. Even though commercial preparations are available for use on a wide range of crops, their use is restricted to conditions in which humidity can be kept at a very high level for a long time, e.g. on chrysanthemum under plastic shelter (Helyer and Wardlow, 1987).

16.2.2. PREDATORS

Among the polyphagous predators able to feed on aphids, some mirids usually colonize greenhouses in the Mediterranean area, limit the populations of different pests and play a part in a conservative kind of biological control. These bugs sting and suck the body contents of soft-bodied insects. They also sometimes feed on the plants, causing cosmetic damage or even stinging developing fruits. This habit is related to shortage of prey, rather than to their own density. *Dicyphus tamaninii* Wagner is widely distributed in northern Spain (Albajes *et al.*, 1996) where it plays a significant role. *Macrolophus caliginosus* Wagner (Malausa *et al.*, 1987; Alvarado *et al.*, 1997) is especially efficient on tomato. It is used in the form of augmentative releases, its main target being whiteflies.

Attention has long been paid to the large aphidophagous predators Chrysopidae and Coccinellidae. *Chrysoperla carnea* (Stephens) is a polyphagous predator accepting aphids as part of its spectrum. Much attention has been paid to this species, which is easy to produce and manipulate. The three larval instars are active predators and the adults feed mainly on pollen. Repeated introductions are necessary. Many young larvae disappear from the plants in the first days after release; older larvae settle more efficiently, but they are expensive to produce. For these reasons, control of aphid populations with *C. carnea* did not become a common practice. This species is considered relatively tolerant to some commonly used insecticides (Kowalska, 1976).

Coccinellids, as a group, are the most common and intensively studied predators of aphids. Both adults and larvae feed on aphids. Only the last instars have very great voracity. Lady beetles are not frequently found in greenhouses. Various attempts to use different species in greenhouses were limited because adults tend to fly to the windows and escape. So, only released larvae, which behave and disappear like *C. carnea* larvae, can be used for control. The idea of using a non-flying strain of a large polyphagous species of Chinese origin, *Harmonia axyridis* (Pallas) could be of interest in the near future (A. Ferran, pers. comm.). Another exotic species, *Hippodamia convergens* (Guérin-Méneville), from North America, is also a potential biological control agent.

In the small group of entomophagous Cecidomyiidae, *Aphidoletes aphidimyza* (Rondani) is the most widespread and polyphagous. The adults are very small (2.5 mm long) and fragile insects; they feed on honeydew. Adults are short-lived; mating and oviposition occur mainly at night. The progeny of one female is either male or female (monogenic reproduction). The female lays orange eggs near the aphids; she is able to detect isolated foci. The larvae pierce aphid bodies with their mandibles and suck their content, leaving shrivelled dark aphid bodies on the plant. The aphids seem anaesthetized during this action. This predator can attack most common aphid species. The full-grown larvae pupate in silk cocoons in the upper layer of the soil. The critical photoperiod for diapause at 20°C is 16 or 17 h (Havelka, 1980). Compared to other predators, larval voracity is limited and they can develop on relatively low numbers of aphids. In the presence of abundant prey, overkill can occur, i.e. they can kill more prey than they are able to feed on (Markkula and Tiittanen, 1985; Harizanova and Ekbom, 1997).

16.2.3. PARASITOIDS

All aphid parasitoids are solitary endoparasitoids and belong to two families of the order Hymenoptera: the Aphidiidae, which are the most important, and the Aphelinidae, which also parasitize other insects, mainly other Homoptera like whiteflies or scales.

The Aphidiidae are all parasitoids of aphids and include many genera, like *Praon, Ephedrus, Aphidius, Lysiphlebus, Monoctonus* or *Trioxys*. The adults are small (2 mm) slender wasps, with orange, brown or dark colours. Reproduction is mostly biparental: unfertilized eggs give rise to males and fertilized eggs to females (haplo–diploidy). When the female parasitoid meets an aphid, she generally bends her abdomen forward between her legs and lays swiftly a minute egg (0.1 mm long) in the host's body cavity. Once inside the aphid, the egg expands to several times its initial size. The larva hatches after a few days and then begins to feed osmotically. It passes through four instars. During the first three, it does not interfere much with the development of its host. In the fourth instar, the parasitoid consumes all the internal tissues of the aphid, completely filling its cuticle. Then, it cuts a slit in the underside of this cuticle and spins its cocoon inside. This is the "mummy", which is attached to the leaf by the silk appearing in the slit. Pupation, and in some cases diapause, occurs inside the mummy. Its colour is yellow, brown and even black in the genus *Ephedrus*. In the genus *Praon*, the cocoon is spun beneath the empty skin of the aphid. At 20°C, the mummy is formed after 8 days and the adult emerges, cutting a circular lid in the top of the mummy 5 days later in *Aphidius matricariae* Haliday. The females do not live longer than 1 or 2 weeks when fed honeydew or nectar (Rabasse and Shalaby, 1980; van Steenis, 1993).

In Aphelinidae, *Aphelinus* is the only genus of importance. The adults are smaller (1 mm) and thicker than Aphidiidae. The female inserts her ovipositor by backing up to her host. She may lay an egg or, in 10% of cases, she turns and feeds from the puncture (this behaviour is called host-feeding). Either operation leads to the death of the aphid. The mummy is not swollen; it is black and the exit hole is ragged. In general, *Aphelinus* has a lower population increase and a longer development time than Aphidiidae (e.g. the development of *Aphelinus abdominalis* (Dalman) until emergence is 21 days at 20°C). They produce fewer eggs per day, but live considerably longer, up to more than a month.

The potential fecundity of both groups of parasitoids can be very high; for instance, van Steenis (1993) observed more than 300 mummies formed per *Aphidius colemani* Viereck female and Monadjemi (1972) observed 250 in *Aphelinus asychis* Walker.

16.3. Successful Cases of Biological Control

16.3.1. PREDATORS

The midge *A. aphidimyza* is the only aphid predator produced on a wide scale. The utility of this species was first demonstrated in Finland (Markkula *et al.*, 1979). The midge larvae are produced and allowed to pupate on moist sand or peat. A key point in production is the control of the diapause, which, however, may facilitate storage. The diapause has to be avoided in the rearings by adjustment of the photoperiod. It can limit the efficiency of the midge in autumn in greenhouses. The use of non-diapausing strains avoids these drawbacks. In soilless cultures, the lack of a suitable pupation substrate can limit the population development of the midge. The cocoons are dispersed in the greenhouse. The adults are very mobile and establish rapidly. However, the release is generally repeated after 2 to 4 weeks. The use of open rearing units containing predatory midges feeding on an aphid unable to establish on the greenhouse crop is an alternative strategy.

Biological control with *C. carnea* involves 2nd instar larvae for a rapid effect (more than 95% of predation occurs after the 1st instar). The eggs can be hatched in cardboard containers and the neonate larvae fed eggs of *Ephestia kuehniella* Zeller. Such containers are released against *M. euphorbiae* and *C. fragaefolii* on strawberries at a rate of 8–10 larvae/m^2 (Celli *et al.*, 1991).

The coccinellid *H. convergens* is introduced in large numbers from the USA at relatively low cost. They are collected from hibernation sites and shipped to Europe. Bags with 10,000 adults are used for eradication of local aphid infestations. This high number is necessary as most of the adults disappear. Sometimes larvae can be found, but establishment in the greenhouse will not occur.

16.3.2. PARASITOIDS

Oligophagous parasitoids, i.e. species that can attack aphids belonging to related genera, are generally chosen for biological control. As they are able to increase rapidly, they are released inoculatively.

Aphis gossypii is the aphid causing most problems at the moment. It is the target of two main parasitoids. *Lysiphlebus testaceipes* (Cresson) is a widely polyphagous species introduced in 1973 into southern France from Cuba and now established in the northern part of the Mediterranean. It can colonize greenhouses spontaneously in that area (Rochat, 1997). *Aphidius colemani* is a complex of species (Rabasse *et al.*, 1985). The north Mediterranean strain does not attack greenhouse aphids. Strains introduced into the laboratory from South America and Africa have been tested. They control *M. persicae* well and parasitize *A. gossypii*. Both species can attack *A. gossypii* very efficiently. At present, *A. colemani* is used on a very large scale in northern Europe (van Steenis, 1992). But inoculative releases of *A. colemani* or *L. testaceipes* have not always been able to prevent outbreaks on cucumber in southern Europe, when the doubling time of the aphid population is around 2 days.

For the control of *M. persicae*, *A. matricariae* has been commercialized for more than 10 years and used on chrysanthemum and Solanaceae. In experiments on aubergine, the aphid populations can be controlled within 3 weeks from observation of the first mummies. *Aphidus matricariae* is ubiquitous (in particular holarctic) and widely polyphagous. It is also able to colonize greenhouses spontaneously (Wyatt, 1965; Rabasse, 1980a). Now, *A. matricariae* has been almost totally replaced by *A. colemani*, which has a slightly broader host range and also attacks *A. gossypii* (van Steenis, 1995).

To control *M. euphorbiae*, *A. abdominalis* has now been used for several years on more than 50 ha of tomato in Europe. *Aphelinus abdominalis* is also indigenous in Europe. An early inoculative release is performed when the aphid population begins to increase. As the adults live longer than those of *Aphidius* and as the females practise host-feeding, *A. abdominalis* can be used successfully for eradicating small foci very early in the season (Rabasse *et al.*, 1989). A recent development in the control of *M. euphorbiae* is the use of *Aphidius ervi* Haliday. This species develops faster than *A. abdominalis* and is a more active searcher.

Aulacorthum solani is a less abundant aphid than the two previous ones; it can be attacked by the parasitoids of *M. persicae* or of *M. euphorbiae*.

Aphid parasitoids are produced on complete food-chains, comprising the host-plant and the aphid. The mummies can be washed off the plants or the adults collected in the rearings. The parasitoids are distributed in both forms. The adult Aphidiidae are very active and need to be released at only a few points per hectare. Aphelinidae disperse less efficiently and have to be released at numerous points (e.g. 250/ha. for *A. abdominalis*). Open rearing systems are used on a very limited commercial scale. The parasitoids are reared in the greenhouse on wheat with grain aphids [e.g. *A. colemani* on *Rhopalosiphum padi* (L.) or *A. ervi* on *Sitobion avenae* (Fabricius)]. Such systems mean that parasitoids are always present in the greenhouse and do not have to be introduced repeatedly. Especially in the Mediterranean area, the indigenous or acclimatized aphid parasitoids are able to colonize greenhouses spontaneously and exert more or less complete control. Because of the rapid growth of an aphid population, it is essential to introduce parasitoids at an early stage of aphid infestation. Therefore, the parasitoids are often introduced before the first aphids are observed (either by standard introductions or an open rearing method).

Aphid parasitoids are regularly attacked by hymenopterous hyperparasitoids, like chalcids, cynipids or ceraphronids. As a precaution, it is necessary to isolate carefully the rearings. In greenhouses, it is not possible to exclude hyperparasitoids and several situations might occur. A single release of parasitoids in a spring population in expansion can result in a wavelike interaction (Ramakers and Rabasse, 1995), even if the parasitoids are destroyed by the hyperparasitoids in the end. The end-result might be a more or less stable interaction. However, the wavelike nature of this interaction can give high aphid densities every now and then. Another way to get lasting control of a lower aphid population can be weekly introduction of low numbers of the parasitoids, which means the interaction is repeated rather than stabilized. At any rate, a stable interaction is not very likely even with repeated introductions. Local aphid infestations can escape parasite pressure and finally develop into rearing units for hyperparasitoids; at this stage standard introductions of parasitoids will not be sufficient. These "hot spots" will have to be suppressed with local chemical treatments.

In practice, a combination of natural enemies can be used. Parasitoids are then used at low aphid densities. As the predators, like midges, need more aphids for successful development, they are introduced when the aphid population is sufficiently large. Ladybirds and chrysopids can then be used additionally to eradicate local aphid infestations. Introduction of midges can also be of use when the parasitoid population is suffering from hyperparasitoids.

An additional help in biological control would be the development of more or less resistant crop varieties. If population development of aphids could be slowed down, it would make biological control much less difficult. Differences in host plant suitability for aphid development and feeding among cultivars have been reported [e.g. in cucumber (van Steenis and El-Khawass, 1995)], but aphid resistance is not used as a criterion when new varieties are screened.

16.4. Conclusion

Research into biological control of aphids continues. However, because of the high growth rates of aphid populations, it is unlikely that they can be fully controlled in all cases. Correct cultural practices (e.g. avoiding nitrogen overfertilization) may reduce the rate of increase of aphid populations and enhance the effectiveness of biological control. Spot treatments with chemicals are still helpful to prevent local outbreaks. The introduction of easy-to-use systemic aphicides implies a serious challenge to biological control. Fortunately, the demand for biological control is still high, both from growers and consumers, and even stronger demand may be expected as resistance to aphicides becomes more common among local aphid populations.

The prospects of using fungi rely mainly on progress made in the production of Entomophthorales. Companies' willingness to diversify their products will probably lead to a wider supply of cosmopolitan oligophagous aphid parasitoids. However, the feasibility and efficiency of multiple-species releases, including or not predators, are to be explored.

References

Albajes, R., Alomar, O., Riudavets, J., Castañé, C., Arno, J. and Gabarra, R. (1996) The mirid bug *Dicyphus tamaninii:* An effective predator for vegetable crops, *IOBC/WPRS Bulletin* **19**(1), 1–6.

Alvarado, P., Balta, O. and Alomar, O. (1997) Efficiency of four Heteroptera as predators of *Aphis gossypii* and *Macrosiphum euphorbiae* (Hom.: Aphididae*), Entomophaga* **42**, 215–226.

Celli, G., Benuzzi, M., Maini, S., Manzaroli, G., Antoniacci, L. and Nicoli, G. (1991) Biological and integrated pest control in protected crops of northern Italy's Po Valley: Overview and outlook, *IOBC/WPRS Bulletin* **14**(5), 2–12.

Furk, C. and Hines, C.M. (1993) Aspects of insecticide resistance in the melon and cotton aphid, *Aphis gossypii* (Hemiptera: Aphididae), *Annals of Applied Biology* **123**, 9–17

Harizanova, V. and Ekbom, B. (1997) An Evaluation of the Parasitoid, *Aphidius colemani* Viereck (Hymenoptera: Braconidae) and the Predator *Aphidoletes aphidimyza* Rondani (Diptera: Cecidomyiidae) for Biological Control of *Aphis gossypii* Glover (Homoptera: Aphididae) on Cucumber, *J. of Entomological Science* **32**, 17–24.

Havelka, J. (1980) Some aspects of photoperiodism of the predacious gall-midge *Aphidoletes aphidimyza* Rond. (Diptera: Cecidomyiidae), *Entomologicheskoe Obozrenie.* **59**, 241–248.

Helyer, N.L. and Wardlow, L.R. (1987) Aphid control on chrysanthemum using frequent, low dose applications of *Verticillium lecanii, IOBC/WPRS Bulletin* **10**(2), 62–65.

Kowalska, T. (1976) Mass rearing and possible uses of Chrysopidae against aphids in glasshouses, *IOBC/WPRS Bulletin* **4**, 80–85.

Malausa, J.C., Drescher, J. and Franco, E. (1987) Prospective for the use of predacious bug *Macrolophus caliginosus* Wagner, (Heteroptera: Miridae) on glasshouse crops, *IOBC/WPRS Bulletin* **10**(2), 106–107.

Markkula, M. and Tiittanen, K. (1985) Biology of the midge *Aphidoletes aphidimyza* and its potential for biological control, in N.W. Hussey and N. Scopes (eds.), *Biological Pest Control. The Glasshouse Experience*, Blandford, Poole, pp. 74–81.

Markkula, M., Tiittanen, K., Hämäläinen, M. and Forsberg, A. (1979) The aphid midge *Aphidoletes aphidimyza* (Diptera: Cecidomyiidae) and its use in biological control of aphids, *Annales Agriculturae Fenniae* **45**, 89–98.

Monadjemi, N. (1972) Sur les variations de la fécondité d'*Aphelinus asychis* (Hym.: Aphelinidae) en fonction des espèces aphidiennes mises à sa disposition à différentes périodes de sa vie imaginale, *Annales de la Société Entomologique de France* **8**, 451–460.

Rabasse, J.M. (1980a) Implantation d'*Aphidius matricariae* dans les populations de *Myzus persicae* en culture d'aubergines sous serre, *IOBC/WPRS Bulletin* **3**(2), 175–185.

Rabasse, J.M. (1980b) Dynamique des populations d'aphides sur aubergine en serre. Considérations générales sur la colonisation et le développement des populations de quatre espèces dans le Sud de la France, *IOBC/WPRS Bulletin* **3**(2), 187–198.

Rabasse, J.M., Lafont, J.P., Guenaoui, Y., Tardieux, I. and Lopin, N. (1989) Potentialités des parasites de pucerons comme agents de lutte biologique en cultures maraîchères protégées, in R. Cavalloro and C. Pelerents (eds.), *Integrated Pest Management in Protected Vegetable Crops*, Balkema, Roterdam, pp. 73–78.

Rabasse, J.M. and Shalaby, F.F. (1980) Laboratory studies on the development of *Myzus persicae* Sulz. (Hom., Aphididae) and its primary parasite *Aphidius matricariae* Hal. (Hym.: Aphidiidae) at constant temperatures, *Acta Oecologica, Oecologia Applicata* **1**, 21–28

Rabasse, J.M., Tardieux, I. and Pintureau, B. (1985) Comparaison de deux populations française et brésilienne d'*Aphidius colemani* Viereck (Hym.: Aphidiidae), *Annales de la Société Entomologique de France* **21**, 41–49.

Ramakers, P.M.J. and Rabasse, J.-M. (1995) IPM in protected cultivation, in R. Reuveni (ed.), *Novel Approaches to Integrated Pest Management*, Lewis Publishers, Boca Raton, pp. 199–229.

Rochat, J. (1997) Delayed effects in aphid-parasitoid systems: consequences for evaluating biological control species and their use in augmentation strategies, *Entomophaga* **42**, 201–213.

van Steenis, M.J. (1992) Biological control of the cotton aphid, *Aphis gossypii* Glover (Hom.: Aphididae): Pre-introduction evaluation of natural enemies, *J. of Applied Entomology* **114**, 362–380.

van Steenis, M.J. (1993) Intrinsic rate of increase of *Aphidius colemani* Vier. (Hym.: Braconidae), a parasitoid of *Aphis gossypii* Glov. (Hom.: Aphididae), at different temperatures, *J. of Applied Entomology* **116,** 192–198

van Steenis, M.J. (1995) Evaluation of four aphidiine parasitoids for biological control of *Aphis gossypii, Entomologia Experimentalis et Applicata,* **75,** 151–157

van Steenis, M.J. and El-Khawass, K.A.M.H. (1995) Life history of *Aphis gossypii:* Influence of temperature, host plant, and parasitism, *Entomologia Experimentalis et Applicata* **76,** 121–131

Wyatt, I.J. (1965) The distribution of *Myzus persicae* (Sulz) on year-round chrysanthemums. I. Summer season, *Annals of Applied Biology* **56,** 439–459.

BIOLOGICAL CONTROL OF THRIPS

Cristina Castañé, Jordi Riudavets and Eizi Yano

17.1. Biology of Major Greenhouse Thrips Pests and Damages

Thrips belong to the order Thysanoptera which includes many herbivorous and predatory species. In greenhouse vegetables three widespread thrips species may be mentioned by their economic importance as pests: *Frankliniella occidentalis* (Pergande), *Thrips palmi* Karny and *Thrips tabaci* Lindeman. These and several other thrips species affect greenhouse-grown vegetable and ornamental plants.

Frankliniella occidentalis, or western flower thrips, is indigenous to the western United States, western Canada and northern Mexico (Bryan and Smith, 1956), and has recently spread to the whole American continent, Europe, Asia, New Zealand and Australia. *Thrips palmi* has been recorded in South and Southeast Asia (Kawai, 1990). After being found in Japan, it invaded Pacific Islands including Hawaii, and it has been reported from northern Australia, Caribbean, Guyana, Florida, Mauritius, Reunion, Sudan and Nigeria (Walker, 1994). *Thrips tabaci* is cosmopolitan with a probable origin in Central Asia (Lewis 1973).

All three species have a high intrinsic rate of increase due to their short developmental time and their high fecundity. In *F. occidentalis* and *T. palmi,* reproduction occurs by facultative parthenogenesis. Fertilized females produce female-biased sex ratios and unmated females produce males parthenogenetically. In greenhouses, *T. tabaci* produces females by parthenogenesis. At temperatures between 25 and 30°C, immature development lasts 10–20 days; females can live up to 30 days and deposit up to 200 eggs, which are inserted in the plant tissue. After hatching, first instar larvae begin to feed almost immediately by piercing and ingesting leaf cell contents. Second-instar larvae are more active and feed more abundantly until they move to the soil or to hidden parts of the plant to pupate. During the two pupal stages, thrips do not feed and move only if disturbed. Emerged adults fly to young leaves, flowers and young fruits, where they feed and lay eggs. *Frankliniella occidentalis* adults are pollenophagous and aggregate in the flowers for feeding and mating while *T. palmi* and *T. tabaci* predominantly aggregate in the leaves.

Frankliniella occidentalis and *T. palmi* remain active and reproduce throughout the winter in the field when temperatures are mild, and in more severe temperatures they only reproduce inside greenhouses.

All three species are highly polyphagous. Their host plants include most vegetables, tree fruits, cereals and ornamentals. Among greenhouse vegetables, peppers, cucumbers, strawberries and tomatoes are the main crops attacked, whilst among ornamentals, chrysanthemums, gerberas, carnations, cyclamens and roses are the most common host. Damage is most serious on cucurbits and solanaceous plants. *Thrips tabaci* prefers host species among the Liliaceae.

R. Albajes et al. (eds.), *Integrated Pest and Disease Management in Greenhouse Crops,* 244-253.
© 1999 *Kluwer Academic Publishers. Printed in the Netherlands.*

Frankliniella occidentalis, T. palmi and *T. tabaci* can cause direct damage due to feeding and egg-laying, and indirect damage due to the transmission of tomato spotted wilt virus (TSWV). When feeding on the epidermal and parenchymatal cells, adults and larvae produce punctures in the plant tissues that cause silver, discoloured and/or necrotic spots. Leaves and terminal shoots become stunted. If the tissue affected is young and still growing, the feeding wounds produce scars that deform the developing organs. Eggs inserted in the plant tissue cause local discoloration and deformity around the site of ovipositor penetration. The virus is persistently transmitted by adults (see Chapter 3) and tomatoes, peppers, aubergines, chrysanthemums, gerberas and cyclamens are the most affected greenhouse crops. *Frankliniella occidentalis* is the predominant vector of recent TSWV epidemics (Cho *et al.*, 1989) and also the species causing the most severe damage due to fruit scarring. When Rosenheim *et al.* (1990) compared damage caused to cucumber by mixed infestations of *F. occidentalis* and *T. palmi* (~6% and 94% respectively) the former was responsible for the most scarring of fruits while the latter mainly produced reductions in total yield. Welter *et al.* (1990) proposed an economic injury level of 16 thrips/leaf for *F. occidentalis* and 94.5 thrips/leaf for *T. palmi* in cucumber. Kawai (1986) estimated economic injury levels of 4.4 adults/leaf, 0.17 adults/leaf and 0.11 adults/flower of *T. palmi* in cucumber, aubergine and sweet pepper, respectively, assuming that 5 % yield loss of undamaged fruits is tolerable.

Thrips can be scored by direct count on the leaves, by tapping flowers, by extraction methods (e.g. Berlese funnels) or by counting adults on coloured sticky traps. *Frankliniella occidentalis* adults and larvae are distributed throughout the plant, although in crops that produce pollen adults tend to aggregate on flowers and larvae on leaves. On strawberries the population develops entirely on flowers and fruits. On cucumbers, adults and larvae are mainly aggregated in the middle leaf strata. On tomatoes, however, adults aggregate on flowers of the upper half and the larvae accumulate on leaves of the lower half of the plant. Presence-absence sampling procedures have been developed for these crops (Steiner 1990; García-Marí *et al.*, 1994; Salguero Navas *et al.*, 1994). On peppers a correlation was found between catches on blue sticky traps and total population, although sampling blossom was the most cost-effective method (Shipp and Zariffa, 1991). *Thrips palmi* adults and larvae are mainly found in the middle leaf strata on aubergine and cucumber, although on sweet pepper both adults and larvae tend to aggregate on flowers, buds and fruits. Random and sequential sampling plans have been developed based on analysis of population dispersion of *T. palmi*, and adult sampling by direct counting is recommended. *Thrips palmi* and *T. tabaci* can also be monitored with white or blue sticky traps (Kawai, 1986, 1990; Brødsgaard, 1993).

17.2. Natural Enemies

As for most insect pests, the list of natural enemies of *F. occidentalis, T. palmi* and *T. tabaci* is long and comprises predators, parasitoids and mycopathogens. Nowadays, predators are the best candidates for biological control of these thrips species (Sabelis and van Rijn, 1997).

Anthocorid bugs are probably major natural enemies in many countries. For *F. occidentalis,* Riudavets (1995) cited seven species of *Orius,* among which *Orius albidipennis* (Reuter), *Orius laevigatus* (Fieber), *Orius majusculus* (Reuter) and *Orius niger* (Wolff) are common in the palearctic region. While *O. laevigatus* and *O. niger* have a wide distribution throughout the whole region, *O. majusculus* is mainly found in the northern part and *O. albidipennis* in the southern Mediterranean (Riudavets and Castañé, 1994; Chyzik *et al.,* 1995) and the Canary islands (Carnero *et al.,* 1993). *Orius insidiosus* (Say) and *Orius tristicolor* (White) are found in the nearctic region, the former being mainly distributed in the eastern part (Kelton, 1963). Most of the *Orius* species cited as predators of *F. occidentalis* also prey upon *T. tabaci.* Yasunaga (1997) recorded seven Japanese species of *Orius,* of which four, *Orius sauteri* (Poppius), *Orius minutus* (L.), *Orius strigicollis* (Poppius) *and Orius tantillus* (Motschulsky), are considered as natural enemies of *T. palmi* in the field. *Orius tantillus* was also recorded in the Philippines and *O. minutus* was found in Thailand. Several Heteroptera have been cited as *F. occidentalis, T. palmi* and *T. tabaci* predators, among which mirid bugs have been found on several vegetable crops (Wang, 1995; Riudavets and Castañé, 1998). Phytoseiid mites have been found preying on thrips. *Neoseiulus (= Amblyseius) cucumeris* (Oudemans), *Amblyseius barkeri* (Hughes) and *Amblyseius degenerans* Berlese are well known natural enemies of *F. occidentalis* and *T. tabaci.* For *T. palmi,* the predatory mites *Amblyseius mckenziei* Schuster & Pritchard, *Amblyseius okinawanus* Ehara (Kajita, 1986), *Amblyseius tsugawai* Ehara and Erythraeidae (Nagai, 1993) have been cited.

Ceranisus menes (Walker) has been reported to parasitize *F. occidentalis, T. palmi* and *T. tabaci* (Hirose *et al.,* 1992; Loomans and van Lenteren, 1995). Among fungi, *Neozygites parvispora* (Macleod and Carl) and *Verticillium lecanii* (A. Zimmerm.) Viégas have been recorded on the three species in vegetable crops (Saito *et al.,* 1989; Saito, 1992; Vacante *et al.,* 1994; Brownbridge, 1995; Montserrat *et al.,* 1998).

17.3. Successful Cases of Biological Control

17.3.1. *Frankliniella occidentalis*

Biological control in vegetable and ornamental crops has been conducted mainly using predators, with varying degrees of success. In peppers, which have high pollen production, inoculative release of several anthocorid bug species established in the crop and controlled thrips populations. *Orius insidiosus* was successful when released at a density of 2 per plant (van den Meiracker and Ramakers, 1991). However, native European *Orius* species such as *O. niger, O. laevigatus* and *O. albidipennis* displaced *O. insidiosus* when released in greenhouses (van de Veire and Degheele, 1992; Tavella *et al.,* 1994; van de Veire 1995). *Orius laevigatus* is abundant and well adapted to greenhouse conditions across continental Europe. *Orius laevigatus* and *O. albidipennis* effectively controlled *F. occidentalis* in peppers when released at a density of 1–2 per plant (Chambers *et al.,* 1993; Rubin *et al.,* 1996). These two *Orius* species do not have diapause, only quiescence at low temperatures (van den Meiracker, 1994; Tommasini and Nicoli, 1996). In Canada, *O. tristicolor* and *O. insidiosus* are native species. When *O. tristicolor* was released in an *O. insidiosus* area the species was replaced naturally by the latter (Shipp *et al.*, 1992). Therefore, there is no single *Orius* species that would be the best biological control agent for all regions. *Neoseiulus cucumeris* and *A. degenerans* are used in inundative releases (Shipp *et al.*, 1991; Jacobson 1995; Ramakers 1995) but they are effective only under specific conditions of temperature and humidity, and when released as preventive introductions.

In cucumber, pollen is virtually absent from most greenhouse varieties and anthocorids do not establish in such crops unless a high density of prey is available, and by then fruits are already damaged (Chambers *et al.*, 1993; Grasselly *et al.*, 1994). The mirid bug *Dicyphus tamaninii* Wagner can effectively control thrips populations and can establish in the crop at low prey densities (Gabarra *et al.*, 1995; Castañé *et al.*, 1996), since it does not feed on pollen. These predators can respond effectively to sudden increases in prey numbers from outside the greenhouse, which are common in the Mediterranean area. In strawberry *O. laevigatus* can keep *F. occidentalis* populations below the economic injury threshold of 10 thrips per flower (Villevielle and Millot 1991; González-Zamora *et al.*, 1994). In ornamentals, good results have been obtained with the release of *O. insidiosus* in chrysanthemum and saintpaulia (Fransen *et al.*, 1993), and with *O. majusculus* in gerbera (Brødsgaard, 1995).

Some crop management practices may help to reduce thrips and virus infestation. Soil sterilization between crops kills immature thrips that are pupating in the soil. Removal of weeds around the greenhouse is advisable if they act as a reservoir for the virus. Screening the greenhouse is another possibility for reducing thrips immigration (Berlinger *et al.*, 1993; Lacasa *et al.*, 1994), although another control method should be used simultaneously.

Cultivars of pepper, cucumber, tomato and chrysanthemum that are resistant to *F. occidentalis* may enhance the application of biological control on these crops (de Kogel, 1997). Tomato varieties resistant to TSWV have recently appeared on the market and have enhanced the use of IPM programmes. The widespread occurrence of the virus and the use of insecticides in an attempt to control the vector has disrupted the application of IPM programmes, as occurs in field tomato crops in the northeast of Spain (Arnó *et al.*, 1995).

17.3.2. *Thrips palmi*

Most studies on the practice of biological control of *T. palmi* have been conducted in Japan with *O. sauteri*. When aubergines in the field were treated with fenthion, which harms *O. sauteri* but does not affect *T. palmi*, the peak density of *T. palmi* was four times as large as that on the treated plants (Nagai, 1990). Further experiments were also conducted using the selective pesticide pyriproxyfen, which has a control effect on the pupal stage of *T. palmi* but is harmless to *O. sauteri*. *Thrips palmi* on aubergines decreased very rapidly after application of this pesticide, due to a combined effect of the pesticide and *O. sauteri* (Nagai, 1996). Since *O. sauteri* has a suppressive effect on *T. palmi* in open fields of aubergines, it is expected to be a biocontrol agent for *T. palmi* on greenhouse aubergines. Different numbers of fifth-instar nymphs of *O. sauteri* were released in four plots in an aubergine greenhouse 12 days after release of *T. palmi*. This was kept effectively below its economic injury level of 0.47 adults per leaf (Kawai, 1986) when more than 2 nymphs of *O. sauteri* per plant were released (Kawai, 1995). *Orius sauteri* showed excellent suppressive effects on populations of *T. palmi* in many greenhouse experiments. The time of release of *O. sauteri* and the initial density of *T. palmi* and *O. sauteri* are important factors affecting the outcome of biological control. It was demonstrated in simulation studies that early and high-density release of *O. sauteri* is crucial for effective control (Yano *et al.*, unpublished).

17.3.3. *Thrips tabaci*

In greenhouse vegetables and ornamentals *T. tabaci* is not considered a threat, probably due to the more serious effects of *F. occidentalis* and *T. palmi*. Most biological control agents are common to the three species, and if the former two species are controlled *T. tabaci* may also be controlled.

The phytoseiid mite *N. cucumeris* is able to control *T. tabaci* populations in peppers and cucumbers (Gillespie, 1989; Brødsgaard and Stengaard, 1992). It establishes in cucumber because spider mites are usually present. They serve as alternative prey until *T. tabaci* populations reach the level for being a preferred prey. Inundative releases with very high numbers (100–300 mites per plant) are performed since only adult mites are predators and they are able to feed only on first instar thrips. Since *N. cucumeris* is a pollenophagous species, it may be preventively introduced in pepper before thrips appear (Ramakers, 1987). In order to be effective, 70–80% of leaves should have 10 or more phytoseiids or their eggs before *T. tabaci* appears (Altena and Ravensberg, 1990).

17.4. Failures and Main Constraints in the Use of Biological Control

17.4.1. *Frankliniella occidentalis*

Biological control of *F. occidentalis* is only possible in crops susceptible to TSWV if the region where they are grown does not have any virus inoculum, since the species is a very efficient vector of the virus.

This new pest has disrupted the IPM programmes that were being applied in some greenhouse crops. The demand for biological solutions in IPM programmes has forced researchers to deliver biological control agents that have not been fully tested and has led to some failures, which are mainly due to lack of information (Steiner, 1995). *Neoseiulus cucumeris*, which is easy and cheap to produce, controlled *T. tabaci* and it was initially used against *F. occidentalis* in greenhouse vegetable and ornamental crops. But it failed since this predator only eat first instar larvae and their rate of increase is lower than that of *F. occidentalis*. They could establish in crops with pollen, like peppers, but not in cucumbers. Other requirements such as high humidity for egg hatching and low tolerance to high temperatures (Cloutier *et al.*, 1995; van Houten *et al.*, 1993) has limited their use in Mediterranean greenhouses (Vacante and Tropea Garzia, 1993).

When crops are grown in winter, *F. occidentalis* continues reproducing and causing problems, but some of the anthocorids (*O. insidiosus, O. majusculus*) and phytoseiid mites (*N. cucumeris*) have diapause. To overcome this problem, non-diapausing species or strains have to be released like *O. laevigatus, O. albidipennis* or *A. degenerans*.

17.4.2. *Thrips palmi*

Although *O. sauteri* is a very effective predator under favourable conditions, predation and oviposition activities of *O. sauteri* are strongly dependent on temperature, day length and host plant. Below 20°C, these activities decrease drastically (Nagai and Yano, unpublished). Induction of reproductive diapause under short day length makes the use of *O. sauteri* more difficult in winter. The most important biotic factor is the effect of the host plant of *T. palmi*. Cucumber is a better host plant for oviposition of *T. palmi* and produces less pollen than aubergine or sweet pepper, so the use of *O. sauteri* is more difficult than on the other two crops. Concerning the release technique of *O. sauteri*, action thresholds of *T. palmi* for deciding the timing of release of *O. sauteri* have not been determined. In *V. lecanii*, high humidity is crucial for spore germination and effective control of *T. palmi* (Saito, 1992).

In Japan registration is required for the commercialization of natural enemies to be used in greenhouses. This takes as much time as for chemical pesticides. Another important factor is the need for the development of integrated control systems for all pests of a crop. If the use of several species of natural enemies is intended, it may take several years before commercialization is possible because of registration. The registration procedure should be simplified for the rapid progress of biocontrol in Japan.

17.5. Conclusions

As with any new pest, the thrips problem has been approached via pesticides, using the whole range of insecticides available in each area. This is currently the principal control measure for *F. occidentalis* and *T. palmi* all over the world. Insecticides do not give a satisfactory solution due to resistance problems. The abuse of the use of insecticides may be the origin of the problem.

Biological control programmes have a good chance of success with thrips problems but, in crops susceptible to TSWV, only after the virus problem has been solved. Virus transmission is very efficient with quite low pest populations, therefore trying to control the vector is not a solution. The recent appearance of virus resistant tomato varieties has increased the possibilities of using biological control strategies in this crop.

Different predators seem to be optimal for thrips control depending on the thrips species, crops and geographic areas considered. Some predators are pollenophagous and are only efficient in crops with abundant pollen production, as is the case of *Orius* spp. in peppers. Native predators are more adapted than exotic introduced species and may replace them, as has been shown with *O. insidiosus* in Europe. Therefore, different successful solutions may be adopted depending on the crop, the area, and the control strategies.

Acknowledgements

Part of this chapter is based on a research project financed by the Instituto Nacional de Investigación y Tecnología Agraria y Alimentaria of Spain (SC95–052).

References

Altena, K. and Ravensberg, W.J. (1990) Integrated pest management in the Netherlands in sweet peppers from 1985–1989, *IOBC/WPRS Bulletin* **13**(5), 10–13.

Arnó, J., Riudavets, J., Moriones, E., Aramburu, J., Laviña, A. and Gabarra, R. (1995). Monitoring western flower thrips as a tomato spotted wilt virus vector in tomato, in B.L. Parker, M. Skinner and T. Lewis (eds.), *Thrips Biology and Management: Proceedings of the 1993 International Conference on Thysanoptera*, Plenum Publishing Co. Ltd, London-New York, pp. 197–200.

Berlinger, M.J., Lebiush-Mordechi, S., Fridja, D. and Mor, N. (1993) The effect of types of greenhouse screens on the presence of western flower thrips, *IOBC/WPRS Bulletin* **16**(2), 13–16.

Brødsgaard, H.F. (1993) Colored sticky traps for thrips (Thysanoptera: Thripidae) monitoring on glasshouse cucumbers, *IOBC/WPRS Bulletin* **16**(2), 19–22.

Brødsgaard, H.F. (1995) "Keep-down" a concept of thrips biological control in ornamental pot plants, in B.L. Parker *et al.* (eds.), *Thrips Biology and Management*, Plenum Press, New York, pp. 221–224.

Brødsgaard, H.F. and Stengaard, L.H. (1992) Effect of *Amblyseius cucumeris* and *Amblyseius barkeri* as biological control agents of *Thrips tabaci* on glasshouse cucumbers, *Biocontrol Science and Technology* **2**, 215–223.

Brownbridge, M. (1995) Prospects for mycopathogens in thrips management, in B.L. Parker *et al.* (eds.), *Thrips Biology and Management*, Plenum Press, New York, pp. 281–295.

Bryan, D.E. and Smith, R.F. (1956) The *Frankliniella occidentalis* (Pergande) complex in California, *University of California Publications in Entomology* **10**(6), 359–410.

Carnero, A., Peña, M.A., Perez-Padron, F., Garrido, C. and Hernández-García, M. (1993) Bionomics of *Orius albidipennis* and *Orius limbatus, IOBC/WPRS Bulletin* **16**(2), 27–30.

Castañé, C., Alomar, O. and Riudavets, J. (1996) Management of western flower thrips on cucumber with *Dicyphus tamaninii* (Heteroptera: Miridae), *Biological Control* **7**, 114–120.

Chambers, R.J., Long, S. and Heyler, N.L. (1993) Effectiveness of *Orius laevigatus* (Hem. Anthocoridae) for the control of *Frankliniella occidentalis* on cucumber and pepper in the UK, *Biocontrol Science and Technology* **3**, 295–307.

Cho, J.J., Mau, R.F.L., German, T.L., Hartmann, R.W., Yudin, L.S., Gonsalves, D. and Provvidenti, R. (1989) A multidisciplinary approach to management of tomato spotted wilt virus in Hawaii, *Plant Disease* **73**(5), 375–383.

Chyzik, R., Klein, M. and Ben-Dov, Y. (1995) Overwintering biology of the predatory bug *Orius albidipennis* (Hemiptera: Anthocoridae) in Israel, *Biocontrol Science and Technology* **5**, 287–296.

Cloutier, C., Arodokum, D., Johnson, S.G. and Gelinas, L. (1995) Thermal dependence of *Amblyseius cucumeris* (Acarina: Phytoseiidae) and *Orius insidiosus* (Heteroptera: Anthocoridae) in greenhouses, in B.L. Parker *et al.* (eds.), *Thrips Biology and Management*, Plenum Press, New York, pp. 231–235.

de Kogel, W.J. (1997). Host plant resistance to western flower thrips: Variable plants and insects, PhD dissertation, University of Amsterdam.

Fransen, J.J., Boogaard, M. and Tolsma, J. (1993) The minute pirate bug, *Orius insidiosus,* as a predator of western flower thrips in chrysanthemum, rose and saintpaulia, *IOBC/WPRS Bulletin* **16**(8), 73–77.

Gabarra, R., Castañé, C. and Albajes, R. (1995) The mirid bug *Dicyphus tamaninii* as a greenhouse whitefly and western flower thrips predator on cucumber, *Biocontrol Science Technology* **5**, 475–488.

García-Marí, F., González-Zamora, J.E., Ribes, A., Benages, E. and Meseguer, A. (1994) Métodos de muestreo binomial y secuencial del trips de las flores *Frankliniella occidentalis* (Thysanoptera: Thripidae) y de antocóridos (Heteroptera, Anthocoridae) en fresón, *Boletín de Sanidad Vegetal – Plagas* **20**, 703–723.

Gillespie, D.R. (1989) Biological control of thrips (Thysanoptera, Thripidae) on greenhouse cucumber by *Amblyseius cucumeris, Entomophaga* **34**(2), 185–192.

González-Zamora, J.E., Ribes, A., Meseguer, A. and García-Marí, F. (1994) Control del trips en fresón: Empleo de plantas de haba como refugio de poblaciones de antocóridos, *Boletín de Sanidad Vegetal – Plagas* **20**, 57–72.

Grasselly, D., Millot, P. and Alauzet, C. (1994) Nuisibilite de *Frankliniella occidentalis* (Pergande) sur concombre, consequences sur la lutte biologique a l'aide d'*Orius majusculus* (Reuter), *IOBC/WPRS Bulletin* **17**(5), 153–157.

Hirose, Y., Takagi, M., and Kajita, H. (1992) Discovery of an indigenous parasitoid *Thrips palmi* Karny (Thysanoptera: Thripidae) in Japan: *Ceranisus menes* (Walker) (Hymenoptera: Eulophidae) on eggplant in home and truck gardens, *Applied Entomology and Zoology* **27**, 465–467.

Jacobson, R.J. (1995) Resources to implement biological control in greenhouses, in B.L. Parker *et al.* (eds.), *Thrips Biology and Management*, Plenum Press, New York, pp. 211–219.

Kajita, H. (1986) Predation by *Amblyseius* spp. (Acarina: Phytoseiidae) and *Orius* sp. (Hemipetra: Anthocoridae) on *Thrips palmi* Karny (Thysanoptera: Thripidae), *Applied Entomology and Zoology* **21**, 482–484.

Kawai, A. (1986). Studies on population ecology and population management of *Thrips palmi* Karny, *Bulletin of the Vegetable and Ornamental Crops Research Station, C, Japan* **9**, 69–135 (in Japanese).

Kawai, A. (1990) Life cycle and population dynamics of *Thrips palmi* Karny, *Japan Agricultural Research Quarterly* **23**, 282–288.

Kawai, A. (1995) Control of *Thrips palmi* Karny (Thysanoptera: Thripidae) by *Orius* spp. (Heteroptera: Anthocoridae) on greenhouse eggplant, *Applied Entomology and Zoology* **30**, 1–7.

Kelton, L.A. (1963) Synopsis of the genus *Orius* Wolff in America north Mexico (Heteroptera: Anthocoridae), *Canadian Entomologist* **95**, 631–633.

Lacasa, A., Contreras, J., Torres, J., González, A., Martínez, M.C., García, F. and Hernández, A. (1994) Utilización de mallas en el control de *Frankliniella occidentalis* (Pergande) y el virus del bronceado del tomate (TSWV) en el pimiento en invernadero, *Boletín de Sanidad Vegetal – Plagas* **20**, 561–580.

Lewis, T. (1973) *Thrips: Their Biology, Ecology and Economic Importance*, Academic Press, London.

Loomans, A.J.M. and van Lenteren, J.C. (1995) Biological control of thrips pests: A review on thrips parasitoids, *Wageningen Agricultural University Papers* **95**(1), 89–201.

Montserrat, M., Castañé, C. and Santamaría, S. (1998) *Neozygites parvispora* (Zygomicotina, Entomophtorales) causing an epizootic in *Frankliniella occidentalis* (Thysanoptera, Thripidae) on cucumber in Spain, *Journal of Invertebrate Pathology* **71**(2), 165–168.

Nagai, K. (1990) Suppressive effect of *Orius* sp. (Hemiptera: Anthocoridae) on the population density of *Thrips palmi* Karny (Thysanoptera: Thripidae), *Japanese J. of Applied Entomology and Zoology* **34**, 109–114 (in Japanese).

Nagai, K. (1993) Studies on integrated pest management of *Thrips palmi* Karny, *Special Bulletin of Okayama Prefectural Agricultural Experiment Station* **82**, 1–55.

Nagai, K. (1996) Integrated pest management of *Thrips palmi* Karny in eggplant fields, in G. Grey *et al.* (eds.), *Biological Pest Control in Systems of Integrated Pest Management*, Food & Fertilizer Technology Center, Taipei, pp. 215–225.

Ramakers, P.M.J. (1987).Control of spider mites and thrips with phytoseiid predators on sweet pepper, *IOBC/WPRS Bulletin* **10**(2), 33–42.

Ramakers, P.M.J. (1995). Biological control using oligophagous predators, in B.L. Parker *et al.* (eds.), *Thrips Biology and Management*, Plenum Press, New York, pp. 225–230.

Riudavets, J. (1995) Predators of *Frankliniella occidentalis* and *Thrips tabaci*: A review, *Wageningen Agricultural University Papers* **95**(1), 43–87.

Riudavets, J. and Castañé, C. (1994) Abundance and host plant preferences for oviposition of *Orius* spp. (Heteroptera: Anthocoridae) along the Mediterranean coast of Spain, *IOBC/WPRS Bulletin* **17**(5), 230–236.

Riudavets, J. and Castañé, C. (1998). Identification and evaluation of native predators of *Frankliniella occidentalis* (Thysanoptera: Tripidae), *Environmental Entomology* **27**(1), 86–93.

Rosenheim, J.A., Welter, S.C., Johnson, M.W., Mau, R.F.L. and Gusukuma-Minuto, L.R. (1990) Direct feeding damage on cucumber by mixed-species infestations of *T. palmi* and *F. occidentalis* (Thysanoptera: Tripidae), *J. of Economic Entomology* **83**(4), 1519–1525.

Rubin, A., Ucko, O., Orr, N. and Offenbach, R. (1996) Efficacy of natural enemies of western flower thrips *Frankliniella occidentalis* in pepper flowers in the Arava valley, Israel, *IOBC/WPRS Bulletin* **19**(1), 139–142.

Sabelis, M.W. and van Rijn, P.C.J. (1997) Predation by insects and mites, in T. Lewis (ed.), *Thrips as Crop Pests*, CAB International, Wallingford, pp. 259–354.

Saito, T. (1992) Control of *Thrips palmi* and *Bemisia tabaci* by a mycoinsecticide preparation of *Verticillium lecanii*, *Proceedings of the Kanto-Tosan Plant Protection Society* **39**, 209–210 (in Japanese).

Saito, T., Kubota, S. and Shimazu, M. (1989) A first record of the entomopathogenic fungus, *Neozygites parvispora* (MacLeod & Carl) Rem. & Kell., on *Thrips palmi* Karny (Thysanoptera: Thripidae) in Japan, *Applied Entomology and Zoology* **24**, 233–235.

Salguero Navas, V.E., Funderburk, J.E., Mack, T.P., Beshear, R.J. and Olson, S.M. (1994) Aggregation indices and sample size curves for binomial sampling of flower-inhabiting *Frankliniella* species (Thysanoptera: Thripidae) on tomato, *J. of Economic Entomology* **87**(6), 1622–1626.

Shipp, J.L., Boland, G.J. and Shaw, L.A. (1991) Integrated pest management of disease and arthropod pests of greenhouse vegetable crops in Ontario: Current status and future possibilities, *Canadian J. of Plant Science* **71**, 887–914.

Shipp, J.L. and Zariffa, N. (1991) Spatial pattern of and sampling methods for western flower thrips (Thysanoptera: Thripidae) on greenhouse sweet pepper, *Canadian Entomologist* **123**(5), 989–1000.

Shipp, J.L., Zariffa, N. and Ferguson, G. (1992) Spatial patterns of and sampling methods for *Orius* spp. (Hemiptera: Anthocoridae) on greenhouse sweet pepper, *Canadian Entomologist* **124**, 887–894.

Steiner, M.Y. (1990) Determining population characteristics and sampling procedures for the western flower thrips (Thysanoptera: Thripidae) and the predatory mite *Amblyseius cucumeris* (Acari: Phytoseiidae) on greenhouse cucumber, *Environmental Entomology* **19**(5), 1605–1613.

Steiner, M.Y. (1995) Marketing considerations for biological control agents, in B.L. Parker *et al.* (eds.), *Thrips Biology and Management*, Plenum Press, New York, pp. 329–335.

Tavella, L., Alma, A. and Arzone, A. (1994) Attivita predatrice de *Orius* spp. (Anthocoridae) su *Frankliniella occidentalis* (Perg.) (Thripidae) in coltura protetta di peperone, *Informatore Fitopatologico* **1**, 40–43.

Tommasini, M.G. and Nicoli, G. (1996) Evaluation of *Orius* spp. as biological control agents of thrips pests: Further experiments on the existence of diapause in *Orius laevigatus*, *IOBC/WPRS Bulletin* **19**(1), 183–186.

Vacante, V., Cacciola, S.O. and Pennisi, A.M. (1994) Epizootiological study of *Neozygites parvispora* (Zygomycota: Entomophthoraceae) in a population of *Frankliniella occidentalis* (Thysanoptera: Thripidae) on pepper in Sicilia, *Entomophaga* **39**(2), 123–130.

Vacante, V. and Tropea Garzia, G. (1993) Impiego programmato di *Amblyseius cucumeris* (Oudemans) contro *Frankliniella occidentalis* (Pergande) su pepperone in serra fredda, *Colture Protette* **1**, 23–32.

van de Veire, M. (1995) Integrated pest management in glasshouse tomatoes, sweet peppers and cucumbers in Belgium, PhD dissertation, University of Gent.

van de Veire, M. and Degheele, D. (1992) Biological control of western flower thrips *Frankliniella occidentalis* (Pergande) (Thysanoptera: Thripidae), in glasshouse sweet pepper with *Orius* spp. (Hemiptera: Anthocoridae). A comparative study between *O. niger* (Wolff) and *O. insidiosus* (Say), *Biocontrol Science and Technology* **2**, 281–283.

van den Meiracker, R.A.F. (1994) Induction and termination of diapause in *Orius* predatory bugs, *Entomologia Experimentalis et Applicata* **73**, 127–137.

van den Meiracker, R.A.F. and Ramakers, P.M.J. (1991) Biological control of western flower thrips *Frankliniella occidentalis* in sweet pepper with the anthocorid predator *Orius insidiosus*, *Mededelingen van de Faculteit Landbouwwetenschappen, Rijksuniversiteit Gent* **56**(2a), 241–249.

van Houten, Y.M., van Rijn, P.C.J., Tanigoshi, L.K. and van Stratum, P. (1993) Potential of phytoseiid predators to control western flower thrips in greenhouse crops, in particular during the winter period, *IOBC/WPRS Bulletin* **16**(8), 98–101.

Villevielle, M. and Millot, P. (1991) Lutte biologique contre *Frankliniella occidentalis* avec *Orius laevigatus* sur fraisier, *IOBC/WPRS Bulletin* **14**(5), 57–64.

Walker, A.K. (1994) A review of the pest status and natural enemies of *Thrips palmi*, *Biocontrol News and Information* **15**(1), 7N–10N.

Wang, C.L. (1995) Predatory capacity of *Campylomma chinensis* Schuh (Hemiptera: Miridae) and *Orius sauteri* (Poppius) (Hemiptera: Anthocoridae) on *Thrips palmi,* in B.L. Parker *et al.* (eds.), *Thrips Biology and Management*, Plenum Press, New York, pp. 259–262.

Welter, S.C., Rosenheim, J.A., Johnson, M.W., Mau, R.F.L. and Gusukuma-Minuto, L.R. (1990) Effects of *Thrips palmi* and western flower thrips (Thysanoptera: Tripidae) on the yield, growth and carbon allocation patterns in cucumbers, *J. of Economic Entomology* **83**(5), 2092–2101.

Yasunaga, T. (1997) The flower bug genus *Orius* Wolff (Heteroptera: Anthocoridae) from Japan and Taiwan, Parts I, II and III, *Applied Entomology and Zoology* **32**, 355–364, 379–386, 387–394.

CHAPTER 18

BIOLOGICAL CONTROL OF LEAFMINERS
Jean-Claude Onillon

18.1. Introduction

Agromyzid leafminers have become very serious pests of north European and Mediterranean ornamental and vegetable crops. Formerly, the agromyzid fauna in greenhouses was restricted to one or two *Chromatomyia* species and one *Liriomyza* species, *Liriomyza bryoniae* (Kaltenbach), the latter also being very common in the open air of southern Europe. Then, as a consequence of crop diversification and importation of infested plants, the *Liriomyza* species were successively joined by *Liriomyza trifolii* (Burgess), *Liriomyza huidobrensis* (Blanchard), and finally *Liriomyza sativae* Blanchard. Due to their polyphagy, these leafminer species can attack many plants such as tomato, cucumber, lettuce, melon, gerbera and beans. *Liriomyza bryoniae* has been reared on host plants belonging to more than thirty botanical families (Spencer, 1973). *Liriomyza trifolii* became a pest of chrysanthemum (Lindquist *et al.*, 1980), celery (Genung and Janes, 1975) and tomato (Zehnder and Trumble, 1984).

When pesticides are not applied, leafmining larvae are usually parasitized by a rich and diversified parasitoid complex. Thus, 21 parasitoid species can attack and develop on *Chromatomyia syngenesiae* Hardy, several dozens of parasitoids have been recorded in *L. trifolii* and *L. sativae* samples (Minkenberg and van Lenteren, 1986) in their original distribution areas; recently, 18 parasitoid species attacking *L. huidobrensis* have been reported (van der Linden, 1990). This rich parasitoid fauna may account for the low leafminer population densities recorded on non-sprayed crops like melon and lettuce (Hills and Taylor, 1951), winter gardens (Harding, 1965) and celery (Trumble, 1981).

Leafminers cause two types of damage. Direct damage occurs as a consequence of larval feeding inside the leaf and adult feeding punctures on the leaf upperside. A high amount of leaf mines causes a significant decrease in photosynthetic assimilate production that may lead to desiccation and premature fall of leaves. Feeding punctures made by adult females can be particularly damaging for seedlings and young plants. Indirect damage takes place when leafminers are able to transmit virus diseases, as is the case of *Liriomyza* on soya (Costa *et al.*, 1988), celery and watermelon (Zitter and Tsai, 1977), or when bacterial and fungal pathogens penetrate into the leaves via female feeding punctures.

There are few data on the relationship between the number of mines per leaf and yield loss. A first economic threshold of 15 mines per leaf was reported by Ledieu and Helyer (1982) for *L. bryoniae* and fruit adjacent leaves in tomato plants. A highly significant correlation between truss yield and the percentage of mines on the six leaves surrounding the tomato truss was obtained by Wyatt *et al.* (1984). Yield loss was

R. Albajes et al. (eds.), Integrated Pest and Disease Management in Greenhouse Crops, 254-264.
© 1999 Kluwer Academic Publishers. Printed in the Netherlands.

closely related to the number of mines (30 mines/leaf led to a 10% loss, 60 mines/leaf to a 20% loss).

18.2. Biology of *Liriomyza* Species

The biology of *Liriomyza* species has been reviewed by Parrella (1987). Only the most relevant features are mentioned here.

18.2.1. FECUNDITY, FEEDING AND LONGEVITY

These three biological parameters are closely related to each other. The reproductive potential of *L. trifolii* is influenced by host plant (Table 18.1) and temperature (Table 18.1 and Fig. 18.1a,b). Fecundity of *L. trifolii* is low at extreme temperatures independently of host plant. A mean fecundity of five eggs on tomato (Minkenberg, 1988), 23 eggs on French beans (Beitia *et al.*, 1998) and celery (Liebee, 1984), and 42 eggs on chrysanthemum (Parrella, 1984) have been recorded at a constant temperature of 15°C. At 32 and 35°C, which are the maximal experimental temperatures at which *L. trifolii* fecundity has been measured, females showed a very high fecundity: 188 and 239 eggs on chrysanthemum and celery respectively. Maximal fecundities have been recorded at 30°C on celery with 405 eggs (Liebee, 1984), at 26.7°C on chrysanthemum with 279 eggs (Parrella, 1984), at 20°C on tomato with 80 eggs (Minkenberg, 1988), and at 25°C on French beans with 117 eggs (Beitia *et al.*, 1998). Temperature also influences fecundity in other *Liriomyza* species on celery (Tryon and Poe, 1981).

TABLE 18.1. Mean total feeding stings, fecundity (no. of eggs) and longevity (days) of *L. trifolii* at different temperatures. 1a: on chrysanthemum (Parrella, 1984). 1b: on tomato (Minkenberg, 1988)

	Temperature (°C)	Number of females	No. of feeding stings per female	Mean fecundity per female (eggs)	Longevity (days)
1a	15.6	19	284	42.4	16.7
	21.1	23	1,253	233.9	14.6
	26.7	19	1,298	278.9	12.8
	32.2	15	1,447	188.5	12.3
	37.8	21	154	0.8	3.1
1b	15	18	339	5	6.5
	20	29	1,406	79	14.4
	25	28	914	59	5.6

Females feed by puncturing the leaf. The number of feeding punctures is maximal at 20°C on tomatoes with a mean of 1406 punctures per female, and at 25°C on beans with a mean value of 2202 punctures per female (Fig. 18.1c), whereas no significant differences among temperatures were found on chrysanthemum. The number of feeding punctures has been positively correlated with female fecundity and longevity. These are also enhanced by supplementing plant diet with pollen and honey; therefore, the presence of aphid or whitefly honeydew on the crop or the occurrence of nectary plants in greenhouse surroundings may cause *L. trifolii* females to increase their reproductive potential. A lower temperature threshold for oviposition at 12.6°C on tomato and at 12.2°C on chrysanthemum have been reported for *L. trifolii* by Minkenberg (1988) and Parrella (1984) respectively. The upper threshold is expected to be near 36°C.

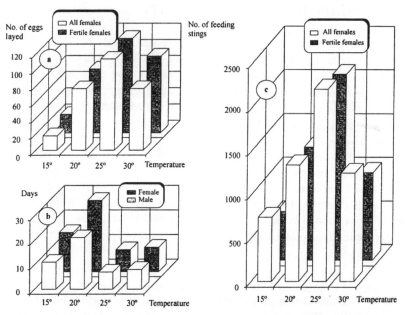

Figure 18.1. Influence of temperature on *L. trifolii* potential biotic parameters: (a) on the fecundity, (b) on the longevity, (c) on the feeding stings (Beitia *et al.*, 1998).

18.2.2. DEVELOPMENT AND MORTALITY OF *Liriomyza* IMMATURES

Eggs
Eggs are inserted into the leaf, just under the epidermis. Egg development takes between two and eight days, depending on temperature. The temperature threshold varies with temperature, host plant and experimental methodology. The lowest temperature threshold for egg development (6.2°C) has been recorded in *L. sativae* on beans (Oatman and Michelbacher, 1959) whereas for *L. trifolii*, the thresholds depend on host plant: 10°C on French beans (Charlton and Allen, 1981), 12.8°C on celery (Leibee, 1984) and 13.4°C on chrysanthemum (Charlton and Allen, 1981).

Larvae
The larva begins feeding immediately after eclosion and it feeds incessantly until ready

to emerge from the leaf. There are four molts and four larval instars. The fourth instar occurs between puparium and pupation formation and is rarely discussed by most authors (Parrella, 1987).

Development time of larvae ranges from 40 days at 12°C to six days at 32°C. Lower threshold in *L. trifolii* varies between 6.1°C on chrysanthemum (Bodri and Oetting, 1985) to 8.5°C on beans (Charlton and Allen, 1981). As for eggs, the lowest developmental threshold has been recorded in *L. sativae* (4.6°C) (Oatman and Michelbacher, 1959) and the highest in *Liriomyza congesta* Becker. When its development is completed, the larva emerges from the mine and falls on the ground to pupate. Occasionally, pupation may occur on the leaf surface, particularly when environmental relative humidity is low.

Pupae
The duration of pupal stage varies inversely with temperature, but at least 50% of the total development time of a *Liriomyza* individual is spent in this stage (Parrella, 1987). The temperature threshold depends on *Liriomyza* species and host plant; it ranges between 8 and 10.3°C.

Preimaginal Mortality
Preimaginal mortality, measured on *Liriomyza* emerging adults, is highest at extreme temperatures. Mortality percentages of 32, 20, 7.5, 25 and 100% were recorded on chrysanthemum at 15.6, 21.2, 26.7, 32.2 and 37.8°C respectively (Parrella *et al.*, 1981). On celery the mortalities recorded at 15, 20, 25, 30 and 35°C were 20, 17, 13, 17 and 91% respectively (Liebee, 1984). Low relative humidity enhances immature mortality. At 11, 51 and 94% of relative humidity the following mortalities were respectively recorded: 94, 36 and 28%.

18.3. Biology of Natural Enemies

Many natural enemies can develop at the expense of the larvae and pupae of the *Liriomyza* miners, among them many species of parasitoid and several species of predators and nematodes.

18.3.1. *Liriomyza* PARASITOIDS

Many species have been identified as parasitoids of *Liriomyza* larvae. Parasitoids recorded on *L. trifolii* and *L. sativae* in the Nearctic Region and on *L. huidobrensis* in the American Continent have been reviewed by Minkenberg and van Lenteren (1986) and van der Linden (1990) respectively. Most parasitoid species belong to two hymenopteran families. Braconidae include *Opius pallipes* Wesmael and *Dacnusa sibirica* Telenga. To Eulophidae belong *Chrysocharis parksi* J.C. Crawford and five *Diglyphus* species: *Diglyphus isaea* (Walker), *Diglyphus begini* (Ashmead), *Diglyphus intermedius* (Girault), *Diglyphus pulchripes* (Crawford) and *Diglyphus websteri* Crawford.

Dacnusa sibirica

Most Braconidae wasps are solitary endoparasitoids. *Dacnusa sibirica* is a parasitoid of *L. bryoniae* and *L. huidobrensis*. The males can be easily distinguished from the females by their pterostigma on the wing, which is black in the former and pale grey in the latter. The female searches a host by drumming the leaf surface with the antennae. Once a mine is located, it is scanned by the antennae and the ovipositor to find the larva, after which the female either rejects the larvae or inserts the ovipositor into it. All the host larval instars are accepted by the female for oviposition. The female is able not only to distinguish if a larva has been already parasitized but also to detect if the leaf has been previously visited by another female (Hendrikse *et al.*, 1980). Usually, a parasitized *Liriomyza* larva can complete its development and pupate. The adult parasitoid emerges from the puparium. At a constant temperature of 22°C, Hendrikse *et al.* (1980) recorded the following values for fecundity, longevity and preimaginal development time: 55.5 eggs, 11.4 days and 15.7 days.

Opius pallipes

It is a solitary endoparasitoid of *L. bryoniae*, *L. trifolii* and *L. huidobrensis* larvae. The female is easily distinguished because of her apparent ovipositor. Its searching behaviour, host instar acceptability to parasitization, and discrimination of already parasitized larvae are the same to those mentioned for *D. sibirica*. Host-feeding has never been recorded. At a constant temperature of 22°C, the values recorded for fecundity, longevity and preimaginal development time are 89.2 eggs, 8.7 days and 18.3 days (Hendrikse, 1983).

Chrysocharis parksii

It is an endoparasitoid of *L. trifolii* and *L. sativae* larvae. The female is blue-green, her legs are pale except for its terminal segments, which are brown. The abdomen of the female is round, whereas it is triangular in the male. The females parasitize third host larvae and the adult emerges from the host pupa (Johnson *et al.*, 1980). The development time is inversely related to temperature; at 26–30°C it is about two weeks. At a mean temperature of 26.7°C, the mean fertility is 134.6 descendants (sex ratio close to 1:1) issued from eggs layed during 11.4 days (Christie and Parrella, 1987). Throughout its lifespan, a *C. parksi* female can kill approximately 56 *L. trifolii* larvae through host-feeding.

Diglyphus begini

This neartic and neotropical species is reported to parasite *L. trifolii* on chrysanthemum (Allen and Charlton, 1981; Parrella *et al.*, 1982), and *L. trifolii* and *L. sativae* on tomato and celery (Zehnder and Trumble, 1984). It is a gregarious ectoparasitoid. The female can lay one or several eggs on the host larva or close to it into the mine. At 25°C, although the mean fecundity recorded is 268 eggs for a longevity of 17 days, most of the host larvae are killed by feeding on them. Allen and Charlton (1981) reported that out of a total of 716 larvae killed, 448 were preyed on.

Diglyphus intermedius

It is a neartic and neotropical parasitoid of larvae of eight leafminer species belonging to three agromyzid genera, such as *L. trifolii* and *L. sativae* (Gordh and Hendrickson, 1979). It is a solitary ectoparasitoid. It can prey on all the three host larval instars but prefers the third one for oviposition. Its mean fecundity at 25.5°C is relatively low, only 40 eggs (Hendrickson and Barth, 1978).

Diglyphus isaea

It is a gregarious ectoparasitoid of *L. bryoniae*, *L. trifolii* and *L. huidobrensis* larvae. Eighteen different species belonging to five agromyzid genera have been reported as hosts. After paralysing the larva, the female usually lays one egg on or near the host. The eggs are cylindrical, translucid and slightly curved. The young larvae are colourless but become green as they develop. Three larval stages can be distinguished (Ibrahim and Madge, 1979). The larva pupates in the leaf and the adult emerges by cutting a hole in the leaf epidermis with the mandibles. *Diglyphus isaea* parasitizes and host feeds in the range of 15–30°C (Franco *et al.*, 1998a). The highest mean fecundity is recorded at 20°C with a value of 212 eggs, whereas it is approximately 125 eggs at 15 and 25°C. Host-feeding is remarkably high and the following numbers of larvae were killed at 15, 20, 25 and 30°C respectively: 355, 511, 358 and 297 larvae. Longevity is quite similar for males and females, with a maximal of 32.3 days at 20°C. The larval instar structure of *L. trifolii* influences the parasitism and host-feeding of *D. isaea*; mortality by host-feeding is highest when host population is composed exclusively of second instar larvae (Franco *et al.*, 1998b). These results are different from those reported by Minkenberg (1989), who recorded the highest mortality at 15°C with a mean of 192 larvae killed per female (Fig. 18.2).

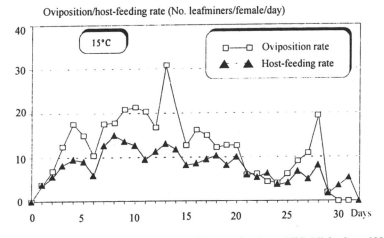

Figure 18.2. Host-feeding and reproduction of *D. isaea* females at 15°C (Minkenberg, 1989).

Diglyphus pulchripes

This neartic species can parasitize seven agromyzid species belonging to five different genera (Gordh and Hendrickson, 1979). This is the most common species in Ohio greenhouses where it can naturally control *L. trifolii* in autumn (Lindquist and Casey, 1983).

Diglyphus websteri
It is a neartic and neotropical species that overwinters on *L. huidobrensis* and *C. syngenesiae.*

18.3.2. PREDATORS AND NEMATODES

Only a few species have been cited as leafminer predators. The mirid bug *Cyrtopeltis modestus* (Distant) has been particularly studied (Parrella and Bethke, 1983). The entomopathogenic nematode *Steinernema carpocapsae* (Weiser) may be effective against *L. trifolii* pupae in the soil.

18.4. Efficacy of Leaf Miner Parasitoids for Biological Control

18.4.1. EFFICACY OF NATURAL CONTROL

In their original distribution areas, leafminer parasitoids are active throughout most of the year and are able to intervene naturally and early in the season in crops infested by several *Liriomyza* species. *Diglyphus isaea* intervenes naturally in spring as greenhouse windows are opened; thus, leafminer control may be achieved within a few generations (Woets and van der Linden, 1983). *Diglyphus begini* is the most important parasitoid of leafminers infesting cantaloups and lettuce (Hills and Taylor, 1951). *Diglyphus begini* was the most common parasitoid of *Liriomyza brassicae* (Riley) infesting cabbage and was responsible for 84% of parasitism recorded in October (Oatman and Platner, 1969). On *L. sativae* and tomato, this species produced 80.7% of all parasitoids reared in summer, which was very high (Oatman and Kennedy, 1976). Finally, this same parasitoid was the responsible for more than 66% of the parasitism recorded on *L. sativae* samples late in the season in Canadian tomato greenhouses, but its natural occurrence was not regular from year to year (McClanahan, 1975). In California, *D. intermedius* is the predominant *Liriomyza* sp. parasitoid in tomatoes. This parasitoid is also predominant on *L. sativae* and tomato and on *L. sativae* and celery in Florida (Tryon and Poe, 1981). *Chrysocharis parksi* is the most abundant parasitoid on *L. sativae* infesting tomatoes in mid to late season (Zehnder and Trumble, 1984), whereas it is a minor parasitoid on *L. sativae* infesting tomatoes in California (Johnson *et al.*, 1980).

18.4.2. EFFICACY OF INOCULATIVE RELEASES

Much effort has been done to standardize parasitoid action in biocontrol of leafminers in greenhouses. Thus, between 10,000 to 15,000 *D. sibirica* adults/ha, i.e. a dose of one adult per two plants, are needed early in the season in The Netherlands to assure a satisfactory leafminer control in greenhouses tomatoes. To ensure this, four weekly releases against the first leafminer generation are needed (Minkenberg and van Lenteren, 1987).

In particularly early infestations, three releases of a total rate of two adults of *D. sibirica* per 150 m^2 are enough to assure 100% parasitism in May, when the initial *L. bryoniae* density is under five larvae per ten plants (Hendrikse *et al.*, 1980). In chrysanthemum, fortnightly releases of *D. intermedius* at a rate of 5000 adults/1000 m^2 along three months resulted in a good control (Parrella *et al.*, 1987).

When *D. isaea* was used at a dose of one adult per 2 m^2 to control an infestation of five *L. trifolii* larvae per plant in French beans, a parasitism near 80% was obtained seven weeks after parasitoid release (Peña, 1988). In trials carried out in commercial greenhouses, early inoculative releases of *D. isaea* at a rate of 2000 adults/ha succeeded in having a parasitism of 100% two months later (Lyon, 1985). Commercial suppliers of *D. isaea* recommend early releases of 1–2 adults/m^2.

Biological control of *L. huidobrensis* can also be achieved using "banker plants" permitting the development of *D. sibirica* and *D. isaea* on an alternative host plant, *Ranunculus asiaticus* L. infested with the leafminer *Phytomyza caulinaris* Hering (van der Linden, 1993).

A foliar application of infective-stage *S. carpocapsae* (5×10^8/ha) to chrysanthemums infested with *L. trifolii* resulted in a mean leafminer mortality of 64.2% (Harris *et al.*, 1990).

18.4.3. MASS REARING OF PARASITOIDS

Mass rearing of *D. isaea* and *D. intermedius* is based on the host-plant, herbivore, and parasitoid food web. *Diglyphus isaea* is reared on *L. trifolii* with young French bean as host-plant. Two chambers climatized at 25±1°C, RH 70% and a minimal daylength of 15 hours are needed. A technique which allows the production of about 700 pupae/day of *L. trifolii* and 400–500 pupae of *L. huidobrensis* has been described (Dalle and Bordat, 1993). Leafminer adults (250) are maintained inside 0.5 m^3 cages during 48 hours for oviposition and then removed. When leafminer larvae reach the third instar, 150 parasitoids adults par cage are released. At the beginning of the parasitoid pupation, bean leaves are cut and placed into emergency boxes under darkness. Attracted by light, the emerging adults are easily collected. In this way, 500–700 *D. isaea* adults per cage are harvested every week.

The same method is used for *D. intermedius* production, but using young chrysanthemum plants. In this case, the initial cage infestation is 2000 leafminers adults and 1000 parasitoid adults per cage. The yield may reach 1000–2000 parasitoid adults per cage and week (Parrella *et al.*, 1987).

Others parasitoids like *Dacnusa* spp. and *Opius* spp., may be produced in similar ways. Recently a parasitoid belonging to Eucoilidae, *Ganaspidium utilis* Beardsley, has been produced on the bean-*Liriomyza* system. A yield of 5000 parasitoids per day may drastically lower the price (US$36 for *G. utilis* vs. US$97 for *D. begini*) (Rathman *et al.*, 1991).

Quality Control of Mass-Produced Parasitoids
The criteria proposed for quality control of mass-produced *D. isaea* and *D. sibirica* adults have been defined in the *OILB Bulletin* (van Lenteren, 1993). The criteria followed cover emerging adult mortality (<5%), sex-ratio (≥45%) and the observation of the number of eggs layed over 4–5 days (≥50 eggs/female).

18.5. Conclusions

Problems caused by agromyzid leafminers in protected vegetables in Europe are spreading as a consequence of the insufficient inspection of plant material imported from other continents. There is an extremely rich fauna of natural enemies which could be used for leafminer control. Many cases of natural control by parasitoid complexes parasitizing non-damaging leafminers that naturally occur in greenhouse surroundings have been reported. However, much research is still needed for developing systems to manage and take benefit from native parasitoid populations.

References

Allen, W.W. and Charlton, C.A. (1981) The biology of *Diglyphus begini* and its performance in caged releases on chrysanthemum, in D.J. Schuster (ed.), *Proceedings of the IFAS Conference on Biological Control of* Liriomyza *Leafminer,* Lake Buena Vista, Fla., pp. 75–81.

Beitia, F., Franco, E. and Onillon, J.C. (1998) Définition des caractéristiques du potentiel biotique de la mineuse serpentine américaine *Liriomyza trifolii* (Burgess) (Dipt., Agromyzidae), *Entomologia Experimentalis et Applicata* (in press).

Bodri, M.S. and Oetting, R.D. (1985) Assimilation of radioactive phosphorus by *Liriomyza trifolii* (Burgess) (Dipt., Agromyzidae) from feeding at different temperatures on different chrysanthemums cultivars, *Proceedings of the Entomological Society of Washington* **87**(4), 770–776.

Charlton, C.A. and Allen, W.W. (1981) The biology of *Liriomyza trifolii* on beans and chrysanthemums, in D.J. Schuster (ed.), *Proceedings of the IFAS Conference on Biological Control of* Liriomyza *Leafminer,* Lake Buena Vista, Fla., pp. 42–49.

Christie, G.D. and Parrella, M.P. (1987) Biological studies with *Chrysocharis parksii* (Hym., Eulophidae), a parasite of *Liriomyza* spp. (Dipt., Agromyzidae), *Entomophaga* **32**(2), 115–126.

Costa, A.S., Silva, D.M. and Duffus, J.E. (1988) Plant virus transmission by a leafminer fly, *Virology* **5**, 145–149.

Dalle, M. and Bordat, D. (1993) Techniques d'élevage de masse permanent de *Liriomyza trifolii* (Burgess) et de *Liriomyza huidobrensis* (Blanchard) (Diptera, Agromyzidae), mouches mineuses des feuilles, in *Proceedings of the CIRAD Conference of Leaf Mining Flies in Cultivated Plants,* Montpellier, France, 24–26 March 1993, CIRAD/ANPP/Ministère de la Recherche, Montpellier, pp. 17–21.

Franco, E., Beitia, F. and Onillon, J.C. (1998a) Predation and parasitism: Two aspects of the efficacy of *Diglyphus isaea* (Hym., *Eulophidae*), parasite of *Liriomyza trifolii* (Dipt., Agromyzidae). I. Influence of the temperature, *Annals of Applied Biology* (in press).

Franco, E., Beitia, F. and Onillon, J.C. (1998b) Predation and parasitism: Two aspects of the efficacy of *Diglyphus isaea* (Hym., *Eulophidae*), parasite of *Liriomyza trifolii* (Dipt., Agromyzidae). II. Influence of host larval instar structure population, *Annals of Applied Biology* (in press).

Genung, W.G. and Janes, M.J. (1975) Host range, wild host significance and in-field spread of *Liriomyza trifolii* and population build-up and effects of its parasites in relation to fall and winter-celery, in M.J. Janes (ed.), *Insecticide Evaluation for Control of Leafminers on Celery,* AREC Research Report No. EV-1975-5, Belle Glade, Fla.

Gordh, G. and Hendrickson, R. Jr. (1979) New species of *Diglyphus*, a world list of the species, taxonomic notes, and a key to New World species of *Diglyphus* and *Diaulinopsis* (Hym., Eulophidae), *Procedings of the Entomological Society of Washington* **81**, 666–684.

Harding, J.A. (1965) Parasitism of the leafminer *Liriomyza munda* in the winter garden area of Texas, *J. of Economic Entomology* **58**, 442–443.

Harris, M.A., Begley, J.W. and Warkentin, D.L. (1990) *Liriomyza trifolii* (Diptera: Agromyzidae) suppression with foliar applications of *Steinernema carpocapsae* (Rhabditida: Steinernematidae) and abamectin, *J. of Economic Entomology* **83**(6), 2380–2384.

Hendrickson, R.M. Jr. and Barth, S.E. (1978) Notes on the biology of *Diglyphus intermedius* (Hym., Eulophidae), a parasite of the alfalfa bloth leafminer, *Agromyza frontella* (Dipt., Agromyzidae), *Proceedings of the Entomological Society of Washington* **80**, 210–215.

Hendrikse, A. (1983) Development of a biological control programme for the tomato leafminer *Liriomyza bryoniae* Kalt., State University of Leiden, Unpublished Internal Report.

Hendrikse, A., Zucchi, R., van Lenteren, J.C. and Woets, J. (1980) *Dacnusa sibirica* Telenga and *Opius pallipes* Wesmael (Hyménopt., Braconidae) in the control of the tomato leafminer *Liriomyza bryoniae* Kalt, *OILB/WPRS Bulletin* **3**, 83–98.

Hills, O.A. and Taylor, E.A. (1951) Parasitization of dipterous leafminers in cantaloups and lettuce in the Salt River Valley, Arizona, *J. of Economic Entomology* **44**, 759–762.

Ibrahim, A.G. and Madge, D.S. (1979) Parasitization of the chrysanthemum leaf-miner *Phytomyza syngenesiae* (Hardy) (Dipt., Agromyzidae) by *Diglyphus isaea* (Walker) (Hym. Eulophidae), *Entomologist's Monthly Magazine* **114**, 71–81.

Johnson, M.W., Oatman, E.R. and Wyman, J.A. (1980) Natural control of *Liriomyza sativae* in pole tomatoes in southern California, *Entomophaga* **25**, 193–198.

Ledieu, M.S. and Helyer, N.L. (1982) Effect of tomato leafminer on yield of tomatoes, in *Annual Report of the Glasshouse Crops Research Institute 1981*, GCRI, Littlehampton, p. 106.

Leibee, G.L. (1984) Influence of temperature on development and fecundity of *Liriomyza trifolii* (Burgess) (Dipt., Agromyzidae) on celery, *Environmental Entomology* **13**(2), 497–501.

Lindquist, R.K. and Casey, M.L. (1983) Introduction of parasites for control of *Liriomyza* leafminers on greenhouse tomato, *IOBC/WPRS Bulletin* **6**(3), 108–115.

Lindquist, R.K., Frost, C. and Wolgamott, M.L. (1980) Integrated control of insect and mites on Ohio greenhouse crops, *IOBC/WPRS Bulletin* **3**(3), 119–126.

Lyon, J.P. (1985) Problèmes particuliers posés par *Liriomyza trifolii* Burgess (Dipt., Agromyzidae) et lutte biologique contre ce nouveau ravageur des cultures protégées, *Colloques de l'INRA* **34**, 85–97.

McClanahan, R.J. (1975) Notes on the vegetable leafminer *Liriomyza sativae* (Dipt., Agromyzidae) in Ontario, *Proceedings of the Entomological Society of Ontario* **105**, 40–44.

Minkenberg, O.P.J.M. (1988) Life history of the agromyzid fly *Liriomyza trifolii* on tomato at different temperatures, *Entomologia Experimentalis et Applicata* **48**, 73–84.

Minkenberg, O.P.J.M. (1989) Temperature effects on the life history of the eulophid wasp *Diglyphus isaea*, an ectoparasitoïd of leafminers (*Liriomyza* spp.) on tomatoes, *Annals of Applied Biology* **115**, 381–397.

Minkenberg, O.P.J.M. and van Lenteren, J.C. (1986) The leafminers *Liriomyza bryoniae* and *L. trifolii* (Diptera: Agromyzidae), their parasites and host plants: A review, *Agricultural University of Wageningen Papers* **86-2**.

Minkenberg, O.P.J.M. and van Lenteren, J.C. (1987) Evaluation of parasitic wasps for the biological control of leafminers, *Liriomyza* spp. in greenhouse tomatoes, *IOBC/WPRS Bulletin* **10**(2), 116–120.

Oatman, E.R. and Kennedy, G.G. (1976) Methomyl induced outbreak of *Liriomyza sativae* on tomato, *J. of Economical Entomology* **69**, 667–668.

Oatman, E.R. and Michelbacher, A.E. (1959) The melon leafminer *Liriomyza pictella* (Thomson) (Dipt., Agromyzidae). II. Ecological studies, *J. of Economical Entomology* **51**(6), 557–66.

Oatman, E.R. and Platner, G.R. (1969) An ecological study of insect populations on cabbage in southern California, *Hilgardia* **40**, 1–40.

Parrella, M.P. (1984) Effect of temperature on oviposition, feeding and longevity of *Liriomyza trifolii* (Dipt., Agromyzidae), *Canadian Entomologist* **16**, 85–92.

Parrella, M.P. (1987) Biology of *Liriomyza*, *Annual Review of Entomology* **32**, 201–224.

Parrella, M.P. and Bethke, J.F. (1983) Biological studies with *Cyrtopeltis modestus* (Hém., Miridae), a facultative predator of *Liriomyza* spp. (Dipt., Agromyzidae), in S.L. Poe (ed.), *Proceedings of the 3rd Annual International Conference on Leafminer,* San Diego, Calif., Soc. Am. Florists, Alexandria, Va., pp. 180–185.

Parrella, M.P., Heinz, K.M. and Ferrentino, G.W. (1987) Biological control of *Liriomyza trifolii* on glasshouse chrysanthemums, *IOBC/WPRS Bulletin* 10(2), 149–151.

Parrella, M.P., Robb, K.L. and Bethke, J.A. (1981) Oviposition and pupation of *Liriomyza trifolii,* in D.J. Schuster (ed.), *Proceedings of the IFAS Conference on Biological Control of* Liriomyza *Leafminer,* Lake Buena Vista, Fla., pp. 50–55.

Parrella, M.P., Robb, K.L., Christie, G.D. and Bethke, J.A. (1982) Control of *Liriomyza trifolii* with biological agents and insect growths regulators, *California Agriculture* 36(11,12), 17–19.

Peña, M.A. (1988) Primeras experiencias de lucha biológica contra *Liriomyza trifolii* (Burgess) (Dipt., Agromyzidae) con *Diglyphus isaea* (Walk) (Hym., Eulophidae) en Islas canarias, *Boletín de Sanidad Vegetal-Plagas* 14, 439–445.

Rathman, R.J., Johnson, M.W. and Tabashnik, B.E. (1991) Production of *Ganaspidium utilis* (Hym., Eucoilidae) for biological control of *Liriomyza* spp. (Dipt., Agromyzidae), *Biological Control* 1(3), 256–260.

Spencer, K.A. (1973) *Agromyzidae (Diptera) of Economic Importance,* Series Entomologica, Vol. 9, Dr. W. Junk, The Hague.

Trumble, J.T. (1981) *Liriomyza trifolii* could become a problem on celery, *California Agriculture* 35(9), 30–31.

Tryon, E.H. and Poe, S.L. (1981) Development rates and emergence of vegetable leafminer pupae and their parasites reared from celery foliage, *Florida Entomologist* 64, 477–483.

van Lenteren, J.C. (1993) Quality control for natural enemies used in greenhouses, *IOBC/WPRS Bulletin* 16(2), 89–92.

van der Linden, A. (1990) Prospects for the biological control of *Liriomyza huidobrensis* (Blanchard), a new leafminer for Europe, *IOBC/WPRS Bulletin* 13(5), 100–103.

van der Linden, A. (1993) Biological control of leafminers in glasshouse lettuce, in *Proceedings of the CIRAD Conference of Leaf Mining Flies in Cultivated Plants,* Montpellier, France, 24–26 March 1993, CIRAD/ANPP/Ministère de la Recherche, Montpellier, pp. 157–162.

Woets, J. and van der Linden, A. (1983) Observation on *Opius pallipes* Wesmael. (Hym., Braconidae) as a potential candidate for biological control of the tomato leafminer *Liriomyza bryoniae* Kalt. (Dipt., Agromyzidae) in Dutch greenhouse tomatoes, *IOBC/WPRS Bulletin* 6, 134–141.

Wyatt, J.J., Ledieu, M.S., Stacey, D.L. and White, P.F. (1984) Crop loss due to pests, in *Annual Report of the Glasshouse Crops Research Institute 1982,* GCRI, Littlehampton, pp. 88–93.

Zehnder, G.W. and Trumble, J.T. (1984) Host selection of *Liriomyza* species (Diptera, Agromyzidae) and associated parasites in adjacent plantings of tomato and celery, *Environmental Entomology* 13, 492–496.

Zitter, T.A. and Tsai, J.H. (1977) Transmission of three potyviruses by the leafminer *Liriomyza sativae* (Dipt., Agromyzidae), *Plant Disease Reporter* 61, 1025–1029.

CURRENT AND POTENTIAL USE OF POLYPHAGOUS PREDATORS

Ramon Albajes and Òscar Alomar

19.1. Introduction: Polyphagous Predators in Plant-Prey-Predator Systems

Many polyphagous or generalist predators, which can feed on a diversity of arthropod families, are found among the arthropods. Many authors use the term when referring to predators or parasitoids which attack different species within a family. This may confuse readers, and we restrict the term to predators of different families. Indeed, monophagy is rather rare in predators and it is probably the exception rather than the rule. It might be better, therefore, to refer to the degree of polyphagy in predators instead of opposing polyphagy to monophagy. Polyphagous predators, however, can display marked preferences for a few prey species and thus have similar characteristics to a more specialized predator. A mixed diet normally enhances their fecundity, longevity, survival or developmental rate. More complex feeding habits are found among omnivorous predators that are able to feed at more than one trophic level. Particularly relevant for this chapter are the omnivorous predators that feed facultatively on plants and other natural enemies.

Polyphagy has often been overlooked, and some polyphagous predators are not considered as such. However, several examples from studies of natural control and biological control by conservation of native natural enemies show that generalist predators, frequently a complex of them, have considerable impact, constitute significant components of prey limitation, and may be responsible for slowing down the rate of increase of potential pests or reducing peak infestations.

Prey or host specificity has frequently been included in the list of positive attributes of effective natural enemies. Arguments for giving priority to specialist natural enemies over more generalist species in biological control have included: (i) the former are more effective at maintaining pest populations at low densities in a stable way; (ii) the use of generalist predators may cause the extinction of non target species (within and outside the agroecosystem) or interference with other natural enemies; and (iii) generalist predators that can facultatively feed on plants may themselves become a pest. Let us consider these three arguments.

It has been argued that the poor response of generalist predators to prey density leads to their inability to regulate prey populations at low densities in a stable way. A stable prey-predator equilibrium that prevents periodic pest outbreaks has been signalled by many authors as necessary for an effective biological control program (e.g. Hassell, 1978). Others (e.g. Murdoch et al., 1985) argue that local extinctions may be common in biological control. The factors that result in stability have been explored by ecologists for many years, but a general conclusion has not yet been reached (Murdoch and Bence, 1987). The stability of prey-predator equilibrium is probably less relevant

R. Albajes et al. (eds.), Integrated Pest and Disease Management in Greenhouse Crops, 265-275.
© 1999 Kluwer Academic Publishers. Printed in the Netherlands.

for biological control of pests in short cycle crops, such as of protected crops. For example, in unheated Mediterranean greenhouses, where crop cycles are rarely longer than 7–8 months, pests and their natural enemies do not usually complete more than 5–6 generations. Over such a short-duration and in such unpredictable environments, attributes other than those directly related to the stability of prey-predator systems should probably be given priority when evaluating candidates for biological control (Wiedenmann and Smith, 1997). When natural control is not efficient enough and seasonal inoculative or augmentative releases for short-term control are needed, a natural enemy should be particularly quick (little or no delay) to respond to sudden and unpredictable increases in prey numbers, possess a high dispersal capacity, switch to the most abundant prey, aggregate in high prey density patches, and also be able to survive at low prey densities. Generalist predators frequently meet several of these requirements due to their behavioural and developmental plasticity to changing environments and prey densities.

Although the ecological risks of releasing non-indigenous species (particularly polyphagous predators) for biological control purposes have been discussed by several authors (see, e.g. Howarth, 1991), there are very few reported examples of the negative impact of biological control on the environment or on non-target species. This, however, do not prove that exotic biological control agents are safe, because monitoring of non-target species is minimal, particularly in sites and habitats far from the point of release (Simberloff and Stiling, 1996). There is more evidence of the negative effects of generalist predators on other natural enemies within the agroecosystem resulting from intraguild predation. This has been defined by Polis et al. (1989) as the cooccurrence of competition of natural enemies for the host and predation on other predators or parasitoids. These authors noted that, in contrast with competition and predation, theoretical work on intraguild predation is scattered and diffuse, and this leads to difficulty in predicting the outcome of adding a generalist predator to a multiprey-multipredator system. In many circumstances, intraguild predation may promote the occurrence of alternative stable states (Polis et al., 1989). Consistent with this, a recent review of intraguild predation concludes that "we feel that it is premature to draw generalizations concerning the effects of intraguild predation on the level of pest suppression achieved by biological control" (Rosenheim et al., 1995). Consequently, the evaluation of a polyphagous predator as a candidate for biological control in greenhouses should ideally include the study of interactions with other released or native natural enemies. This kind of knowledge, which is lacking for most of the natural enemies currently supplied, would help us to determine when and how a polyphagous predator could be released or how it can be managed to optimize its control efficiency and to minimize its possible negative effects on other biocontrol agents.

A case of omnivory that is particularly relevant for biological control is when the predator can feed occasionally or even regularly on plants, from which it may derive nutrients and water. This is a relatively common phenomenon among polyphagous predators. The plant-feeding habits of biological control agents may be regarded as negative if feeding risks damage to the crop. But it may be positive if it allows predator numbers to be maintained when prey densities are low, or if predator development and reproduction are enhanced. The preference of predators for particular plants may condition their preference for the prey that feeds on those plants – habitat preference leads to prey preference – and even their usefulness in biological control for specific crop plants or cultivars.

19.2. Native Polyphagous Predators in Natural and Biological Control in Greenhouses

In warm areas, pests may develop and reproduce both inside and outside greenhouse. The overlapping occurrence throughout the year of vegetable or ornamental crops in the open air and in greenhouses allows pests to build up high populations. In Fig. 19.1, the migrating cycle of the greenhouse whitefly exemplifies the case of several polyphagous pests affecting greenhouse crops in the Mediterranean areas. Within such a framework, conservation of native natural enemies may be useful for lowering pest pressure on greenhouse crops and incorporating beneficial fauna into the outside-inside greenhouse cycle of the pest-natural enemy complex.

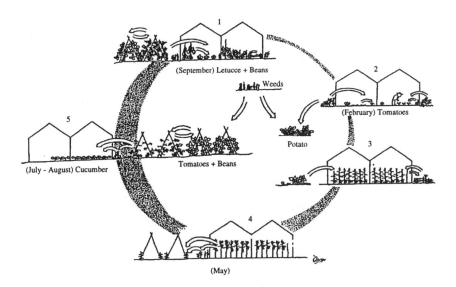

Figure 19.1. Year-round carryover of polyphagous pests and biocontrol agents among greenhouse, field crops and weeds; (1–2), warm winters allow pest survival outdoors, and intensive crop rotation allows carryover of pests within the greenhouse; (2–3), the Mediterranean greenhouse is not an isolated ecosystem: pests enter continuously from surrounding crop and non-crop vegetation; (4), pests leave protected crops early in the summer when they are harvested, and hence sudden increases in pests are recorded in field crops; (5), coexistence of field crops aids build-up of pests during summer; (5–1), autumn crops are then heavily infested from the onset.

With such a landscape approach, natural enemies that are able to track moving pests may act as ecological buffers by limiting or delaying the population growth of phytophages. In this respect, cyclic colonizers, mostly general predators, may be particularly useful (Wissinger, 1997). On this basis, a management system for two predatory mirid bug populations was developed for pest control in field tomatoes (Alomar and Albajes, 1996). Several wild plants were identified as major winter refuges from which the mirid adults invade protected crops, where they feed on whiteflies and other pests, and reproduce. At tomato senescence, the mirids leave the greenhouses and colonize outdoor crops until they return to wild plants for overwintering. Conservation programmes for these predators reduce the number of chemical applications in the field and complement the natural enemies that are released in greenhouses. However, further research is needed to determine which stimuli cause mirids to successively stay or leave crops during the year, and thus to improve the predictability of natural control.

The natural occurrence of predators and parasitoids in unsprayed greenhouse crops is quite common in warm areas and it is frequently referred to in the literature, although their ability to prevent pests reaching damaging densities has rarely been assessed. Frequently the establishment of native fauna of natural enemies is too late, when pests have already built up high populations that are difficult to control. Understanding the mechanisms that influence the abundance and phenology of natural enemy immigration into greenhouses would help us to favour the earlier establishment of native predators and parasitoids.

19.3. Uses of Polyphagous Predators in Greenhouse Crops

Amongst the predators that spontaneously enter and establish in greenhouses or that are released, two heteropteran families may be emphasized by their polyphagy: Anthocoridae and Miridae.

19.3.1. ANTHOCORIDS

The Anthocoridae contains many predatory genera, among which *Orius* is the most relevant for biological control in greenhouses. The genus *Orius* includes many species, most of which are polyphagous predators (Péricart, 1972). They have been reported to prey on soft-bodied arthropods such as spider mites, aphids, psyllids, whiteflies, thrips and lepidopteran eggs and small larvae (Riudavets, 1995); cannibalism by adults of young nymphs is common even in the presence of other food. For biological control of greenhouse pests, several *Orius* spp. are commercially supplied for thrips control (see Chapter 17), but few data support the alleged preference of *Orius* for Thysanoptera over other prey. Aphids, mites and noctuid eggs have been observed to be preyed upon in greenhouses by *Orius* species. There is no evidence of switching-prey preference with prey abundance in the field in the literature.

Anthocorids may also feed on plants – including pollen and other plant material – but benefits derived from such feeding habits for their development and reproduction are species-specific and depend on the quality of prey and plant components in the diet (see Naranjo and Gibson, 1996, for a recent review). Although *Orius* species are found on a variety of wild and agricultural plants and naturally enter IPM greenhouses in warm and cold areas, plant characteristics by themselves may also affect the establishment on the crop (Coll, 1996) and subsequent population dynamics. For example, *Orius* is rarely found on tomatoes. On sweet pepper, *Orius insidiosus* (Say) can persist on the crop despite the very low western flower thrips density because of the availability of pollen (van den Meiracker and Ramakers, 1991). Conversely, as cucumber flowers do not produce pollen, higher thrips densities are required to ensure settlement of *Orius laevigatus* (Fieber) on that crop (Chambers *et al.*, 1993).

When present in greenhouses, *Orius* may interfere with other natural enemies but also brings benefits from intraguild predation. For example, Gillespie and Quiring (1992), Ramakers (1993) and Brødsgaard and Enkegaard (1995) reported that by preying on phytoseiids, *Orius* can remain in the crop and complete its development; Cloutier and Johnson (1993) showed that *Orius tristicolor* (White) killed high numbers of *Phytoseiulus persimilis* Athias-Henriot, even when prey was available, but this was not considered as necessarily negative because it can prevent population crashes of *P. persimilis* toward the end of a spider mite infestation. These examples, however, should not mask the risks of reduced control of target pests.

19.3.2. MIRIDS

Traditionally considered as plant feeders, Miridae include many species that are also predaceous to a great extent. Several species, mostly belonging to Dicyphinae [*Macrolophus caliginosus* Wagner, *Macrolophus costalis* Fieber, *Dicyphus tamaninii* Wagner, *Dicyphus errans* (Wolff) and *Nesidiocoris tenuis* (Reuter)], have been or are being considered as potentially useful for release or conservation in biological control programmes of greenhouse and outdoor vegetable pests. Recently, Carapezza (1995) suggested the name *Macrolophus melanotoma* (Costa) for *M. caliginosus* but the authors have preferred to keep the older nomenclature.

Most of the studies on these species have focused on their predatory activity on *Trialeurodes vaporariorum* (Westwood) or *Bemisia* spp. eggs and nymphs, on which they actively prey, but they also feed on aphids, spider mites, thrips, leafminer larvae, and eggs and young larval instars of lepidoptera (e.g. Salamero *et al.*, 1987; Foglar *et al.*, 1990; Riudavets, 1995; Albajes *et al.*, 1996; Malausa and Trottin-Caudal, 1996; Alvarado *et al.*, 1997). *Dicyphus tamaninii* is able to control *Frankliniella occidentalis* (Pergande) on cucumber (Gabarra *et al.*, 1995; Castañé *et al.*, 1996). Nymphal development rates and fecundity – and therefore control efficacy – are dependent on the type or species of prey (e.g. Fauvel *et al.*, 1987). Little is known about prey preferences of predatory mirids in greenhouse conditions.

Predatory mirids feed facultatively on plants. The intake of plant material may enhance the developmental and reproductive rates of facultative mirids. Feeding on plant juices, however, may have negative implications for biological control if feeding results in commercial loss. There is some controversy as to the extent of the risk posed by *M. caliginosus*. Malausa and Trottin-Caudal (1996) reported that *M. caliginosus* only causes feeding marks on tomato fruit at extreme densities, and so they consider it to be harmless to the crop. Sampson (1996) and van Schelt *et al.* (1996) report damage by *M. caliginosus* in cherry tomatoes and gerbera flower buds when prey is depleted at the end of the crop. Feeding marks on tomato fruits by *D. tamaninii* and *D. errans* have also been related to a shortage of prey (Salamero *et al.*, 1987; Gabarra *et al.*, 1988; Alomar and Albajes, 1996). However, we have never observed effects caused by *D. tamaninii* or *M. caliginosus* on the vegetative growth of tomatoes after eight years of field use of an IPM program based on the conservation of the two mirids. Furthermore, no damage on cucumber fruit by *D. tamaninii* was observed in cage and field trials despite high predator-to-prey ratios (Gabarra *et al.*, 1995; Castañé *et al.*, 1996).

Many references report that several Dicyphinae species spontaneously colonize vegetable and ornamental greenhouses in the Mediterranean basin, where they contribute to the control of *T. vaporariorum* and also *Bemisia tabaci* (Gennadius) (see, e.g. Alomar and Albajes, 1996, and references within). Only one predatory mirid, *M. caliginosus*, is currently commercially supplied for inoculative releases. It is mostly used in northern Europe and France for whitefly control in greenhouse tomatoes (van Schelt *et al.*, 1996; R. GreatRex, pers. com.). Its ability to control pests other than whitefly has been linked to full establishment of the mirid in the crop (Trottin-Caudal and Millot, 1994). Lack of establishment is one of the main causes of predator failure (van Schelt *et al.*, 1996) particularly on cucumbers (R. GreatRex, pers. com.). This contrasts with multiple observations of natural establishment of the predatory mirid on several greenhouse crops throughout the Mediterranean region. Factors other than host plant and prey density are probably involved in predatory mirid establishment on the crop and they would merit investigation in order to optimize their use for biological control in greenhouses.

Defining management strategies for such facultative predators is not straightforward, and the risks of some economic loss have to be balanced with the tangible benefits. Risks are probably maximum when high predator density occurs with low prey numbers at susceptible periods of crop growth. To minimize the risks, elements and relationships within the tri-trophic interaction (plant-multiprey-predator) should be determined in order to define a safe strategy for using mirids that are released or that naturally enter the greenhouses. Questions such as what is the relative susceptibility of crops or cultivars to mirid attack? what crop growth stages are the most susceptible to mirid feeding punctures? what pests may be preyed upon by mirids and what prey preference may be expected from the predator? how does the predator respond numerically to different prey densities? what predator-to-prey ratio may be critical for causing damage to the crop? have no reliable answer with the present state of knowledge. Alomar and Albajes (1996) attempted to answer to some of these questions by developing a mirid population management strategy for biological control in field tomatoes. In greenhouses, the release strategy for mirids only focuses on whitefly control, and early predator releases – alone or in combination with *Encarsia formosa* Gahan – are recommended. This strategy may lead to a more or less rapid whitefly population decline with, in the case of a rapid numerical response in the predator, the consequent increase in mirid density. This leads to high damage risk if no suitable alternative prey is present at numbers high enough to prevent the predator from switching to feeding on the plant. In such a complex system, mirid management strategies should be taken into account, considering all potential prey and not just whitefly.

19.3.3.OTHER GROUPS OF POLYPHAGOUS PREDATORS FOR GREENHOUSE USE

The Phytoseiidae include specialized and generalist predators, which prey mostly on mites (McMurtry and Croft, 1997). Some species may feed on Thysanoptera and hence they have been used for biological control of thrips pests. *Phytoseiulus persimilis* is the most widely used species for biological control in greenhouses, but its maintenance depends strictly on spider mites. Other Phytoseiids, however, are more generalist, and are used for biocontrol of thrips or have been tested. for other pests such as *Polyphagotarsonemus latus* (Banks) (Fan and Petitt, 1994) or *Bemisia* (Meyerdirk and Coudriet, 1986; Nawar and El-Sherif, 1993). Joint use with *Orius* has been discussed previously. At low thrips densities, competition among phytoseiids for spider mites is likely, and the polyphagous *Amblyseius barkeri* (Hughes) may also attack the monophagous *P. persimilis* (Kabicek and Horka, 1994). Pollen is known to facilitate the establishment of these predators but it may also reduce predation rate (van Rijn and Sabelis, 1993). Evaluation of phytoseiids should also consider predation on non-tetranychid mites such as *P. latus* and *Aculops lycopersici* (Massee), which are becoming important pests in warm areas.

The Hypoaspididae are used for the control of soil stages of pests. *Hypoaspis miles* (Berlese) is polyphagous, but prefers sciarids over thrips pupae and collembola (Brødsgaard *et al.*, 1996, and references therein). *Hypoaspis miles* and *Hypoaspis aculeifer* (Canestrini) occur naturally in European greenhouses and are also marketed, mainly for soil-inhabiting stages of different crop pests.

The Chrysopidae or green lacewings are prominent among generalist predators associated with many agroecosystems (Tulisalo, 1984; Wang and Nordlund, 1994). In greenhouses they can spontaneously colonize crops, commonly at low numbers. The adults feed on honeydew, nectar and pollen but the adults of some species, and all larvae, are predaceous, mainly on soft-bodied, slow-moving prey (New, 1988). Most chrysopid species are observed in association with aphids and these have been the principal target of chrysopid releases (Bondarenko, 1987), but some green lacewings have been tested against other greenhouse pests such as *B. tabaci* (Breene *et al.*, 1992). *Chrysoperla rufilabris* (Burmeister) prefers lepidoptera eggs or larvae over aphids in the laboratory (Nordlund and Morrison, 1990; Legaspi *et al.*, 1994). One of the features that most limits the use of chrysopids in greenhouses is their lack of reproduction and establishment on crops, and costs resulting from high release rates.

Some predaceous coccinellids have a wide range of essential prey, and adults may feed on nectar, pollen and even green leaves, these probably for providing fluid (Hodek, 1967). However, most coccinellids that are released in greenhouses are fairly specific for scales, whiteflies or aphids and have little effect on other pests. Coccinellids seldom occur spontaneously in greenhouses. When released for aphid control, they tend to leave the greenhouse, especially when aphid densities are low (van Steenis, 1995). *Harmonia axyridis* (Pallas) has been imported for aphid and coccid control in greenhouses and orchards.

Several other groups of polyphagous predators, such as muscid and empidid flies (von Kühne and Schrameyer, 1994), spiders, carabids, staphylinids, some pentatomid and lygaeid bugs, and thrips, are known to contribute to lower pest populations in field crops and, when present, may reduce pest pressure on neighbouring greenhouses. The development of systems for their management may prevent pest outbreaks outside and save many interventions in protected crops.

19.4. Conclusions

Little is known about the role of polyphagous predators in agricultural systems in general and greenhouses in particular. Of those data presented here, several points can be highlighted to indicate areas for further research, and also to challenge some current views on biological control practice in greenhouses.

Polyphagous or generalist predators have been regarded as largely inefficient and their use risky in biological control, mainly because they cannot provide stability in prey-predator systems and they can prey on non target prey, causing negative environmental and control impacts. As biological control in greenhouses does not need to establish a lasting prey-predator interaction, attributes other than giving stability should be prioritized in screening candidates for release or conservation in protected crops. Some of these attributes – such as being quick to respond to sudden and unpredictable increases in prey numbers, a high capacity to disperse and to aggregate in high prey density patches, the ability to switch to the most abundant prey, a high survival capacity at low prey densities – are frequently found among polyphagous predators. Very few environmental effects of using generalist predators have been soundly documented in the literature. There are risks from releasing non-specific predators, as these can prey on other natural enemies and diminish the global efficacy of biological control. However, no general outcome can be predicted from intraguild predation or resource competition.

Many predators naturally colonize the crops from surrounding crops or natural vegetation. This is important whenever pest problems originate from immigrating populations, when greenhouses are dispersed as a mosaic in the landscape, either in alternation with outdoor vegetable crops or non-agricultural vegetation, or in large areas of protected cultivation. In such landscapes, models for encouraging natural enemies should focus on annual dispersal cycles between winter refuges and patchy-work of fields (among which greenhouses are located) as suggested by Wissinger (1997). IPM programmes developed for field tomatoes create reservoirs for the predators, and also considerably reduce the scale of whitefly immigration into autumn greenhouses. Inclusion of polyphagous predators that are able to prey and survive on several prey and to successively colonize different crops may be crucial for success in such a scenario. The creation of refuges adjacent to greenhouses or the use of banker-plant systems within them, could assist the process of crop colonization. Understanding factors that determine the movements of predators out of refuges and between crops would selectively enhance natural control and reduce the need for localized interventions.

Polyphagous predators are currently released for biocontrol in greenhouses, but mostly against one target prey. Polyphagy is not always considered when discussing results, nor are interactions between natural enemies. It is therefore difficult to evaluate their impact on biocontrol from literature searches. Some data are available on the effect of prey species on predator life history traits, but more work is needed to establish the control efficacy on a complex of prey.

For biocontrol to be successful, sufficient predator numbers should be present at the correct time, in some cases even before the pest arrives. Such synchronization can best be ensured if the predator establishes on other prey in advance of the target prey. The response may thus not show lag-times to sudden immigrations of other prey. Moreover, already established predators may be able to exert control at low prey densities. Not being dependent on one prey species, predators can remain in the crop and maintain themselves on other prey if the target prey is at a low density, and so prevent reinfestation by immigrating individuals of the target prey. However, the outcome depends on the degree of switching shown by the predator.

Particularly critical is the determination of the role of the plant in the ecology of those predators that may feed on plants. Intake of plant products will affect the predator directly, independently of the prey. Plant characteristics will therefore be more important for these predators than is normal for beneficials. Plant-feeding may allow predators to establish on the crop in advance of pests or their maintenance on the crop when prey is scarce. In fact, the plant should be considered as another prey, which is unlimited, and different crops may alter insect prey preference. However, benefits should be balanced with the risks of damage for certain plant feeding habits (e.g. when the plant tissue attacked relates to crop yield). In these circumstances, predator management systems must avoid high predator-to-prey ratios on susceptible crops and cultivars.

In summary, a better understanding of polyphagous predator ecology within and outside the greenhouse environment may allow us to derive benefits from their potential for biological control in protected crops.

References

Albajes, R., Alomar, O., Riudavets, J., Castañé, C., Arnó, J. and Gabarra, R. (1996) The mirid bug *Dicyphus tamaninii*: An effective predator for vegetable crops, *IOBC/WPRS Bulletin* **19**(1), 1–4.

Alomar, O. and Albajes, R. (1996) Greenhouse whitefly (Homoptera: Aleyrodidae) predation and tomato fruit injury by the zoo-phytophagous predator *Dicyphus tamaninii* (Heteroptera: Miridae), in O. Alomar and R. Wiedenmann (eds.), *Zoophytophagous Heteroptera: Implications for Life History and IPM*, Thomas Say Publications in Entomology: Proceedings, Entomological Society of America, Lanham, Md., pp. 155–177.

Alvarado, P., Baltà, O. and Alomar, O. (1997) Efficiency of four heteroptera as predators of *Aphis gossypii* Glover and *Macrosiphum euphorbiae* (Thomas) (Hom.: Aphididae), *Entomophaga* **42**(1/2), 215–226.

Bondarenko, N.V. (1987) The experience of biological and integrated control of pests on glasshouse crops in the USSR, *IOBC/WPRS Bulletin* **10**(2), 33–36.

Breene, R.G., Meagher, R.L., Nordlund, D.A. and Yin-Tung Wang (1992) Biological control of *Bemisia tabaci* (Homoptera: Aleyrodidae) in a greenhouse using *Chrysoperla rufilabris* (Neuroptera: Chrysopidae), *Biological Control* **2**, 9–14.

Brødsgaard, H.F. and Enkegaard, A. (1995) Biological control of aphids and thrips on pot Gerbera, *12th Danish Plant Protection Conference, Pests and Diseases, SP Rapport – Statens Planteavlsforsog* **4**, 257–266.

Brødsgaard, H.F., Sardar, M.A. and Enkegaard, A. (1996) Prey preference of *Hypoaspis miles* (Berlese)

(Acarina: Hypoaspididae): Non-interference with other beneficials in glasshouse crops, *IOBC/WPRS Bulletin* **19**(1), 23–26.

Carapezza, A. (1995) The specific identities of *Macrolophus melanotoma* (A. Costa, 1853) and *Stenodema curticolle* (A. Costa, 1853) (Insecta Heteroptera, Miridae), *Naturalista Sicil*iano, S.IV, **XIX** (3–4), 295–298).

Castañé, C., Alomar, O. and Riudavets, J. (1996) Management of western flower thrips on cucumber with *Dicyphus tamaninii* (Hetroptera: Miridae), *Biological Control* **7**, 114–120.

Chambers, R.J., Long, S. and Helyer, N.L. (1993) Effectiveness of *Orius laevigatus* for the control of *Frankliniella occidentalis* on cucumber and pepper in the UK, *Biocontrol Science and Technology* **3**, 295–307.

Cloutier, C. and Johnson, S.G. (1993) Predation by *Orius tristicolor* (Hemiptera: Anthocoridae) on *Phytoseiulus persimilis* (Acarina: Phytoseiidae): Testing for compatibility between biological control agents, *Environmental Entomology* **22**, 477–482.

Coll, M. (1996) Feeding and ovipositing on plants by an omnivorous insect predator, *Oecologia* **105**, 214–220.

Fan, Y. and Petitt, F.L. (1994) Biological control of broad mite, *Polyphagotarsonemus latus* (Banks), by *Neoseiulus barkeri* Hughes on pepper, *Biological Control* **4**, 390–395.

Fauvel, G., Malausa, J.C. and Kaspar, B. (1987) Etude en laboratoire des principales caractéristiques biologiques de *Macrolophus caliginosus* (Heteroptera: Miridae), *Entomophaga* **32**, 529–543.

Foglar, H., Malausa, J.C. and Wajnberg, J. (1990) The functional response and preference of *Macrolophus caliginosus* (Heteroptera: Miridae) for two of its preys: *Myzus persicae* and *Tetranychus urticae*, *Entomophaga* **35**, 465–474.

Gabarra, R., Castañé, C. and Albajes, R. (1995) The mirid bug *Dicyphus tamaninii* as a greenhouse whitefly and western flower thrips predator on cucumber, *Biocontrol Science and Technology* **5**, 475–488.

Gabarra, R., Castañé, C., Bordas, E. and Albajes, R. (1988) *Dicyphus tamaninii* Wagner as a beneficial insect and pest of tomato crops in Catalonia, *Entomophaga* **33**, 219–228.

Gillespie, D.R. and Quiring, D.J.M. (1992) Competition between *Orius tristicolor* (White) (Hemiptera: Anthocoridae) and *Amblyseius cucumeris* (Oudemans) (Acari: Phytoseiidae) feeding on *Frankliniella occidentalis* (Pergande) (Thysanoptera: Thripidae), *Canadian Entomologist* **124**, 1123–1128.

Hassell, M.P. (1978) *The Dynamics of Arthropod Predator-Prey Systems*, Princeton University Press, Princeton, NJ.

Hodek, I. (1967) Bionomics and ecology of predaceous coccinellidae, *Annual Review of Entomology* **12**, 7–104.

Howarth, F.G. (1991) Environmental impacts of classical biological control, *Annual Review of Entomology* **36**, 485–509.

Kabicek, J. and Horka, D. (1994) Development of the predatory mite *Amblyseius barkeri* (Acarina: Phytoseidae) on two food types, *Sbornik Vysoke Skoly Zemedelske v Praze, Fakulta Agronomicka, Rada A, Rostlinná Výroba* **56**, 135–140.

Legaspi, J.C., Carruthers, I. and Nordlund, D.A. (1994) Life history of *Chrysoperla rufilabris* (Neuroptera: Chrysopidae) provided sweetpotato whitefly *Bemisia tabaci* (Homoptera: Aleyrodidae) and other food, *Biological Control* **4**, 178–184.

Malausa, J.-C. and Trottin-Caudal, Y. (1996) Advances in the strategy of use of the predaceous bug *Macrolophus caliginosus* (Heteroptera: Miridae) in glasshouse crops, in O. Alomar and R. Wiedenmann (eds.), *Zoophytophagous Heteroptera: Implications for Life History and IPM*, Thomas Say Publications in Entomology: Proceedings, Entomological Society of America, Lanham, Md., pp. 178–189.

Meyerdirk, D.E. and Coudriet, D.L. (1986) Evaluation of two biotypes of *Euseius scutalis* (Acari: Phytoseidae) as predators of *Bemisia tabaci* (Homoptera: Aleyrodidae), *J. of Economic Entomology* **79**, 659–663.

McMurtry, J.A. and Croft, B.A. (1997) Life-styles of phytoseiid mites and their roles in biological control, *Annual Review of Entomology* **42**, 291–321.

Murdoch, W.W. and Bence, J. (1987) General predators and unstable prey populations, in W.C. Kerfoot and A. Sih (eds.), *Predation. Direct and Indirect Impacts on Aquatic Communities*, University Press of New England, Hannover, pp.17–30.

Murdoch, W.W., Chesson, J. and Chesson, P.L. (1985) Biological control in theory and practice, *American Naturalist* **125**, 344–366.

Naranjo, S.E. and Gibson, R.L. (1996) Phytophagy in predaceous Heteroptera: Effects on life history and population dynamics, in O. Alomar and R. Wiedenmann (eds.), *Zoophytophagous Heteroptera:*

Implications for Life History and IPM, Thomas Say Publications in Entomology: Proceedings, Entomological Society of America, Lanham, Md., pp. 57–93.

Nawar, M.S. and El-Sherif, A.A. (1993) *Neoseiulus cucumeris* (Oudemans), a predator of whitefly *Bemisia tabaci* (Gennadius), *Bulletin of Entomological Society of Egypt* **71**, 9–17.

New, T.R. (1988) Neuroptera, in A.K. Minks and P. Harrewijn (eds.), *Aphids; Their Biology, Natural Enemies and Control*, Vol. B, Elsevier, Amsterdam, pp. 249–258.

Nordlund, D.A. and Morrison, R.K. (1990) Handling time, prey preference, and functional response for *Chrysoperla rufilabris* in the laboratory, *Entomologia Experimentalis et Applicata* **57**, 237–242.

Péricart, J. (1972) *Hemiptères Anthocoridae, Cimicidae et Microphysidae de l'Ouest-Paléarctique*, Masson et Cie, Paris.

Polis, G.A., Myers, C.A. and Holt, R.D. (1989) The ecology and evolution of intraguild predation: Potential competitors that eat each other, *Annual Review of Ecology and Systematics* **20**, 297–330.

Ramakers, P.M.J. (1993) Coexistence of two thrips predators, the anthocorid *Orius insidiosus* and the phytoseiid *Amblyseius cucumeris* on sweet pepper, *IOBC/WPRS Bulletin* **16**(2), 133–136.

Riudavets, J. (1995) Predators of *Frankliniella occidentalis* and *Thrips tabaci*: A review, *Wageningen Agricultural University Papers* **95**(1), 43–87.

Rosenheim, J.A., Kaya, H.K., Ehler, L.E., Marois, J.J. and Jaffee, A. (1995) Intraguild predation among biological control agents: Theory and evidence, *Biological Control* **5**, 303–335.

Salamero, A., Gabarra, R. and Albajes, R. (1987) Observations on the predatory and phytophagous habits of *Dicyphus tamaninii* Wagner (Heteroptera: Miridae), *IOBC/WPRS Bulletin* **10**(2), 165–169.

Sampson, C. (1996) Macrolophus pros and cons, *Grower* **26**, 9.

Simberloff, D. and Stiling, P. (1996) How risky is biological control? *Ecology* **77**, 1965–1974.

Trottin-Caudal, Y. and Millot, P. (1994) Lutte integrée contre les ravageurs sur tomate sous abri. Situation et perspectives en France, *IOBC/WPRS Bulletin* **17**(5), 5–13.

Tulisalo, U. (1984) Biological and integrated control by Chrysopids, in M. Canard, Y. Semeria and T.R. New (eds.), *Biology of Chrysopidae*, Dr. W. Junk, The Hague, pp. 228–233.

van den Meiracker, R.A.F. and Ramakers, P.M.J. (1991) Biological control of the western flower thrips *Frankliniella occidentalis*, in sweet pepper with the anthocorid *Orius insidiosus*, *Mededelingen van de Faculteit Landbouwwetenschappen, Rijksuniversiteit Gent* **56/2a**, 241–249.

van Rijn, P.C.J. and Sabelis, M.W. (1993) Does alternative food always enhance biological control? The effect of pollen on the interaction between western flower thrips and its predators, *IOBC/WPRS Bulletin* **16**(8), 123–125.

van Schelt, J., Klapwijk, J., Letard, M. and Aucouturier, C. (1996) The use of *Macrolophus caliginosus* as a whitefly predator in protected crops, in D. Gerling and R.T. Mayer (eds.), *Bemisia 1995. Taxonomy, Biology, Damages, Control and Management*, Intercept, Andover, Hants, pp. 515–521.

van Steenis, M. (1995) *Evaluation and Application of Parasitoids for Biological Control of Aphis gossypii in Glasshouse Cucumber Crops*, PhD Thesis, Wageningen Agricultural University.

von Kühne, K. and Schrameyer, K. (1994) On the occurrence of predatory flies of the family Hybotidae (Dipt. Empidoidea) in greenhouses and on the predatory efficiency of two species of the genus *Platypalpus* Marquart, *J. of Applied Entomology* **118**, 209–216.

Wang, R. and Nordlund, D.A. (1994) Use of *Chrysoperla* spp. (Neuroptera: Chrysopidae) in augmentative release programmes for control of arthropod pests, *Biocontrol News and Information* **15**, 51–57.

Wiedenmann, R.N. and Smith, J.W. Jr. (1997) Attributes of natural enemies in ephemeral crop habitats, *Biological Control* **10**, 16–22.

Wissinger, S.A. (1997) Cyclic colonization in predictably ephemeral habitat: A template for biological control in annual crop systems, *Biological Control* **10**, 4–15.

MASS PRODUCTION, STORAGE, SHIPMENT AND QUALITY CONTROL OF NATURAL ENEMIES
Joop C. van Lenteren and Maria Grazia Tommasini

20.1. Introduction

Since the beginning of this century, mass production of natural enemies has been considered as a means of improving biological control programmes, especially those based on inundative and seasonal inoculative releases. For general information on mass production and quality control of insects and other arthropods, we refer to Morrison and King (1977), King and Morrison (1984), Singh (1984) and van Lenteren (1986a); for mass production and quality control related to commercially produced natural enemies for greenhouse use, we refer to Nicoli *et al.* (1994). We will not discuss the question on how to obtain a good stock colony to start a mass production, because this issue is addressed in Chapter 13. In this section we will briefly summarize developments in mass rearing during the 20th century.

Mass production of beneficials is a "skilful and highly defined processing of an entomophagous species through insectary procedures, which results in economical production of millions of beneficial insects" (Finney and Fisher, 1964). This is true for most of the mass-rearing programmes, but there are important exceptions where mass production seems to be a fairly simple process.

The first step in a mass-rearing programme is a trial to rear the natural enemy on a natural host (the pest organism) in an economical way. Most of the natural enemies are reared in this way. However, several natural enemies are not mass reared on their natural host because it is either too expensive, or undesirable due to the risk of infection with the pest organism or concurrent infection with other pests or diseases when natural enemies are released on their natural substrate. In these cases a search is made for an opportunity to rear the natural enemy on alternative host (and often an alternative host plant).

A subsequent step in making mass rearing more economical is to change from a natural host medium (host plant) to an artificial medium for rearing the host. Rearing insects on artificial diets was developed earlier this century and considerable progress has been made recently. Rearing on artificial diets is considerably cheaper as less expensively climatized space is needed, but artificial rearing may create serious quality problems which will be discussed later in this chapter. Singh (1984) summarizes the historical development, recent advances and future prospects for insect diets as follows: (i) some 750 species, mainly phytophagous insects can be reared successfully on (semi-) artificial diets; (ii) only about two dozen species have been successfully reared for several generations on completely artificial diets; (iii) large-scale mass rearing on artificial media has been developed for less than twenty species of insects; (iv) quality control is essential, as there can be dietary effects on all critical performance traits of the mass-reared insect and also on the natural enemy produced on a host that was mass reared on an artificial medium; and

R. Albajes et al. (eds.), Integrated Pest and Disease Management in Greenhouse Crops, 276-294.
© 1999 *Kluwer Academic Publishers. Printed in the Netherlands.*

(v) suitable bioassays are important for answering the question "what is the ultimate effect of the diet on the reared insect?"

A final step when trying to minimize rearing costs is the search for ways to rear the natural enemy on an artificial diet. This has been attained for several endo- and ectoparasitoids (e.g. *Trichogramma*) but is not yet commercially applied, and has been commercialized for a few predators (e.g. *Chrysoperla*). The technology for rearing natural enemies on diets is, however, far less developed than that for rearing of pest species (Grenier *et al.*, 1994).

Although biological control of arthropod pests has been used in protected crops since 1926, large-scale production of natural enemies in Europe emerged in the 70s (van Lenteren and Woets, 1988). Initially mass rearing involved the production of several thousands of individuals per week of a few beneficial species, nowadays millions of individuals are produced per week, and the number of natural enemies available for commercial use has increased very fast during the past 30 years (Fig. 20.1). None of the early publications on commercial biocontrol in greenhouses mention the topic of quality control of natural enemies (e.g. Hussey and Bravenboer, 1971). Quality control of mass-reared beneficial insects is mentioned in the mid 80s for the first time in relation to greenhouse biocontrol. From then on the topic was frequently raised in publications (e.g. van Lenteren, 1986b; Nicoli *et al.*, 1994).

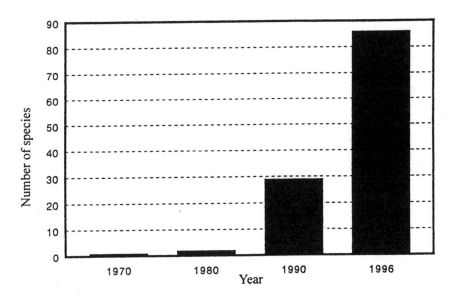

Figure 20.1. Number of natural enemy species commercially available for biological control in greenhouses.

20.2. Obstacles Encountered in Mass Production

The main problems encountered in mass production of entomophagous insects are summarized in Table 20.1. For a detailed discussion of these obstacles, we refer to van Lenteren (1986a). Many problems relate to the artificial nature of the situation under which the mass production takes place. The best advice one can give a new commercial producer is that during the first stage of developing the mass production of a natural enemy, every effort should be made: (i) to rear the natural enemy on the target pest; (ii) to rear the target pest on the plant that is to be protected; and (iii) to rear under normal climate conditions. This is often a realistic option for natural enemies which are to be used in greenhouses.

TABLE 20.1. Problems related to mass rearing of natural enemies

1. Difficult to produce biocontrol agents of good quality at low costs
2. Lack of artificial diets for rearing host and/or natural enemy
3. Lack of techniques that prevent selection pressure, leading to genetic deterioration of natural enemies
4. Occurrence of cannibalism (predators), host feeding and/or superparasitism (parasitoids), leading to inefficient production and loss of quality
5. Behavioural changes as a result of rearing on alternative host/prey or alternative conditions (e.g. conditioning and learning)
6. Reduced vigour, e.g. when reared on alternative host and/or alternative host diet
7. Contamination with pathogens (nematodes, micro-organisms)
8. Colonization by other arthropod species

Anyone starting mass production must overcome the above mentioned obstacles and should also realize the conflicting requirements for natural enemies in mass rearing programmes and field performance (Table 20.2). The main conflicts relate to searching, migration and learning, which are not important or appreciated in mass rearing, while these are very important for functioning well in the greenhouse or field (for a detailed discussion of this problem, see Vet and Dicke, 1992). Due to these different demands, artificial selection in the laboratory may lead to reduced field performance (Bartlett, 1984b).

TABLE 20.2. Conflicting requirements for natural enemies in mass rearing and field performance

Appreciated in mass-rearing units		Important for field performance	
1.	Polyphagy, makes rearing on alternative host easier	1.	Mono-, oliphagy, more specific, greater pest reduction
2.	Good host attack at high host densities	2.	Good host attack at low densities
3.	No strong dispersal as a result of direct interference or competition for hosts	3.	Strong dispersal as a result of direct interference or competition for hosts
4.	Migration behaviour unnecessary and unwanted, minimal ability to disperse	4.	Migration behaviour essential
5.	Associative learning not desirable	5.	Associative learning desirable
6.	Diapausing capacity appreciated	6.	Diapause not appreciated

An additional problem is the risk of inbreeding (homozygosis). Typically, an insect colony for mass rearing is started from small populations (van Lenteren and Woets, 1988). To reduce the risk of a too small genetic pool, it is important to start with a large population collected at different locations. This is not always possible, however, and bottleneck effects may occur, resulting in populations which have lost certain qualities that are essential for greenhouse or field performance (see Section 20.7.3). To correct for these problems, the producers should regularly restart their mass rearing with a new genetic pool by collecting individuals from the field or greenhouse. Existing laboratory populations should not be mixed with the field-collected material, but the old population should be replaced with the new one.

20.3. Mass Production of Natural Enemies

About 150 species of natural enemies have been imported and released into Europe during the 20th century to control about 55 mite and insect pest species. Until 1970 this mainly concerned inoculative (classical) biological control. After 1970 many developments took place in greenhouses, and commercial biological control programmes for *circa* 50 pest species were developed by importing 60 species of natural enemies. In addition, 40 endemic species of natural enemies were employed in commercial biological control. Our experience with the development of new biological control programmes has shown that dogmatism is useless when selecting natural enemies; this contrasts with the approach of earlier biocontrol workers (see e.g. DeBach, 1964). We have, for example, had excellent control results by releasing endemic natural enemies against exotic pests and vice versa: all combinations are worth trying (Table 20.3).

TABLE 20.3. Commercial biological control of endemic or exotic arthropods with endemic or exotic natural enemies in Europe (the numbers reflect the number of combinations in which a certain natural enemy is used for control of a certain pest)

Different combinations of natural enemy use in northwest Europe	
Use of endemic natural enemies for the control of endemic pests. Example: *Chrysoperla carnea* for control of endemic aphid species	56
Use of endemic natural enemies for the control of exotic pests. Example: *Diglyphus isaea* for control of exotic *Liriomyza* species	24
Use of exotic natural enemies for the control of endemic pests. Example: *Harmonia axyridis* for control of endemic aphid species	43
Use of exotic natural enemies for the control of exotic pests. Example: *Encarsia formosa* for control of exotic whitefly species	49

Greenhouse pests are presently managed through biological control on some 14,000 ha of the about 250,000 ha of protected cultivation world-wide, compared to 200 ha under biological control in 1970 (van Lenteren, 1995). In 1968, when commercial biological control in greenhouses started in Europe, two small commercial producers were active.

Today, Europe has 26 natural enemy producers including the world's three largest, whereas there are 65 producers world-wide. These three largest companies serve more than 75% of the greenhouse biological control market. Of the *circa* 100 biological control agents that are marketed today, about thirty make up 90% of the total sales (Table 20.4).

TABLE 20.4. Most commonly sold biological control agents in Europe

Biological control agent	No. of producers	Shipped as
Neoseiulus (= Amblyseius) californicus	3	Mixed life stages
Neoseiulus (= Amblyseius) cucumeris	5	Mixed life stages
Amblyseius degenerans	5	Mixed life stages
Aphelinus abdominalis	3	Adult
Aphidius colemani	5	Mummy
Aphidius ervi	3	Mummy
Aphidoletes aphidimyza	4	Pupa
Chrysoperla carnea	2	Egg/larva
Cryptolaemus montrouzieri	5	Adult
Dacnusa sibirica	3	Adult
Delphastus pusillus	3	Adult
Diglyphus isaea	5	Adult
Encarsia formosa	7	Pupa
Eretmocerus eremicus (= Eretmocerus californicus)	3	Pupa
Eretmocerus mundus	2	Pupa
Harmonia axyridis	2	Adult
Heterorhabditis megidis	2	Juveniles
Hypoaspis aculeifer	3	Mixed life stages
Hypoaspis miles	4	Mixed life stages
Leptomastidea abnormis	3	Adult
Leptomastix dactylopii	5	Adult
Leptomastix epona	4	Adult
Lysiphlebus testaceipes	2	Mummy
Macrolophus caliginosus	5	Adult
Orius insidiosus	4	Adult
Orius laevigatus	5	Adult
Orius majusculus	5	Adult
Phytoseiulus persimilis	9	Mixed life stages
Steinernema feltiae	2	Juveniles
Trichogramma brassicae	2	Parasitized host egg
Trichogramma evanescens	2	Parasitized host egg

Very little is published about prices of commercially produced organisms (e.g. van Lenteren and Woets, 1988), but recently two comprehensive reviews were written, one for the North American market (Cranshaw *et al.*, 1996) and one for the European market (van Lenteren *et al.*, 1997). From these reviews, it appears that many more species of biocontrol agents are available in Europe than in North America. This is caused by the much larger European greenhouse industry and a longer history of research in greenhouse biocontrol in Europe.

Although on farm production of natural enemies is possible, most growers purchase them from commercial suppliers. Many of the mass production companies are, understandably, reluctant to provide information on many aspects of mass production. Our experience is that many of the natural enemies produced for biocontrol in protected cultivation are reared on their natural hosts (the pests) and host plants. Rearing on purely artificial media (without organic additives) is very rare, primarily because this technology is insufficiently developed for mass production and because this way of production may lead to poor performance of natural enemies when exposed to their target hosts (for details, see van Lenteren 1986a). Rearing conditions should be as similar as possible to the conditions under which the natural enemies will have to function in commercial greenhouses. Two examples of mass production schemes, one for the predator *Orius* and the other for the parasitoid *Encarsia*, are presented in Figs 20.2 and 20.3.

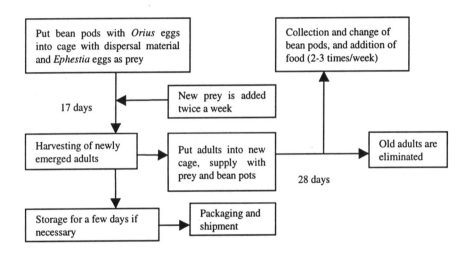

Figure 20.2. Production scheme for the thrips predator *Orius* sp.

20.4. Storage of Natural Enemies

It is necessary to have storage methods and facilities available to meet the requirements for good planning for a mass production unit and because of the difficulty of accurately predicting demand from clients (both delivery dates and quantities). This is relatively simple for microbial biocontrol agents like fungi, viruses and bacteria because they can often be stored in a resting stage for months or even years. Many predators and parasitoids can only be stored for a short time. This usually involves placing the natural enemies as immatures at temperatures between 4 and 15°C. Normally, storage only lasts several weeks, but even then reduction in fitness is the rule (Posthuma-Doodeman *et al.*, 1996). Storage during the adult stage leads to even higher and faster reduction in fitness than with storage of immatures. The pupal stage seems to be most suitable for short-term storage.

Data on long-term storage of natural enemies or their hosts are limited. Host material [e.g. eggs of *Sitotroga cerealella* (Olivier) and *Graphosoma lineatum* (L.)] stored for long periods (in the case of *Graphosoma* for up to five years) in liquid nitrogen could still be used for production of *Trichogramma* and *Trissolcus simoni* (Mayr) respectively (Gennadiev and Khilistovskii, 1980). Eggs of *Ephestia kuehniella* Zeller can be sterilized by UV radiation or freezing, and then be stored at low temperature for several months without losing their value as alternative food for mass production of predators such as *Chrysoperla* and *Orius*.

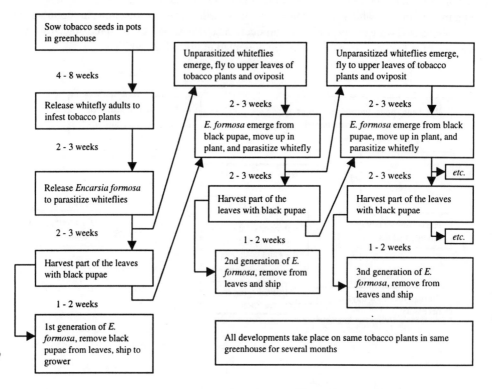

Figure 20.3. "Continuous" production scheme for the whitefly parasitoid *Encarsia formosa* Gahan (see Fig. 14.3, Chapter 14, for the development of distribution patterns of whitefly and parasitoid over the host plant).

The parasitoid *Diglyphus isaea* (Walker) can be stored at a low temperature for at least two months, during which time mortality does not increase and fecundity remains the same (Burgio and Nicoli, 1994). Hagvar and Hofsvang (1991) reported that some species of Aphidiidae (e.g. *Aphidius matricariae* Haliday) can be stored at low temperatures for several weeks.

The possibility of storing beneficials in the diapausing stage has been studied, but most of this work has not yet led to practical application, because unacceptably high mortality occurred during the artificially induced diapause. There are, however, some positive exceptions. Diapausing adults of the predator *Chrysoperla carnea* (Stephens) can be stored at a low temperature for about 30 weeks while maintaining an acceptable level of survival and reproduction activity (Tauber *et al.*, 1993). Also the predator *Orius insidiosus* (Say) maintains good longevity and reproduction rate after storage in diapause for up to 8 weeks (Ruberson *et al.*, 1998). The predator *Aphidoletes aphidimyza* (Rondani) can survive periods of three to eight months when stored at 10°C (Tiitanen, 1988). Long-term storage of the diapausing stage of the parasitoid *Trichogramma* has been successful for periods up to a year, and is now commercially exploited (J. Frandon, Biotop, France, pers. com.).

Long-term storage capability is very desirable for production companies, because: (i) continuous production of the same quantity of beneficial insects is often economically more attractive than seasonal production of very large numbers; (ii) storage facilities enable them to build up reserve supplies of entomophages to compensate for periods of low production or periods of unexpected high demands; and (iii) storage makes rearing possible at the best period of the year, e.g. at a period that host plants can be grown under optimal conditions.

20.5. Collection and Shipment of Natural Enemies

After production, the beneficials should be delivered to the growers as soon as possible. If delivery is looked after by the producer and occurs within 48 hours after harvesting the organisms, no special shipment procedures are normally needed for parasitoids and non-cannibalistic predators other than protection against excessive heat, cold or rough handling. When transport takes several days, climatized containers should be used and it may be necessary to add food (e.g. honey in the case of parasitoids and pollen/prey for predators). To overcome high mortality rates in predators due to long transportation time, young stages can be packaged with food, so that further development takes place during transportation. Packaging of predators demands special attention when cannibalism is a common phenomenon. Many of the commercially available predators are generalists and exhibit cannibalism when kept at high densities, even if food is available in the containers for shipment. To reduce the risk of cannibalism, it is common to provide hiding places for the natural enemy by using paper, buckwheat, vermiculite or wheat bran in the container (see Table 20.5). In the early days of mass production, the biological control agents were often collected and shipped on the host plant on which they were reared. With the internationalization of biocontrol, shipment on or in inert media became a necessity. Ingenious collection and shipping procedures have been developed. Poor shipping conditions frequently led to natural enemies arriving either dead or in poor condition. Difficulties in shipping can be considerable in countries where greenhouses are not concentrated together and where distances are large. Most transport is still by truck, although an increasing quantity is sent by aircraft. With intercontinental transport, problems are caused less by containerization than by the sometimes excessively long handling time at customs, which leads to high mortality or decrease in fitness. Logistics of shipments remains one of the main problems for the commercialization of biological control. Examples of the different techniques for collecting, counting, packaging and shipping of the natural enemies are given in Table 20.5.

TABLE 20.5. Examples of how natural enemies are collected, shipped and released

Natural enemy	Stage at which collected	Special handling	Counting	Packaging	Shipping	Mode of introduction	Quality control
Amblyseius spp.	All stages		Estimate	In wheat bran	Plastic container	Sprinkling on plants	Yes
Aphidus spp.	Mummy	Removal from the leaf	Weight		Plastic container	Small numbers in sheltered locations	Yes
Aphidoletes aphidimyza	Pupa		Weight	Vermiculate	Plastic container	Small numbers in sheltered locations	Yes
Chrysoperla carnea	Egg	Removal from the leaf	Volumetr.	Buckwheat	Plastic container	Sprinkling on plants	Yes
Cryptolaemus montrouzieri	Adult		Counting	Paper strips	Plastic container	Small numbers on plants	Yes
Dacnusa sibirica	Adult		Counting		Plastic container	Tapped from container	Yes
Diglyphus isaea	Adult		Counting		Plastic container	Tapped from container	Yes
Encarsia formosa	Pupa on leaves	Removal from the leaf	Volumetr.	Glued on cardboard	Paper box	Cards hung in plants	Yes
Eretmocerus mundus	Pupa on leaves	Removal from the leaf	Counting	Glued on cardboard	Paper box	Cards hung in plants	Yes
Harmonia axyridis	Larvae		Counting	Pop corn	Plastic container	Tapped from container on plants	Yes
Leptomastix dactylopii	Adult		Counting		Plastic container	Tapped from container	Yes
Lysiphlebus testaceipes	Mummy	Removal from the leaf	Counting		Plastic container	Small numbers in sheltered locations	Yes
Macrolophus caliginosus	Nymphs and adult		Counting	Buckwheat + vermic.	Plastic container	Small numbers on plants	Yes
Orius spp.	Nymphs and adult		Volumetr.	Buckwheat + vermic.	Plastic container	Small numbers on plants	Yes
Phytoseiulus persimilis	All moving stages		Volumetr.	In wheat bran	Plastic container	Sprinkling on plants	Yes
Trichogramma spp.	Pupa		Volumetr.	Glued on card	Paper box	Cards hung in plants or by sprinkling on plants	Yes

20.6. Release of Natural Enemies

20.6.1. DEVELOPMENTAL STAGE AT WHICH ORGANISM IS RELEASED

Entomophagous insects can be brought into commercial greenhouses in different stages of their development (Table 20.5): (i) eggs (e.g. *Chrysoperla*); (ii) larvae or nymphs (e.g. *Chrysoperla, Phytoseiulus, Amblyseius, Orius*); (iii) pupae or mummies (e.g. *Aphidius, Trichogramma, Encarsia*); (iv) adults (e.g. *Dacnusa, Diglyphus, Orius, Phytoseiulus*); and (v) all stages together (e.g. *Phytoseiulus, Amblyseius*).

The stage in which the beneficials are introduced depends mainly on the ease of transport and manipulation in the field, but it is – of course – also important to release the natural enemy at a stage which is most active at killing the pest. Usually, the stage which is least vulnerable to mechanical handling is chosen, and therefore a none-mobile stage, often the egg or pupa, is most suited for transport and release. In situations where it is difficult, but essential, to distinguish the natural enemy from the pest, the only solution is to introduce adults. Adult releases for parasitoids are advised only when younger natural enemy stages cannot be distinguished or separated from the pest insect: handling and releasing of delicate adult parasitoids is very difficult, and often a large reduction of fertility is observed compared to the fertility of parasitoids when released as immatures. When the natural enemy is released in one of the developmental stages which do not predate or parasitize the host, the timing should be such that the active stage emerges at the right moment of pest population development. For some natural enemies the stage of release depends on pest development: when pest density is low, release of first instar *C. carnea* suffices, when the infestation with the pest organisms is already relatively high, it is better to release second instar larvae, which have a much higher predation capacity.

20.6.2. METHODS OF INTRODUCTION

Beneficials are introduced into the field in many ways (Table 20.5). Eggs and pupae are either distributed over the field on their normal substrate (leaves of the host plant, e.g. *Chrysoperla* and *Encarsia*) or glued on paper/cardboard cards (e.g. *Encarsia, Trichogramma*). These stages of the natural enemies can also be collected, and put into containers which are then brought into the field (e.g. *Trichogramma*).

The mobile stages of natural enemies, larvae or nymphs and adults, can be put into the field in containers from which they emerge (e.g. many adult parasitoids and predators) or the grower can distribute natural enemies in these stages over the crop, for example, by "sprinkling" them onto the plant. In this case, the use of dispersal material (e.g. buckwheat, vermiculite) is often necessary in order to obtain a homogeneous distribution of small natural enemies. When natural substrates (e.g. buckwheat or wheat bran) are used as dispersal materials, they must be free from pesticides.

Instead of introducing the predator or parasitoid by itself, one can also introduce a whole "production unit": e.g. "banker-plants" containing the host insect and its natural enemy can be brought into a crop. When the introduced host population is almost exterminated, the natural enemies invade the surrounding crop (van Steenis, 1995).

20.6.3. THE MOMENT OF INTRODUCTION

In many cases, the natural enemies are released when the pest organism has been observed, although it is not unusual to apply "blind releases" when sampling of the pest is difficult (e.g. whiteflies) or when pest populations develop very quickly, like those of aphids and thrips. When pest generations are not yet overlapping early in the growing season, proper timing of the release(s) is essential, so that the beneficials are available when the preferred host stages are present.

Determining the dosage, the distribution and the frequency of the releases are very difficult problems which are encountered in both inundative and seasonal inoculative release programmes. Release ratios are not critical in inundative release programmes as long as it is possible to release a (super)abundance of natural enemies. This, however, may be limited by the cost of mass production. In seasonal inoculative programmes, release ratios are more critical: if too few beneficials are released, effective control will be obtained after the pest has caused economic damage. If too many are released, there is a risk of exterminating the pest and thus, eventually, also of the natural enemy. This is a practical problem in small tunnels and greenhouses. In the latter situation, resurgence of the pest is likely and a serious threat. In these seasonal inoculative release programmes, the release ratios are usually determined by trial and error, but the first simulation programmes are appearing for a more scientific estimate of release rates (number of releases, spacing between release points and timing of releases) (see e.g. van Roermund, 1995, for seasonal inoculative releases and Suverkropp, 1997, for inundative releases).

20.7. Quality Control

20.7.1. WHY QUALITY CONTROL?

The past 30 years of commercial biological control in greenhouses are characterized by appearance and disappearance of natural enemy producers. Only a few producers active in the 70s are still in the market. In Europe there are now more than 20 small producers, and three large facilities (i.e. those having more than 50 persons employed). The number of beneficials produced by these three large companies is often more than 5–10 million per agent per week (van Lenteren and Woets, 1988), and they provide the full spectrum of natural enemies needed for an entire IPM programme in a specific commodity. As biological control is a rapidly developing market influenced by small competing companies, product quality and prices are continuously under pressure. In the short-term this may be profitable for greenhouse growers, but in the long run it could lead to pest control failures. Some 20 years ago, natural enemies were properly evaluated before commercial use, nowadays some species of natural enemies are sold without testing under practical cropping conditions to see if they are effective against the target pest. The rise and fall of so many producers resulted in a negative attitude to the concept of practical biological control due to pest control failures, which were partly because beneficials of poor quality were provided and because insufficient guidance was provided.

The quantity of each natural enemy species produced and the number of biocontrol agents which are commercially available have increased dramatically over the past 25 years (see Fig. 20.1). Nowadays, there are almost 90 natural enemy species on the market for greenhouse biocontrol, and 30 of these are produced in commercial insectaries in very large quantities (Table 20.4). The three large natural enemy producers and a few of the smaller ones can now be considered professional, with research facilities, some application of quality control, an international distribution network, public relation activities and an advisory service. They are well respected for their work and their market will certainly increase with the increasing demand for unsprayed food, a cleaner environment and because of the growing pesticide resistance problem, which is particularly a problem for crop protection in greenhouses.

The problems addressed in the previous sections on mass production and releases mean that good quality-control programmes are essential. Quality control should ensure that natural-enemy numbers but also natural-enemy performance in the greenhouse. Simple, representative and reliable quality control programmes for natural enemies are now emerging as a result of intensive co-operation between researchers and the biocontrol practitioners, and it is expected that these developments will result in a quick improvement of the biocontrol industry. In an EC/IOBC funded European project, quality control guidelines were designed for the 20 species of natural enemies that are most widely used in greenhouses. In addition, fact sheets about natural enemies and pests are being composed for training. At present, the guidelines comprise mainly characteristics which are relatively easy to determine in the laboratory (e.g. emergence, sex ratio, life span, fecundity, adult size, predation/parasitization rate). Future work will be focused at development of: (i) flight tests; and (ii) a test relating laboratory and semi-field tests to field efficiency.

20.7.2. WHAT IS QUALITY CONTROL?

Quality control is applied to mass-reared organisms to maintain the performance of the population; in this case the performance of a natural enemy in its intended role after release into the greenhouse. The aim of quality control is then to check whether the overall quality is maintained, but that is too general a statement to be manageable. Characteristics which are quantifiable and relevant for greenhouse performance have to be identified. This is a straightforward statement, but very difficult to make concrete. The aim of releases of mass-produced natural enemies is to control a pest. In this context, the aim of quality control should be to determine whether a natural enemy is still in a condition to properly control the pest. So we deal with something like *acceptable quality*, and not necessarily with maximal or optimal quality. An important consideration is that quality control is not applied for the sake of the scientist, but as a necessity. Leppla and Fisher (1989) formulated this dilemma as: "Information is expensive, so it is important to separate *need* to know from *nice* to know". Characteristics to be measured should be few in number, but directly linked to greenhouse performance, if companies producing natural enemies are going to apply quality control on a regular basis.

20.7.3. BASIC CONSIDERATIONS FOR QUALITY CONTROL

The problem of quality control of beneficial insects can be approached from many sides. Theoretically, the best approach would be to list what changes can be expected when a mass rearing is started, measure these, and if the changes are undesirable, improve the rearing method. A practical disadvantage of this approach is that too many measurements have to be performed, and it assumes that potential problems are foreseeable and that corrections can be made in time. Bartlett (1984a) states that remedial measures have been proposed for assumed genetic deterioration, but that causes for deterioration are not so easily identified, demand detailed genetic studies and that it is difficult to define and measure detrimental genetic traits. He continuous with: "I believe an unappreciated element of this problem is that the genetic changes taking place when an insect colony is started are natural ones that occur whenever any biological organism goes from one environment to another (in the authors' case from the greenhouse or field to the mass production situation). These processes have been very well studied as evolutionary events and involve such concepts as colonization, selection, genetic drift, effective population numbers, migration, genetic revolutions, and domestication theory". In two other articles (Bartlett, 1984b, 1985), he discusses what happens to genetic variability in the process of domestication, what factors might change variability and which ones might be expected to have little or no effect. In laboratory domestications, those insects are selected that have suitable genotypes to survive in this new environment: a process called winnowing by Spurway (1955) or, less appropriately but widely used, forcing insects through a bottleneck (e.g. Boller, 1979). In Table 20.6, the changes that a field population may undergo when introduced into the laboratory are listed.

TABLE 20.6. Factors influencing changes in field populations when introduced into the laboratory

1. Laboratory populations are kept at constant environments with stable abiotic factors (light, temperature, wind, humidity) and constant biotic factors (food, no predation or parasitism), there is no selection to overcome unexpected stresses. The result is a change of the criteria that determine fitness, and a modification of the whole genetic system
2. There is no interspecific competition in laboratory populations with as result a possible change in genetic variability
3. Laboratory conditions are made suitable for the average, sometimes for the poorest, genotype. No choice of environment is possible as all individuals are confined to the same environment. The result is a possible decrease in genetic variability
4. Density-dependent behaviours (e.g. searching efficiency) may be affected in laboratory situations
5. Mate-selection processes may be changed because unmated or previously mated females will have restricted means of escape
6. Dispersal characteristics, specifically adult flight behaviour and larval dispersal, may be severely restricted by laboratory conditions

Variability in performance traits is usually abundantly present in natural populations (Prakash, 1973), and can remain large even in inbred populations (Yamazaki, 1972). But the differences between field and laboratory conditions will result in differences in variability. When part of the "open population", where gene migration can occur and environmental diversity is large, is brought into the laboratory, becomes a "closed population" and all the genetic changes will be made from the limited genetic variation in the original founders (Bartlett, 1984b, 1985). The size of the founder population will directly affect how much variation will be taken from the native gene pool. Although there is no agreement on the size of founder populations for starting a mass production colony, a minimum number of 1000 individuals is mentioned in the literature (Bartlett, 1985). Founder populations for a number of natural enemies were much smaller and sometimes less than 20 individuals (see above and van Lenterten and Woets 1988 for examples). Fitness characteristics for the greenhouse will be different from those for the laboratory [e.g. difference in importance of ability to diapause, or the ability to locate hosts/prey or mates (role of infochemicals)], so laboratory selection forces may produce a genetic revolution (Mayr, 1970) and new balanced gene systems will be selected (Lopez-Fanjul and Hill, 1973).

One of the cures often suggested to overcome or correct for genetic revolutions is the regular introduction of wild individuals from the field or greenhouse, and several mass producers actually do this. But if the rearing conditions remain the same in the laboratory, the introduced wild individuals will be subjected to the same process of genetic revolution. If genetic differentiation has occurred between laboratory and field population which has led to genetic isolation (Oliver, 1972) – and positive correlations have been found between the incompatibility of races and the difference between the environments where the races occur (e.g. Jansson, 1978; Jaenson, 1978), and for the length of time two populations have been isolated – and if incompatibility is complete, then introduction of native individuals seems to be useless. If the mass producer wants to introduce wild genes, it should be done regularly, based on good methodology and from the start of a laboratory colony. It should not be delayed until problems occur. However, introducing native insects into an insect colony has its own risks from the introduction of parasitoids, predators or pathogens (Bartlett, 1984b).

Another problem with insect colonies can be inbreeding: mating of relatives and production of progeny more homozygous than when random mating occurs in large populations. Homozygous individuals often contain harmful traits. The inbreeding coefficient is directly related to the size of the founder population and, because of artificial selection in the laboratory which results in an even smaller population size, the rate of inbreeding will increase and the result is often a definite and rapid effect on the genetic composition of the laboratory population (Bartlett, 1984b). Inbreeding can be prevented by several procedures to maintain genetic variability. Joslyn (1984) proposes the following methods: (i) precolonization methods; and (ii) postcolonization methods.

Precolonization methods involve selection and pooling of founder insects from throughout the range of the species to provide a wide representation of the gene pool, resulting in greater fitness of the laboratory material.

Postcolonization methods may involve: (i) variable laboratory environments (variation over time and space); although the concept is simple, putting it into practice is difficult; consider, for example, the investment for rearing facilities with varying temperatures, humidities, and light regimes, the creation of the possibility to choose from diets (hosts), the provision of space for dispersal, etc.; and (ii) gene infusion, the regular rejuvenation of the gene pool with wild insects.

A fundamental question to this inbreeding problem is: what is the effective population size to keep genetic variation sufficiently large? Joslyn (1984) says that to maintain sufficient heterogeneity, the size of the laboratory colony should not decline below the number of founder insects. The larger the colony the better. Very few data are available about effective population size. Joslyn mentions a minimum number of 500 individuals.

The major natural enemy producers always keep much larger colonies throughout the year.

Based on the way of rearing, the degree of artificial selection can be estimated. A natural enemy reared in a glasshouse on the same host plant and same pest organism as in the commercial greenhouses, will undergo less artificial selection than a beneficial reared under climate room conditions on an alternative food or artificial diets. Furthermore, results of artificial selection will show earlier in natural enemies with many generations per year than in beneficials with a few generations only.

20.7.4. HOW QUALITY CONTROL?

Natural enemies for greenhouse biological control are often mass-produced under greenhouse situations which are similar to cropping conditions. The main exception is that pest densities are much higher, so most of the points listed in Table 20.6 are applicable and, therefore, quality control is required. The practical development of quality control for greenhouse natural enemies in Europe was approached rather pragmatically and it is expected that in the coming decade quality control will mature. Quality control guidelines have been developed for 20 of the 30 natural enemy species listed in Table 20.4. Full descriptions of the tests can be found in van Lenteren (1996). The elements of the quality control test are given in Table 20.7 and an overview of the recently developed quality control guidelines is given in Table 20.8.

TABLE 20.7. General quality control criteria

Elements of all quality control tests will be:	
Quantity	Predators: number of live predators in container
	Parasitoids: if delivered as adults, number of live parasitoids
	if delivered as immatures, number of emerging adults over a defined period
Sex ratio	Minimum percentage females; male biased ratio may indicate poor rearing conditions
Fecundity	Number of offspring produced during a defined period; fecundity of parasitoids is also an indication of the host kill rate
Longevity	Minimum longevity in days
Predation	Number of prey eaten during a defined period
Adult size	Hind tibia length, sometimes pupal size (size is often a good indication for longevity, fecundity and predation capacity if natural enemy is not manipulated during harvest, packaging, shipment and release)
Flight	Short-range flight (natural enemy can still fly)
	Long-range flight + predation/parasitization capacity (can fly and perform)
Field performance	Capacity to locate and consume or parasitize prey/host in crop under field conditions
Quality control is done under standardized test conditions (temperature, relative humidity and light regime are specified for each test)	
Fecundity, longevity and predation capacity tests can often be combined	
Expiry date for shipment is given on packaging material	
All numbers/ratios/sizes, etc. are mentioned on the container or packaging material	

TABLE 20.8. Summary of quality control aspects for the most important natural enemies used in greenhouses[1]. Environmental conditions: RH 75±10%; photoperiod 16L+8D. The quantity of natural enemies shipped in the container should be specified

Natural enemy	Temperature (°C)	Sex-ratio (% female) n = sample size	Fecundity (eggs/fem.) n = number of fem. tested	Mortality (%) n = sample size	Emergence
Neoseiulus (= Amblyseius) cucumeris	22±2	≥50; n = 100 ♦	≥7 in 7 d; n = 30 ♦		
Aphelinus abdominalis	22±2	≥45; n = 300	>60 in 8 d; n = 30 ♦	≤10; n = 300	80% (in 15 d)
Aphidius spp.	25±2	≥45; n = 200	≥60 or 35 at 1st d; n = 30 ♦		≥45% Δ
Aphidoletes aphidimyza	22±2 (emergence at 25)	≥45; n = 500	>40 in 4 d; n = 30 Δ		70% (in 7 d)
Chrysoperla carnea	25±2				Hatching rate: >65% in 5 d
Dacnusa sibirica	22±2	≥45; n = 500 Δ	≥60 in 3 d; n = 10 ♦	≤5; n = 500	
Diglyphus isaea	25±2	≥45; n = 500 Δ	≥70% fertile fem.; n = 30 ♦	≤8; n = 500	
Encarsia formosa	22±2	≥98; n = 500 Δ	≥7/day (2–4 d); n = 30 ♦		≥number specified
Eretmocerus mundus	25±2	≥45; n = 500 •	≥10; n = 20 ♦		
Leptomastix dactylopii	25±2	≥45 (c.); n = 500	≥40 in 14 d; n = 30 ♦	≤10; n = 500	
Macrolophus caliginosus	22±2	≥45; n = 100 •	≥7 in 3 d; n = 30 ♦	≤5; n = 500	
Orius spp.	25±2	≥45; n = 100 •	≥30 in 14 d; n = 30 ♦		
Phytoseiulus persimilis	22–25	>70; n = 100 ♦	>10 in 5 d; n = 30 •		
Trichogramma brassicae	23±2	≥50; n = 500	≥40 in 7 d; n = 30 Δ	<20 after 7 d; n = 30 Δ	>80%

[1] ♦ annual test; • seasonal test; Δ monthly test; where it is not specified, the tests are done weekly or batchwise
Note: Flight tests are available for E. formosa, Aphidius spp. and A. aphidimyza

The guidelines developed up to now refer to *product control* procedures, not to production or process control. They were designed to be as uniform as possible, so they can be used in a standardized manner by many producers, distributors, pest management advisory personnel and even by greenhouse growers. The tests should preferably be carried out by the producer *after all handling procedures and just before shipment*. However, additional testing after shipment will often be important as well, although it is expected that the grower will only perform a few of the quality tests, e.g. percent emergence or number of live adults in the package. Some tests should be carried out frequently by the producer, i.e. on a daily, weekly or batch-wise basis. Others will be done less frequently, i.e. annually or seasonally, or when rearing procedures are changed. In the near future, flight tests and field performance tests will be added to these guidelines. Such tests are needed to show the relevance of the laboratory measurements. Laboratory tests are only adequate when a good correlation has been established between the laboratory measurements, flight tests and field performance. In addition to the quality control tests, fact sheets on natural enemies and pests are needed to inform new quality control personnel and plant protection services on biological details.

20.8. Conclusions

Mass production of natural enemies has seen a very fast development during the past three decades: the numbers produced have greatly increased, the spectrum of species available has widened dramatically, and mass production methods clearly have evolved. Developments in the area of mass production, quality control, storage, shipment and release of natural enemies have decreased production costs and led to better product quality, but much more can be done. Innovations in long-term storage (e.g. through diapause), shipment and release methods may lead to a further increase in natural enemy quality with a concurrent reduction in costs of biological control, thereby making it easier and more economical to apply.

Companies starting the production of natural enemies usually have very little knowledge about the obstacles and complications related to mass rearing. They are even more ignorant about the development and application of quality control. A special point of concern is the lack of knowledge about the sources of variability of natural enemy behaviour and methods to prevent genetic deterioration of natural enemies. Mass-rearing of natural enemies often takes place in small companies with little know-how and understanding of conditions influencing performance, which may result in natural enemies of bad quality and failures of biological control programmes. But even when the natural enemies leave the insectary in top condition, it does not mean that they are in top shape when released in the greenhouse. Shipment and handling by the producers, distributors and growers may result in deterioration of the biocontrol agents. This makes robust quality control programmes a necessity. The larger companies apply quality control, but methods differ widely.

In the 90s, commercial producers of biological control agents and scientists started to work on development and standardization of quality control methods. Quality control procedures for natural enemies are presently being developed for commercially applied natural enemies in greenhouses. Quality control criteria relate to product control and are based on laboratory measurements which are often easy to carry out. The criteria will soon be complemented with flight tests and field performance tests.

If the biological control industry is to survive and florish, the production of highly reliable natural enemies is essential.

Acknowledgement

Dr. N. Martin (New Zealand Institute for Crop and Food Research, Auckland, New Zealand) is thanked for editing the text carefully and for suggesting a number of improvements.

References

Bartlett, A.C. (1984a) Establishment and maintenance of insect colonies through genetic control, in E.G. King and N.C. Leppla (eds.), *Advances and Challenges in Insect Rearing*, USDA/ARS, New Orleans, p. 1.

Bartlett, A.C. (1984b) Genetic changes during insect-domestication, in E.G. King and N.C. Leppla (eds.), *Advances and Challenges in Insect Rearing*, USDA/ARS, New Orleans, pp. 2–8.

Bartlett, A.C. (1985) Guidelines for genetic diversity in laboratory colony establishment and maintenance, in P. Singh and R.F. Moore (eds.), *Handbook of Insect Rearing*, Vol. 1, Elsevier, Amsterdam, pp. 7–17.

Boller, E.F. (1979) Behavioral aspects of quality in insectary production, in M.A. Hoy and J.J. McKelvey (eds.), *Genetics in Relation to Insect Management*, Rockefeller Foundation, New York, pp. 153–160.

Burgio, C. and Nicoli, G. (1994) Cold storage of *Diglyphus isaea*, in G. Nicoli, M. Benuzzi and N.C. Leppla (eds.), *Proceedings 7th Global IOBC Workshop "Quality Control of Mass Reared Arthropods"*, Rimini, Italy, 13–16 September 1993, Studiostampa, Cesena, pp. 171–178.

Cranshaw, W., Sclar, D.C. and Cooper, D. (1996) A review of 1994 pricing and marketing by suppliers of organisms for biological control of arthropods in the United States, *Biological Control* 6, 291–296.

DeBach, P. (ed.) (1964) *Biological Control of Insect Pests and Weeds*, Chapman & Hall, London.

Finney, G.L. and Fisher, T.W. (1964) Culture of entomophagous insects and their hosts, in P. DeBach (ed.), *Biological Control of Insect Pests and Weeds*, Chapman & Hall, London, pp. 329–355.

Gennadiev, V.G. and Khlistovskii, E.D. (1980) Long-term cold storage of host eggs and reproduction in them of egg-parasites of insect pests, *Zhurnal Obshchei Biologii* 41, 314–319.

Grenier, F., Greany, P. and Cohen, A.C. (1994) Potential for mass release of insect parasitoids and predators through development of artificial culture techniques, in D. Rosen, F.D. Bennett and J.L. Capinera (eds.), *Pest Management in the Subtropics – A Florida Perspective*, Intercept, Andover, Hants, pp. 181–205.

Hagvar, E.B. and Hofsvang, T. (1991) Aphid parasitoids (Hymenoptera: Aphidiidae): Biology, host selection and use in biological control, *Biocontrol News and Information* 12, 13–41.

Hussey, N.W. and Bravenboer, L. (1971) Control of pests in glasshouse culture by the introduction of natural enemies, in C.B. Huffaker (ed.), *Biological Control*, Plenum, New York, pp. 195–216.

Jaenson, T.G.T. (1978) Mating behavior of *Glossina pallides* Austen (Diptera, Glossinidae): Genetic differences in copulation time between allopatic populations, *Entomologia Experimentalis et Applicata* 24, 100–108.

Jansson, A. (1978) Viability of progeny in experimental crosses between geographically isolated populations of *Arctocorisa carinata* (C. Sahlberg) (Heteroptera, Corixidae), *Annales Zoologici Fennici* 15, 77–83.

Joslyn, D.J. (1984) Maintenance of genetic variability in reared insects, in E.G. King and N.C. Leppla (eds.), *Advances and Challenges in Insect Rearing*, USDA/ARS, New Orleans, pp. 20–29. .

King, E.G. and Morrison, R.K. (1984) Some systems for production of eight entomophagous arthropods, in E.G. King and N.C. Leppla (eds.), *Advances and Challenges in Insect Rearing*, USDA/ARS, New Orleans, pp. 206–222.

Leppla, N.C. and Fisher, W.R. (1989) Total quality control in insect mass production for insect pest management, *J. of Applied Entomology* 108, 452–461.

Lopez-Fanjul, C. and Hill, W.G. (1973) Genetic differences between populations of *Drosophila melanogaster* for quantitative trait. II. Wild and laboratory populations, *Genetical Research* 22, 60–78.

Mayr, E. (1970) *Populations, Species, and Evolution,* Harvard University Press, Cambridge, Mass.

Morrison, R.K. and King, E.G. (1977) Mass production of natural enemies, in R.L. Ridgway and S.B. Vinson (eds.), *Biological Control by Aumentation of Natural Enemies*, Plenum, New York, pp. 183–217.

Nicoli, G., Benuzzi, M. and Leppla, N.C. (eds.) (1994) *Proceedings 7th Global IOBC Workshop "Quality Control of Mass Reared Arthropods"*, Rimini, Italy, 13–16 September 1993, Studiostampa, Cesena.

Oliver, C.G. (1972) Genetic and phenotypic differentiation and geographic distance in four species of Lepidoptera, *Evolution* 26, 221–241.

Posthuma-Doodeman, C.J.A.M., van Lenteren, J.C., Sebestyen, I. and Ilovai, Z. (1996) Short-range flight test for quality control of *Encarsia formosa, Proceedings of Experimental and Applied Entomology* 7, 153–158.

Prakash, S. (1973) Patterns of gene variation in central and marginal populations of *Drosophila robusta, Genetics* 75, 347–369.

Ruberson, J.R., Kring, T.J. and Elkassabany, N. (1998) Overwintering and the diapause syndrome of predatory Heteroptera, in *Predatory Heteroptera in Agroecosystems: Their Biology and Use in Biological Control*, Thomas Say Publications in Entomology, ESA, Lamham (in press).

Singh, P. (1984) Insect diets. Historical developments, recent advances, and future prospects, in E.G. King and N.C. Leppla (eds.), *Advances and Challenges in Insect Rearing*, USDA/ARS, New Orleans, pp. 32–44.

Spurway, H. (1955) The causes of domestication: An attempt to integrate some ideas of Konrad Lorenz with evolution theory, *J. of Genetics* **53**, 325–362.

Suverkropp, B.P. (1997) *Host-Finding Behaviour of Trichogramma brassicae in Maize*, PhD Thesis, Agricultural University Wageningen.

Tauber, M.J., Tauber, C.A. and Gardescu, S. (1993) Prolonged storage of *Chrysoperla carnea* (Neuroptera: Chrysopidae), *Environmental Entomology* **22**, 843–848.

Tiitanen, K. (1988) Utilization of diapause in mass production of *Aphidoletes aphidimyza* (Rond.) (Dipt., Cecidomyiidae), *Annales Agriculturae Fenniae* **27**, 339–343.

van Lenteren, J.C. (1986a) Evaluation, mass production, quality control and release of entomophagous insects, in J.M. Franz (ed.), *Biological Plant and Health Protection*, Fischer, Stuttgart, pp. 31–56.

van Lenteren, J.C. (1986b) Parasitoids in the greenhouse: Successes with seasonal inoculative release systems, in J.K. Waage and D.J. Greathead (eds.), *Insect Parasitoids*, Academic Press, London, pp. 341–374.

van Lenteren, J.C. (1995) Integrated pest management in protected crops, in D. Dent (ed.), *Integrated Pest Management*, Chapman & Hall, London, pp. 311–343.

van Lenteren, J.C. (1996) Quality control tests for natural enemies used in greenhouse biological control, *IOBC/WPRS Bulletin* **19**(1), 83–86.

van Lenteren, J.C., Roskam, M. and Timmer, R. (1997) Commercial mass production of organisms for biological control in Europe, *Biological Control* **10**, 143–149.

van Lenteren, J.C. and Woets, J. (1988) Biological and integrated control in greenhouses, *Annual Review of Entomology* **33**, 239–269.

van Roermund, H.J.W. (1995) *Understanding Biological Control of Greenhouse Whitefly with the Parasitoid Encarsia formosa: From Individual Behaviour to Population Dynamics*, PhD Thesis, Agricultural University Wageningen.

van Steenis, M. (1995) *Evaluation and Application of Parasitoids for Biological Control of Aphis gossypii in Glasshouse Cucumber Crops*, PhD Thesis, Agricultural University, Wageningen.

Vet, L.E.M. and Dicke, M. (1992) Ecology of infochemical use by natural enemies in a tritrophic context, *Annual Review of Entomology* **37**, 141–172.

Yamazaki, T. (1972) Detection of single gene effect by inbreeding, *Nature* **240**, 53–54.

MICROBIAL CONTROL OF PESTS IN GREENHOUSES
Jerzy. J. Lipa and Peter H. Smits

21.1. Introduction

There are several reasons why microbial control of arthropods and nematodes in greenhouses attracts the attention of researchers, extension workers and greenhouse owners (Hussey and Scopes, 1985; van Lenteren and Woets, 1988): (i) glass houses or plastic tunnels are closed premises inside which it is much easier to use and manipulate the occurrence of micro-organisms than in open fields; (ii) widespread use of bumble bees (*Bombus* spp.) as flower pollinators in greenhouse crops limits the use of many chemical pesticides, and the use of biopesticides may be an excellent alternative; and (iii) biopesticides normally do not require a preharvest waiting period and there are no residue restrictions so they can be used when needed without interference with the procedures of healthy or organic food production.

Presently, commercially available biopesticides can be effectively used against lepidopterans, whiteflies, aphids and nematodes. However, for important pests such as leafminers or spider mites no effective and reliable biopesticides are available.

21.2. Summary of Characteristics of Insect Pathogens

21.2.1. BACTERIA

Although hundreds of genera and thousands of species of bacteria are known from insects, only sporeforming bacteria belonging to the Bacillaceae family have found practical use in insect control. Essentially three species of sporeforming *Bacillus* genus are of prime interest: *Bacillus thuringiensis* Berliner, *Bacillus sphaericus* Meyer & Neide, and *Bacillus popilliae* Dutky.

Bacillus thuringiensis (Bt), which occurs in several subspecies or varieties and in over forty serological forms, is a rod-shaped 2.5 by 1 μm bacterium which, during growth and sporulation, produces oval spores and several toxins. The most important group of toxins are the delta-endotoxins that are present in the form of a crystalline inclusion body, which is easily visible with a light microscope. Numerous strains and subspecies of Bt are known and they are characterized by flagellar antigens and gene types that produce different delta-endotoxins and a beta-exotoxin. Since the beta-exotoxin also has broad toxic activity against vertebrates, such Bt strains are avoided in commercial products, with some exceptions in Finland and in Russia where Bitoxybacillin (contains endo- and exotoxin) is widely used with no reported negative side-effects.

Three well known Bt subspecies are widely used in commercial products: *Bacillus*

R. Albajes et al. (eds.), Integrated Pest and Disease Management in Greenhouse Crops, 295-309.
© 1999 *Kluwer Academic Publishers. Printed in the Netherlands.*

thuringiensis Berliner ssp. *kurstaki* Dulmage in products that control larvae of Lepidoptera, *Bacillus thuringiensis* Berliner ssp. *israelensis* (Goldberg & Margalit) de Barjac in products against larvae of Diptera, and *Bacillus thuringiensis* Berliner ssp. *tenebrionis* Krieg & Huger in products against larvae and adults of Coleoptera. Recently, some Bt strains producing crystal toxins with activity against Nematoda were found, and this opens up prospects for new products (Shevtsov *et al.*, 1996, Zuckerman *et al.*, 1993).

The quick action and high effectiveness of Bt products is due to the activity of the crystalline endotoxins. When ingested by a susceptible insect, the toxin chrystal dissolves in the gut under alkaline conditions (pH>9) and the activated delta-endotoxin damages the midgut epithelium causing gut paralysis and general shock. The affected insects stop feeding in minutes or hours so Bt biopesticides provide quick crop protection similar to the effect of chemical pesticides. Mass production is done in large scale fermenters on a simple medium and is relatively easy and cheap. This contributes much to the commercial success of Bt as a biopesticide.

21.2.2. VIRUSES

There are more than 1600 viruses known from over 1100 insect and mite species which are assigned to several virus families. Most of the viruses studied and used in biological control come from the family Baculoviridae. Not only are they very effective control agents, but viruses from this family are reported exclusively from arthropods and are therefore considered absolutely safe for vertebrates and mammals.

The Baculoviridae family includes large, rod-shaped DNA-containing viruses. Three sub-groups are recognized based on the type and morphology of the virus inclusion bodies (VIB): (i) nuclear polyhedrosis viruses (NPVs); (ii) granulosis viruses (GVs); and (iii) the small group of non-occluded baculoviruses.

A characteristic of the NPVs is that several hundreds of the rod-shaped virus particles are occluded into a polyhedral inclusion body (PIB), also often called the occlusion body (OB). These range in size from 1 to15 μm and therefore can easily be detected with a light microscope. In the case of GVs, a single rodshaped virus particle is occluded into a small ellipsoidal inclusion body (granule/capsule) having the average size of about 0.5 μm. The protein matrix of inclusion bodies protects the virus particles and enables them to persist in the environment.

According to the most recent virus taxonomy, the NPVs belong to the genus *Nucleopolyhedrovirus* while GVs belong to the genus *Granulovirus*. Both genera are mainly found in Lepidoptera and Hymenoptera. Although most baculoviruses are host specific and infect only a single host, viruses such as *Autographa californica* (Speyer) NPV and *Mamestra brassicae* (L.) NPV infect different insect species and are most extensively used in microbial control.

Virus particles are liberated from OBs under alkaline conditions in the midgut of insects. Virus particles bind to the gut wall, infect cells and replication starts. It takes several days before most cells of the body are infected and produce free and occluded virus particles. The insect dies and often liquifies releasing billions of OBs into the environment.

Baculoviruses often cause epizootics, almost wiping out complete populations of caterpillars, e.g. *Spodoptera exigua* (Hübner) (Smits, 1987; Caballero *et al.*, 1992), and this makes them very useful biopesticides. Mass production is done on mass-reared insects. Production in cell culture is possible but not on a large commercial scale yet.

21.2.3. FUNGI

Fungi were the first micro-organisms to be recognized as the cause of an insect disease and used for microbial pest control (Rombach and Gillespie, 1988). At present we know about 800 species of fungi that are pathogenic to insects and mites, belonging to the Mastigomycotina, Zygomycotina, Ascomycotina, Basidiomycotina and Deuteromycotina. However, most of the entomopathogenic fungi that show use for microbial control are in the Zygomycotina (Zygomycetes: Entomophthorales), and Deuteromycotina (Hyphomycetes: Moniliales). Although several species belonging to various families and genera play an important role in reducing insect abundance, only some are being used in microbial control due to technical difficulties in their mass production as well as their great dependence on the correct combination of high humidity and temperatures.

The most important groups of entomopathogenic fungi are found in the Deuteromycotina, which is comprised of species known only from asexual forms. These include the genera with wide host ranges such as: *Beauveria* [*Beauveria bassiana* (Balsamo) Vuillemin], *Metarhizium* [*Metarhizium anisopliae* (Metschnikoff) Sorokin], *Paecilomyces* [*Paecilomyces fumosoroseus* (Wize) Brown & Smith], *Nomuraea* [*Nomuraea rileyi* (Farlow)], *Verticillium* [*Verticillium lecanii* (A. Zimmerm.) Viégas] and *Aschersonia* (*Aschersonia aleyrodis* Webber). The other important group is the Zygomycotina, consisting of the order Entomophthorales, with important genera such as: *Entomophthora* [*Entomophthora muscae* (Cohn) Fress], *Conidiobolus* [*Conidiobolus coronatus* (Constantin) Batko], *Erynia* [*Erynia neoaphidis* Remaudière & Hennebert (= *Entomophthora aphidis* Hoffmann)], *Zoophthora* [*Zoophthora radicans* (Brefeld)] and *Tarichium* (*Tarichium gammae* Weiser).

The infective stage is the spore (conidiospore) that attaches to the insect host cuticle, and germinates and penetrates into the hemocoel as a germ tube. In the hemocoel, fungal growth takes place as mycelium; hyphal bodies or protoplasts gradually fill the body cavity and destroy all the tissues. Immediately after, the host dies and, at the correct humidity, the mycelium spreads outside and covers the dead insect body with a thick layer of hyphae and spores, often giving the cadaver a characteristic colour, e.g. white (*B. bassiana*), green (*M. anisopliae*) or pink (*P. fumosoroseus*). Spores present outside on the cadaver's body are the source of infection for healthy insects, either by direct contact or through wind distribution.

Since, as a rule, fungi infect their hosts through the cuticle, it makes them extremely useful for control of mites and those insects such as whiteflies, aphids and thrips, which have sucking mouthparts and do not ingest bacteria spores or virus particles as do the many foliage eating Lepidoptera or Coleoptera. Most fungi can be easily grown on artificial solid media, which makes large scale commercial production of fungus biopesticides feasible. However, in the case of *Beauveria*, *Metarhizium* and *Paecilomyces,* two-phase growing is necessary to obtain resistant conidiospores.

21.2.4. PROTOZOA

Of the approximately 15,000 known species of Protozoa, about 1200 are associated with insects (Lipa, 1974). This number includes symbionts, commensals and parasites. Entomopathogenic protozoa occur in six phyla: Zoomastigina (flagellates), Rhizopoda (amoebas), Apicomplexa (eugregarines, neogregarines, coccidia), Microspora (microsporidia), Haplosporidia and Ciliophora (ciliates). Although species belonging to all phyla cause mortality of insects, research on the use of protozoa for biological control focuses only on Microspora and Neogregarinida. Both groups develop intracellularly, infect and destroy various tissues, including the fat body, and multiply by asexual and sexual reproduction, producing a large number of spores which are released in the excrement or from cadavers and which easily infect healthy insects. Vertical transmission from parent to offspring of the host occurs especially in microsporidian and neogregarine infections. For this reason several species from these groups are responsible for the collapse of natural populations or laboratory cultures of many insect species. Natural epizootics caused by *Nosema laphygmae* Weiser in *S. exigua* populations and by *Nosema heliothidis* (Lutz & Splendore) in *Helicoverpa* (= *Heliothis*) spp. populations have been reported in several countries.

Microsporidian infection typically begins when an infective spore is ingested from which an infective sporoplasm is released into the gut and penetrates into cells of the midgut epithelium. Some genera (i.e. *Pleistophora*) develop in the gut, while others (i.e. *Nosema*) cause a general infection including in the gonads and for this reason are also transmitted vertically.

The limiting factor in the wide use of microsporidia or neogregarines is that as obligatory parasites they cannot be grown on media but have to multiplied in living hosts. For this reason so far only one protozoan biopesticide (Nolock® against grasshoppers) has been developed and registered in the USA.

21.2.5. NEMATODES

From over 30 families of nematodes associated with insects, only nine have members with potential as biological control agents: Tetradonematidae, Mermithidae, Steinernematidae, Heterorhabditidae, Phaenopsitylenchidae, Itonchiidae, Allantonematidae, Parasitylenchidae and Sphaerulariidae (Georgis, 1992). However, research on the use of nematodes for biocontrol focuses only on two families, the Steinernematidae and Heterorhabditidae. These are associated with pathogenic symbiotic bacteria that enable them to rapidly kill a wide range of hosts. In addition, they can be mass produced in fermenters.

The infective stage of the nematode is a third instar juvenile that is free-living in the soil. The juveniles of most species have a size of 500–800 μm and carry the symbiotic bacteria of the genus *Xenorhabdus* or *Photorhabdus* in their gut. The juveniles actively search for the insect host and penetrate it through the cuticle or natural openings, i.e. the mouth, the anus or the spiracula. Once inside the hemocoel, the nematodes release their symbionts and excrete metabolites that repress the immune system of the host, so that the symbiotic bacteria can develop. Since the bacteria produce toxins, the insect is killed within 24–48 hours and filled with bacteria on which the nematodes feed and develop. After two or three weeks, the whole cadaver is exploited and two to three generations of nematodes have developed. In a normal sized caterpillar, about 100,000 juveniles are produced, which in turn search for new hosts and are able to survive in the soil for up to six months. Due to the above features, several commercial biopesticides based on *Steinernema* spp. or *Heterorhabditis* spp. are produced and used on a large scale.

21.3. Greenhouse Environment and Microbial Control

As pointed out above, the effectiveness of many biopesticides, especially those based on fungi, greatly depends on the correct combination of temperature and humidity. Temperature is always a very important factor in biological as well as in microbial control of pests, as it affects the activity and development of pests and of their natural enemies as well as the development of crop plants (Hussey and Scopes, 1985).

Of all crops grown in greenhouses, cucumber and sweet pepper require the highest temperatures and humidity. In the period before fruit development, both crops require 22–24°C during the day and 17–18°C during the night. During fruit bearing, 24–28°C and 18–20°C is required respectively. Humidity during the pre-fruiting period should be 75–80% RH, and during the fruiting period 85–90% RH. Tomato plants require a lower temperature regime: 22–24°C during the day and 16–18°C during the night. This explains why the pathogenic fungi *Paecilomyces farinosus* (Holmsk.) A.H.S. Brown & G. Sm. and *V. lecanii* are always more effective against whiteflies [*Trialeurodes vaporariorum* (Westwood), *Bemisia tabaci* (Gennadius)] in cucumber than in tomato crops. Conidia sporulate best at temperatures of about 25°C and RH above 80%, while at low temperatures and at a RH below 50% sporulation is inactivated (Ekbom, 1981). Since in ornamental crops temperature and humidity are relatively low, when fungal biopesticides are used, both parameters must be changed to favour the infection process and fungus development. This is mostly achieved by performing treatments in the late afternoon, so that the germination of fungi on the insect cuticle will take place in the evening or during the night, when humidity reaches the highest levels.

A major benefit for the use of pathogens as biological control agents in the greenhouse environment is the reduced level of ultraviolet (UV) radiation. Under field conditions, UV radiation is responsible for inactivation of bacteria, viruses or fungal pathogens used as biopesticides within hours after application. In greenhouses and tunnels, UV radiation is in most cases negligible, due to absorption by glass and plastic covers or white paint usually used to reduce insolation.

21.4. Epizootiology of Pathogens

21.4.1. METHODS OF PATHOGENS USE

Knowledge of the principles of epizootics caused by pathogens in host populations is essential for the success of microbial control attempts. In general pathogens can be used in three ways (Fuxa, 1987):

(i) Inundative releases resulting in an immediate effect. This approach is especially used for *B. thuringiensis* and viral biopesticides. Several treatments during the growing season are necessary to protect crops in greenhouses because of the constant migration of insects from fields and weeds around the greenhouses and growth of unprotected new foliage.

(ii) Inoculation of pathogens which results in more or less permanent suppression of pest populations. This approach can be used for nematodes (*Steinernema, Heterorhabditis*) or nematophagous fungi (*Arthrobotrys*), which, once introduced to the soil in greenhouses, survive for a prolonged time and provide long lasting reduction of soil pests or those pests which spend part of their life cycle (i.e. thrips) in the soil (Smits, 1996). This approach can also be used for baculoviruses (*Nucleopolyhedrovirus, Granulovirus*) and fungi (*Verticillium, Aschersonia*). Permanent establishment of fungi and viruses is due to the so called recycling, which involves multiplication during infection or on cadavers and subsequent infection of healthy insects in the population. The best recent example of such an approach is given by Vestergaard *et al.* (1996), who incorporated the entomopathogenic fungus *M. anisopliae* into the growth medium of Gerbera in order to disrupt the part of life cycle of *Frankliniella occidentalis* (Pergande) that takes place in the soil substrate.

(iii) Environmental manipulations involving enhancement of naturally occurring pest control by means other than direct addition of pathogen units to those already present. This method needs to be explored with respect to greenhouse pests and one possibility is to release virus-infected larvae of *S. exigua* or *M. brassicae* which would spread infection among offspring of present or migrating moths and start the epizootic.

21.4.2. HORIZONTAL TRANSMISSION

Pathogens have the ability to spread through pest populations. Horizontal transmission includes all modes of transmission, except transovarial and transovum infection of offspring from parents. The most common method of horizontal transmission of viruses, bacteria and protozoa is by feeding on foliage contaminated with pathogens arising from infected insects or cadavers. Parasitoids can also transfer pathogens from host to host. Fungi usually infect through spores that attach themselves to the cuticle. The spores are either picked up by the insect, when it moves on foliage or soil, or transported by air or water to the cuticle. Nematodes actively move from host to host.

21.4.3. VERTICAL TRANSMISSION

Vertical transmission is the direct transfer of pathogens from parents to their offspring. The two main infection routes are transovum and transovarial transmission. In the first case the pathogen is transferred inside the egg. With transovarial transmission the pathogen is often present on the outside of the egg, and infection takes place when the larvae hatch and eat contaminated parts of the eggshell. Vertical transmission in particular occurs commonly in protozoa and viruses.

21.4.4. INUNDATIVE AND INOCULATIVE CONTROL EXAMPLES

Most pathogens in greenhouses are used as inundative biopesticides. A total of 500 g of Bt is mixed with 500 l of water and sprayed on the crop with conventional spraying equipment. Virus biopesticides are sprayed at dosages of 10^{11}–10^{12} VIB/ha in large volumes of water. Entomopathogenic nematodes are applied in large volumes of water at a dosage of 0.5–1×10^6 juveniles/m^2 to control *Otiorhynchus sulcatus* (Fabricius) or Sciaridae. Fungi are sprayed at dosages of 10^{12}–10^{13} spores/ha in 1000–3000 lof water.

Repetitive treatments are required as the young shoots, often preferred by insects, are not treated, and horizontal and vertical transmission is usually insufficient to cause rapid and efficient control of the pest population. De Moed *et al.* (1990) simulated in a computer model the possibilities for inundative and inoculative use of SeNPV for control of *S. exigua* in greenhouse chrysanthemums. Their simulations predicted that a single high dose inoculative spray would give insufficient long-term control. The main reasons for not finding sufficient long-term damping of the population was the removal of much of the virus inoculum from the greenhouse by harvesting of plants, in combination with the preference of adults to lay eggs on newly planted young cuttings and the preference of larvae to feed on young shoots. Their simulations also showed that when the chrysanthemum crop was replaced by more long-term crops, such as roses, long-term damping of the pest population could occur.

The computer simulations also predicted that multiple low dose applications at weekly intervals would be more effective than a single high dose application. The model predictions were proven to be valid during later field trials. These results may be similar for the inundative use of other insect pathogens as well.

Except for one paper by Tverdyukov *et al.* (1993), who recommends the use of *E. neoaphidis* by introduction in aphid infested greenhouse crops of diseased mummified aphids in a ratio from 1:25 to 1:75, no examples can be found in the literature of successful inoculative use of insect pathogens in greenhouses, whereas many can be found for predators and parasitoids. Research has focused on the inundative use of pathogens, but there must be potential for inoculative use of nematodes, fungi, protozoa and viruses to manage populations of aphids, thrips, mites and caterpillars in greenhouses.

21.5. Practical and Experimental Use of Pathogens in Greenhouses

21.5.1. MICROBIAL CONTROL OF LEPIDOPTERA

Use of Bacteria
A large number of commercial biopesticides based on Bt [*B. thuringiensis* ssp. *kurstaki*, *Bacillus thuringiensis* Berliner ssp. *aizawai* de Barjac & Bonnefoi, and *Bacillus thuringiensis* Berliner ssp. *thuringiensis* Heimpel & Angus] are widely used against different moth species in various countries (Table 21.1). In order to obtain the best control results, it is necessary to follow the producer label instructions with respect to doses. Timing is critical and treatments should be done during egg hatching, as first and second instar larvae are most susceptible to Bt toxins. Old instar larvae are relatively resistant and

no satisfactory control results will be obtained. Van de Vrie (1991) gives information on the use of *Bacillus thruringiensis* ssp. *kurstaki* against tortricids in greenhouse ornamental crops. As Bt has antifeedant activity, it is important to make sure that a lethal dose is acquired before the larvae stop feeding, otherwise recovery may occur. A low or medium volume application may prove more successful than a high volume spraying.

Use of Viruses

Only a few virus biopesticides are available to be used against moth caterpillars occurring in greenhouses (Smits, 1987). Spod-X is specifically recommended against *S. exigua*, which is the principal lepidopterous pest in sweet pepper and other vegetable and ornamentals in greenhouses in Europe and the USA (Smits and Vlak, 1994). When other noctuid species (such as *Helicoverpa, Mamestra, Lacanobia, Plusia* and *Diaparopis*) are present, a broad spectrum virus must be used, as indicated in Table 21.1.

TABLE 21.1. Biopesticides based on viruses, bacteria, fungi and nematodes used against pests of greenhouse crops in Europe, the USA and Japan

Bioagents and Biopesticides	Target Pests
Viral	
NPV of *Mamestra brassicae*	Lepidoptera
Mamestrin (France), Virin EKS L (Russia),	
Virin EKS WP (Russia)	
NPV of *Spodoptera exigua*	*Spodoptera exigua*
Spod-X (USA)	
NPV of *Spodoptera littoralis*	*Spodoptera littoralis*
Spodopterin (France)	
GV of *Agrotis segetum*	*Agrotis segetum*
Virin OS (Uzbekistan)	
NPV of *Helicoverpa* (= *Heliothis*) *armigera*	*Helicoverpa* spp., *Heliothis* spp.
Virin KHS (Russia), Elcar (USA)	
Bacterial	
Bacillus thuringiensis ssp. *kurstaki*	Lepidoptera
Bactospeine (France), Dipel (USA), Agree (USA), Thuricide	
(USA), Astur (Russia)	
Bacillus thuringiensis ssp. *dendrolimus*	Lepidoptera
Dendrobacillin (Russia), Baksin (Russia)	
Bacillus thuringiensis ssp. *kurstaki* +	Lepidoptera
Bacillus thuringiensis ssp. *aizawai*	
Turex (Switzerland)	
Bacillus thuringiensis ssp. *israelensis*	Diptera
Baktoculicid (Russia)	
Bacillus thuringiensis ssp. *thuringiensis*	*Tetranychus* spp., Lepidoptera
Bitoxybacillin (Russia), Turingin-1 (Russia)	
Fungal	
Aschersonia spp.	*Trialeurodes vaporariorum, Bemisia*
Aschersonin (Russia)	spp.
Beauveria bassiana	*Thrips tabaci, Trialeurodes*
Boverin K-BL (Russia), Boverin (Ukraine),	*vaporariorum*
Boverin ZH (Russia), Naturalis-O (USA)	

TABLE 21.1. Biopesticides based on viruses, bacteria, fungi and nematodes used against pests of greenhouse crops in Europe, the USA and Japan (cont.)

Bioagents and Biopesticides	Target Pests
Fungal (cont.)	
Paecilomyces fumosoroseus	*Bemisia* spp., *Trialeurodes*
PFR-20 GW (USA)	*vaporariorum*
Verticillium lecanii	*Bemisia* spp., *Trialeurodes*
Mycotal (UK), Vertalec (UK), Verticilin (Russia), Verticilin Zh (Russia), Verticillin Z-BL (Moldova), Verticillin M (Russia), Verticillin K (Russia), Cefalosporin (Ukraine)	*vaporariorum*, aphids
Entomophthora thaxteriana	Aphids
Mikoafidin T (Russia), Entox (Russia)	
Entomophthora pyriformis	Aphids
Piriformin (Russia)	
Arthrobotrys oligospora	*Meloidogyne* spp.
Nematofagin-BL (Russia)	
Metarhizium anisopliae	*Otiorhynchus sulcatus, Trialeurodes*
Bio 1020 (Germany), Metarizin (Russia)	*vaporariorum*
Streptomyces avermitilis	*Meloidogyne* spp., *Tetranychus* spp.,
Avertron (Russia), Vertimex (Russia), Fitoverm (Russia)	aphids
Streptomyces aurantiacus	*Tetranychus* spp., *Bemisia* spp.,
Aleicid (Russia)	*Trialeurodes vaporariorum*
Nematode	
Heterorhabditis bacteriophora	*Otiorhynchus sulcatus*
Otinem (Switzerland)	
Heterorhabditis megidis	*Otiorhynchus sulcatus*
Larvanem (The Netherlands), Nemasys (Belgium), NovoNem (Germany)	
Steinernema carpocapsae	*Bradysia* spp., *Lycoriella* spp.,
Biosafe (Ireland), Exhibit (Switzerland), Nemabakt (Russia)	Sciaridae
Steinernema feltiae	*Bradysia* spp., *Lycoriella* spp.,
Entonem (The Netherlands), Nemalogi (Sweden), Owinema SC (Poland), Nemasys (Belgia), Nemalogi (Sweden)	Sciaridae
Phasmarhabditis hermaphrodita	Slugs: *Arion* spp., *Deroceras* spp.,
Nemaslug (UK), Bioslug (Ireland)	*Tandonia* spp. and others

21.5.2. MICROBIAL CONTROL OF APHIDS AND WHITEFLIES

Whiteflies and aphids have sucking mouthparts, and therefore cannot be controlled by sprayed bacteria and viruses that must be ingested with the food for infection to occur. Fungi infect through the cuticle and for this reason can be used effectively against sucking insects. Fungi belonging to the genera *Beauveria*, *Verticillium*, *Paecilomyces* and *Aschersonia* are particularly useful control agents. Rombach and Gillespie (1988) and Fransen (1990) provide comprehensive reviews on the use of fungi against arthropod pests on greenhouse crops.

Use of V. lecanii

The fungus *V. lecanii* is the best studied pathogen of whiteflies and aphids, and commercial biopesticides based on this fungus are available in several countries in western (Schuler *et al.*, 1991) as well as in eastern Europe (Lipa, 1985, 1996). When using *V. lecanii* biopesticides, the most effective control is obtained if the temperature after application is 15–28°C and RH is above 90%. Under the right conditions, in some crops, i.e. year round chrysanthemums, one application provides aphid control as well as whitefly control through repeated cycles for up to three months.

Use of Aschersonia
Out of over 30 known *Aschersonia* species, the most extensively studied and used are *A. aleyrodis, Aschersonia confluens* Henn., *Aschersonia flava* Petch and *Aschersonia placenta* Berkeley & Broome (Solovei and Koltsov, 1976; Ramakers and Samson, 1984; Fransen, 1990). Only larvae are infected while pupae and adults are resistant, which limits the effectiveness of these fungi. Although *A. aleyrodis* was experimentally used against *T. vaporariorum* in western Europe (Ramakers and Samson, 1984), no commercial products are available. On the other hand, in Russia, Belarus and the Ukraine *Aschersonia* spp. are cultured in regional biolaboratories, and liquid formulations of Aschersonin are distributed with a titer of 20–50×10^6 spores/ml (Izhevskii and Prilepskaya, 1977; Tverdyukov *et al.*, 1993).

Use of B. bassiana
This fungus is recommended against nymphs of *T. vaporariorum* which die 5–7 days after treatment, and 10–12 days later become covered with a white layer of mycelium and spores which serve as source of infection for other individuals.

Use of P. fumosoroseus
This fungus is infective to all stages (eggs, nymphs and adults) of *Trialeurodes* and *Bemisia* whiteflies, so is more effective than other fungi mentioned above (Smith, 1993; Lindquist, 1996; Sosnowska and Piatkowski, 1996).

Use of Metarhizium *spp.*
Although it is generally admitted that *Metarhizium* species have good potential in insect control, surprisingly only few reports refer to their use against greenhouse pests: *M. anisopliae* against *O. sulcatus* on greenhouse pot plants and against *F. occidentalis* on gerbera; *Metarhizium flavoviridae* Gams & Rozsypal against *Pemphigus bursarius* (L.) on lettuce; *Metarhizium album* Petch against whiteflies, thrips, spider mites and leafminers.

Use of Entomophthora *spp.*
Entomophthora thaxteriana Petch and *Entomophthora pyriformis* Thoizon have been recommended to control several aphid species, *Tetranychus* spp. and *Thrips tabaci* Lindeman. Tverdyukov *et al.* (1993) recommend the use of *E. neoaphidis* by introduction on aphid infested greenhouse crops of diseased mummified aphids in a ratio from 1:25 to 1:75.

21.5.3. MICROBIAL CONTROL OF THRIPS

Gillespie (1986) discussed the potential of entomogenous fungi as control agents for *T.*

tabaci, while Ravensberg *et al.* (1990) demonstrated the effectiveness of *V. lecanii* against whiteflies and thrips. There is potential to use entomopathogenic nematodes such as *Steinernema carpocapsae* Weiser and *Heterorhabditis bacteriophora* Poinar for the control of thrips.

21.5.4. MICROBIAL CONTROL OF DIPTERA

In some regions, i.e. England or Russia, dipterans such as *Bradysia* spp. and *Sciara* spp. appear in up to eight generations per year. Feeding in the roots or stem base damages various crops. Nedstam and Burman (1990) and Harris *et al.* (1995) demonstrated the effectiveness of *Steinernema* and *Heterorhabditis* nematodes in the control of *Bradysia brunnipes* (Meigen), *Bradysia coprophila* (Lintner) and other similar noxious dipterans in greenhouse crops. In a recent study, Broadbent and Olthof (1995) evaluated foliar application of *S. carpocapsae* against a leafminer, *Liriomyza trifolii* (Burgess), on chrysanthemums. Nematodes as well as *Bacillus thuringiensis* ssp. *israelensis* are quite effective against *Lycoriella solani* (Winnertz) and *Lycoriella auripila* (Winnertz) in mushroom cultivations, so they also have good potential in greenhouses (Grewal and Richardson, 1993; Rinker *et al.*, 1995).

21.5.5. MICROBIAL CONTROL OF MITES

Tverdyukov *et al.* (1993) reports on the effective use of *Bacillus thuringiensis* ssp. *thuringiensis* against *Tetranychus urticae* Koch and *Tetranychus cinnabarinus* (Boisduval). Andreeva and Shternshis (1995) report results of laboratory and field tests with *Streptomyces avermitilis* Burg *et al.*, *E. thaxteriana* and *V. lecanii* used to control *T. urticae* and *T. vaporariorum* attacking greenhouse crops in the Novosibirsk region of Siberia.

21.5.6. MICROBIAL CONTROL OF NEMATODES

Several species of the genus *Meloidogyne* cause serious damage in greenhouse crops and are very difficult to control with chemical or physical methods. An intensive search and screening has led to the production of some microbial nematicides or the discovery of strains of fungi and bacteria with good potential against nematodes. Stirling (1991) provides a general overview of previous attempts of biocontrol at plant parasitic nematodes. More recent publications refer to the potential and experimental use of several micro-organisms against *Meloidogyne* species: *Arthrobotrys oligospora* Fresen., *S. avermitilis*, *B. thuringiensis* (this also against *Pratylenchus penetrans* Cobb and *Rotylenchus fragaricus* Maqbool & Shahina), *Bacillus cereus* Frankland & Frankland, *Pasteuria penetrans* (Thorne) Sayre & Starr, *Hirsutella rhossiliensis* Minter & Brady, *Monacrosporium cionopagum* (Drechsler), *Monacrosporium ellipsosporium* (Grove) Cooke & Dickinson, *Verticillium chlamydosporium* Goddard, *Pseudomonas chitynolytica* Spiegel *et al.*, *Paecilomyces lilacinus* (Thom) R.A. Samson and *V. lecanii*.

21.5.7. MICROBIAL CONTROL OF GASTEROPODA (SLUGS)

Several slugs species can cause damage in greenhouse crops. *Deroceras reticulatum* Müller and *Deroceras agreste* L. are the most common and can feed on over 140 plant species. Wilson *et al.* (1993) reported on the development of a biopesticide based on a nematode, *Phasmarhabditis hermaphrodita* (Schneider) (Rhabditidae), that is effective against many slug species occurring in open field crops as well as in protected crops. In a recent publication, Wilson *et al.* (1995) reported the effective protection of lettuce grown in plastic tunnels against the following slugs: *Arion ater* (L.), *Arion distinctus* Mabille, *Arion intermedius* Normand, *Arion silvaticus* Lohmander, *Deroceras panormitamum* (Lessona & Pollonera) [= *Deroceras caruanae* (Pollonera)], *D. reticulatum*, *Milax gagates* (Draparnaud), *Tandonia budapestensis* (Hazay) and *Tandonia sowerbyi* (Férussac).

21.6. Pathogens as Part of an IPM System in Greenhouses

21.6.1. COMPATIBILITY WITH CHEMICAL PESTICIDES

Bacterial and viral biopesticides are highly compatible with chemical pesticides, and the producer of Mamestrin even recommends tank mixtures of that product with any pyrethroid (Anonymous, 1995). Problems arise when fungal biopesticides are used in combination with fungicides. It should be emphasized that the IOBC/WPRS Working Group on Pesticide Side-Effects to Beneficial Organisms evaluates and regularly publishes information on the compatibility or non-compatibility of various pesticides and biocontrol agents (see Chapter 11). Distributors publish an extensive list of chemical pesticides which can be used in greenhouses in the presence of bioagents.

21.6.2. COMPATIBILITY WITH PARASITOIDS, PREDATORS AND POLLINATORS

Bacterial and viral biopesticides are harmless to parasitoids and predators used in greenhouses, and no published records of their side-effects to that group of beneficial arthropods are known. But fungi such as *Beauveria, Metarhizium, Verticillium* and *Paecilomyces* have a wide infectivity spectrum and do create a real threat to adults or larvae of most predators and to parasitoids. Pavlyushin (1996) reports on some deleterious effect of *V. lecanii, P. fumosoroseus,* and *B. bassiana* on larvae of *Chrysoperla carnea* (Stephens) and *Chrysoperla sinica* (Tjeder), as well as on the coccinnellid *Cycloneda limbifer* Casey. However, in general, biopesticides can be considered as fully compatible with other biological control measures in greenhouses. This conclusion may also be generalized with respect to the use of bumblebees and bees as flower pollinators in greenhouses.

21.6.3. SAFETY FOR USE

Biopesticides based on micro-organisms such as bacteria, fungi, viruses and protozoa are generally registered through similar procedures as those used for chemical pesticides. This procedure ensures that only biopesticides that are safe to man and are not phytotoxic to plants can be put on the market. All biopesticides are exempted from residue level and pre-harvest waiting periods and can be used even during the harvest period. However, because of general hygienic precautions, product label instructions must be read and followed.

21.7. Expected Developments

Ravensberg (1994) discussed general aspects of the current state of biological protection of greenhouse crops and pointed out limiting as well as promoting factors of the future developments in this area. Although the availability and use of bumblebees for pollination promotes the application of selective microbial insecticides, their number is still very limited. However, there is noticeable interest by companies in the development and production of microbial biopesticides, and this offers some optimism (Smits, 1997a,b).

Feitelson *et al.* (1992), Gelerntner (1994) and Marrone (1994) present possible developments in the area of *B. thuringiensis* and transgenic plants which offer several opportunities to control various greenhouse pests. Genetic improvement of strains of insect pathogens and entomopathogenic nematodes is an area to be intensively explored. Tomalak (1994a,b) discussed ways to genetically improve *Steinernema feltiae* (Filipjev) for improved efficacy in the control of *F. occidentalis* and *L. solani*.

There is much potential in the use of pathogens to control greenhouse pests. The number of commercial products in the near future will however remain rather limited due to the fact that all insect pathogens, except for entomopathogenic nematodes, have to go through registration. The costs involved in the process of registration seriously hamper the development of commercial products as the greenhouse market, cash intensive as it may be, is often too limited in size to enable the company to get a fair return on their investments.

References

Andreeva, I.V and Shternshis, M.V (1995) Microbiological formulations against spider mites in greenhouses, *Zashchita Rastenii* **11**, 41–42.

Anonymous (1995) *Mamestrin*, NPP–Natural Plant Protection, Nogueres.

Broadbent, A.B. and Olthof, T.H.A. (1995) Foliar application of *Steinernema carpocapsae* (Rhabditida: Steinernematida) to control *Liriomyza trifolii* (Diptera: Agromyzidae) larvae in chrysanthemus, *Environmental Entomology* **24**, 201–206.

Caballero, P., Aldebis, H.K., Vargas-Osuna, E. and Santiago-Alvarez, C. (1992) Epizootics caused by a nuclear polyhedrosis virus in populations of *Spodoptera exigua* in Southern Spain, *Biocontrol Science and Technology* **2**, 35–38.

de Moed, G.H., van der Werf, W. and Smits, P.H. (1990) Modelling the epozootiology of *Spodoptera exigua* nuclear polyhedrosis virus in a spatially distributed population of *Spodoptera exigua* in greenhouse chrysanthemus, *IOBC/WPRS Bulletin* **12**(5), 135–141.

Ekbom, B.S. (1981) Humidity requirements and storage of the entomopathogenic fungus *Verticillium lecanii* for use in greenhouses, *Annales Entomologica Fennica* **47**, 61–62.

Feitelson, J.S., Payne, J. and Kim, L. (1992) *Bacillus thuringiensis*: Insects and beyond, *Biotechnology* **10**, 271–275.

Fransen, J.J. (1990) Natural enemies of whiteflies: Fungi, in D. Gerling (ed.), *Whiteflies: Their Bionomics, Pests Status and Management,* Intercept, London, pp. 187–210.

Fuxa, J.R. (1987) Ecological considerations for the use of entomopathogens in IPM, *Annual Review of Entomology* **32**, 225–251.

Gelerntner, W.D. (1992) *Bacillus thuringiensis*, bioengineering and the future of bioinsecticdes, in *Brighton Crop Protection Conference – Pests and Diseases,* Vol. 2, BCPC, Farnham, pp. 617–624.

Georgis, R. (1992) Present and future prospects for entomopathogenic nematode products, *Biocontrol Science and Technology* **2**, 83–99.

Gillespie, A.T. (1986) The potential of entomogenous fungi as control agents for onion thrips, *Thrips tabaci, BCPC Monograph: Biotechnlogy and Crop Improvement and Protection* **34**, 237–243.

Grewal, P.S. and Richardson, P.N. (1993) Effects of application rates of *Steinernema feltiae* (Nematoda: Steinernematidae) on biological control of the mushroom fly *Lycoriella auripila* (Diptera: Sciaridae), *Biocontrol Science and Technology* **3**, 29–40.

Harris, M.A., Oetting, R.D. and Gardner, W.A. (1995) Use of entomopathogenic nematodes an a new monitoring technique for control of fungus gnats, *Bradysia coprophila* (Diptera: Sciaridae), in floriculture, *Biological Control* **5**, 412–418.

Hussey, N.W. and Scopes, N. (1985) *Biological Pest Control, The Glasshouse Experience*, Blandford Press, Poole.

Izhevskii, S.A. and Prilepskaya, N.A. (1977). [Ashersonia against greenhouse whitefly], Ministry of Agriculture, Moscow (in Russian).

Lindquist, R. (1996) Microbial control of greenhouse pests using entomopathogenic fungi in the USA, *IOBC/WPRS Bulletin* **19**(9), 153–156.

Lipa, J.J. (1974) *An Outline of Insect Pathology,* USDA/NSF/NCSTEI, Warsaw.

Lipa J.J. (1985) History of biological control in protected cultures. 2. Eastern Europe, in N.W. Hussey and N. Scopes (eds.), *Biological Pest Control, The Glasshouse Experience*, Blandford Press, Poole, pp. 23–29.

Lipa, J.J. (1996) Insect pathology and microbial control in the EPRS region and in Poland, *IOBC/WPRS Bulletin* **19**(9), 1–11.

Marrone, P.G. (1994) Present and future use of *Bacillus thuringiensis* in integrated pest management systems: An industrial perspective, *Biocontrol Science and Technology* **4**, 517–526.

Nedstam, B. and Burman, M. (1990) The use of nematodes against sciarids in Swedish greenhouses, *IOBC/WPRS Bulletin* **13**(5), 147–148.

Pavlyushin, V.A. (1996) Effect of entomopathogenic fungi on entomophagous arthropods, *IOBC/WPRS Bulletin* **19**(9), 247–249.

Ramakers, P.M.J. and Samson, R.A. (1984) *Aschersonia aleyrodis*, a fungal pathogen of whitefly. II. Application as biological insecticide in glasshouses, *Zeitschrift für angewandte Entomologie* **97**, 1–8.

Ravensberg, W.J. (1994) Biological control of pests: Current trends and future prospects, in *Brighton Crop Protection Conference – Pests and Diseases,* Vol. 2, BCPC, Farnham, pp. 591–600.

Ravensberg, W.J., Malais, M. and van der Schaaf, D.A. (1990) Applications of *Verticillium lecanii* in tomatoes and cucumbers to control whitefly and thrips, *IOBC/WPRS Bulletin* **13**(5), 173–178.

Rinker, D.L., Olthoff, H.A. and Dano, G.A. (1995) Effects of entomopathogenic nematodes on control of a mushroom-infesting sciarid fly and on mushroom production, *Biocontrol Science and Technology* **5**, 109–119.

Rombach, M.C. and Gillespie, A.T. (1988) Entomogenous hyphomycetes for insect and mite control on greenhouse crops, *Biocontrol News and Information* **9**, 7–18.

Schuler, T., Hommes, M., Plate, H. and Zimmermann, G. (1991) *Verticillium lecanii* (Zimmermann) Viegas (Hyphomycetes: Moniliales) Geschichte, Systematik, Verbraitung, Biologie und Anwendung im Pflanzenchutz, *Mitteilungen aus der Biologischen Bundesanstalt für Land- und Forstwirtschaft* **269**, 1–154.

Shevtsov, V., Schyolokova, E., Krainova, O., Jigletsova, S. and Ichtchenko, V. (1996) Application horizons of crystalliferous bacilli for control of pest insects, nematodes and mosquitoes, *IOBC/WPRS Bulletin* **19**(9), 289–292.

Smith, P. (1993) Control of *Bemisia tabaci* and the potential of *Paecilomyces fumosoroseus* as a biopesticide, *Biocontrol News and Information* **14**, 71N–78N.

Smits, P.H. (1987) *Nuclear Polyhedrosis Virus as Biological Control Agent of* Spodoptera exigua, Ph.D. Thesis, Agricultural University, Wageningen.

Smits, P.H. (1996) Post-application persistence of entomopathogenic nematodes, *Biocontrol Science and Technology* 6, 379–387.

Smits, P.H. (1997a) Insect pathogens, their suitability as biopesticides, in *Microbial Insecticides: Novelty or Neccessity*, BCPC Symposium Proceedings No. 68, BCPC, Farnham, pp. 21–28.

Smits, P.H. (1997b) Microbial control of insect pests, in J.C. Zadoks (ed.), *Modern Crop Protection: Developments and Perspectives*, Wageningen Press, Wageningen, pp. 189–198.

Smits, P.H. and Vlak, J.M. (1994) Registration of the first viral insecticide in the Netherland: The development of Spod-X, based on *Spodoptera exigua* nuclear polyhedrosis virus, *Mededelingen van de Faculteit Landbouwwetenschappen Rijksuniversiteit Gent* 59/2a, 385–392.

Solovei, E.F. and Koltsov, P.D. (1976) [The action of entomogenous fungi of the genus *Aschersonia* on the whitefly], *Mikologia i Fitopatologia* 10, 425–429 (in Russian).

Sosnowska, D. and Piatkowski, J. (1996) Efficacy of entomapthogenic fungus *Paecilomyces fumosoroseus* against whitefly (*Trialeurodes vaporariorum*) in greenhouse tomato cultures, *IOBC/WPRS Bulletin* 19(9), 179–182.

Stirling, G.R. (1991) *Biological Control of Plant Parasitic Nematodes*, CAB International, Wallingford.

Tomalak, M. (1994a) Genetic improvement of *Steinernema feltiae* for integrated control of the Western flower thrips, *Frankliniella occidentalis*, *IOBC/WPRS Bulletin* 17, 17–20.

Tomalak, M. (1994b) Selective breeding of *Steinernema feltiae* (Filipjev) (Nematoda: Steinemematidae) for improved efficacy in control of a mushroom fly, *Lycoriella solani* Winnertz (Diptera: Sciaridae), *Biocontrol Science and Technology* 4, 187–198.

Tverdyukov, A.P., Nikonov, P.V. and Yushchenko, N.P. (1993) *[Biological Control Methods against Pests and Diseases in Protected Crops]*, Kolos, Moscow (in Russian).

van de Vrie, M. (1991) Tortricids in ornamental crops in greenhouses, in L.P.S. van der Geest and H.H. Evenhuis (eds.), *Tortricid Pests, Their Biology, Natural Enemies and Control*, Elsevier, Amsterdam, pp. 515–539.

van Lenteren, J.C. and Woets, J. (1988) Biological and integrated pest control in greenhouses, *Annual Review of Entomology* 33, 239–269.

Vestergaard, S., Gillespie, A.T. and Eilenberg, J. (1996) Control of western flower thrips, *Frankliniella occidentalis* (Thysanoptera: Thripidae) in gerbera by incorporating the entomopathogenic fungus *Metarhizium anisopliae* into the growth medium, *IOBC/WPRS Bulletin* 19(9), 240–246.

Wilson, M.J., Glen, D.M. and George, S.K. (1993) The rhabditid nematode *Phasmorhabditis hermaphrodita* as a potential biological control agent for slugs, *Biocontrol Science and Technology* 3, 503–511.

Wilson, M.J., Glen, D.M., George, S.K. and Hughes, L.A. (1995) Biocontrol of slugs in protected lettuce using the rhabditid nematode *Phasmarhabditis hermaphrodita*, *Biocontrol Science and Technology* 5, 233–242.

Zuckerman, B.M., Dicklow, M.B. and Acosta, N. (1993) A strain of *Bacillus thuringiensis* for the control of plant-parasitic nematodes, *Biocontrol Science and Technology* 3, 41–46.

CHAPTER 22

COMMERCIAL ASPECTS OF BIOLOGICAL PEST
CONTROL IN GREENHOUSES
Karel J.F. Bolckmans

22.1. Introduction

Hussey (1985) gives a detailed description of the first two decades of the history of commercial biological pest control in greenhouses starting in 1967. During these years, biocontrol grew from a scientific activity to a commercially interesting venture and a technically reliable way of pest management. The discovery of "new" beneficial arthropods and the use of bumble-bees for pollination of greenhouse vegetables have led to an explosive development of a real biological control industry. Today about 30 companies world-wide specialize in the production of beneficial arthropods for the biological control of pests in greenhouses. More than 30 beneficial arthropods are currently commercially available on the market for biological pest control in greenhouses. In total, about 90 species of natural enemies are commercially available for application under field conditions in Europe (van Lenteren, 1997).

22.2. Why Biocontrol?

In a commercial context it is important to understand the reasons why greenhouse growers want to use biological pest control. Probably the most important motive for growers to switch from chemical control to integrated pest managment was and still is pesticide resistance. Additionally, growers of greenhouse vegetables which use bumble-bees for pollination are very restricted in the number of pesticides they can use. The introduction of bumble-bees in 1987 has clearly created a bigger and more sensitive market for biological control. But nowadays, demands from retailers and consumers also stimulate use of biocontrol agents.

Biological control involves similar costs or can even be cheaper than chemical control. A major cost component of chemical control is labour. Generally speaking, biological control requires less labour. However, for some beneficials the time needed to introduce the beneficials in the greenhouse is still significant. Mechanization of release methods could provide a solution. The time which growers spend on scouting and monitoring for biological control should also be spent when using chemical pest control in order to determine whether sprays are needed. The close involvement of the grower in pest management activities and the closer follow-up of pests when using biological control are possibly major reasons why, in many cases, integrated pest managment is even more reliable than chemical control.

Many growers also appreciate the absence of visible spray residue on the crop or

310

R. Albajes et al. (eds.), Integrated Pest and Disease Management in Greenhouse Crops, 310-318.
© 1999 Kluwer Academic Publishers. Printed in the Netherlands.

fruits. Furthermore, there are no phytotoxic effects with biological control. Growth inhibition, commonly caused by repeated application of chemical pesticides, often results in yield reduction. The use of beneficials requires no safety or re-entry periods allowing continued harvesting without danger to the health of personnel working in the greenhouse. Consumers have become increasingly interested in safer and healthier food with less pesticide residues. In some countries, e.g. the UK and The Netherlands, supermarkets and auctions require growers-suppliers to produce their vegetables according to strict guidelines and standards, only allowing minimal use of pesticides from a restricted list. Breaking the rules is punished by a (temporary) prohibition to supply. Independent certification companies are increasingly being used to audit growers. Horticultural products are certified with various labels. In The Netherlands and Belgium, the auctions have had a strong influence on the implementation of biological pest control. In The Netherlands, the authorities exert strong pressure on the horticulture industry to reduce the use of pesticides. The government encourages and even subsidizes the development of biological pest control strategies in glasshouse grown crops. Growers and growers associations in western Europe and North America also rely heavily on biological control as a marketing tool, especially in an attempt to distinguish themselves from cheaper imported products from southern countries with lower production costs and higher residue loads.

22.3. The Market for Biological Pest Control in Greenhouses

22.3.1. AREA OF PROTECTED CROPS

The world area of protected crops (glasshouses, plastic houses, walk-in plastic tunnels and multispan) is estimated at around 300,000 ha (Wittwer and Castilla, 1995) with more than 220,000 ha of greenhouse vegetables and almost 80,000 hectares of greenhouse ornamentals. Europe has an estimated 120,000 ha of protected crops. About 60% of this area is located around the Mediterranean Sea, mainly in Spain and in Italy. The most important greenhouse vegetables are tomatoes, peppers, cucumbers, melons and eggplants, with tomatoes accounting for about 30% of the European greenhouse vegetables (see Chapter 1).

22.3.2. PRODUCTS

Over 30 different beneficial arthropods have been developed for greenhouse use (see Chapter 20). In an integrated pest management programme, beneficial arthropods are used in conjunction with monitoring tools such as: (i) pheromone traps and coloured sticky traps; (ii) microbial insecticides (bacteria, fungi and viruses); (iii) botanical insecticides (e.g. pyrethrum, azadirachtin); (iv) selective chemical pesticides; and (v) mechanical and cultural methods. All these tools together form a finely balanced pest management programme which, in order to maintain itself, requires effective solutions for each individual pest without endangering the total programme. Regularly "new" pests appear and in such cases solutions need to be found quickly. If not biological, the solution needs to be at least compatible with the rest of the system. The large number of available beneficial arthropods, often with several natural enemy species for each pest, has made biological control programmes more stable and reliable. Table 22.1 gives an overview of the beneficial arthropods currently available for pest control in greenhouses.

TABLE 22.1. Commercially produced beneficial arthropods for biological pest control in greenhouses (after van Lenteren, 1992)

Beneficial arthropod	Pest	In use since
Phytoseiulus persimilis	Spider mites	1968
Encarsia formosa	Whiteflies	1970 (1926)
Opius pallipes	Leafminers	1980–1983, 1996
Amblyseius barkeri	Tobacco thrips	1981–1990
	Western flower thrips	1986–1990
Dacnusa sibirica	Leafminers	1981
Diglyphus isaea	Leafminers	1984
Heterorhabditis spp.	Vine weevil	1984
Steinernema spp.	Fungus gnats	1984
Neoseiulus (= *Amblyseius*) *cucumeris*	Thrips	1985
Chrysoperla carnea	Aphids	1987
Bombus terrestris (pollination)	–	1987
Aphidoletes aphidimyza	Aphids	1989
Aphidius matricariae	Green peach aphid	1990
Orius spp.	Thrips	1991
Aphidius colemani	Green peach aphid, cotton aphid	1992
Aphelinus abdominalis	Potato aphid	1992
Hippodamia convergens	Aphids	1993
Delphastus pusillus	Whiteflies	1993
Amblyseius degenerans	Thrips	1994
Macrolophus caliginosus	Whiteflies	1994
Anagrus atomus	Leafhoppers	1994
Hypoaspis miles	Fungus gnats, shore flies	1994
Neoseiulus (= *Amblyseius*) *californicus*	Spider mites	1994
Phasmarhabditis hermaphrodita	Slugs	1994
Eretmocerus eremicus (= *Eretmocerus californicus*)	Whiteflies	1994
Harmonia axyridis	Aphids	1995
Feltiella acarisuga	Spider mites	1995
Episyrphus balteatus	Aphids	1995
Aphidius ervi	Potato aphid, glasshouse aphid	1996
Podisus maculiventris	Caterpillars	1996
Eretmocerus mundus	Whiteflies	1996
Coenosia spp.	Fungus gnats	1996

Natural enemies cannot be patented. Therefore production techniques are usually kept as closely guarded secrets. When a producer has developed a mass-rearing system for a "new" beneficial and has introduced its use among growers, other suppliers are able to copy the technology in a very short time. Since growers need to be provided with a full package of natural enemies and technical service, suppliers should be able to provide all beneficial arthropods which are available on the market as well as bumble-bees.

22.3.3. ACTUAL MARKET

The total market for natural enemies at end-user level for greenhouses in 1997 is estimated at more than US$30 million. The most important markets are The Netherlands, the UK and France, followed by the USA. Together, these countries account for about two thirds of the total market. Most biocontrol producers are located in north Europe and North America. A large potential but still undeveloped market for biological pest control is located in south Europe around the Mediterranean Sea, in Asia and in Latin America. Bumble-bees are already widely used in south Europe, Japan and Korea.

Currently, greenhouse vegetables account for more than 90% of the market for beneficial arthropods. Since the early 90s, the market in greenhouse ornamentals is continuing to grow steadily. Biological control is growing fastest in potplants like poinsettias and in cut flowers such as roses, gerberas and chrysanthemums.

Figure 22.1 shows the distribution of the total market for natural enemies divided over the respective greenhouse pests against which they are used. Beneficial arthropods against whiteflies (33%), thrips (22%), spider mites (16%) and aphids (13%) account for 84% of the total market for biological pest control.

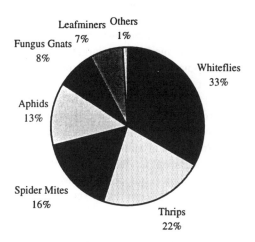

Figure 22.1. Distribution of the total market in 1997 for natural enemies applied in greenhouses divided over the respective greenhouse pests.

The most important beneficial arthropods are *Encarsia formosa* Gahan, which accounts for 25% of the total market for natural enemies, *Phytoseiulus persimilis* Athias-Henriot (12%) and *Neoseiulus* (= *Amblyseius*) *cucumeris* (Oudemans) (12%).

22.4. Producers and Producer Associations

To date, of more than 30 companies which are involved in the production of beneficial insects and mites for greenhouse crops, 20 are in Europe (van Lenteren *et al.*, 1997). A list of producers as well as distributors from North and Central America has been compiled by Hunter (1994). Besides these companies, others are involved in the production of beneficials for pest control in citrus, olives, interior plantscapes and botanical gardens, corn, stables (flies), etc.

The number of producers has grown almost exponentially after the first company started in 1967. Most companies are rather small and employ less than ten people. A few are much larger companies with subsidiaries in different parts of the world. Larger companies try to produce most of the beneficials themselves, except for some smaller specialized products. Other companies produce some beneficials themselves and purchase other species from other producers.

Recently, producers of biocontrol agents have started to organize themselves in different associations. The oldest is the Association of Natural Biocontrol Producers (ANBP) which serves the producers from the USA and Canada. The producers from Australia have created Australasian Biological Control (ABC). In Europe the producers are associated as the International Biocontrol Manufacturers Association (IBMA). The most important reasons for producers to found these associations are the upcoming new regulations for importation and release of exotic beneficial arthropods.

22.5. Marketing, Distribution and Logistics

Most producers of beneficials rely on their own technical personnel to distribute their products to growers in their home market. The larger, internationally active producers have either created local subsidiaries in important export countries, or, in most cases, have developed a network of distribution companies to serve their clients in foreign countries. These distributors are usually local private companies which are already involved in horticulture through the sales of fertilizers, pesticides and equipment or services for greenhouse growers. Some of them are only involved in the sales of natural enemies and bumble-bees. Producers of natural enemies train their distributors. Usually they have specially assigned technical advisors who visit the distributors on a regular basis to give them additional technical advice and to assist them in developing locally adapted introduction programmes. A continuous exchange of information between the producer and the distributor is essential to keep both partners up to date.

Marketing is done through advertisements and articles in technical grower magazines, horticulture fairs and more recently through the Internet. Most new customers are found through direct visits by technical advisors. Customers usually are signed up for an entire growing season. The supplier provides the grower with all biocontrol agents and monitoring tools (pheromones and sticky traps) as well as technical support. Usually beneficials and bumble-bees are both supplied by the same company.

The products, packed in plastic bottles, sachets or stuck on cards, are shipped by truck, train or aeroplane in polystyrene boxes with icepacks (see Chapter 20). They should not be in transit longer than 36 hours since after that time the temperature within the polystyrene boxes may start to rise. Upon arrival, the distributor will keep the products in cool storage until forwarded, while cooled, to his customers. Each distributor is supplied at least weekly with fresh products.

22.6. Biological Pest Control: How Much Does It Cost?

The total cost of biological control will depend on the crop, the country, the pest pressure, the experience of the grower and his advisor, and on how far the grower wants to commit himself. A grower can choose to rely on a combination of beneficials and selective chemicals or he can decide to try to control all pests as much as possible with beneficials. Biological pest control generally does not have to be more expensive than chemical pest control. Spraying one hectare of greenhouse tomatoes with a vertical spray boom requires about 8 hours of heavy work, whereas biological control, generally speaking, requires less labour.

Table 22.2 gives an estimate of the cost per square metre and per year of biological pest control for different crops in The Netherlands in 1997 (see also Cranshaw *et al.*, 1996; Jacobson, 1997; van Lenteren *et al.*, 1997). A difference has been made in Table 22.2 in the case of an "IPM" programme (biological control combined with selective chemical insecticides when needed) and a "maximal biocontrol" programme where the use of chemical insecticides is avoided.

TABLE 22.2. Average cost (US$/m^2/year) of a biological pest control programme in different crops in 1997 in The Netherlands

Crop	IPM	Maximal biocontrol
Tomatoes	0.25	0.55
Peppers	0.22	0.65
Cucumbers	0.32	0.90
Eggplants	0.50	0.65
Roses	1.00	–
Gerberas	0.40	–
Poinsettias	0.25	–

22.7. Technical Support: Essential but Expensive

For IPM to be successful, regular and reliable technical support is essential. To date, biological pest control without technical support from a specialist very seldom works. Technical support is still included in the price of the purchased beneficials. In case of large coops, there is a close cooperation between the supplier's technical staff (supplier is not always the same as producer) and the coop's technicians. Growers hence do not really buy only products but a service. Each client is generally visited once every week although visits may be less frequent for small growers or those in remote areas. Technical advisors inspect old and new hot spots, check the level of parasitism and population development of predators and pests, discuss the situation with the growers, and advise on the quantities of beneficials to introduce, which pesticides to use (or not to use) and how and where, among other tasks. Regular visits are crucial for success, especially during the early phase of a biocontrol programme. Growers who want to make the switch from chemical control to biological pest management require extra attention. These growers and their personnel first need to be trained in recognizing pests and natural enemies, as well as in scouting and monitoring techniques. Next they need to be assisted in the transition from the use of broad spectrum persistent pesticides to the use of more selective materials and those with a short residual activity. After this transition period, which can take several months before all pesticide residues have disappeared, introductions of beneficials can be started.

Technical advisors need to be able to recognize all pests and beneficials and have a thorough knowledge of their biology, ecology and behaviour, the introduction programmes for all natural enemies in the different greenhouse crops, the effects and side-effects of chemical pesticides, the greenhouse crops and growing systems, the different diseases, scouting and monitoring techniques and bumble-bee pollination. They need to be able to interpret the interactions between pest and natural enemies, and in case of problems, find the reason and give acceptable solutions. Most technical advisors have a technical degree and build up experience during on-the-job training within biocontrol companies. Such training takes a minimum of one year, and it usually takes two entire growing seasons for a technical advisor to be able to give reliable technical advice independently.

Prices of natural enemies and bumble-bees have come down drastically and therefore giving technical advice has become relatively more expensive for biocontrol producers and distributors. Some companies are putting a lot of effort now into training the growers and in developing detailed technical literature and even computer programs which have to help growers in their scouting and monitoring activities and in interpreting the collected data. There are even attempts to give advice through intranets, via telephone helplines and via e-mail (see Chapter 12). Besides, more technical information is becoming available through the Internet. Through these efforts, companies try to reduce their costs for technical advice and to give better technical support to growers in remote areas. In some countries, growers are increasingly ready to hire independent scouts to inspect their crops.

22.8. Regulatory Issues

Most countries have no regulations for the importation and release of beneficial arthropods, except Sweden, Austria, France, Poland and Hungary in Europe where it is required to submit a technical file for each beneficial and a fee for each beneficial must be paid prior to the issuing of an authorization for the import and release of natural enemies. Such fees can be very high and will thus prohibit use of biological control, especially if they are widely introduced at this level. The European Plant Protection

Organization (EPPO) is currently developing guidelines for the importation and release of non-indigenous natural enemies in Europe.

Overregulation of biological control with beneficial arthropods could severely slow down the development of "new" beneficials and increase the reaction time in case of an outbreak of a new exotic pest. Smaller companies are not able to bear the costs involved in putting together all the data required to obtain approval for marketing a beneficial.

Commercial importation of natural enemies and bumble-bees is not allowed in Israel which has its own local producers. The USA and Canada request the submission of a technical file for each beneficial. A risk assessment study is conducted by the authorities (USDA-APHIS-PPQ, Agriculture Canada) after which a formal decision is taken to issue an import permit or not. Both countries exchange information on the permitting of beneficials. Japan has very stringent regulations for both the import and use of beneficials which have to go through a real registration procedure including efficacy trials over two growing seasons. Because of these strict regulations, biocontrol is still in its infancy in Japan. The strict quarantine regulations of New Zealand and Australia do not allow regular commercial importation of natural enemies. Growers are supplied by local producers. Australian growers do not have access to bumble-bees either as they are not endemic to Australia and therefore import is forbidden.

22.9. Opportunities and Threats for Biological Pest Control

High quality products and technical advice are basic requirements in business. Scientists and producers in Europe have developed quality control guidelines and standards for most natural enemies within the IOBC working group for "Quality Control of Mass-reared Arthropods" (see Chapter 20). Also the North American producers have developed a set of quality control procedures. Both European and North American producers are currently working on the implementation of these guidelines through the development of a certification scheme for producers of natural enemies. Some also plead for the certification of technical advisors in order to assure the quality of the advice.

Several companies label their products with information about the product, storage, numbers to introduce, target pests, etc. Other companies just mention the scientific name or a brand name. In Europe the producers are currently developing guidelines for uniform labelling of natural enemies. To solve the problem of having to print labels in many languages one company has developed pictograms which will also be adopted by other producers.

The inevitable development of more restrictive regulations for the import and release of natural enemies and for the introduction of non-indigenous species in Europe, the most important market for greenhouse biological control, will certainly harmonize the current situation and prevent the premature introduction of "new" natural enemies. On the other hand, regulations can potentially be a threat for the industry if large, expensive studies were to be required prior to the introduction of new beneficials. The main mechanisms through which biocontrol companies compete are: (i) the quality of the products; (ii) the quality, quantity and frequency of the technical support; (iii) the development of cost-effective production systems; (iv) the development of new products and strategies; and (v) unfortunately, the price. Fierce competition, mainly in Europe, has led in the mid 90s to considerable reductions in the end-user prices of beneficials and bumble-bees. The reduced profit margins have undoubtedly slowed down the development of new beneficials and improvement of IPM programmes by the producers of natural enemies. Another risk is that producers and their distributors can no longer afford to give sufficient technical support to the growers.

In a large part of the greenhouse area, mainly located around the Mediterranean Sea, in Central and South America and in Asia, biological control is still in its early days. Nevertheless, in many cases biological control has already been implemented successfully in these areas. Adapted introduction programmes with suitable beneficials are being developed. Significant research efforts and the development of local technical support and training are conditional for furthering biological control successfully in these new areas.

References

Cranshaw, W., Sclar, D.C. and Cooper, D. (1996) A review of 1994 pricing and marketing by suppliers of organisms for biological control of arthropods in the United States, *Biological Control* 6, 291–296.

Hunter, C.D. (1994) *Suppliers of Beneficial Organisms in North America*, California Protection Agency, Sacramento, Calif.

Hussey, N.W. (1985) History of biological control in protected crops, western Europe, in N.W. Hussey and N. Scopes (eds.), *Biological Pest Control, the Glasshouse Experience*, Blandford Press, Poole, Dorset, pp. 11–22.

Jacobson, R.J. (1997) Integrated Pest Management (IPM) in glasshouses, in T. Lewis (ed.), *Thrips as Crop Pests*, CAB International, Wallingford, pp. 639–666.

van Lenteren, J.C. (1992) Biological control in protected crops: Where do we go? *Pesticide Science* 36, 321–327.

van Lenteren, J.C. (1997) Benefits and risks of introducing exotic macro-biological control agents in Europe, OEPP/EPPO *Bulletin* 27, 15–27.

van Lenteren, J.C., Roskam, M.M. and Timmer, R. (1997) Commercial mass production and pricing of organisms for biological control of pests in Europe, *Biological Control* 10, 143–149.

Wittwer, S.H. and Castilla, N. (1995) Protected cultivation of horticultural crops worldwide, *HortTech* 5(1), 6–23.

CHAPTER 23

BIOLOGICAL CONTROL OF SOILBORNE PATHOGENS

Dan Funck Jensen and Robert D. Lumsden

23.1. Introduction

Soilborne pathogens cause severe disease problems on plants grown under greenhouse conditions. This is the case both in greenhouses based on low technology and in advanced high technology houses. Chemical control is often unreliable because it can be difficult to reach the targeted soilborne pathogen with the pesticide and, because of development of fungicide resistance by some pathogens (Dekker, 1976). When soils are treated with chemicals it is not uncommon to have problems with pathogens which have the ability to quickly recolonize the disinfested soil (Jarvis, 1989). There is also an increasing concern for the environment and how it is influenced by pesticide use. Thus, both for environmental protection reasons, and since the possibilities for chemical control of soilborne pathogens are limited, growers are forced to seek new disease control measures. Development of IPM programmes which include the control of soilborne diseases is being considered as an approach for reducing the use of pesticides in greenhouses (Rattink, 1992; van Steekelenburg, 1992). It seems obvious, that biological disease control will form an important part of such IPM programmes. This will, however, require an intensified research effort in plant pathology and microbial ecology in greenhouse cultivation systems, and there will be a need for training growers and extension officers in new bio-intensive IPM procedures before they can be implemented in commercial plant production.

There are several examples of successful biological control of plant diseases in greenhouses at the experimental stages as reviewed by Whipps and Lumsden (1991) and Whipps (1997). Research in biocontrol has mainly been concentrated on antagonistic fungi belonging to the genera *Gliocladium* and *Trichoderma* (Papavizas, 1985; Lumsden and Locke, 1989; Jensen and Wolffhechel, 1995), mycoparasitic *Pythium* spp. (Benhamou *et al.*, 1997), non-pathogenic *Fusarium* spp. (Postma and Rattink, 1992; Eparvier and Alabouvette, 1994; Minuto *et al.*, 1995a; Larkin *et al.*, 1996) binucleate *Rhizoctonia* spp. (Herr, 1995) and *Laetisaria* spp. (Lewis and Papavizas, 1992); and antagonistic bacteria belonging to the genus *Bacillus* (Pleban *et al.*, 1995; Bochow *et al.*, 1996), fluorescent *Pseudomonas* (Weller, 1988) and *Streptomyces* (Tahvonen, 1982a). Potential antagonists for use in greenhouses might also be found among other species if the right isolation and screening methods were employed as discussed by Jensen (1996). More than 30 biocontrol agents (BCAs) are now being commercialized for the control of plant diseases [according to the latest information from the Biocontrol of Plant Diseases Laboratory, Beltsville, MD, USA, http://www.barc.usda.gov/psi/bpdl/bioprod.htm (Chapter 26)], and several of these BCAs are developed for the control of soilborne diseases under greenhouse conditions. There is considerable interest from growers to use biocontrol methods in greenhouse production systems. BCAs for use in greenhouses on a commercial scale are based largely

319

R. Albajes et al. (eds.), Integrated Pest and Disease Management in Greenhouse Crops, 319-337.
© 1999 *Kluwer Academic Publishers. Printed in the Netherlands.*

on antagonistic *Trichoderma* spp. and *Gliocladium virens* J.H. Miller, J.E. Giddens & A.A. Foster. The latter which has been determined to be synonymous with *Trichoderma virens* (J.H. Miller, J.E. Giddens & A.A. Foster) von Arx (Rehner and Samuels, 1994). For purposes of clarity and to avoid confusion in this review, the name *Gliocladium virens* will be retained, instead of *Trichoderma virens*, fully recognizing the taxonomic correctness and reasons for the name change. Acceptance of commercial preparations in the USA has been favourable so far (Harman and Lumsden, 1990; Mintz and Walter, 1993). In the past, experience often showed that the results were inconsistent, sometimes with good results and sometimes ineffective results (Postma and Rattink, 1992). Successful antagonists must be active against pathogen(s) at the right place and at the right time in the greenhouse. For employing biological control in greenhouses it will be required that detailed information be available about the ecology of the antagonist and the pathogens and their behaviour in the plant growing system. In this paper, this will be illustrated by case studies of biocontrol of *Pythium* spp. causing damping-off and root rot diseases in greenhouse crops. Especially examples where antagonistic *Trichoderma* spp. and *Gliocladium* spp. have been used for biocontrol of *Pythium* spp. will be considered. The possibilities for implementing biological control in an IPM strategy will also be discussed.

23.2. Greenhouses, Growth Systems and Disease Problems

In the subtropics, in East Central Europe and in Mediterranean areas most protected plants are grown in simple houses or tunnels where wooden or metal frames are covered with plastic film. Plants are normally grown in soil in the bottom of the houses, using very simple technology, and disease control is based on traditional methods (Gullino, 1992). In the USA, Canada, Northern Europe, North Italy and France, plant production is mainly in greenhouses or houses covered with hard plastic in which highly advanced technologies are in use. In some cases, soil is still used on raised beds (Gullino, 1992), but in most cases soilless culture systems are used. Rattink (1996) has described the different principles and methods for growing plants in soilless cultures. In brief, the systems are as follows: the plants are supported by growing in soilless media (peat, perlite, vermiculite, rockwool, expanded clay, polyurethane foam, etc.) placed in different ways in the greenhouse (in plastic bags, enveloped in plastic, in pots, trenches and on bench tops, etc.). In all cases, the growing media are not in direct contact with the soil. The plants are supplied with nutrient solution continuously or discontinuously either by overhead irrigation or irrigation from below. There are different types of irrigation systems, of which drip irrigation and the ebb and flow systems are used the most. The nutrient solution is either drained away (open systems) or recirculated (closed systems). For environmental protection reasons, some countries require that the recirculating irrigation systems are used (van Oosten, 1992).

23.2.1. DISEASE PROBLEMS

The different disease problems in greenhouses have been reviewed elsewhere (Stanghellini and Rasmussen, 1994; Besri, 1997) and this is also the subject of Chapters 2 and 3. It is highly relevant to consider the diseases which will be of importance in the different cultivation systems and then carry out a thorough investigation of how the pathogens behave in the systems before a successful biological control strategy can be developed. Most of the same pathogens cause problems in all cultivation systems. Soilborne diseases may be influenced by the type of the irrigation system and whether production is with or without soil. When soil is present, problems with pathogens such as *Pyrenochaeta lycopersici* R. Schneider & Gerlach in tomato, *Phomopsis sclerotioides* van Kestern in cucumber and *Sclerotinia sclerotiorum* (Lib.) de Bary on lettuce (Ebben, 1987), may be serious because resting structures of the pathogen can build up in the soil and might resist disinfestation between crops. In advanced soilless houses with recirculating irrigation systems, most problems encountered are with pathogens which can be spread by zoospores (Bates and Stanghellini, 1984; Paulitz, 1997) or bacteria (Nieves-Brun, 1985; Schuerger and Batzer, 1993). Diseases caused by zoosporic *Pythium* spp. and *Phytophthora* spp. can be very troublesome in soilless cultivation (Thinggaard and Middelboe, 1989). Also, some pathogens with airborne spores such as *Botrytis cinerea* Pers.:Fr. (Elad, 1996), *Fusarium* spp. causing crown rot (Couteaudier and Alabouvette, 1989; Rattink, 1993), and *Thielaviopsis basicola* (Berk. & Broome) Ferraris causing root rot (Stanghellini and Rasmussen, 1990), can give substantial disease problems in soilless systems. These pathogens have all been the target for biological control strategies in greenhouses.

23.3. Greenhouses Are Well Suited for Biological Control

Greenhouse conditions offer good opportunities for using biological control (Baker, 1992; Lumsden *et al.*, 1996; Paulitz, 1997). It is often said that greenhouses should be the first place for successful use of biocontrol strategies because the environment is more uniform in greenhouses compared to the often extreme fluctuations in field environments, and because of the possibilities for regulating the environment in favour of antagonists. Establishment of antagonists by incorporation in soil or soilless media in greenhouses requires less volume of BCAs than in the field, and BCAs can be introduced not only at seeding, but also throughout the cultivation period. Cultural practices and amendments inducing disease suppressiveness in soil or soilless media might also be regarded as more effective in greenhouses than in field soils. Different antagonists have different requirements in the environment. For example, the growth of *Trichoderma* spp. and *Gliocladium* spp. was found to be favoured at low pH (Chet and Baker, 1980; Harman and Taylor, 1988) whereas the pH optimum for antagonistic bacteria normally would be higher (Bochow, 1989). Differences were also found among isolates of *Trichoderma* and *Gliocladium* in their ability to control diseases at different temperatures (Lifshitz *et al.*, 1986; Tronsmo, 1989; Knudsen *et al.*, 1995) and at different matric potential (Wolffhechel and Jensen, 1991). Although there are some possibilities for regulating the environment in favour of the antagonists, it is, however, primarily the requirements of the plant that direct how the environmental conditions will be regulated. Thus, antagonists for use in greenhouses should be selected to fit the set of conditions used for cultivation of the crop plant and the conditions which favour disease outbreaks. It might be mentioned that *Trichoderma* spp. are found naturally as a part of the indigenous microflora in soilless systems and, therefore, seem to fit such environments in greenhouses (Loschenkohl, 1994).

It has also been argued that bacteria would be well adapted to soilless cultures due to the higher water content in these systems (Paulitz, 1997). Thus, the use of antagonistic bacterial products is also possible.

23.3.1. DISEASE SUPPRESSIVE SOILS AND SOILLESS GROWING MEDIA

Soils can be disease suppressive. Examples are the soils suppressive to disease caused by *Fusarium oxysporum* Schlechtend.:Fr. (Alabouvette *et al.*, 1985), *Pythium* spp. (Lumsden *et al.*, 1987), *T. basicola* (Stutz *et al.*, 1986) and *Gaeumannomyces graminis* (Sacc.) Arx & D. Oliver (Weller, 1988). Suppressiveness has also been obtained by transferring small amounts of suppressive soil to both conducive soil and peat (Alabouvette, 1987). The main factor for suppressiveness seems to be the ability of the soil to support an active microflora which can have a suppressive effect on the pathogen populations or possibly more importantly, suppressing the disease-causing activities of the pathogens (Schneider, 1982).

Suppressiveness can be due to: (i) fungistasis exerted by the general microflora, due mainly to competition for substrates (Alabouvette *et al.*, 1985; Lockwood, 1988); (ii) antagonistic activities such as antibiosis (Fravel, 1988; Cook *et al.*, 1995); (iii) mycoparasitism (Lumsden, 1992); and (iv) induction of disease resistance in plants (Sticher *et al.*, 1997). Soilless growing media are normally considered to be conducive to soilborne diseases since they normally do not support an actively suppressive microflora (Hoitink and Fahy, 1986; Gullino and Garibaldi, 1994). There are, however, some exceptions to this. Hoitink (1980) found that composted bark could be used for growing ornamentals and that this was highly suppressive to root rot caused by *Phytophthora* spp. and, by amending conducive peat growing media with composted bark, the peat became suppressive to several soilborne diseases (Hoitink, 1980). The disease control is believed to be due to the biomass and the general microbial activity in the amended medium (Chen *et al.*, 1988; Boehm and Hoitink, 1992) and, possibly, the synergistic effect of several antagonistic micro-organisms (Hoitink *et al.*, 1991). Other types of composted organic material can exert similar effects, for example, against damping-off caused by *Pythium ultimum* Trow (Lewis *et al.*, 1992; Grebus *et al.*, 1994). Methods to predict the quality of peat and compost as it relates to suppression of disease caused by, for example, *Fusarium* spp., *Pythium* spp. and *Rhizoctonia solani* Kühn has been developed (Boehm and Hoitink, 1992; Hoitink and Grebus, 1994). For improving compost quality, procedures for amending the compost with beneficial antagonists at the end of the composting process are described (Hoitink and Grebus, 1994). Pathogen suppressive compost prepared in this way is now in use in commercial greenhouses in the USA (Harman and Björkman, 1998). Light coloured sphagnum peat lots have in some cases been shown to be suppressive to diseases caused by seed- and soilborne pathogens (Tahvonen, 1982a; Wolffhechel, 1988). Similar results were later obtained by Hoitink *et al.* (1991). They found that the microbial activity, as measured by the fluorescein diacetate technique, was higher in the least decomposed light peat, demonstrating that this type of peat was able to support a high activity of microbial biomass (Boehm and Hoitink, 1992). Tahvonen (1982b) isolated antagonistic *Trichoderma* spp. and *Gliocladium* spp. from the suppressive peat lots and from some of them also *Streptomyces* spp. The BCA Mycostop™ is based on these *Streptomyces* spp. (Chapter 26). Wolffhechel (1989) found antagonistic *Penicillium* spp., *Gliocladium* spp.

and *Trichoderma* spp., which contributed to the suppressiveness in different sources of sphagnum peat. It was also demonstrated that disease suppression could be restored by inoculating conducive steamed sphagnum peat with isolates of *Trichoderma harzianum* Rifai and *G. virens* isolated from disease suppressive peat lots. Fortunately, despite the belief that most soilless growing media are disease conducive and do not support the activity of antagonists, there are several other examples of successful biological control by introducing antagonists to conducive soilless growing medium (Lumsden *et al.*, 1996; Whipps, 1997).

23.3.2. ESTABLISHMENT OF ANTAGONISTS

When an antagonist is introduced to soil or soilless medium, it is important that it establishes and becomes active in suppressing the diseases it is meant to control. Micro-organisms have evolved different strategies for survival, nutrient exploitation, growth and reproduction. Such ecological strategies are more complex for fungi than bacteria. Thus, micro-organisms differ in their ability to exploit different sources of organic substrate. The natural indigenous microflora will, due to varied composition, respond quickly and colonize organic substrates, if available. Competition for substrates in the soil and rhizosphere result in microbial buffering. Therefore, it is difficult for a single micro-organism, whether it is indigenous or introduced, to increase its population and activity as a response to excess substrate in soil.

Soilless growing media normally have a low microbial buffering capacity, because they are composed of either inert materials or the nutritional status is such that an active biomass cannot be supported (Hoitink and Fahy, 1986; Gullino, 1992). The organic substrate needed for the activity of introduced antagonists could be deliberately introduced with the antagonist (Lumsden and Locke, 1989) or used to amend the growing medium such as has been done with composted bark (Hoitink, 1980). It should be mentioned, however, that there could be a risk of stimulating pathogens by amending with organic substrates, as indicated by Harman *et al.* (1981).

Micro-organisms have different niches to which they are adapted. Antagonists, which possess rhizosphere competence are of considerable interest (Ahmad and Baker, 1987; Nemec *et al.*, 1996) because they are adapted to live in the rhizosphere and thereby able to interact with the pathogens at the infection sites on the root surface. Likewise, non-pathogenic strains of plant pathogenic fungi, which are adapted to the same ecological niches as the pathogenic strains, are of special interest. Examples of these are non-pathogenic *F. oxysporum* (Eparvier and Alabouvette, 1994; Larkin *et al.*, 1996), non-pathogenic *Pythium* spp. (Paulitz and Baker, 1987) and the binucleate *Rhizoctonia* spp. (Herr, 1995). Recently it has also been demonstrated that endophytic micro-organisms can be potential biocontrol organisms (Benhamou *et al.*, 1996). Antagonists which are adapted to colonize sclerotia or other resting propagules of pathogens also have had some attention in relation to biocontrol in greenhouse crops (Budge *et al.*, 1995). This type of antagonist might be of special importance for controlling diseases in soil-grown plants in low technology houses.

The Use of Genetically Marked Antagonists in Autecology Studies
A fungus, *T. harzianum* T3, which was isolated by Wolffhechel (1989) from a *Pythium*

suppressive peat lot, has been used in several experiments with biological control and for studying the ecology of the antagonist following its introduction in sphagnum peat. This isolate has, in pot experiments, been shown to protect cucumber roots from primary infections by germinating oospores (Green and Jensen, submitted). The secondary spread from plant to plant by growing mycelium of *P. ultimum* has also been significantly reduced by incorporation of *T. harzianum* T3 in the growing medium (Green and Jensen, submitted). Autecology studies of an introduced antagonist are very difficult to carry out in the rhizosphere where a complex and active microflora and fauna is present. The introduced antagonist is difficult to distinguish from other related organisms and methods for specific measurements of the activity of the antagonist in the complex environment found in the rhizosphere are few. One approach to overcome these problems is to use genetically marked strains of the antagonist. *Trichoderma harzianum* T3, has been transformed with the GUS gene (Thrane *et al.*, 1995). GUS activity in the transformant, *T. harzianum* T3a, correlates with the general activity of the antagonist. Results have shown that if *T. harzianum* T3a is incorporated in pasteurized sphagnum peat it will be active for only a few days before conidia or chlamydospores are formed (Green and Jensen, 1995). Even in the rhizosphere on the surface of healthy roots, *T. harzianum* T3a remained as inactive conidia without any GUS activity (Green *et al.*, submitted). The antagonist, however, was very active in and around wounds on the roots (Green and Jensen, 1995). These results may also apply to other *Trichoderma* strains. Strains of *Trichoderma* which possess rhizosphere competence might, however, express higher activity even around healthy roots. The activity the first days after application of *T. harzianum* T3 is probably due to accompanying organic substrate delivered with the antagonists or nutrient released by pasteurization of the growing media. It has also been demonstrated, that *T. harzianum* T3a is actively colonizing dead or dying roots infected by *P. ultimum* (Green *et al.*, submitted). This is one way that the secondary spread of *Pythium* by hyphal growth from diseased plants is significantly restricted (Green and Jensen, submitted). *Trichoderma harzianum* strain T3a may also be able to restrict the formation of zoosporangia when it is colonizing the infected root tissue. Thus, secondary spread of *Pythium* by zoospores may also in this way be restricted by using antagonistic *Trichoderma* spp. *Trichoderma* conidia and chlamydospores can survive in the greenhouse in soilless growing media such as peat for several months (Heiberg, Green and Jensen, unpublished). Whether the propagules of *Trichoderma* can be reactivated by cultural practices such as amending organic substrate into the soilless system or if they can be activated due to interactions with the pathogen in the bulk growing medium should be studied more closely. *Trichoderma* spp. are fast growing organisms with a high competitive saprophytic ability in colonizing organic substrates in soil compared to weaker competitors such as *Pythium* spp. Thus, the antagonistic activity of *Trichoderma* spp. in the bulk growing medium might also be of importance in controlling *Pythium* spp. in soilless systems.

23.3.3. MECHANISMS OF ANTAGONISTIC ACTIVITY

Several antagonistic mechanisms could be involved in biological control of soilborne diseases, including competition, antibiosis, mycoparasitism and induced resistance.

Competition

Nutrients can be a limiting factor in the rhizosphere and growing medium, resulting in starvation and death of micro-organisms. Therefore, competition for nutrients such as nitrogen, carbon and iron can be a mechanism in biological control of soilborne pathogens. Competition for sites, for example on the root surface, may also be important in biocontrol.

Lifshitz *et al.* (1986) found that antibiosis rather than competition for nutrients was the important mechanism in the control of Pythium damping-off in pea by *T. harzianum*, because there still was a biocontrol effect following amendment to seeds with excess nitrogen and carbon. The results with the GUS marked *T. harzianum* (strain T3a) described above (Green *et al.*, submitted), indicate that this strain has difficulty competing for nutrients at the surface of healthy roots. However, competition for nutrients in wounds, or dead or dying root tissue seems to be involved in the control of *P. ultimum*. Other mechanisms could also be involved. It could, in fact, be a prerequisite that the antagonist is able to compete for and gain access to substrates before antibiosis and mycoparasitism can be effected. It could, thus, be a concerted action of more than one mechanism, resulting in the control of Pythium root rot. In the example from Lifshitz *et al.* (1986) with pea, a combined effect of competition for nutrients and antibiosis might have been involved in the biocontrol. Induced rhizosphere competent strains of *T. harzianum* is an example of an antagonist having improved biocontrol effect, probably because it was able to compete for cellulose in the mucilage layer at the root surface (Ahmad and Baker, 1987).

Zhou and Paulitz (1993) investigated whether bacteria could interfere with pre-infection events taking place when zoospores were released from sporangia of *Pythium*, but before root infection occurred. Following screening of more than 600 bacteria isolates, isolates of *Pseudomonas* spp. were found which reduced root rot in greenhouse experiments, leading to yield increases in cucumber grown on rockwool (Rankin and Paulitz, 1994; Paulitz, 1997). The isolates inhibited the chemotaxis response of the zoospores towards root exudates and cyst germination *in vitro*. The distribution of sporecysts of *Pythium* was different on roots treated with the bacteria compared to non-treated roots. Therefore, it was assumed that reduced chemotaxis could be due to competition for root exudates by the bacteria, and for micro-sites on the root. It would be of interest if these mechanisms were studied further to reveal more details about the interference with pre-infection events at the cellular level.

Antibiosis

Fungi belonging to the genera *Trichoderma* and *Gliocladium* have the potential for producing many different secondary metabolites (Ghisalberti and Sivasithamparam, 1991). Some of these compounds function as antibiotics inhibiting other microbes in *in vitro* tests. The action of antibiotics *in vivo* in a soilless system has been described (Harris and Lumsden, 1997). The production of secondary metabolites are to a high degree substrate dependent. Thus, it is difficult to predict their actual importance in biocontrol within *in vivo* systems based on experience from *in vitro* experiments. From experiments with antagonistic bacteria in soil, clear evidence is also accumulating that antibiosis is an important mechanism in biological control. Mutants lacking the ability to produce an antibiotic have been constructed and methods for detecting the metabolite in soil and rhizospheres have been developed. Mutants lost part of the biocontrol ability but this ability could be regained by restoring the ability to produce the antibiotic by complementation (Cook *et al.*, 1995). Results with fungal antagonists are less conclusive as it is more difficult to construct antibiotic minus mutants of fungi. However, good indications for the involvement of antibiosis in the interaction between *Trichoderma* and soilborne pathogens including *Pythium* spp. are coming from the work with *G. virens*. The fungus is able to produce a range of secondary metabolites of which gliovirin (Howell and Stipanovic, 1983) and gliotoxin (Lumsden *et al.*, 1992) are believed to play important roles in biological control of *Pythium* spp. Lumsden *et al.* (1992) incorporated *G. virens* in different soilless media as an alginate prill formulation containing wheat bran. They found by chloroform extraction that gliotoxin was produced in soilless medium in connection with mycelium which had grown up to 4 cm from the food base (the alginate prill). The disease suppression correlated with the biomass of *G. virens* or the amount of prill mixed in the medium. In this case, gliotoxin was the primary antibiotic active against *Pythium* spp. at sites where the pathogen was in close contact with *G. virens* and sufficient organic substrate was available for gliotoxin production. In further studies, using gliotoxin minus mutants generated by UV radiation, about 50% of the disease suppressive effect was lost compared to the wild type of *G. virens* (Wilhite *et al.*, 1994). The results do not, however, exclude the involvement of other secondary metabolites from *G. virens* in biological control of *Pythium* spp. Howell and Stipanovic (1983) for example, reported a similar study showing that gliovirin was important for the control of *Pythium* spp.

Mycoparasitism and Cell Wall Degrading Enzymes
Mycoparasitism of plant pathogens has been thoroughly reviewed by Lumsden (1992). Examples of mycoparasitism are known from all groups of fungi and it appears to be of widespread occurrence in natural systems. Cell wall degrading enzymes are considered to be important in mycoparasitic interactions (Sivan and Chet, 1989). Here we will only briefly consider mycoparasitism as a possible mechanism in the interaction between *Trichoderma* spp. or *Gliocladium* spp. and pathogenic *Pythium* spp. in soilless growing media. Evidence for mycoparasitism between *Trichoderma* spp. and *Pythium* spp. is derived from *in vitro* studies (Chet *et al.*, 1981). The break down of glucans in the cell wall of *Pythium* in mycoparasitic interactions between *Pythium* spp. and *T. harzianum* has also been shown *in vitro*, indicating the involvement of hydrolytic enzymes produced by *Trichoderma* in mycoparasitism (Benhamou and Chet, 1997). Inhibition of germination and germ-tube elongation of *Pythium* zoospore cysts by ß-1,3-glucanase has been demonstrated *in vitro* and evidence for the production of this enzyme by *T. harzianum* in the interaction with *P. ultimum* in sphagnum peat has been obtained (Thrane *et al.*, 1997). From experiments with pathogens with chitin in the cell walls, evidence has been obtained indicating that a synergistic effect between chitinase and the secondary metabolite gliotoxin might be important in the control of this type of pathogens by *G. virens* (Di Pietro *et al.*, 1993). Whether a similar synergistic effect is important between ß-1,3-glucanase and secondary metabolites from *Trichoderma* in the control of *Pythium* spp. is not known. Mycoparasitism, however, of *P. ultimum* in soil and rhizosphere by *T. harzianum* has not been demonstrated. In a time course study Lifshitz *et al.* (1986) showed that it is unlikely that *Trichoderma* spp. will have the time to establish a mycoparasitic relationship with *P. ultimum* before the primary infection of the plants takes place if germination of oospores

or sporangia is triggered by plant seed exudates. It might, however, be expected that mycoparasitism is an important mechanism in the interaction between *Pythium* and *Trichoderma* in infected root tissue and in competition for decaying roots (Green *et al.*, submitted). Thus, mycoparasitism might be important in reducing the secondary spread of the pathogen by *Trichoderma* from infected plant roots. This is now under further investigation. An attempt to demonstrate mycoparasitism between *G. virens* and *P. ultimum* in soilless media was not successful. Colonization of mycelia already presumably killed by secondary metabolites was observed (Harris and Lumsden, 1997).

Induced Resistance

Recently, data was obtained indicating that *T. harzianum* strain T39 (the antagonist in the BCA Trichodex™) can induce systemic resistance against *B. cinerea* in bean (De Meyer *et al.*, 1998). Less disease symptoms were observed on the leaves when *T. harzianum* was incorporated in the soil. To our knowledge this is the first time it is repported that *Trichoderma* can induce disease resistance in plants.

Zhou and Paulitz (1994) found evidence for the involvement of induced resistance in their experiments with antagonistic bacteria against Pythium root rot in soilless systems. New evidence has also been obtained indicating that mycoparasitic *Pythium* spp. are able to induce resistance responses in plants (Benhamou *et al.*, 1997). Thus, this mechanism might also be interesting for further studies.

Combining Antagonists

Alabouvette *et al.* (1993) clearly demonstrated that a synergistic effect can be obtained in controlling *Fusarium oxysporum* Schlechtend.:Fr. f. sp. *radicis-lycopersici* W.R. Jarvis & Shoemaker by combining a fluorescent *Pseudomonas* sp. with the non-pathogenic *F. oxysporum* (Fo47). The strategy of Hoitink *et al.* (1991) is to incorporate several antagonists in combinations in peat substrates and in this way render them disease suppressive. Whether a synergistic effect can be obtained in the control of *Pythium* by combining different antagonists is not clarified. However, an additive effect might be expected if the organisms are interfering with different events in the disease cycle.

23.4. Selection, Production, Formulation and Delivery Systems

The antagonistic micro-organisms contained in the BCAs used in protected crops often originate from quite different ecological niches than those found in greenhouse production. In spite of this, antagonists have been selected which are very efficient as BCAs especially in soilless systems. The use of nonpathogenic *Fusarium* spp. (FusaClean™), *Streptomyces* spp. (Mycostop™), *T. harzianum* (RootShield™), *Pseudomonas cepacia* (ex *Burkholderia*) Palleroni & Holmes type Wisconsin (DENY™) and *G. virens* (SoilGard™) are examples (Chapter 26). Before a new BCA can be commercialized it usually must be registered. Thus, a part of a screening procedure should consider whether the selected organisms would blatently fail toxicology tests and risk assessment required by the authorities. Screening systems for selecting antagonists for use in protected crops have been discussed elsewhere (Lumsden and Lewis, 1989; Jensen, 1996). Improvement of selected antagonists using DNA technology is the subject of Chapter 25. Methods used for production, formulation, testing efficacy and delivery are very important in the development of commercial BCAs. Those subjects have also been discussed elsewhere (Harman and Lumsden, 1990; Taylor and Harman, 1990; Lumsden *et al.*, 1993; Mintz and Walter, 1993; Lumsden *et al.*, 1995; Lumsden *et al.*, 1996) and will also be discussed in Chapter 26.

Many methods of delivery can be visualized. Newly prepared soilless media generally have a low microbial buffering capacity and thus, are disease conducive (Paulitz, 1997). Therefore amendment of these media with antagonists before seeding crops is a proven successful method of application (Lumsden *et al.*, 1996). Treatment of seeds and cuttings and delivery with nutrient solutions are also possible means of delivery. The antagonists can also be incorporated successfully in growth media before delivering to growers. Sphagnum peat, either raw or pasteurized, which has been treated with *Trichoderma* spp. by the manufacturer, is now for sale in Denmark (Pindstrup Mosebrug A/S, Denmark). Antagonists can be applied at different time intervals later in the growing period more readily than with field grown plants. This may be done either with transplants, as top spray, drenches or with the nutrient solution in closed systems before it is recirculated to the plants. Preferred delivery systems depend on the BCA and the plant cultivation system and decisions must be on a case to case basis. It is, for example, recommended that the BCA SoilGard is incorporated in the growing medium 1–3 days before planting in order to protect the plants against damping-off and root rot (Walter and Lumsden, 1997). The recommended timing and the need for an external food base (Lumsden and Locke, 1989) applied with the antagonist is based on knowledge about the ecology of *G. virens* and that antibiosis is an important mechanism involved in the biocontrol by this organism. In this way, the antagonists will be active against the pathogens in the growing medium from the time of planting and it does not depend on an organic substrate originating from the plants. RootShield is based on the strain *T. harzianum* T-22 (Harman and Björkman, 1998). For use in greenhouses it is formulated in two ways. A granular product (i.e. the fungus colonized on clay particles) can be incorporated in the growing medium as with SoilGard, and a product can be suspended in water and used as a drench. The latter consist mainly of conidia. Both methods of application are based on the fact that T-22 is a rhizosphere competent strain, and is expected to colonize the roots and be active against the pathogens in the rhizosphere. DENY is based on an aggressive root colonizing bacterium, effective against *Pythium* spp. The organism is formulated either in peat for seed or hopper box treatment or as a liquid formulation for treating transplants and for application with irrigation water. For all three BCAs and also for the use of suppressive compost improved with antagonists, mentioned above, the delivery systems have been developed based on a comprehensive knowledge of the ecology of the antagonists and the mechanisms of antagonism active in soilless growing systems. There is, however, a general lack of information about the ecology of most BCAs in soil and soilless media.

23.5. Implementation of Biological Disease Control in IPM Strategies

Integrated Pest Management as it relates to plant pathology has recently been reviewed (Jacobsen, 1997). The term, IPM, was first introduced in relation to insect pest control in the sense of integrating the use of pesticides and biological control organisms in agriculture. In plant pathology, attempts to reduce the use of chemicals for disease control relies on several approaches such as banning the use of certain chemicals, use of certified healthy plant material, sanitation, host resistance, agricultural cultivation practices and biological control. Thus, the original definition of IPM was too narrow to be useful in plant pathology. A definition of IPM which includes these other approaches to disease control, and which defines ecologically based or biointensive IPM strategies, has proven more useful in relation to disease control (Jacobsen, 1997). With this definition, IPM aims at minimizing economic, health and environmental risks (Jacobsen, 1997).

In the following we will discuss how other strategies for disease control can be integrated with biological control measures, in order to reduce the use of pesticides and how they can be used for enhancing biological control in cultivation systems.

23.5.1. HEALTHY PROPAGATION MATERIAL

Several pathogens can be brought into the greenhouse with propagation material. Quarantine programmes restricting the spread of pathogens with propagation material between different regions have been legislated in many countries. In this way, some pathogens can be excluded. In some countries, certification schemes for the production of healthy plant material for the greenhouse industry have been established and rules for the production of healthy propagation material is determined by law (i.e. in the EU). Thus, these measures are an attempt to reduce initial inoculum by eradication or exclusion. Mass production of certified plant material for flower production is often based on *in vitro* tissue culture propagation. BCAs might be used successfully when such plants are transferred from sterile conditions and are established in soilless media at a time when plants are considered to be vulnerable to attack by soilborne pathogens. A new approach under consideration is to inoculate with endophytic micro-organisms (J. van Wurde, pers. comm.). In spite of all the attempts to produce healthy propagation material, it is not always free of pathogens. Seeds, for example, with a certain percentage of infection with seedborne pathogens sometimes contaminate greenhouse operations. Coating seeds with BCAs has been shown to control seedborne pathogens (Jensen *et al.*, 1996). This method can also be effective in the control of damping-off caused by soilborne pathogens (Taylor and Harman, 1990). Combining seed priming and seed treatment with BCAs is a special integrated strategy where avoidance or disease escape is combined with biological control. This has been successfully demonstrated in the control of *Pythium* spp. with isolates of *T. harzianum* (Harman and Taylor, 1988; Harman *et al.*, 1989).

23.5.2. SANITATION

Soil disinfestation is commonly used for sanitation in low technology houses as reviewed by Garibaldi and Gullino (1995), and fumigation with chemicals, heating, steaming and solarization are the main methods in use. These methods have also been used for disinfestation of soilless growing media (Garibaldi and Gullino, 1995). Although the targets are soilborne pathogens and the methods aim at reducing the initial inoculum, soil disinfestation can also have deleterious effects on beneficial soil micro-organisms. This results in a reduction of the microbial buffering capacity and thus, leaves the soil or soilless medium as a microbial vacuum, ready for invasion by nearby micro-organisms, including plant pathogens. The sources for recolonization of the soil after disinfestation can be airborne micro-organisms, organisms from the untreated subsoil, organisms introduced with the planting material or organisms resistant to the disinfestation treatment. Some of the indigenous micro-organisms can, thus, be less sensitive to heat than the pathogens, leading to a recolonization by beneficial antagonists such as *Aspergillus terreus* Thom in Thom & Church and *Talaromyces flavus* (Klöcker) A.C. Stolk & R.A. Samson (Tjamos, 1992) and *Bacillus* spp., *Pseudomonas* spp. and *Streptomyces* spp. (Antoniou *et al.*, 1995). *Penicillium* spp., *Trichoderma* spp. and *Gliocladium* spp. are often less sensitive to fumigants and other chemicals used in disinfestation, resulting in recolonization of the soil by these organisms (Garibaldi and Gullino, 1995). Disinfestation of recirculated water in a soilless system either with a disinfectant or a fungicide has proven effective (Loschenkohl, 1994). Following such disinfestation, *Trichoderma* became the dominant fungus and disease caused by *Chalara elegans* Nag Raj & Kendrick (synanamorph of *T. basicola*) was suppressed. Disinfestation with chemicals or heat treatments might lead to increased disease suppressiveness (i.e. biological control) due to a change in the microflora both in soil and soilless systems. Resting propagules of pathogens such as microsclerotia can also be resistant to disinfestation, although it has been hypothesized that these propagules can be predisposed to infection by, for example, *Gliocladium* spp. as a result of the treatment (Tjamos, 1992). There is, however, a high risk following disinfestation that the soil and soilless media will be recolonized with pathogens instead, leading to more serious disease problems than before treatment (Jarvis, 1989). Incorporation of antagonists following disinfestation of the soil may have varying results (Tamietti and Garibaldi, 1989; Minuto *et al.*, 1995b). Problems may be due to uneven distribution of the antagonist in the medium, or the introduced organisms do not compete well with other organisms in the recolonization of the soil. Different methods for controlling the spread of pathogens with nutrient solutions in closed systems have been tested, such as slow sand filtration, biofiltration, UV exposure, or heat treatment of the nutrient solution. Many of these methods are now in use in commercial greenhouses (Runia, 1995). In most cases, the pathogen propagules are destroyed (eradicated) before the nutrient solution is recirculated. BCAs can spread readily through the nutrient solution, but might also be affected by these treatments, although this should be of less importance if the right delivery system is used. Addition of surfactants to the nutrient solution controls the spread of motile zoospores, but seems to have no effect on propagules protected by a cell wall (Stanghellini *et al.*, 1996). Thus, these approaches can be combined with most biocontrol strategies for controlling soilborne diseases.

23.5.3. CULTIVATION PRACTICES

Manipulation of greenhouse environmental conditions can be used for controlling diseases caused by soilborne pathogens. The control of *Phytophthora cryptogea* Pethybr. & Lafferty in tomato (Kennedy *et al.*, 1993) and *Pythium* spp. in cucumber (Paternotte, 1992) was successfully implemented by changing the temperature or changing the frequency in watering. In low technology houses and in houses with organic production systems (without any agrochemical input), crop rotation and organic amendments are common practices. Disease control based on such practices is probably, in most cases, a result of a stimulation of antagonistic micro-organisms in the soil and, thus, a biocontrol measure. Crop rotation with suitable non-host plants may also be a method for the control of some pathogens.

23.5.4. RESISTANCE

Host plant resistance to plant diseases is in itself considered to be a biological control measure (Cook and Baker, 1983). Thus, resistant cultivars can be used against specialized pathogens such as *F. oxysporum* in several crops, for example, in tomatoes (Jones *et al.*, 1993). Strong resistance to pathogens having a broad host range, such as *Pythium* spp., is unusual, although some differences in susceptibility between different cultivars exist (Paternotte and de Kreij, 1993). Susceptible or moderately resistant cultivars can be combined with other suitable characteristics (Zinnen, 1988). For example, carnation cultivars with differences in susceptibility to *Fusarium oxysporum* Schlechtend.:Fr. f. sp. *dianthi* (Prill. & Delacr.) W.C. Snyder & H.N. Hans. (van Peer *et al.*, 1989) showed the best biocontrol effect by treating the least susceptible cultivar with antagonistic pseudomonads. In addition, Bird (1982) showed that plants can be selected which will support an antagonistic microflora in the rhizosphere. Coupling biocontrol with resistant cultivars as a control strategy in greenhouses needs further study as does control by induced resistance elicited by antagonistic rhizosphere micro-organisms (Sticher *et al.*, 1997).

23.5.5. CHEMICAL CONTROL

Tolerance to fungicides, and incorporating fungicide resistance into antagonists, has made it possible for fungicides to be used in integrated control along with antagonists. There are several examples at the experimental stage showing a synergistic or an additive effect in the control of different diseases by combining low doses of fungicides with fungicide resistant antagonists (Locke *et al.*, 1985; Howell, 1991). Results from *in vitro* experiments indicate that hydrolytic enzymes produced by the antagonists may be important for the synergistic effect with fungicides (Lorito *et al.*, 1994; Thrane *et al.*, 1997). Seed treatments with BCAs which are insensitive to chemicals (i.e. pesticides) used in traditional seed technology are used in commercial delivery systems based on, for example, antagonistic *Trichoderma* spp. (Harman and Björkman, 1998). Also, in the use of RootShield in greenhouses, where severe disease pressure of *Pythium* spp. or other soilborne pathogens is known to exist, it is recommended to use a compatible fungicide together with the BCA in order to ensure effective disease control (Harman and Björkman, 1998). However, if fungicides with a broad spectrum are used, the antagonists might be inhibited, resulting in an insufficient biocontrol (De and Mukhopadhyay, 1994).

23.6. Conclusion

The use of BCAs has just recently been adapted on a commercial scale in greenhouse operations. There is a great interest from growers for using this new technology, both for reducing the use of chemical control and for obtaining better control of soilborne diseases. For being effective the BCAs must be applied at the right time and in the right way to the growing system. This requires comprehensive information about the ecology of pathogens and antagonists in the greenhouse, which in most cases must be much more detailed than that required for using more traditional chemical control measures. Some BCAs which are being commercialized are developed on the basis of such detailed information. For other BCAs more information is still needed before they can be used successfully in greenhouses. Given that the needed information is available, it must be passed on to extension service personnel and growers for BCAs to be used correctly and effectively. Other methods used for disease control are, in most cases, compatible with the use of biological control measures. If it was better understood how cultural practices and alternative control measures influence not only pathogens, but also antagonists, it might be possible to obtain even better biological control of soilborne diseases in greenhouses. An increased research effort in these directions can form the basis for developing biointensive IPM strategies, in which disease control includes an intensive use of biological control measures.

References

Ahmad, J.S. and Baker, R. (1987) Rhizosphere competence of *Trichoderma harzianum*, *Phytopathology* **77**, 182–189.

Alabouvette, C. (1987) Biological control of soil-borne diseases, especially Fusarium-wilts in protected crops, in R. Cavalloro (ed.), *Integrated and Biological Control in Protected Crops*, A.A. Balkema, Rotterdam, Brookfield, pp. 209–222.

Alabouvette, C., Couteaudier, Y. and Louvet, J. (1985) Soils suppressive to Fusarium wilt: Mechanisms and management of suppressiveness, in C.A. Parker, A.D. Rovira, K.J. Moore, P.T.W. Wong and J.F. Kollmorgen (eds.), *Ecology and Management of Soilborne Plant Pathogens*, The American Phytopathological Society, St Paul, Minn., pp. 101–106.

Alabouvette, C., Lemanceau, P. and Steinberg, C. (1993) Recent advances in the biological control of Fusarium wilts, *Pesticide Science* **37**, 365–373.

Antoniou, P.P., Tjamos, E.C., Andreou, M.T. and Panagopoulos, C.G. (1995) Effectiveness, modes of action and commercial application of soil solarization for control of *Clavibacter michiganensis* subsp. *michiganensis* of tomatoes, *Acta Horticulturae* **382**, 119–128.

Baker, R. (1992) Biological control of diseases of crops grown in covered and environmentally controlled structures, in E.C. Tjamos, G.C. Papavizas and R.J. Cook (eds.), *Biological Control of Plant Diseases: Progress and Challenges for the Future*, Plenum Press, New York, pp. 231–241.

Bates, M.L. and Stanghellini, M.E. (1984) Root rot of hydroponically grown spinach caused by *Pythium aphanidermatum* and *P. dissotocum*, *Plant Disease* **68**, 989–991.

Benhamou, N. and Chet, I. (1997) Cellular and molecular mechanisms involved in the interaction between *Trichoderma harzianum* and *Pythium ultimum*, *Applied and Environmental Microbiology* **63**, 2095–2099.

Benhamou, N., Kloepper, J.W., Quadt-Hallman, A. and Tuzun, S. (1996) Induction of defense-related ultrastructural modifications in pea root tissues inoculated with endophytic bacteria, *Plant Physiology* **112**, 919–929.

Benhamou, N., Rey, P., Chérif, M., Hockenhull, J. and Tirilly, Y. (1997) Treatment with the mycoparasite *Pythium oligandrum* triggers induction of defence-related reactions in tomato roots when challenged with *Fusarium oxysporum* f.sp. *radicis-lycopersici*, *Phytopathology* **87**, 108–122.

Besri, M. (1997) Integrated management of soil-borne diseases in the Mediterranean protected vegetable cultivation, *IOBC/WPRS Bulletin* **20**, 45–57.

Bird, L.S. (1982) The MAR (Multi-Adversity Resistance) system genetic improvement of cotton, *Plant Disease* **66**, 172–176.

Bochow, H. (1989) Use of microbial antagonists to control soil borne pathogens in greenhouse crops, *Acta Horticulturae* **255**, 271–280.

Bochow, H., Dolej, S., Fischer, I. and Alemayehu, M. (1996) Modes of action in biocontrolling fungal seedling and root rot by *Bacillus subtilis*, in C. Alabouvette (ed.), *Biological and Integrated Control of Root Diseases in Soilless Cultures, IOBC/WPRS Bulletin* **19**, pp. 99–108.

Boehm, M.J. and Hoitink, H.A.J. (1992) Sustenance of microbial activity in potting mixes and its impact on severity of Pythium root rot of Poinsettia, *Phytopathology* **82**, 259–264.

Budge, S.P., McQuilken, M.P., Finlon, J.S. and Whipps, J.M. (1995) Use of *Coniothyrium minitans* and *Gliocladium virens* for biological control of *Sclerotinia sclerotiorum* in glasshouse lettuce, *Biological Control* **5**, 513–522.

Chen, W., Hoitink, H.A.J. and Madden, L.V. (1988) Microbial activity and biomass in container media for predicting suppressiveness to damping-off caused by *Pythium ultimum*, *Phytopathology* **78**, 1447–1450.

Chet, I. and Baker, R. (1980) Induction of suppressiveness to *Rhizoctonia solani* in soil, *Phytopathology* **70**, 994–998.

Chet, I., Harman, G.E. and Baker, R. (1981) *Trichoderma hamatum*: Its hyphal interactions with *Rhizoctonia solani* and *Pythium* spp., *Microbial Ecology* **7**, 29–38.

Cook, R.J. and Baker, K.F. (1983) *The Nature and Practice of Biological Control of Plant Pathogens*, The American Phytopathological Society, St Paul, Minn.

Cook, R.J., Thomashow, L.S., Weller, D.M., Fujimoti, D., Mazzola, M., Bangera, G. and Kim, D. (1995) Molecular mechanisms of defense by rhizobacteria against root disease, *Proceedings of National Academic Sciences of the United States of America* **92**, 4197–4201.

Couteaudier, Y. and Alabouvette, C. (1989) Essais de lutte contre la pourriture racinaire de la tomate due à *Fusarium oxysporum* f.sp. *radicis lycopersici*, in R. Cavalloro and C. Pelerents (eds.), *Integrated Pest Management in Protected Vegetable Crops*, A.A. Balkema, Rotterdam, Brookfield, pp. 225–232.

De, R.K. and Mukhopadhyay, A.N. (1994) Biological control of tomato damping-off by *Gliocladium virens*, *J. of Biological Control* **8**, 34–40.

De Meyer, G., Bigirimana, J., Elad, Y. and Hofte, M. (1998) Induced systemic resistance in *Trichoderma harzianum* T39 biocontrol of *Botrytis cinerea*, *European J. of Plant Pathology* (in press).

Dekker, J. (1976) Acquired resistance to fungicides, *Annual Review of Phytopathology* **14**, 405–428.

Di Pietro, A., Lorito, M., Hayes, C.K., Broadway, R.M. and Harman, G.E. (1993) Endochitinase from *Gliocladium virens*: Isolation, characterization, and synergistic antifungal activity in combination with gliotoxin, *Phytopathology* **83**, 308–313.

Ebben, M.H. (1987) Observations on the role of biological control methods within integrated system, with reference to three constraining diseases of protected crops, in R. Cavalloro (ed.), *Integrated and Biological Control in Protected Crops*, A.A. Balkema, Rotterdam, Brookfield, pp. 197–208.

Elad, Y. (1996) Mechanisms involved in the biological control of *Botrytis cinerea* incited diseases, *European J. of Plant Pathology* **102**, 719–732.

Eparvier, A. and Alabouvette, C. (1994) Use of ELISA and GUS-transformed strains to study competition between pathogenic and non-pathogenic *Fusarium oxysporum* for root colonization, *Biocontrol Science and Technology* **4**, 35–47.

Fravel, D.R. (1988) Role of antibiosis in the biocontrol of plant diseases, *Annual Review of Phytopathology* **26**, 75–91.

Garibaldi, A. and Gullino, M.L. (1995) Focus on critical issues in soil and substrate disinfestation towards the year 2000, *Acta Horticulturae* **382**, 21–36.

Ghisalberti, E.L. and Sivasithamparam, K. (1991) Antifungal antibiotics produced by *Trichoderma* spp., *Soil Biology and Biochemistry* **23**, 1011–1020.

Grebus, M.E., Watson, M.E. and Hoitink, H.A.J. (1994) Biological, chemical and physical properties of composted yard trimmings as indicators of maturity and plant disease suppression, *Compost Science & Utilization* **2**, 57–71.

Green, H., Heiberg, N. and Jensen, D.F. (1998) The use of a GUS transformed strain of *Trichoderma harzianum* to study spermosphere and rhizosphere activity in relation to biocontrol of *Pythium* damping-off and root rot, *European J. of Plant Pathology* (submitted).

Green, H. and Jensen, D.F. (1995) A tool for monitoring *Trichoderma harzianum*: II. The use of a GUS transformant for ecological studies in the rhizosphere, *Phytopathology* **85**, 1436–1440.

Green, H. and Jensen, D.F. (1998) Disease progression and biocontrol of *Pythium ultimum* var. *ultimum*. The use of a rhizobox system to study the spread by active mycelial growth, *Phytopathology* (submitted)

Gullino, M.L. (1992) Integrated control of diseases in closed systems in the sub-tropics, *Pesticide Science* **36**, 335–340.

Gullino, M.L. and Garibaldi, A. (1994) Influence of soilless cultivation on soilborne diseases, *Acta Horticulturae* **361**, 341–354.

Harman, G.E. and Björkman, T. (1998) Potential and existing uses of *Trichoderma* and *Gliocladium* for plant disease control and plant growth enhancement, in G.E. Harman and C.P. Kubicek (eds.), *Trichoderma and Gliocladium: Enzymes, Biological Control, and other Commercial Applications*, Taylor and Francis, London, (in press).

Harman, G.E., Chet, I. and Baker, R. (1981) Factors affecting *Trichoderma hamatum* applied to seeds as a biocontrol agent, *Phytopathology* **71**, 569–572.

Harman, G.E. and Lumsden, R.D. (1990) Biological disease control, in J.M. Lynch (ed.), *The Rhizosphere*, John Wiley & Sons, Chichester, pp. 259–280.

Harman, G.E. and Taylor, A.G. (1988) Improved seedling performance by integration of biological control agents at favorable pH levels with solid matrix priming, *Phytopathology* **78**, 520–525.

Harman, G.E., Taylor, A.G. and Stasz, T.E. (1989) Combining effective strains of *Trichoderma harzianum* and solid matrix priming to improve biological seed treatments, *Plant Disease* **73**, 631–637.

Harris, A.R. and Lumsden, R.D. (1997) Interactions of *Gliocladium virens* with *Rhizoctonia solani* and *Pythium ultimum* in non-sterile potting medium, *Biocontrol Science and Technology* **7**, 37–47.

Herr, L.J. (1995) Biological control of *Rhizoctonia solani* by binucleate *Rhizoctonia* spp. and hypovirulent *R. solani* agents, *Crop Protection* **14**, 179–186.

Hoitink, H.A.J. (1980) Composted bark, a lightweight growth medium, *Plant Disease* **64**, 142–147.

Hoitink, H.A.J. and Fahy, P.C. (1986) Basis for the control of soilborne plant pathogens with compost, *Annual Review of Phytopathology* **24**, 93–114.

Hoitink, H.A.J. and Grebus, M.E. (1994) Nurseries find new value in composted products, *BioCycle* **35**, 51–52.

Hoitink, H.A.J., Inbar, Y. and Boehm, M.J. (1991) Status of compost-amended potting mixes naturally suppressive to soilborne diseases of floricultural crops, *Plant Disease* **75**, 869–873.

Howell, C.R. (1991) Biological control of Pythium damping-off of cotton with seed-coating preparations of *Gliocladium virens*, *Phytopathology* **81**, 738–741.

Howell, C.R. and Stipanovic, R.D. (1983) Gliovirin, a new antibiotic from *Gliocladium virens*, and its role in the control of *Pythium ultimum*, *Canadian J. of Microbiology* **29**, 321–324.

Jacobsen, B.J. (1997) Role of plant pathology in integrated pest management, *Annual Review of Phytopathology* **35**, 373–391.

Jarvis, W.R. (1989) Epidemiology of *Fusarium oxysporum* f.sp. *radicis-lycopersici*, in E.C. Tjamos and C. Beckman (eds.), *Vascular Wilt Diseases of Plants*, NATO ASI Series H28, Springer-Verlag, Berlin, Heidelberg, pp. 397–411.

Jensen, B., Knudsen, I.M.B., Jensen, D.F. and Hockenhull, J. (1996) Application of antagonistic microorganisms to seeds to control fungal pathogens, *Combined Proceedings of the International Plant Propagators' Society* **46**, 256–262.

Jensen, D.F. (1996) Screening for effective antagonists to control root diseases in soilless cultures, in C. Alabouvette (ed.), *Biological and Integrated Control of Root Diseases in Soilless Cultures, IOBC/WPRS Bulletin* **19**, pp. 27–35.

Jensen, D.F. and Wolffhechel, H. (1995) The use of fungi, particularly *Trichoderma* spp. and *Gliocladium* spp., to control root rot and damping-off diseases, in H.M.T. Hokkanen and J.M. Lynch (eds.), *Biological Control Benefits and Risks*, Cambridge University Press, Cambridge, pp. 177–189.

Jones, J.B., Stall, R.E. and Zitter, T.A. (1993) *Compendium of Tomato Diseases*, 2nd edn, APS Press, St Paul, Minn.

Kennedy, R., Pegg, G.F. and Welham, S.J. (1993) *Phytophthora cryptogea* root rot of tomato in rockwool nutrient culture. III. Effect of root zone temperature on growth and yield of winter-grown plants, *Annual Applied Biology* **123**, 563–578.

Knudsen, I.M.B., Hockenhull, J. and Jensen, D.F. (1995) Biocontrol of seedling diseases of barley and wheat caused by *Fusarium culmorum* and *Bipolaris sorokiniana*: Effects of selected fungal antagonists on growths and yield components, *Plant Pathology* **44**, 467–477.

Larkin, R.P., Hopkins, D.L. and Martin, F.N. (1996) Suppression of Fusarium wilt of watermelon by nonpathogenic *Fusarium oxysporum* and other microorganisms recovered from a disease-suppressive soil, *Phytopathology* **86**, 812–819.

Lewis, J.A., Lumsden, R.D., Millner, P.D. and Keinath, A.P. (1992) Suppression of damping-off of peas and cotton in the field with composted sewage sludge, *Crop Protection* **11**, 260–266.

Lewis, J.A. and Papavizas, G.C. (1992) Potential of *Laetisaria arvalis* for the biocontrol of *Rhizoctonia solani*, *Soil Biology and Biochemistry* **24**, 1075–1079.

Lifshitz, R., Windham, M.T. and Baker, R. (1986) Mechanism of biological control of preemergence damping-off of pea by seed treatment with *Trichoderma* spp., *Phytopathology* **76**, 720–725.

Locke, J.C., Marois, J.J. and Papavizas, G.C. (1985) Biological control of Fusarium wilt of greenhouse-grown chrysanthemums, *Plant Disease* **69**, 167–169.

Lockwood, J.L. (1988) Evolution of concepts associated with soilborne plant pathogens, *Annual Review of Phytopathology* **26**, 93–121.

Lorito, M., Peterbauer, C., Hayes, C.K. and Harman, G.E. (1994) Synergistic interaction between fungal cell wall degrading enzymes and different antifungal compounds enhances inhibition of spore germination, *Microbiology* **140**, 623–629.

Loschenkohl, B. (1994) Eftervirkning af kemisk bekaempelse i recirkulerende vandingsvand (After-effects of chemical control of fungi in recirculating watering systems), *SP Rapport* **7**, 347–353.

Lumsden, R.D. (1992) Mycoparasitism of soilborne plant pathogens, in G.C. Carroll and D.T. Wicklow (eds.), *The Fungal Community, Its Organization and Role in the Ecosystem*, Marcel Dekker, Inc., New York, pp. 275–293.

Lumsden, R.D., Garcia-E., R., Lewis, J.A. and Frías-T., G.A. (1987) Suppression of damping-off caused by *Pythium* spp. in soil from the indigenous Mexican Chinampa agricultural system, *Soil Biology and Biochemistry* **19**, 501–508.

Lumsden, R.D. and Lewis, J.A. (1989) Selection, production, formulation and commercial use of plant disease biocontrol fungi: Problems and progress, in J.M. Whipps and R.D. Lumsden (eds.), *Biotechnology of Fungi for Improving Plant Growth*, Cambridge University Press, Cambridge, pp. 171–190.

Lumsden, R.D., Lewis, J.A. and Fravel, D.R. (1995) Formulation and delivery of biocontrol agents for use against soilborne plant pathogens, in F.R. Hall and J.W. Barry (eds.), *Biorational Pest Control Agents: Formulation and Delivery*, American Chemical Society, Washington, DC, pp. 166–182.

Lumsden, R.D., Lewis, J.A. and Locke, J.C. (1993) Managing soilborne pathogens with fungal antagonists, in R.D. Lumsden and J.L. Vaughn (eds.), *Pest Management: Biologically Based Technologies*, American Chemical Society, Washington, DC, pp. 196–203.

Lumsden, R.D. and Locke, J.C. (1989) Biological control of damping-off caused by *Pythium ultimum* and *Rhizoctonia solani* with *Gliocladium virens* in soilless mix, *Phytopathology* **79**, 361–366.

Lumsden, R.D., Locke, J.C., Adkins, S.T., Walter, J.F. and Ridout, C.J. (1992) Isolation and localization of the antibiotic gliotoxin produced by *Gliocladium virens* from alginate prill in soil and soilless media, *Phytopathology* **82**, 230–235.

Lumsden, R.D., Walter, J.F. and Baker, C.F. (1996) Development of *Gliocladium virens* for damping-off disease control, *Canadian J. of Plant Pathology* **18**, 463–468.

Mintz, A.S. and Walter, J.F. (1993) A private industry approach: Development of GlioGard™ for disease control in horticulture, in R.D. Lumsden and J.L. Vaughn (eds.), *Pest Management: Biologically Based Technologies*, American Chemical Society, Washington, DC, pp. 398–403.

Minuto, A., Migheli, Q. and Garibaldi, A. (1995a) Evaluation of antagonistic strains of *Fusarium* spp. in the biological and integrated control of Fusarium wilt of cyclamen, *Crop Protection* **14**, 221–226.

Minuto, A., Migheli, Q. and Garibaldi, A. (1995b) Integrated control of soil-borne plant pathogens by solar heating and antagonistic microorganisms, *Acta Horticulturae* **382**, 173–182.

Nemec, S., Datnoff, L.E. and Strandberg, J. (1996) Efficacy of biocontrol agents in planting mixes to colonize plant roots and control root diseases of vegetables and citrus, *Crop Protection* **15**, 735–742.

Nieves-Brun, C. (1985) Infection of roots of *Dieffenbachia maculata* by the foliar blight and soft rot pathogen, *Erwinia chrysanthemi*, *Plant Pathology* **34**, 139–145.

Papavizas, G.C. (1985) *Trichoderma* and *Gliocladium*: Biology, ecology, and potential for biocontrol, *Annual Review of Phytopathology* **23**, 23–54.

Paternotte, S.J. (1992) Influence of growing conditions on disease development of *Pythium* in glasshouse cucumbers on rockwool, *Mededelingen – Faculteit Landbouwkundige en Toegepaste Biologische Wetenschappen, Universiteit Gent* **57/2b**, 373–379.

Paternotte, S.J. and de Kreij, C. (1993) The influence of humic substances and pH on *Pythium* spp., *Mededelingen – Faculteit Landbouwkundige en Toegepaste Biologische Wetenschappen, Universiteit Gent* **58/3b**, 1223–1227.

Paulitz, T.C. (1997) Biological control of root pathogens in soilless and hydroponic systems, *HortScience* **32**, 193–196.

Paulitz, T.C. and Baker, R. (1987) Biological control of *Pythium* damping-off of cucumbers with *Pythium nunn*: Population dynamics and disease suppression, *Phytopathology* **77**, 335–340.

Pleban, S., Ingel, F. and Chet, I. (1995) Control of *Rhizoctonia solani* and *Sclerotium rolfsii* in the greenhouse using endophytic *Bacillus* spp., *European J. of Plant Pathology* **101**, 665–672.

Postma, J. and Rattink, H. (1992) Biological control of Fusarium wilt of carnation with a nonpathogenic isolate of *Fusarium oxysporum*, *Canadian J. of Botany* **70**, 1199–1205.

Rankin, L. and Paulitz, T.C. (1994) Evaluation of rhizosphere bacteria for biological control of Pythium root rot of greenhouse cucumbers in hydroponic culture, *Plant Disease* **78**, 447–451.

Rattink, H. (1992) Targets for pathology research in protected crops, *Pesticide Science* **36**, 385–388.

Rattink, H. (1993) Biological control of Fusarium crown and root rot of tomato on a recirculation substrate system, *Mededelingen – Faculteit Landbouwkundige en Toegepaste Biologische Wetenschappen, Universiteit Gent* **58/3b**, 1329–1336.

Rattink, H. (1996) Root pathogens in modern cultural systems – Assessment of risks and suggestions for integrated control, in C. Alabouvette (ed.), *Biological and Integrated Control of Root Diseases in Soilless Cultures*, *IOBC/WPRS Bulletin* **19**, pp. 1–10.

Rehner, S.A. and Samuels, G.J. (1994) Taxonomy and phylogeny of *Gliocladium* analysed from nuclear large subunit ribosomal DNA sequences, *Mycological Research* **98**, 625–634.

Runia, W.T. (1995) A review of possibilities for disinfection of recirculation water from soilless cultures, *Acta Horticulturae* **382**, 221–229.

Schneider, R.W. (1982) *Suppressive Soils and Plant Disease*, APS Press, St Paul, Minn.

Schuerger, A.C. and Batzer, J.C. (1993) Identification and host range of an *Erwinia* pathogen causing stem rots on hydroponically grown plants, *Plant Disease* **77**, 472–477.

Sivan, A. and Chet, I. (1989) Degradation of fungal cell walls by lytic enzymes of *Trichoderma harzianum*, *J. of General Microbiology* **135**, 675–682.

Stanghellini, M.E. and Rasmussen, S.L. (1990) Thielaviopsis root rot of corn-salad, *Plant Disease* **74**, 81.

Stanghellini, M.E. and Rasmussen, S.L. (1994) Hydroponics. A solution for zoosporic pathogens, *Plant Disease* **78**, 1129–1138.

Stanghellini, M.E., Rasmussen, S.L., Kim, D.H. and Rorabaugh, P.A. (1996) Efficacy of nonionic surfactants in the control of zoospore spread of *Pythium aphanidermatum* in a recirculating hydroponic system, *Plant Disease* **80**, 422–428.

Sticher, L., Mauch-Mani, B. and Métraux, J.P. (1997) Systemic acquired resistance, *Annual Review of Phytopathology* **35**, 235–270.

Stutz, E.W., Défago, G. and Kern, H. (1986) Naturally occurring fluorescent pseudomonads involved in suppression of black root rot of tobacco, *Phytopathology* **76**, 181–185.

Tahvonen, R. (1982a) The suppressiveness of Finnish light coloured *Sphagnum* peat, *J. of the Scientific Agricultural Society of Finland* **54**, 345–356.

Tahvonen, R. (1982b) Preliminary experiments into the use of *Streptomyces* spp. isolated from peat in the biological control of soil and seedborne diseases in peat culture, *J. of the Scientific Agricultural Society of Finland* **54**, 357–369.

Tamietti, G. and Garibaldi, A. (1989) Effectiveness of soil solarisation against *Rhizoctonia solani* in northern Italy, in R. Cavalloro and C. Pelerents (eds.), *Integrated Pest Management in Protected Vegetable Crops*, A.A. Balkema, Rotterdam, Brookfield, pp. 193–197.

Taylor, A.G. and Harman, G.A. (1990) Concepts and technologies of selected seed treatments, *Annual Review of Phytopathology* **28**, 321–339.

Thinggaard, K. and Middelboe, A.L. (1989) *Phytophthora* and *Pythium* in pot plant cultures grown on ebb and flow bench with recirculating nutrient solution, *J. of Phytopathology* **125**, 343–352.

Thrane, C., Lübeck, M., Green, H., Degefu, Y., Allerup, S., Thrane, U. and Jensen, D.F. (1995) A tool for monitoring *Trichoderma harzianum*. I. Transformation with the GUS gene by protoplast technology, *Phytopathology* **85**, 1428–1435.

Thrane, C., Tronsmo, A. and Jensen, D.F. (1997) Endo-1,3-β-glucanase and cellulase from *Trichoderma harzianum*: Purification and partial characterization, induction of and biological activity against plant pathogenic *Pythium* spp., *European J. of Plant Pathology* **103**, 331–344.

Tjamos, E.C. (1992) Selective elimination of soilborne plant pathogens and enhancement of antagonists by steaming, sublethal fumigation and soil solarization, in E.C. Tjamos, G.C. Papavizas and R.J. Cook (eds.), *Biological Control of Plant Diseases: Progress and Challenges for the Future*, Plenum Press, New York, pp. 1–15.

Tronsmo, A. (1989) *Trichoderma harzianum* used for biological control of storage rot on carrots, *Norwegian J. of Agricultural Science* **3**, 157–161.

van Oosten, H.J. (1992) IPM in protected crops: Concerns, challenges and opportunities, *Pesticide Science* **36**, 365–371.

van Peer, R., Xu, T., Rattink, H. and Schippers, B. (1989) Biological control of carnation wilt caused by *Fusarium oxysporum* f.sp. *dianthi* in hydroponic systems, in *Proceedings of the Seventh International Congress on Soilless Culture*, Flevohof, The Netherlands, 1988, ISOSC, Wageningen, pp. 361–373.

van Steekelenburg, N.A.M. (1992) Novel approaches to integrated pest and disease control in glasshouse vegetables in The Netherlands, *Pesticide Science* **36**, 359–362.

Walter, J.F. and Lumsden, R.D. (1997) *Gliocladium* and biological control of damping-off complex, in D.A. Andow, D.W. Ragsdale and R.F. Nyvall (eds.), *Ecological Interactions and Biological Control*, Westview Press, A Division of Harper Collins Publishers, Oxford, pp. 254–267.

Weller, D.M. (1988) Biological control of soilborne pathogens in the rhizosphere with bacteria, *Annual Review of Phytopathology* **26**, 379–407.

Whipps, J.M. (1997) Developments in the biological control of soil-borne plant pathogens, *Advances in Botanical Research* **26** *Incorporating Advances in Plant Pathology*, 1–134.

Whipps, J.M. and Lumsden, R.D. (1991) Biological control of *Pythium* species, *Biocontrol Science and Technology* **1**, 75–90.

Wilhite, S.E., Lumsden, R.D. and Straney, D.C. (1994) Mutational analysis of gliotoxin production by the biocontrol fungus *Gliocladium virens* in relation to suppression of Pythium damping-off, *Phytopathology* **84**, 816–821.

Wolffhechel, H. (1988) The suppressiveness of *Sphagnum* peat to *Pythium* spp., *Acta Horticulturae* **221**, 217–222.

Wolffhechel, H. (1989) Fungal antagonists of *Pythium ultimum* isolated from a disease-suppressive sphagnum peat, *Växtskyddsnotiser* **53**, 7–11.

Wolffhechel, H. and Jensen, D.F. (1991) Influence of the water potential of peat on the ability of *Trichoderma harzianum* and *Gliocladium virens* to control *Pythium ultimum*, *Development in Agricultural and Managed-Forest Ecology* **23**, 392–397.

Zhou, T. and Paulitz, T.C. (1993) *In vitro* and *in vivo* effects of *Pseudomonas* spp. on *Pythium aphanidermatum*: Zoospore behavior in exudates and on the rhizoplane of bacteria-treated cucumber roots, *Phytopathology* **83**, 872–876.

Zhou, T. and Paulitz, T.C. (1994) Induced resistance in the biocontrol of *Pythium aphanidermatum* by *Pseudomonas* spp. on cucumber, *J. of Phytopathology* **142**, 51–63.

Zinnen, T.M. (1988) Assessment of plant diseases in hydroponic culture, *Plant Disease* **72**, 96–99.

BIOLOGICAL CONTROL OF DISEASES IN THE PHYLLOSPHERE
Yigal Elad, Richard R. Bélanger and Jürgen Köhl

24.1. Introduction

Restricted ventilation and low light intensity, especially on the lower parts of the plant, are common in the greenhouse environment. As a result, the greenhouse atmosphere is saturated with water for long periods each day, which in turn reduces plant growth. This environment favours the development of pathogens, and most of the vegetables and ornamental plants grown in greenhouses can be affected by a variety of diseases. However, biocontrol agents (BCAs) are also encouraged by greenhouse conditions, which can be manipulated to favour them. Currently several BCAs are being developed or have already been registered for use in greenhouses (Elad *et al.*, 1995a,b, 1997; Bélanger *et al.*, 1997). The effect of BCAs on greenhouse plant diseases, with particular emphasis on those already or soon to be put on the market, will be discussed in this chapter.

24.1.1. THE PATHOGENS

Very few examples of foliar diseases will be described here (see also Chapter 3). Powdery mildews are certainly among the most common greenhouse parasites. Their conidia are self-sufficient in water and nutrients and, although germination is favoured by low vapour pressure deficit (VPD), free water on the plant surface can reduce spore viability. Powdery mildew fungi are obligate parasites; they grow on the surface of the host and obtain nutrients through haustoria which penetrate the epidermal cells. The temperature requirements of the fungi overlap with the conditions generally prevailing in greenhouses, up to a maximum of about 30–35°C. Therefore, climate control in greenhouses is generally not effective against these diseases.

Diseases caused by *Botrytis cinerea* Pers.:Fr. (grey mould) and *Sclerotinia sclerotiorum* (Lib.) de Bary (white mould) on various crops, *Fulvia fulva* (Cooke) Cif. (= *Cladosporium fulvum* Cooke) (tomato leaf mould) and downy mildews [e.g. *Pseudoperonospora cubensis* (Berk. & M.A. Curtis) Rostovzev on cucurbits and *Peronospora sparsa* Berk. on roses] are very common in greenhouses under a variety of growing conditions. However, they all share conditions of low VPD for optimal development. *Botrytis cinerea* can infect the leaves, stems, flowers and fruits of most greenhouse vegetables and flower crops. During severe epidemics the entire foliage may be destroyed. *Fulvia fulva* only infects leaves, whereas *S. sclerotiorum* can infect all plant parts and cause plant death due to stem infection. Downy mildews may devastate the leaves of the plants. *Botrytis cinerea* can sporulate profusely on necrotic tissues such as senescent leaves and flower parts in the greenhouse crop. Conidia

R. Albajes et al. (eds.), Integrated Pest and Disease Management in Greenhouse Crops, 338-352.
© 1999 *Kluwer Academic Publishers. Printed in the Netherlands.*

produced on such substrates can be a major cause of disease outbreaks (Hausbeck *et al.*, 1996) and are very common in the greenhouse air. Low VPD, free moisture on plant surfaces and cool weather are considered the environmental factors which most influence infection by these pathogens. Optimum temperatures for infection are between 10 and 20°C, but infection could occur even at 2°C and above 25°C. Conidia of *B. cinerea* and ascospores of *S. sclerotiorum* require significantly more nutrients for germination and for subsequent germ tube growth on the host surface than the two other pathogens. Fewer nutrients reduce the infection rate (Blakeman, 1993).

Disease-resistant cultivars are not commonly used or widely developed for greenhouse crops. One can find resistant cultivars of melon against strains 1, 2 and 3 of powdery mildew {*Sphaerotheca fusca* (Fr.) Blumer. [= *Sphaerotheca fuliginea* (Schlechtend.:Fr.) Pollacci]}, and there are also tolerant cucumber cultivars of the long English variety available on the market (e.g. 'Flamingo' cv.). In addition there are gradients of powdery mildew resistance among rose cultivars. There are tomato cvs. resistant to leaf mould, but genetic resistance to *B. cinerea* has yet to be reported among commonly infected greenhouse crops. So far, only recourse to chemical fungicides before the logarithmic phase of the epidemic has offered satisfactory control of the above-mentioned diseases. As a result, pathogen strains which are resistant to several systemic fungicides have developed rapidly (see Chapters 3 and 11). One alternative way of controlling these diseases is intensive heating and ventilation of the greenhouse to prevent canopy wetness. This is often effective against infection of leaves, flowers and fruits by *B. cinerea*, but not against stem infections. However, heating in some countries is very energy-intensive and expensive, and may even increase deaths of plants with infected stems (A.J. Dik, pers. com.).

24.2. Biological Control

24.2.1. BIOCONTROL MECHANISMS

The use of BCAs to prevent infection is based on competition for nutrients and space, production of antibiotics, hyperparasitism and/or induced resistance in the host plant. There are numerous reports of attempts to prevent infection by means of antagonist micro-organisms (e.g. Andrews, 1992; Tronsmo, 1992). Species of leaf bacteria, yeasts and filamentous fungi can inhibit pathogens by competing for nutrients (Blakeman, 1993). Mycoparasitism of fungi by filamentous fungi has been recorded in many systems (Kranz, 1981). A few bacteria are also capable of direct parasitism (Scherff, 1973). Antibiosis is a feature of many bacteria and fungi (Andrews, 1985).

Antagonistic interactions can also be exploited during the saprophytical stage of necrotrophic pathogens such as *B. cinerea* (Köhl *et al.*, 1995b,c). Thus, biocontrol can be the result of multi-mechanism actions. For instance, *Trichoderma harzianum* Rifai T39 competes for nutrients and also interferes with the production of pathogenicity enzymes by the pathogen; thus, in addition to slowing the germination of the pathogen's conidia, T39 also prevents the penetration and maceration of the host tissue (Zimand *et al.*, 1996). As well as this, T39 induces resistance in host plants (de Meyer *et al.*, 1998). The antagonism of *Ulocladium atrum* G. Preuss to *Botrytis* spp. seems to be based on competition in necrotic tissues, since no effect of toxins or cell-wall degrading enzymes was found (Köhl *et al.*, 1997). Microbial suppression of sporulation on necrotic tissues reduces the spore load in the crop, which leads to a slower progression of disease epidemics, as shown for *Botrytis* spp. in field-grown onions (Fokkema, 1993; Köhl and Fokkema, 1993; Köhl *et al.*, 1995a).

Reducing spore dissemination, can be a successful biocontrol strategy against biotrophs. The behaviour of foliar pathogens during their pre-penetration phase on the healthy leaf and the relative importance of necrotic tissues within the crop as inoculum source determine the mechanism which may be effective in biocontrol. Biotrophic pathogens such as powdery mildews and rusts, which are independent of exogenous nutrients during germination and penetration, can still establish infection in a nutrient-depleted phyllosphere (Staples *et al.*, 1962). However, on the leaf surface the conidia or germ tubes of the biotrophs are exposed to antibiotics and lytic enzymes which are produced by micro-organisms (mainly bacteria such as *Bacillus* spp. and *Pseudomonas* spp.) and which can inhibit germination and lyse germ tubes (Doherty and Preece, 1978).

Unllike biotrophs, necrotrophic pathogens, such as the necrotrophs *B. cinerea* and *S. sclerotiorum*, use exogenous nutrients in many circumstances during germination of their spores and during superficial growth of the mycelium on the plant surface before penetration. The necrotrophs may have to compete with the phylloplane microflora for limited nutrient resources. Reduction in the concentration of nutrients generally results in a reduced rate of spore germination and in slower germ tube growth, thereby reducing the number of infection courts and the extent of subsequent necrosis incited by the pathogen (reviewed in Blakeman, 1993; Elad *et al.*, 1994a,b).

24.2.2. BIOCONTROL OF BIOTROPHS

Biotrophs such as powdery mildews are in principle easy targets for BCAs because of their ectotrophic growth. Since powdery mildew conidia are self-sufficient and have no saprophytic phase, one can reasonably assume that BCAs will rely on parasitism and/or antibiosis rather than competition to antagonize them. Incidentally, all micro-organisms reported as BCAs of powdery mildew fungi have been found to act by hyperparasitism or antibiosis (Elad *et al.*, 1995b; Bélanger *et al.*, 1997).

Hyperparasitism
The most common hyperparasite of powdery mildews is the coelomycete, *Ampelomyces quisqualis* Cesati:Schltdl. (Sztejnberg *et al.*, 1989). The species does not appear to be confined to one host since, for instance, an isolate from an *Oidium* sp. infecting *Catha edulis* (Vahl.) Forsk. in Israel, was able to antagonize several powdery mildew fungi belonging to the genera *Oidium*, *Erysiphe*, *Sphaerotheca*, *Podosphaera*, *Uncinula* and *Leveillula* (Sztejnberg *et al.*, 1989). Several authors reported biocontrol of *S. fusca* in cucumbers by *A. quisqualis* (e.g. Jarvis and Slingsby, 1977; Sundheim, 1982; Sztejnberg *et al.*, 1989). *Ampelomyces quisqualis* penetrates from cell to cell through the septal pores of the fungus and continues to grow during the gradual degeneration of

the infected cells (Hashioka and Nakai, 1980). In experiments, the hyperparasite *A. quisqualis* has been applied at regular intervals to cover new growth of the host plant and to prevent the rapid dissemination of powdery mildews. Recovery of the antagonist on non-sprayed control plots is indicative of the spread of airborne *A. quisqualis* inoculum (Falk *et al.*, 1995). One isolate of *A. quisqualis* developed in the Hebrew University of Jerusalem (Sztejnberg *et al.*, 1989) was developed into a commercial product named AQ10 (Ecogen, Jerusalem, Israel) and is currently being tested under greenhouse conditions for the control of powdery mildew on cucurbits (A. Sztejnberg and Y. Elad, pers. com.) and roses (Pasini *et al.*, 1997). *Verticillium lecanii* (A. Zimmerm.) Viégas is another hyperparasite of several pathogenic fungi including *S. fusca*. In greenhouse experiments it was reported to reduce the incidence of powdery mildew when certain conditions were respected, mainly low VPD (Spencer and Ebben, 1983; Verhaar *et al.*, 1996). Considered for a long time a strict hyperparasite, recent findings have suggested that early degradation of powdery mildew cells in interaction with *V. lecanii* was mediated by the production of antibiotic substances (Askary *et al.*, 1998). A few other fungi have been reported in the literature as parasites on powdery mildew fungi: for example, both *Acremonium alternatum* Link:Fr. and *Cladosporium cladosporioides* (Fresen.) G.A. De Vries antagonized and destroyed the thallus of *S. fusca* under conditions of low VPD (Malathrakis and Klironomou, 1992b).

Antibiosis

Recently, the antagonistic relationship between the epiphytic yeast-like fungi, *Sporothrix flocculosa* Traquair, Shaw & Jarvis and *Sporothrix rugulosa* Traquair, Shaw & Jarvis, and the powdery mildew pathogens of rose [*Sphaerotheca pannosa* (Wallr.:Fr.) Lév. var. *rosae* Woronichin] and cucumber (*S. fusca*) has been described (Jarvis *et al.*, 1989; Hajlaoui and Bélanger, 1991). *Sporothrix flocculosa* colonized powdery mildew colonies faster than *S. rugulosa* and it was reported that *S. flocculosa* was able to severely alter the mycelial growth and spore production of the pathogen in less than 48 h. Investigations into the mode of action of *S. flocculosa* revealed that the antagonist did not penetrate its host, but rather induced rapid plasmolysis in powdery mildew cells (Hajlaoui *et al.*, 1992; Hajlaoui *et al.*, 1994). These cell reactions are presumably the result of the production of fatty acids by the antagonist (Choudhury *et al.*, 1994; Benyagoub *et al.*, 1996a). The fatty acids appear to interfere with the integrity of the plasmalemma, the composition of which determines the specificity of the antagonist. Indeed, fungi with a low sterol content and a high level of unsaturated fatty acids on their membrane are extremely susceptible to *S. flocculosa* (Benyagoub *et al.*, 1996b). Bélanger and Deacon (1996) recently showed that there was a correlation between sensitivity to fatty acids and susceptibility to the antagonist. In efficacy trials, *S. flocculosa* was tested on a commercial scale against powdery mildew in roses and was found to be as effective as a fungicide, provided that the VPD was favourable to the antagonist (Bélanger *et al.*, 1994). In large-scale comparative trials on cucumber, *S. flocculosa* was found to be more efficient at controlling *S. fusca* than other BCAs of powdery mildew including *A. quisqualis*, *V. lecanii* and *Tilletiopsis washingtonensis* Nyland (Hajlaoui and Bélanger, 1991; Askary *et al.*, 1997; Dik *et al.*, 1998). The fungus is currently being targeted for marketing under the name Sporodex™.

Other examples of BCAs presumably acting by antibiosis against powdery mildews are *Tilletiopsis* spp. They are common phyllosphere yeasts belonging to the Sporobolomycetaceae. Hoch and Provvidenti (1979) found strong antagonism between a *Tilletiopsis* sp. isolate and the cucumber powdery mildew pathogen *S. fusca*. Inoculation with a *Tilletiopsis*-cell suspension on detached, mildew-infected cucumber leaves destroyed the superficial thallus of the powdery mildew. Like *S. flocculosa*, *Tilletiopsis* spp. do not seem to penetrate the powdery mildew fungus. Microscopic studies and the use of culture filtrates suggest that antibiosis is the main mode of action (Hijwegen, 1989). Hijwegen (1986) found practically no effect of the application of *Tilletiopsis minor* Nyland on small cucumber plants one day before inoculation with *S. fusca*. When applied twice after inoculation with powdery mildew, *T. minor* was very effective in controlling the disease on cucumber plants under controlled conditions. However, under greenhouse conditions, the effect was disappointing, probably because the VPD was too high (Hijwegen, 1992). In comparative experiments on small cucumber plants under controlled conditions, *Tilletiopsis albescens* Gokhale and *A. quisqualis* controlled *S. fusca* better than *T. minor* (Hijwegen, 1986). Urquhart *et al.* (1994) found that two other *Tilletiopsis* spp., *T. washingtonensis* and *Tilletiopsis pallescens* Gokhale, both reduced the density of powdery mildew spores on greenhouse-grown cucumbers when sprayed at a concentration of 1×10^8 cfu/ml.

24.2.3. BIOCONTROL OF NECROTROPHS

Infection by necrotrophic pathogens, including *B. cinerea*, can be reduced under controlled and field conditions by pre-inoculation of the phylloplane with epiphytic filamentous fungi, bacteria or yeasts (Blakeman, 1993). Newhook (1951, cited in Blakeman, 1993) and Wood (1951, cited in Blakeman, 1993) inoculated senescent lettuce leaves with antagonists such as *Fusarium* sp. and *Penicillium claviforme* Bainier isolated from the same crop in order to prevent primary establishment of *B. cinerea*. Later, *Cladosporium herbarum* (Pers.:Fr.) Link effectively controlled grey mould on strawberries by protecting the flowers under field conditions (Bhatt and Vaughan, 1962, cited in Blakeman, 1993). Similarly, several fungi were found effective against *S. sclerotiorum* on various crops (Boland and Inglis, 1988). Control of *B. cinerea* by *Trichoderma* spp. has been reported for several crops. *Trichoderma hamatum* (Bonord.) Bainier reduced pod infection by 94% in snap beans when applied to the blossom before or at the same time as the pathogen (Nelson and Powelson, 1988). *Exophiala jeanselmei* (Langeron) McGinnis & Padhye was also effective in controlling *B. cinerea* (Redmond *et al.*, 1987). *Bacillus brevis* Migula reduced disease on protected Chinese cabbage by 64–71% shortening the period of leaf wetness (Edwards and Seddon, 1992). Li and Leiffert (1994) found that an antibiotic-producing isolate of *Bacillus subtilis* (Ehrenberg) Cohn was capable of protecting *Astilbe* microplants as long as resistance to the antibiotic in the *B. cinerea* population did not develop.

Isolates of the yeasts *Rhodotorula glutinis* (Fresenius) Harrison and *Cryptococcus albidus* (K. Saito) C.E. Skinner, of the bacteria *Xanthomonas maltophilia* Swings *et al.*, *Bacillus pumilus* Meyer & Gottheil, *Lactobacillus* sp. and *Pseudomonas* sp., and of the filamentous fungus *Gliocladium catenulatum* Gilman & E. Abbott were found to control grey mould in bean and tomato plants (Elad *et al.*, 1994a,b). They reduced germination of conidia of *B. cinerea* and the severity of rot symptoms on detached leaves, and were able to control the disease on plants under controlled conditions. The BCAs competed with *B. cinerea* for nutrients and possibly induced host resistance to the fungus. Establishment of yeast populations on healthy and *Botrytis*-infected leaves and on bean and tomato flowers was successful. Most of the above-mentioned isolates and those of the *Penicillium* sp., *Apiospora montagnei* Sacc., *Arthrinium phaeospermum* (Corda) M.B. Ellis, *Sesquicillium candelabrum* (Bonord.) W. Gams, *Chaetomium globosum* Kunze:Fr., *Alternaria alternata* (Fr.:Fr.) Keissl., *U. atrum* and *Trichoderma viride* Pers.:Fr. reduced sporulation of the pathogen in previously established lesions and lesion expansion (Elad *et al.*, 1994a,b). Köhl *et al.* (1995a) screened saprophytic fungi for their ability to suppress sporulation of *Botrytis* spp. on necrotic leaf tissues. The most efficient antagonist, *U. atrum*, consistently reduced by over 90% sporulation of *B. cinerea* on dead lily leaves under field conditions in a series of experiments in various microclimatic conditions. In recent experiments, this antagonist controlled *Botrytis* spp. in open-field crops of strawberry and onion as efficiently as fungicides. The antagonist was also tested in greenhouse-grown ornamentals. Naturally senescing leaves within the dense canopy of cyclamen plants are the initial substrate for *B. cinerea*. From such necrotic leaves, *B. cinerea* attacks the neighbouring petioles and leaves. Conidia of *U. atrum* sprayed on green cyclamen survived on the leaves for at least ten weeks. The antagonist was able to colonize such leaves when they grew old and so protect them from colonization by *B. cinerea*. The antagonist was also tested in commercial greenhouses (Köhl *et al.*, 1998). The disease incidence and the disease severity expressed as number of diseased petioles per cyclamen plant was significantly reduced by five applications of *U. atrum* sprayed every two to three weeks as compared to an untreated control (Table 24.1). The biological control was as efficient as a chemical control consisting of five fungicide applications [dichlofluanid (2×), prochloraz-manganese (2×) and iprodione sprayed in alternation]. A second antagonist, *Gliocladium roseum* Bainier, applied with the same frequency, was less efficient.

TABLE 24.1. Effect of *U. atrum*, *G. roseum* and a fungicide programme on the incidence and severity of grey mould on cyclamen caused by *B. cinerea*

	Disease incidence (%)	Diseased petioles (No./plant)
Control	84 a	3.5 a
Fungicide programme	46 b	0.7 b
Ulocladium atrum	40 b	0.9 b
Gliocladium roseum	69 a	1.8 a

[a,b]Figures of the same column with a common letter do not differ statistically; P ≤ 0.05

Peng and Sutton (1991) tested various antagonists for their ability to control sporulation of *B. cinerea* on strawberry leaflets and found that isolates of *Trichoderma* and *Gliocladium* were most effective. *Gliocladium roseum* and *Myrothecium verrucaria* (Albertini & Schwein.) Ditmar:Fr. effectively suppressed sporulation of the pathogen on black spruce seedlings (Zhang and Sutton, 1994). A conidial preparation of *G. roseum* was transferred to strawberry flowers by bees and the transferred conidia of the BCA suppressed *B. cinerea* on the flowers and fruits (Peng *et al.*, 1992). An isolate of *Streptomyces griseovirides* Anderson, Ehrlich, Sun & Burkholder is the core of another preparation, Mycostop, that has been developed commercially (White *et al.*, 1990). It is aimed at lettuce grey mould and applied by drenching the plantlets with the suspension.

Intensive biocontrol work with *Trichoderma* spp. under commercial conditions has been carried out on greenhouse crops (Garibaldi and Corte, 1987; Gullino *et al.*, 1990). Isolate T39 of *T. harzianum* from Israel has effectively controlled *Botrytis* diseases in greenhouse crops in various countries (Elad *et al.*, 1993; O'Neill *et al.*, 1996a). A commercial preparation developed from T39 (Trichodex®, 20P, Makhteshim Ltd, Be'er Sheva, Israel), registered for agricultural use in Israel and other countries, is the first such product to be introduced commercially to greenhouses. An isolate of *Penicillium* sp. and Trichodex controlled grey mould of greenhouse tomato effectively; in this case, a tank mix with iprodione controlled no better than iprodione on its own (Malathrakis and Klironomou, 1992a). In other cases, the mixture of the commercial preparation with iprodione was superior to either of the components on its own (Elad *et al.*, 1994c). Trichodex mixed with Phyton 27 (a copper-based pesticide) was most effective (97% control) on greenhouse tomato (Bourbos and Skoudridakis, 1994). Studying the effect of *T. viride* on chocolate spot of broad beans, Bennett and Lane (1992) found that combining the biocontrol with sulphur did not improve the degree of control but ensured control in time when conditions in the field were unfavourable for the BCA. *Trichoderma harzianum* T39 established high populations (10^3 propagules/cm^2) on leaves and fruits of greenhouse cucumber plants treated with Trichodex (Elad and Kirshner, 1993; Elad *et al.*, 1993). The BCA also established itself on non-treated plants in the greenhouse: a drawback in the experiment. However, the secondary dispersal of the BCA can be advantageous in practice, in that it protects new plant parts (Elad *et al.*, 1993). Isolates of the yeasts *C. albidus* and *Aureobasidium pullulans* (de Bary) G. Arnaud and the fungus *T. harzianum* T39 were compared for their efficacy against *B. cinerea* in cucumber and tomato in large-scale glasshouse trials with different climate regimes. In cucumber, the most consistent control was achieved by *A. pullulans* and *T. harzianum*, which maintained sufficiently high population densities and controlled stem infections as well as, or significantly better than, the broad-spectrum fungicide tolylfluanid under all climate regimes tested (Dik and Elad, unpublished). In tomato, control was better when relative humidity was high during part of the night (Dik and Elad, unpublished).

The biocontrol of Fulvia and Sclerotinia diseases has been studied much less than grey mould, which has been researched in detail, as mentioned above. The reason for this may be the availability of effective chemicals and cultivars resistant to *F. fulva* in some countries. One BCA tested against *F. fulva* is the fungus *Dicyma pulvinata* (Berk. & M.A. Curtis) Arx [= *Hansfordia pulvinata* (Berk. & M.A. Curtis) S.J. Hughes] which is a mycoparasite of the plant pathogen and was found to secrete a compound (sesquiterpene) which inhibits the pathogen (Tirilly *et al.*, 1987). We found that *T. harzianum* T39 reduces the rate of leaf mould on greenhouse tomatoes by *c.* 50–60%, and of *S. sclerotiorum* on cucumber by *c.* 70–90% (Elad *et al.*, 1997).

24.3. Improved Control and Integration

BCAs may be combined to control more than one disease in the greenhouse, either with additives to improve their efficacy or survival, or with other means of control in order to achieve an acceptable level of disease suppression.

Since powdery mildew and grey mould may simultaneously occur on the same cucumber crop, Sztejnberg and Elad (unpublished) tested the control of both diseases by the respective BCAs, AQ10 (0.04% w/v, applied with 0.3% supplement ADDQ vegetable oil) and Trichodex (0.2% w/v). Each of the BCAs on its own significantly reduced the severity of powdery mildew. Mixing them did not improve control of this disease. Grey mould on the fruits was reduced by AQ10, Trichodex and by their combination: each treatment resulted in a better control than the previous one, respectively (Table 24.2). The effect on grey mould of AQ10 supplemented by oil may be attributed to the added oil.

TABLE 24.2. Control of cucumber powdery mildew (*S. fusca*) and of grey mould (*B. cinerea*) by Trichodex (*T. harzianum* T39), AQ10 (*A. quisqualis*) supplemented with 0.3% ADDQ and their combination (Elad and Sztejnberg, unpublished)

	Powdery mildew (% leaf coverage)		Grey mould (infected fruits/plant)	
	14 January	19 February	6 March	9 April
(i) Control	62 a	88 a	4.1 a	11.6 a
(ii) Trichodex	36 b	65 b	1.7 c	5.2 c
(iii) AQ10 + ADDQ	19 c	51 c	3.0 b	7.8 b
(iv) = (ii) + (iii)	19 c	67 bc	1.6 c	3.0 d

[a,b,c,d]Figures of the same column with a common letter do not differ significantly; $P \leq 0.05$

24.3.1. BIOTROPHS

Without exception, all known powdery mildew antagonists are dependent on conditions of low VPD for maximum efficiency. Some hyperparasites colonize powdery mildews only when there is free water on plant surfaces; others need VPD below 2.5 mbar for a fast and complete colonization. While some BCAs such as *S. flocculosa* appear to be more active under a wider range of humidity conditions than most powdery mildew antagonists (Hajlaoui and Bélanger, 1991; Bélanger *et al.*, 1997; Dik *et al.*, 1998), VPD remains the most critical factor affecting the colonization and destruction of powdery mildew structures on plant surfaces. In this context, the greenhouse environment is far more suitable for the application of hyperparasites against powdery mildew. The fact that, in most cases, the inconsistency noticed in the control of powdery mildews with hyperparasites in field experiments is due to differences in the VPD, indicates that BCAs should be integrated with other products or approaches to ensure optimal control of the disease.

To reduce the dependence of hyperparasites on low VPD, researchers have used several additives. Spencer and Ebben (1983) mixed spores of *V. lecanii* with 2% glycerol and 1% gelatin and so increased their longevity on cucumber leaves. Malathrakis and Klironomou (1992b) found that in greenhouse experiments, glycerol at 0.2% and *A. alternatum* at 10^6 or 10^7 spores/cm^2 were equally effective, either applied separately or as a mixture against *S. fusca*. However, Jarvis (1992) was unable to increase natural populations of *A. quisqualis* using sugars, peptone, gelatin or glycerol. In addition, several types of oils mixed with hyperparasites increased the effectiveness of the biocontrol agents at high VPD. Philipp *et al.* (1990), by using 1% paraffin oil, alleviated *A. quisqualis'* need for low VPD to control powdery mildew on cucumber. Hijwegen (1992) counteracted the loss of activity of *T. minor* at RH below 80% by using a formulation of paraffin oil (Hora Oleo 11E) and coffee cream mixed with a spore suspension of the hyperparasite. Both these additives were also effective in the absence of a BCA. Bélanger *et al.* (1994) improved the effectiveness of *S. flocculosa* against rose powdery mildew by using 1% paraffin oil mixed with the spore suspension of the antagonist and, finally, Dik *et al.* (1998) found that 0.3% oil improved the control of cucumber powdery mildew by the same antagonist.

Integration of biological control with genetic resistance will usually improve disease control: Verhaar *et al.* (1996) and Dik *et al.* (1998) obtained better results on tolerant than on susceptible cultivars when they used *V. lecanii* against cucumber powdery mildew. Control was further improved by the addition of silicon to the nutrient solution, which renders the plants less susceptible to powdery mildew (Dik *et al.*, 1998). Integration of fungicides with BCAs also offered the opportunity of reducing the amount of fungicide applied. In this case, knowledge of the compatibility between fungicides and biocontrol agents is essential. Data reported by several researchers indicate that some strains of *A. quisqualis* are compatible not only with fungicides but also with other pesticides used on crops on which the antagonist might be used (Sundheim, 1986). Hijwegen (1986) tested the compatibility of *T. minor* with various fungicides. Sztejnberg *et al.* (1989) studied the effects of *A. quisqualis* on its own, the fungicide pyrazophos on its own, and both treatments in alternation. Pyrazophos on its own and in alternation with *A. quisqualis* gave better results than *A. quisqualis* on its own. Finally, Benyagoub and Bélanger (1995) were able to isolate a strain of *S. flocculosa* resistant to dodemorph-acetate, fungicide used in the control of *S. pannosa* var. *rosae*.

24.3.2. NECROTROPHS

Integration of biological and chemical control into disease suppression in *B. cinerea* pathosystems has been investigated (Elad *et al.*, 1994c, 1995a). In most trials conducted, the BCAs were treated as biological fungicides, i.e. they were applied at predetermined intervals. These spraying schedules were arranged either on a weekly basis (in greenhouses) or according to the stage of development of the crop, as occurs in grapes (O'Neill *et al.*, 1996a). The biocontrol preparation, Trichodex, was tested on its own and compared with the standard chemical fungicides, and was tested in combinations of the BCA and chemical fungicides. The combinations consisted of either tank mix or weekly alternation. Tank mixing was effective, but is not desirable because it does not reduce the use of chemicals. Alternation of biocontrol treatment with the chemical treatments on a calendar basis was more effective than biocontrol alone on tomato, cucumber and strawberry crops. Furthermore, this treatment was more reliable than the other treatments (Elad *et al.*, 1994c). The variation in disease control was lower in the alternation treatments than in the treatments with either *T. harzianum* or the fungicides alone. Disease suppression achieved by the alternation treatment was sometimes insufficient if the BCA was applied in weather which was particularly favourable to grey mould (Shtienberg and Elad, 1997). The microclimate in the greenhouse can affect the efficacy of the BCA. For instance, *T. harzianum* T39 was found effective at temperatures above 15°C, whereas at low temperature it is not effective while the pathogen is still significantly harmful (Elad *et al.*, 1993; O'Neill *et al.*, 1996b). However, in heated glasshouses, alternation of Trichodex with iprodione did not improve control of *B. cinerea* in cucumber over weekly application with Trichodex. It did improve control compared to iprodione. In the heated glasshouses, no adverse conditions for the BCA occur and so biocontrol as a stand-alone treatment is as good as or better than chemical control or alternation (Dik and Elad, unpublished).

The integration of biological and chemical controls aided by the use of a forecaster to predict *Botrytis* outbreaks was examined recently. The overall goal is to develop an integrated control programme in which BCA is the most important tool and chemical control is implemented on a need-basis only (Elad *et al.*, 1994c). A decision support system named BOTMAN (Botrytis manager) was developed for integration of chemical and biological controls in unheated or partially heated greenhouses (Shtienberg and Elad, 1997). This system was further broadened by Elad and Shtienberg (1997) to also solve problems of other diseases and was named GREENMAN (greenhouse disease manager). Decisions on whether to apply biological or chemical measures are taken before each spray according to a 4-day weather microclimate within greenhouses. Therefore, a conversion from the conditions forecast for the outside environment is employed, and not for the conditions within the greenhouse. Parameters concern occurrence and severity of foliar diseases, resistance of pathogens in the greenhouse to fungicides, and conduciveness of the greenhouse and crop to disease development. The decision follows these lines: when weather, crop and greenhouse are extremely unfavourable to disease development, then no spraying at all is recommended, but when conditions are extremely favourable to *B. cinerea* development (e.g. forecast for humid and rainy weather, high disease pressure and conducive crop situation), then a chemical fungicide is suggested; in all other cases, a BCA (Trichodex) is recommended. GREENMAN was tested in twenty experiments carried out in tomato and cucumber greenhouses during the winter seasons of 1993–1997. On average ten sprays were applied in each experiment in the standard treatments (alternation of various fungicides on a calendar basis). In plots treated according to GREENMAN four chemical sprays and six biological sprays were applied. Control of white mould, leaf mould and grey mould were 55, 65 and 70%, respectively, in the standard treatment, and 63, 60 and 64% in treatments when GREENMAN recommendations were followed. The difference between these treatments was not significant. Trichodex significantly contributed to the control achieved by the fungicides in BOTMAN. Relying on this system for management of diseases reduced chemical spraying by 60%.

24.4. Future Perspectives

From this review, it is clear that some significant progress has been made toward biological and integrated control of greenhouse diseases in the phylloplane. For instance, some biofungicides are already on the market in a few countries, and these products are likely to become more widely available as they are registred in more areas. Other BCAs should reach the market soon and one can reasonably assume that biological alternatives will be available for most important greenhouse diseases in a near future. These encouraging developments should not divert attention from the fact that BCAs are living organisms, the activity of which greatly depends on environmental conditions. In this context, greenhouse operations provide an ideal niche for BCAs since they are less subject to sudden climate changes, and are often equipped with a sophisticated system of climate control. Nonetheless, it is unrealistic to assume that perfect conditions for the development of BCAs will always prevail in the greenhouse, and as a result, biofungicide will rarely stand alone as a complete measure of disease control under all conditions. For this reason, scientists and growers alike must accept the fact that BCAs are usually not as effective as pesticides. Rather, biological control should be viewed as an important if not essential component of an integrated disease management scheme if we are to achieve a significant and permanent reduction of pesticide use.

References

Andrews, J.H. (1985) Strategies for selecting antagonistic microorganisms from the phylloplane, in C.E. Windels and S.E. Lindow (eds.), *Biological Control on the Phylloplane*, American Phytopathological Society, St Paul, Minn., pp. 31–44.

Andrews, J.H. (1992) Biological control in the phyllosphere, *Annual Review of Phytopathology* **30**, 603–635.

Askary, H., Benhamou, N. and Brodeur, J. (1997) Ultrastructural and phytochemical investigations of the antagonistic effect of *Verticillium lecanii* on cucumber powdery mildew, *Phytopathology* **87**, 359–368.

Askary, H., Carrire, Y., Bélanger, R.R. and Brodeur, J. (1998) Pathogenicity of the fungus *Verticillium lecanii* to aphids and powdery mildew, *Biocontrol Science and Technology* (in press).

Bélanger, R.R. and Deacon, J.W. (1996) Interaction specificity of the biocontrol agent *Sporothrix flocculosa*: A video microscopy study, *Phytopathology* **86**, 1317–1323.

Bélanger, R.R., Dik, A.J. and Menzies, J.G. (1997) Powdery mildews: Recent advances toward integrated control, in G.J. Boland and L.D. Kuykendall (eds.), *Plant-Microbe Interactions and Biological Control*, Marcel Dekker, Inc., New York, pp. 89–109.

Bélanger, R.R., Labbè, C. and Jarvis, W.R. (1994) Commercial-scale control of rose powdery mildew with a fungal antagonist, *Plant Disease* **78**, 420–424.

Bennett, A.W. and Lane, S.D. (1992) The potential role of *Trichoderma viride* in the integrated control of *Botrytis fabae*, *Mycologist* **6**, 199–201.

Benyagoub, M., Bel Rhlid, R. and Bélanger, R.R. (1996a) Purification and characterization of new fatty acids with antibiotic activity produced by *Sporothrix flocculosa*, *J. of Chemical Ecology* **22**, 405–413.

Benyagoub, M. and Bélanger, R.R. (1995) Development of a mutant strain of *Sporothrix flocculosa* with resistance to dodemorph-acetate, *Phytopathology* **85**, 766–769.

Benyagoub, M., Willemot, C. and Bélanger, R.R. (1996b) Influence of a subinhibitory dose of antifungal fatty acids from *Sporothrix flocculosa* on cellular lipid composition in fungi, *Lipids* **31**, 1077–1082.

Blakeman, J.P. (1993) Pathogens in the foliar environment, *Plant Pathology* **42**, 479–493.

Boland, G.J. and Inglis, G.D. (1988) Antagonism of white mold (*Sclerotinia sclerotiorum*) of bean by fungi from bean and rapeseed flowers, *Canadian J. of Botany* **67**, 1775–1781.

Bourbos, V.A. and Skoudridakis, M.T. (1994) Integrated control of *Botrytis cinerea* in non-heated greenhouse tomatoes, *9th Congress Medit. Phytopathol. Union*, Rabat, Morocco, pp. 327–328.

Choudhury, S.R., Traquair, J.A. and Jarvis, W.R. (1994) 4-methyl 7,11-heptadecanoic acid: A new antibiotic from *Stephanoascus flocculosus* and *St. rugulosus*, *J. of Natural Products* **57**, 700–704.

de Meyer, G., Bigirimana, J., Elad, Y. and Hofte, M. (1998) Induced systemic resistance in *Trichoderma harzianum* T39 biocontrol of *Botrytis cinerea*, *European J. of Plant Pathology* (in press).

Dik, A.J., Verhaar, M.A. and Bélanger, R.R. (1998) Comparison of three biocontrol agents against cucumber powdery mildew (*Sphaerotheca fuliginea*) in commercial-scale greenhouse trials, *European J. of Plant Pathology* (in press).

Doherty, M.A. and Preece, T.F. (1978) *Bacillus cereus* prevents germination of uredospores of *Puccinia allii* and the development of rust disease of leek, *Allium porrum*, in controlled environment, *Physiological Plant Pathology* **12**, 123–132.

Edwards, S.G. and Seddon, B. (1992) *Bacillus brevis* as biocontrol agent against *Botrytis cinerea* on protected Chinese cabbage, in K. Verhoeff, N.E. Malathrakis and B. Williamson (eds.), *Recent Advances in Botrytis Research*, Pudoc Scientific Publishers, Wageningen, pp. 267–271.

Elad, Y., Cohen, A. and Abir, H. (1997) Control of foliar pathogens by the biocontrol preparation Trichodex (*Trichoderma harzianum* T39), Proceedings of the International Symposium on Production and Protection of Horticultural Crops, Agadir, 6–9 May 1997 (in press).

Elad, Y., Gullino, L.M., Shtienberg, D. and Aloi, C. (1995a) Coping with tomato grey mould under Mediterranean conditions, *Crop Protection* **14**, 105–109.

Elad, Y. and Kirshner, B. (1993) Survival in the phylloplane of an introduced biocontrol agent (*Trichoderma harzianum*) and populations of the plant pathogen *Botrytis cinerea* as modified by abiotic conditions, *Phytoparasitica* **21**, 303–313.

Elad, Y., Köhl, J. and Fokkema, N.J. (1994a) Control of infection and sporulation of *Botrytis cinerea* on bean and tomato by saprophytic yeasts, *Phytopathology* **84**, 1193–1200.

Elad, Y., Köhl, J. and Fokkema, N.J. (1994b) Control of infection and sporulation of *Botrytis cinerea* on bean and tomato by saprophytic bacteria and fungi, *European J. of Plant Pathology* **100**, 315–336.

Elad, Y., Malathrakis, N.E. and Dik, A.J. (1995b) Biological control of Botrytis incited diseases and powdery mildews in greenhouse crops, *Crop Protection* **15**, 224–240.

Elad, Y. and Shtienberg, D. (1997) Integrated management of foliar diseases in greenhouse vegetables according to principles of a decision support system – GREENMAN, *IOBC/WPRS Bulletin* **20**(4), 71–76.

Elad, Y., Shtienberg, D. and Niv, A. (1994c) *Trichoderma harzianum* T39 integrated with fungicides, Improved biocontrol of grey mould, in *Brighton Crop Protection Conference – Pests and Diseases*, BCPC, Thornton Heath, pp. 1109–1113.

Elad, Y., Zimand, G., Zaqs, Y., Zuriel, S. and Chet, I. (1993) Use of *Trichoderma harzianum* in combination or alternation with fungicides to control cucumber grey mould (*Botrytis cinerea*) under commercial greenhouse conditions, *Plant Pathology* **42**, 324–332.

Falk, S.P., Gadoury, D.M., Cortesi, P., Pearson, R.C. and Seem, R.C. (1995) Parasitism of *Uncinula necator* cleistothecia by the mycoparasite *Ampelomyces quisqualis*, *Phytopathology* **85**, 794–800.

Fokkema, N.J. (1993) Opportunities and problems of control of foliar pathogens with microorganisms, *Pesticide Science* **37**, 411–416.

Garibaldi, A. and Corte, A. (1987) Integrated control of the most severe diseases of some protected vegetables, in R. Caval (ed.), *Integrated and Biological Control in Protected Crops*, A.A. Balkema, Rotherdam, pp. 85–91.

Gullino, M.L., Aloi, L. and Garibaldi, A. (1990) Chemical and biological control of grey mould of strawberry, *Mededelingen van de Faculteit Landbouwwetenschappen Rijkuniversiteit Gent* **55**, 967–970.

Hajlaoui, M.R. and Bélanger, R.R. (1991) Comparative effects of temperature and humidity on the activity of three potential antagonists of rose powdery mildew, *Netherlands J. of Plant Pathology* **97**, 203–208.

Hajlaoui, M.R., Benhamou, N. and Bélanger, R.R. (1992) Cytochemical study of the antagonistic activity of *Sporothrix flocculosa* on rose powdery mildew, *Sphaerotheca pannosa* var. *rosae*, *Phytopathology* **82**, 583–589.

Hajlaoui, M.R., Traquair, J.A., Jarvis, W.R. and Bélanger, R.R. (1994) Antifungal activity of extracellular metabolites produced by *Sporothrix flocculosa*, *Biocontrol Science and Technology* **4**, 229–237.

Hashioka, Y. and Nakai, Y. (1980) Ultrastructure of pycnidial development and mycoparasitism of *Ampelomyces quisqualis* parasitic on *Erysiphales*, *Transactions of the Mycological Society of Japan.* **21**, 329–338.

Hausbeck, M.K., Pennypacker, S.P. and Stevenson, R.E. (1996) The effect of plastic mulch and forced heated air on *Botrytis cinerea* on geranium stock plants in a research greenhouse, *Plant Disease* **80**, 170–173.

Hijwegen, T. (1986) Biological control of cucumber powdery mildew by *Tilletiopsis minor*, *Netherlands J. of Plant Pathology* **92**, 93–95.

Hijwegen, T. (1989) Effect of culture filtrate of seventeen fungicolous fungi on sporulation of cucumber powdery mildew, *Netherlands J. of Plant Pathology* **95**, 95–98.

Hijwegen, T. (1992) Biological control of cucumber powdery mildew with *Tilletiopsis minor* under greenhouse conditions, *Netherlands J. of Plant Pathology* **98**, 211–215.

Hoch, H.C. and Provvidenti, R. (1979) Mycoparasitic relationships: Cytology of the *Sphaerotheca fuliginea-Tilletiopsis* sp. interaction, *Phytopathology* **69**, 359–362.

Jarvis, W.R. (1992) *Managing Diseases in Greenhouse Crops*, APS Press, St Paul, Minn.

Jarvis, W.R., Shaw, L.A. and Traquair, J.A. (1989) Factors affecting antagonism of cucumber powdery mildew by *Stephanoascus flocculosus* and *S. rugulosus*, *Mycological Research* **92**, 162–165.

Jarvis, W.R. and Slingsby, K. (1977) The control of powdery mildew of greenhouse cucumber by water sprays and *Ampelomyces quisqualis*, *Plant Disease Reporter.* **61**, 728–730.

Köhl, J., Bélanger, R.R. and Fokkema, N.J. (1997) Interaction of four antagonistic fungi with *Botrytis aclada* in dead onion leaves: A comparative microscopic and ultrastructural study, *Phytopathology* **87**, 634–642.

Köhl, J. and Fokkema, N.J. (1993) Fungal interactions on living and necrotic leaves, in J.P. Blakeman and B. Williamson (eds.), *Ecology of Plant Pathogens*, CAB International, Wallingford, pp. 321–334.

Köhl, J., Gerlagh, M., de Haas, B.H. and Krijger, M.C. (1998) Biological control of *Botrytis cinerea* in cyclamen with *Ulocladium atrum* and *Gliocladium roseum* under commercial conditions, *Phytopathology* **88**, 568–575.

Köhl, J., Molhoek, W.M.L., van der Plas, C.H. and Fokkema, N.J. (1995a) Suppression of sporulation of *Botrytis* spp. as a valid biocontrol strategy, *European J. of Plant Pathology* **101**, 251–259.

Köhl, J., Molhoek, W.M.L., van der Plas, C.H. and Fokkema, N.J. (1995b) Effect of *Ulocladium atrum* and other antagonists on sporulation of *Botrytis cinerea* on dead lily leaves exposed to field conditions, *Phytopathology* **85**, 393–401.

Köhl, J., van der Plas, C.H., Molhoek, W.M.L. and Fokkema, N.J. (1995c) Effect of interrupted leaf wetness periods on suppression of sporulation of *Botrytis allii* and *B. cinerea* by antagonists on dead onion leaves, *European J. of Plant Pathology* **101**, 627–637.

Kranz, J. (1981) Hyperparasitism of biotrophic fungi, in J.P. Blakeman (ed.), *Microbiology of the Phylloplane*, Academic Press, London, pp. 327–352.

Li, H. and Leiffert, C. (1994) Development of resistance in *Botryotinia fuckeliana* (de Bary) Whetzel against the biological control agent *Bacillus subtilis*, *J. of Plant Diseases and Protection* **101**, 414–418.

Malathrakis, N.E. and Klironomou, E.J. (1992a) Control of grey mould of tomatoes in greenhouses with fungicides and antagonists, in K. Verhoeff, N.E. Malathrakis and B. Williamson (eds.), *Recent Advances in Botrytis Research*, Pudoc Scientific Publishers, Wageningen, pp. 282–286.

Malathrakis, N.E. and Klironomou, E.J. (1992b) Effectiveness of *Acremonium alternatum* and glycerol against cucumber powdery mildew (*Sphaerotheca fuliginea*), in E.C. Tjamos, G.C. Papavizas and R.J. Cook (eds.), *Biological Control of Plant Diseases: Progress and Challenges for the Future*, Plenum Press, New York, pp. 443–446.

Nelson, M.E. and Powelson, M.L. (1988) Biological control of gray mold of snap beans by *Trichoderma hamatum*, *Plant Disease* **72**, 727–729.

O'Neill, T.M., Elad, Y., Shtienberg, D. and Cohen, A. (1996a) Control of grapevine grey mould with *Trichoderma harzianum* T39, *Biocontrol Science and Technology* **6**, 139–146.

O'Neill, T.M., Niv, A., Elad, Y. and Shtienberg, D. (1996b) Biological control of *Botrytis cinerea* on tomato stem wounds with *Trichoderma harzianum*, *European J. of Plant Pathology* **102**, 635–643.

Pasini, C., D'Aquilla, F., Curir, P. and Gullino, M.L. (1997) Effectiveness of antifungal compounds against rose powdery mildew (*Sphaerotheca pannosa* var. *rosae*) in greenhouses, *Crop Protection* **16**, 251–256.

Peng, G. and Sutton, J.C. (1991) Evaluation of microorganisms for biocontrol of *Botrytis cinerea* in strawberry, *Canadian J. of Plant Pathology* **13**, 247–257.

Peng, G., Sutton, J.C. and Kevan, P.G. (1992) Effectiveness of honey bees for applying the biocontrol agent *Gliocladium roseum* to strawberry flowers to suppress *Botrytis cinerea*, *Canadian J. of Plant Pathology* **14**, 117–129.

Philipp, W.-D., Beuther, E., Hermann, D., Klinkert, F., Oberwalter, C., Schmidtke, M. and Straub, B. (1990) Zur Formulierung des Mehltauhyperparasiten *Ampelomyces quisqualis* Ces., *Zeitschrift fur Pflanzenkrankheiten und Pflanzenschutz* **97**, 120–132.

Redmond, J.C., Marois, J.J. and MacDonald, J.D. (1987) Biological control of *Botrytis cinerea* with epiphytic microorganisms, *Plant Disease* **71**, 799–802.

Scherff, R.H. (1973) Control of bacterial blight of soybean by *Bdellovibrio bacteriovorous*, *Phytopathology* **63**, 400–402.

Shtienberg, D. and Elad, Y. (1997) Incorporation of weather forecasting in integrated, biological-chemical management of *Botrytis cinerea*, *Phytopathology* **87**, 332–340.

Spencer, D.M. and Ebben, M.H. (1983) Biological control of cucumber powdery mildew, in *Annual Report of the Glasshouse Crops Research Institute 1981*, Littlehampton, pp. 128–129.

Staples, R.C., Syamanda, R., Kao, V. and Bloch, R.J. (1962) Comparative biochemistry of obligately parasitic and saprophytic fungi. II. Assimilation of ^{14}C-labelled substrates by germinating spores, *Contributions Boyce Thompson Institute* **21**, 345–362.

Sundheim, L. (1982) Control of cucumber powdery mildew by the hyperparasite *Ampelomyces quisqualis* and fungicides, *Plant Pathology* **31**, 209–214.

Sundheim, L. (1986) Use of hyperparasites in biological control of biotrophic plant pathogens, in N.J. Fokkema and J. van den Heuvel (eds.), *Microbiology of the Phylloplane*, Cambridge University Press, London, pp. 333–347.

Sztejnberg, A., Galper, S., Mazar, S. and Lisker, N. (1989) *Ampelomyces quisqualis* for biological and integrated control of powdery mildew in Israel, *J. of Phytopathology* **124**, 285–295.

Tirilly, Y., Trique, B. and Maisonneuve, J. (1987) Perspectives d'utilisation de *Hansfordia pulvinata* contre la cladosporiose de la tomate, *Bulletin OEPP* **17**, 39–643.

Tronsmo, A. (1992) Leaf and blossom epiphytes and endophytes as biological control agents, in E.S. Tjamos, G.C. Papavizas and R.J. Cook (eds.), *Biological Control of Plant Diseases: Progress and Challenges for the Future*, Plenum Press, New York, pp. 43–54.

Urquhart, E.J., Menzies, J.G. and Punja, Z.K. (1994) Growth and biological control activity of *Tilletiopsis* species against powdery mildew (*Sphaerotheca fuliginea*) on greenhouse cucumber, *Phytopathology* **84**, 341–351.

Verhaar, M.A., Hijwegen, T. and Zadoks, J.C. (1996) Glasshouse experiments on biocontrol of cucumber powdery mildew (*Sphaerotheca fuliginea*) by the mycoparasites *Verticillium lecanii* and *Sporothrix rugulosa, Biological Control* 6, 353–360.

White, J.G., Linfield, C.A., Lahdenpera, M.L. and Voti, J. (1990) Mycostop – A novel biofungicide based on *Streptomyces griseoviridis*, in *Brighton Crop Protection Conference – Pests and Diseases*, BCPC, Thornton Heath, pp. 221–226.

Zhang, P.G. and Sutton, J.C. (1994) Evaluation of microorganisms for biocontrol of *Botrytis cinerea* in container-grown black spruce seedlings, *Canadian J. of Forest Research* 24, 1312–1316.

Zimand, G., Elad, Y. and Chet, I. (1996) Effect of *Trichoderma harzianum* on *Botrytis cinerea* pathogenicity, *Phytopathology* 86, 1255–1260.

CHAPTER 25

GENETIC MANIPULATION FOR IMPROVEMENT OF MICROBIAL
BIOCONTROL AGENTS

Sonja Sletner Klemsdal and Arne Tronsmo

25.1. Introduction

In the future, biological control agents are expected to become an important component
in plant-disease management. Fungi and bacteria with antagonistic activity, or which
are able to promote plant growth by either controlling minor pathogens or producing
growth-stimulating compounds, can be considered as potential biocontrol agents.
Molecular techniques allow modification of wild type strains to improve their ability to
suppress plant disease. Genetic modifications could result in new biocontrol strains with
increased production of antifungal or antibacterial compounds, strains with improved
ability to compete for limited nutrients, strains with expanded host range compared to
the wild type strains, or strains better adjusted to colder temperatures or other climatic
factors than the original biocontrol strains. Unwanted genes might be deleted, or one or
more genes could be added. During the last few years there has been a revolution in the
field of genetic modification of biocontrol agents.

25.2. Methods for Genetic Modification of Biocontrol Agents

25.2.1. CHEMICAL OR UV-INDUCED MUTAGENESIS

Mutants of bacterial and fungal biocontrol agents can be induced by chemicals or by
ultraviolet (UV) light. Chemicals like ethyl methane sulphonate (EMS) (Miller, 1972)
or N-methyl-N'-nitro-N-nitrosoguanidine (NTG) (Carlton and Brown, 1981) can be
used to alter a base that is already incorporated into the double-stranded DNA molecule,
causing specific mispairing. UV light stimulates formation of dimers from two adjacent
pyrimidines, causing insertion of an incorrect base at the position of the dimer. Both
kinds of mutagens can also produce frameshifts, duplications or deletions.

25.2.2. PROTOPLAST FUSION

Protoplast fusion has been used as a method to produce heterokaryons between two
fungal strains that can not form progeny by sexual crosses. In the resulting recombinant
progeny strains, traits from both parents will be combined as a result of asexual genetic
recombination. Hybridizations can be done between strains of the same species. Stasz *et
al.* (1988) fused protoplasts of two biocontrol strains of *Trichoderma harzianum* Rifai
and were able to obtain an isolate with improved biocontrol abilities (Harman and

R. Albajes et al. (eds.), Integrated Pest and Disease Management in Greenhouse Crops, 353-364.
© 1999 *Kluwer Academic Publishers. Printed in the Netherlands.*

Taylor, 1989). Intergeneric crosses fusing protoplasts from different fungal species have also been reported (Kirimura *et al.*, 1989).

25.2.3. TRANSFORMATION

Several methods have been developed that allow transformation of bacterial or fungal cells. In theory, DNA fragments from any organism can be integrated and expressed in the transformed organism, if the right regulatory regions are used. Integration could be random or site-specific. Fungal protoplasts can be transformed in the presence of polyethylene glycol (PEG) and calcium (Penttilä *et al.*, 1987; Thomas and Kenerley, 1989; Herrera-Estrella *et al.*, 1990). Using lithium acetate (LiAc) and PEG, Dickman (1988) was able to transform whole cells, not protoplasts. Transformation of bacteria and fungi can be performed by electroporation using a high-voltage electric pulse (Goldman *et al.*, 1990). In the technique called biolistic transformation, fungi can be transformed by bombardment of conidia using particles covered by DNA (Armaleo *et al.*, 1990; Lorito *et al.*, 1993). Plasmids or cosmids containing the genes of interest can be transferred from *Escherichia coli* Castellani & Chalmers to *Pseudomonas* by conjugation, using special helper plasmids and triparental matings (Voisard *et al.*, 1988).

25.2.4. TRANSPOSON MUTAGENESIS

Insertion of transposable elements like Tn5 or Tn3 randomly in the genome can be used to make series of mutants (Simon *et al.*, 1983). Genes in which the transposon is inserted will not give a functional gene product. Genes of interest can be cloned using the transposon as a tag.

25.3. Approaches to Improve Biocontrol Agents Using Genetic Modifications

25.3.1. MUTATION TO FUNGICIDE RESISTANCE

When disease pressure is high, fungal biological control agents often become less effective, and the integration of other approaches to control disease is necessary. Introduction of fungicide tolerance in the antagonistic fungi would allow the combined application of a biocontrol agent and a chemical fungicide, either simultaneously or in rotation. Benomyl-resistant strains of *Gliocladium virens* J.H. Miller, J.E. Giddens & A.A. Foster and *Trichoderma* spp. were developed using protoplast fusion or mutagenesis by UV and EMS (Ahmad and Baker, 1988; Papavizas *et al.*, 1990). Some mutants showed improved rhizosphere competence and retained their biocontrol ability, while in other mutants the biocontrol ability was lost (Ahmad and Baker, 1988; Papavizas *et al.*, 1990). A dodemorph-acetate resistant mutant of *Sporothrix flocculosa* Traquair, Shaw & Jarvis was selected (Belanger and Benyagoub, 1997). The new strain was found to be as effective in controlling rose powdery mildew as the wild type, alone or in combination with this fungicide. UV-induced benomyl-resistant mutants of nonpathogenic *Fusarium*-isolates antagonistic to Fusarium wilt in carnation have been developed (Postma and Luttikholt, 1993). Genes conferring fungicide resistant phenotypes have been cloned and characterized (Orbach *et al.*, 1986; Cooley and Caten, 1993; Benyaacov *et al.*, 1994), which should soon allow genetic transformation of such genes into biocontrol agents.

25.3.2. MUTATION TO HYPOVIRULENCE

The best studied pathosystem where mycovirus infection has been shown to result in hypovirulence, involves *Cryphonectria parasitica* (Murrill) Barr, the fungal pathogen causing chestnut blight and double stranded RNA molecules shown to reduce the virulence of this pathogen (Nuss, 1992). Hypovirulent strains of several other phytopathogens have been described, including some causing disease in greenhouse crops (Boland, 1992; Kousik *et al.*, 1994; Howitt *et al.*, 1995; Juan-Abgona *et al.*, 1996). Recently research has been done to genetically modify strains of *C. parasitica* to hypovirulence (Choi and Nuss, 1992; Chen *et al.*, 1993; Chen *et al.*, 1994; Choi *et al.*, 1995; Monteiro-Vitorello *et al.*, 1995). A hypovirulent phenotype can theoretically be genetically engineered by any of three strategies: introduction of synthetic viral transcripts, or mutations induced in mitochondrial or nuclear DNA. The results obtained on hypovirulence in the *C. parasitica* system provide the background for future research in other pathosystems including systems important in greenhouse crops.

25.3.3. PRODUCTION OF BACTERIOCINS

Crown gall, a tumorous disease of pome, stone, and several small fruits and ornamentals (rose and euonymus), is caused by the soil bacterium, *Agrobacterium tumefaciens* Conn. In many plants, the disease can be controlled by the related but nonpathogenic bacterium *Agrobacterium radiobacter* (Beijerinck & van Delden) Conn strain K84 (Kerr and Htay, 1974). Production of the bacteriocin agrocin 84 is an important part of the mechanism of this biocontrol (Kerr, 1980). Agrocin 84-encoding genes (Ellis *et al.*, 1979; Wang *et al.*, 1994), genes giving immunity to the bacteriocin, and genes responsible for conjugal transfer (*tra*) (Farrand *et al.*, 1985; Ryder *et al.*, 1987) are all located on the plasmid pAgK84. Natural transfer of pAgK84 to pathogenic bacteria resulted in strains of *A. tumefaciens* insensitive to agrocin 84 (Ellis and Kerr, 1979). By recombinant DNA techniques, a new strain of *A. radiobacter* (strain K1026) was constructed in which a 5.9 kb region overlapping the transfer (*tra*) region of plasmid pAgK84 was deleted (Jones *et al.*, 1988). The purpose of this modification was to decrease the risk of breakdown of biological control of crown gall. The resulting Tra⁻ strain was unable to transfer its genes giving immunity to agrocin 84 (Vicedo *et al.*, 1993). The genetically modified strain is currently in commercial use on stone fruits. It was found to be as effective to colonize roots and control crown gall as the parent strain K84 (Jones and Kerr, 1989).

Another approach to improve the effectivity of bacteriocin-producing biological control agents might be to increase the number of bacteriocins secreted, or to construct a biocontrol strain which produces more effective bacteriocins. Agrocin 434, a second agrocin produced by *A. radiobacter* strains K84 and K1026, inhibits a broader range of *Agrobacterium* than agrocin 84 (Donner *et al.*, 1993). When more genes involved in the biosynthesis of agrocins or other bacteriocins have been cloned, genetic modification may be used to construct new and improved biocontrol agents.

25.3.4. PRODUCTION OF SIDEROPHORES

The role of siderophore production by biological control rhizosphere bacteria in antagonism against phytopathogens has been studied by molecular methods for many years. However, only recently has the effect of siderophores in a fungal biocontrol agent, *T. harzianum*, been reported (Ratto *et al.*, 1996). No fungal gene involved in siderophore production has yet been cloned.

Siderophores are iron-chelating compounds. With limited amounts of iron present, iron depletion of the competing phytopathogens will result, if they lack receptors for the siderophore of a given biocontrol strain (Loper and Ishimaru, 1991). Recent research has focused on cloning of outer membrane receptors (Marugg *et al.*, 1989; Morris *et al.*, 1992) and the identification of genes regulating their expression (Leong *et al.*, 1991; Laville *et al.*, 1992). Morris *et al.* (1992) found that *Pseudomonas* sp. strain M114 contained a receptor for pseudobactin MT3A, a siderophore that this strain does not produce. The authors suggested that utilization of heterologous siderophores can provide a competitive advantage for rhizosphere bacteria. Recently a fluorescent *Pseudomonas* sp. B24 was genetically modified to utilize additional ferric siderophores (Moenne Loccoz *et al.*, 1996). The plasmid pCUP2 containing the gene *pbuA*, encoding the membrane receptor of ferric pseudobactin M114, enabled the modified *Pseudomonas* to utilize the pseudobactin of *Pseudomonas fluorescens* (Trevisan) Migula M114 in addition to its own siderophore. However this ability did not improve the ecological fitness of the modified *Pseudomonas* strain as compared to the wild type.

25.3.5. PRODUCTION OF ANTIBIOTICS

In iron-rich conditions, many bacterial or fungal strains used in biocontrol of plant pathogens produce antibiotics that inhibit the growth of other fungi. DNA fragments containing structural or regulatory genes involved in the synthesis of antibiotics have recently been identified from several strains effective in biocontrol of plant pathogens in greenhouse crops (Keel *et al.*, 1992; Laville *et al.*, 1992; Gaffney *et al.*, 1994; Hill *et al.*, 1994; Asaka and Shoda, 1996; Hammer *et al.*, 1997). Complementation of antibiotic-negative mutants with these fragments restores the ability to produce antibiotics and the ability to control disease. The availability of such genes can, in principle, be used to improve biocontrol strains by genetic engineering. Some strains of *Erwinia herbicola* (Löhnis) Dye produce peptides with antifungal activities (Winkelmann *et al.*, 1980). The component produced by *E. herbicola* CHS1065 is highly related to herbicolin A and encoded by genes located on a plasmid, pHER1065 (Tenning *et al.*, 1993). After some modification, the plasmid was stably introduced into *E. coli*, which then expressed the antifungal activity quantitatively and qualitatively comparable to *E. herbicola* CHS1065 (Tenning *et al.*, 1993). The authors suggest that the herbicolin-synthesizing genes can be introduced into a root-colonizing bacterium and thus improve biocontrol activity by the production of herbicolin along the roots of the plants.

After UV-induction of *T. harzianum*, mutants were found with high or low production of antibiotics compared to the wild type strain, at least one of which was reported to give slightly better control of *Pythium ultimum* Trow than its parent (Faull and Graeme-Cook, 1992).

Introduction of cosmid pME3090 with an insert containing the gene encoding the housekeeping sigma factor into a *P. fluorescens* CHA0 background increased production of the antibiotics pyoluteorin (Plt) and 2,4-diacetylphloroglucinol (Phl) (Maurhofer *et al.*, 1992; Schnider *et al.*, 1995). The overproducing strain showed improved protection of cucumber against *P. ultimum, Phomopsis sclerotioides* van Kestern, and *Fusarium oxysporum* Schlechtend.:Fr. f. sp. *cucumerinum* J.H. Owen compared to the wild-type CHA0 strain (Maurhofer *et al.*, 1992). The pME3090-containing strain also showed improved protection of tobacco from *Thielaviopsis basicola* (Berk. & Broome) Ferraris, and cress and sweet corn from *P. ultimum*. But for these crops, the treatment results in a significant reduction of the fresh weights of the plants. The authors conclude that whether cosmid pME3090 should result in toxic effects on the plants or improved disease control, is determined by the plant species rather than the pathogen.

Pseudomonas fluorescens strain BL915 is an effective biocontrol agent for the control of seedling disease induced by *Rhizoctonia solani* Kühn. Ligon *et al.* (1996) describe an impressive collection of strains originating from BL915, genetically modified for increased production of the antibiotic pyrrolnitrin (Prn). Four constructs were made: (i) introduction of an extra copy of the wild type global regulator gene, *gacA* (Hill *et al.*, 1994); (ii) replacement of the TTG translation initiation codon in the wild type *gacA* gene with an ATG codon to increase the translational efficiency; (iii) control of the *gacA* gene by the *tac* promoter; and (iv) the introduction of an extra copy of the four genes involved in the biosynthesis of Prn, *prnA, prnB, prnC* and *prnD* controlled by the *tac* promoter. The *tac* promoter is a constitutive promoter found to be expressed at high levels in pseudomonads. The production of Prn in the modified strains increased by 2- to 4-fold compared to the parent strain. With one exception, all the constructs resulted in increased biocontrol of *R. solani* on cucumber and impatiens. The construct containing an extra copy of the wild type *gacA* gene had increased activity in impatiens, but lower biocontrol activity in cucumber.

25.3.6. IMPROVED ROOT COLONIZATION ABILITY

Some soil bacteria are found to be beneficial to the plant without establishing a symbiotic relationship. These plant growth promoting rhizobacteria (PGPR) can be found free-living in soil, or close to and even within the roots of plants. Plant growth promotion may be accomplished by inhibition of the growth of plant pathogenic microorganisms. The status of research aiming to genetically modify PGPR to enhance the biocontrol of phytopathogens has recently been reviewed (Glick, 1995; Glick and Bashan, 1997). Increasing the ability to compete for limiting amounts of nutrients (i.e.

carbon, nitrogen, iron) might improve the root colonization ability of a PGPR strain. Genetic modification to extend the range of ferric-siderophores that can be used in competition for iron has been discussed previously in this chapter (Section 25.3.3). Strains that can utilize unusual carbon or nitrogen compounds will benefit when such compounds are present. A biocontrol strain transformed with genes encoding enzymes capable of degrading naphthalene and salicylate increased the persistence of the host bacterium in the presence of salicylate (Colbert *et al.*, 1993). Introduction of a gene encoding an antifreeze protein into a PGPR might enable the bacterium to persist and proliferate at colder temperatures and so increase its biocontrol activity under these conditions (Glick, 1995). So far no bacterial antifreeze protein coding gene has been cloned, but attempts to isolate and characterize the gene product are in progress (Sun *et al.*, 1995).

Site competition as a mechanism to control plant pathogens has been investigated. Application of an Ice⁻ mutant of *Pseudomonas syringae* van Hall reduced subsequent colonization of an Ice⁺ *P. syringae,* resulting in reduced frost injury (Lindemann and Suslow, 1987). The two isogenic strains differed only in ice nucleation activity. Avirulent mutants of *Pseudomonas solanacearum* (Smith) Smith have been used in the biocontrol of virulent *P. solanacearum* strains. Hrp⁻ mutants, Tn5-induced mutants containing insertions in the *hrp* gene (Boucher *et al.*, 1985; Frey *et al.*, 1994), were found to be nonpathogenic to the plant. Hrp⁻ mutants were able to invade the plant, and to survive and multiply within the plant (Trigalet and Demery, 1986). The Hrp⁻ mutants were found to exclude pathogenic strains from susceptible tomato plants (Trigalet and Trigalet-Demery, 1990; Frey *et al.*, 1994), and thus to give improved biological control of bacterial wilt.

25.3.7. PRODUCTION OF LYTIC ENZYMES

One mechanism involved in the biocontrol of plant pathogenic fungi is mycoparasitism. Lytic enzymes such as chitinases, proteases and glucanases produced by bacterial and fungal biocontrol agents cause degradation of the fungal cell walls, resulting in death or inhibition of growth of the attacked fungus (Chet, 1987). During the last years, extensive research has focused on the cloning and characterization of genes encoding lytic enzymes. This information can be used in an attempt to improve existing biocontrol agents. Genes encoding chitinolytic enzymes have been cloned from a wide variety of organisms, including the bacteria *Enterobacter agglomerans* (Beijerinck) Ewing & Fife (Chernin *et al.*, 1997) and *Serratia marcescens* Bizio (Sundheim *et al.*, 1988), and different strains of the filamentous fungus *T. harzianum* (García *et al.*, 1994; Hayes *et al.*, 1994; Draborg *et al.*, 1995; Limón *et al.*, 1995; Peterbauer *et al.*, 1996). These microorganisms are all found to be antagonistic to plant pathogenic fungi. From *T. harzianum* the gene encoding an endo-β-1,3-glucanase has also been cloned (de la Cruz *et al.*, 1995). After transformation of *E. coli*, the *E. agglomerans* (Chernin *et al.*, 1997) and *S. marcescens* (Shapira *et al.*, 1989) chitinase genes were expressed, and the enzyme produced and secreted. The transformants containing the *E. agglomerans* gene were found to inhibit *R. solani* (Chernin *et al.*, 1997). When the chitinase gene from *S. marcescens* was introduced into *P. fluorescens* strains, the genetically engineered strains were able to suppress disease caused by *Fusarium redolens* Wollenweb (Sundheim *et al.*, 1988). The plasmid was, however, found to be very unstable in *Pseudomonas*. The *Serratia* chitinase gene has also been introduced into *Pseudomonas putida* (Trevisan) Migula (Chet *et al.*, 1993), *Rhizobium meliloti* Dangeard (Sitrit *et al.*, 1993) and *T. harzianum* strain T-35 (Haran *et al.*, 1993). Reduction of disease caused by *Sclerotium rolfsii* Sacc. and *R. solani* were observed using the modified *Pseudomonas* strain, but the plasmid proved again to be rapidly lost when grown without selective pressure. The transformed *R. meliloti* and *T. harzianum* were stable, but their effect on biological control of plant pathogenic fungi has to be further investigated.

The endochitinase production in *T. harzianum* was increased several fold when *T. harzianum* was transformed with the *T. harzianum* endochitinase gene, *ThEn-42* (Hayes *et al.*, 1994), linked to the high expression promoter, *cbh1*, from *Trichoderma reesei* E. Simmons in H.E. Bigelow & E. Simmons (Margolles-Clark *et al.*, 1996). Further studies are needed to see whether the increased chitinase production has resulted in improved biocontrol ability of the transformed strains.

25.4. Risks of Releasing Genetically Modified Biocontrol Organisms

In order to use genetically modified organisms as biocontrol agents, these organisms have to be released into the environment (see Chapter 27). The potential risks of releasing genetically modified biocontrol organisms have been thoroughly discussed (Fry and Day, 1992; Schroth, 1992; Ryder, 1994; Cook *et al.*, 1996). Cook *et al.* (1996) summarize the potential hazards associated with the use of microorganisms for the biological control of plant diseases: displacement of nontarget microorganisms, allergenicity to consumers, and toxigenicity or pathogenicity to nontarget organisms. We agree with the authors that regardless of whether the organisms to be used in disease control are unmodified but released into an ecosystem where they normally do not exist, or are genetically modified by traditional or recombinant DNA techniques, the safety issues are the same. Schroth (1992) finds no scientific evidence to support the notion that genetic modifications may make saprophytic biocontrol agents themselves phytopathogenic or more ecologically competent than the wild type. Generally we find it easier to accept the use of genetically modified organisms originating from nonpathogenic species than a modification of a pathogen to a nonpathogenic antagonistic isolate. Any genetically engineered organism should, however, be thoroughly tested regarding its ability to cause disease on all relevant hosts.

As presented, there is a broad range of bacterial and fungal strains and species used as the parent organism for genetic modification to improve biological control activities. There is also a large variation in the kind of genetic modification that has been performed, spanning from deletion of a fragment from the wild type genome of *Agrobacterium* strain K1026 (Jones *et al.*, 1988), to modification or amplification of a global regulation gene, *gacA*, in *P. fluorescens* BL915 (Ligon *et al.*, 1996). Different organisms and different modifications necessitate the case to case procedures to determine whether or how the genetically engineered organism should be tested or released in the environment.

25.5. Conclusions

During the last few years genetic modification of microorganisms to improve their biological control of plant pathogens has created some very interesting results, which indicates that molecular biology can be used not only to understand the mechanism behind biological control, but also to improve biocontrol ability. It is, however, important to remember that in most strains, the ability to control disease will be explained by more than one mechanism, and the host for genetic modification has to be ecologically fit in the environment where the biocontrol is to be performed. Modification of a biocontrol mechanism may result in an improved biocontrol ability, but may also give unexpected effects. The effect of a new or modified biocontrol agent must therefore be carefully determined empirically for each host-pathogen system. There is also in the public a resistance against introduction of genetically modified organisms into the environment, which probably will slow down the introduction of genetically modified biocontrol agents in many countries. So far only a few modified organisms are on the market, e.g. the deletion mutants *A. radiobacter* strain K1026 (Section 25.3.3) and the Ice⁻ mutant of *P. syringae* (Section 25.3.6), plus the protoplast fusant *T. harzianum* strain 1295-22 (Bio-Trek 22) (Section 25.2.2), but with the great demand for alternatives to chemicals in plant protection it is expected that more modified organisms soon will become commercially available.

References

Ahmad, J.S. and Baker, R. (1988) Rhizosphere competence of benomyl-tolerant mutants of *Trichoderma* spp., *Canadian J. of Microbiology* **34,** 694–696.

Armaleo, D., Ye, G., Klein, T.M., Shark, K.B., Sanford, J.C. and Johnston, S.A. (1990) Biolistic nuclear transformation of *Saccharomyces cerevisiae* and other fungi, *Current Genetics* **17,** 97–103.

Asaka, O. and Shoda, M. (1996) Biocontrol of *Rhizoctonia solani* damping-off of tomato with *Bacillus subtilis* RB14, *Applied and Environmental Microbiology* **62,** 4081–4085.

Belanger, R.R. and Benyagoub, M. (1997) Challenges and prospects for integrated control of powdery mildews in the greenhouse, *Canadian J. of Plant Pathology* **19,** 310–314.

Benyaacov, R., Knoller, S., Calwell, G.A., Becker, J.M. and Koltin, Y. (1994) *Candida albicans* gene encoding resistance to benomyl and methotrexate is a multidrug-resistance gene, *Antimicrobial Agents and Chemotherapy* **38,** 648–652.

Boland, G.J. (1992) Hypovirulence and double-stranded RNA in *Sclerotinia sclerotiorum, Canadian J. of Plant Pathology* **14,** 10–17.

Boucher, C.A., Barberis, P.A., Trigalet, A.P. and Demery, D.A. (1985) Transposon mutagenesis of *Pseudomonas solanacearum*: Isolation of Tn5-induced avirulent mutants, *J. of General Microbiology* **131,** 2449–2457.

Carlton, B.C. and Brown, B.J. (1981) Gene mutation, in P. Gerhardt (ed.), *Manual of Methods for General Bacteriology*, American Society for Microbiology, Washington, DC, pp. 222–242.

Chen, B.S., Choi, G.H. and Nuss, D.L. (1993) Mitotic stability and nuclear inheritance of integrated viral cDNA in engineered hypovirulent strains of the chestnut blight fungus, *EMBO J.* **12,** 2991–2998.

Chen, B.S., Choi, G.H. and Nuss, D.L. (1994) Attenuation of fungal virulence by synthetic infectious hypovirus transcripts, *Science* **264,** 1762–1764.

Chernin, L.S., de la Fuente, L., Sobolev, V., Haran, S., Vorgias, C.E., Oppenheim, A.B. and Chet, I. (1997) Molecular cloning, structural analysis, and expression in *Escherichia coli* of a chitinase gene from *Enterobacter agglomerans*, *Applied and Environmental Microbiology* **63,** 834–839.

Chet, I. (1987). *Trichoderma*-application, mode of action, and potential as a biocontrol agent of soilborne plant pathogenic fungi, in I. Chet (ed.), *Innovative Approaches to Plant Disease Control,* John Wiley & Sons, Inc., New York, pp.137–160.

Chet, I., Barak, Z. and Oppenheim, A. (1993) Genetic engineering of microorganisms for improved biocontrol activity, in I. Chet (ed.), *Biotechnology in Plant Disease Control,* John Wiley & Sons, Inc., New York, pp. 211–235.

Choi, G.H., Chen, B.S. and Nuss, D.L. (1995) Virus-mediated or transgenic suppression of a G-protein alpha-subunit and attenuation of fungal virulence, *Proceedings of the National Academy of Sciences of the United States of America* **92**, 305–309.

Choi, G.H. and Nuss, D.L. (1992) Hypovirulence of chestnut blight fungus conferred by an infectious viral cDNA, *Science* **257**, 800–803.

Colbert, S.F., Hendson, M., Ferri, M. and Schroth, M.N. (1993) Enhanced growth and activity of a biocontrol bacterium genetically engineered to utilize salicylate, *Applied and Environmental Microbiology* **59**, 2071–2076.

Cook, R.J., Bruckart, W.L., Coulson, J.R., Goettel, M.S., Humber, R.A., Lumsden, R.D., Maddox, J.V., McManus, M.L., Moore, L., Meyer, S.F., Quimby, P.C., Stack, J.P. and Vaughn, J.L. (1996) Safety of microorganisms intended for pest and plant disease control: A framework for scientific evaluation, *Biological Control* **7**, 333–351.

Cooley, R.N. and Caten, C.E. (1993) Molecular analysis of the *Septoria nodorum* beta-tubulin gene and characterization of a benomyl resistance mutation, *Molecular and General Genetics* **237**, 58–64.

de la Cruz, J., Pintor-Toro, J.A., Benítez, T., Llobell, A. and Romero, L.C. (1995) A novel endo-1,3-glucanase, BGN13.1, involved in the mycoparasitism of *Trichoderma harzianum, J. of Bacteriology* **177**, 6937–6945.

Dickman, M.B. (1988) Whole cell transformation of the alfalfa fungal pathogen *Colletotrichum trifolii, Current Genetics* **14**, 241–246.

Donner, S.C., Jones, D.A., McClure, N.C., Rosewarne, G.M., Tate, M.E., Kerr, A., Fajardo, N.N. and Clare, B.G. (1993) Agrocin 434, a new plasmid encoded agrocin from the biocontrol *Agrobacterium* strains K84 and K1026, which inhibits biovar 2 agrobacteria, *Physiological and Molecular Plant Pathology* **42**, 185–194.

Draborg, H., Kauppinen, S., Dalbøke, H. and Christgau, S. (1995) Molecular cloning and expression in *S. cerevisiae* of two exochitinases from *Trichoderma harzianum, Biochemistry and Molecular Biology International* **36**, 781–791.

Ellis, J.G. and Kerr, A. (1979) Transfer of agrocin production from strain 84 to pathogenic recipients: A comment on the previous paper, in B. Schippers and W. Gams (eds.), *Soil Borne Plant Pathogens,* Academic Press, London, pp. 579–583.

Ellis, J.G., Kerr, A., van Montagu, M. and Schell, J. (1979). *Agrobacterium*: Genetic studies on agrocin 84 production and the biological control of crown gall, *Physiological Plant Pathology* **15**, 311–319.

Farrand, S.K., Slota, J.E., Shim, J.S. and Kerr, A. (1985) Tn5 insertions in the agrocin 84 plasmid: The conjugal nature of pAgK84 and the locations of determinants for tranfer- and agrocin 84 production, *Plasmid* **13**, 106–117.

Faull, J.L. and Graeme-Cook, K. (1992) Characterization of mutants of *Trichoderma harzianum* with altered antibiotic production characteristics, in E.C. Tjamos, G.C. Papavizas and R.J. Cook (eds.), *Biological Control of Plant Diseases,* Plenum Press, New York, pp. 345–351.

Frey, P., Prior, P., Marie, C., Kotoujansky, A., Trigalet-Demery, D. and Trigalet, A. (1994) Hrp⁻ mutants of *Pseudomonas solanacearum* as potential biocontrol agents of tomato bacterial wilt, *Applied and Environmental Microbiology* **60**, 3175–3181.

Fry, J.C. and Day, M.J. (1992) *Release of Genetically Engineered & Other Micro-organisms,* Cambridge University Press, Cambridge.

Gaffney, T., Lam, S., Ligon, J., Gates, K., Frazelle, A., Di Maio, J., Hill, S., Goodwin, S., Torkewitz, N., Allshouse, A., Kempf, H.-J. and Becker, J. (1994) Global regulation of expression of antifungal factors by a *Pseudomonas fluorescens* biological control strain, *Molecular Plant-Microbe Interaction* **7**, 455–463.

García, I., Lora, J.M., de la Cruz, J., Benítez, T., Llobell, A. and Pintor-Toro, J.A. (1994) Cloning and characterization of a chitinase (CHIT42) cDNA from the mycoparasitic fungus *Trichoderma harzianum, Current Genetics* **27**, 83–89.

Glick, B.R. (1995) The enhancement of plant growth by free-living bacteria, *Canadian J. of Microbiology* **41**, 109–117.

Glick, B.R. and Bashan, Y. (1997) Genetic manipulation of plant growth-promoting bacteria to enhance biocontrol of phytopathogens, *Biotechnology Advances* **15**, 353–378.

Goldman, G.H., van Montagu, M. and Herrera-Estrella, A. (1990) Transformation of *Trichoderma harzianum* by high-voltage electric pulse, *Current Genetics* **17**, 169–174.

Hammer, P.E., Hill, D.S., Lam, S.T., van Pée, K.-H. and Ligon, J.M. (1997) Four genes from *Pseudomonas fluorescens* that encode the biosynthesis of pyrrolnitrin, *Applied and Environmental Microbiology* **63**, 2147–2154.

Haran, S., Schickler, H., Pe'er, S., Logemann, S., Oppenheim, A. and Chet, I. (1993) Increased constitutive chitinase activity in transformed *Trichoderma harzianum*, *Biological Control* **3**, 101–108.

Harman, G.E. and Taylor, A. (1989) Cloning effective strains of *Trichoderma harzianum* and solid matrix priming to improve biological seed treatment, *Plant Disease* **73**, 631–637.

Hayes, C.K., Klemsdal, S., Lorito, M., Di Pietro, A., Peterbauer, C., Nakas, J.P., Tronsmo, A. and Harman, G.E. (1994) Isolation and sequence of an endochitinase-encoding gene from a cDNA library of *Trichoderma harzianum*, *Gene* **138**, 143–148.

Herrera-Estrella, A., Goldman, G.H. and van Montagu, M. (1990) High-efficiency transformation system for the biocontrol agents, *Trichoderma* spp., *Molecular Microbiology* **4**, 839–843.

Hill, D.S., Stein, J.I., Torkewitz, N.R., Morse, A.M., Howell, C.R., Pachlatko, J.P., Becker, J.O. and Ligon, J.M. (1994) Cloning of genes involved in the synthesis of pyrrolnitrin from *Pseudomonas fluorescens* and role of pyrrolnitrin synthesis in biological control of plant disease, *Applied and Environmental Microbiology* **60**, 78–85.

Howitt, R.L.J., Beever, R.E., Pearson, M.N. and Forster, R.L.S. (1995) Presence of double-stranded RNA and virus-like particles in *Botrytis cinerea*, *Mycological Research* **99**, 1472–1478.

Jones, D.A. and Kerr, A. (1989) *Agrobacterium radiobacter* K1026, a genetically engineered derivative of strain K84, for biological control of crown gall, *Plant Disease* **73**, 15–18.

Jones, D.A., Ryder, M.H., Clare, B.G., Farrand, S.K. and Kerr, A. (1988) Construction of a Tra- deletion mutant of pAgK84 to safeguard the biological control of crown gall, *Molecular & General Genetics* **212**, 207–214.

Juan-Abgona, R.V., Katsuno, N., Kageyama, K. and Hyakumachi, M. (1996) Isolation and identification of hypovirulent *Rhizoctonia* spp. from soil, *Plant Pathology* **45**, 896–904.

Keel, C., Schnider, U., Maurhofer, M., Boisard, C., Laville, J., Burger, U., Wirthner, P., Haas, D. and Défago, G. (1992) Suppression of root diseases by *Pseudomonas fluorescens* CHA0: Importance of the bacterial secondary metabolite 2,4-diacetylphloroglucinol, *Molecular Plant-Microbe Interactions* **5**, 4–13.

Kerr, A. (1980) Biological control of crown gall through production of agrocin 84, *Plant Disease* **64**, 25–30.

Kerr, A. and Htay, K. (1974) Biological control of crown gall through bacteriocin production, *Physiological Plant Pathology* **4**, 37–44.

Kirimura, K., Imura, M., Lee, S.P., Kato, Y. and Usami, S. (1989) Intergeneric hybridization between *Aspergillus niger* and *Trichoderma viride* by protoplast fusion, *Agricultural and Biological Chemistry* **53**, 1589–1596.

Kousik, C.S., Snow, J.P. and Valverde, R.A. (1994) Comparison of double-stranded-RNA components and virulence among isolates of *Rhizoctonia solani* AG-1 IA and AG-1 IB, *Phytopathology* **84**, 44–49.

Laville, J., Voisard, C., Keel, C., Maurhofer, M., Défago, G. and Haas, D. (1992) Global control in *Pseudomonas fluorescens* mediating antibiotic synthesis and suppression of black root rot of tobacco, *Proceedings of the National Academy of Sciences of the United States of America* **89**, 1562–1566.

Leong, J., Bitter, W., Koster, M., Marugg, J.D., Venturi, V. and Weisbeck, P.J. (1991) Molecular analysis of iron transport in plant growth-promoting *Pseudomonas putida* WC358, in C. Keel, B. Koller and G. Défago (eds.), *Plant Growth-Promoting Rhizobacteria – Progress and Prospects*, IOBC/WPRS Bulletin **14**(8), 127–135.

Ligon, J.M., Lam, S.T., Gaffney, T.D., Hill, D.S., Hammer, P.E. and Torkewitz, N. (1996) Biocontrol: Genetic modifications for enhanced antifungal activity, in G. Stacey, B. Mullin and P.M. Gresshoff (eds.), *Biology of Plant-Microbe Interactions*, Int. Soc. Plant-Microbe Interactions, St. Paul, Minn., pp. 457–462.

Limón, M.C., Lora, J.M., García, I., de la Cruz, J., Llobell, A., Benítez, T. and Pintor-Toro, J.A. (1995) Primary structure and expression pattern of the 33-kDa chitinase gene from the mycoparasitic fungus *Trichoderma harzianum*, *Current Genetics* **28**, 478–483.

Lindemann, J. and Suslow, T.V. (1987) Competition between ice nucleation-active wild type and ice nucleation-deficient deletion strains of *Pseudomonas syringae* and *P. fluorescens* biovar I and biological control of frost injury on strawberry blossoms, *Phytopathology* **77**, 882–886.

Loper, J.E. and Ishimaru, C.A. (1991) Factors influencing siderophore-mediated biocontrol activity of rhizosphere *Pseudomonas* spp., in D.L. Keister and P.B. Cregan (eds.), *The Rhizosphere and Plant Growth*, Kluwer Academic Publishers, Dordrecht, pp. 253–261.

Lorito, M., Hayes, C.K., Di Pietro, A. and Harman, G.E. (1993) Biolistic transformation of *Trichoderma harzianum* and *Gliocladium virens* using plasmid and genomic DNA, *Current Genetics* **24**, 349–356.

Margolles-Clark, E., Harman, G.E. and Penttilä, M. (1996) Enhanced expression of endochitinase in *Trichoderma harzianum* with the *chb1* promoter of *Trichoderma reesei*, *Applied and Environmental Microbiology* **62**, 2152–2155.

Marugg, D., Weger, L.A. de, Neilander, H.B., Oorthuizen, M., Recourt, K., Lugtenberg, B., van der Hofstad, G.A.J.M. and Weisbeek, P.J. (1989) Cloning and characterization of a gene encoding an outer membrane protein required for siderophore-mediated uptake of Fe^{3+} in *Pseudomonas putida* WCS358, *J. of Bacteriology* **171**, 2819–2826.

Maurhofer, M., Keel, C., Schnider, U., Voisard, C., Haas, D. and Défago, G. (1992) Influence of enhanced antibiotic production in *Pseudomonas fluorescens* strain CHA0 on its disease suppressive capacity, *Phytopathology* **82**, 190–195.

Miller, J.F. (1972) *Experiments in Molecular Genetics*, Cold Spring Laboratory, New York.

Moenne Loccoz, Y., McHugh, B., Stephens, P.M., McConnel, F.I., Glennon, J.D., Dowling, D.N. and Ogara, F. (1996) Rhizosphere competence of fluorescent *Pseudomonas* sp. B24 genetically modified to utilise additional ferric siderophores, *FEMS Microbiology Ecology* **19**, 215–225.

Monteiro-Vitorello, C.B., Bell, J.A., Fulbright, D.W. and Bertrand, H. (1995) A cytoplasmically transmissible hypovirulence phenotype associated with mitochondrial-DNA mutations in the chestnut blight fungus *Cryphonectria parasitica*, *Proceedings of the National Academy of Sciences of the United States of America* **92**, 5935–5939.

Morris, J., O'Sullivan, D.J., Koster, M., Leong, J., Weisbeck, P.J. and O'Gara, F. (1992) Characterization of fluorescent siderophore-mediated iron uptake in *Pseudomonas* sp. strain M114: Evidence for the existence of an additional ferric receptor, *Applied and Environmental Microbiology* **58**, 630–635.

Nuss, D.L. (1992) Biological control of chestnut blight: An example of virus-mediated attenuation of fungal pathogenesis, *Microbiological Reviews* **56**, 561–576.

Orbach, M.J., Porro, E.B. and Yanovsky, C. (1986) Cloning and characterization of the gene for β-tubulin and its use as a dominant marker, *Molecular and Cellular Biology* **6**, 2452–2461.

Papavizas, G.C., Roberts, D.P. and Kim, K.K. (1990) Development of mutants of *Gliocladium virens* tolerant to benomyl, *Canadian J. of Microbiology* **36**, 484–489.

Penttilä, M., Nevalainen, H., Rättö, M., Salminen, E. and Knowles, J. (1987) A versatile transformation system for the cellulytic filamentous fungus *Trichoderma reesei*, *Gene* **61**, 155–164.

Peterbauer, C.K., Lorito, M., Hayes, C.K., Harman, G.E. and Kubicek, C.P. (1996) Molecular cloning and expression of the *nag1* gene (N-acetyl-β-D-glucosaminidase-encoding gene) from *Trichoderma harzanum* P1, *Current Genetics* **30**, 325–331.

Postma, J. and Luttikholt, A.J.G. (1993) Benomyl-resistant *Fusarium*-isolates in ecological studies on the biological control of *Fusarium* wilt in carnation, *Netherlands J. of Plant Pathology* **99**, 175–188.

Ratto, M., Niku Paavola, M.L., Raaska, L., Mattila Sandholm, T. and Viikari, L. (1996) The effect of *Trichoderma harzianum* siderophores on yeasts and wood-rotting fungi, *Material und Organismen* **30**, 279–292.

Ryder, M. (1994) Key issues in the deliberate release of genetically-manipulated bacteria, *FEMS Microbiology Ecology* **15**, 139–146.

Ryder, M.H., Slota, J.E., Scarim, A. and Farrand, S.K. (1987) Genetic analysis of agrocin 84 production and immunity in *Agrobacterium* spp., *J. of Bacteriology* **169**, 4184–4189.

Schnider, U., Keel, C., Blumer, C., Troxler, J., Défago, G. and Haas, D. (1995) Amplification of the housekeeping sigma factor in *Pseudomonas fluorescens* CHA0 enhances antibiotic production and improves biocontrol abilities, *J. of Bacteriology* **177**, 5387–5392.

Schroth, M.N. (1992) Risks of releasing wild-type and genetically engineered biocontrol organisms into the ecosystem, in E.C. Tjamos, G.C. Papavizas and R.J. Cook (eds.), *Biological Control of Plant Diseases*, Plenum Press, New York, pp. 371–379.

Shapira, R., Ordentlich, A., Chet, I. and Oppenheim, A.B. (1989) Control of plant diseases by chitinase expressed from cloned DNA in *Escherichia coli*, *Phytopathology* **79**, 1246–1249.

Simon, R., Priefer, U. and Puhler, A. (1983) A broad host range mobilization system for *in vivo* genetic engineering: Transposon mutagenesis in gram negative bacteria, *Bio/Technology* **1**, 784–791.

Sitrit, Y., Barak, Z., Kapulnik, Y., Oppenheim, A.B. and Chet, I. (1993) Expression of *Serratia marcescens* chitinase gene in *Rhizobium meliloti* during symbiosis on alfalfa roots, *Molecular Plant-Microbe Interactions* **6**, 293–298.

Stasz, T.E., Harman, G.E. and Weeden, N.F. (1988) Protoplast preparation and fusion in two biocontrol strains of *Trichoderma harzianum*, *Mycologia* **80**, 141–150.

Sun, X.Y., Griffith, M., Pasternak, J.J. and Glick, B.R. (1995) Low-temperature growth, freezing survival, and production of antifreeze protein by the plant-growth promoting rhizobacterium *Pseudomonas putida* GR12-2, *Canadian J. of Microbiology* **41**, 776–784.

Sundheim, L., Poplawsky, A. and Elllingboe, H. (1988) Molecular cloning of two chitinase genes from *Serratia marcescens* and their expression in Pseudomonas species, *Physiological and Molecular Plant Pathology* **33**, 483–491.

Tenning, P., van Rijsbergen, R., Zhao, Y. and Joos, H. (1993) Cloning and transfer of genes for antifungal compounds from *Erwinia herbicola* to *Escherichia coli*, *Molecular Plant-Microbe Interactions* **6**, 474–480.

Thomas, M.D. and Kenerley, C.M. (1989) Transformation of the mycoparasite *Gliocladium*, *Current Genetics* **15**, 415–420.

Trigalet, A. and Demery, D. (1986) Invasiveness in tomato plants of Tn5-induced avirulent mutants of *Pseudomonas solanacearum*, *Physiological and Molecular Plant Pathology* **28**, 423–430.

Trigalet, A. and Trigalet-Demery, D. (1990) Use of avirulent mutants of *Pseudomonas solanacearum* for the biological control of bacterial wilt of tomato plants, *Physiological and Molecular Plant Pathology* **36**, 27–38.

Vicedo, B., Penalver, R., Asins, M.J. and López, M.M. (1993) Biological control of *Agrobacterium tumefaciens*, colonization, and pAgK84 transfer with *Agrobacterium radiobacter* K84 and the Tra⁻ mutant strain K1026, *Applied and Environmental Microbiology* **59**, 309–315.

Voisard, C., Rella, M. and Haas, D. (1988) Conjugative transfer of plasmid RP1 to soil isolates of *Pseudomonas fluorescens* is facilitated by certain large RP1 deletions, *FEMS Microbiology Letters* **55**, 9–14.

Wang, C.L., Farrand, S.K. and Hwang, I. (1994) Organization and expression of the genes on pAgK84 that encode production of agrocin 84, *Molecular Plant-Microbe Interactions* **7**, 472–481.

Winkelmann, G., Lupp, R. and Jung, G. (1980) Herbicolins – New peptide antibiotics from *Erwinina herbicola*, *J. of Antibiotics* **33**, 353–358.

CHAPTER 26

PRODUCTION AND COMMERCIALIZATION OF
BIOCONTROL PRODUCTS

Deborah R. Fravel, David J. Rhodes and Robert P. Larkin

26.1. Introduction

Commercialization of a biocontrol product is likely to conjure up somewhat different images in the minds of the research plant pathologist and the industry representative. The researcher may view the process as a series of challenges from designing rational screening procedures to find microbes with biocontrol potential, initial discovery of a microbe with potential, studies of the mechanisms of action of the organism, knowledge of the ecological and biological requirements for efficacy of the biocontrol agent, compatibility with common pesticides and agricultural practices and testing under production conditions. In addition to acquisition of these basic data, there are several pragmatic considerations to producing this microbe on a large scale and formulating it so that it has acceptable shelf life and other desired characteristics imparted by formulation. The industry representative's first thoughts about commercialization of a biocontrol agent are likely to be about market size, cost effectiveness, ability to produce, formulate and distribute the product and the ability to patent or otherwise protect the investment. This chapter addresses the scientific challenges of producing and formulating biocontrol agents on a commercial scale. The chapter focuses on living microbes for control of plant diseases, although some microbial insecticides and herbicides are included to illustrate particular points. Previous reviews on related topics include those by Lisansky (1985, 1997), Jutsum (1988), Rhodes (1990, 1996), Newton *et al.* (1996) and Fravel *et al.* (1998).

26.2. Production and Scale up

Perhaps the most significant reason for the limited commercial acceptance of microbial control agents is the high cost of production, resulting in low profit margins. This may be due to the inherently high cost of substrate, low biomass productivity, or limited economies of scale (Rhodes, 1996). The latter problem is related to low market volume, which may not justify the use of large, dedicated fermentors.

Two methods are commonly used for producing inoculum of biocontrol microbes – liquid and solid fermentation. Because industry has developed equipment and methods for large-scale liquid fermentation for production of microbial products such as antibiotics, enzymes and organic acids, this expertise provides a starting point for production of biocontrol microbes. Although information developed for production of microbial products can be used as a guideline, each organism is different and specific schedules for aeration, pH control, nutrients and other requirements must be developed for each organism (Rhodes, 1990, 1993; Slininger and Shea-Wilbut, 1995).

R. Albajes et al. (eds.), Integrated Pest and Disease Management in Greenhouse Crops, 365-376.
© 1999 *Kluwer Academic Publishers. Printed in the Netherlands.*

Many biocontrol microbes are easily produced in the laboratory in liquid culture, but do not produce the expected quantity or quality of propagule when produced in large scale. One of the most important differences between small- and large-scale liquid fermentation is gas exchange. Small-scale fermentation flasks are often placed on a shaker to increase the amount of oxygen available to the biocontrol agent. In large-scale fermentation, the amount of air or oxygen introduced into the fermentor can be carefully controlled. Both the amount of inoculum produced and the type of propagule produced can be affected by the amount of oxygen during fermentation. For example, the number of colony-forming units of the mycoherbicide *Fusarium oxysporum* Schlechtend.:Fr. was significantly greater when dissolved oxygen (DO) was high than when DO was low (Hebbar *et al.*, 1997). However, the percentage of chlamydospores was significantly higher when DO was low. The pH of the medium also affected the amount and type of propagule produced as well as the percentage of chlamydospores.

Altering the medium to produce the desired propagules of the antagonist is also important. The survival stage structure of the organism is generally considered to be the preferred propagule for formulation since it is the most likely to provide adequate shelf life, particularly under the adverse environmental conditions that may be encountered during shipping and storage. Consequently, chlamydospores rather than conidia are preferred for *Trichoderma* (Lewis and Papavizas, 1983), while ascospore rather than conidia are preferred for *Talaromyces flavus* (Klöcker) A.C. Stolk & R.A. Samson (Fravel *et al.*, 1985). Similarly, bacteria such as *Bacillus* that produce endospores survive well in formulation. Some bacteria, such as *Agrobacterium*, are easily dried (Kerr, 1980) and can be provided as dry cells or formulated further. Because gram-negative bacteria, such as the genera *Pseudomonas* and *Burkholderia*, do not form specific survival structures during fermentation, they are difficult to formulate into long-term viable formulations.

The production medium affects not only the type of propagule formed, but also the efficacy of these propagules. For example, pH, temperature and carbon source regulate the phenazine activity of the biocontrol bacterium *Pseudomonas fluorescens* (Trevisan) Migula 2–79 (Slininger and Shea-Wilbur, 1995). Nutrition during production can affect the efficacy of biocontrol without affecting the type of propagule produced. For example, carbon and nitrogen sources that slightly increased ascospore production of *T. flavus* reduced efficacy of biocontrol of Verticillium wilt compared with ascospores produced on potato dextrose agar (Engelkes *et al.*, 1997).

If the biocontrol organism has been produced by liquid fermentation, will it be shipped as is, or will it require further formulation? If shipped as a liquid, is it necessary to reduce the volume of liquid (i.e. concentrate the inoculum)? Is the final product a dry formulation? If so, then microbes produced in liquid fermentation need to be dried. Usually drying is accomplished by first separating the propagules from the production medium by filtration and/or centrifugation. The resulting dry biomass will likely be milled for further formulation. Although there is general agreement that drying is often a very critical step in production, there is little published data on the effect of drying on shelf life or biocontrol efficacy. Some biocontrol agents begin to germinate if the drying is too slow while a longer drying time increases the chances for microbial contamination. Likewise, rapid drying may cause cell membrane damage, particularly if heating is used to speed drying. A fluid-bed dryer has been used to successfully dry atoxigenic strains of *Aspergillus flavus* Link:Fr. and *Aspergillus parasiticus* Speare to the desired level of water activity (Daigle *et al.*, 1997).

Although we often automatically think of liquid fermentation when we think of large-scale production, the capability for large-scale solid fermentation has also been developed. For ease of shipping and handling, it may be desirable to have the final form of the product as a solid. Thus, producing the inoculum by solid fermentation, rather than liquid, may save the labour and technical difficulties in separation of the inoculum from the substrate and drying the material. For example, Sylvan Spawn Laboratories (Cabot, PA) routinely produces mushroom spawn in solid fermentation. They have used this same technology to produce inoculum of the mycoparasite *Sporidesmium sclerotivorum* Uecker, Ayers & Adams (Ayers and Adams, 1983). Large-scale fermentation has also been developed for *Trichoderma* spp. (Roussos *et al.*, 1991; Durand *et al.*, 1993) and other biocontrol fungi (Durand *et al.*, 1993).

In addition to the biological considerations in choosing liquid or solid fermentation and in the selection of the growing media and other manipulations of the system, the cost of these materials, the amount of time the fermentation system will be tied up and the labour involved must be considered also. In an effort to reduce costs, inexpensive waste products such as molasses, peanut hulls, corn cobs, fish meal, various chitin sources, yeast extracts, soy bean hulls and others have been used for production of biocontrol agents. Although inexpensive waste materials used for fermentation may lower the apparent cost, they may increase the variability in the product produced.

26.3. Formulation

After propagules of the biocontrol agent are produced, they generally must be further formulated before use. The formulation can affect many aspects in the biocontrol success of the product. Some of the obvious benefits include greater efficacy, increased shelf life, ease of handling, increased safety (i.e. reduced inhalation or skin permeability), proper coverage of the target area, compatibility with agricultural equipment and practices and lower production costs.

Formulation of microbial agents presents several fundamental problems (Rhodes, 1993). First, microbial cells must typically be maintained in a stable form during long periods of exposure to uncontrolled temperature fluctuations as the product passes through the distribution chain. Limited shelf life, or a requirement for refrigeration is only likely to be compatible with highly specialized distribution systems. Products based on *Bacillus thuringiensis* Berliner are relatively stable, since the organism contains both a resistant spore and a proteinaceous parasporal crystal, but other products based on non-sporing bacteria, fungi and nematodes are likely to be much more difficult to formulate. Secondly, biological agents are exposed to hostile and fluctuating environmental conditions, particularly water activity and ultraviolet light during and after application, resulting in a rapid reduction in population density. This is less of an issue with greenhouse crops than in open field situations and may be alleviated to some extent by formulation, but it is still a significant barrier to commercialization. Unfortunately, the wealth of formulation technology available within the chemical industry is typically directed toward problems of distribution, uptake and safety rather than stabilization of labile active ingredients such as microbial cells.

The type of formulation desired depends on the intended use. For example, a granular material would be more appropriate for combining with potting mix, while a wettable powder would be more appropriate for root dips or sprays. Although many formulations for microbial pesticides are modifications of those used for chemical pesticides, others have been developed specifically for biocontrol agents. Many biocontrol agents can be supplied in clays such as talc, pyrophyllite or kaolinite, or in other carriers such as peat, vermiculite or lignite (Kloepper and Schroth, 1981; Vidhyasekaran et al., 1997). Alginate formulations have been used for a variety of biocontrol agents (Fravel et al., 1985; Lewis and Papavizas, 1985; Bashan, 1986; Magan and Whipps, 1988; DeLucca et al., 1990; Knudsen et al., 1991; Mintz and Walter, 1993), but because of the cost involved in the production of alginate formulations, other technologies are being explored. "Pesta" uses pasta-making technology to produce granules containing biocontrol organisms (Connick et al., 1991a, 1993; Daigle et al., 1997). Similar extruded formulations have been used successfully to formulate mycoherbicides and the biocontrol fungi Gliocladium virens J.H. Miller, J.E. Giddens & A.A. Foster and Trichoderma spp. (Hebbar et al., 1996, 1998; Lewis and Larkin, 1997). Pregelatinized starches, modified by the addition of 10 ml isopropanol/100 g corn flour, are inexpensive and easy to prepare and have been used to formulate G. virens and Trichoderma hamatum (Bonord.) Bainier for the control of damping-off induced by Rhizoctonia solani Kühn (Lewis et al., 1995). Similarly, biopolymers such as rice flour have been used to entrap microbial insecticides (Bok et al., 1996). Invert emulsions (water in dispersed phase surrounded by oil in continuous phase) have also received attention because of their ability to protect living organisms (Daigle et al., 1990; Connick et al., 1991b; Womack et al., 1996).

Various compounds can be added to formulations to improve efficacy, shelf life, environmental tolerance, or ease of handling. Biocontrol efficacy of the product can often be improved by altering the nutritional status of the formulation. Some biocontrol agents, such as Trichoderma and Gliocladium spp., perform better when a food base such as wheat bran is present (Elad et al., 1980; Elad and Hadar, 1981; Lewis and Papavizas, 1985). Others, such as T. flavus, provide better control when the formulation is poor in nutrients (Fravel et al., 1995). In addition, in some cases, specialized nutrient sources added to the formulation that can be utilized by the biocontrol agent, but not by the pathogen, may also enhance efficacy (Chun et al., 1997).

Shelf life can usually be extended by reducing the oxygen content of the formulation through vacuum packaging or addition of oxygen scavengers. Lowering the water content of the formulation can also extend shelf life (Connick et al., 1996). Similarly, addition of compounds to regulate osmotic tolerance may also extend shelf life. Refrigeration usually extends shelf life, but is not practical in most cases. Optical brighteners have been added to formulations to protect entomopathogenic viruses from ultraviolet light (UV) (Shapiro, 1992; Shapiro and Robertson, 1992). These or other UV protectants may be useful in formulation of biocontrol fungi and bacteria.

26.4. Registration

One of the major hurdles to be overcome in the commercialization process is registration of the organism. Registration of biocontrol agents in the European Union (EU) has historically been the responsibility of the EU member states. Under national legislation, a number of bacterial, fungal and viral agents have been registered, particularly in France, Spain, the UK and the Netherlands. In order to provide a consistent legal framework throughout the EU, microbial agents were specifically addressed by Directive 91/414/EEC "Concerning the placement of plant protection products on the market". There has been considerable subsequent debate surrounding this issue, however, since the Directive was felt by many producers of biocontrol agents to draw too heavily on guidelines devised for chemicals. The issue of registration of microbial agents in Europe has still not been fully resolved, creating considerable uncertainty in the industry.

Regulations concerning registration vary from country to country. The regulations may even vary within a country if a particular state or province decides to enforce more stringent guidelines than the national standard, as is the case with the state of California in the US. Generally, one of the first facts that needs to be established is that the particular strain of the microbe can be uniquely identified so that one can determine with certainty the fate of the organism released into the environment. Depending on the type of microbe involved, this can be provided by a DNA fingerprint, antibiotic profile, fatty acid profile, or identification of another unique physiological or morphological characteristic. The ability to identify with certainty the particular strain is likely to be required prior to receiving permission for large-scale (>2 ha) field testing and is necessary to establish purity during manufacturing.

Toxicology data will be required for registration to establish that the microbe, the process for producing and formulating the microbe and the formulated product are not toxic to humans. The types of tests required may depend on the microbial product and its intended use. In the US, toxicology tests are divided into three tiers of increasing toxicological concern. Tier I consists of a battery of short-term tests designed to identify any potential for toxicity, infectivity or pathogenicity. A microbe that shows any adverse data in Tier I testing would be an unlikely candidate for registration. Toxicology data submitted to the US Environmental Protection Agency (EPA) to support registration of a microbe must be the product of an independent laboratory and not those from the group seeking registration. In some cases, the EPA may grant waivers from selected toxicity tests based on published reports in peer-reviewed journals. Data are also required to determine how much residue from the microbe or microbial toxins might appear on or in food and whether this residue constitutes a health hazard.

The EPA has four tiers of testing for adverse environmental effects that might occur from the microbial pesticide. These include non-target toxic or pathogenic effects on both plants and animals. The tests required may vary according to the product and its intended use.

Related to toxicology and environmental fate issues is the question of risk assessment. For *G. virens*, Lumsden and Walter (1995) have described the mammalian toxicity risks, as well as risks for other parts of the environment. Gullino *et al.* (1995) provided a case study for the use of antagonistic Fusaria. They compared naturally occurring saprophytic *F. oxysporum*, UV-induced mutant hybrids derived from protoplast fusions and transformed strains for persistence and survival, effects on indigenous microbial communities and pathogenicity and toxicity. They concluded that saprophytic *F.*

oxysporum did not pose a risk to natural or agricultural ecosystems. Similar studies have been conducted with strains of *Pseudomonas* (De Leij *et al.*, 1995; Weller *et al.*, 1995; Défago *et al.*, 1996). These studies demonstrated that data needed to make rational decisions about the release of microbes can be collected. It is important that the scientific community communicates to the public the soundness of these data so that the decisions can be based on actual, rather than perceived, risk. This need for communication appears to be particularly important in the case of genetically altered organisms.

26.5. Barriers to Commercialization

Despite at least 25 years of research into biological control of plant pests and diseases, microbial agents currently comprise less than 1% of the total world market of crop protection products, accounting for some US$75–200 million of sales (Newton *et al.*, 1996; Lisansky, 1997). Although the proportion in the greenhouse disease control sector may be somewhat higher, these products still represent no more than a specialized market niche. This limited market penetration is despite a consensus view among the public, growers and regulators that such products are, in theory, a valuable and desirable component of integrated pest management strategies.

There may be several explanations for this apparent anomaly. Biological agents tend to be highly specific in their activity. While this may be desirable from an environmental standpoint, extreme specificity also has the effect of restricting market potential to situations where it is economic to control one or two species only. A related factor is that large companies tend to be characterized by large sales forces and substantial fixed cost bases, both of which may be difficult to support on the basis of niche products. Conversely, small specialist companies are unlikely to have sufficient resources and market presence to ensure rapid uptake of their products.

Despite lower regulatory costs associated with biocontrol products, much of the cost of product development is made up of field testing, manufacturing and formulation research. These costs are largely independent of whether the active ingredient is chemical or biological in nature. While the cost of registration of biocontrol products is, in most cases, significantly lower than for their chemical counterparts, specific regulatory guidelines for biological products in many parts of the world are incompletely formulated and subject to frequent change. Not all countries adopt a favourable registration track for biologicals. This uncertainty continues to discourage investment in the area.

26.6. Commercially Available Products

Although still clearly in the minority of available pesticides, approximately 40 commercial products containing live microbes for the control of plant diseases are now available (Table 26.1). Most commercial products contain beneficial fungi or bacteria, although actinomycetes and bacteriophages are also represented. The products are targeted against a wide variety of plant diseases including soilborne, foliar and post-harvest diseases. The products are applied to a wide range of crops including flowers, ornamentals, vegetables, field and row crops, turf, trees and fruit. The Biocontrol of Plant Diseases Laboratory, ARS, USDA, maintains a current list of commercially available biocontrol products, including addresses and phone or fax numbers for the manufacturers or distributors of these products (http://www.barc.usda.gov/psi/bpdl/bpdl.html). Anyone with updates to the products list is encouraged to send the information to the e-mail address indicated at the web site.

Diseases in greenhouse settings make logical targets for the use of biocontrol because the low biological diversity of soilless potting mixes and control over environmental parameters, such as water and temperature, make it easier to establish biocontrol agents than in field situations. Consequently, some products, such as SoilGard™, were developed specifically with greenhouse diseases in mind and other uses for the product were subsequently found (Lumsden *et al.*, 1996).

TABLE 26.1. Commercial products containing live microbes for control of plant pathogens

Product/biocontrol organism	Target pathogen/crop disease	Crop	Source
AQ10 Biofungicide *Ampelomyces quisqualis*	Powdery mildew	Apples, cucurbits, grapes, ornamentals, strawberries, tomatoes	Ecogen, Inc., USA, Israel
Aspire *Candida oleophila*	*Botrytis* spp., *Penicillium* spp.	Citrus, pome fruit, post-harvest	Ecogen, Inc., USA, Israel
Binab T *Trichoderma harzianum* *Trichoderma polysporum*	Pathogenic fungi causing wilt, take-all, root rot, internal decay of wood products and decay in tree wounds	Flowers, fruit, ornamentals, turf, vegetables	Bio-Innovation AB, Sweden, UK
Biofox C *Fusarium oxysporum* (non-pathogenic)	*Fusarium oxysporum* *Fusarium moniliforme*	Basil, carnation, cyclamen, tomato	ȘIAPA, Italy
Bio-Fungus (formerly Anti-Fungus) *Trichoderma* spp.	*Sclerotinia*, *Phytophthora*, *Rhizoctonia solani*, *Pythium* spp., *Fusarium, Verticillium*	Flowers, strawberries, trees, vegetables	Grondortsmettingen DeCuester, Belgium
Bio-Save 10 *Pseudomonas syringae*	*Botrytis cinerea*, *Penicillium* spp., *Mucor piriformis, Geotrichum candidum*	Citrus and pome fruit, post-harvest	EcoScience Corp., USA
Bio-Save 11 *Pseudomonas syringae*	*Botrytis cinerea*, *Penicillium* spp., *Mucor piriformis, Geotrichum candidum*	Citrus and pome fruit, post-harvest	EcoScience Corp., USA
BlightBan A506 *Pseudomonas fluorescens*	Frost, *Erwinia amylovora*	Almond, apple, cherry, peach, pear, potato, strawberry, tomato	Plant Health Technologies, USA

TABLE 26.1. Commercial products containing live microbes for control of plant pathogens (cont.)

Product/biocontrol organism	Target pathogen/crop disease	Crop	Source
Blue Circle *Pseudomonas cepacia* (ex *Burkholderia*)	*Fusarium, Pythium,* lesion, spiral, lance and sting nematodes	Vegetables	CTT Corp., USA
Conquer *Pseudomonas fluorescens*	*Pseudomonas tolaasii*	Mushrooms	Mauri Foods, Australia, USA
Contans *Coniothyrium minitans*	*Sclerotinia sclerotiorum* and *Sclerotinia minor*	Canola, sunflower, peanut, soybeans, vegetables (lettuce, bean, tomato)	Prophyta Biologischer Pflanzenschutz GmbH, Germany
Deny (formerly PRECEP) *Pseudomonas cepacia* (ex *Burkholderia*) (see Blue Circle)	*Rhizoctonia, Pythium, Fusarium,* lesion, spiral, lance and sting nematodes	Alfalfa, barley, beans, clover, cotton, peas, grain sorghum, vegetable crops, wheat	CTT Corp., USA
DiTera *Myrothecium verrucaria*	Root-knot, citrus, cyst, stubby root, sting, lesion and burrowing nematodes	Fruit, vegetable and ornamental crops, turf	Abbott Laboratories, USA
Epic *Bacillus subtilis*	*Rhizoctonia solani, Fusarium* spp., *Alternaria* spp. and *Aspergillus* spp.	Cotton, legumes	Gustafson, Inc., USA
Fusaclean *Fusarium oxysporum* (non-pathogenic)	*Fusarium oxysporum*	Asparagus, basil, carnation, cyclamen, gerbera, tomato	Natural Plant Protection, France
Galltrol-A *Agrobacterium radiobacter*	Crown gall disease, *Agrobacterium tumefaciens*	Fruit, nut, ornamental nursery stock	AgBioChem, Inc., USA
Intercept *Pseudomonas cepacia* (ex *Burkholderia*)	*Rhizoctonia solani, Fusarium* spp., *Pythium* sp.	Maize, vegetables, cotton	Soil Technologies Corp., USA
Kodiak, Kodiak HB, Kodiak A·T *Bacillus subtilis*	*Rhizoctonia solani, Fusarium* spp., *Alternaria* spp. and *Aspergillus* spp.	Cotton, legumes	Gustafson, Inc., USA
Mycostop *Streptomyces griseoviridis*	*Fusarium* spp., *Alternaria brassicicola, Phomopsis* spp., *Botrytis* spp., *Pythium* spp. and *Phytophthora* spp.	Field, ornamental, vegetable crops	Kemira Agro Oy, Finland
Nogall, Diegall *Agrobacterium radiobacter*	*Agrobacterium tumefaciens*	Trees	Bio-Care Technology Pty. Ltd., Australia
Norbac 84C *Agrobacterium radiobacter*	Crown gall disease, *Agrobacterium tumefaciens*	Fruit, nut, ornamental nursery stock	New BioProducts, Inc., USA
Phagus Bacteriophage	*Pseudomonas tolaasii*	*Agaricus* spp. *Pleurotus* spp.	Natural Plant Protection, France

TABLE 26.1. Commercial products containing live microbes for control of plant pathogens (cont.)

Product/biocontrol organism	Target pathogen/crop disease	Crop	Source
Polygandron *Pythium oligandrum*	*Pythium ultimum*	Sugarbeet	Vyskumny ustav rastlinnej [Plant Production Institute], Slovak Republic

PSSOL
  Pseudomonas solanacearum
  (non-pathogenic)
	Pseudomonas solanacearum	Vegetables	Natural Plant Protection, France

Rotstop, P. g. Suspension
  Phanerochaete gigantea
  (= *Phlebia gigantea*)
	Heterobasidium annosum	Trees	Kemira Agro Oy, Finland

SoilGard (formerly GlioGard) *Gliocladium virens*	*Rhizoctonia solani* and *Pythium* spp.	Ornamental and food plants in greenhouses, nurseries, homes and interiorscapes	ThermoEcotek (formerly W. R. Grace & Co.), USA
Supresivit *Trichoderma harzianum*	Various fungi		Produced in Czech Republic

System 3
  Bacillus subtilis and
  chemical pesticides
	Seedling pathogens	Barley, beans, cotton, peanut, pea, rice, soybean	Helena Chemical Co., USA

T-22G and T-22HB
  (RootShield)
  Trichoderma harzianum T-22
	Pythium spp., *Rhizoctonia solani*, *Fusarium* spp. and *Sclerotinia homoeocarpa*	Bean, cabbage, corn, cotton, cucumber, peanut, shrubs, sorghum, soybean, sugar beet, tomato, transplants, trees, all ornamentals	BioWorks, Inc. (formerly TGT, Inc.), USA

Trichodex *Trichoderma harzianum* T-39	Primarily *Botrytis cinerea, Sclerotinia sclerotiorum*, also *Colletotrichum* spp., *Fulvia fulva, Monilia laxa, Plasmopara viticola, Pseudoperonospora cubensis, Rhizopus stolonifer*	Cucumber, grape, nectarine, soybean, strawberry, sunflower, tomato	Makhteshim Chemical Works Ltd, Israel

Trichopel, Trichoject, Trichodowels, Trichoseal
  Trichoderma harzianum and
  Trichoderma viride
	Armillaria, Botryosphaeria, Chondrostereum, Fusarium, Nectria, Phytophthora, Pythium, Rhizoctonia		Agrimm Technologies Ltd., New Zealand

Trichoderma 2000 *Trichoderma* sp.	*Rhizoctonia solani, Sclerotium rolfsii, Pythium* sp.	Nursery and field crops	Mycontrol Ltd, Israel
Victus *Pseudomonas fluorescens*	*Pseudomonas tolaasii*	Mushrooms	Sylvan Spawn Laboratory, USA

26.7. Outlook

Greenhouse crops present, in many ways, an ideal opportunity for the adoption of biological control strategies, being characterized by relatively high profit margins and the ability to exert a high level of environmental control. However, the economic constraints presented by restricted market opportunities, high production costs and complications with formulation and application represent fundamental barriers to the commercial success of microbial control in the industry and will need to be further addressed. The fact that so many microbes have survived the journey from discovery to the market shelf is encouraging for the future of biological pesticides. How well these products are accepted by the growers and profit earned by the products will help to determine the future of biocontrol products.

There are numerous differences between biological and chemical pesticides. These differences are apparent at all stages of discovery, development, commercialization, sale and use of the pesticides. Thus, everyone involved in the process, from researchers to industry personnel, to registering agencies, to pesticide producers, to sales people, to the growers and the consumers, must play roles somewhat different from those they played with chemical pesticides. It will take some time to sort out these roles and develop new paradigms for biocontrol agents. Communication across these groups is vital for the process to succeed.

References

Ayers, W.A. and Adams, P.B. (1983) Improved media for growth and sporulation of *Sporidesmium sclerotivorum, Canadian J. of Microbiology* **29**, 325–330.

Bashan, Y. (1986) Alginate beads as synthetic inoculant carriers for slow release of bacteria that affect plant growth, *Applied and Environmental Microbiology* **51**, 1089–1098.

Bok, S.H., Son, K.H., Lee, H.W., Choi, D. and Kim, S.U. (1996) Bioencapsulated biopesticides, in T. Wenhua, R. J. Cook and A. Rovira, (eds.), *Advances in Biological Control of Plant Diseases,* China Agricultural University Press, Beijing, pp. 303–309.

Chun, S-.C., Schneider, R.W., Groth, D.E. and Giles, C.G. (1997) Effect of different carbon sources on efficacy of biological control of seedling diseases in water-seeded rice, *Phytopathology* **87**, S19.

Connick, W.J., Jr., Boyette, C.D. and McAlpine, J.R. (1991a) Formulations of mycoherbicides using a pasta–like process, *Biological Control* **1**, 281–287.

Connick, W.J., Jr., Daigle, D.J., Boyette, C.D., Williams, K.S., Vinyard, B.T. and Quimby, P.C., Jr. (1996) Water activity and other factors that affect the viability of *Colletotrichum truncatum* conidia in wheat flour-kaolin granules ("Pesta"), *Biocontrol Science and Technology* **6**, 277–284.

Connick, W.J., Jr., Daigle, D.J. and Quimby, P.C., Jr. (1991b) An improved invert emulsion with high water retention for mycoherbicide delivery, *Weed Technology* **5**, 442–444.

Connick, W.J., Jr., Nickle, W.R. and Vinyeard, B.T. (1993) "Pesta": New granular formulations for *Steinernema carpocapsae, J. of Nematology* **25**, 198–203.

Daigle, D.J., Connick, W.J., Jr., Boyette, C.D., Lovisa, M.P., Williams, K.S. and Watson, M. (1997) Twin–screw extrusion of "Pesta"-encapsulated biocontrol agents, *World J. of Microbiology and Biotechnology* **13**(6), 671–676.

Daigle, D.J., Connick, W.J., Jr., Quimby, P.C., Jr., Evans, J., Trask-Morrell, B. and Fulgham, F.E. (1990) Invert emulsions: Carrier and water source for the mycoherbicide *Alternaria cassiae, Weed Technology* **4**, 327–331.

De Leij, F.A.M., Sutton, E.J., Whipps, J.M., Fenlon, J.S. and Lynch, J.M. (1995) Impact of field release of genetically modified *Pseudomonas fluorescens* on indigenous microbial populations of wheat, *Applied and Environmental Microbiology* **61**, 3443–3453.

Défago, G., Keel, C. and Moënne-Loccoz, Y. (1996) Fate of introduced biocontrol agent *Pseudomonas fluorescens* CHA0 in soil: Biosafety considerations, in T. Wenhua, R.J. Cook and A. Rovira (eds.), *Advances in Biological Control of Plant Diseases*, China Agricultural University Press, Beijing, pp. 241–245.

DeLucca, A.J., II, Connick, W.J., Jr., Fravel, D.R., Lewis, J.A. and Bland, J.M. (1990) The use of bacterial alginates to prepare biocontrol formulations, *J. of Industrial Microbiology* 6, 129–134.

Durand, A., Renaud, R., Almanza, S., Maratray, J., Diez, M. and Desgranges, C. (1993) Solid state fermentation reactors: From lab scale to pilot plant, *Biotechnology Advances* 11, 591–597.

Elad, Y., Chet, I. and Katan, J. (1980) *Trichoderma harzianum*: A biocontrol agent effective against *Sclerotium rolfsii* and *Rhizoctonia solani*, *Phytopathology* 70, 119–121.

Elad, Y. and Hadar, Y. (1981) Biological control of *Rhizoctonia solani* by *Trichoderma harzianum* in carnation, *Plant Disease* 65, 675–677.

Engelkes, C.A., Nuclo, R.L. and Fravel, D.R. (1997) Effect of carbon, nitrogen and C:N ratio on growth, sporulation and biocontrol efficacy of *Talaromyces flavus*, *Phytopathology* 87, 500–505.

Fravel, D.R., Connick, W.J., Jr. and Lewis, J.A. (1998) Formulation of microorganisms to control plant diseases, in H.D. Burgess (ed.), *Formulation of Microbial Biopesticides, Beneficial Microorganisms and Nematodes*, Chapman and Hall, London, (in press).

Fravel, D.R., Lewis, J.A. and Chittams, J.L. (1995) Alginate prill formulations of *Talaromyces flavus* with organic carriers for biocontrol of *Verticillium dahliae*, *Phytopathology* 85, 165–168.

Fravel, D.R., Marois, J.J., Lumsden, R.D. and Connick, W.J., Jr. (1985) Encapsulation of potential biocontrol agents in an alginate-clay matrix, *Phytopathology* 75, 774–777.

Gullino, M.L., Migheli, Q. and Mezzalama, M. (1995) Risk analysis in the release of biological control agents, *Plant Disease* 79, 1193–1201.

Hebbar, K.P., Lewis, J.A., Poch, S.M. and Lumsden, R.D. (1996) Agricultural by-products as substrates for growth, conidiation and chlamydospore formation by a potential mycoherbicide, *Fusarium oxysporum* strain EN-4, *Biocontrol Science and Technology* 6, 263–275.

Hebbar, K.P., Lumsden, R.D., Lewis, J.A., Poch, S.M. and Bailey, B.A. (1998) Formulation of mycoherbicidal strains of *Fusarium oxysporum*, *Weed Science* 46, 501–507.

Hebbar, K.P., Lumsden, R.D., Poch, S.M. and Lewis, J.A. (1997) Liquid fermentation to produce biomass of mycoherbicidal strains of *Fusarium oxysporum*, *Applied Microbiology and Biotechnology* 48, 714–719.

Jutsum, A.R. (1988) Commercial application of biological control: Status and prospectus, *Philosophical Transactions of the Royal Society of London. Series B* 318, 357–373.

Kerr, A. (1980) Biological control through production of Agrocin 84, *Plant Disease* 64, 25–30.

Kloepper, J.W. and Schroth, M.N. (1981) Development of a powder formulation of Rhizobacteria for inoculation of potato seedpieces, *Phytopathology* 71, 590–592.

Knudsen, G.R., Eschen, D.J., Dandurand, L.M. and Wang, Z.G. (1991) Method to enhance growth and sporulation of pelletized biocontrol fungi, *Applied and Environmental Microbiology* 57, 2864–2867.

Lewis, J.A., Fravel, D.R., Lumsden, R.D. and Shasha, B.S. (1995) Application of fungi in granular formulations of pregelatinized starch–flour to control damping-off diseases caused by *Rhizoctonia solani*, *Biological Control* 5, 397–404.

Lewis, J.A. and Larkin, R.P. (1997) Extruded granular formulation with biomass of biocontrol *Gliocladium virens* and *Trichoderma* spp. to reduce damping-off of eggplant caused by *Rhizoctonia solani* and saprophytic growth of the pathogen in soil-less mix, *Biocontrol Science and Technology* 7, 49–60.

Lewis, J.A. and Papavizas, G.C. (1983) Production of chlamydospores and conidia by *Trichoderma* spp. in liquid and solid growth media, *Soil Biology & Biochemistry* 15, 351–357.

Lewis, J.A. and Papavizas, G.C. (1985) Characteristics of alginate pellets formulated with *Trichoderma* and *Gliocladium* and their effect on the proliferation of the fungi in soil, *Plant Pathology* 34, 571–577.

Lisansky, S.G. (1985) Production and commercialization of pathogens, in N.W. Hussey and N. Scopes (eds.), *Biological Pest Control*, Blanford Press, Poole, pp. 210–218.

Lisansky, S. (1997) Microbial biopesticides, in H.F. Evans (ed.), *Microbial insecticides: Novelty or necessity*, BCPC Symp. Proc. No. 68, Warwick, Coventry, 16–18 April 1997, PCPC, Farnham, pp. 3–10.

Lumsden, R.D. and Walter, J.F. (1995) Development of the biocontrol fungus *Gliocladium virens*: Risk assessment and approval for horticultural use, in H.M.T. Hokkanen and J.M. Lynch (eds.), *Biological Control: Benefits and Risks*, Cambridge Univ. Press, Cambridge, pp. 263–269.

Lumsden, R.D., Walter, J.F. and Baker, C.P. (1996) Development of *Gliocladium virens* for damping–off disease control, *Canadian J. of Plant Pathology* **18**, 463–468.

Magan, N. and Whipps, J.M. (1988) Growth of *Coniothyrium minitans, Gliocladium roseum, Trichoderma harzianum* and *T. viride* from alginate pellets and interaction with water availability, *Bulletin OEPP* **18**, 37–45.

Newton, P.J., Neale, M.C., Arslan-Bir, M., Brandl, M., Figett, M.J. and Greatrex, R.M. (1996) Full–range pest management with IPM systems – An industry view of the options for non-indigenous biopesticides, in J.K. Waage (ed.), *Biological control introductions – Opportunities for improved crop production*, BCPC Symp. Proc. No. 67, Brighton, 18 Nov 1996, BCPC, Farnham, pp. 77–97.

Rhodes, D.J. (1990) Formulation requirements for biological control agents, *Aspects of Applied Biology* **24**, 145–153.

Rhodes, D.J. (1993) Formulation of biological control agents, in D.G. Jones (ed.), *Exploitation of Microorganisms*, Chapman & Hall, London, pp. 411–439.

Rhodes, D.J. (1996) Economics of baculovirus–Insect cell production systems, *Cytotechnology* **20**, 291–297.

Roussos, S., Olmos, A., Raimbault, M., Saucedo-Castañeda, G. and Lonsane, B.K. (1991) Strategies for large scale inoculum development for solid state fermentation system: Conidiospores of *Trichoderma harzianum, Biotechnology Techniques* **5**, 415–240.

Shapiro, M. (1992) Use of optical brighteners as radiation protectants for gypsy moth (Lepidoptera: Lymantriidae) nuclear polyhedrosis virus, *J. of Economic Entomology* **85**, 1682–1686.

Shapiro, M. and Robertson, J.L. (1992) Enhancement of gypsy moth (Lepidoptera: Lymantriidae) baculovirus activity by optical brighteners, *J. of Economic Entomology* **85**, 1120–1124.

Slininger, P.J. and Shea-Wilbut, M.A. (1995) Liquid-culture pH, temperature and carbon (not nitrogen) source regulate phenazine productivity of the take-all biocontrol agent *Pseudomonas fluorescens* 2-79, *Applied Microbiology and Biotechnology* **43**, 794–800.

Vidhyasekaran, P., Sethuraman, K., Rajappan, K. and Vasumathi, K. (1997) Powder formulations of *Pseudomonas fluorescens* to control pigeonpea wilt, *Biological Control* **8**, 166–171.

Weller, D.M., Thomashow, L.S. and Cook, R.J. (1995) Biological control of soil-borne pathogens of wheat: Benefits, risks and current challenges, in H.M.T. Hokkanen and J.M. Lynch (eds.), *Biological Control: Benefits and Risks*, Cambridge Univ. Press, Cambridge, pp. 149–160.

Womack, J.G., Eccleston, G.M. and Burge, M.N. (1996) A vegetable oil-based invert emulsion for mycoherbicide delivery, *Biological Control* **6**, 23–28.

CHAPTER 27

EVALUATION OF RISKS RELATED TO THE RELEASE OF BIOCONTROL
AGENTS ACTIVE AGAINST PLANT PATHOGENS

Jan Dirk van Elsas and Quirico Migheli

27.1. Introduction

Biological control of plant pathogens in greenhouse and field cropping greatly relies on
the use of antagonistic micro-organisms. These organisms can provide protection against
an array of foliar, soilborne or post-harvest pathogens, with presumably no negative
effects on the ecosystem. When released into soil, irrigation water, or onto plant organs,
biocontrol agents may protect the plants from attack by plant pathogens through
mechanisms such as antibiosis, competition, or parasitism by lytic enzymes (Chet, 1993).
In order to compete with conventional plant disease management, biological control has to
be effective, reliable, consistent and economical. This can be accomplished only by
developing superior antagonists (Chet *et al.*, 1993) and suitable delivery systems (Harman,
1992; Trevors *et al.*, 1992). One important approach is the genetic boosting of the
biocontrol potential via molecular means. The biocontrol efficacy of antagonistic micro-
organisms, such as *Pseudomonas* spp., *Agrobacterium* spp., *Trichoderma* spp.,
Gliocladium virens J.H. Miller, J.E. Giddens & A.A. Foster, or saprophytic *Fusarium* spp.,
has been enhanced by improving wild strains using conventional mutagenesis (Papavizas,
1987), protoplast fusion (Migheli *et al.*, 1992, 1995; Harman and Hayes, 1993) or genetic
modification (Lindow *et al.*, 1989; van Elsas *et al.*, 1991b, 1994; Chet *et al.*, 1993; Flores
et al., 1997). These efforts led to the production of new antagonistic strains, some of
which are now in the process of registration.

The application of novel biotechnology products has been suggested to pose potential
hazard to the environment as well as to sensitive organisms. Initial fears originated from
the notion that ecosystems and organisms might be disturbed by exposure to novel genetic
combinations. Hence, considerable effort has been put into investigating factors
determining hazard and into designing safe ways of releasing candidate organisms for
application. The EU BAP, BRIDGE and BIOTECH research programmes have provided
ample support for these endeavours. In spite of the fact that not a single case of clear
adverse effects due to the release of a genetically modified micro-organism has occurred
to date, there is a consensus now that releases of "novel" organisms should be preceded by
a careful examination of their threat to the environment (OTA, 1988; Tiedje *et al.*, 1989;
Cairns and Orvos, 1992; Tzotzos, 1995). This attitude has even pervaded the area of
unmodified biocontrol agents (Défago *et al.*, 1997). Hence, before commercialization of
micro-organisms in agricultural environments, their behaviour and impact in ecosystems
should be evaluated (van Elsas *et al.*, 1998). Regulations in many countries now require an
analysis of environmental impact as part of an application for registration and commercial
development of genetically modified as well as unmodified biocontrol agents.

This review will focus on crucial events in the release of biocontrol micro-organisms

R. Albajes et al. (eds.), Integrated Pest and Disease Management in Greenhouse Crops, 377-393.
© 1999 *Kluwer Academic Publishers. Printed in the Netherlands.*

that determine their fate and potential adverse effects. It will be argued that factors that affect biosafety often also affect the efficacy of the application.

27.2. Factors for Consideration in Biosafety Studies

After its deliberate or accidental release, an antagonistic micro-organism may affect the physical or biological environment in different ways. Among the main factors to be considered for an adequate assessment of adverse effects we will refer to: (i) establishment and survival of released biocontrol agents; (ii) dispersal of released biocontrol agents; (iii) genetic stability of, and horizontal genetic transfer from, the introduced micro-organisms; (iv) effects of the introduced micro-organism on the resident microflora and fauna (e.g. pathogenicity, virulence, allergenicity and toxicity towards humans, animals and plants); and (v) availability and applicability of effective containment systems.

27.3. Establishment and Survival of Released Biocontrol Agents

An effective biocontrol strain should be able to persist at high population density for adequate activity after introduction into soil or phyllosphere. Soil is a complex and heterogeneous environment consisting of solid, liquid and gaseous phases, which all affect the fate of the micro-organisms present (van Elsas *et al.*, 1991a). Main factors in soil controlling the proliferation of micro-organisms are limitations of substrate and soil water. In addition, soil structure and texture, pH and temperature affect bacterial and fungal fate. Introduced organisms will face the often harsh soil conditions in relatively open (unprotected) soil sites. In addition, stress conditions may be posed upon introduced micro-organisms by competing, antagonistic and predatory indigenous soil organisms. The role of predation by soil protozoa in limiting the population size of introduced bacteria has been firmly established (e.g. Heijnen *et al.*, 1988). Thus, populations of non-pathogenic, non-adapted micro-organisms usually decline once introduced into soil and may reach low levels (van Elsas *et al.*, 1994). The possible conversion into viable but non-culturable forms, which escape cultivation-based detection, is a point of concern, as such forms conceptually form an environmental reservoir (van Elsas *et al.*, 1998). Van Veen *et al.* (1997) have recently reviewed the factors that cause declines of biocontrol agents in soil as well as their physiological response upon introduction. Survival of biocontrol agents (such as antagonistic *Trichoderma* spp.) in the phylloplane is also often limited by fluctuating and adverse environmental conditions, which are not buffered like in soil (Elad, 1990; Migheli *et al.*, 1994).

Pre-release survival studies should be conducted in contained environments, such as soil microcosms, growth chambers or glasshouses, where natural conditions can be simulated in a limited space (Teuben and Verhoef, 1992). To achieve this goal, unambiguous detection of the target organism is necessary. This can be accomplished through the use of selectable markers (e.g. antibiotic or fungicide resistance, nutritional complementation, induced mutations which confer particular colony morphology) or via molecular techniques [PCR/hybridization, restriction fragment length polymorphisms (RFLP), REP/ERIC/BOX-PCR fingerprinting, electrophoretic karyotyping, nucleic acid hybridization techniques]. The use of these methods has been described in the Molecular Microbial Ecology Manual (Akkermans *et al.*, 1995) and in Trevors and van Elsas (1995). Several other reviews have also discussed the use of different marker systems in environmental monitoring protocols (Prosser, 1994; Jansson, 1995; Smalla and van Elsas, 1996; van Elsas *et al.*, 1998). An array of different markers, mainly for bacteria, is nowadays available for monitoring, and markers for fungi are emerging. In the case of antagonistic *Fusarium* spp. that are active against phytopathogenic formae speciales of *F. oxysporum* Schlechtend.:Fr., the use of colour and fungicide resistance markers in combination with electrophoretic karyotyping (Migheli *et al.*, 1993), ERIC/REP-PCR, RFLP (Edel *et al.*, 1995), Southern analysis of transforming sequences (Migheli *et al.*, 1996) and random amplification of polymorphic DNA (RAPD) (Migheli and Cavallarin, 1994) allowed the recognition of selected antagonistic strains, even several months after their introduction into both disinfested and non-disinfested soils and in the plant rhizosphere (Mezzalama *et al.*, 1994). RAPD fingerprinting proved effective for typing of antagonistic *Trichoderma* spp. (Zimand *et al.*, 1993; Arisan-Atac *et al.*, 1995) and for assignment of strains to species (Turner *et al.*, 1997). RFLP analysis of the internal transcribed spacer (ITS) region of nuclear ribosomal DNA was recently carried out on a world-wide collection of *Ampelomyces* spp. isolates, hyperparasites of powdery mildew fungi; seven RFLP groups were detected, but no correlation between geographical origin and genetic similarity was found. With respect to this, the role of agricultural commerce in the spread of *Ampelomyces* hyperparasites cannot be excluded (Kiss, 1997).

Establishment and survival of introduced biocontrol agents is determined by the presence, either naturally or in an added form, of specific (protective) habitats or substrates which provide the organisms with a survival or growth advantage in comparison with the indigenous soil microflora. The presence and accessibility of either a protective microhabitat or a specific degradable substrate at the moment of introduction of the agent probably determines the population levels at which the introduced organisms establish initially and survive in later stages (van Veen *et al.*, 1997). Clay minerals such as bentonite added to soil can improve inoculant survival and blooming of predatory protozoa has been shown to be concomitantly reduced (Heijnen *et al.*, 1988). This reduced protozoan proliferation was attributed to a change in soil pore size class distribution. Bentonite was assumed to increase the volume of pores with a relatively small neck, which are accessible to inoculant bacteria and inaccessible to protozoa (van Elsas *et al.*, 1991a; Heijnen *et al.*, 1992). Alternatively, a protective and/or nutritional matrix can be provided around cells to be released, in which these can survive and establish (Trevors *et al.*, 1992; Rhodes, 1993). Traditionally, undefined matrices (carriers) such as peat have been employed for *Rhizobium* inoculants. However, due to inconsistencies encountered with peat, the use of more defined matrices has been increasingly explored. Powdered or slurried bentonite clay has been suggested for this purpose (Heijnen *et al.*, 1992). Moreover, polymeric matrices such as alginate have also been examined (Trevors *et al.*, 1992). Survival of the potential biocontrol agent *Pseudomonas fluorescens* (Trevisan) Migula cells in alginate beads (2–3 mm) in a loamy sand soil was significantly improved over survival of unencapsulated cells in soil (Trevors *et al.*, 1992; van Elsas *et al.*, 1992). Furthermore, colonization of wheat roots by the alginate-encapsulated cells was not inhibited (van Elsas *et al.*, 1992).

The use of defined protective matrices such as alginate for inoculation thus offers great potential for enhancing the persistence of biocontrol agents in soil, and it can easily be adapted to allow the inoculation of plant parts or seeds. At the same time, the potential for adverse effects may be enhanced by the use of these carriers, and this aspect needs careful scrutiny.

27.4. Dispersal of Released Biocontrol Agents

The dispersal of biocontrol agents has important biosafety implications (Dighton *et al.*, 1997). In contrast, spread can be desirable in cases where a better spread of a "point" inoculum is required, for instance when uncolonized parts of root systems have to be reached. Active motility of biocontrol agents does not play a great role for translocation over greater distances (Madsen and Alexander, 1982; Trevors *et al.*, 1990). Such transport is mainly passive and can be brought about by biological (soil animals, developing plant roots) or physical (wind and water) factors.

Soil animals (moles, earthworms, insects) as well as plant roots can move around micro-organisms. Insects such as the cutworm *Peridroma saucia* (Hübner) have been implicated in the spread of genetically-marked gram-negative bacteria able to colonize their intestinal tracts (Armstrong *et al.*, 1989). Burrowing earthworms (*Lumbricus* spp.) can also translocate inoculant bacteria (Madsen and Alexander, 1982; Henschke *et al.*, 1989), and introduced bacteria might survive in the gut for a period of up to 50 days. Growing plant roots also can translocate bacteria present on seeds or root parts over cm-scale distances (Parke *et al.*, 1986; Bahme and Schroth, 1987). However, Madsen and Alexander (1982) and Trevors *et al.* (1990) found that some organisms, e.g. fluorescent pseudomonads, apparently have a limited capacity of hitch-hiking along with growing roots. On the other hand, antagonistic *Trichoderma* and *Fusarium* spp. presented a high level of rhizosphere competence, which allows them to establish along developing roots (Ahmad and Baker, 1987; Garibaldi *et al.*, 1990). The mode of inoculant application, either directly on the root/seed or mixed within the soil, will affect the degree of plant root-induced translocation, as micro-organisms will contact roots to different extents.

Wind-induced transport is a factor to be dealt with in any large-scale release of biocontrol agents (Dighton *et al.*, 1997). Shearing of the soil surface leading to aerosolization of soil particles is the main process involved. Wind speed, duration and direction, soil texture, plant leaf condition and age, inoculum age, spore concentration, cultural practices and resident fauna or microfauna can affect the dispersal (Cairns and Orvos, 1992). The effects of formulation, soil texture, cultural practices and resident microflora have been evaluated for several bacteria (Donegan *et al.*, 1992; Kluepfel, 1992; Hekman *et al.*, 1994). Aerosolized *Pseudomonas syringae* van Hall (ice⁻) cells sprayed over an experimental potato plot (Lindow and Panopoulos 1988) were found, in low numbers, meters away from the release spot. Bacteria introduced into soil top layers also translocated over meter-scale distances due to wind movement of soil particles (Knudsen 1989), and their numbers decreased drastically at increasing distances from the release spot. Mathematical models such as that developed by Knudsen (1989) adequately described the distribution of the introduced cells. Epidemiological models were also developed in the case of *Chondrostereum purpureum* (Pers.:Fr.) Pouzar, a fungal biocontrol agent for the wild blackcherry *Prunus serotina* Ehrh., to determine low-risk use areas in the Netherlands (De Jong *et al.*, 1990).

Another physical factor that can transport inoculant cells through soil is water movement. Such transport depends on inoculum cell properties (e.g. cell size, type of capsular material, hydrophobicity and/or surface charge), on physical soil properties (texture, pH, temperature, clay mineral content, pore size distribution or soil structure) and on biological soil properties, i.e. the activity of the indigenous community. Cellular properties determine the extent to which inoculants interact with soil particles (Peterson and Ward, 1989). Introduced bacteria may largely starve in soil (van Overbeek *et al.*, 1995), but some cells might divide, causing the production of "ultramicrocells", which are presumably easily translocated by water. Harvey *et al.* (1989) found transport of bacteria to be even faster than that of small latex beads, and an effect of surface charge. Moreover, cells with high surface hydrophobicity can adhere more strongly to a solid surface than cells of low hydrophobicity (van Loosdrecht *et al.*, 1987).

Physical soil properties also play a role in water-induced transport, since they affect the degree of inoculum adsorption. Bacterial cells adsorb onto clay minerals such as kaolinite and montmorillonite (Breitenbeck *et al.*, 1988), whereas they may adsorb poorly in sandy soil. Clay minerals are therefore important in the adsorption of bacteria to soil, causing decreased dispersal. The finding of a greater degree of inoculant transport in a loamy sand as compared to a loam soil corroborates this (Trevors *et al.*, 1990). Furthermore, the presence of irregularities in soil structure such as cracks, macropores or root channels influences bacterial dispersal by affecting water flow patterns (Smith *et al.*, 1985).

The mode of application also affects the extent of dispersal of biocontrol agents. Genetically-marked *P. fluorescens* applied to soil in alginate beads showed reduced transport as compared to unencapsulated cells (Hekman *et al.*, 1994). Over 95% of the cells in alginate beads remained in the inoculated layer of the column, whereas in the control this percentage was only about 70. This low transport rate, coupled with the high survival rate and good rhizosphere and rhizoplane colonization, suggests that alginate can promote the *in situ* persistence of biocontrol agents. The type of soil and formulation used also influenced dispersal of antagonistic *F. oxysporum* propagules through different soils in vertical column microcosms (Mezzalama *et al.*, 1994; Gullino *et al.*, 1995). In a sandy loam, recovery of *F. oxysporum* from percolating water was observed up to 60 days from release both as a talc powder and an alginate pellet formulation. In a soil-based potting mix, dispersal through water beyond 14 days from release was observed only when the antagonists were formulated in alginate pellets. Encapsulation of *F. oxysporum* in alginate beads enhanced their survival capability (Mezzalama *et al.*, 1994).

In summary, inoculant dispersal through soil via biological factors (soil animals and roots) may be fairly limited in scale (Madsen and Alexander, 1982; Parke *et al.*, 1986; Bahme and Schroth, 1987), and may only affect a relatively small proportion of the bacterial population. Transport with wind or percolating water on the other hand will be less localized (scale of meters or more) and affect a large part of the bacterial population. Indeed, Madsen and Alexander (1982) showed that percolating water caused a 100-fold greater dispersal than a developing plant root or an earthworm. Combinations of water and plant or earthworm respectively did not differ from water alone. Dispersal of bacteria applied to plant roots through the rhizosphere was also stimulated by percolating water (Parke *et al.*, 1986; Bahme and Schroth, 1987). Hence, there is ample and strong evidence that pinpoints water as the major dispersing agent for inoculant cells in soil.

27.5. Genetic Stability and Transfer of Genes to Indigenous Micro-organisms

Genetically modified inoculant bacteria are commonly extensively screened for the stability of the construct as well as their ecological fitness (e.g. van Elsas *et al.*, 1991b, 1994). Chromosomal insertions, the preferred strategy for modification, have generally been found to be stable, and a slight reduction of growth rate and fitness has been noted in, for instance, modified fluorescent pseudomonads (van Elsas *et al.*, 1991b, 1994). These constructs have been proposed as ecologically acceptable, as they presumably could perform well, and at the same time did not outcompete the parent organism. Genetic stability of transformed biocontrol fungi is also a key factor in their safety. Any evidence for genetic instability in contained, pre-release experiments should prompt a re-evaluation of the antagonist and its release, as changes in the transformant sequences may lead to erroneous biosafety assessments (Leslie and Dickman, 1991). This is particularly true for fungi used as weed control agents. Many genetically altered plant pathogenic fungi, such as *Glomerella cingulata* (Stoneman) Spauld. & H. Schrenk (Rodriguez and Yoder, 1987), *F. oxysporum* (Kistler and Benny, 1988) and *Cochliobolus heterostrophus* (Drechs.) Drechs. (Keller *et al.*, 1991) can be mitotically unstable when passaged on a plant host or cultured under non-selective conditions (Kistler, 1991). A change in the genetic structure may eventually lead to a modification of the host range, thus representing a serious environmental hazard.

The mitotic stability of antagonistic fungi can be determined through comparative Southern analysis of the introduced DNA sequences, by using DNA extracted from cultures derived from cells obtained before and after a release into the environment. Southern analysis of nine hygromycin B resistant transformants of antagonistic *F. oxysporum* showed that all but one underwent loss of plasmids during *in vitro* growth without selective pressure or after release in soil microcosms (Migheli *et al.*, 1996). This high instability is not surprising in the case of multicopy transformants, where homologous recombination events between plasmid copies scattered around the genome might be responsible for a rearrangement of transforming DNA. In contrast, with four hygromycin B resistant transformants of antagonistic *Trichoderma harzianum* Rifai no modification in the restriction pattern of the introduced DNA and in the plasmid copy number per haploid genome were observed both after *in vitro* growth without selective pressure and after release onto the phylloplane of tomato plants (Migheli *et al.*, 1994). This high stability of transforming DNA was probably due to the presence of homologous DNA sequences in the plasmid used, pHATa (Herrera-Estrella *et al.*, 1990), a derivative of pAN7-1 which carries a 2.4 kb fragment of a *T. harzianum* putative α-amylase gene inserted into its unique *Hind*III site. Also, different transformed strains of antagonistic *T. harzianum* were shown to be mitotically stable after release into soil (Pe'er *et al.*, 1991; Flores *et al.*, 1997).

Methods to enhance the stability of transforming DNA are particularly needed in the case of biocontrol fungi. Promising developments are the identification of trapping DNA sequences and their use in constructing new plasmid vectors, useful to increase the frequency of targeted and stable integration events, electroporation, restriction enzyme-mediated integration (REMI) (Schliestl and Petes, 1991) of plasmid DNA, which might be effective in generating single insertions into genomic restriction sites, transposon-based gene tagging and transformation systems, which proved effective in the case of *F. oxysporum* (Daboussi and Langin, 1994).

Transfer of heterologous genes to indigenous micro-organisms is a major biosafety concern. The three mechanisms of gene transfer between bacteria, i.e. transduction, transformation and conjugation, might all function in soil and the phylloplane. While there is emerging evidence for the occurrence of transformation (Nielsen *et al.*, 1997a,b) and transduction (Zeph *et al.*, 1988) in soil, most experimental evidence has been provided for the occurrence of conjugation (Smit *et al.*, 1991, 1993, 1995; Wellington and van Elsas, 1992; Day and Fry, 1992b).

A prerequisite for conjugal transfer in soil is cell-to-cell contact. The chance that a donor meets a potential recipient is dependent on the *in situ* donor and recipient population densities. Postma and van Veen (1990) suggested that a large part of the accessible pore space may be hostile to bacterial cells, as less than 0.5% of the total pore space was occupied by bacterial cells. Thus, the "diluted" distribution of microbial cells through soil may limit conjugation between them. Most initial experiments on bacterial gene transfer in soil have focused on detection of transconjugants upon co-introduction of donor (usually containing a selftransmissible plasmid) and recipient strains (Stotzky, 1989; Wellington and van Elsas, 1992). The effect on gene transfer of various soil parameters, such as sterile versus non-sterile soil, addition of nutrients, soil temperature and moisture content, pH, clay content, chemical pollution and the presence of plants has been studied (van Elsas *et al.*, 1988a,b). Addition of nutrients increased the frequency of plasmid transfer in soil between introduced bacteria (van Elsas *et al.*, 1988b). The presence of plant (wheat) roots has also been shown to enhance plasmid transfer frequencies (van Elsas *et al.*, 1988a; Smit *et al.*, 1991). Bacteria that are metabolically activated by nutrients, e.g. in root exudates, apparently form mating pairs more readily than bacteria in an energy-depleted state. In sterilized soil, gene transfer occurred more frequently than in non-sterilized soil (Wellington and van Elsas, 1992). This confirmed that microbial activation enhances transfer frequencies, since bacteria in sterile soil are able to grow, whereas those in non-sterile soil may convert to a state of starvation (van Overbeek *et al.*, 1995).

In the last 6 years, gene transfer from soil-introduced to indigenous bacteria has become increasingly known (Henschke and Schmidt, 1990; Smit *et al.*, 1991, 1993, 1995). Henschke and Schmidt (1990) detected plasmid mobilization from an introduced donor to indigenous pseudomonads in soil by using an ecologically unfit donor (*Escherichia coli* Castellani & Chalmers). Smit *et al.* (1991), using a *P. fluorescens* donor strain and bacteriophage-based donor counterselection, found that the broad-host-range plasmid RP4 was transferred to various indigenous soil bacteria in the rhizosphere of wheat. This work was later extended to include mobilization of an IncQ plasmid, pSKTG (Smit *et al.*, 1993). On the other hand, no evidence was found for transfer of a chromosomally inserted gene cassette to the indigenous community (Smit *et al.*, 1995). Kluepfel *et al.* (1991) released a root-colonizing strain of *Pseudomonas aureofaciens* Kluyver modified by the insertion of *E. coli lacZY* genes into the rhizosphere. More than 10,000 bacterial rhizosphere isolates were screened for the presence of these sequences. No transfer into any microbe of the rhizosphere was ever detected (Kluepfel *et al.*, 1991). Hence, plasmids are likely to be transferred under rhizosphere conditions, whereas chromosomal inserts indeed can be shown to be mostly confined to the host. Moreover, the potential of plant rhizospheres to provide gene mobilizing capacity to incoming bacteria has been indicated to be low (van Elsas *et al.*, in preparation).

In filamentous fungi, genetic recombination may occur through several mechanisms.

Mitotic and meiotic recombination were studied extensively in *Aspergillus* and *Neurospora*, while parasexual genetic exchange via hyphal anastomosis is still poorly characterized. Reassortment of characters through parasexual recombination can be artificially induced by protoplast fusion, which may help to overcome vegetative incompatibility. However, post-fusion vegetative incompatibility may occur even if protoplast fusion is utilized, resulting in low levels of recovery of progeny and in slow growth of heterokaryons. Thus, parasexuality in filamentous fungi should be considered as a rare event in nature, which does not play a significant role in the horizontal transfer of characters.

In considering the biosafety of biocontrol agents that are able to differentiate into a sexual stage in nature, the presence of related phytopathogenic species in the introduction area, as well as the existence of common hosts that might support hybridization should be carefully evaluated (Weidemann, 1991). Cereal rusts were shown to be able to hybridize on wild grasses by producing progeny with altered host range (Eshed and Dinoor, 1981) and similar results have been obtained with related *Cochliobolus* species (Kline and Nelson, 1971). The uncertainty should be more acceptable for micro-organisms that lack a sexual reproductive system, as the frequency of genetic exchanges would be reduced.

An assessment of the location of inserted DNA in the genome of fungal inoculants is important (OTA, 1988; Comeaux *et al.*, 1990). In this respect, engineered fungal antagonists may have a lower potential for horizontal gene transfer, as transforming DNA sequences are usually integrated into chromosomes, although the existence of extrachromosomal autonomously replicating plasmids in some filamentous fungi has been demonstrated (Kistler, 1991; Timberlake, 1992). The transfer of an autonomously replicating vector carrying hygromycin B resistance was shown to occur between vegetatively incompatible biotypes of *Colletotrichum gloeosporioides* (Penz.) Penz. & Sacc. in Penz., a pathogenic fungus infecting *Stylosanthes* spp. in Australia (Poplawski *et al.*, 1997).

Moreover, different classes of transposable elements (TE) have been characterized in many fungal species (Daboussi, 1996). Some elements affected gene structure and function through gene inactivation upon insertion, modification of nucleotide sequence through excision, and chromosome rearrangement. Increasing data suggest that TE may be horizontally transmitted. The uneven distribution of *Tad* in *Neurospora* sp. (Kinsey, 1990) and of *grh* in *Magnaporthe grisea* (T.T. Hebert) Yaegashi & Udagawa [telomorph of *Pyricularia grisea* (Cooke) Sacc.] (Dobinson *et al.*, 1993) may reflect recent acquisition of these TE through interspecific transfer. Moreover, the high sequence similarity between *Fot1* elements in two *Fusarium* spp. as compared to divergence measured for non-transposable sequences, and the unequal distribution of this element among *Fusarium* species and *F. oxysporum* formae speciales indicate that *Fot1* may have transferred horizontally (Daboussi and Langin, 1994). TE could also act as carriers for non-transposable sequences, as in the case of the *Ant* element in *Aspergillus niger* Tiegh., shown to host genomic sequences (Glayzer *et al.*, 1995). If horizontal transmission of such a transposon occurs, it is possible that host genes are transferred between different organisms (Daboussi, 1996).

The genetic distance between potential donor and recipient species can negatively affect horizontal gene transfer. The potential for genetic exchanges in bacterium-plant and fungus-plant interactions is now being investigated. Initial results have indicated that barriers to plant-bacterium transfers are enormous (e.g. Nielsen *et al.*, 1997a). Intra-specific or intra-generic DNA transfer occurs most readily. For instance, agrocin-encoding plasmids can be mobilized from antagonistic to pathogenic *Agrobacterium* strains, and tumour-inducing sequences can be transferred from pathogenic to antagonistic ones. The resulting pathogenic strains would not be subject to biological control, thus representing an environmental threat (Thomson, 1987). Concern has been expressed about the potential of *Phytophthora palmivora* (E.J. Butler) E.J. Butler, registered in the United States as a bioherbicide for strangler vine (*Morrenia odorata* Lindl.) in citrus (Ridings, 1986), to hybridize with other *Phytophthora* species that were pathogenic on citrus, and the potential to adapt to citrus as a host (Weidemann, 1991). Oospores were produced in some crosses with related *Phytophthora* species but these were not able to germinate. The potential for genetic adaptation to citrus was evaluated, but no change in host range or virulence was demonstrated, and the commercial use of this pathogenic fungus as biocontrol agent was permitted.

In summary, the potential for gene spread from inoculant organisms to members of the indigenous community clearly exists, and both cellular (localization of the heterologous gene, spread to related versus distant organisms) and environmental factors can affect the outcome. The expected gene transfer frequencies, even in worst-case scenarios, are probably low, but forces of selection might boost these frequencies, and should be considered.

27.6. Effects of Released Biocontrol Agents

Released micro-organisms may cause both qualitative and quantitative alterations in microbial community structure. This is probably the most difficult aspect of the determination of the biosafety of a release, since a variety of assessments are required to estimate the effects of releases (Cairns and Orvos, 1992; Kluepfel, 1992; van Elsas *et al.*, 1998). Displacement of indigenous microbial groups can be important if introduced organisms possess high fitness. Often, such an effect is envisaged for biological control agents, like in the case of ice⁻ *Pseudomonas* spp. to be used against ice-nucleating *P. syringae* (Lindow and Panopoulos, 1988). The population size of ice⁺ *P. syringae* strains coinoculated on potato leaves with the corresponding ice⁻ strain was reduced over 300-fold on treated plants compared to plants treated only with the ice⁺ strain. Introduction of genetically modified *P. aureofaciens* on wheat seeds caused large perturbations in the total microbial population (up to 2 log units) at the seedling stage on seeds and roots, while, as the inoculated plants matured, perturbations were not significant (De Leij *et al.*, 1994). In contrast, spray applications of the marked *P. aureofaciens* strain onto the leaf surface of wheat caused no perturbations of the indigenous microbial populations present on the phylloplane (De Leij *et al.*, 1994). Antagonistic strains of *G. virens* and *Pseudomonas* spp. showed no negative effect on the colonization of tests plants by the vesicular arbuscular mycorrhizal fungi *Glomus intraradices* Schenck & Smith and *Glomus etunicatum* Becker & Gerdemann (Paulitz and Linderman, 1989, 1991). Similar results were obtained with antagonistic strains of *F. oxysporum* on the colonization of basil by *Glomus versiforme* (Daniels & Trappe) Berch. (Migheli, unpublished results).

However, unwanted effects may occur when the displaced microflora plays a key role in the geochemical cycling of nutrients, thus having broad consequences for the ecosystem. Given the rapid decrease of introduced microbial populations in the environment, the probability of adverse effects may be small, but this potential has to be carefully assessed (van Elsas *et al.*, 1998). To estimate the effects on microbial communities, the population dynamics of specific functional groups of fungi or bacteria should be evaluated, e.g. those linked to the processing of nitrogen, sulphur and phosphorous (Fenchel and Blackburn, 1979) and mycorrhizae. Soil DNA based PCR with universal or specific (bacterial or fungal) primers followed by denaturing gradient gel electrophoresis can provide insight in shifts of populations of the respective microbial groups (Heuer and Smalla, 1997; van Elsas *et al.*, 1998).

Other important parameters considered for evaluating the effects of newly introduced microbial antagonists are total microbial biomass, nutrient and enzyme concentrations, production/respiration ratios, oxidation-reduction potentials and pH. Finally, soil faunal processes, such as predation, grazing, propagule dissemination and microbial dynamics in animal guts, may lead to interactions with both the newly introduced and the resident microbial communities (Couteaux and Bottner, 1994).

Furthermore, a biocontrol agent has to be harmless towards the plant on which it should be applied (unless this represents the specific target, as in the case of mycoherbicides) but also towards other plants which could be exposed to its propagules. It is therefore necessary to carry out pathogenicity tests on the widest range of potential hosts, by testing many cultivars and biotypes of the same plant species (Weidemann, 1991; TeBeest, 1991). Host range determination is critical in the case of agents for weed control, as the safety of non-target plants must be ensured. In this respect, the use as mycoherbicides of plant pathogens with a broad host spectrum, such as *P. palmivora*, *F. oxysporum* or *Fusarium solani* (Mart.) Sacc. should be considered with extreme care. Also, release into an area where the pathogen did not occur previously should be avoided.

The antagonist should also not affect humans and animals. *In vitro* growth at 35–38°C, pathogenicity in immunocompromised patients and allergenicity should be absent. A number of *in vitro* toxicological tests have been developed to evaluate the presence of toxic or mutagenic metabolites in culture filtrates. Testing the inhibition of root development in tomato germlings, radial growth of the fungus *Geotrichum candidum* Link, toxicity towards larvae of the crustacean *Artemia salina* L., or genotoxicity on germ cells of locust are some available means to characterize the toxic potential of biocontrol agents. Large-scale release of microbial antagonists which produce toxic metabolites may represent an environmental threat and should be discouraged. Thus, the selection of antagonistic micro-organisms should not be merely based on *in vitro* tests, which emphasize biological activity linked to the production of toxic metabolites in the substrate. Antibiotic production (Fravel, 1988) should be critically considered when the antagonist acts against foliar or post-harvest pathogens, since the presence of toxic substances on edible fruit and vegetables implies a potential hazard. Examples of "safe" antagonists are antagonistic *Metschnikowia pulcherrima* P.I. Pitt & M.W. Miller isolate 4.4, active against Botrytis postharvest rot on apple fruit, and *Ulocladium atrum* G. Preuss isolate 385, active against *Botrytis aclada* Fresen. on onion leaves. Both antagonists may act as competitors in the fruit/plant tissues, since no effect of toxins or cell wall degrading enzymes could be found (Köhl *et al.*, 1997; Piano *et al.*, 1997). On the other side, the early interaction between *Verticillium lecanii* (A. Zimmerm.) Viégas, considered for a long time to be a strict hyperparasite, and cucumber powdery mildew was recently found to be mediated by the production of antibiotics (Askary *et al.*, 1997). Also, antibiotic production by antagonists which are released into soil should not be underestimated, and the fate of toxic metabolites in the environment carefully assessed (Thomashow *et al.*, 1990; Lumsden *et al.*, 1992; Raaijmakers *et al.*, 1997).

27.7. Concluding Remarks

The understanding and prediction of behaviour of both genetically modified and unmodified micro-organisms is a prerequisite for any deliberate release into the environment. Only this knowledge will allow to take adequate risk management decisions. It is important to avoid irrational concerns that ignore the significant body of cases that all indicate a lack of adverse effects on the environment. However, there is a clear need to recognize and test those organisms that have the potential for an unwanted environmental effect (Day and Fry, 1992a). Once this potential has been estimated, a quantification of the uncertainty in that estimate has to be made (Cairns and Orvos, 1992), and this can be achieved through ecological studies carried out both in contained environments and in small-scale field releases. Based on the data from such studies presented here, it is clear that all environmental factors that enhance inoculant survival and activity, also enhance the putative adverse effects. Although the hazards can potentially be minimized by a careful choice of the genetic construction, e.g. by opting for a chromosomal insertion instead of a plasmid-borne gene, and by the delivery system, e.g. using a matrix carrier which minimizes water-induced translocation like in the case of alginate, the potential for adverse effects can obviously not be reduced to nil. Therefore, minimal potential biohazard will always be inherent to any application of biocontrol agents. Before any release is cogitated, small-scale trials might be run. In these trials, both physical and biological containment systems might be used. Physical containments can be accomplished through physical barriers (growth chambers, enclosures, sealed windows, air locks, filters), in order to prevent the escape and dispersal of the organism. In addition, organisms neutralized at the end of the experiment. Since decontamination via burning and biocide application, alone or in combination with tillage, may not provide satisfactory control of introduced bacteria, effective and selective mitigation methods need to be developed (Donegan *et al.*, 1992). In contrast, bacterial populations released into the field may reach the detection limit within one growing season, suggesting that such releases can be conducted in a safe and responsible manner.

Finally, biological containment systems have been developed (Cuskey, 1992), which are based on the insertion in the host genome of lethal genes (e.g. *hok*, host killing, or *kilA*, killing factor A) under the control of promoters, which are activated under certain environmental conditions and trigger cell death when those conditions are met. The populations of such strains would therefore decrease until reaching undetectable levels after a given time. However, problems still exist in controlling manipulated organisms deliberately released to the environment; population killing is often incomplete due to genetic instability of the constructs, and some containment systems may not be efficiently expressed outside *E. coli* (Cuskey, 1992). Containment of eukaryotic micro-organisms, such as filamentous fungi, is less developed. A system using non-engineering techniques has been developed for release of an endemic broad host-range pathogen, *Sclerotinia sclerotiorum* (Lib.) de Bary used as a biocontrol agent of weeds. Mutants of *S. sclerotiorum* lacking survival capabilities were generated; some failed to overwinter in limited field trials. One mutant lacked the ability to differentiate sclerotia (Miller *et al.*, 1989a), whereas another one was a cytosine auxotroph and required an exogenous source of pyrimidine to cause infection (Miller *et al.*, 1989b). This mutant was able to differentiate apothecia and ascospores, but these were unable to infect plants in the absence of exogenous cytosine, thus limiting the spread beyond the area of application.

Taken together, only a multidisciplinar approach, which involves plant pathology, microbiology, toxicology, molecular biology and ecology, will allow the construction of a risk assessment paradigm to predict the fate and effects of biocontrol agents released in agricultural environments.

Acknowledgements

Studies carried out by the authors were partially supported by the following grants: EC BRIDGE, CNR RAISA (Subproject No. 2), MIRAAF (P.N. "Biotecnologie vegetali", Area 10 – I diagnostici, Program No. 451) and EU-BIOTECH BIO2-CT92-0491.

References

Ahmad, J.S. and Baker, R. (1987) Competitive saprophytic ability and cellulolytic activity of rhizosphere competent mutants of *Trichoderma harzianum, Phytopathology* **77**, 358–362.

Akkermans, A.D.L., van Elsas, J.D. and De Bruijn, F.J. (1995) *Molecular Microbial Ecology Manual,* Kluwer Academic Publishers, Dordrecht.

Arisan-Atac, I., Heidenreich, E. and Kubicek, C.P. (1995) Randomly amplified polymorphic DNA fingerprinting identifies subgroups of *Trichoderma viride* and other *Trichoderma* sp. capable of chestnut blight biocontrol, *FEMS Microbiology Letters* **126**, 249–256.

Armstrong, J.L., Porteous, L.A. and Wood, N.D. (1989) The cutworm *Peridroma saucia* (Lepidoptera: Noctuidae) supports the growth and transport of pBR322-bearing bacteria, *Applied and Environmental Microbiology* **55**, 2200–2205.

Askary, H., Benhamou, N. and Brodeur, J. (1997) Ultrastructural and cytochemical investigations of the antagonistic effect of *Verticillium lecanii* on cucumber powdery mildew, *Phytopathology* **87**, 359–368.

Bahme, J.B. and Schroth, M.N. (1987) Spatial-temporal colonization patterns of a rhizobacterium on underground organs of potato, *Phytopathology* **77**, 1093–1100.

Breitenbeck, G.A., Yang, H. and Dunigan, E.P. (1988) Water-facilitated dispersal of inoculant *Bradyrhizobium japonicum* in soils, *Biology and Fertility of Soils* **7**, 58–62.

Cairns, J. and Orvos, D.R. (1992) Establishing environmental hazards of genetically engineered microorganisms, *Reviews of Environmental Contamination and Toxicology* **124**, 19–39.

Chet, I. (ed.) (1993) *Biotechnology in Plant Disease Control,* Wiley-Liss, Inc., New York.

Chet, I., Barak, Z. and Oppenheim, A. (1993) Genetic engineering of microorganisms for improved biocontrol activity, in I. Chet (ed.), *Biotechnology in Plant Disease Control,* Wiley-Liss, Inc., New York, pp. 211–235.

Comeaux, J.L., Pooranampillai, C.D., Lacy, G.H. and Stromberg, V.K. (1990) Transfer of genes to other populations, and analysis of associated potential risks, in J.J. Marois and G. Bruening (eds.), *Risk Assessment in Agricultural Biotechnology,* Regents of the University of California, Oakland, Calif., pp. 132–145.

Coûteaux, M.M. and Bottner, P. (1994) Biological interactions between fauna and the microbial community in soils, in K. Ritz, J. Dighton and K.E. Giller (eds.), *Beyond the Biomass – Compositional and Functional Analysis of Soil Microbial Communities,* John Wiley & Sons, Chichester, pp. 159–172.

Cuskey, S.M. (1992) Lethal genes in biological containment of released micro-organisms, in J.C. Fry and M.J. Day (eds.), *Release of Genetically Engineered and Other Microorganisms*, Cambridge University Press, Cambridge, pp. 94–99.

Daboussi, M.J. (1996) Fungal transposable elements: Generators of diversity and genetic tools, *J. of Genetics* **75**, 325–339.

Daboussi, M.J. and Langin, T. (1994) Transposable elements in the fungal plant pathogen *Fusarium oxysporum, Genetica* **93**, 49–59.

Day, M.J. and Fry, J.C. (1992a) Microbial ecology, genetics and risk assessment, in J.C. Fry and M.J. Day (eds.), *Release of Genetically Engineered and Other Microorganisms*, Cambridge University Press, Cambridge, pp. 160–167.

Day, M.J. and Fry, J.C. (1992b) Gene transfer in the environment: Conjugation, in J.C. Fry and M.J. Day (eds.), *Release of Genetically Engineered and Other Microorganisms*, Cambridge University Press, Cambridge, pp. 40–53.

De Jong, M.D., Scheepens, P.C. and Zadoks, J.C. (1990) Risk analysis for biological control: A Dutch case study in biocontrol of *Prunus serotina* by the fungus *Chondrostereum purpureum, Plant Disease* **74**, 189–194.

De Leij, F.A.A.M., Sutton, E.J., Whipps, J.M. and Lynch, J.M. (1994) Effect of a genetically modified *Pseudomonas aureofaciens* on indigenous microbial populations of wheat, *FEMS Microbiology Ecology* **13**, 249–258.

Défago, G., Keel, C. and Moenne-Loccoz, Y. (1997) Fate of released *Pseudomonas* bacteria in the soil profile: Implications for the use of genetically-modified microbial inoculants, in J.T. Zelnikoff (ed.), *Ecotoxicology: Responses, Biomarkers and Risk Assessment*, SOS Publishers, Fair Haven, NJ, pp. 403–418.

Dighton, J., Jones, H.E., Robinson, C.H. and Beckett, J. (1997) The role of abiotic factors, cultivation practices and soil fauna in the dispersal of genetically modified micoorganisms in soils, *Applied Soil Ecology* **5**, 109–132.

Dobinson, K.F., Harris, R.E. and Hamer, J.F. (1993) Grasshopper, a long terminal repeat (LTR) retroelement in the phytopathogenic fungus *Magnaporthe grisea, Molecular Plant-Microbe Interactions* **6**, 114–126.

Donegan, K., Fieland, V., Fowles, N., Ganio, L. and Seidler, R. (1992) Efficacy of burning, tillage, and biocides in controlling bacteria released at field sites and effects on indigenous bacteria and fungi, *Applied and Environmental Microbiology* **58**, 1207–1214.

Edel, V., Steinberg, C., Avelange, I., Laguerre, G. and Alabouvette, C. (1995) Comparison of three molecular methods for the characterization of *Fusarium oxysporum* strains, *Phytopathology* **85**, 579–585.

Elad, Y. (1990) Reasons for the delay in development of biological control of foliar pathogens, *Phytoparasitica* **18**, 99–105.

Eshed, N. and Dinoor, A. (1981) Genetics of pathogenicity in *Puccinia coronata*: The host range among grasses, *Phytopathology* **71**, 156–163.

Fenchel, T. and Blackburn, T.H. (1979) *Bacteria and Mineral Cycling,* Academic Press, New York.

Flores, A., Chet, I. and Herrera-Estrella, A. (1997) Improved biocontrol activity of *Trichoderma harzianum* by over-expression of the proteinase-encoding gene *prb1, Current Genetics* **31**, 30–37.

Fravel, D.R. (1988) Role of antibiosis in the biocontrol of plant diseases, *Annual Review of Phytopathology* **26**, 75–91.

Garibaldi, A., Guglielmone, L. and Gullino, M.L. (1990) Rhizosphere competence of antagonistic *Fusaria* isolated from suppressive soils, *Symbiosis* **9**, 401–404.

Glayzer, D.C., Roberts. I.N., Archer. D.B. and Oliver, R.P. (1995) The isolation of *Ant1*, a transposable element from *Aspergillus niger, Molecular and General Genetics* **249**, 432–438.

Gullino, M.L., Migheli Q., Mezzalama, M. and Garibaldi, A. (1995) Risk analysis for biological control agents: Antagonistic *Fusarium* spp. as a case study, *Plant Disease* **79**, 1193–1201.

Harman, G.E. (1992) Production and delivery systems for biocontrol agents, *IOBC/WPRS Bulletin* **15**(1), 201–205.

Harman, G.E. and Hayes, C.K. (1993) The genetic nature and biocontrol ability of progeny from protoplast fusion in *Trichoderma*, in I. Chet (ed.), *Biotechnology in Plant Disease Control*, Wiley-Liss, Inc., New York, pp. 237–255.

Harvey, R.W., George, L.H., Smith, R.L. and LeBlanc, D.R. (1989) Transport of microspheres and indigenous bacteria through a sandy aquifer: Results of natural- and forced-gradient tracer experiments, *Environmental Science and Technology* **23**, 51–56.

Heijnen, C.E., Hok-A-Hin, C.H. and van Veen, J.A. (1992) Improvements to the use of bentonite clay as a protective agent, increasing survival levels of bacteria introduced into soil, *Soil Biology and Biochemistry* **24**, 533–538.

Heijnen, C.E., van Elsas, J.D., Kuikman, P.J. and van Veen, J.A. (1988) Dynamics of *Rhizobium leguminosarum* biovar *trifolii* introduced into soil; the effect of bentonite clay on predation by protozoa, *Soil Biology and Biochemistry* **20**, 483–488.

Hekman, W.E., Heijnen, C.E., Trevors, J.T. and van Elsas, J.D. (1994) Water flow induced transport of *Pseudomonas fluorescens* cells through soil columns as affected by inoculant treatment, *FEMS Microbiology Ecology* **13**, 313–324.

Henschke, R.B., Nücken, E. and Schmidt, F.R.J. (1989) Fate and dispersal of recombinant bacteria in a soil microcosm containing the earthworm *Lumbricus terrestris*, *Biology and Fertility of Soils* **7**, 374–376.

Henschke, R.B. and Schmidt, F.R.J. (1990) Plasmid mobilization from genetically engineered bacteria to members of the indigenous soil microflora *in situ*, *Current Microbiology* **20**, 105–110.

Herrera-Estrella, A., Goldman, G.H. and van Montagu, M. (1990) High-efficiency transformation system for the biocontrol agents, *Trichoderma* spp, *Molecular Microbiology* **4**, 839–843.

Heuer, H. and Smalla, K. (1997) Application of denaturing gradient gel electrophoresis and temperature gradient gel electrophoresis for studying soil microbial communities, in J.D. van Elsas, E.M.H. Wellington and J.T. Trevors (eds.), *Modern Soil Microbiology*, Marcel Dekker, Inc., New York, pp. 353–373.

Jansson, J.K. (1995) Tracking genetically engineered microorganisms in nature, *Current Opinions in Biotechnology* **6**, 275–283.

Keller, N.P., Bergstrom, G.C. and Yoder, O.C. (1991) Mitotic stability of transforming DNA is determined by its chromosomal configuration in the fungus *Cochliobolus heterostrophus*, *Current Genetics* **19**, 227–233.

Kinsey, J.A. (1990) Restricted distribution of the Tad transposon in strains of *Neurospora*, *Current Genetics* **15**, 271–275.

Kiss, L. (1997) Genetic diversity in *Ampelomyces* isolates, hyperparasites of powdery mildew fungi, inferred from RFLP analysis of the rDNA ITS region, *Mycological Research* **101**, 1073–1080.

Kistler, H.C. (1991) Genetic manipulation of plant pathogenic fungi, in D.O. TeBeest (ed.), *Microbial Control of Weeds*, Chapman & Hall, London, pp. 152–170.

Kistler, H.C. and Benny, U.K. (1988) Genetic transformation of the fungal wilt pathogen *Fusarium oxysporum*, *Current Genetics* **13**, 145–149.

Kline, D.M. and Nelson, R.R. (1971) The inheritance of factors in *Cochliobolus sativus* conditioning lesion induction on gramineous hosts, *Phytopathology* **61**, 1052–1054.

Kluepfel, D.A. (1992) The behavior of nonengineered bacteria in the environment: What can we learn from them? in R. Casper and J. Landsmann (eds.), *Proceedings Second International Symposium on the Biosafety Results of Field Tests of Genetically Modified Plants and Microorganisms*, BBA, Braunschweig, pp. 37–42.

Kluepfel, D.A., Kline, E.L., Skipper, H.D., Hughes, T.A. and Gooden, T.A. (1991) The release and tracking of genetically engineered bacteria in the environment, *Phytopathology* **81**, 348–352.

Knudsen, G.R. (1989) Model to predict aerial dispersal of bacteria during environmental release, *Applied and Environmental Microbiology* **55**, 2641–2647.

Köhl, J., Bélanger, R.R. and Fokkema, N.J. (1997) Interaction between four antagonistic fungi with *Botrytis aclada* in dead onion leaves: A comparative microscopic and ultrastructural study, *Phytopathology* **87**, 634–642.

Leslie, J.F. and Dickman, M.B. (1991) Fate of DNA encoding hygromycin resistance after meiosis in transformed strains of *Gibberella fujikuroi* (*Fusarium moniliforme*), *Applied and Environmental Microbiology* **57**, 1423–1429.

Lindow, S.E. and Panopoulos, N.J. (1988) Field tests of recombinant ice⁻ *Pseudomonas syringae* for biological frost control in potato, in M. Sussman, C.H. Collins, F.A. Skinner and D.E. Stewart-Tull (eds.), *The Release of Genetically-Engineered Micro-organisms*, Academic Press, London, pp. 121–138.

Lindow, S.E., Panopoulos, N.J. and McFarland, B.L. (1989) Genetic engineering of bacteria from managed and natural habitats, *Science* **244**, 1300–1307.

Lumsden, R.D., Locke, J.C., Adkins, S.T., Walter, J.F. and Ridout, C.J. (1992) Isolation and localization of the antibiotic gliotoxin produced by *Gliocladium virens* from alginate prill in soil and soilless media, *Phytopathology* **82**, 230–235.

Madsen, E.L. and Alexander, M. (1982) Transport of *Rhizobium* and *Pseudomonas* through soil, *Soil Science Society of America J.* **46**, 557–560.

Mezzalama, M., Mocioni, M. and Gullino, M.L. (1994) Survival of antagonistic *Fusarium* spp. in soil, *Microbial Releases* **2**, 255–259.

Migheli, Q., Berio, T. and Gullino, M.L. (1993) Electrophoretic karyotypes of *Fusarium* spp., *Experimental Mycology* **17**, 329–337.

Migheli, Q. and Cavallarin, L. (1994) Characterization of antagonistic and pathogenic *Fusarium oxysporum* isolates by random amplification of polymorphic DNA, *Molecular Biotechnology* **2**, 197–200.

Migheli, Q., Friard, O., Del Tedesco, D., Musso, M.R. and Gullino, M.L. (1996) Stability of transformed antagonistic *Fusarium oxysporum* strains *in vitro* and in soil microcosms, *Molecular Ecology* **5**, 641–649.

Migheli, Q., Herrera-Estrella, A., Avataneo, M. and Gullino, M.L. (1994) Fate of transformed *Trichoderma harzianum* in the phylloplane of tomato plants, *Molecular Ecology* **3**, 153–159.

Migheli, Q., Piano, S., Enrietti, S. and Gullino, M.L. (1992) Protoplast fusion in antagonistic *Fusarium* spp., *IOBC/WPRS Bulletin* **15**(1), 196–198.

Migheli, Q., Whipps, J.M., Budge, S.P. and Lynch, J.M. (1995) Production of inter- and intra-strain hybrids of *Trichoderma* spp. by protoplast fusion and evaluation of their biocontrol activity against soil-borne and foliar pathogens, *J. of Phytopathology* **143**, 91–97.

Miller, R.V., Ford, E.J. and Sands, D.C. (1989a) A non-sclerotial pathogenic mutant of *Sclerotinia sclerotiorum, Canadian J. of Microbiology* **35**, 517–520.

Miller, R.V., Ford, E.J., Zidack, N.J. and Sands, D.C. (1989b) A pyrimidine auxotroph of *Sclerotinia sclerotiorum* for use in biological weed control, *J. of General Microbiology* **135**, 2085–2091.

Nielsen, K.M., Gebhard, F., Smalla, K., Bones, A.M. and van Elsas, J.D. (1997a) Evaluation of possible horizontal gene transfer from transgenic plants to the soil bacterium *Acinetobacter calcoaceticus* BD413, *Theoretical and Applied Genetics* **95**, 815–821.

Nielsen, K.M., van Weerelt, M.D.M., Berg, T.N., Bones, A.M., Hagler, A.N. and van Elsas, J.D. (1997b) Natural transformation and availability of transforming DNA to *Acinetobacter calcoaceticus* in soil microcosms, *Applied and Environmental Microbiology* **63**, 1945–1952.

OTA, Office of Technology Assessment (1988) *New Developments in Biotechnology. Field-testing Engineered Organisms: Genetic and Ecological Issues,* OTA–BA-350, Washington DC.

Papavizas, G.C. (1987) Genetic manipulation to improve the effectiveness of biocontrol fungi for plant disease control, in I. Chet (ed.), *Innovative Approaches to Plant Disease Control*, John Wiley & Sons, New York, pp. 193–212.

Parke, J.L., Moen, R., Rovira, A.D. and Bowen, G.D. (1986) Soil water flow affects the rhizosphere distribution of a seed-borne biological control agent, *Pseudomonas fluorescens, Soil Biology and Biochemistry* **18**, 583–588.

Paulitz, T.C. and Linderman, R.G. (1989) Interactions between fluorescent pseudomonads and VA mycorrhizal fungi, *New Phytologist* **113**, 37–45.

Paulitz, T.C. and Linderman, R.G. (1991) lack of antagonism between the biocontrol agent *Gliocladium virens* and vesicular arbuscular mycorrhizal fungi, *New Phytologist* **117**, 303–308.

Pe'er, S., Barak, Z., Yarden, O. and Chet, I. (1991) Stability of *Trichoderma harzianum amdS* transformants in soil and rhizosphere, *Soil Biology and Biochemistry* **23**, 1043–1046.

Peterson, T.C. and Ward, R.C. (1989) Development of a bacterial transport model for coarse soils, *Water Resources Bulletin* **25**, 349–357.

Piano, S., Neyrotti, V., Migheli, Q. and Gullino, M.L. (1997) Biocontrol capability of *Metschnikowia pulcherrima* against Botrytis postharvest rot of apple, *Postharvest Biology and Technology* **11**, 131–140.

Poplawski, A.M., He, C., Irwin, J.A.G. and Manners, M. (1997) Transfer of an autonomously replicating vector between vegetatively incompatible biotypes of *Colletotrichum gloeosporioides, Current Genetics* **32**, 66–72.

Postma, J. and van Veen, J.A. (1990) Habitable pore space and survival of *Rhizobium leguminosarum* biovar *trifolii* introduced into soil, *Microbial Ecology* **19**, 149–161.

Prosser, J.I. (1994) Molecular marker systems for detection of genetically engineered microorganisms in the environment, *Microbiology* **140**, 5–17.

Raaijmakers, J.M., Thomashow, L.S. and Weller, D.M. (1997) Plant defense by antibiotic-producing bacteria in natural disease-suppressive soils, in *Proceedings of the IOBC/EFPP Workshop on Molecular Approaches in Plant Disease Control*, Délemont, September 15–18 1997, (abs.).

Rhodes, D.J. (1993) Formulation of biological control agents, in D.J. Jones (ed.), *Exploitation of Microorganisms*, Chapman & Hall, London, pp. 411–439.

Ridings, W.J. (1986) Biological control of stranglervine in citrus – A researcher's view, *Weed Science* 34, 31–32.

Rodriguez, R.J. and Yoder, O.C. (1987) Selectable genes for transformation of the fungal plant pathogen *Glomerella cingulata* f.sp. *phaseoli* (*Colletotrichum lindemuthianum*), *Gene* 54, 73–81.

Schliestl, R.H. and Petes, T.D. (1991) Integration of DNA fragments by illegitimate recombination in *Saccharomyces cerevisiae*, *Proceedings of the National Academy of Sciences USA* 88, 7585–7589.

Smalla, K. and van Elsas, J.D. (1996) Monitoring genetically modified organisms and their recombinant DNA in soil environments, in J. Tomiuk, K. Wöhrmann and A. Sentker (eds.), *Transgenic Organisms – Biological and Social Implications*, Birkhäuser Verlag, Basel, pp. 127–146.

Smit, E., van Elsas, J.D., van Veen, J.A. and de Vos, W.M. (1991) Detection of plasmid transfer from *Pseudomonas fluorescens* to indigenous bacteria in soil by using phage ΦR2f for donor counterselection, *Applied and Environmental Microbiology* 57, 3482–3488.

Smit, E., Venne, D. and van Elsas, J.D. (1993) Mobilization of a recombinant IncQ plasmid between bacteria on agar and in soil via co-transfer or retrotransfer, *Applied and Environmental Microbiology* 59, 2257–2263.

Smit, E., Wolters, A. and van Elsas, J.D. (1995) Genetic stability, conjugal transfer and expression of heterologous DNA inserted into different plasmids and the genome of *Pseudomonas fluorescens* in soil, *Revista de Microbiologia* 26, 169–179.

Smith, M.S., Thomas, G.W., White, R.E. and Ritonga, D. (1985) Transport of *Escherichia coli* through intact and disturbed soil columns, *J. of Environmental Quality* 14, 87–91.

Stotzky, G. (1989) Gene transfer among bacteria in soil, in S.B. Levy and R.V. Miller (eds.), *Gene Transfer in the Environment*, McGraw–Hill, New York, pp. 165–222.

TeBeest, D.O. (1991) Ecology and epidemiology of fungal plant pathogens studied as biological control agents of weeds, in D.O. TeBeest (ed.), *Microbial Control of Weeds*, Chapman & Hall, London, pp. 97–114.

Teuben, A. and Verhoef, H.A. (1992) Relevance of micro- and mesocosm experiments for studying soil ecosystem processes, *Soil Biology and Biochemistry* 24, 1179–1183.

Thomashow, L.S., Weller, D.M., Bonsall, R.F. and Pierson, L.S. (1990) Production of the antibiotic phenazine-1-carboxylic acid by fluorescent *Pseudomonas* species in the rhizosphere of wheat, *Applied and Environmental Microbiology* 56, 908–912.

Thomson, J.A. (1987) The use of agrocin-producing bacteria in the biological control of crown gall, in I. Chet (ed.), *Innovative Approaches to Plant Disease Control*, John Wiley and Sons, New York, pp. 213–228.

Tiedje, J.M., Colwell, R.K., Grossman, Y.L., Hodson, R.E. and Lenski, R.E. (1989) The planned introduction of genetically engineered organisms: Ecological considerations and recommendations, *Ecology* 70, 298–315.

Timberlake, W.E. (1992) Cloning and analysis of fungal genes, in J.W. Bennett and L.L. Lasure (eds.), *More Gene Manipulations in Fungi*, Academic Press, San Diego, Calif., pp. 51–85.

Trevors, J.T. and van Elsas, J.D. (1995) *Nucleic Acids in the Environment; Methods and Applications*, Springer Verlag, Heidelberg.

Trevors, J.T., van Elsas, J.D., Lee, H. and van Overbeek, L.S. (1992) Use of alginate and other carriers for encapsulation of microbial cells for use in soil, *Microbial Releases* 1, 61–69.

Trevors, J.T., van Elsas, J.D., van Overbeek, L.S. and Starodub, M.-E. (1990) Transport of a genetically engineered *Pseudomonas fluorescens* strain through a soil microcosm, *Applied and Environmental Microbiology* 56, 401–408.

Turner, D., Kovacs, W., Kuhls, K., Lieckfeldt, E., Peter, B., Arisan-Atac, I., Strauss, J., Samuels, G.J., Börner, T. and Kubicek, C.P. (1997). Biogeography and phenotypic variation in *Trichoderma* sect. *Longibrachiatum* and associated *Hypocrea* species, *Mycological Research* 101, 449–459.

Tzotzos, G.T. (1995). *Genetically Modified Organisms. A Guide to Biosafety*, CAB International, Wallingford.

van Elsas, J.D., Duarte, G.F., Rosado, A.S. and Smalla, K. (1998) Microbiological and Molecular Biological methods for monitoring microbial inoculants and their effects in the soil environment, *J. of Microbiological Methods* - Special Issue 32, 133–154.

van Elsas, J.D., Heijnen, C.E. and van Veen, J.A. (1991a) The fate of introduced genetically engineered microorganisms (GEMs) in soil, in microcosms and the field: Impact of soil textural aspects, in D.R. McKenzie and S.C. Henry (eds.), *Biological Monitoring of Genetically Engineered Plants and Microbes*, Agricultural Research Institute, Bethesda, Md., pp 67–79.

van Elsas, J.D., Trevors, J.T., Jain, D., Wolters, A.C., Heijnen, C.E. and van Overbeek, L.S. (1992) Survival of, and root colonization by, alginate-encapsulated *Pseudomonas fluorescens* cells following introduction into soil, *Biology and Fertility of Soils* **14**, 14–22.

van Elsas, J.D., Trevors, J.T. and Starodub, M.-E. (1988a) Bacterial conjugation between pseudomonads in the rhizosphere of wheat, *FEMS Microbiology Ecology* **53**, 299–306.

van Elsas, J.D., Trevors, J.T. and Starodub, M.-E. (1988b) Plasmid transfer in soil and rhizosphere, in W. Klingmüller (ed.), *Risk Assessment for Deliberate Releases*, Springer Verlag, Berlin, pp 89–99.

van Elsas, J.D., van Overbeek, L.S., Feldmann, A.M., Dullemans, A.M. and de Leeuw, O. (1991b) Survival of genetically engineered *Pseudomonas fluorescens* in soil in competition with the parent strain, *FEMS Microbiology Ecology* **85**, 53–64.

van Elsas, J.D., Wolters, A.C., Clegg, C.D., Lappin-Scott, H.M. and Anderson, J.M. (1994) Fitness of genetically modified *Pseudomonas fluorescens* in competition for soil and root colonization, *FEMS Microbiology Ecology* **13**, 259–272.

van Loosdrecht, M.C.M., Lylkema, J., Norde, W., Schraa, G. and Zehnder, A.J.B. (1987) The role of bacterial cell wall hydrophobicity in adhesion, *Applied and Environmental Microbiology* **53**, 1893–1897.

van Overbeek, L.S., Eberl, L., Givskov, M., Molin, S. and van Elsas, J.D. (1995) Survival of, and induced stress resistance in, carbon-starved *Pseudomonas fluorescens* cells residing in soil, *Applied and Environmental Microbiology* **61**, 4202–4208.

van Veen, J.A., van Overbeek, L.S and van Elsas, J.D. (1997) Fate and activity of microorganisms introduced into soil, *Microbiology and Molecular Biology Reviews* **61**, 121–135.

Weidemann, G.J. (1991) Host-range testing: Safety and science, in D.O. TeBeest (ed.), *Microbial Control of Weeds*, Chapman & Hall, London, pp. 83–96.

Wellington, E.H.M. and van Elsas, J.D. (1992) *Genetic Interactions among Microorganisms in the Natural Environment*, Pergamon Press, Oxford.

Zeph, L.R., Onaga, M.A. and Stotzky, G. (1988) Transduction of *Escherichia coli* by bacteriophage P1 in soil, *Applied and Environmental Microbiology* **54**, 190–193.

Zimand, G., Valinsky, L., Elad, Y., Chet, I. and Manulis, S. (1993) DNA fingerprinting of a Trichoderma isolate by the RAPD procedure, *IOBC/WPRS Bulletin* **16**(11), 173–176.

CHAPTER 28

THE ROLE OF THE HOST IN BIOLOGICAL CONTROL OF DISEASES
Timothy C. Paulitz and Alberto Matta

28.1. Introduction

In the past 30 years, great progress has been made in understanding the mechanisms of biological control. Most of this research has focused on a direct interaction between the biocontrol agent and the pathogen. For example, the biocontrol agent produces antibiotics or other antifungal compounds that antagonize the pathogen, either on the root or leaf surface (Fravel, 1988). Parasites directly attack the hyphae of the pathogenic fungus, using it as a food source (Adams, 1990). Bacteria produce lytic enzymes which dissolve the hyphal tips of the fungus (Chernin et al., 1995). Biocontrol agents can compete with the pathogen for limiting resources such as carbon or iron (Loper and Buyer, 1991; Paulitz, 1991).

However, until recently little research has focused on indirect effects of biocontrol agents mediated through the plant itself. To have a proper understanding of biocontrol, just as to understand the disease process, all the players in the interaction must be considered. In the same way that disease can be described by a disease triangle, with the pathogen, host, and environment interacting; so too the process of biocontrol can be described by a pyramid. The critical players are the plant, the pathogen, the biocontrol agent, and the environment (Fig. 28.1). The direct effects of biocontrol agents on pathogens are represented by the solid arrow labelled (A) on the left side of the pyramid. The sign in parentheses indicates whether the interaction is positive or negative. This chapter will concentrate on the effect of the host on the outcome of this complex set of interactions that leads to biocontrol, that is the suppression of the pathogen or its harmful effects.

The biocontrol agent itself can have an indirect effect on the pathogen via host-mediated defence responses, by inducing higher levels of resistance in the plant. Within the last few years, this area of research has received increased attention. Its findings may have practical implications for the greenhouse industry. This interaction is represented by the solid and dashed arrow labelled (B) in Fig. 28.1.

The host may also directly influence the biocontrol agents by offering hospitality and stimulating antagonism. "Host hospitality", a term coined by Smith et al. (1997), provides favourable environmental niches and nutrients that influence biocontrol by favouring growth and survival of the antagonist. The outcome of the antagonist-pathogen interaction may also depend on the plant exudation of nutrients needed for the production of antibiotics and on the level of nutrients that may affect competition. All these factors are determined by the genetic background of the plant, which can be manipulated to favour growth and activity of the antagonist, thus effecting biocontrol.

Genetic background also affects the resistance or susceptibility of the plant to the

394

R. Albajes et al. (eds.), Integrated Pest and Disease Management in Greenhouse Crops, 394-410.
© 1999 Kluwer Academic Publishers. Printed in the Netherlands.

pathogen. The inherent resistance of the host to the pathogen may also influence the outcome of the biocontrol reaction, a factor not considered in most biocontrol research. The direct effect of the host on the biocontrol agent is represented by the solid arrow (C) in Fig. 28.1.

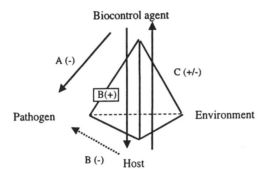

Figure 28.1. Schematic representation of interactions between the biocontrol agent, pathogen, host, and environment. (A) represents direct antagonistic effects of the biocontrol agent on the pathogen. The biocontrol agent can have a beneficial effect on the plant (B+) by having an indirect negative effect on the pathogen (B–) via inducing a higher level of resistance in the host (induced resistance). The host can also have beneficial or limiting effects on the biocontrol agent (C) by providing nutrients and environmental niches for the microbes on and near plant surfaces.

28.2. Ability of the Biocontrol Agent to Indirectly Affect the Pathogen by Inducing Resistance in the Host Plant

The phenomenon of induced resistance has been described for over 70 years (Chester, 1933), but has received increased attention in the last 10 to 20 years (Matta, 1971; Kuć, 1982; Madamanchi and Kuć, 1991). It is defined as "the process of active resistance dependent on the host plant's physical or chemical barriers, activated by biotic or abiotic agents (inducing agents)" (Kloepper *et al.*, 1992). The pioneering work of J. Kuć in the 70s and 80s has further advanced this area of study. The field of induced resistance is summarized by Hammerschmidt and Kuć (1995). Sticher *et al.* (1997) have summarized the current molecular and biochemical knowledge about the mechanisms behind systemic acquired resistance (SAR). Induced systemic resistance (ISR) is a term coined by Kloepper *et al.* (1992) to distinguish resistance induced by Plant Growth-Promoting Rhizobacteria (PGPR), which may have different underlying mechanisms than SAR.

The phenomenon of induced resistance has been demonstrated in many crops, both monocots and dicots, especially cucurbits and solanaceous plants (Kuć and Strobel, 1992). It is non-specific and broad spectrum: inoculation of a plant with an inducing organism leads to plant resistance against many pathogens, related and unrelated to the inducing organism. Besides the pathogen itself, non-pathogenic organisms such as saprophytes, avirulent races, strains, or formae speciales of the pathogen can also induce resistance, a phenomenon which biocontrol practitioners hope to exploit. However, classical SAR induced by necrogenic pathogens may be quite different from ISR induced by non-necrogenic nonpathogens such as PGPR, in terms of persistence, mechanism of induction, expression of pathogenesis-related (PR) proteins and role of salicylic acid (van Loon *et al.*, 1997). In addition to biotic elicitors of resistance, toxic and non-toxic chemicals, UV, and even compost-derived products have also been shown to induce resistance.

Much of this experimentation has been done in the greenhouse, and could offer potential technologies for greenhouse crops. However, to our knowledge, few of these technologies or inducers are marketed for commercial use, and those biocontrol agents that are on the market use other mechanisms that are not plant mediated. One technology that has been commercially applied is the use of avirulent strains of a virus to protect against more virulent strains, commonly known as cross protection. The mechanisms of cross protection are not the same as classical SAR, but in both cases, biocontrol is mediated via a host response. This strategy has been successfully used in the field for over 20 years in Brazil to protect against citrus tristeza virus, where trees are grafted onto rootstocks infected with a mild form of the virus (Fulton, 1986). Cross protection is also being developed for field use in Hawaii to control papaya mosaic (Gonsalves and Garnsey, 1989) and has been used in the past world-wide to protect tomatoes against tomato mosaic virus (Fulton, 1986). However, this chapter will concentrate on how the host influences the biological control of fungal and bacterial diseases. The next section will summarize some of the major types of inducers and the types of diseases they protect against.

28.2.1. PATHOGENIC ORGANISMS AS INDUCERS

In the classical SAR, the pathogen induces resistance to itself. That is, a pathogen such as *Colletotrichum orbiculare* (Berk. & Mont.) Arx can be applied to the cotyledon or first true leaf of a cucumber plant, producing lesions in a few days. This is called the inducing treatment. However, if leaves formed after the induction treatment are challenged with the pathogen, the lesion size and number on the challenged leaves are reduced compared to those on plants not induced with the pathogen. Thus, some signal has been transported from the induced leaves to the challenged leaves to make them more resistant. A lag phase is usually needed for this resistance to develop (Tuzun and Kuć, 1985), but protection can persist for a long time, sometimes the life of the plant, in the case of tobacco. However, cucumbers cannot be immunized once they have started to flower and set fruit (Guedes *et al.*, 1980). The induced resistance response may result in rapid lignification at the infection site that limits the development of the pathogen (Hammerschmidt and Kuć, 1982) and increases in enzymes involved in the defence reaction such as peroxidase (Hammerschmidt *et al.*, 1982) and chitinase (Métraux and Boller, 1986).

The literature contains many other examples of pathogens being inducers, although most examples have been on field crops, such as pre-inoculation of barley with *Bipolaris maydis* (Nisikado & Miyake) Shoemaker and *Stagonospora nodorum* (Berk.) Castellani & E.G. Germano [= *Septoria nodorum* (Berk.) Berk. in Berk. & Broome], pathogens of corn and wheat, respectively, to reduce net blotch caused by *Drechslera teres* (Sacc.) Shoemaker (Jørgensen *et al.*, 1996). There are few examples of induced resistance on greenhouse or ornamental plants. Some examples of induced vegetable crops include bacterial wilt of tomatoes [*Pseudomonas solanacearum* (Smith) Smith] (Sequeira, 1983) and *Phytophthora infestans* (Mont.) de Bary on tomatoes (Heller and Gessler, 1986). Foliar pathogens on cucumbers also induce protection against vascular pathogens (Gessler and Kuć, 1982).

Some pathogens have even been used as inducers under field conditions. Tobacco plants in the field were protected against blue mould of tobacco by injecting stems with sporangiospores of the pathogen itself, *Peronospora tabacina* D.B. Adam, giving control comparable to the fungicide metalaxyl (Tuzun *et al.*, 1986). However, the major drawback to this strategy in the field is the intensive labour involved in immunizing the plants and the possibility of the pathogen inoculum exacerbating disease problems on the crop in the field or surrounding areas. This is also a risk for many greenhouse diseases that are polycyclic and can produce secondary inoculum. Wei *et al.* (1996) demonstrated in a field trial that the classic treatments with the pathogen as the inducer were more diseased than the diseased check, probably due to secondary spread. There is a psychological factor of introducing a potentially harmful agent, in terms of grower acceptance and regulatory approval. However, non-pathogenic organisms do not have this problem of psychological acceptance.

28.2.2. NON-PATHOGENIC ORGANISMS AS INDUCERS

Non-pathogens and avirulent races and formae speciales of pathogens have been used to induce resistance (Kuć and Strobel, 1992). Some examples, mainly regarding vegetable or ornamental plants, are given in the following section. Systemic resistance in cucumber can also be induced by saprophytic fungi which also promote plant growth (Meera *et al.*, 1994).

Mycorrhizal fungi have been implicated in some induced resistance-like reactions, although, to our knowledge, no studies have shown this definitively, by ruling out other indirect effects such as improved plant nutrition. Morandi *et al.* (1984) found an increase in the concentration of a phytoalexin-like isoflavonoid in vesicular-arbuscular mycorrhizae (VAM) inoculated soybean. VAM colonization of marigold roots reduced infection by *Pythium ultimum* Trow (St. Arnaud *et al.*, 1994). St. Arnaud *et al.* (1994) ruled out a nutritional response because phosphorus levels were similar in VAM and non-VAM plants. However, since the inducer (VAM fungus) was not spatially separated from the challenge pathogen, the question of induced resistance remains to be proven.

A great deal of research has been done on induced resistance to vascular diseases. The first successful attempts were made with *Acremonium*-like fungi, whose ability to act as resistance inducers has been more recently confirmed with Fusarium wilt of tomato (Bargmann and Schoenbeck, 1992). Soilborne fungi and bacteria nonpathogenic or avirulent on the target plant have been repeatedly proven to protect plants such as tomato, cucumber, melon, carnation, from Fusarium or Verticillium wilt (Matta, 1989). Even foliar infections, at least on cucurbitaceous plants, seem to be able to induce protection against vascular pathogens (Gessler and Kuć, 1982).

Different avirulent inducers can be effective against the same disease and an exchange of roles between the inducer and challenger is not unusual, i.e. avirulent isolates of *Verticillium* spp. and formae speciales of *Fusarium oxysporum* Schlechtend.:Fr. can protect tomato plants from Fusarium and Verticillium wilt respectively; *Fusarium oxysporum* Schlechtend.:Fr. f. sp. *melonis* W.C. Snyder & H.N. Hans. protects tomato from *Fusarium oxysporum* Schlechtend.:Fr. f. sp. *lycopersici* (Sacc.) W.C. Snyder & H.N. Hans. and *F. oxysporum* f. sp. *lycopersici* protects melon from *F. oxysporum* f. sp. *melonis* (see Matta, 1989). Thus, the induction as well as the resulting protective effect appear to be largely aspecific. An interval of time is required between induction and challenge. With few exceptions, 1–3 day intervals have been reported to be optimal for the control of Fusarium and Verticillium wilts (Matta, 1966; Price and Sackston, 1983). Absence of antagonism *in vitro*, possible exchange of role between inducer and challenger on different plant species and especially expression of maximal activity after one or a few days after the induction inoculation suggest that protection probably relies on plant-mediated mechanisms. Separation of inducer and challenger, in the case of systemic induced resistance, will definitely exclude the occurrence of fungus-fungus competition for sites or nutrients.

Protection from vascular pathogens can be localized or systemic depending on the plant and the experimental conditions. Full evidence of systemicity has been obtained on cucumber (Mandeel and Baker, 1991; Liu *et al.*, 1995a), watermelon (Larkin *et al.*, 1996), tomato (Olivain *et al.*, 1995; Fuchs *et al.*, 1997), and carnation (van Peer and Schippers, 1991). The systemicity of the effect should be particularly important in practice if obtainable by foliar application of the inducer, a feature up to now reported on cucumber only (Gessler and Kuć, 1982).

Although induced resistance to wilt diseases with avirulent fungi has been repeatedly demonstrated in laboratory conditions, its practical applicability in the field has been shown in rare instances. Protection of sweet potato from *Fusarium oxysporum* Schlechtend.:Fr. f. sp. *batatas* (Wollenweb.) W.C. Snyder & H:N. Hans. can be systemically induced by nonpathogenic formae speciales of *F. oxysporum* (Ogawa and Komada, 1986). Four different varieties of watermelon have been partially protected from *Fusarium oxysporum* Schlechtend.:Fr. f. sp. *niveum* (E.F. Sm.) W.C. Snyder & H.N. Hans. in a highly infested field by previous inoculation of the seedlings with a spore and mycelial suspension of *Bipolaris zeicola* (G.L. Stout) Shoemaker (= *Helminthosporium carbonum* Ullstrup) (Shimotsuma *et al.*, 1972). The severity of Verticillium wilt of tomato under commercial greenhouse conditions was reduced by 88% in artificially inoculated soil by root inoculation of the seedlings at transplanting time with conidial and mycelial suspensions of strains of *Verticillium albo-atrum* Reinke & Berthier and formae speciales of *F. oxysporum* avirulent on tomato (Matta and Garibaldi, 1977). Results (unpublished) even more spectacular were obtained against Verticillium wilt of tomato in the same cultural conditions when the fungal inducers were substituted with heat treatments. Previous laboratory experiments had shown that heat treatments increased resistance to subsequent vascular infections causing slight injury on the roots without impairing the vitality of the entire plant (Anchisi *et al.*, 1985). Considering that induced resistance to wilt diseases in tomato is generally short lasting, the success obtained on Verticillium wilt is explained by the fact that the critical period for the infection of the pathogen does not last more than 15 days after transplanting while subsequent infections are decreasingly important for symptom expression and plant productivity (Matta and Garibaldi, 1972). As frequently happens in biological control, high incidence of other pathogens (*Pyrenochaeta lycopersici* R. Schneider & Gerlach and *Fusarium oxysporum* Schlechtend.:Fr. f. sp. *radicis-lycopersici* W.R. Jarvis & Shoemaker) prevented a practical development of the method.

28.2.3. INDUCED RESISTANCE IN FUSARIUM SUPPRESSIVE SOILS

The existence of soils naturally suppressive to Fusarium wilt diseases has been reported in different areas of the world (Alabouvette *et al.,* 1996). Physico-chemical factors might have some influence in determining Fusarium wilt-suppressiveness. However the experimental evidence accumulated up to now (Alabouvette and Couteaudier, 1992) indicates that its origin is mainly biological, but does not rule out the involvement of abiotic factors. Suppressiveness to soilborne pathogens in general can be due to the activity of the total microbial biomass depleting the nutrients available to the pathogen (general suppression) or to the antagonism of specific groups of micro-organisms (specific suppression) (Cook and Baker, 1983). Fusarium wilt suppressiveness appears to be specific, depending on the activity of populations of *Fusarium,* mainly *F. oxysporum.* Besides being specifically induced, Fusarium wilt suppressiveness is also specifically effective in that it operates against diseases incited by formae speciales of *F. oxysporum* but not against diseases caused by other groups of plant pathogens. It appears then as a case of natural cross protection whose actual mechanisms are not completely understood. Selected *F. oxysporum* isolates induce suppressiveness in naturally conducive soils and in soils made conducive by physical or chemical disinfestation (Paulitz *et al.,* 1987) or increase suppressiveness possibly acting on a background of general suppression. The mode of action of *F. oxysporum* populations has been attributed to saprophytic competition with the pathogen for the nutrients in the soil (Alabouvette *et al.,* 1983) and/or on the root surface (Garibaldi *et al.,* 1991) as well as to parasitic competition for the infection sites (Schneider, 1984). The hypothesis that induced resistance also contributes to the overall protection effect of suppressive soils has been emerging more recently. Nonpathogenic *F. oxysporum* can colonize the rhizosphere permanently and the rhizosphere competence of selected strains is related to their Fusarium-suppressive activity in soil (Garibaldi *et al.,* 1991). Moreover, the potential capacity of such strains to interact with the plant root, is shown by the fact that at high inoculum level they can irreversibly injure the plant rootlets. Thus preliminary conditions for the induction of resistance appear to be satisfied.

Fusarium strains selected for their high suppressive activity in soil are at the same time good inducers of resistance when tested with methods specifically developed to investigate induction of local (Tamietti and Matta, 1991) or systemic (Olivain *et al.,* 1995) resistance to *F. lycopersici* in tomato. On the other hand, formae speciales of *F. oxysporum* avirulent on tomato that had been found quite active as resistance inducers were also effective in inducing suppressiveness in a *Fusarium* conducive soil.

Induced resistance to wilt diseases is characterized by marked biochemical changes in the plant (Madamanchi and Kuć, 1991). The resistance induced in tomato plants by avirulent fungi as well as by abiotic stresses is constantly associated with the activation of the defence related enzymes chitinase, 1,3-ß-glucanase, peroxidase, polyphenoloxidase (Abbattista Gentile *et al.*, 1988; Matta *et al.*, 1988; Fuchs *et al.*, 1997). If induced resistance had a role in suppressiveness, analogous changes should also arise from the interaction between Fusarium suppressive micro-organisms and the plant in the soil. Evidence that the suppressiveness of a soil influences the physiology of tomato plants was given by the enhanced activity of defence-related enzymes in roots, stem and leaves of tomato plants grown in a suppressive soil compared to plants grown in a conducive soil. Among five different isolates from the soil having different capacity to restore suppressiveness, the strain most suppressive was also the most active in stimulating defence related enzymes (Tamietti *et al.*, 1993).

It is difficult at the moment to determine what part of Fusarium suppressiveness of the soils depends on induced resistance. Induced resistance can be differently important on different plants, in different soils and for different Fusarium isolates. However, research should still look for antagonists to be employed to restore Fusarium-suppressive conditions in the soil for biological control. Future research could be aimed at developing strains with the capacity to induce resistance, to compete for nutrients in the soil, and with rhizosphere competence.

28.2.4. PLANT GROWTH-PROMOTING RHIZOBACTERIA (PGPR) AS INDUCING AGENTS

Since the late 70s, this group of bacteria have been the subject of extensive research. Initially, studies focused on their ability to directly promote plant growth (Schroth and Becker, 1990) through such mechanisms as the production of plant growth hormones (Frankenberger and Arshad, 1995) and their capacity to antagonize pathogens and reduce plant disease (Kloepper *et al.*, 1991). However, in 1991, van Peer *et al.* provided the first published evidence that PGPR could also induce systemic resistance to a disease, Fusarium wilt of carnations. Using a system of spatially separating the inoculation of the inducer (*Pseudomonas* sp. strain WCS417) from the challenger [*Fusarium oxysporum* Schlechtend.:Fr. f. sp. *dianthi* (Prill. & Delacr.) W.C. Snyder & H.N. Hans.] on the stem, disease was reduced in carnations by the PGPR treatment.

Resistance to bacterial diseases was demonstrated by treating bean seeds with *Pseudomonas fluorescens* (Trevisan) Migula and showing a reduction in lesions caused by later inoculation with *Pseudomonas syringae* van Hall pv. *phaseolicola* (Burkholder) Young *et al.* (Alström, 1991). Treatment of seeds of cucumber with PGPR strains reduced infection by a foliar anthracnose pathogen *C. orbiculare* (Wei *et al.*, 1991). Subsequent work by this group has also demonstrated that seed treatments with PGPR protected against cucumber mosaic virus (CMV) (Raupach *et al.*, 1996), *Pseudomonas syringae* van Hall pv. *lachrimans* (Smith & Bryan) Young *et al.* (Liu *et al.*, 1995b), Fusarium wilt (Liu *et al.*, 1995a), and *Erwinia tracheiphila* (Smith) Bergey *et al.* (Wei *et al.*, 1996). Protection against *E. tracheiphila* may be due to reduced feeding on induced plants by cucumber beetles that vector this wilt pathogen (Zehnder *et al.*, 1997).

Until recently, there were no reports of induced resistance to *Botrytis cinerea* Pers.:Fr., an important necrotrophic greenhouse pathogen. Elad *et al.* (1994) suggested that saprophytic bacteria and fungi may reduce the severity of grey mould by inducing a localized resistance. However, de Meyer and Höfte (1997) demonstrated that the rhizobacterium *Pseudomonas aeruginosa* (Schroeter) Migula induced systemic resistance to leaf infection by *B. cinerea* and provided evidence that salicylic acid production by the bacterium was important in this response.

PGPR have also been shown to induce protection against a root-rotting pathogen (Zhou and Paulitz, 1994). They used a split root system of cucumber to treat spatially-separated roots with zoospores of the challenger [*Pythium aphanidermatum* (Edson) Fitzp.] either simultaneously or one week after treatment with the inducers *Pseudomonas corrugata* (ex Scarlett *et al.*) Roberts & Scarlett or *P. fluorescens*. Plants induced with the bacteria showed increased root and shoot dry weights and a reduction in disease severity. Further work showed that these strains stimulated levels of phenylalanine ammonium lyase, peroxidase and polyphenoloxidase in cucumber roots, before they were challenged with *P. aphanidermatum* (Chen *et al.*, 1996). In later studies, when the opposite side of the root system was induced with bacteria prior to inoculation with the pathogen, the movement of *P. aphanidermatum* up the root system from a root tip was delayed and reduced (Chen *et al.*, 1997). Benhamou *et al.* (1996) also showed ultrastructural evidence of root defence reactions to *Fusarium oxysporum* Schlechtend.:Fr. f. sp. *pisi* (J.C. Hall) W.C. Snyder & Hanna induced by PGPR applied to pea roots. Systemic induced resistance to Pythium root rot and anthracnose in cucumber has also been demonstrated using a bark compost (Zhang *et al.*, 1996).

Exactly how PGPR induce resistance is still a mystery. Research is presently focused on the role of siderophores and salicylic acid produced by bacteria, often with contradictory results. One more hypothesis is that PGPR have indirect effects on the plant by changing the microbial communities on the surface of and inside the plant. Cucumbers inoculated with several resistance-inducing PGPR strains showed an altered composition of endophytes within the plant (Press *et al.*, 1995). The composition of bacterial communities in the rhizosphere of soybean were altered by inoculation with the biocontrol agent *Bacillus cereus* Frankland & Frankland UW85n1 (Gilbert *et al.*, 1996).

28.3. Direct Effects of the Plant on the Biocontrol Agent

Many biocontrol agents are developed to be used in a protective strategy to be applied at the infection site and become established before the pathogen can infect. These bacteria and fungi are often natural rhizosphere or phyllosphere colonizers and depend on the plant for carbon, nitrogen and other essential elements. These soluble and volatile nutrients arise from root, seed, flower and leaf exudates. Sloughed-off root cells shed by roots growing in soil and mucigel from the root cap are other important sources of nutrients for rhizosphere microbes. The list of compounds provided by the plant is extensive, including sugars, lipids, fatty acids, amino acids, organic acids, enzymes, hormones, vitamins and phenolic compounds. Despite our knowledge about what the plant gives off, we still know very little about the qualitative and quantitative effects of these sources of plant nutrients on the microbial community, including biocontrol agents.

Besides having an effect on the growth and reproduction of the biocontrol agent, this nutrient flux may also directly affect the mechanisms involved in the antagonism of the pathogen. Antibiotics such as phenazine 1-6 carboxylic acid and 2,4-diacetylphloroglucinol have been detected *in planta* on the surface of roots, and have been shown to be crucial for biocontrol in many systems (Thomashow and Weller, 1988; Keel *et al.*, 1992). These compounds are secondary metabolites typically produced in the stationary phase of bacterial growth. Nutrients quantitatively and qualitatively affect antibiotic production during fermentation, and may have a similar role *in planta*. Strains of *P. fluorescens* CHAO were engineered to overproduce 2,4 diacetylphloroglucinol and pyoluteorin (Maurhofer *et al.*, 1995). The increase in biocontrol efficiency of these overproducer strains was much greater on cucumber than on wheat, suggesting that the plant species affects the antibiotic production due to differences in root exudates between the two plant species. Further work with this strain has revealed that if a global regulator gene that controls production of these antibiotics is knocked out, the strain no longer controls Pythium damping-off on cucumber and cress, but continues to protect against the same pathogens on wheat and maize (Schmidli-Sacherer *et al.*, 1997).

Work with reporter gene systems has also given insights into the effect of the plant on antibiotic production. Georgakopoulos *et al.* (1994) used an ice nucleation gene reporter system to look at the expression of phenazine production on the surface of seeds. When the promoter regulating phenazine production is expressed, the bacterium produces an ice nucleation protein which can be quantified. They found high levels of expression of the reporter gene on wheat seeds and low levels on cotton. Inoculum levels, matric potential and soil did not appear to influence expression. Kraus and Loper (1995) observed temporal differences in expression of pyoluteorin biosynthesis genes by *P. fluorescens* on cucumber and cotton seed. Although expression was similar in the rhizosphere of these plants, expression differed in the spermosphere within 24 hours after seeds were planted. This suggests that the nutrient profile of the seed exudates may be important.

In the same way, the plant sets the stage for the drama of competition to unfold. In some biocontrol systems, competition for iron or carbon is critical for pathogen antagonism and the plant influences this. For example, competition for iron is mediated by siderophores produced by bacteria. Siderophores are low molecular weight compounds produced by microbes that chelate iron. Siderophores produced by fluorescent *Pseudomonas* spp. have a higher affinity for iron than the siderophores produced by pathogens such as Fusarium whose growth can be suppressed under iron-limiting conditions. However, is iron limiting enough on the plant surface for this competition to occur? Using an ice nucleation reporter gene fused to an Fe-regulator, *P. fluorescens* Pf-5 was shown to encounter iron-limiting conditions immediately after inoculation onto bean roots in field soil (Loper and Henkels, 1997). But 1 to 2 days later, iron became more available to the bacterium. pH also influenced the expression of the ice nucleation reporter gene. This shows that competition for iron can occur in the environment that the bacteria are encountering.

Many of these compounds emanating from the plant could also act as signals to stimulate the germination of the pathogen. Bacteria may interfere with this signal by catabolizing the substrate before it reaches the pathogen surrounding the plant. Some of these signals are volatile. Hyphal growth of *P. ultimum* was stimulated by volatiles, such as ethanol and acetaldehyde, produced by germinating pea and soybean seeds. Treatment of seeds with *Psudomonas putida* (Trevisan) Migula N1R reduced ethanol concentrations in the spermosphere and the volatile stimulation of *P. ultimum* (Paulitz, 1991). The fatty acid linoleic acid stimulated sporangial germination of *P. ultimum* and the biocontrol bacterium *Enterobacter cloacae* (Jordan) Hormaeche & Edwards can catabolize this fatty acid (van Dijk and Nelson, 1994). Mutants unable to utilize the fatty acid no longer protected cotton seeds from Pythium damping-off (van Dijk and Nelson, 1997).

In the future, the biocontrol agent may be engineered with a plant-derived competitive advantage, by giving the bacterium the ability to use a carbon source that other microbes cannot use, and by providing the plant with the genes to produce that carbon source. Fukui *et al.* (1994) compared two isogenic lines of *P. fluorescens-putida* ML5, one with a plasmid to utilize salicylate as a carbon source. When applied to the spermosphere of sugar beet with exogenous salicylate added, the strain with the plasmid outcompeted the other strain, even when applied at a lower inoculum density than the strain without the plasmid. If the production of this unique carbon source could be engineered into the plant, the plant could "select for" the biocontrol agent, a concept termed "biased rhizosphere". Tobacco plants were engineered with the ability to produce a unique class of compounds called opines (Savka and Farrand, 1997). These compounds are normally produced by tumours in plants infected with *Agrobacterium tumefaciens* (Smith & Townsend) Conn, which incorporates the genes for opine synthesis into the plant genome via T-DNA. They also constructed two strains of *P. fluorescens*, one with a plasmid-containing genes for the utilization of the opine, which *P. fluorescens* normally does not utilize. When inoculated alone, both bacteria reached similar levels on both normal and transgenic plants. However, when co-inoculated together, the catabolizing strain reached higher levels on the transgenic plant. A similar study was done with transgenic opine-producing tobacco plants and *P. syringae*, a phyllosphere colonizing bacterium (Oger *et al.*, 1997). When the bacterium was inoculated on a transgenic plant, it reached higher populations compared to a near-isogenic non-utilizer. These results suggest that plants could be "custom engineered" to fit a particular biocontrol agent, but commercialization is still far in the future.

The plant can exert a profound influence on the populations of fluorescent pseudomonads on the roots (Lemanceau *et al.*, 1995; Latour *et al.*, 1996). Bacteria were isolated and characterized from the roots of flax and tomato grown in the same soil. The bacteria were phenotypically characterized based on utilization of organic compounds and on PCR fingerprinting, and groups were clustered based on similarity. Some of the clusters were unique to either flax or tomato, demonstrating that different plant species select for different bacteria. The effect of bacteria may vary according to plant species. *Bacillus subtilis* (Ehrenberg) Cohn prevented Phytophthora and Pythium damping-off of *Astilbe, Photinia*, and *Hemerocallis* microplants, but not on *Daphne* plants (Berger *et al.*, 1996).

Different plant cultivars may have different degrees of "host hospitality" to biocontrol agents, a term coined by Smith *et al.* (1997) to indicate the amenability of different host genotypes to biocontrol. Different cultivars of wheat responded differently to PGPR strains (Chanway *et al.*, 1988) and biocontrol strains (Weller, 1986). *Pseudomonas cepacia* (ex *Burkholderia*) Palleroni & Holmes was not equally effective in suppressing Aphanomyces root rot of each of four cultivars of pea (King and Parke, 1993) and different cultivars of cucumber with different levels of resistance showed differing responses to ISR-inducing PGPR (Liu *et al.*, 1995c). Smith *et al.* (1997) attempted to use modelling to factor out and separate the specific response of the host to the pathogen and the biocontrol agent. They used six inbred tomato lines, different doses of *Pythium torulosum* Coker & F. Patterson and different doses of the biocontrol bacterium *B. cereus* UW85. The lowest dose of the pathogen revealed the greatest differences in seedling mortality among the inbred lines, inoculated with just the pathogen, while the highest pathogen dose revealed the greatest differences in biocontrol efficacy. Two of the inbreds, NIJ8 and 66, did not show any response to increasing concentrations of the biocontrol agents. Also, there was no correlation between the host response to the pathogen and the inoculum level of the biocontrol agent on this set of inbred lines. This indicates that our assumptions about the uniformity of cultivar response to biocontrol agents may be erroneous. Our idea of "one size fits all" in expecting a biocontrol agent to show equal levels of control on all cultivars or species attacked by a pathogen must be reconsidered. Differences in "host hospitality" to biocontrol agents could be exploited through plant breeding and selection.

The plant may influence the biocontrol agent indirectly via the physical and biological environment. Plants provide the niche for the biocontrol agent, and regulate the temperature, pH, osmotic potential and water potential of the microenvironment. The plant may also play a role in the distribution of the biocontrol agent on the plant surface by providing the microniches where the biocontrol agent and pathogen can interact together.

28.4. Conclusions

Control of diseases by biocontrol agents mediated through the plant (induced resistance) may offer many advantages as a disease control strategy for greenhouse diseases (Kuć, 1987). With some plant hosts, such as cucumber, induced resistance is broad-spectrum, effective against viral, bacterial, and fungal diseases. It is systemic and persistent, and utilizes the natural defences of the plant. This overcomes three of the major constraints of the protective strategy of using the biocontrol agent in contact with the pathogen at the infection site. These constraints are applying the biocontrol agent so it is well distributed and contacts all susceptible infection sites, getting the biocontrol agent to establish in high enough numbers to antagonize, and getting the biocontrol agents to persist over the period when the plant is susceptible. With induced resistance, treatment of one part of the plant can protect the whole plant. One of the major constraints in the use of induced resistance in the field is the labour involved in immunizing the plant. However, such treatments may be economical in the greenhouse, and immunized plants could be transplanted into the field already protected. Induced resistance is also graft-transmissible, which may be useful for woody perennials propagated by grafting. The finding that plants can be immunized by seed treatment with PGPR has also overcome a major barrier in the realistic use of this technology. The mechanisms involved in induced resistance may even be transferred to transgenic plants, making the application even easier. This technology has as much potential to become commercial reality as other biocontrol technologies. However, there is still a question of low persistence and systemicity in hosts other than cucumber. This obstacle will have to be overcome before the technology can be feasible.

Our understanding of how the plant exerts an influence on the biocontrol interaction will also lead to applications in disease management. Plants from breeding programs can be selected for "hospitality" to the biocontrol agent, in much the same way that plants are selected for resistance to diseases. This may be especially useful for pathogens that have been traditionally difficult to breed resistance to, for example Pythium damping-off and Botrytis grey mould.

Another potential future technology mediated through the plant could be to genetically engineer a "biased rhizosphere" (O'Connell et al., 1996). The idea is to enhance the effects of beneficial organisms in the rhizosphere by engineering plants to produce inducers or signals that would turn on microbial genes, such as antibiotic production. Or the population of beneficial organisms could be enhanced by engineering the plant to produce unusual nutrients and engineering the microbes to catabolize these nutrients, thus giving the microbe a competitive advantage. The paradigm of biological control is undergoing a major shift. Clearly, biological control is not just a one-on-one battle between the pathogen and the biocontrol agent, but rather the plant is an active ally in the struggle.

References

Abbattista Gentile, I., Ferraris, L. and Matta, A. (1988) Variations of polyphenoloxidase activities as a consequence of stresses that induce resistance to Fusarium wilt, J. of Phytopathology 122, 45–53.

Adams, P.B. (1990) The potential of mycoparasites for biological control of plant diseases, Annual Review of Plant Pathology 28, 59–72.

Alabouvette, C. and Couteaudier, Y. (1992) Biological control of fusarium wilts with nonpathogenic fusaria, in E.C. Tjamos, G.C. Papavizas and R.J. Cook (eds.), Biological Control of Plant Diseases, Progress and Challenges for the Future, Plenum Press, New York, pp. 415–426.

Alabouvette, C., Couteaudier, Y. and Louvet, J. (1983) Importance des phénomènes de compétition nutritive dand l'antagonisme entre microorganismes, Les Colloques de l'INRA 18, 7–16.

Alabouvette, C., Lemanceau, P. and Steinberg, C. (1996) Biological control of Fusarium wilts: Opportunities for developing a commercial product, in R. Hall (ed.) Principles and Practice of Managing Soilborne Plant Pathogens, American Phytopathological Press, St. Paul, Minn., pp. 192–212.

Alström, S. (1991) Induction of disease resistance in common bean susceptible to halo blight bacterial pathogen after seed bacterization with rhizosphere pseudomonads, J. of General and Applied Microbiology 37, 495–501.

Anchisi, M., Gennari, M. and Matta, A. (1985) Retardation of Fusarium wilt symptoms in tomato by pre- and post-inoculation treatments of the roots and aerial parts of the host in hot water, Physiological and Plant Pathology 26, 175–183.

Bargmann, C. and Schonbeck, F. (1992) *Acremonium kiliense* as inducer of resistance to wilt diseases on tomatoes, *Zeitschrift fur Pflanzenkrankheiten und Pflanzenschutz* **99**, 266–272.

Benhamou, N., Bélanger, R. and Paulitz, T. (1996) Induction of defence reactions by *Pseudomonas fluorescens* in Ri-TDNA transformed pea roots: Differential effect after challenge with *Fusarium oxysporum* f. sp. *pisi* or *Pythium ultimum*, *Phytopathology* **86**, 1174–1185.

Berger, F., Hong Li, D., White, D., Frazer, R. and Leifert, C. (1996) Effect of pathogen inoculum, antagonist density, and plant species on biological control of *Phytophthora* and *Pythium* damping-off by *Bacillus subtilis* Cot1 in high- humidity fogging glasshouses, *Phytopathology* **86**, 428–433.

Chanway, C.P., Nelson, L.M. and Holl, F.B. (1988) Cultivar-specific growth promotion of spring wheat (*Triticum aestivum* L.) by coexistent *Bacillus* species, *Canadian J. of Microbiology* **34**, 925–929.

Chen, C., Paulitz, T., Bélanger, R. and Benhamou, N. (1997) Inhibition of growth of *Pythium aphanidermatum* and stimulation of plant defense enzymes in split roots of cucumber systemically induced with *Pseudomonas* spp., *Phytopathology* **87**, S18 (abs.)

Chen, C., Paulitz, T., Benhamou, N. and Bélanger, R. (1996) Resistance to *Pythium aphanidermatum* in cucumber induced by *Pseudomonas* spp., *Phytopathology* **86**, S52 (abs.).

Chernin, L., Ismailov, Z., Haran, S. and Chet, I. (1995) Chitinolytic *Enterobacter agglomerans* antagonistic to fungal plant pathogens, *Applied and Environmental Microbiology* **61**, 1720–1726.

Chester K.S. (1933) The problem of acquired physiological immunity in plants, *Quarterly Review of Biology* **8**, 129–154, 275–324.

Cook, R.J. and Baker, K.F. (1983) *The Nature and Practice of Biological Control of Plant Pathogens,* American Phytopathological Society, St Paul, Minn.

de Meyer, G. and Höfte, M. (1997) Salicylic acid produced by the rhizobacterium *Pseudomonas aeruginosa* 7NSK2 induces resistance to leaf infection by *Botrytis cinerea* on bean, *Phytopathology* **87**, 588–593.

Elad, Y., Köhl, J. and Fokkema, N.J. (1994) Control of infection and sporulation of *Botrytis cinerea* on bean and tomato by saprophytic bacteria and fungi, *European J. of Plant Pathology* **100**, 315–336.

Frankenberger, W.J. and Arshad, M. (1995) *Phytohormones in Soils: Microbial Production and Function,* Marcel Dekker, New York.

Fravel, D.R. (1988) Role of antibiotics in the biocontrol of plant diseases, *Annual Review of Plant Pathology* **26**, 75–91.

Fuchs, J.G., Moënne-Loccoz, Y. and Défago, G. (1997) Nonpathogenic *Fusarium oxysporum* strain Fo47 induces resistance to Fusarium wilt in tomato, *Plant Disease* **81**, 492–496.

Fukui, R., Schroth, M.N., Hendson, M. and Hancock, J.G. (1994) Interactions between strains of pseudomonads in sugar beet spermospheres and their relationship to pericarp colonization by *Pythium ultimum* in soil, *Phytopathology* **84**, 1322–1330.

Fulton, R.W. (1986) Practices and precautions in the use of cross protection for plant virus disease control, *Annual Review of Phytopathology* **26**, 67–81.

Garibaldi, A., Guglielmone, L., Cugudda, L. and Gullino, M.L. (1991) Osservazioni sulla "rhizosphere competence" di alcuni isolati di *Fusarium* antagonisti provenienti da terreni repressivi verso le trachofusariosi, *Petria* **1**, 51–60.

Georgakopoulos, D.G., Hendson, M., Panopoulos, N.J. and Schroth, M.N. (1994) Analysis of expression of a phenazine biosynthesis locus of *Pseudomonas aureofaciens* PGS12 on seeds with a mutant carrying a phenazine biosynthesis locus-ice nucleation reporter gene fusion, *Applied and Environmental Microbiology* **60**, 4573– 4579.

Gessler, C. and Kuć, J. (1982) Induction of resistance to Fusarium wilt in cucumber by root and foliar pathogens, *Phytopathology* **72**, 1439–1441.

Gilbert, G.S., Clayton, M.K., Handelsman, J. and Parke, J.L. (1996) Use of cluster and discriminant analyses to compare rhizosphere bacterial communities biological perturbation, *Microbial Ecology* **32**, 123–147.

Gonsalves, D. and Garnsey, S.M. (1989) Cross-protection techniques for control of plant virus diseases in the tropics, *Plant Disease* **73**, 592–597.

Guedes, M.E., Richmond, S. and Kuć, J. (1980) Induced systemic resistance to anthracnose in cucumber as influenced by the location of the inducer inoculation with *Colletotrichum lagenarium* on the onset of flowering and fruiting, *Physiological Plant Pathology* **17**, 229–233.

Hammerschmidt, R. and Kuć, J. (1982) Lignification as a mechanism for induced systemic resistance in cucumber, *Physiological Plant Pathology* **20**, 73–82.

Hammerschmidt, R. and Kuć, J. (1995) *Induced Resistance to Disease in Plants,* Kluwer Academic Publishers, Dordrecht.

Hammerschmidt, R., Nuckles, E.M. and Kuć, J. (1982) Association of enhanced peroxidase activity with induced systemic resistance of cucumber to *Colletotrichum lagenarium, Physiological Plant Pathology* **20**, 73–82.

Heller, W.E. and Gessler, C. (1986) Induced systemic resistance in tomato plants against *Phytophthora infestans, J. of Phytopathology* **116**, 323–328.

Jørgensen, H.J.L., Andresen, H. and Smedegaard-Petersen, H. (1996) Control of *Drechslera teres* and other barley pathogens by preinoculation with *Bipolaris maydis* and *Septoria nodorum, Phytopathology* **86**, 602-607.

Keel, C., Schnider, U., Maurhofer, M., Voisard, C., Laville, J., Burger, U., Wirthner, P., Hass, D. and Défago, G. (1992) Suppression of root diseases by *Pseudomonas fluorescens* CHAO: Importance of the bacterial secondary metabolite 2,4-diacetylphloroglucinol, *Molecular Plant-Microbe Interactions* **5**, 4–13.

King, E.B. and Parke, J.L. (1993) Biocontrol of Aphanomyces root rot and Pythium damping-off by *Pseudomonas cepacia* AMDD on four pea cultivars, *Plant Disease* **77**, 1185–1188.

Kloepper, J.W., Tuzun, S., and Kuć, J. (1992) Proposed definitions related to induced resistance, *Biocontrol Science and Technology* **2**, 349–351.

Kloepper, J.W., Zablotowicz, R.M., Tipping, E.M. and Lifshitz, R. (1991) Plant growth promotion mediated by bacterial rhizosphere colonizers, in D.L. Keister and P.B. Cregan (eds.), *The Rhizosphere and Plant Growth,* Kluwer Academic Publishers, Dordrecht, pp. 315–326.

Kraus, J. and Loper, J.E. (1995) Characterization of a genomic region required for production of the antibiotic pyoluteorin by the biological control agent *Pseudomonas fluorescens* Pf-5, *Applied and Environmental Microbiology* **61**, 849–854.

Kuć, J. (1982) Plant immunization-mechanisms and practical implications, in R.K.S. Wood (ed.), *Active Defense Mechanisms in Plants,* Plenum Press, New York, pp. 157–178.

Kuć, J. (1987) Plant immunization and its applicability for disease control, in I. Chet (ed.), *Innovative Approaches to Plant Disease Control,* John Wiley and Sons, New York, pp. 255–274.

Kuć, J. and Strobel, N.E. (1992) Induced resistance using pathogens and non pathogens, in E.C. Tjamos, G.C. Papavizas and R.J. Cook (eds.), *Biological Control of Plant Diseases, Progress and Challenges for the Future,* Plenum Press, New York, pp. 295– 303.

Larkin, R.P., Hopkins, D.L. and Martin, F.N. (1996) Suppression of Fusarium wilt of watermelon by nonpathogenic *Fusarium oxysporum* and other microorganisms recovered from a disease-suppressive soil, *Phytopathology* **86**, 812–819.

Latour, X., Corberand, T., Laguerre, G., Allard, F. and Lemanceau, P. (1996) The composition of fluorescent pseudomonad populations associated with roots is influenced by plant and soil type, *Applied and Environmental Microbiology* **62**, 2449–2456.

Lemanceau, P., Corebrand, T., Gardan, L., Latour, X., Laguerre, G., Boeufgras, J.-M., and Alabouvette, C. (1995) Effect of two plant species, flax (*Linum usitatissimum* L.) and tomato (*Lycopersicon esculentum* Mill.), on the diversity of soilborne populations of fluorescent pseudomonads, *Applied and Environmental Microbiology* **61**, 1004–1012.

Liu, L., Kloepper, J.W. and Tuzun, S. (1995a) Induction of systemic resistance in cucumber against Fusarium wilt by plant growth-promoting rhizobacteria, *Phytopathology* **85**, 695–698.

Liu, L., Kloepper, J.W. and Tuzun, S. (1995b) Induction of systemic resistance in cucumber against bacterial angular leaf spot by plant growth-promoting rhizobacteria, *Phytopathology* **85**, 843–847.

Liu, L., Kloepper, J.W. and Tuzun, S. (1995c) Induction of systemic resistance in cucumber by plant growth-promoting rhizobacteria: Duration of protection and effect of host resistance on protection and root colonization, *Phytopathology* **85**, 1064–1068.

Loper, J.E. and Buyer, J.S. (1991) Siderophores in microbial interactions on plant surfaces, *Molecular Plant-Microbe Interactions* **4**, 5–13.

Loper, J. and Henkels, M.D. (1997) Availability of iron to *Pseudomonas fluorescens* in rhizosphere and bulk soil evaluated with an ice nucleation reporter gene, *Applied and Environmental Microbiology* **63**, 99–105.

Madamanchi, N.R. and Kuć, J. (1991) Induced systemic resistance in plant, in G.T. Cole and H.C. Hoch (eds.), *The Fungal Spore and Disease Initiation in Plants and Animals*, Plenum Press, New York, pp. 347–362.

Mandeel, Q. and Baker, R. (1991) Mechanisms involved in biological control of Fusarium wilt of cucumber with strains of non-pathogenic *Fusarium oxysporum, Phytopathology* **81**, 462–469.

Matta, A. (1966) Effetto immunizzante di alcuni micromiceti verso le infezioni di *Fusarium oxysporum* f. sp. *lycopersici* su pomodoro, *Annali della Facolta di Scienze Agrarie della Universita di Torino* **3**, 85–98.

Matta, A. (1971) Microbial penetration and immunization of uncongenial host plants, *Annual Review of Phytopathology* **9**, 387–410.

Matta, A. (1989) Induced resistance to Fusarium wilt diseases, in E.C. Tjamos and C.H. Beckman (eds.), *Vascular Wilt Diseases of Plants*, Springer, Berlin, pp. 175–196.

Matta, A. and Garibaldi, A. (1972) Ricerca dei periodi critici per le infezioni di *Verticillium dahliae* sul pomodoro coltivato in serra, *Annali della Facolta di Scienze Agrarie della Universita di Torino* **8**, 1–12.

Matta, A. and Garibaldi, A. (1977) Control of Verticillium wilt of tomato by preinoculation with avirulent fungi, *Netherlands J. of Plant Pathology* **83**(1), 457–462.

Matta, A., Gentile, I. and Ferraris, L. (1988) Stimulation of 1,3-ß-glucanase and chitinase by stresses that induce resistance to Fusarium wilt in tomato, *Phytopathologia Mediterranea* **2**, 45–50.

Maurhofer, M., Keel, C., Hass, D. and Défago, G. (1995) Influence of plant species on disease suppression by *Pseudomonas fluorescens* strain CHAO with enhanced antibiotic production, *Plant Pathology* **44**, 40–50.

Meera, M.S., Shivanna, M.B., Kageyama, K. and Hyakumachi, M. (1994) Plant-growth promoting fungi from zoysiagrass rhizosphere as potential inducers of systemic resistance in cucumbers, *Phytopathology* **84**, 1399–1486.

Métraux, J.P. and Boller, T. (1986) Local and systemic induction of chitinase in cucumber plants in response to viral, bacterial and fungal infection, *Physiological and Molecular Plant Pathology* **28**, 161–169.

Morandi, D., Bailey, J.A. and Gianinazzi-Pearson, V. (1984) Isoflavonoid accumulation in soybean roots infected with vesicular-arbuscular mycorrhizal fungi, *Physiological and Plant Pathology* **24**, 357–364.

O'Connell, K.P., Goodman, R.M. and Handelsman, J. (1996) Engineering the rhizosphere: Expressing a bias, *Trends in Biotechnology* **14**, 83–88.

Ogawa, K. and Komada, H. (1986) Induction of systemic resistance against Fusarium wilt of sweet potato, *Annals of the Phytopathological Society of Japan* **52**, 15–21.

Oger, P., Petit, A. and Dessaux, Y. (1997) Genetically engineered plants producing opines alter their biological environment, *Nature Biotechnology* **15**, 369–372.

Olivain, C., Steinberg, C. and Alabouvette, C. (1995) Evidence of induced resistance in tomato inoculated by nonpathogenic strains of *Fusarium oxysporum*, in M. Manka (ed.), in *Environmental Biotic Factors in Integrated Plant Disease Control, Proceedings of the 3rd Conference of the European Foundation for Plant Pathology,* Poznan, 5–9 September 1994, pp. 427–430.

Paulitz, T.C. (1991) Effect of *Pseudomonas putida* on the stimulation of *Pythium ultimum* by seed volatiles of pea and soybean, *Phytopathology* **81**, 1282–1287.

Paulitz, T.C., Park, C.S. and Baker, R. (1987) Biological control of Fusarium wilt of cucumber with nonpathogenic isolates of *Fusarium oxysporum, Canadian J. of Microbiology* **33**, 349–353.

Press, C.M., Wilson, M., Kloepper, J.W. and Tuzun, S. (1995) Salicylate production by plant growth-promoting rhizobacteria which induce systemic disease resistance in cucumber, *Phytopathology* **85**, 1154 (abs.)

Price, D. and Sackston, W.E. (1983) Cross protection in sunflower against Verticillium wilt, *Canadian J. of Plant Pathology* **5**, 210.

Raupach, G.S., Liu, L., Murphy, J.F., Tuzun, S. and Kloepper, J.W. (1996) Induced systemic resistance in cucumber and tomato against cucumber mosaic cucumovirus using plant growth-promoting rhizobacteria (PGPR), *Plant Disease* **80**, 891–894.

Savka, M. and Farrand, S.K. (1997) Modification of rhizobacterial populations by engineering bacterium utilization of a novel plant-produced resource, *Nature Biotechnology* **15**, 363–368.

Schmidli-Sacherer, P., Keel, C. and Défago, G. (1997) The global regulator GacA of *Pseudomonas fluorescens* CHAO is required for suppression of root diseases in dicotyledons but not in Gramineae, *Plant Pathology* **46**, 80–90.

Schneider, R.W. (1984) Effects of nonpathogenic strains of *Fusarium oxyporum* on celery root infection by *F. oxysporum* f. sp. *apii* and a novel use of the Lineweaver Burke double reciprocal plot technique, *Phytopathology* **74**, 646– 653.

Schroth, M.N. and Becker, J.O. (1990) Concepts of ecological and physiological activities of rhizobacteria related to biological control and plant growth promotion, in D. Hornby (ed.), *Biological Control of Soil-Borne Plant Pathogens,* CAB International, Wallingford, pp. 389–414.

Sequeira, L. (1983). Mechanisms of induced resistance in plants, *Annual Review of Microbiology* **37**, 51–79.

Shimotsuma, M., Kuć, J. and Jones, C.M. (1972) The effect of prior inoculations of non-pathogenic fungi on Fusarium wilt of watermelon, *HortScience* **7**, 72–73.

Smith, K.P., Handelsman, J. and Goodman, R.M. (1997) Modeling dose-response relationships in biological control: Partitioning host responses to the pathogen and biocontrol agent, *Phytopathology* **87**, 720–729.

St. Arnaud, M., Hamel, C., Caron, M. and Fortin, J.A. (1994) Inhibition of *Pythium ultimum* in roots and growth substrate of mycorrhizal *Tagetes patula* colonized with *Glomus intraradices, Canadian J. of Plant Pathology* **16**, 187–194.

Sticher, L., Mauch-Mani, B. and Métraux, J.P. (1997) Systemic acquired resistance, *Annual Review of Phytopathology* **35**, 235–270.

Tamietti, G., Ferraris, L., Matta, A. and Abbattista Gentile, I. (1993) Physiological responses of tomato plants grown in Fusarium suppressive soil, *J. of Phytopathology* **138**, 66–76.

Tamietti, G. and Matta, A. (1991) The effect of inoculation methods on biocontrol of Fusarium wilt of tomato with non-pathogenic *Fusarium* isolates, in A.B.R. Beemster, G.J. Bollen, M. Gerlagh, M.A. Ruissen, B. Schippers and A. Tempel (eds.), *Biotic Interactions and Soil-borne Diseases*, Elsevier, Amsterdam, pp. 329–334.

Thomashow, L.S. and Weller, D.M. (1988) Role of a phenazine antibiotic from *Pseudomonas fluorescens* in biological control of *Gaeumannomyces graminis* var. *tritici, J. of Bacteriology* **170**, 3499–3508.

Tuzun, S. and Kuć, J. (1985) A modified technique for inducing systemic resistance to blue mold and increasing growth of tobacco, *Phytopathology* **75**, 1127–1129.

Tuzun, S., Nesmith, W., Ferriss, R.S. and Kuć, J. (1986) Effects of stem injections with *Peronospora tabacina* on growth of tobacco and protection against blue mold in the field, *Phytopathology* **76**, 938–941.

van Dijk, K.V. and Nelson, E.B. (1994) Loss of biological control ability in an *Enterobacter cloacae* mutant unable to catabolize lineoleic acid, *Phytopathology* **84**, 1091 (abs.)

van Dijk, K.V and Nelson, E.B. (1997) Fatty acid uptake and beta-oxidation by *Enterobacter cloacae* is necessary for seed rot suppression of *Pythium ultimum, Phytopathology* **87**, S100 (abs.)

van Loon, L.C., Bakker, P.A.H.M. and Pieterse, C.M.J (1997) Mechanisms of PGPR-induced resistance against pathogens, in A. Ogoshi, K. Kobayashi, Y. Homma, F. Kodama, N. Kondo and S. Akino (eds.), *Plant Growth-Promoting Rhizobacteria – Present Status and Future Prospects*, OECD, Paris, pp. 50–57.

van Peer, R., Niemann, G.J. and Schippers, B. (1991) Induced resistance and phytoalexin accumulation in biological control of Fusarium wilt of carnation by *Pseudomonas* sp. strain WCS417r, *Phytopathology* **81**, 728–734.

van Peer, R. and Schippers, B. (1991) Biocontrol of Fusarium wilt by *Pseudomonas* sp. strain WCS4172: induced resistance and phytoalexin accumulation, in A.B.R. Beemster, G.J. Bollen, M. Gerlagh, M.A. Ruissen, B. Schippers and A. Tempel (eds.), *Biotic Interactions and Soil-borne Diseases*, Elsevier, Amsterdam, pp. 274–280.

Wei, G., Kloepper, J.W. and Tuzun, S. (1991) Induction of systemic resistance of cucumber to *Colletotrichum orbiculare* by select strains of plant growth-promoting rhizobacteria, *Phytopathology* **81**, 1508–1512.

Wei, G., Kloepper, J.W. and Tuzun, S. (1996) Induced systemic resistance to cucumber diseases and increased plant growth-promoting rhizobacteria under field conditions, *Phytopathology* **86**, 221–224.

Weller, D.M. (1986) Effects of wheat genotype on root colonization by a take-all suppressive strain of *Pseudomonas fluorescens*, *Phytopathology* **76**, 1059 (abs.).

Zehnder, G., Kloepper, J. Tuzun, S., Yao, C., Wei, G., Chambliss, O. and Shelby, R. (1997) Insect feeding on cucumber mediated by rhizobacteria-induced plant resistance, *Entomologia Experimentalis et Applicata* **83**, 81–85.

Zhang, W., Dick, W.A. and Hoitink, H.A.J. (1996) Compost-induced systemic acquired resistance in cucumber to Pythium root rot and anthracnose, *Phytopathology* **86**, 1066–1070.

Zhou, T. and Paulitz, T.C. (1994) Induced resistance in the biocontrol of *Pythium aphanidermatum* by *Pseudomonas* spp. on cucumber, *J. of Phytopathology* **142**, 51–63.

IMPLEMENTATION OF IPM: FROM RESEARCH TO THE CONSUMER
Jean-Claude Onillon and M. Lodovica Gullino

29.1. Introduction

Several chapters in this book provide a complete view of the feasibility of Integrated Pest and Disease Management (IPM) on most protected crops. However, there is still room for further improvement in such crops and for development of new systems in other crops. Successful IPM systems start with appropriate planning of the necessary research; only well-oriented research will lead to results that will find practical application. However, the fact that an IPM programme has been developed by researchers does not necessarily mean that it will automatically be implemented by farmers. In this chapter, the entire process to transfer the innovative knowledge from research to application is summarized. Special emphasis has been laid on the research, development and application of biological control, as IPM relies strongly on biological control agents (BCAs). The subject of biological control has already been covered in Chapters 13 to 22 for pests, and 23 to 28 for diseases. A few more considerations, all related to the use of biocontrol in IPM, are summarized here.

29.2. Research on BCAs and their Development in the Framework of IPM Programmes

In order to be effective, IPM research must not take place in isolation. Although it usually starts in a University Department, the development, adaptation and implementation of IPM systems requires, in contrast to that of chemical control, an interdisciplinary approach. Very close co-operation among scientists, extension services, growers, producers of BCAs and policy makers is necessary and all participants in an IPM programme must be willing to help its implementation. Moreover, IPM research should be primarily oriented towards the solutions of problems that are not easily managed through presently available control measures.

In the case of pest, and less frequently, disease control, it appears relatively easy to find the BCAs able to replace pesticides. However, years of fundamental and applied research, together with considerable investments in the private sector, may be necessary before the BCA can be used commercially.

29.2.1. RESEARCH UNDER CONTROLLED CONDITIONS

Knowledge of the biology and ecology of the components of the tri-trophic system, the host plant-pest/pathogen-antagonist, must be fully gained under laboratory and

411

R. Albajes et al. (eds.), Integrated Pest and Disease Management in Greenhouse Crops, 411-419.
© 1999 *Kluwer Academic Publishers. Printed in the Netherlands.*

greenhouse conditions. In the case of insect pests, the effect of temperature on fecundity, duration of preimaginal development and mortality of immatures under laboratory conditions is used to determine the thermal interval at which the natural enemy population will grow faster than the pest population and, therefore, the interval within which the BCA will be useful for biological control with inoculative releases. Such information is used to adopt the most favourable thermal regime for the BCA or, alternatively, to choose the BCA best adapted to particular greenhouse thermal regimes. These data are available for the most common greenhouse pests and natural enemies and are given in earlier chapters. For instance, *Encarsia formosa* Gahan has long been the only available BCA for greenhouse whitefly control since the thermal range in which the net rate of increase of the parasitoid is higher than that of the pest is 20–30°C, which is the common range of temperature in heated glasshouses. However, in Mediterranean conditions, where summer and winter temperatures in plastic greenhouses are outside the optimal interval mentioned, its effectiveness is much lower. In these conditions, native Mediterranean fauna including better adapted and polyphagous predators like *Macrolophus caliginosus* Wagner has been evaluated (Alomar *et al.*, 1991). In the case of greenhouse plant pathogens and their antagonists, the effect of temperature and also of relative humidity must be determined.

The host plant may also influence the performance of a BCA and its use. It is well known, for example, that host plant species and even cultivars greatly affect the rate of increase of the greenhouse whitefly and the behaviour of its parasitoid *E. formosa* (Chapter 14, Malausa *et al.*, 1983). When such information was obtained from careful and sometimes sophisticated research, the use and doses of the parasitoid in greenhouses were defined more precisely. Particularly relevant for biological control are the studies of pest-BCA interactions. Knowledge of the host larval instars that are the most susceptible to parasitoid attack or the influence of whitefly larval instar proportion on the crop is necessary for the correct parasitoid release strategy. Whilst not very important for parthenogenetic parasitoids such as *E. formosa*, the latter is crucial in the case of bisexual species such as *Encarsia tricolor* Foerster and *Encarsia pergandiella* Howard.

At a more advanced research stage, greenhouse experiments are used to study the ecology of BCAs under conditions closer to practice. The effects of the whole crop plant, influence of fluctuating temperatures and greenhouse structure are some of the issues analysed on the greenhouse scale. Moreover, heterogeneity in the greenhouse environment, particularly in plastic-houses, may favour the development of pest and disease foci that will affect the strategy for the use of BCAs in practice.

29.2.2. APPLICATION OF BCAs UNDER PRACTICAL CONDITIONS

BCAs are just one component of IPM (Gullino, 1995). The possibility of their application under practical conditions must be evaluated in the early phase of their development. Such studies must take place first in experimental greenhouses, where it is possible to optimize the application of the BCAs, and later in commercial greenhouses, where it is possible to evaluate the effect of several factors on the success of the BCA.

In experimental greenhouses, evaluation of the application methodology of *E. formosa* against *Trialeurodes vaporariorum* (Westwood), or other parasitoids against *Bemisia tabaci* (Gennadius), enabled definition of the best application techniques (Onillon, 1990). The aspects to be taken into consideration are: (i) the application of the right strain; (ii) the best timing of application (Parr *et al.,* 1976; Stacey, 1977; Hussey and Scopes, 1985); (iii) the dosage of the parasitoid to be released in relation to the surface area; (iv) the number of release points (Foster, 1980); (v) the pest population (Helgesen and Tauber, 1974; Onillon *et al.*, 1976); (vi) the number of treatments; and (vii) the most susceptible pest stage.

Another important aspect is to assess the compatibility between the introduced BCAs and the other components of IPM, particularly pesticides (Delorme and Angot, 1983) still applied to combat other pathogens or pests. This is covered in Chapter 11.

In commercial greenhouses, other aspects must be taken into account, such as external constraints related to the sensitivity of the grower and the personal motivation which moved him to offer his greenhouse for the experiment or, at least, to accept hosting it. Such motivation may be based on economic or technical issues. Internal constraints, sometimes difficult for farmers to accept, particularly the level of contamination (especially in the glasshouse or nurseries) and the adaptation to local uses, must be taken into account. Finally, economic constraints, such as carrying out the experiment at no cost, may be important.

29.2.3. COMMERCIAL DEVELOPMENT OF BCAs

This very crucial aspect, considered in Chapters 20 and 26, is basic for the practical application of BCAs. It requires full co-operation between the public and private sector. It includes several steps aimed at obtaining effective BCAs at a sufficient quantitative and qualitative level and at transferring their application into practice, in co-operation with the extension service. Practical application of most BCAs, particularly those active against plant pathogens, cannot take place without their registration (Cook, 1993). The many problems encountered at registration are considered in Chapter 26. The importance of quality control is stressed in Chapters 20 and 26.

We wish to emphasize once more the importance of shelf life of BCAs, as these must compete with chemicals in terms of their storability (Gullino, 1993). In the case of biocontrol of pests, BCAs are not always available all year and a balance must be maintained between production, cold storage (Scopes *et al.*, 1973) and commercialization.

29.3. Transfer of the New Technology to Extension Services and Growers

The final objective of an IPM programme is to bring it to the growers. However, before direct, practical advice is given to growers, they first have to be convinced that the proposed new technology is reliable. Therefore, the basic technology has to be further adapted to regional or even local conditions, according to particular environment/crop/pest conditions.

29.3.1. DEMONSTRATION TRIALS

Once a BCA is commercially developed and produced, larger scale demonstration trials must take place. This step must be linked to the research level since it allows studying the reliability of the treatment. Demonstration trials imply co-operation among researchers, the extension service and interested growers. Other growers visit such demonstration trials, so that implementation can start. The cultural environment of the crop, not only at the single greenhouse level, but at the level of a group of greenhouses, belonging or not to the same grower, must be considered. This implies knowing the climate and the flora, since the approach will be different if the group of greenhouses is isolated inside a complex of greenhouses (with the possibility of fast recontamination with resistant adults of *T. vaporariorum*, lack of indigenous predators) or if exchanges with the exterior are possible (with possible activity of native natural enemies when environment is favourable).

Internal constraints in trials are similar to those in commercial greenhouses. One of the common technical constraints for demonstration trials deals with the availability of the sufficient number of BCAs. When this type of trials are carried out in commercial greenhouses, there is not necessarily and industrial procedure to mass rear BCAs, at least in the case of natural enemies of insect pests. For example, the only *E. formosa* available for the first demonstration trials were produced in the laboratory and it was necessary to produce the sufficient amount of parasitoids to treat the area required. At the beginning, an overdosage is suggested, in order to obtain good results. At the level of psychological constraints, beside those already mentioned, it is helpful to make the grower understand that the strategy proposed is as simple as a chemical spray. Finally, the economic constraints, linked to the need to maintain the cost of biological control similar to that of chemical control, should also be taken into account.

Trials allow demonstration of the efficacy of biological control under different environmental, cultural and technical situations. Geographic and climatic diversity, as well as cultural conditions and varietal diversity should be taken into account (Onillon et al., 1983). Demonstration trials also permit careful evaluation of the technical limitations that can make IPM impossible under practical conditions (van Lenteren, 1995).

29.3.2. ROLE OF THE EXTENSION SERVICE

The use of the most sophisticated technologies (i.e. the application of BCAs, and the choice of the most suitable resistant cultivar) requires in depth knowledge, which is not always available everywhere among growers.

It is evident that extension services play a major role (Wearing, 1988; Wardlow, 1992). First of all, extension services provide the necessary link between application and research. When setting up IPM programmes, the participation of the extension service is essential during the planning, as well as during the implementation phase. When an IPM programme for a certain crop is introduced, the extension service must provide growers with a period of very intensive attention. Not only public extension services, present in most countries, but also private consultants are important in assisting growers in making complex control decisions, and informing them of constantly changing regulatory issues (Rogers, 1996). Advisers should maintain a high profile, in order to be credible and valued. In particular, they need to be independent.

29.3.3. TRAINING AND EDUCATION

Training is essential to the success of IPM (Wardlow, 1992). A well-instructed and motivated grower is crucial for effective IPM. The basic instruction of the young grower starts at agricultural schools and continues with refreshment courses organized by the extension service and by local growers' associations. Expert systems (see Chapter 12) may also be useful in this respect. Technicians working in the extension service should be regularly updated, in order to make sure that they will be able to provide growers with the most advanced and recent technologies. It is also important to take care of the education of professors teaching scientific subjects in secondary schools. Students are future consumers and an early knowledge of how food is produced will make them more considerate when choosing produce. Finally, the consumers should be clearly informed about the availability of food produced under IPM (see Section 29.4).

A comprehensive review of the role of education and training in IPM has been given by Jeger (1995), who stresses the importance of all formal instruction, ranging from short courses to research training for PhD degrees, including continuing education programmes. The teaching of crop protection has drastically changed at all levels (from vocational schools to University) during the last 20 years. Pure technical information on how to spray and with what chemical has been replaced with information on other forms of pest and disease control (van Lenteren, 1995).

29.3.4. ROLE OF THE GOVERNMENTS

Legislation and political decisions can positively influence implementation of IPM. This has been shown in several countries such as The Netherlands, Italy (Gullino and Kuijpers, 1994) and, more recently, the USA (Jacobsen, 1997). Governments may require manufacturers or food distributors to include accurate information on product labels. This enables consumers to adjust their buying behaviour appropriately (Marsh, 1995).

Such policies should be accompanied by strong support of research and demonstration projects and by farmer incentives. In some cases, growers' participation in funding IPM research has resulted in faster results, as shown in Canada in the case of chrysanthemum and poinsettia IPM programmes (Murphy and Broadbent, 1993).

29.3.5. ROLE OF THE AGROCHEMICAL COMPANIES

Although any complication in a simple, straight chemical control programme has been viewed in the past as a negative development by the large agrochemical industries (van Lenteren, 1995), a much more sustainable approach is now pursued by industry (Urech, 1996). Some new chemicals with a novel mode of action can be used together with

BCAs. Moreover, large companies have started investing by acquiring small companies interested in developing BCAs. This will, hopefully, eventually help in filling the gap between chemical and biological control. Although BCAs represent niche markets that are too small for large companies, their interest in their development could speed up their usage in practice.

29.3.6. EXTENSION ACTIVITIES

IPM implementation is also strongly supported by information given throughout specialized growers journals. Researchers should pay more attention to making the results of their research available not only through scientific, peer reviewed journals, but also, in a more simple form, through technical journals. Advisers also need to publish relevant information on new developments in IPM by several means, such as articles for local and national press, lectures to growers' groups, seminars and conferences, radio and television interviews and talks, and specialized growers' notes and leaflets (Wardlow, 1992). The Internet can now help in such a task (Biggs and Grove, 1998). In some countries, technical booklets and posters are frequently used to assist growers in the recognition of pests and diseases, their damage and their natural enemies.

29.4. Reaching the Consumer

Today's consumers have started looking not only at the appearance but also at factors such as flavour, nutritional value, method of cultivation and absence of pesticide residues in the produce they buy. They show interest in the way food is produced: for some this stems from concern for health. An important group of consumers are critical of contemporary farming because it strongly relies on pesticides. IPM programmes seem to provide a good method for anticipating consumer requirements. IPM production systems should be and frequently are advertised in a way that consumers are also reached.

In the future, more care should be taken to provide consumers with adequate information. Formal and informal education play a major role in determining behaviour. An informed public is better placed to judge and make choices. In a largely urban community, people have little first hand knowledge of agriculture and are very vulnerable to propaganda and romantic notions about how it actually operates (Marsh, 1995). Education, which includes both the underlying science involved in food production and the role of the agricultural and food industries in food supply, will be essential.

29.4.1. SPECIAL LABELING OF PRODUCTS GROWN UNDER IPM

The general "green" attitude in public life has widened interest in decreased pesticide usage and this has favoured IPM by increasing its popularity. Food retailers are finding it useful to have sections in their stores devoted to IPM grown produce and food.

Several supermarket chains and canning companies already market fresh and processed produce grown under their own requirements and regulations. A special label identifies these products. Several regional European administrations have also taken the initiative in defining and regulating Integrated Production (IP) and IPM labels. Many examples of products marketed under IPM labels exist in most European countries (Anonymous, 1988). One well-documented IPM success story is that of the Campbell Soup Company's IPM implementation with growers in several States of the USA and in Mexico. In early 1989 the company made IPM implementation a priority and in 1994 pesticide usage on tomato, celery and carrot by their growers had been reduced by at least 50%, with no loss of yield or quality (Bolkan and Rienert, 1994). The key to this success was the use of epidemiological models for some key pathogens, the use of resistant varieties, understanding the epidemiology of gemini viruses and the biology of the vector of whitefly populations. The Campbell Soup Company IPM programme addressed weeds, arthropods and pathogens by emphasizing cultural practices, environmental and pest monitoring, and economically/ecologically based pesticide application (Bolkan and Rienert, 1994). Sometimes growers themselves take the initiative for such production and labelling systems. Growers participating in those programmes must comply with rules concerning IPM. Under these regulations, pesticide reduction is achieved at levels that often surpass the government requirements. This has been observed in many cases in Europe and, most recently, in the USA, where the Wegman's grocery chain is selling IPM labelled produce: the IPM definition agreed to by growers is more restrictive than that proposed by the IPM team at Cornell University (Jacobsen, 1997). Market demand for IPM grown products already exists. Such demand may increase if proper advertisement is done and if a larger number of consumers become more educated.

29.5. Conclusions

Successful IPM programmes for greenhouses share a number of characteristics (van Lenteren, 1995): (i) their use was promoted only after a complete programme had been developed covering all aspects of pest and disease control for a certain crop; (ii) they have been intensively supported by the extension service during the first few years; (iii) the total cost of crop protection in the IPM programme was not higher than that of the standard (chemical control) programme; and (iv) all their components (particularly BCAs) were easily available to growers.

Successful IPM programmes start with well-defined, goal-oriented research projects, carried out in a very interdisciplinary framework. Success in implementing IPM requires partnerships with economists, sociologists, ecologists, horticulturists, agronomists, agricultural engineers, geographic information specialists, food processors, crop consultants, pesticide applicators, regulatory agencies, computer scientists, consumers, public policy interest groups, and, most important, farmers (Jacobsen, 1997).

In European greenhouses, many success stories are already available (see the many examples in this book). Large investments in research and appropriate policies have been crucial (Marsh, 1995). Lack of funding for both disciplinary and interdisciplinary developmental research and implementation is considered responsible for the paucity of comprehensive IPM programmes for most crops in the USA (Jacobsen, 1997). Actually, IPM policies (the Clinton Administration IPM Initiative) have only very recently been adopted in the USA (Jacobsen, 1997). New, stricter policies, an increasing growers' acceptance, the public trend in favour of food produced with less pesticides, and the standardization of the production technology for BCAs will greatly help in expanding the application of IPM in new areas and on new crops.

References

Alomar, O., Castane, C., Gabarra, R., Arno, J., Arino, J. and Albajes, R. (1991) Conservation of native mirid bugs for biological control in protected and outdoor tomato crops, *IOBC/WPRS Bulletin* **14**(5), 33–42.

Anonymous (1998) *Integrated Crop Management Fruit and Vegetable*, EuroHandelsinstitut e. V., Köln.

Biggs, A.R. and Grove, G.G. (1998) Role of the world wide web in extension plant pathology: Case studies in tree fruits and grapes, *Plant Disease* **82**, 452–464.

Bolkan, H.A. and Reinert, W.R. (1994) Developing and implementing IPM strategies to assist farmers: An industry approach, *Plant Disease* **78**, 545–550.

Cook, R.J. (1993) Making greater use of introduced microorganisms for biological control of plant pathogens, *Annual Review of Phytopathology* **31**, 53–80.

Delorme, R. and Angot, A. (1983) Toxicités relatives de divers pesticides pour *Encarsia formosa* Gahan (Hyménopt., *Aphelinidae*) et pour son hôte *Trialeurodes vaporariorum* West. (Homopt., Aleurodidae), *Agronomie* **3**(6), 577–584.

Foster, G.N. (1980) Biological control of whitefly, initial pest density as the most important factor governing success, *IOBC/WPRS Bulletin* **3**, 41–44.

Gullino, M.L. (1993) Commercialization of biocontrol agents, *IOBC/WPRS Bulletin* **16**(11), 219–224.

Gullino, M.L. (1995) Use of biocontrol agents against fungal diseases, in *Proceedings of the International Conference on Microbial Control Agents in Sustainable Agriculture*, St. Vincent, 18–19 October 1995, pp. 50–59.

Gullino, M.L. and Kuijpers, L.A.M. (1994) Social and political implications of managing plant diseases with restricted fungicides in Europe, *Annual Review of Phytopathology* **32**, 559–579.

Helgesen, R.G. and Tauber, M.J. (1974) Biological control of greenhouse whitefly *Trialeurodes vaporariorum (Aleyrodidae*, Homopt.) on short-term crops by manipulating biotic and abiotic crops, *Canadian Entomologist* **106**, 1175–1188.

Hussey, N.W. and Scopes, N. (1985) *Biological Pest Control: The Glasshouse Experience,* Blandford Press, Poole, Dorset.

Jacobsen, B.J. (1997) Role of plant pathology in integrated pest management, *Annual Review of Phytopathology* **35**, 373–391.

Jeger, M.J. (1995) The implication of integrated crop protection approaches for education and training, in *Integrated Crop Protection: Towards Sustainability?* BCPC Symposium Proceedings No. 63, BCPC, Farnham, pp. 457–468.

Malausa, J.C., Sicard, J.C. and Daunay, M.C. (1983) Fecundity and survival of *Trialeurodes vaporariorum* West. on different eggplant varieties, *IOBC/WPRS Bulletin* **4**, 41–42.

Marsh, J.S. (1995) The policy approach to sustainable farming systems in the EU, in *Integrated crop Protection: Towards Sustainability?* BCPC Symposium Proceedings No. 63, BCPC, Farnham, pp. 13–23.

Murphy, G.D. and Broadbent, A.B. (1993) Development and implementation of IPM in greenhouse floriculture in Ontario, Canada, *IOBC/WPRS Bulletin* **16**(2), 113–116.

Onillon, J.C. (1990) The use of natural enemies for the biological control of whiteflies, in D. Gerling (ed.), *Whiteflies: Their bionomics, Pest Status and Management*, Intercept Ltd, Andover, Hants, pp. 287–313.

Onillon, J.C., Maisonneuve, J.C. and Miquel, J. (1983) Influence des conditions de milieu sur l'emploi et l'efficacité des auxiliaires en cultures maraîchères, in *Faune et Flore Auxiliaires en Agriculture*, ACTA éd., Paris, pp. 251–254.

Onillon, J.C., Onillon, J. and Di Pietro, J.P. (1976) Résultats préliminaires du contrôle biologique de l'aleurode des serres *T. vaporariorum* (Homopt., *Aleurodidae*) par *E. formosa* (Hyménopt., *Aphelinidae*) en serres d'aubergine, *IOBC/WPRS Bulletin* **4**, 138–150.

Parr, W.J., Gould, H.J., Jessop, N.H. and Ludlam, A.B. (1976) Progress towards a biological control program for glasshouse whitefly *Trialeurodes vaporariorum* on tomatoes, *Annals of Applied Biology* **83**, 349–363.

Rogers, S. (1996) The plant pathologist in private practice: Opportunities for the present and challenges for the future, *Plant Disease* **80**, 5–13.

Scopes, N.E.A., Biggerstaff, S.M. and Goodal, D.E. (1973) Cool storage of some parasites used for pest control in glasshouses, *Plant Pathology* **22**, 189–193.

Stacey, D.L. (1977) Banker plant production of *Encarsia formosa* and its use in the control of glasshouse whitefly on tomatoes, *Plant Pathology* **26**, 63–66.

Urech, P.A. (1996) Is more legislation and regulation needed to control crop protection products in Europe? in *Brighton Crop Protection Conference: Pests and Diseases*, Vol. 2, Proceedings of an International Conference, Brighton, 18-21 November 1996, BCPC, Farnham, pp. 549–558.

van Lenteren, J.C. (1995) Integrated pest management in protected crops, in D. Dent (ed.), *Integrated Pest Management*, Chapman & Hall, London, pp. 311–343.

Wardlow, L.R. (1992) The role of extension services in integrated pest management in glasshouse crops in England and Wales, in J.C. van Lenteren, A.K. Minks and O.M.B. de Ponti (eds.), *Biological Control and Integrated Crop Protection: Towards Environmentally Safer Agriculture*, PUDOC, Wageningen, pp. 193–199.

Wearing, C.H. (1988) Evaluating the IPM implementation process, *Annual Review of Entomology* **33**, 17–38.

CHAPTER 30

TOMATOES

Rosa Gabarra and Mohamed Besri

30.1. Introduction

Integrated Pest and Disease Management (IPM) in greenhouse tomato (*Lycopersicon esculentum* Mill.) crops is closely related to climatic areas. In cold areas (e.g. northern Europe) tomatoes are grown in glasshouses, whereas in the warmer areas (e.g. Mediterranean basin and Middle East) plastic houses predominate. In the EU most tomato production (78%) is in the warmer areas (Aldanondo, 1995).

In the last thirty years, protected cultivation of tomato has increased greatly in the warm regions. Only a few years ago, a yield of 60 t/ha of tomato under plastic was considered a fair production. Now, yields of more than 200 t/ha can be harvested in many Mediterranean countries (van Alebeek and van Lenteren, 1990). There has been a revolution in greenhouse production technology: type of greenhouse, quality of the plastic cover, fertirrigation, plastic mulch, new high-yield hybrids and varieties, specific pesticides, soil fumigation, etc. However, the intensification of protected tomato production has created optimal conditions for many pests. For example, several pathogens {i.e. *Leveillula taurica* (Lév.) G. Arnaud, *Botrytis cinerea* Pers.:Fr., *Clavibacter michiganensis* (Smith) Davis *et al.* ssp. *michiganensis* (Smith) Davis *et al.* [= *Corynebacterium michiganense* (Smith) Jensen ssp. *michiganense* (Smith) Jensen], *Meloidogyne* spp.} were relatively easy to control in the early years, but wreaked more damage as cultivation became more intense (van Alebeek and van Lenteren, 1990; Besri, 1991a).

30.2. Major Pests and Diseases

30.2.1. INSECTS AND MITES

The most harmful pests on greenhouse tomato crops are polyphagous. Their relative importance varies with the climatological area and type of greenhouse (Table 30.1). The key pests on this crop are whiteflies and, to a lesser extent, leafminers. Aphids, lepidopteran larvae and mites may cause severe economic damage, but their incidence is variable. In the cold area and in part of the warm area (milder subarea), *Trialeurodes vaporariorum* (Westwood) is the major whitefly species on greenhouse tomato (Onillon, 1990; van Lenteren *et al.*, 1992). *Trialeurodes vaporariorum* and *Bemisia tabaci* (Gennadius) coexist in the transition subarea (Arnó and Gabarra, 1994) and only *B. tabaci* causes damage in the warmer subarea (Traboulsi, 1994; Gerling, 1996). Different *Bemisia* species and/or *B. tabaci* biotypes coexist in several parts of the world (Markham *et al.*, 1996).

R. Albajes et al. (eds.), Integrated Pest and Disease Management in Greenhouse Crops, 420-434.
© 1999 *Kluwer Academic Publishers. Printed in the Netherlands.*

TABLE 30.1. Control methods, action thresholds and rates used for pest control in IPM programmes for greenhouse tomato crops[1]

Pest	Control agent/technique or method	Climatic area²/rate³/action threshold and general remarks
Whiteflies		
Trialeurodes vaporariorum (Tv)	Encarsia formosa (Ef)	(C) 2–28 Ef/m², at least 4 releases or until black scales are found or 80–90% of parasitized whitefly scales. (W) 16–40 Ef/m², 4–10 releases
		All the time less than 1 adult/plant or Tv adults in chromo-attractive traps
	Encarsia formosa + Macrolophus caliginosus (Mc)	(C) 8–16 Ef/m², 4 releases, + 1–2 Mc/m², 2–4 releases. (W) 12–24 Ef/m², 4–6 releases, + 4 Mc/m², 2–3 releases
		All the time less than 1 adult/plant or Tv adults in chromo-attractive traps
	Macrolophus caliginosus	(W) 4 Mc/m², 2–3 releases
		Less than 0.1 Tv/plant
Trialeurodes vaporariorum and Bemisia tabaci	Eretmocerus mundus (Em)	3 Em/m², minimum of 3 weekly introductions (curative light) Action threshold not determined
Leafminers		
Liriomyza trifolii, Liriomyza bryoniae, Liriomyza huidobrensis	Diglyphus isaea (Di), Dacnusa sibirica (Dac), Opius pallipes (Op)	(C) Low temperatures 0.25 Dac/m², until sufficient parasitism. From March 1st onwards 0.3 Di/m², 3 weekly releases. (W) 0.2–0.4 Di/m², 2–3 releases
		Presence of first miners or adults in chromo-attractive traps or check for less than 25% natural parasitism
Aphids		
Macrosiphum euphorbiae, Myzus persicae	Pirimicarb	Treat foci if not widespread in the greenhouse. Labelled rates
		When aphids or damage are first seen, check for natural enemies
	Aphidoletes aphidimyza (Aa), Aphelinus abdominalis (Ab), Aphidius ervi (Ae)	2 (Ab and Ae)/m² in hot spots or 0.5–1 Aa/m²/week or 0.5 Ab/m², in 2–3 releases
		Presence of first foci
Spider mites		
Tetranychus urticae, Tetranychus cinnabarinus	Specific and selective acaricides	Labelled rates
	Phytoseiulus persimilis T strain (Pp)	Treat at first sign of damaged plants
		25 Pp/m² in infested areas and 3 Pp/m², 2 releases over the total area

TABLE 30.1. Control methods, action thresholds and rates used for pest control in IPM programmes for greenhouse tomato crops[1] (cont.)

Pest	Control agent/technique or method	Climatic area[2]/rate[3]/action threshold and general remarks
Tomato russet mite *Aculops lycopersici*	Specific and selective acaricides	Treat foci if not widespread in the greenhouse. Labelled rates Treat at first sign of damaged plants
Lepidoptera *Chrysodeixis chalcites, Autographa gamma, Lacanobia oleracea, Spodoptera littoralis, Helicoverpa* (= *Heliothis*) *armigera* (Ha)	*Bacillus thuringiensis* (Bt), *Trichogramma* spp.	Labelled rates For Bt: Presence of 2nd instar caterpillars or check for more than two young larvae per plant; specifically for Ha: when eggs or small caterpillars are first seen
Viral diseases Tomato yellow leaf curl virus (TYLCV), cucumber mosaic virus (CMV), tobacco mosaic virus (TMV), alfalfa mosaic virus (AMV), tomato spotted wilt virus (TSWV)	Resistant varieties, use of healthy plants and of pathogen-free seeds, eradication of alternative hosts, sowing date, cross protection, control of vectors (aphids, whitefly) by insecticides and mineral oil sprays, use of reflective mulches	No tomato cultivars resistant to or tolerant of AMV, CMV are available Vector control is not always effective for all viruses (TSWV)
Bacterial diseases *Clavibacter michiganensis* ssp. *michiganensis* (Cmm) (bacterial canker), *Pseudomonas syringae* pv. *tomato* (Pst) (bacterial speck), *Pseudomonas corrugata* (pith necrosis), *Pseudomonas solanacearum* (bacterial wilt), *Xanthomonas vesicatoria* (Xv) (bacterial spot)	Rogueing, pathogen-free seeds and transplants, destruction of crop residues, eradication of alternative hosts, disinfection of pruning tools and stakes, chemical sprays, soil disinfection, space solarization	No tomato cultivars resistant to Cmm, Pst and Xv are available
Fungal diseases *Verticillium dahliae, Verticillium albo-atrum* (Verticillium wilt), *Fusarium oxysporum* f. sp. *lycopersici* (Fol) (Fusarium wilt), *Fusarium oxysporum* f. sp. *radicis-lycopersici* (Fusarium crown and root rot)	Resistant cultivars, destruction of crop residues, irrigation with non-saline water, weeding, solarization, chemical soil disinfection, nematode control, pathogen-free seeds (Fol), grafting, soilless culture	No tomato cultivar resistant to *Verticillium* strain 2 and to Fol strain 3 is available. Irrigation with saline water increases plant susceptibility and breaks down cultivar resistance

TABLE 30.1. Control methods, action thresholds and rates used for pest control in IPM programmes for greenhouse tomato crops[1] (cont.)

Pest	Control agent/technique or method	Climatic area[2]/rate[3]/action threshold and general remarks
Fungal diseases (cont.)		
Botrytis cinerea (Bc) (grey mould), Phytophthora infestans (late blight), Phytophthora nicotianae var. parasitica (Phytophthora root rot and buckeye rot), Sclerotinia spp. (white mould)	Greenhouse ventilation, wound protection, plant pruning, destruction of crop residues, lower plant densities, chemicals	Fungicide resistance of Bc to benzimidazoles, thiophanate methyl and dithiocarbamates, and resistance of Phytophthora spp. to anilides. No resistant cultivars are available. Forecasting systems should be developed
Alternaria solani (early blight), Alternaria alternata f. sp. lycopersici (Alternaria stem canker), Fulvia fulva (Fv) (leaf mould), Colletotrichum spp. (anthracnose), Stemphylium spp. (Stemphylium leaf spot), Septoria lycopersici (Septoria leaf spot)	Greenhouse ventilation, destruction of crop residues, chemical sprays, resistant cultivars, pathogen-free seeds, eradication of alternative hosts	Fv is mainly important in glasshouses. No resistant cultivars are available to all the races of the pathogen
Didymella lycopersici (Didymella stem canker)	Pathogen-free seeds, destruction of crops residues, chemical spray, stake disinfection	No Didymella resistant cultivars are available
Pyrenochaeta lycopersici (corky root)	Resistant cultivars, destruction of crop residues, solarization, grafting, soilless culture	Pyrenochaeta is often associated to other pathogens
Erysiphe sp., Leveillula taurica (powdery mildew)	Weed control, sanitation, chemicals	No resistant cultivar is available
Nematodes		
Meloidogyne spp.	Resistant cultivars, destruction of crop residues, fumigation, solarization	Many high yielding cultivars (i.e. 'Daniella') are susceptible

[1]References: Gould, 1987; Bues et al., 1989; Minkenberg and van Lenteren, 1990; Onillon, 1990; van Alebeek and van Lenteren, 1990; Besri, 1991a,b; Celli et al., 1991; Albajes et al., 1994; Trottin-Caudal et al., 1995; van Schelt et al., 1996

[2]Climatic areas: (C) cold area, (W) warm area

[3]Rates recommended by suppliers of natural enemies (Biobest B.S., Koppert B.V., Novartis BCM, Biolab)

In this chapter, the name *B. tabaci* will be used to refer to all *Bemisia* biotypes/species. *Bemisia tabaci* transmits the tomato yellow leaf curl virus (TYLCV), which causes big economic losses in tomato crops. The virus is present in the transition and in the warmer subareas (Credi *et al.*, 1989; Moriones *et al.*, 1993).

30.2.2. DISEASES

The major diseases of tomato grown under protected cultivation are reported in Table 30.1 (van Alebeek and van Lenteren, 1990; Jones *et al.*, 1991). The disease distribution, incidence and severity vary from one region to another, according to many factors such as the cultivars, the climatic conditions, the greenhouse type, the cultural practices and the control methods used (van Alebeek and van Lenteren, 1990). *Phytophthora infestans* (Mont.) de Bary is more severe in polyethylene houses than in glasshouses, but the opposite is true in the case of *Fulvia fulva* (Cooke) Cif. (= *Cladosporium fulvum* Cooke) (Garibaldi and Corte, 1987). This latter pathogen is more frequent and more severe in the northern and eastern Mediterranean countries than in the southern ones. In the cold areas Verticillium wilt is due to *Verticillium albo-atrum* Reinke & Berthier, and in the warm areas to *Verticillium dahliae* Kleb. (Jones *et al.*, 1991). *Pyrenochaeta lycopersici* R. Scheneider & Gerlach, *B. cinerea* and *Rhizoctonia solani* Kühn are severe in all cropping areas, but their severity varies with the region and the cultural practices. Viruses, nematodes and bacterial diseases cause the greatest economic losses in warm countries (van Lenteren, 1987; van Alebeek and van Lenteren, 1990). Incidence of *Orobanche* spp. is increasing and may constitute a threat to the tomato production in some Mediterranean countries (Fer and Thalouarn, 1997).

30.3. Components of IPM

30.3.1. CULTURAL CONTROL

Sanitation is a very important component of IPM. This control method includes all actions designed to eliminate or reduce the inoculum present in a plant or plot, and so prevent the spread of the pest (Agrios, 1988). Thus, ploughing under or removal and proper disposal of infected plant debris that may harbour the pest reduces the amount of inoculum. Washing the soil off farm equipment before movement from one plot to another may also help to avoid the spread of pathogens present in the soil. Since weeds can be infected by the same pest as the primary host, it is important to control their growth (Besri, 1991a) and eliminate weeds from the edges of greenhouses at least two weeks before planting. This practice delays whitefly crop infestation in warm areas (Alomar *et al.*, 1989). Many tomato pathogens, such as *Alternaria solani* Sorauer, *Didymella lycopersici* Kleb. [teleomorph of *Phoma lycopersici* Cooke (= *Diplodina lycopersici* Hollós)], *Fusarium oxysporum* Schlechtend.:Fr. f. sp. *lycopersici* (Sacc.) W.C. Snyder & H.N. Hans., *C. michiganensis* ssp. *michiganensis*, *Pseudomonas syringae* van Hall pv. *tomato* (Okabe) Young *et al.* and tomato mosaic virus (ToMV), are seed-transmitted (Jones *et al.*, 1991). The use of pathogen-free seeds should, therefore, be an important component of any tomato IPM programme (Besri, 1978; van Alebeek and van Lenteren, 1990).

Some cultural practices strengthen tomato plants and consequently increase their resistance to pest attacks. Thus, proper fertilization, drainage, proper spacing of plants, weed control, etc. improve the plants' growth and may have a direct or indirect effect on the control of a particular pest (Agrios, 1988; Besri, 1991a).

Soil and water salinity increase the susceptibility of tomato plants to many diseases and particularly to Fusarium and Verticillium wilts. Resistant varieties become susceptible when the irrigation water has a high salt content. Reducing the water salt content by mixing non-salty water from dams with salty water pumped from wells decreases the incidence and severity of these two pathogens (Besri, 1981).

In the prevailing greenhouse type in the Mediterranean and Middle East, prevention of airborne pathogens such as *A. solani*, *B. cinerea*, *L. taurica* and *P. infestans* by keeping greenhouse vents closed is not always possible, because this may lead to insufficient ventilation and a consequent increase in humidity-promoted diseases (van Alebeek and van Lenteren, 1990; Nicot and Allex, 1991; Nicot and Baille, 1996). When *B. tabaci* and TYLCV are present in the area, screening the greenhouse is an important IPM tool (see Chapter 8 for greenhouse climate management and the use of screens).

Tomato plants are pruned to remove axillary buds and leaves. Plant pruning creates a drier microclimate in the lower plant, but also provides numerous points of entry for many pathogens such as *D. lycopersici* and *B. cinerea*. Pruning wounds on tomato plants are less likely to become infected by these pathogens if the leaves are cut close to the stem than if a fragment of petiole is left on the stem (Besri and Diatta, 1992). However, some natural enemies, like parasitoids of whitefly, develop on the older leaves. In this case pruning too early should be avoided and pruned leaves should be kept in the greenhouse until the parasitoids have emerged.

Effective rotation is not feasible in protected tomato cultivation and soil disinfestation may be prohibitively expensive (steaming) or toxicologically and environmentally undesirable (methyl bromide, MBr). Soilless culture is a technique originally developed to reduce the severity of soilborne pathogens (Zinnen, 1988; Braun and Supkoff, 1994). However, although soilless media are initially pathogen-free, their infestation by pathogens such as *Phytophthora*, *Pythium*, *Rhizoctonia* and *Fusarium* may occur in greenhouses (Jenkins and Averre, 1983; Zinnen, 1988). For some pathogens such as *Pythium* and *Phytophthora*, incidence and severity of the disease may be higher in soilless culture than in the traditional soil cultivation system (Zinnen, 1988). Exclusion of inoculum has proved impractical and sanitation to reduce inoculum load appears futile, once fungi are established in a recycling hydroponic facility (Zinnen, 1988). At present, except in a few cases such as in Holland, most tomato growers use a non-circulating system (Steinberg et al., 1995). However, this system is not profitable, and nor does it protect the environment. It was reported that for one hectare of tomato, recycling the nutrient solution will save 20–30% of water and 60% of fertilizers. It was also estimated that for one hectare of tomato, 3000 m^3 of nutrient solution containing eight tonnes of fertilizers are released back into the environment. Recycling the nutrient solution could be a solution to these problems, but an economically acceptable technique of nutrient solution disinfection is not yet available. Irradiation, ozonation, ultrafiltration and thermo-disinfection of the nutrient solution gave remarkable control of various soilborne pathogens in small experimental systems, but they have been rarely implemented on a commercial scale.

30.3.2. RESISTANT VARIETIES

Many tomato cultivars are resistant to various soil and airborne pathogens. However, there is no resistance to some pathogens, such as alfalfa mosaic virus (AMV), cucumber mosaic virus (CMV), *C. michiganensis* ssp. *michiganensis*, *P. syringae* pv. *tomato*, *Xanthomonas vesicatoria* (ex Doidge) Vauterin *et al.*, *Sclerotinia sclerotiorum* (Lib.) de Bary, *L. taurica* and *D. lycopersici* (Jones *et al.*, 1991). Even in the case of the resistant cultivars, the rise of new strains, particularly of *Fusarium* and *Verticillium*, is a threat to tomato production (Besri *et al.*, 1984; Jones *et al.*, 1991).

Resistant rootstocks control excellently many soilborne tomato pathogens, particularly *F. oxysporum* f. sp. *lycopersici*, *Fusarium oxysporum* Schlechtend.:Fr. f. sp. *radicis-lycopersici* W.R. Jarvis & Shoemaker, *P. lycopersici* and *Meloidogyne* spp. (Jones *et al.*, 1991). This technique, which used to be considered too expensive, is now widely used in many countries (van Alebeek and van Lenteren, 1990). In Morocco, grafting is used commercially to control root-knot nematodes. Plant grafting is a control method that could decrease the frequency of soil disinfestation with MBr (Ristaino and Thomas, 1997).

30.3.3. PHYSICAL DISINFECTION

Soil solarization is a promising technique, which may have an important future in many countries, particularly when the use of MBr is stopped under the Montreal protocol (Ristaino and Thomas, 1997). Though initially used only in hot regions during the summer, technological advances are extending the range of solarization to cooler areas and cooler seasons. Soil solarization controls many tomato pathogens such as *Colletotrichum coccodes* (Wallr.) S.J. Hughes, *F. oxysporum* f. sp. *lycopersici*, *V. dahliae*, *P. lycopersici* and *R. solani* (Katan, 1996). Solarization also decreases the population of *Meloidogyne* spp. in tomato greenhouse (Calabretta *et al.*, 1991a). In Sicily, soil solarization, in alternation with MBr, decreases by 50% the use of fumigants (Calabretta *et al.*, 1991b). Solarization of tomato stakes is a successful control method of some diseases such as Didymella stem canker, and could easily be achieved by storing this agricultural material in empty plastic greenhouses during the hot months of the year (Besri, 1982,1991b). In warmer areas, pathogens are killed by closing the greenhouse in the off-season (space solarization) and allowing sunlight to disinfest the greenhouse (Shlevin *et al.*, 1994).

30.3.4. BIOLOGICAL CONTROL

Insects and Mites
Biological control, which includes seasonal inoculative releases and techniques of augmentation and conservation of natural enemies, is used against major greenhouse tomato pests. The decision threshold and the rate of natural enemies required for the control of different pests are variable (see Table 30.1).

Biological control of *T. vaporariorum* in greenhouse tomatoes with seasonal inoculative releases of the parasitoid *Encarsia formosa* Gahan is widely used in the temperate greenhouse area and to a lesser extent in the milder subarea (Onillon, 1990).

However, *E. formosa* is not very efficient under cool, cloudy conditions and, in Europe, inoculative releases of the predator *Macrolophus caliginosus* Wagner are also used, either on its own or with *E. formosa* (Malézieux *et al.*, 1995; Trottin-Caudal *et al.*, 1995; van Schelt *et al.*, 1996). In fact, the inoculation of both natural enemies is now used in many tomato greenhouses where formerly only *E. formosa* was released (K. Bolckmans and R. GreatRex, pers. com.). In the warm area, initial populations of *T. vaporariorum* are usually higher than in the cold area. Whitefly migration between crops also occurs. Thus, higher densities of natural enemies are required, but for shorter growing seasons (Albajes *et al.*, 1994; Manzaroli and Benuzzi, 1995; Trottin-Caudal *et al.*, 1995). *Encarsia formosa* does not control *B. tabaci* sufficiently in winter greenhouse tomato crops (Arnó and Gabarra, 1996). At present, *Eretmocerus mundus* Mercet is released to control *T. vaporariorum* and *B. tabaci* in some tomato greenhouses (Novartis BCM and Koppert B.V., pers. com.). Studies on the efficiency of other natural enemies are being carried out (see Chapter 14).

Inoculative releases of *Diglyphus isaea* (Walker) are commercially used for biological control of leafminers in tomato. In the cold area this parasitoid is applied together with *Dacnusa sibirica* Telenga or *Opius pallipes* Wesmael (van Lenteren and Woets, 1988). In the warm area, natural populations of leafminer parasitoids are abundant all year and natural parasitism controls leafminers in the crop (Nicoli and Burgio, 1997). Augmentative releases of *D. isaea* are used only when natural parasitism is low.

No suitable parasitoid is yet available to control all the species of aphid that attack tomatoes. In fact, aphids are not yet controlled biologically in the majority of greenhouses. However, in the milder area, aphids in greenhouses which do not use broad-spectrum insecticides do not normally reach economic thresholds, due to the presence of indigenous populations of their natural enemies (Manzaroli and Benuzzi, 1995; Alomar *et al.*, 1997).

Biological control of spider mites with *Phytoseiulus persimilis*. Athias-Henriot on tomato crops has been largely ineffective and is not widely employed (van Lenteren and Woets, 1988). However, a new strain of *P. persimilis* (called the T strain) has given better results on tomatoes (Kielkewiicz, 1995).

Lacanobia oleracea (L.), *Chrysodeixis chalcites* (Esper), *Autographa gamma* (L.) and *Spodoptera littoralis* (Boisduval) are kept well under control by *Bacillus thuringiensis* Berliner treatments. *Helicoverpa* (= *Heliothis*) *armigera* (Hübner) is also well controlled if the treatment is applied when eggs or young larvae are present. However, intensive sampling is necessary to locate *H. armigera* eggs or larvae (Bues *et al.*, 1989). Inoculative releases of *Trichogramma evanescens* Westwood are also used for biological control of *C. chalcites* in some greenhouse tomato crops (W. Ravensberg, pers. com.).

Diseases

Biological control of tomato soil- and airborne pathogens is gaining an increasing interest (Malathrakis, 1991). *Penicillium oxalicum* Currie & Thom reduces the incidence of *F. oxysporum* f. sp. *lycopersici* both in hydroponic and soil systems (De Cal *et al.*, 1997), and *Trichoderma harzianum* Rifai and *Trichoderma koningii* Oudem.

control Fusarium root and crown rot (Bourbos *et al.*, 1997). Commercially, the
interpolation of a lettuce or dandelion (*Taraxacum officinale* Weber) crop between
successive tomato crops is an efficient form of control of *F. oxysporum* f. sp. *radicis-
lycopersici* (Jarvis, 1988). The hyperparasites *Dicyma pulvinata* (Berk. & M.A. Curtis)
Arx [= *Hansfordia pulvinata* (Berk. & M.A. Curtis) S.J. Hughes] and *Acremonium
sclerotigenum* (Valenta) W. Gams, alone or with phosethyl-Al, control *F. fulva*
(Bourbos and Skoudriakis, 1994). The introduction into the soil of non-pathogenic
strains of *Fusarium oxysporum* Schlechtend.:Fr. (F.o. 74), obtained from suppressive
soil, controls Fusarium wilts (Alabouvette, 1988). Many of these biological control
agents, however, are still being tested and are not commercially available.

Several biocontrol agents were reported to be effective against *B. cinerea*, such as
some isolates of *T. harzianum*. A commercial preparation developed from isolate T39
of *T. harzianum* (Trichodex) is at the moment registered for agricultural use in Israel
and other countries (Shtienberg and Elad, 1997). The combination of the antagonist
with a fungicide (iprodione) reduces fungicide use and consequently minimizes
pesticide residues on fruits and in the soil (Elad and Zimand, 1991). Many other cases
of biological control of tomato soil- and airborne pathogens have been reported
(Malathrakis, 1991).

30.3.5. SELECTIVE CHEMICAL CONTROL

Insect and Mite Pests
The majority of broad-spectrum insecticides are highly toxic to the natural enemies and
bumble-bees [*Bombus terrestris* (L.)] which are used in greenhouse tomato crops (see
Chapter 11). Buprofezin and pyriproxyfen are used in registered countries to control
whitefly when the initial population is very high, or where natural enemies are not well
installed. In the majority of IPM programmes, the control of aphids, red spider mites
and the tomato russet mite is achieved with selective pesticides applied only at specific
points of outbreak rather than on the entire crop (Table 30.1).

Diseases
MBr is widely used for soil disinfection in tomato. MBr controls soilborne pathogens,
insects and weeds. Fumigation before planting enables the soil to be replanted after a
short waiting period. Chloropicrin (CP) is a very effective fungicide for the control of
tomato soilborne fungi, but not for weed and nematode control. 1,3-dichloropropene
(1,3 D) is as efficacious as MBr in controlling nematodes, but does not control fungi or
insects. At high rates, 1,3 D has some efficacy against a few weeds. Dazomet and
metham-sodium added to moist soil decompose to methyl isothiocyanate, which is the
biocidal agent. These chemicals do not provide as consistent control of soilborne
pathogens as MBr (Braun and Supkoff, 1994). In the soilless culture system, chemical
control is greatly limited by the lack of registered products (Zinnen, 1988).
Propamocarbe and metalaxyl control Pythium and Phytophthora root rots, although
risks of metalaxyl resistance in Pythium have been noticed (Gold and Stanghellini,
1985).

Control of tomato airborne pathogens still relies largely on chemical control.

However, sprays with wrong fungicides (non-specific ones or ones to which resistance has developed in the greenhouse) could lead to a failure of chemical control. Tomato growers very often apply fungicide as paste over an individual stem lesion of *B. cinerea* in an attempt to prevent or delay the development of the canker on the stem. This fungicide application technique does not control the disease, and in addition leads to a greater number of resistant spores than fungicide sprays do (Besri and Diatta, 1992).

In general, tomato greenhouses receive routine fungicide application on a regular schedule during the growing season. The number of fungicide applications may be reduced by good cultural practices, the right time of application and the appropriate active ingredient. The establishment of regional disease-warning systems could also lead to a drop in fungicide pressure on the environment. Though such systems exist for some perennial crops, they unfortunately have not yet been set up for tomato or adapted to different climates and greenhouse types (Shtienberg and Elad, 1997). Some commonly used fungicides are dangerous for natural enemies and their use is not recommended in greenhouses where IPM is applied.

30.4. IPM Programmes

IPM is used in greenhouse tomato crops in many countries (van Lenteren and Woets, 1988; van Lenteren, 1995). Programmes are based on the biological control of the main insect pests and several secondary pests, the use of selective pesticides for the remaining pests (Table 30.1), and the use of host plant resistance and fungicides with low toxicity against natural enemies. The total area of greenhouses that use IPM is not known. However, if the area of tomato crops in which *E. formosa* is used is taken as an indicator, the area in Europe is 1410 ha in the cold area, and 697 ha in the warm area (Onillon, 1990). In Japan the area is 150 ha (E. Yano, pers. com.) and in Canada it is 183 ha (Elliot, 1996). In recent years, there has been an increase ·in the use of IPM following the introduction of bumble-bees for crop pollination, a method which is seen as cheap and effective by growers.

IPM programmes should be implemented at the various stages of tomato production: (i) in the field, prior planting (choice of the field, nematodes analysis, soil disinfection, etc.); (ii) in the seedbeds (choice of the cultivar, seed quality, chemical control, etc.); and (iii) in the production field (plant spacing, proper irrigation and nutrition, pruning, chemical control, etc.) (van Alebeek and van Lenteren, 1990; Besri, 1991a). An example of such a programme for tomato is shown in Table 30.1.

30.5. Factors Limiting Wider Application

Various factors limit the wider use of IPM in greenhouse tomato crops, some of which are common to all climatic areas, whilst others are specific to particular areas. Widespread problems include the lack of specialist technical supervision during IPM application. The service that growers obtain from the producers of natural enemies or/and from advisory personnel may be insufficient. For example, growers may lack information on the best moment to introduce a natural enemy or on pesticides suitable for integration. If a grower begins to use IPM, the quality and quantity of the initial guidance determines the success of the programme. The quantity and quality of the natural enemies to be released are also important. If natural enemies are in poor condition on arrival at the greenhouse, they will not provide effective control. This problem is exacerbated when the growers are far from the supplier and so need to apply more natural enemies (e.g. in most Mediterranean countries). In addition, the full system of integrated control may become too complicated for a grower. Biological control in tomato crops is based on the introduction of five to nine natural enemies (Table 30.1). The low number of selective pesticides is also a problem; for example, at present, only one selective aphicide is used to control aphids in tomato crop under IPM. New compounds in the pesticide market create imbalances in biological control. Usually, the negative effects of new pesticides on natural enemies are not evaluated before these pesticides replace older ones. Furthermore, growers and consumers are not fully aware of the advantages (lower residues and protection of the environment) of crops grown in IPM programmes.

In the warm areas, *B. tabaci* and TYLCV are very harmful. As yet, no TYLCV-resistant or tolerant tomato varieties have been developed. Populations of *B. tabaci* and virus inoculum are present in both indoor and outdoor crops, and the pest moves between crops, increasing the incidence of the virus. In addition, *B. tabaci* is resistant to all main groups of insecticides and even some of the newer compounds, including buprofezin and pyriproxyfen (Denholm *et al.*, 1996). Although many natural enemies of *B. tabaci* have been identified (see Chapter 14), commercial introduction is slow.

Many components of IPM for disease control are not yet available or need to be made more efficient. Because of the availability of broad-spectrum and effective fumigants such as MBr for the control of soilborne diseases, the need for host resistance diminished and plant breeders devoted more time and effort to the improvement of yield and quality. Many high-yield tomato varieties used at present,. such as 'Daniella', are susceptible to nematodes. No tomato variety is resistant to *Verticillium* strain 2. Commercial cultivars of tomato which are resistant to *Fusarium oxysporum* f. sp. *radicis-lycopersici*, *P. lycopersici* and to *F. oxysporum* f. sp. *lycopersici* strain 3 are not available (Besri *et al.*, 1984; van Alebeek and van Lenteren, 1990; Besri, 1991a). All the other control methods need to be improved and adapted to each ecological condition. In addition, techniques for population monitoring to estimate pest population densities should be improved or developed. New techniques for the detection of insecticide and fungicide resistance should also be researched so that the incidence of pest resistance can be rapidly detected and appropriate counter-measures taken.

30.6. Future of IPM in Greenhouse Tomatoes

In recent years, concerns expressed by environmentalists about pesticides contaminating water, soil and air have convinced politicians to launch ambitious programmes for reducing the amount of pesticides used (see Chapter 1). Furthermore, the successful introduction of bumble-bees for pollinating tomatoes also meant that more biocontrol had to be introduced.

In the Mediterranean region, one of the largest tomato-growing areas, research into biological control and IPM in greenhouse crops has greatly expanded in recent years (IOBC/WPRS Working Group "Integrated Control in Protected Crops in Mediterranean Climate" from 1985 until now). As a result, new natural enemies have been identified (for example, *M. caliginosus* and *D. isaea*) which control effectively the main pest species in tomato (Nicoli and Burgio, 1997). The introduction of techniques for the conservation of indigenous populations of natural enemies (Alomar *et al.*, 1991; Benuzzi and Nicoli, 1993) and the identification of cultural methods that can be used in greenhouses in this area have also led to improvements in IPM programmes.

IPM is still not used in most of the area dominated by *B. tabaci*. However, studies on different natural enemies are being conducted. In any case, placing mesh over ventilation openings to prevent entry of adult *B. tabaci* and the use of selective insecticides has led to a 50% reduction in pesticide use in Israel, and enabled bumble-bees to be introduced (Ausher, 1996).

In the Mediterranean, the most abundant and effective indigenous natural enemies for control of *T. vaporariorum* on tomatoes are the predatory mirid bugs *M. caliginosus*, *Dicyphus tamaninii* Wagner and *Dicyphus errans* (Wolff) (see Chapter 19). *Macrolophus caliginosus* and *D. tamaninii* also feed on *B. tabaci* (Barnadas *et al.*, 1998) and so may be good candidates for biological control by conservation and/or inoculation. In Israel, populations of *B. tabaci* in open-air fields have been decreasing for the last 20 years, partly because of the reduction in pesticide use, which has enabled the conservation of natural enemies. It is possible that candidates for inoculative or augmentative releases in greenhouses can be chosen from amongst these species (Gerling, 1996).

In some countries, farmers producing tomato mainly for the local market do not apply most components of integrated tomato disease management, both because technology is unavailable and because of the high costs of some of the available technology (e.g. soilless culture, grafting, etc.). However, farmers who are producing tomato for export are more technically advanced and are also on the lookout for any new technology regardless of its cost. MBr is the most widely used fumigant in the world, but all countries will have to replace MBr with sustainable economic alternatives in the near future. Research and demonstration projects are underway in many countries to implement the existing alternatives and to develop new ones (Ristaino and Thomas, 1997).

More research should be conducted into varieties resistant to diseases which continue to be a threat to tomato production both in developing and developed countries (i.e. TYLCV, CMV, AMV, root-knot nematodes and bacterial canker). Warning systems for the most important airborne diseases (i.e. grey mould and late blight) should also be introduced.

References

Agrios, G.N. (1988) *Plant Pathology*, Academic Press, New York.
Alabouvette, C. (1988) Manipulation of soil environment to create suppressiveness in soil, in E.C. Tjamos and C.H. Beckman (eds.), *Vascular Wilt Diseases of Plants*, Springer-Verlag, New York, pp. 457–478.

Albajes, R., Gabarra, R., Castañé, C., Alomar, O., Arnó, J., Riudavets, J., Ariño, J., Bellavista, J., Martí, M., Moliner, J. and Ramírez, M. (1994) Implementation of an IPM program for spring tomatoes in Mediterranean greenhouses, *IOBC/WPRS Bulletin* **17**(5), 14–21.

Aldanondo, A.M. (1995) Cultivo y producción de tomate en la Unión Europea y en España, in F. Nuez (ed.), *El Cultivo del Tomate*, Ediciones Mundi-Prensa, Madrid, pp. 696–740.

Alomar, O., Castañé, C., Gabarra, R., Arnó, J., Ariño, J. and Albajes, R. (1991)Conservation of mirid bugs for biological control in protected and outdoor tomato crops, *IOBC/WPRS Bulletin* **14**(5), 33–42.

Alomar, O., Castañé, C., Gabarra, R., Bordas, E., Adillon, J. and Albajes, R. (1989) Cultural practices for IPM in protected crops in Catalonia, in R. Cavalloro and C. Pelerents (eds.), *Integrated Pest Management in Protected Vegetable Crops*, A.A. Balkema, Rotterdam, pp. 347–354.

Alomar, O., Gabarra, R. and Castañé, C. (1997) The aphid parasitoid *Aphelinus abdominalis* (Hym.: Aphelinidae) for biological control of *Macrosiphum euphorbiae* on tomatoes grown in unheated plastic greenhouses, *IOBC/WPRS Bulletin* **20**(4), 203–206.

Arnó, J. and Gabarra, R. (1994) Whitefly species composition in winter tomato greenhouses, *IOBC/WPRS Bulletin* **17**(5), 104–109.

Arnó, J. and Gabarra, R. (1996) Potential for biological control of mixed *Trialeurodes vaporariorum* and *Bemisia tabaci* populations in winter tomato crops grown in greenhouses, in D. Gerling and R.T. Mayer (eds.), *Bemisia 1995. Taxonomy, Biology, Damage, Control and Management*, Intercept Ltd, Andover, pp. 523–526.

Ausher, R. (1996) Implementation of Integrated Pest Management in Israel, in D. Gerling and R.T. Mayer (eds.), *Bemisia 1995. Taxonomy, Biology, Damage, Control and Management*, Intercept Ltd, Andover, pp. 659–667.

Barnadas, I., Gabarra, R. and Albajes, R. (1998) Predatory capacity of two mirid bugs preying on *Bemisia tabaci*, *Entomologia Experimentalis et Applicata* **86**, 215–219.

Benuzzi, M. and Nicoli, G. (1993) Outlook for IPM in protected crops in Italy, *IOBC/WPRS Bulletin* **16**, 9–12.

Besri, M. (1978) Phases de la transmission de *F. oxysporum* f.sp. *lycopersici* et de *V. dahliae* par les semences de quelques variétés de tomate, *Phytopathologische Zeitschrift* **93**, 148–163.

Besri, M. (1981) Qualité des sols et des eaux d'irrigation et manifestation des trachéomycoses de la tomate au Maroc, *Phytopathologia Mediterranea* **20**, 107–111.

Besri, M. (1982) Conservation de *Didymella lycopersici* dans les cultures de tomate par les tuteurs, *Phytopathologische Zeitschrift* **105**, 1–10.

Besri, M. (1991a) Lutte intégrée contre les maladies cryptogamiques de la tomate au Maroc, *IOBC/WPRS Bulletin* **14**(5), 187–191.

Besri, M. (1991b) Solarization of soil and agricultural materials for control of *Verticillium* wilt and *Didymella* stem canker in Morocco, in J. Katan and J.E. Devay (eds.), *Soil Solarization*, CRC Press, Boca Raton, Fla., pp. 237–243.

Besri, M. and Diatta, F. (1992) Effect of fungicide application techniques on the control of *Botrytis cinerea* and development of fungal resistance, in K. Verhoeff, N.E. Malathrakis and B. Williamson (eds.), *Recent Advances in Botrytis Research*, Pudoc, Wageningen, pp. 248–251.

Besri, M., Zrouri, M. and Beye, I. (1984) Appartenance raciale et pathogénie comparée de quelques isolats de *Verticillium dahliae* (Kleb.) obtenus à partir de tomates résistantes au Maroc, *Phytopathologische Zeitschrift* **109**, 289–294.

Bourbos, V.A., Michalopoulos, G. and Skoudriakis, M.T. (1997) Lutte biologique contre *Fusarium* f.sp. *radicis-lycopersici* chez la tomate en serre non chauffée, *IOBC/WPRS Bulletin* **20**(4), 58–62.

Bourbos, V.A. and Skoudriakis, M.T. (1994) Possibilité d'une lutte intégrée contre la Cladosporiose de tomate en serre, *IOBC/WPRS Bulletin* **17**(5), 39–42.

Braun, A.L. and Supkoff, D.M. (1994) *Options to Methyl bromide for the Control of Soil-borne Diseases and Pests in California with Reference to the Netherlands*, Pest Management Analysis and Planning Program, PM 94-02, Environmental Protection Agency, Sacramento, Cal..

Bues, R., Toubon, J.F. and Boudinhon, L. (1989) Dynamique des populations et lutte biologique contre *Heliothis armigera* en culture de tomate sous serre dans le sud de la France, in R. Cavalloro and C. Pelerents (eds.), *Integrated Pest Management in Protected Vegetable Crops*, A.A. Balkema, Rotterdam, pp. 91–98.

Calabretta, C., Colombo, A., Cosentino, S., Schiliro, E. and Sortino, O. (1991a) Effects of soil solarization on larvae of *Meloidogyne* spp. in soil and root rot of tomato 'Novi F1' in cold greenhouse, *IOBC/WPRS Bulletin* **14**(5), 164–171.

Calabretta, C., Colombo, A., Nucifora, S. and Privetera, S. (1991b) "Soil solarization" and Methyl-Bromide alternated treatments as new method in the disinfestation of soil in protected crops, *IOBC/WPRS Bulletin* **14**(5), 153–163.

Celli, G., Benuzzi, M., Maini, S., Manzaroli, G., Antoniacci, L. and Nicoli, G. (1991) Biological and integrated pest control in protected crops of Northern Italy's Po Valley: Overview and outlook, *IOBC/WPRS Bulletin* **14**(5), 2–12.

Credi, R., Betti, L. and Canova, A. (1989) Association of a geminivirus with severe disease of tomato in Sicily, *Phytopthologia Mediterranea* **28**(3), 223–226.

De Cal, A., Pascual, S., García-Lepe, R. and Melgarejo, P. (1997) Biological control of Fusarium wilt of tomato, *IOBC/WPRS Bulletin* **20**(4), 63–70.

Denholm, I., Byrne, F.J., Cahill, M. and Devonshire, A.L. (1996) Progress with documenting and combating insecticide resistance in *Bemisia*, in D. Gerling and R.T. Mayer (eds.), *Bemisia 1995. Taxonomy, Biology, Damage, Control and Management*, Intercept Ltd, Andover, pp. 577–604.

Elad, Y. and Zimand, G. (1991) Experience in integrated chemical-biological control of grey mould (*Botrytis cinerea*), *IOBC/WPRS Bulletin* **14**(5), 195–199.

Elliot, D. (1996) Biological control in Canadian vegetable greenhouses, *Sting – Newsletter on Biological Control in Greenhouses* **16**, 7–8.

Fer, A. and Thalouarn, P. (1997) L'Orobanche: Une ménace pour nos cultures, *Phytoma – La Défense des Végétaux* **499**, 34–40.

Garibaldi, A. and Corte, A. (1987) Integrated control of the most severe diseases of some protected vegetable crops in Italy, in R. Cavallaro and C. Pelerents (eds.), *Integrated and Biological Control in Protected Crops*, A.A. Balkema, Rotterdam, pp. 35–43.

Gerling, D. (1996) Status of *Bemisia tabaci* in the Mediterranean countries: Opportunities for biological control, *Biological Control* **6**, 11–22.

Gold, S.E. and Stanghellini, M.E. (1985) Effects of temperature on *Pythium* root rot of spinach grown under hydroponic conditions, *Phytopathology* **75**, 333–337.

Gould, H.J. (1987) Protected crops, in A.J. Burn, T.H. Coaker and P.C. Jepson (eds.), *Integrated Pest Management*, Academic Press, London, pp. 403–424.

Jarvis, W.R. (1988) Allelopathic control of *Fusarium oxysporum* f.sp. *radicis-lycopersici*, in E.C. Tjamos and C.H. Beckman (eds.), *Vascular Wilt Diseases of Plants*, Springer-Verlag, New York, pp. 479–786.

Jenkins, S.F. Jr. and Averre, C.W. (1983) Root diseases of vegetables in hydroponic culture systems in North Carolina greenhouses, *Plant Disease* **67**, 968–970.

Jones, J.B., Jones, J.P., Stall, R.E. and Zitter, T.A. (1991) *Compendium of Tomato Diseases*, APS Press, St Paul, Minn.

Katan, J. (1996) Soil solarization: Integrated control aspects, in R. Hall (eds.), *Principles and Practice of Managing Soilborne Plant Pathogens*, APS Press, St Paul, Minn., pp. 250–278.

Kielkewiicz, M. (1995) Effectiveness of *Phytoseilus persimilis* Athias-Henriot (Acari: Phytoseiidae) against the carmine spider mite, *Tetranychus cinnabarinus* Boisduval (Acari: Tetranychidae) on tomato crops, in *Proceedings of the Conference on "Actual and Potential Use of Biological Pest Control on Plants"*, Skierniewice, Poland, 1995, pp. 68–75.

Malathrakis, N. (1991) Biological control of diseases of protected crops: Present status and prospects, *Acta Horticulturae* **287**, 335–347.

Malézieux, S., Girardet, C., Navez, B. and Cheyrias, J.M. (1995) Contre l'aleurode des serres en cultures de tomates sous abris. Utilisation et développement de *Macrolophus caliginosus* associé a *Encarsia formosa*, *Phytoma – La Défense des Végétaux* **471**, 29–32.

Manzaroli, G. and Benuzzi, M. (1995) Pomodoro in serra, lotta biologica e integrata, *Culture Protette* **1**, 41–47.

Markham, P.G., Bedford, I.D., Liu, S., Frolich, D.R., Rosell, R. and Brown, J.K. (1996) The transmission of geminiviruses by biotypes of *Bemisia tabaci* (Gennadius), in D. Gerling and R.T. Mayer (eds.), *Bemisia 1995. Taxonomy, Biology, Damage, Control and Management*, Intercept Ltd, Andover, pp. 69–75.

Minkenberg, O.P.J.M. and van Lenteren, J.C. (1990) Evaluation of parasitoids for the biological control of leafminers on glasshouse tomatoes: Development of a preintroduction selection procedure, *IOBC/WPRS Bulletin* **13**(5), 124–128.

Moriones, E., Arnó, J., Accotto, G.P., Noris, E. and Cavallarin, L. (1993) First report of tomato yellow leaf curl virus in Spain, *Plant Disease* **77**, 953.

Nicoli, G. and Burgio, G. (1997) Mediterranean biodiversity as source of new entomophagous species for biological control in protected crops, *IOBC/WPRS Bulletin* **20**(4), 27–38.

Nicot, P. and Allex, D. (1991) Grey mould of greenhouse-grown tomatoes: Disease control by climate management, *IOBC/WPRS Bulletin*, **14**(5), 200–210.

Nicot, C. and Baille, A. (1996) Integrated control of *Botrytis cinerea* on greenhouse tomatoes, in C.E. Morris, P.C. Nicot and C. Nguyen (eds.), *Aerial Plant Surface Microbiology*, Plenum Press, New York, pp. 169–189.

Onillon, J.C. (1990) The use of natural enemies for the biological control of whiteflies, in D. Gerling (ed.), *Whiteflies: Their Bionomics, Pest Status and Management*, Intercept Ltd, Andover, pp. 287–314.

Ristaino, J.B. and Thomas, W. (1997) Agriculture, methyl bromide and the ozone hole. Can we fill the gaps? *Plant Disease* **9**, 964–977.

Shlevin, E., Katan, J., Mahrer, Y. and Kritzman, G. (1994) Sanitation of inocula plant pathogens in the greenhouse structure by space solarization, *Phytoparasitica* **22**(1), 156.

Shtienberg, D. and Elad, Y. (1997) Incorporation of weather forecasting in integrated, biological-chemical management of *Botrytis cinerea*, *Phytopathology* **87**, 332–340.

Steiberg, C., Gautheron, N., Gaillard, P. (1995) Désinfection des eaux, *Fruits et Légumes* **126**, 85–86.

Traboulsi, R. (1994) *Bemisia tabaci*: A report on its pest status with particular reference to the Near East, *FAO Plant Protection Bulletin* **42**(1–2), 33–58.

Trottin-Caudal, Y., Grasselly, D., Millot, P. and Veschambre, D. (1995) *Maîtrise de la Protection Sanitaire. Tomate sous Serre et Abris*, Centre Technique Interprofessionnel des Fruits et Légumes, Paris.

van Alebeek, F.A.N. and van Lenteren, J.C. (1990) *Integrated Pest Management for Vegetables Grown Under Protected Cultivation in the Near East*, FAO Consultation Report, Parts I and II, FAO, Rome.

van Lenteren, J.C. (1987) Integrated pest management in protected crops in the Netherlands, in R. Cavallaro and C. Pelerents (eds.), *Integrated and Biological Control in Protected Crops*, A.A. Balkema, Rotterdam, pp. 95–104.

van Lenteren, J.C. (1995) Integrated Pest Management in protected crops, in D. Dent (ed.), *Integrated Pest Management*, Chapman & Hall, London, pp. 311–343.

van Lenteren, J.C., Benuzzi, M., Nicoli, G. and Maini, S. (1992) Biological control in protected crops in Europe, in J.C. van Lenteren, A.K. Minks and O.M.B. de Ponti (eds.), *Biological Control and Integrated Crop Protection: Towards Environmentally Safer Agriculture*, Pudoc, Wageningen, pp. 77–89.

van Lenteren, J.C. and Woets, J. (1988) Biological and integrated pest control in greenhouses, *Annual Review Entomology* **33**, 239–269.

van Schelt, J., Klapwijk, J., Letard, M. and Aucouturier, C. (1996) The use of *Macrolophus caliginosus* as a whitefly predator in protected crops, in D. Gerling and R.T. Mayer (eds.), *Bemisia 1995. Taxonomy, Biology, Damage, Control and Management*, Intercept Ltd, Andover, pp. 515–521.

Zinnen, T.M. (1988) Assessment of plant diseases in hydroponic culture, *Plant Disease* **72**, 96–99.

CHAPTER 31

CUCURBITS

Pierre M.J. Ramakers and Timothy M. O'Neill

31.1. Cucumber Production

Cucumber crops are grown in most regions of the world (Table 31.1), especially in northern Europe, the Mediterranean region, North Africa, Canada and Japan. As the plant responds to warm temperatures and high humidities, cultivation in glasshouses or plastic-covered structures is widely practised. Air humidity in a greenhouse may be sub-optimal on winter nights because of the intense heating required, and on summer days because of strong solar irradiation. Cultivation is therefore particularly successful in temperate maritime climates, where it is easier to maintain conditions optimal for the crop: temperatures between 16 and 30°C and air humidity around 80%. Such conditions allow year-round production of high quality fruits. In many countries, hydroponic cropping has become commonplace because of the increased yields attainable. Yields of between 600 and 700 t/ha are normal for hydroponic crops in heated glasshouses, and with artificial light yields in excess of 1000 t/ha are possible.

TABLE 31.1. World production of cucumbers and gherkins in 1995 (adapted from FAO, 1996)

Region	Area (ha)	Production (t)
Africa	23,000	388,000
North and Central America	106,000	1,462,000
South America	4,000	67,000
USSR (1989–91)	167,000	1,384,000
Asia	780,000	13,372,000
Europe	90,000	2,434,000
Oceania	1,000	21,000
World	1,200,000	19,353,000

Cucumbers are extremely fast growing plants and a crop can reach the roof of a greenhouse within three weeks of planting. The main stem is then stopped and production of fruits is continued on lateral shoots. Unlike other fruiting vegetables, it is difficult to maintain one cucumber crop all year without sacrificing both yield and fruit quality. Cucumbers are therefore replanted, between one and three times a year. Most growers will

435

R. Albajes et al. (eds.), Integrated Pest and Disease Management in Greenhouse Crops, 435-453.
© 1999 Kluwer Academic Publishers. Printed in the Netherlands.

replant the whole crop at once, others practice interplanting in order to maintain some continuity of fruit production. Recently, some growers have tried to avoid replanting by layering, a system originally developed in the UK for tomato growing. In this system, the top of the plant is kept at the same height throughout the season, by laying horizontal the lower part of the stem at the same speed as the plant grows upwards. The productive part of the plant thereby continues to receive maximal benefit from light allowing high yields of quality fruit. The final length of layered plants may exceed 20 metres. Layering is applicable only in tall glasshouses, and systematic leaf trimming is required to keep the horizontal part of the stem free from diseases.

Intensive production methods pose specific problems for integrated pest and disease management, particularly with regard to reliability and cost of the control programme. Typical difficulties associated with the different cropping practices described above are: (i) replanting [disturbance of the balance between pest organisms and beneficials; persistence of root and stem base pathogens (e.g. *Pythium*, *Mycosphaerella*)]; (ii) replanting in summer (high rates of airborne whiteflies, aphids and powdery mildew spores attacking young crop); (iii) frequent replanting (high costs of repeated introductions of beneficials); (iv) interplanting (too many remaining insects with possible preference for young crop; immediate reinfection with powdery mildew); and (v) layering (stem diseases; inadvertent removal of parasitized whitefly scales before hatching of parasitoids, because of leaf trimming).

31.2. Major Pests and Diseases and Methods Employed for their Control

Cucumber crops are subject to attack by some 20 pests (Shipp, in preparation), at least 40 fungal pathogens and several bacteria and viruses (Blancard *et al.*, 1995; Zitter *et al.*, 1996) causing often considerable reductions in yield and fruit quality. The more common pests and diseases are listed in Table 31.2. Where there is intensive production in a concentrated area with a long growing season (December to October), dispersal of airborne pathogens and pests between greenhouses is a real problem. Powdery and downy mildews, for example, can both become epidemic very quickly in a particular locality. Practices used for pest and disease control may involve utilization of host resistance, reduction of infection sources (e.g. disease-free young plants; scrupulous nursery hygiene between crops), manipulation of the greenhouse climate, appropriate cultural practices (e.g. regular deleafing), monitoring for pests and diseases and the rational use of pesticides and biological control agents (Table 31.3).

Host resistance has provided a very effective method of control against several important fungal pathogens of cucumber (Fletcher, 1992). Unfortunately, it is not available for all pathogens and most crops are treated with fungicides, in northern Europe usually to control powdery mildew, *Botrytis* or *Pythium*. Resistance against arthropods is usually only partial; present breeding programmes include resistance against spider mites and thrips in cucumber (Mollema, 1992) and against cotton aphid in melon (Klingler *et al.*, 1998).

TABLE 31.2. Major pests and diseases of cucurbits (A: cucumber, B: courgette, C: melon)

Common name	Scientific name	Transmission	A	B	C
Pests					
Twospotted mite	*Tetranychus urticae*		*	*	*
Aphids, mainly cotton aphid (= melon aphid)	*Aphis gossypii*		*	*	*
Greenhouse whitefly	*Trialeurodes vaporariorum*		*	*	*
Tobacco whitefly	*Bemisia tabaci*		*	*	*
Thrips, mainly Western flower thrips	*Frankliniella occidentalis*		*	*	*
Semiloopers	*Chrysodeixis chalcites, Autographa gamma*		*	*	*
Other caterpillars	Mainly *Spodoptera* spp.		*	*	*
Fungus gnats	*Bradysia* spp.		*	*	*
Leaf miners	*Liriomyza* spp.		*	*	*
"French fly"	*Tyrophagus* spp.		*	*	
Root-knot nematodes	*Meloidogyne* spp.		*	*	*
Viruses					
Cucumber mosaic virus	CMV	Aphids	*	*	*
Watermelon mosaic virus 2	WMV2	Aphids			*
Zucchini yellow mosaic virus	ZYMV	Aphids	*	*	*
Cucurbit aphid-borne yellows virus	CABYV	Aphids	*	*	*
Beet pseudo yellows virus	BPYV	*Trialeurodes vaporariorum*	*		*
Cucumber vein yellowing virus	CVYV	*Bemisia tabaci*	*		*
Cucurbit yellow stunting disorder virus	CYSDV	*Bemisia tabaci*	*		*
Melon necrotic spot virus	MNSV	Fungus *Olpidium* sp. through water irrigation	*		*
Tobacco necrosis virus	TNV	Fungus *Olpidium* sp. through water irrigation	*		
Cucumber green mottle mosaic virus	CGMMV	Mechanically, seeds	*		*
Squash mosaic virus	SqMV	Beetles, seeds, mechanically		*	*
Bacteria					
Angular leaf spot	*Pseudomonas syringae* pv. *lachrymans*	Seeds, insects, rain splash	*		*
Bacterial leaf spot	*Xanthomonas campestris*	Seeds	*		
Bacterial soft rot	*Erwinia carotovora* ssp. *carotovora*		*		*
Bacterial wilt	*Erwinia tracheiphila*	Beetles	*		*
Root mat	*Agrobacterium* sp.		*		
Fungi					
Alternaria leaf spot	*Alternaria alternata* f. sp. *cucurbitae*		*	*	*
Anthracnose	*Colletotrichum orbiculare* (= *Colletotrichum lagenarium*)		*		*
Black root rot	*Phomopsis sclerotioides*		*		

TABLE 31.2. Major pests and diseases of cucurbits (A: cucumber, B: courgette, C: melon) (cont.)

Common name	Scientific name	Transmission	A	B	C
Fungi (cont.)					
Cercospora leaf spot	*Cercospora citrullina*		*		*
Charcoal rot	*Macrophomina phaseolina*				*
Corynespora blight	*Corynespora cassiicola*		*		
Downy mildew	*Pseudoperonospora cubensis*		*	*	*
Fusarium crown and root rot	*Fusarium solani* f. sp. *cucurbitae*			*	
Fusarium wilt	*Fusarium oxysporum* f. sp. *cucumerinum*		*		
Fusarium wilt	*Fusarium oxysporum* f. sp. *melonis*				*
Grey mould	*Botrytis cinerea*		*	*	*
Gummosis	*Cladosporium cucumerinum*		*	*	*
Gummy stem blight	*Didymella bryoniae*		*	*	*
Penicillium stem rot	*Penicillium oxalicum*		*		
Phytophtora crown and root rot	*Phytophthora capsici*				*
Powdery mildew	*Sphaerotheca fusca* (= *Sphaerotheca fuliginea*)		*	*	*
Powdery mildew	*Erysiphe orontii* (= *Erysiphe cichoracearum*)		*	*	*
Powdery mildew	*Leveillula taurica*		*	*	*
Pythium root and stem base rot	*Pythium* spp.		*	*	*
Root rot and vine decline	*Monosporascus cannonballus*				*
Root rot	*Thielaviopsis basicola*			*	
Sclerotinia stem rot	*Sclerotinia sclerotiorum*		*	*	*
Septoria leaf spot	*Septoria cucurbitacearum*		*	*	*
Southern blight	*Sclerotium rolfsii*				*
Ulocladium leaf spot	*Ulocladium cucurbitae*		*	*	*
Verticillium wilt	*Verticillium albo-atrum* and *Verticillium dahliae*		*	*	*

Manipulation of the greenhouse climate can be very effective in the control of some fungal diseases. Crop management practices can also influence disease development (O'Neill *et al.*, 1991), but are rarely adequate as control methods alone.

The trend towards production in hydroponic systems, primarily to increase yields, has significantly reduced the importance of root diseases and soilborne pests. However, even in hydroponic systems, Pythium root and stem base rot can be a serious problem. This is especially so when a subsequent crop or a second crop in the same season is grown without sterilizing the growing medium.

The very rapid growth of greenhouse crops in general and cucumbers in particular, with continued proliferation of new leaves, makes foliar pest and disease control a particular challenge. This is one of the additional reasons (apart from pesticide resistance) for using biological pest control, since natural enemies have the potential to disperse and protect new foliage as it develops. Biological control of cucumber pests became popular during the 70s and is discussed further in Section 31.4.

TABLE 31.3. Methods commonly employed for the control of major pests and diseases in protected cucumber crops (*: seldom used, of minor importance; **: used in many crop, considerable contribution to control; ***: used in most crops, a major component of control)

Target	Method of control					
	Host plant resistance	Chemical	Biological	Climate	Hygiene and sanitation	Nutrition
Two-spotted spider mite	*[2]	***	***	*Lime shading; misting		
Greenhouse whitefly		***	***			
Tobacco whitefly		**	*			
Thrips	*[2]	***	***		**Soil disinfection *New or sterilized slabs + new soil covers	
Aphids		***	**			
Noctuids		**	**			
Leaf miners		*	*		**Soil disinfection *New or sterilized slabs + new soil covers *Picking off (nursery)	
Fungus gnats		*	*		**Soil disinfection; new or sterilized slabs	
Root-knot nematodes		*			**Soil disinfection; inert growing media	
CMV	*	**Aphid control			*Weed control; removal of affected plants	
CGMMV					**Removal of affected plants; general hygiene; milk	
Root mat					*General hygiene; new or sterilized slabs	
Powdery mildew	**Partial resistant cvs[1]	***Frequent in summer			*Picking off; avoid draughts and open doors	*Additional silicon
Downy mildew	*	**		**Extra heat and ventilation	*Picking off	
Grey mould		**	*	**Extra heat and ventilation	**Remove affected tissues	
Sclerotinia stem rot		*		**Extra heat and ventilation	***Prompt removal of affected plants or plant parts; cover soil floors	
Gummy stem blight	*Resistance claimed	*Moderately effective		**Extra heat and ventilation	**Remove wounded leaves	*Increase conductivity
Penicillium stem rot	*[2]	*Moderately effective			*Removal of affected plants	

TABLE 31.3. Methods commonly employed for the control of major pests and diseases in protected cucumber crops (*: seldom used, of minor importance; **: used in many crop, considerable contribution to control; ***: used in most crops, a major component of control) (cont.)

Target	Method of control					
	Host plant resistance	Chemical	Biological	Climate	Hygiene and sanitation	Nutrition
Pythium root and stem base rot	*Grafting onto resistant rootstock	**[3]			***Soil disinfection; inert growing media; general hygiene	
Black root rot	*Grafting onto resistant rootstock	**			***Soil disinfection; inert growing media; general hygiene	
Fusarium wilt	**	*			***Soil disinfection; inert growing media; general hygiene	
Verticillium wilt	**	*			***Soil disinfection; inert growing media; general hygiene	
Leaf spot	***					
Gummosis	***					

[1]In all but earliest plantings; greenhouse perimeter or throughout
[2]Differences in varietal susceptibility demonstrated
[3]At (re)-planting in mid-summer

31.3. Integrated Control of Diseases

31.3.1. POWDERY MILDEW {Sphaerotheca fusca (Fr.) Blumer. [= Sphaerotheca fuliginea (Schlechtend.:Fr.) Pollacci]} AND OTHER SPECIES

Of the three powdery mildews occurring on cucumber, *S. fusca* is generally the most important in greenhouse crops. *Erysiphe orontii* Cast (= *Erysiphe cichoracearum* DC.) and *Leveillula taurica* (Lév.) G. Arnaud occur occasionally, especially in the Mediterranean region. Fungicides remain the principal method of control with a wide range of fungicides used including bupirimate, carbendazim, chlorothalonil, dimethirimol, fenarimol, imazalil, triforine, pyrazophos and sulphur. Growers may apply ten or more sprays during a season to susceptible cultivars and fungicides from different chemical groups are alternated to prevent selection of resistant pathotypes. Selection of fungicide-resistant strains and loss of disease control has occurred with bupirimate, methyl-benzimidazole (MBC), pyrazophos and the ergosterol biosynthesis inhibitors (EBIs) such as dimethirimol (e.g. McGrath, 1996). However, the level of resistance may decline when use of the fungicide ceases, allowing successful reintroduction into a spray programme the following season, providing they are not overused. Bupirimate and EBI fungicides are still used successfully, many years after the first reports of resistance to these fungicides. Where conventional fungicides are unavailable or their use is not permitted (e.g. in organic crop production), sodium bicarbonate and plant extracts [e.g. Milsana, based on an extract from *Reynoutria sachalinensis* (F. Schmidt) Nakai] may be used to provide some control. If mildew is at a low incidence, prompt removal of affected leaves delays the need for chemical treatment and this is practised by some growers as a first step in integrated control strategies.

Where IPM is practised, fungicide choice is limited by the harmful side-effect of some of them on predators and parasitoids. For this reason, MBC fungicides and pyrazophos are now rarely used, while bupirimate, which is moderately harmful to introduced beneficial insects, may only be used when other treatments are not sufficiently effective.

Application of fungicides for powdery mildew control is now frequently done as a low volume mist (25–50 l/ha). Treatment can be automated and is done at night in a closed and unoccupied greenhouse with strategically placed fans to assist spray distribution. The ease of this treatment method allows more timely application of fungicides in response to increasing mildew pressure. Unfortunately, the efficacy when mildew is well established in a crop is relatively poor and a high volume spray is more likely to be used in this situation.

Mildew-resistant cultivars have significantly reduced the need for prophylactic spray programmes in summer-planted crops, but they tend to show leaf chlorosis under low light conditions, and therefore are rarely used in northern Europe for plantings before mid-February. As early crops are less likely to be seriously affected by powdery mildew, there is less need to use mildew-tolerant varieties at this time. Resistant varieties are sometimes planted around the crop if the main cultivar is susceptible, as mildew often first develops at greenhouse perimeters and row ends.

Manipulation of crop nutrition can be used to control powdery mildew, notably in hydroponic crops (O'Neill, 1991). Rockwool slabs impregnated with supplementary silicon or addition of potassium metasilicate to the feed solution were used in the UK and The Netherlands in the early 90s, although this practice has declined in recent years with the advent of mildew-tolerant cultivars.

Although several antagonistic fungi (e.g. *Ampelomyces quisqualis* Cesati:Schltdl., *Sporothrix flocculosa* Traquair, Shaw & Jarvis, *Stephanoascus* spp., *Tilletiopsis* spp.) have been shown capable of providing partial control of cucumber powdery mildew, none has yet been successfully developed into a commercial product for use on this crop. *Ampelomyces quisqualis* was recently registered in the USA as a product (AQ10) for use in grapes, and *S. flocculosa* is in process of commercialization (see Chapter 24). Registration costs and reliability are major constraints in their development as biological control agents.

31.3.2. DOWNY MILDEW [*Pseudoperonospora cubensis* (Berk. & M.A. Curtis) Rostovzev]

Downy mildew is controlled in heated crops by manipulation of the greenhouse environment. Research in Austria (Bedlan, 1987) has shown that the disease is strongly influenced by temperature, humidity and leaf wetness, with the disease cycle completed in just four days under optimum conditions. A critical stage of 60 degree-hours of leaf wetness is needed for sporangial germination and infection. The use of radiant heat and ventilation to prevent prolonged leaf wetness has proved a very simple and effective control strategy. In unheated or partially heated crops the disease can be very difficult to control even with fungicides. These include chlorothalonil, copper oxychloride, cymoxanil, fosetyl aluminium, mancozeb and metalaxyl (Cohen and Grinberger, 1987). Metalaxyl was very effective until the occurrence of resistant pathotypes, now present both in the Mediterranean area and the USA (Cohen and Grinberger, 1987; Moss, 1987).

Resistance to powdery mildew is often correlated with resistance to downy mildew in commercial cucumber varieties, and cultivars resistant against downy mildew are available in the USA.

As *P. cubensis* has a narrow host-range, confined to members of the Cucurbitaceae family, effective disposal of infected crops can help prevent carryover to new crops. In the UK, the lack of alternative hosts for overwintering supported statutory measures taken from the mid 70s to 1991 to try and prevent establishment of the disease, believed to have been introduced with imported symptomless plants.

31.3.3. GREY MOULD (*Botrytis cinerea* Pers.:Fr.)

While grey mould can be found in many cucurbit crops, it is usually most troublesome in unheated or partially heated crops, in long-season crops and in crops grown in polythene structures with restricted ventilation. Yield loss occurs directly from fruit rot and indirectly from stem infections leading to shoot or plant death. Rotting of the main stem is particularly damaging. Senescent leaves can be important sites for Botrytis development and removal from the main stem minimizes stem rot. Weekly leaf trimming in cucumbers was shown to be more effective than fortnightly or monthly removal (O'Neill *et al.*, 1996). If lateral shoots are insufficiently thinned, grey mould may develop in the upper canopy, associated with reduced air movement. In unheated greenhouses, Botrytis rotting of aborted and young fruit commonly occurs following cool nights. Picking off the affected fruits limits subsequent shoot and stem rot. Spectrally-modified polyethylenes which suppress sporulation of *B. cinerea* have been used to clad tunnels growing cucumber with the aim of reducing grey mould. Although they are currently little used in commercial practice, there is increasing interest in this novel control method.

Fungicides are widely used to reduce losses by grey mould, the intensity of spraying varying with the disease risk and the availability of alternative control methods (e.g. heating and ventilation). They include chlorothalonil, iprodione and other dicarboximide fungicides, diethofencarb, carbendazim and dichlofluanid. Attempts to control the disease by spraying fungicides alone, rather than as part of an integrated programme, can lead to intensive fungicide use and the selection of resistant pathotypes. Resistance of *B. cinerea* has occurred to iprodione, carbendazim and diethofencarb. New fungicides in the anilinopyrimidine and strobilurin groups show promise for improved control of grey mould. Some of the strobilurin fungicides have a very broad spectrum of activity (including control of powdery and downy mildews) but are safe to predators, and thus appear to have good potential for use in integrated programmes.

Grey mould is one of the few diseases where a biological control product is available. *Trichoderma harzianum* Rifai T39 (Trichodex) is registered for use on cucumber in several countries and is used by some growers as part of an integrated strategy (Elad and Shtienberg, 1995). It is primarily used at times when the disease risk is not high.

31.3.4. GUMMY STEM BLIGHT [*Didymella bryoniae* (Auersw.) Rehm]

Gummy stem blight can affect all the aerial parts of cucurbits. In replanted crops, lesions tend to be particularly common at the stem base. Partial resistance to gummy stem blight has been described (Wyszogrodzka *et al.*, 1985) and recently introduced in some cucumber varieties.

As *D. bryoniae* infection is favoured by high humidity and leaf wetness, disease incidence can be reduced by growing under drier conditions (van Steekelenburg, 1985). In The Netherlands, greenhouse environment control with appropriate ventilation was found more effective than sanitation or application of fungicides (van Steekelenburg, 1986). Use of heat and ventilation early in the day, at least one hour before sunrise, is widely practised in cucumber growing in northern Europe and Canada to prevent condensation on fruit thus reducing the incidence of fruit rot. Appropriate staff training can further help to minimize fruit rot, as careful handling during picking, grading and packing, to minimize wounding, reduces the incidence of external fruit rot. Regular washing of hands and knives minimizes disease spread.

Fungicides used to control *D. bryoniae* include benomyl, carbendazim, chlorothalonil, iprodione and triforine, but protection by spraying is difficult because of the numerous stem wounds and the dense crop canopy. Although weekly sprays may be necessary to exert reasonable control in cucumber (van Steekelenburg, 1978), for practical reasons a less intensive schedule is usually used, with sprays applied at key growth stages (e.g. a stem base spray soon after replanting) and when the disease begins to increase. Isolates of *D. bryoniae* resistant to carbendazim and iprodione were detected in some cucumber crops in England in surveys in 1983 and 1986.

31.3.5. PYTHIUM ROOT AND STEM BASE ROT [*Pythium aphanidermatum* (Edson) Fitzp.] AND OTHER *Pythium* spp.

Pythium root rot can seriously affect both crops grown in the soil and those grown in inert media, with the latter appearing at greater risk when replanted in mid-summer, possibly associated with high substrate temperatures at this time. Good hygiene is of paramount importance in preventing Pythium. The greenhouse structure and pathways are washed down with a disinfectant at the end of a cropping season and footbaths may be used. Inert growing media (principally rockwool) is increasingly reused, sometimes for up to three seasons, with annual disinfection usually by steaming. For soil-grown crops, disinfection by steaming is preferred to methyl bromide treatment. The latter method was found to be ineffective in controlling black root rot.

Where there is a known soilborne problem, plants may be grafted onto rootstocks (usually *Cucurbita ficifolia* Boucé) resistant to *Pythium* spp. and a range of other common root diseases to avoid the need for fungicide root drenches. Grafted plants are more expensive, but for many years grafting proved an effective strategy for root disease control. It has decreased markedly in recent years following the widespread adoption of hydroponic production systems.

Chemicals widely used for control of Pythium root diseases include etridiazole and propamocarb hydrochloride. In the UK, soil-grown crops are usually drenched with a mixture of etridiazole and carbendazim (for control of black root rot) at monthly intervals. Hydroponic crops are treated with propamocarb hydrochloride, often prophylactically.

Following experimental work indicating spread of *Pythium* in water films on the polyethylene floor covering (McPherson, unpublished), some growers with profiled floors leave occasional gaps between slabs to minimize the risk of extensive spread along a row. Fungus gnat and shore fly larvae have recently been recognized as vectors of *Pythium* in cucumber crops (Jarvis *et al.*, 1993). As poor drainage increases the risk of both *Pythium* and fungus gnats, growers will improve drainage to prevent flooding, and take other measures if necessary to control the pests.

31.3.6. CUCUMBER MOSAIC VIRUS (CMV)

CMV is occasionally damaging to cucurbits, with outbreaks almost always associated with an obvious aphid (*Aphis gossypii* Glover) attack. The virus is not readily contact-transmitted, the extent of damage depending on the level of aphid infestation, age of plant at infection and strain of the virus. Good control of CMV can be readily achieved by effective control of aphids and by ensuring prompt removal of any infected plants and weeds in and around the greenhouse. It is also important to ensure good control of root diseases, especially *Pythium*, as plants affected by both CMV and Pythium wilt die quickly.

31.3.7. CUCUMBER GREEN MOTTLE MOSAIC VIRUS (CGMMV)

CGMMV is an occasional problem but it is highly infectious spreading throughout a crop in 6–8 weeks. It is limited by prompt and careful removal of affected plants. Many growers will remove all apparently healthy plants for 1–2 m immediately surrounding the infected area. Rockwool slabs are also removed, and replaced with new slabs after the whole area (plastic sheeting, support wires, etc.) has been treated with a disinfectant. Some growers also use an ultra-heat treated (UHT) milk suspension to limit spread of the virus, either as a hand and knife dip between working on each plant or to spray plants. If the outbreak is limited to a particular area, restrictions on staff access can also reduce the risk of further spread. At the end of cropping after an outbreak of CGMMV, an intensive disinfection procedure is adopted. For soil-grown crops, steaming is usually used to reduce the risk of carryover in the soil. Rockwool slabs to be reused are steamed at 100°C for at least ten minutes.

31.3.8. LESS COMMON DISEASES

Sclerotinia stem rot [*Sclerotinia sclerotiorum* (Lib.) de Bary] occurs on stems, shoots and fruit in the dense upper canopy of a crop. It is rarely a widespread problem in crops where the floor has been covered with polythene sheeting, and careful removal of affected parts provides effective control. Penicillium stem rot (*Penicillium oxalicum* Currie & Thom) is occasionally very damaging (O'Neill *et al.*, 1991). The disease is partially controlled by prompt removal of affected plants and by treatment with iprodione or an MBC fungicide. *Trichoderma harzianum*, the active ingredient of Trichodex (see Section 31.3.3), is known to be an antagonist of this pathogen.

Black root rot (*Phomopsis sclerotioides* van Kestern), a damaging disease in UK soil-grown cucumbers, is controlled by soil steaming, grafting onto a resistant rootstock or by MBC fungicide root drenches. It is uncommon at damaging levels in hydroponic crops.

With the use of resistant varieties, both Fusarium and Verticillium wilts were rare in the UK but there have been some damaging attacks of both diseases in recent years. MBC fungicide root drenches give partial control. Leaf spot [*Corynespora cassiicola* (Berk. & M.A. Curtis) C.T. Wei] and gummosis (*Cladosporium cucumerinum* Ellis & Arth.) are now extremely rare as a result of the use of resistant varieties.

Two *Olpidium*-transmitted virus diseases, tobacco necrosis virus (TNV) and melon necrotic spot virus (MNSV), occur occasionally in cucumber but are rarely damaging. However, a mixed TNV and CGMMV infection can result in severe crop damage. Beet pseudo yellows virus (BPYV), transmitted by the greenhouse whitefly [*Trialeurodes vaporariorum* (Westwood)], occasionally has caused significant damage in the UK and The Netherlands. Zucchini yellow mosaic virus (ZYMV), transmitted by aphids, may affect cucurbits in the autumn months resulting in severely deformed fruit. Damage by these viruses is minimized by effective control of the vector insects. Cucurbit yellows virus transmitted by aphids has recently caused significant losses in southern France and several Mediterranean countries while cucumber vein yellowing virus (CVYV), transmitted by the tobacco whitefly [*Bemisia tabaci* (Gennadius)], has caused losses in Israel. Resistant cultivars have been identified to both these diseases.

Root mat, caused by a rhizogenic plasmid carried by *Agrobacterium* bv. 1 (Weller et al., in preparation), has affected an increasing number of hydroponic crops in the UK since 1991. In 1997, over 17 crops were affected and losses on just two nurseries were estimated at over £100,000. A control strategy for this disease has still be developed.

31.4. Integrated Control of Pests

31.4.1. RED SPIDER MITE

The most devastating pest on cucumber is the two-spotted or red spider mite, *Tetranychus urticae* Koch. Uncontrolled, it will not just damage but can quickly kill plants. The rapid development of pesticide resistance in this mite was the main incentive for researching biological control methods. While resistance development in insects is often a regional or even global phenomenon, the resistance status of local spider mite populations might differ considerably, even between adjacent greenhouses. This is the result of the combination of a high genetic plasticity, a low migration capacity and the ability to hibernate in an empty greenhouse. Thus an individual grower is likely to get immediate benefit from a sound resistance prevention strategy. Indeed, using integrated rather than only chemical control, many growers report that the same acaricides seem to work better.

Biological control with the predatory mite *Phytoseiulus persimilis* Athias-Henriot is common practice. The number of mites introduced should be adapted to the abundance of the spider mites. Since the density of mites within a colony at the beginning of an infestation is host-plant specific and reasonably constant, it has been suggested to estimate the infested leaf area rather than the number of mites (Sabelis, 1983). IPM advisors, however, find this approach too time-consuming and recommend fixed rates of predatory mites per plant or per area, such as 20 predators/m^2 in hot spots and 1/m^2 elsewhere. With advanced infestations, the introduction of sufficient numbers of predators is often impracticable and uneconomic, and use of selective acaricides is required. It is debatable whether the acaricide or the predator should be applied first in such situations. If time allows and available acaricides are sufficiently selective, it is probably better to permit the predator to complete one or two generations before applying the acaricide.

If introduced in time, *P. persimilis* will control spider mites under most conditions except during prolonged periods of hot and dry weather. The predator, a species of low vegetation, tends to abandon the upper regions of the crop canopy then, whereas the spider mites thrive particularly well. In such situations, growers need to apply acaricides in order to restore the balance between pest and predator. Overhead misting has been suggested for repairing or (probably better) preventing such situations (Lindquist *et al.*, 1987), but is not widely applied due to the risk of encouraging diseases.

Selective acaricides available for integrated control include fenbutatin oxide, hexythiazox and clofentezine. Abamectin can also be integrated as at a very low rate it is more toxic to spider mites than to predatory mites; however, this chemical is usually applied at a much higher rate, meant for insect control, and thus conflicts with integrated control.

Control of spider mites without any use of acaricides is unlikely to succeed. In order to avoid acaricides as much as possible, the best strategy to follow would be the pest-in-first method (Hussey, 1967) with frequent monitoring of the artificially established colonies of spider mites.

31.4.2. WHITEFLY

Biological control with *Encarsia formosa* Gahan in cucumber has long been obstructed by chemical control of other insects, particularly thrips. Also, the frequent application of fungicides against powdery mildew is suspected to affect this delicate wasp. Although most fungicides are classified as harmless in the standard IOBC/WPRS tests (see Chapter 11), chemicals in this category still cause up to 50% mortality of *E. formosa* adults under laboratory conditions. Wider adoption of mildew-tolerant cultivars could therefore further improve biocontrol of whiteflies.

Cucumbers are among the best host plants for the greenhouse whitefly, *T. vaporariorum*, as evidenced by the high fecundity and longevity of adults on this plant (van Lenteren and Noldus, 1990). It has been a subject of discussion between researchers as to whether or not biocontrol for whitefly is feasible on cucumber, and various strategies have been devised for tackling this problem (Hussey, 1985). The dispute was finally settled "on the battlefield": following the example of Humber Growers in Hull, most growers have now adopted the so-called dribble method. Assuming that somewhere in the crop some whiteflies are present, introductions of *E. formosa* are started soon after planting once the residues of any recently applied insecticides have worn off. Introductions are repeated weekly and continued until the parasitoid is found to be sufficiently established or even until the end of the season. In recent years, some biocontrol companies have introduced a related Aphelinid parasitoid, *Eretmocerus eremicus* Rose & Zolnerowich (= *Eretmocerus californicus* Howard), for the control of whiteflies. It is claimed to perform better at extremely high temperatures and to be somewhat more tolerant to pesticides. It is, however, recommended to be used as an addition to rather than as a substitute for *E. formosa*.

For restoring unbalanced situations, insect growth regulators (IGRs) such as buprofezin or teflubenzuron may be used. The whitefly fungus *Aschersonia aleyrodis* Webber might serve the same purpose (Ramakers and Samson, 1984) and would be preferable to an IGR because it is a highly-specific pathogen and unlikely to affect any natural enemy. Some progress has been made in producing and formulating *A. aleyrodis*, but it has not yet been commercialized. Whiteflies are also potential hosts for various less-specific fungal pathogens, of which only *Verticillium lecanii* (A. Zimmerm.) Viégas is registered, but this is not very popular.

Both the IGRs and *A. aleyrodis* affect the immature instars and are thus competing with *E. formosa*. It would be preferable to have a selective agent aimed at the adults for bringing local outbreaks quickly under control. Broad-spectrum insecticides, vacuum cleaners and even gas flames have been used for this purpose, indicating that there is a desperate need rather than a solution. A selective (chemical or microbial) control agent against adult whiteflies would make IPM more attractive on crops with a high whitefly risk, such as cucumbers.

Systematic leaf trimming as required in the layering system (see Section 31.1) might remove parasitized scales before the parasitoids were able to hatch. As a consequence, a balance between whiteflies and parasitoids may never occur, and inundative introductions have to be repeated continuously. Factors that may disturb an already established balance include application of pesticides, mass immigration of whiteflies from outdoors and, particularly in cucumber, replanting of the crop. Replantings in spring and summer are carried out very quickly, with the greenhouse only empty for a few days. Survival of some adult whiteflies is very likely, even after intensive chemical treatment of the old crop. For the parasitoids, however, replanting results in a difficult situation: on the new crop there are only some whitefly adults and maybe a few eggs, but no suitable instar for parasitization. Restarting the whole system after every replanting would make biocontrol rather costly, especially in summer when the initial population densities are higher.

The recent advance of (a new strain of) *B. tabaci* has revived the search for new natural enemies of whiteflies, but is as yet an increasing impediment for commercial growers on biological whitefly control in cucurbits, especially in the Mediterranean area.

31.4.3. THRIPS

In chemical-orientated pest control, thrips was considered a minor, though omnipresent pest. Usually it remained suppressed by chemical control of other pests or by its immediate food-competitor, the red spider mite, because of the much shorter generation time of the latter. In the late 60s when cucumber growers started to control spider mites with predatory mites, thrips filled up the newly created niche and soon became a dominating problem in IPM.

When onion thrips, *Thrips tabaci* Lindeman, was still the major thrips species, different strategies were used seeking integration with biocontrol of the main pests. Dutch cucumber growers applied foliar treatments of organophosphorus compounds (OPs), compatible with

P. persimilis but with few other biocontrol organisms. In the UK, preventing pupation of thrips in the ground was tried (Pickford, 1984). This strategy required aggressive and inconvenient soil treatments, but allowed greater application of biocontrol methods against leaf-borne pests.

During the 80s, *T. tabaci* was superseded by western flower thrips (WFT), *Frankliniella occidentalis* (Pergande). The advance of this new thrips species coincided with and was a further stimulus for the introduction of *Amblyseius* spp. as predators of thrips larvae (Ramakers *et al.*, 1989). *Neoseiulus* (*Amblyseius*) *cucumeris* (Oudemans) is now used in most cucumber growing countries around the world. Because of the insecticide resistance of *F. occidentalis*, no insecticides are available that would fit, to any extent, in an integrated programme. Biocontrol of thrips should therefore be based on an all-or-nothing approach. The predator should be established prior to pest attack ("predator-in-first") (see Ramakers and Rabasse, 1995). Modern cucumber cultivars, however, are fully gynoecious, so there is no pollen for the predators to feed on. Therefore they are not just released, but introduced in small bags containing a living culture in miniature. Growers or their IPM advisors need to check these cultures regularly to determine when they should be replaced. If such expertise is not available, it is recommended that cultures are replaced monthly.

Anthocorid predators, particularly *Orius majusculus* (Reuter) in northern Europe, may occur spontaneously in association with thrips. They are more effective predators of thrips than phytoseiid mites but cucumber is not their favourite host plant because of the absence of pollen. These predators are usually observed in late spring or summer, when the population density of thrips is high (Schreuder and Ramakers, 1989). Artificial introduction of anthocorids for an earlier effect on thrips is worthwhile considering, but has not become as popular as on pepper crops. While other *Orius* spp. hunt in flowers, *O. majusculus* is considered more of a leaf-dwelling species. It is therefore preferred to other *Orius* spp. for cucumber, where even with flower thrips the vast majority of the larval population is found on the leaves.

Abamectin, methiocarb in The Netherlands and endosulfan in Canada, are used around replanting to prepare for IPM. It is uncertain, however, whether these chemicals help or hinder subsequent biological pest control. Dichlorvos is to be preferred because of its short persistence against biological control agents, but cucumber growers are reluctant to use it after planting because of its phytotoxicity.

31.4.4. APHIDS

Aphis gossypii is the most usual aphid on cucurbits. Unlike the pests mentioned above, it does not occur throughout the year in the greenhouse. Attack usually starts with winged aphids originating from surrounding vegetation or nearby greenhouses. Some years ago *A. gossypii* was suppressed by chemicals used at that time for control of whitefly. Hydrogen cyanide was very effective in controlling both pests. When growers converted to biological control of whitefly, the selective chemical pirimicarb proved a satisfactory alternative for control of aphids. However, resistance of *A. gossypii* against pirimicarb in the late 80s led to a renewed interest in natural enemies of aphids.

Aphis gossypii has a wide host plant range, but its reproductive capacity is particularly high on cucurbits, and parasitoids with a similar or higher intrinsic rate of increase are unknown. *Aphis gossypii* is a poor host for *Aphidius matricariae* Haliday, the most widely used aphid parasitoid in the 70s and 80s. When IPM advisors found that in some cucumber houses aphids became highly parasitized, it was first thought that *A. matricariae* had adapted to this host. This so-called cucumber-strain was later found to be another *Aphidius* species, *Aphidius colemani* Viereck. After it was confirmed that this species is effective against both *A. gossypii* and *Myzus persicae* (Sulzer), most biocontrol companies switched from *A. matricariae* to *A. colemani* (van Steenis, 1995). In recent years, *A. colemani* has been introduced into cucumber crops on cereal aphids as substitute hosts using barley seedlings as "banker plants", a technique originally developed for introducing the aphid predator *Aphidoletes aphidimyza* (Rondani) (Kuo-Sell, 1989). However, *A. colemani* is more often used in crops other than cucumber because with *A. gossypii* on cucumber the interval between detecting the aphid and reaching the damage threshold is often very short.

The new systemic insecticide imidacloprid, if applied to the roots, has little effect on natural enemies except for those which also feed on the plant (e.g. *Orius*, *Macrolophus*). It is very effective against aphids on fast growing plants like cucumber. Although a welcome addition to the IPM programme as a whole, it is also a serious competitor for biological aphid control agents. Where there is an immediate risk of an aphid-transmitted virus (see Section 31.3.6), a grower will probably not consider biological aphid control at all.

31.4.5. MINOR PESTS

Lepidopteran pests are a problem in many IPM programmes, but few species are common pests on cucurbits. An exception are some Plusiinae species [*Chrysodeixis chalcites* (Esper), *Autographa gamma* (L.)], that, unlike most noctuids, scatters their eggs rather than producing egg clusters. Local sprays with broad-spectrum insecticides against the hatching larvae are therefore not an option. Larvae are susceptible to *Bacillus thuringiensis* Berliner, but application of this microbial agent has to be repeated frequently. For this reason, a grower might prefer an IGR like teflubenzuron.

Capsid bugs might intrude into the greenhouse in summer. Control is often not necessary but imidacloprid, where permitted, may be added to the irrigation system to control these and other phloem feeders.

Leaf miners are often present but rarely a problem. If natural biological control is not sufficient, artificial introduction of larval parasitoids may be considered. Additional control with the systemic oxamyl is possible against some species but is seldom required.

31.5. Integrated Control Programmes

Cucumber has the honour of being the first crop on which integrated control in the narrow sense (harmonizing chemical and biological control) was used commercially. The nucleus of this early IPM programme was control of spider mite with the predatory mite *P. persimilis* and control of powdery mildew with dimethirimol, the first systemic fungicide. It was actually not the availability of the predator, but the registration of the fungicide that triggered large-scale application of the IPM programme (Bartlett, 1987). Dimethirimol was soon withdrawn from the market because of resistance problems, but IPM was continued as other fungicides could be safely used in a IPM programme providing a few restrictions were observed (e.g. not spraying benomyl, avoiding frequent use of pyrazophos). Also, most acaricides were found to be sufficiently selective. Integration with insect control was more problematic, though this problem was reduced by the development of OP-resistance in *P. persimilis* (Schulten *et al.*, 1976). Key insecticides in this initial IPM programme were hydrogen cyanide against whiteflies, pirimicarb against aphids and a number of broad-spectrum insecticides including nicotine, sulfotep and diazinon against miscellaneous insects.

A modern IPM programme for cucumber would include at least biological control of spider mites, thrips and whiteflies and sometimes also of aphids and noctuids. Key pesticides used in The Netherlands are: (i) fenbutatin oxide and hexythiazox against spider mites; (ii) buprofezin against whiteflies; (iii) teflubenzuron against whiteflies and noctuids; (iv) imidacloprid or pirimicarb against aphids; and (v) abamectin (propagation) and methiocarb (end of season) against thrips.

The adoption of hydroponic cropping systems has considerably reduced the risk of root diseases and soilborne pests like nematodes, "French flies" (*Tyrophagus* spp.) and symphilids. Similarly, the move to replanting once or twice a year has lessened the potential yield loss which might occur following a damaging attack of Botrytis, Mycosphaerella or Penicillium stem rot. Fungicides are available for these diseases when they do occur, but with only occasional use they are unlikely to disrupt the biological pest control.

While entomologists took the lead in applying biological control methods, plant pathologists have been more successful in applying other non-chemical methods such as plant breeding and climate control. The drier aerial environment created by cropping over polythene sheeting, as commonly practised with hydroponic production, has lessened the risk of damaging leaf and fruit infection by *Botrytis* and *Mycosphaerella*. Disease risk can be further reduced by use of automated greenhouse climate control, especially to control downy mildew and to minimize leaf and fruit infection by *Mycosphaerella*. The introduction of mildew-tolerant cultivars and the continued effectiveness of certain mildew fungicides which are safe to introduced beneficial insects has assisted the development of current IPM programmes.

IPM is now widely applied on cucumbers in northern Europe and Canada, mostly in winter-plantings. Reasons why IPM is less popular in other parts of the season include higher initial pest populations, higher temperatures, more pest species and more interference with outdoor populations. For similar reasons, IPM is more complicated in subtropical areas than in the temperate zone. Researchers in the Mediterranean therefore focus on exploiting native natural enemies rather than artificial establishment of populations originating from mass-rearings (Ramakers and Rabasse, 1995).

Pest and disease problems on related crops such as melon and courgette are similar but not identical (Table 31.2). Growers of these crops can tolerate somewhat more thrips than cucumber growers, which facilitates the application of IPM. Another advantage is the fact that these crops require pollination and that thus the use of broad-spectrum insecticides should better be avoided. However, since these crops are usually grown in subtropical climates, the abundance of field populations of *B. tabaci*, *A. gossypii* and viruses associated with these insects constitutes an important impediment.

31.6. The Future of IPM

IPM in cucurbits will continue to develop to prevent pesticide resistance, to overcome new pest and disease problems and to respond better to public pressure on minimizing pesticide use. There is an immediate need for selective insecticides effective against thrips and adult whiteflies. Development of (partial) pest resistance and a wider range of mildew-tolerant cultivars would facilitate more widespread use of delicate natural enemies such as hymenopterous parasitoids and *Aphidoletes*. In the longer term, microbial control might become more important, especially against fungal diseases, providing the level and reliability of control can be improved and registration costs are not prohibitive. The unpopularity of applying biocontrol in the spring and summer plantings, and the trend to replant cucumbers more frequently, is a matter of concern. Promoting the layering system rather than replanting might help to increase the attractiveness of biological methods in summer, but would at the same time create a stronger need for applying occasional treatments with selective chemicals to restore pest/predator balances. It may also increase the need for more fungicide treatments to control stem diseases. Depending on the availability of selective pesticides, growers in subtropical areas will obtain more opportunity to benefit from predators and parasitoids resident in their area.

Growing cucurbits without any pesticides is not possible without suffering yield losses. It is feasible, however, to try and avoid the use of pesticides for considerable parts of the growing season. Whether an individual grower chooses for a low or high input of biological control agents depends not only on the availability of these agents, but much more on local climatical and epidemiological conditions and the demands from the market. Labour and energy costs will also affect the extent to which diseases will be controlled by non-chemical methods. IPM is not only part of a sound resistance prevention strategy, but also a response to the increasing demand of consumers for "green" products (Wardlow and O'Neill, 1992). Associations of growers now make agreements with their retailer(s) about produce quality and production methods, including restrictions on pesticide use. Such self-imposed restrictions, which are additional to the state regulations, are rapidly gaining in importance. Dutch auctions are using a "green label" for a number of greenhouse crops including cucumber, melon and courgette. Growers carrying this label have to demonstrate the presence of a minimum amount of a natural enemy; for cucumber the demand is 50% "black scales" in the whitefly population, or three quarters of the leaves being colonized by phytoseiids (Ramakers, 1996). In Canada, marketing organizations like BC Hothouse exert considerable pressure – on top of the already very restrictive governmental policy – on their members to minimize pesticide use. UK growers are required to comply with the retailers' Integrated Crop Management protocols, which include the maximum use of non-chemical control methods and minimal use of pesticides. Further developments will depend on the willingness of the market, or a market segment, to pay a premium for such "green" fruits.

Acknowledgements

We are grateful to David Ann, Jude Bennison and Dan Drakes of ADAS Consulting Ltd,
UK, for their helpful comments.

References

Bartlett, K. (1987) Partnership: Dutch and British; Partnership: Biology and chemistry, *Innovation ICI* **8**, 7–11.
Bedlan, G. (1987) Studies for optimisation of spraying dates to control *Pseudoperonospora cubensis* in cucumbers
    in Austria, *Pflanzenschutzberichte* **48**, 1–11.
Blancard, D., Lecoq, H. and Pitrat, M. (1995) *A Colour Atlas of Cucurbit Diseases: Observation, Identification
    and Control*, Manson Publishing Ltd, London.
Cohen, Y. and Grinberger, M. (1987) Control of metalaxyl-resistant casual agents of late blight in potato and
    tomato and downy mildew in cucumber by cymoxanil, *Plant Disease* **77**, 1283–1288.
Elad, Y. and Shtienberg, D. (1995) *Botrytis cinerea* in greenhouse vegetables: Chemical, cultural, physiological
    and biological controls and their integration, *Integrated Pest Management Reviews* **1**, 15–25.
FAO (1996) *FAO Production Yearbook*, Vol. 49–1995, FAO Statistics Series No. 130, FAO, Rome.
Fletcher, J.T. (1992) Disease resistance in protected crops and mushrooms, *Euphytica* **63**, 33–49.
Hussey, N.W. (1967) *Provisional Programme for the Use of the Predatory Mite* Phytoseiulus riegeli *to Control
    Red Spider Mite (*Tetranychus urticae*) on Cucumbers*, in Glasshouse Crops Research Institute Annual Report
    1966, GCRI, Littlehampton, pp. 140–143.
Hussey, N.W. (1985) Whitefly control by parasites, in N.W. Hussey and N. Scopes (eds.), *Biological Pest Control:
    The Glasshouse Experience*, Blandford Press, Poole, Dorset, pp. 104–107.
Jarvis, W.R., Shipp, J.L. and Gardiner, R.B. (1993) Transmission of *Pythium aphanidermatum* in greenhouse
    cucumber by the fungus gnat *Bradysia impatiens* (Diptera: Sciaridae), *Annals of Applied Biology* **122**, 23–39.
Klingler, J., Powell, G., Thompson, G.A. and Isaacs, R. (1998) Phloem specific aphid resistance in *Cucumis melo*
    line AR 5: Effects on feeding behaviour and performance of *Aphis gossypii*, *Entomologia Experimentalis et
    Applicata* **86**(1), 79–88.
Kuo-Sell, H.L. (1989) Getreideblattläuse als Grundlage zur biologischen Bekämpfung der Pfirsichblattlaus, *Myzus
    persicae* (Sulz.), mit *Aphidoletes aphidimyza* (Rond.) (Dipt., Cecidomyiidae) in Gewächshäusern, *J. of Applied
    Entomology* **107**, 58–64.
Lindquist, R.K., Casey, M.L., Bauerle, W.L. and Short, T.L. (1987) Effects of an overhead misting system on
    thrips populations and spider mite-predators interactions on greenhouse cucumber, *IOBC/WPRS Bulletin* **10**(2),
    97–100.
McGrath, M.T. (1996) Increased resistance to triadimefon and to benomyl in *Sphaerotheca fuliginea* populations
    following fungicide use over one season, *Plant Disease* **80**, 633–639.
Mollema, C. (1992) Western flower thrips in cucumber: Resistant genotypes keep pest under control, *Prophyta*
    46(2), 22–23.
Moss, M.A. (1987) Resistance to metalaxyl in the *Pseudoperonospora cubensis* population causing downy mildew
    of cucumber in South Florida, *Plant Disease* **71**, 1045.
O'Neill, T.M. (1991) Investigation of glasshouse structure, growing medium and silicon nutrition as factors
    affecting disease incidence in cucumber crops, *Mededelingen van de Faculteit Landbouwwetenschappen
    Rijksuniversiteit Gent* **56**, 359–367.
O'Neill, T.M., Bagabe, M. and Ann, D.M. (1991) Aspects of biology and control of stem rot of cucumber caused
    by *Penicillium oxalicum*, *Plant Pathology* **40**, 78–84.
O'Neill, T.M., Hand, D., Harriman, M. and McPherson, G.M. (1996) Influence of some cultural practices on the
    development of cucumber stem rot caused by *Botytis cinerea*, in *XI International Botrytis Symposium*,
    Wageningen, June 1996, p. 106.
Pickford, R.J. (1984) Evaluation of soil treatment for control of *Thrips tabaci* on cucumbers, *Annals of Applied
    Biology* **5**, 18–19.
Ramakers, P.M.J. (1996) Use of natural enemies as "indicators" for obtaining an IPM label, *IOBC/WPRS Bulletin*
    19(1), 119–122.

Ramakers, P.M.J., Dissevelt, M. and Peeters, K. (1989) Large scale introductions of phytoseiid predators to control thrips on cucumber, *Mededelingen van de Faculteit Landbouwwetenschappen Rijksuniversiteit Gent* **54/3a**, 923–929.

Ramakers, P.M.J. and Rabasse, J.M. (1995) Integrated pest management in protected cultivation, in R. Reuveni (ed.), *Novel Approaches to Integrated Pest Management*, Lewis Publishers, Boca Raton, Fla., pp. 199–229.

Ramakers, P.M.J. and Samson, R.A. (1984) *Aschersonia aleyrodis*, a fungal pathogen of whitefly. II. Application as a biological insecticide in glasshouses, *Zeitschrift für Angewandte Entomologie* **97**, 1–8.

Sabelis, M.W. (1983) The dose of predatory mites required for spider-mite control on greenhouse cucumber: Computer simulations, presented at a meeting of the IOBC Working Group Crop Protection & Modelling, Leuven, March 1983.

Schreuder, R.G. and Ramakers, P.M.J. (1989) Onverwachte hulp bij geïntegreerde bestrijding, *Groenten & Fruit* **45**(5), 28–29.

Schulten, G.G.M., van de Klashorst, G. and Russell, V.M. (1976) Resistance of *Phytoseiulus persimilis* A.H. (Acari: Phytoseiidae) to some insecticides, *Zeitschrift für Angewandte Entomologie* **80**(4), 337–341.

Shipp, J.L. (in preparation) IPM program for greenhouse cucumber, in M.P. Parrella and K.M. Heinz (eds.), *Control of Arthropod Pests in Protected Culture*, Ball Publishing, Batavia, Ill.

van Lenteren, J.C. and Noldus, L.P.J.J. (1990) Behavioural and ecological aspects of whitefly-plant relationships, in D. Gerling (ed.), *Whiteflies: Their Bionomics, Pest Status and Management*, Intercept Ltd, Andover, Hants, pp. 47–89.

van Steekelenburg, N.A.M. (1978) Chemical control of *Didymella bryoniae* in cucumbers, *Netherlands J. of Plant Pathology* **84**, 27–34.

van Steekelenburg, N.A.M. (1985) Influence of humidity on incidence of *Didymella bryoniae* on cucumber leaves and growing tips under controlled environment conditions, *Netherlands J. of Plant Pathology* **91**, 277–283.

van Steekelenburg, N.A.M. (1986) Factors influencing internal fruit rot of cucumber caused by *Didymella bryoniae*, *Netherlands J. of Plant Pathology* **92**, 81–91.

van Steenis, M.J. (1995) *Evaluation and Application of Parasitoids for Biological Control of* Aphis gossypii *in Glasshouse Cucumber Crops*, PhD Thesis, Wageningen Agricultural University.

Wardlow, L.R. and O'Neill, T.M. (1992) Management strategies for controlling pests and diseases in glasshouse crops, *Pesticide Science* **36**, 341–347.

Weller, S.A., Stead, E.D., O'Neill, T.M., Hargreaves, D. and McPherson, G.M. (in preparation) Rhizogenic *Agrobacterium* biovar 1 strains and cucumber root mat in the UK.

Wyszogrodzka, A.J., Williams, P.H. and Peterson, C.E. (1985) Search for resistance to gummy stem blight (*Didymella bryoniae*) in cucumber (*Cucumis sativus*), *Euphytica* **35**, 603–613.

Zitter, T.A., Hopkins, D.L. and Thomas, C.E. (1996) *Compendium of Cucurbit Diseases*, APS Press, St Paul, Minn.

CHAPTER 32

STRAWBERRIES
Stanley Freeman and Giorgio Nicoli

32.1. Strawberry Cultivation

The world surface cultivation of strawberry, *Fragaria x ananassa* (Duchesne), has been estimated at *c.* 165,000 ha by FAO in 1994, with a total production of 2.3 million tonnes. Europe is the main producer (106,150 ha with a total yield of *c.* 1.0 million t), followed by North and Central America (729,000 t) and Asia (419,000 t). The main European producers are Spain (219,400 t), Italy (191,600 t), Poland (170,000 t), France (81,500 t) and Germany (54,600 t). Production in California (USA) was calculated as 511,660 t in 1992 (Strand, 1994).

Strawberries can be grown both in open fields and in greenhouses: the cultivation under protection is frequent in the cold regions, such as northern Europe (where hydroponic culture in heated glasshouses is rather common) while outdoor cultivation is predominant in warmer areas, such as California and Florida. In the Mediterranean region, cultivation under plastic is frequently adopted, although its incidence varies among countries. Strawberry planting and fruit harvesting take place in Europe over two growing seasons, winter and summer, respectively. In single growing seasons, transplants for production fields, originating from disease-free material, are propagated from mother plants in nurseries during the summer months. Winter plantings from mid to late summer bear fruit during the winter and spring months. In Europe however, summer production is predominant: in the Mediterranean, cultivation is generally designed for a single season whereas in central Europe a biannual season may be planned, but the use of this growing technique has been significantly reduced in the past years.

Integrated Pest Management (IPM) relies on optimization of production with minimal use of chemicals, in order to reduce hazards to humans, animals, plants and the environment. Successful methods begin prior to crop cultivation and include a thorough knowledge of cropping history, soil and water sampling for nutritional purposes and influence of weather on the crop. Additional factors include soil disinfestation, use of certified clean plant material and biocontrol of pests. The use of IPM can provide new commercial opportunities for growers thanks to "green labels" certifying environment-safe production techniques and absence of pesticide residues on fruits (Celli, 1987).

32.2. Management Methods

In general, monitoring of both pests and natural enemies should be carried out on a weekly basis for a large part of the crop cycle, because plants are particularly susceptible and infestation/infection can increase rapidly. It is recommended to examine at least 50 plants per unit area (1000–2000 m^2), sampling populations and looking for "hot spots" which may lead to disease epidemics.

Numbers of pest species and the damage they cause generally increase from cold

R. Albajes et al. (eds.), Integrated Pest and Disease Management in Greenhouse Crops, 454-472.
© 1999 *Kluwer Academic Publishers. Printed in the Netherlands.*

regions to the warmer temperate and sub-tropical ones, where IPM programmes need to deal with a wider range of problems. Outbreaks of some arthropod pests can vary yearly and they appear dependent mainly on three parameters: growing techniques, climate and latitude. For instance, when the crop is grown in a permanent greenhouse or the plastic covers are set up before winter, the western flower thrips, *Frankliniella occidentalis* (Pergande), can overwinter in quiescence, being active and reproducing early in the season, causing more severe outbreaks than in winter-covered tunnels. With a hot and dry climate the red spider mite, *Tetranychus urticae* Koch, is favoured, the sensitivity of the crop increases while the predatory mite, *Phytoseiulus persimilis* Athias-Henriot, is affected particularly by the low relative humidity. At the southern European latitudes, strawberries are frequently exposed to destructive migrations of noctuid moths from Africa (Cayrol, 1972).

32.2.1. NATURAL AND BIOLOGICAL CONTROL

Natural control of arthropod pests can be very important for IPM, mainly in the warmer regions, where predators can colonize both greenhouse and open-field strawberries. Some species of predatory mites and insect predators can often control the red spider mite, the western flower thrips and other arthropod pests.

On strawberries, biological control by the release of mass-reared arthropods is generally applied by two methods: (i) the "seasonal inoculative" method by introducing natural enemies that can establish on plants and multiply up to the end of the crop cycle, as *P. persimilis* and *Orius laevigatus* (Fieber); and (ii) the "inundative" method used for the larvae of two predators, *Chrysoperla carnea* (Stephens) and *Harmonia axyridis* (Pallas), released to control aphids and lasting on the plants up to the emergence of adults, that generally leave the crop without reproducing on it. Microbial agents can also be used: formulations of *Bacillus thuringiensis* Berliner ssp. *kurstaki* Dulmage are sprayed on plants or used as active compounds for the preparation of baits to control some lepidopteran pests. Entomopathogenic nematodes are used in soil, mainly in northern Europe, against weevils feeding on roots.

32.2.2. AGROTECHNICAL PRACTICES

Healthy Propagation Material and Cultivar Selection
Healthy propagation material is a prerequisite for disease control. Heating of bare transplants provides a non-chemical treatment to ensure healthy plant stock. Nurseries are established in disinfestated soils, therefore, plant material must be disease free. Likewise, transplants from nurseries to production fields should also be void of pathogens. Some strawberry cultivars vary in their resistance to certain pests and diseases: it is therefore important to chose the cultivars best suited for the area planned for production.

Crop Rotation
Crop rotation is another way of reducing pests specific to strawberry, such as root-knot nematodes and soilborne insect pests. Strawberries plantings are generally designed for only one season of fruit production which reduces the risk of heavy outbreaks of root weevil larvae, more frequent when a polyannual crop cycle is adopted. By rotating strawberry cultivation with cover crops, weeds can be limited, inoculum levels of certain diseases (*Colletotrichum, Verticillium*) are reduced and populations of beneficial micro-organisms are preserved.

Plasticulture
The use of plastic products features predominantly in strawberry cultivation at both the nursery and field production stages and is an asset to any IPM programme. It is generally a prerequisite for productive yields. Splash dispersal of fungal spore inoculum is reduced when drip irrigation is implemented early on, in establishing mother plants in nurseries and transplants in production fields.

Strawberries are generally grown on beds, with plastic mulch or directly on the soil, constituting a large part of the Californian cultivation practice. In Europe, plastic mulch is used for strawberry crops grown in soil, both in greenhouses and open fields, for reducing decay problems (by preventing fruit contact with soil and limiting inoculum build-up of some diseases) and for reducing the need of weed control when black polyethylene is used.

Hydroponic culture prevents several problems of root diseases and pests. Strawberry fruiting fields may be completely protected by plasticulture, whether grown under low tunnels or walk-in greenhouses, to reduce splash dispersal of inoculum and to protect the crop from adverse weather conditions. The ability to aerate under plasticulture also alleviates accumulation of moisture, which is important for reduction of inoculum levels of most pathogens, as in the case of grey mould caused by *Botrytis cinerea* Pers.:Fr.

Pruning and Sanitation
Removal of senescent leaves in late winter limits the inoculum of some diseases and reduce populations of some overwintering pests (such as noctuid larvae, aphids and mites), although some natural enemies (such predatory mites) can be removed as well.

Sanitation is an elementary component in all IPM programmes. All infected material (runners, whole plants and fruits) should be removed from the cultivation area and not left in rows since this serves as an additional inoculum source.

Pesticide Usage
Pesticides should be correctly used for controlling certain pests and diseases. Soil disinfestation for the eradication of soilborne pests (such as root weevils and root-knot nematodes), diseases (e.g. Phytophthora crown, root and fruit rots, Anthracnose, Verticillium wilt,) and weeds is an integral part of any IPM programme for the production of disease free plant material.

Pesticide resistance is one of the main reasons leading growers to move to biocontrol and IPM. Two of the most harmful pests of strawberries, the western flower thrips and the red spider mite, are resistant to a large number of pesticides, including many active ingredients selective for beneficial arthropod species. Additional difficulties arise in the control of secondary pest outbreaks caused by the scarcity of natural enemies following a non-selective treatment against a primary pest. The "red list" of insecticides/acaricides to avoid is long and only a few of them can be considered selective for the beneficial arthropods colonizing (or released on) strawberries. Even some fungicides show negative side-effects: in a greenhouse experiment, pyrazophos affected the wild populations of *P. persimilis*, whereas other fungicides (sulphur and barium polysulphide) were harmless (Benuzzi *et al.*, 1989). Sulphur appears to be selective for predatory mites naturally occurring in the Mediterranean, probably due to its lasting wide use in this area. In general, inorganic fungicides (including copper compounds) should be preferred for their selectivity.

32.3. IPM for Key Pests and Diseases

32.3.1. MAIN ARTHROPOD PESTS

Red Spider Mite
The red spider mite (also called "two-spotted spider mite"), *T. urticae*, is a serious world-wide pest in all strawberry growing areas. Feeding by mites can reduce plant vigour and decrease fruit size and yield; young plants may die if mites remain uncontrolled. *Tetranychus urticae* is often introduced on planting material; the mite can overwinter on the crop as diapausing adult females characterized by an orange-coloured body (similar in colour to the predatory mite *P. persimilis*), very different from the two-spotted livrea, typical of non-diapausing mites.

Being endemic along the Mediterranean sea (Galazzi and Nicoli, 1996), *P. persimilis* can colonize plants after transplanting and overwintering on them, both in greenhouses and in open fields, as recorded in 73% of infested strawberries at the end of winter, near the northern Adriatic coast (Celli *et al.*, 1988). Other predatory mites play an effective role in the natural control of *T. urticae* (Cross *et al.*, 1994; Strand, 1994), as well as the Coccinellidae *Stethorus picipes* Casey (Nearctic region), *Stethorus punctillum* (Weise) (Palaearctic region) and other polyphagous insects.

Biocontrol of *T. urticae* is feasible by releasing mass-reared predatory mites (particularly *P. persimilis*), as reported for the main cultivation areas: North America (Strand, 1994), northern Europe (Scopes, 1985) and Mediterranean basin (Vacante and Nucifora, 1987; Benuzzi and Nicoli, 1990; Benuzzi *et al.*, 1992). Biocontrol by *P. persimilis* has to be carried out following the "seasonal inoculative" criteria (see Chapter 13), and the amount of predators to release is generally commensurate with severity of the outbreak. The absence/presence sampling can facilitate the estimation of the pest population, maintaining a good degree of accuracy up to 60–70% of infested leaves (Fig. 32.1). The number of red spider mites per plant can be easily calculated according to the number of leaves per plant, as well as the number of predatory mites to release to obtain a favourable prey:predator ratio. Figure 32.2 shows the results of a greenhouse experiment where both *T. urticae* and *P. persimilis* females were artificially introduced on plants to obtain six different initial prey:predator ratios. The initial ratio influenced the severity of the outbreaks, but not the final balance of the populations, because high densities of preys induced a quick multiplication of the predatory mite.

The release ratio should be decided taking into account susceptibility of plants, climatic conditions, need for a rapid response and natural presence of other predators. Generally, the ratios adopted vary from 15:1 to 25:1 adult females (prey:predator); using the highest amounts of predators in the warmer areas. Predatory mites should be released as soon as the pest is detected after transplanting and generally also when plants renew their vegetation (Table 32.1). In northern Europe, another predatory mite, *Neoseiulus* (= *Amblyseius*) *californicus* (McGregor), has been tested and, in California, a mixture of species [*P. persimilis*, *N. californicus* and *Metaseiulus occidentalis* (Nesbitt)] is under evaluation to control the pest over a wider range of conditions than the release of *P. persimilis* alone (Strand, 1994).

Western Flower Thrips
Although some species were recorded on strawberry in Europe (Gremo *et al.*, 1997), phytophagous Thysanoptera were generally considered harmless until the introduction of the western flower thrips, *F. occidentalis*, from America. Damage to leaves is rarely

recorded whereas fruits are frequently affected: thrips feed on flowers (pollen is very important) and a large number of insects may cause blossoms to drop and fruit to remain small and hard.

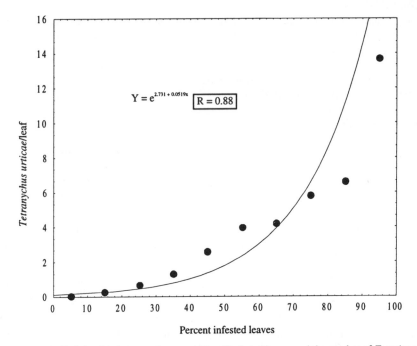

Figure 32.1. Relationship between the percentage of infested leaves and the number of *T. urticae* adult females per trifoliate leaf (from Benuzzi and Nicoli, 1990).

In open field-grown strawberries, intensity of *F. occidentalis* outbreaks depend mainly on climatic conditions: in the warm areas, the pest can overwinter outdoors in quiescence, being active for a large part of the winter and multiplying quickly as soon as spring temperatures rise. For protected cultivations, damages are generally more severe in artificially heated glasshouses and in plastic tunnels covered in autumn compared to those covered at the end of winter. Outbreaks are favoured also by broad-spectrum pesticides, because thrips is resistant to many insecticides whereas naturally-occurring predators are eliminated. Many native predators adapted to this exotic pest providing effective new associations, mainly in the Mediterranean region (Riudavets, 1995, Chapter 17).

Particularly in the areas (or greenhouses) where *F. occidentalis* can overwinter in quiescence, thrips should be sampled weekly on flowers, as soon they open, and blue sticky traps can help for early detection of the pest. The release of predators generally starts after the heating system is activated, or at the first opening of flowers. In the warm areas, *Orius* spp. are released adopting the threshold of 1–2 thrips per flower (Table 32.1). Some species of *Orius*, both native and exotic, have been evaluated for controlling *F. occidentalis* in Europe, and the palaearctic *O. laevigatus* has been finally selected. This species tolerates high temperatures (Alauzet *et al.*, 1994) and shows a

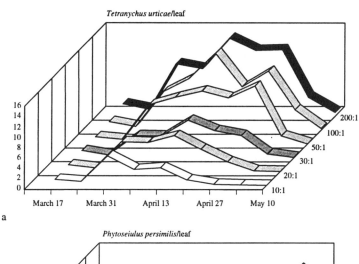

a

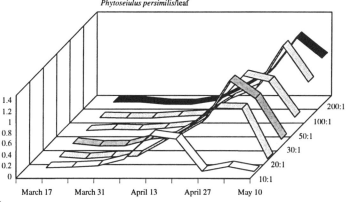

b

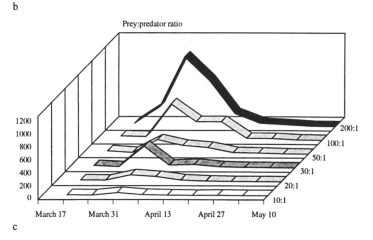

c

Figure 32.2. Trends of *T. urticae* (a) and *P. persimilis* (b) populations (adult females), and prey:predator ratios (c) in a protected strawberry experiment. Each plot was artificially infested by the same number of *T. urticae* females and various amounts of *P. persimilis* females to originate six different initial prey:predator ratios (10:1, 20:1, 30:1, 50:1, 100:1 and 200:1) (from Benuzzi and Nicoli, 1990).

TABLE 32.1. General guidelines for Integrated Pest Management in protected strawberry

Target pest/pathogen	Control agent/technique	Release rate/active ingredient	Threshold and general remarks
Arthropod pests			
Tetranychus urticae (Red spider mite)	*Phytoseiulus persimilis*	Multiple releases of predatory mites: from a total of 4-6 predators/m² (cold areas) to 10-15 predators/m² (warm areas)	Inorganic fungicides (i.e. sulphur, copper compounds) should be preferred for controlling plant diseases for their selectivity for predatory mites and other natural enemies
	Neoseiulus (= *Amblyseius*) *californicus*	One or two releases with a total of 2-4 predators/m²	
Frankliniella occidentalis (Western flower thrips)	*Orius laevigatus* (other *Orius* spp.)	Multiple releases of predatory bugs: a total of 2-4 predators/m²	Release of *Orius laevigatus* (0.5-1 predators/m² per release) should start at the first evidence of infestation, particularly at the beginning of flowering; other diapausing species of *Orius* can be released during springtime. Highest rates are used in southern Europe. Predatory mites are generally used in northern Europe
	Neoseiulus (= *Amblyseius*) *cucumeris*	Release 50-200 predators/m² in shaker bottles or sachets	
Macrosiphum euphorbiae *Chaetosiphon fragaefolii* *Aphis gossypii* *Myzus persicae* *Myzus ascalonicus* (Aphids)	*Chrysoperla carnea* *Harmonia axyridis* Aphid parasitoids	Release 20 second-instar larvae/m² Release 5-10 young larvae/m² Release 0.25 parasitoids/m² (weekly releases)	Threshold for predators' release: >20% of young leaves infested
			Harmonia axyridis is generally used for hot spot releases on heavily infested plants
			Parasitoids are generally used in northern Europe; the species must be chosen in relation to the aphid species to control
	Natural pyrethrum or other selective insecticides	Hot spot treatments or general treatments	Selective insecticides can be used also to reduce the outbreaks before the release of natural enemies
Lepidopteran larvae (Several species)	*Bacillus thuringiensis* ssp. *kurstaki* Neurotoxic insecticides *Chrysoperla carnea*	General treatments (sprays) or baits Baits Two releases with >50 second instar larvae/m² on hot spots	*Bacillus thuringiensis* is effective mainly for the control of young larvae. Baits activated by neurotoxic insecticides should be used to control the species non-susceptible to *Bacillus thuringiensis* *Chrysoperla carnea* larvae are currently advised in northern Europe
Otiorhynchus sulcatus *Otiorhynchus rugosostriatus* (Root weevils)	*Steinernema carpocapsae* or *Heterorhabditis* spp.	500,000 entomopathogenic nematodes/m²	Different species of root-weevils have been recorded in northern America. Entomopathogenic nematodes can control the larvae in the soil. The soil must be moist (not waterlogged) both before and after application. Nematodes are used mainly in northern Europe

TABLE 32.1. General guidelines for Integrated Pest Management in protected strawberry (cont.)

Target pest/pathogen	Control agent/technique	Release rate/active ingredient	Threshold and general remarks
Steneotarsonemus pallidus (Cyclamen mite)	*Neoseiulus* (= *Amblyseius*) *cucumeris* Sulphur	Release 50 predators/m² (prevention on second year crops) Spray or powder treatments	*Neoseiulus* (= *Amblyseius*) *cucumeris* is used particularly in northern Europe, mainly for hot spot control: 30 predators/infested plant and 50 predators/m² surrounding the hot spot. Sulphur is effective also for control of the powdery mildew; it is used mainly in the Mediterranean area
Viruses and mycoplasma diseases	Virus and mycoplasma indexing, propagation of pathogen-free material		
Fungal diseases			
Phytophthora spp. (Crown root and leather rot)	Soil disinfestation, fungicides, resistant varieties, plastic mulch, improve soil drainage		
Colletotrichum spp. (Anthracnose)	Soil disinfestation, fungicides, resistant varieties, avoid overhead irrigation, aeration, sanitation		
Verticillium spp. (Verticillium wilt)	Soil disinfestation, solarization, grow cover crops		
Botrytis cinerea (Grey mould)	Fungicides, aeration, plastic mulch, sanitation, biological control using *Trichoderma* spp.		
Sphaerotheca macularis f. sp. *fragariae* (Powdery mildew)	Timely fungicide sprays (sulphur), sanitation		
Nematodes			
Ditylenchus dipsaci Aphelenchoides fragariae Aphelenchoides ritzemabosi (Foliar nematodes)	Annual planting Nematode-free plantings		Allowing foliage to dry out between irrigations may help reduce foliar nematode activity
Meloidogyne spp. (Root-knot nematodes)	Annual planting and crop rotation Nematode-free plantings Soil solarization/disinfestation		Solarization requires additional practices to prevent infestations, such as field selection, sanitation and crop rotation

wide natural distribution on several Mediterranean crops (Riudavets and Castañé, 1994). In addition, the overwintering females of a southern European strain showed a high oviposition propensity, indicating that at least a part of the population can overwinter in quiescence (Tommasini and Nicoli, 1996). These biological traits indicate the possibility of releasing this species over a wide range of climatic/growing conditions, even during the short-daylength periods, when other *Orius* spp. undergo diapause (van den Meiracker, 1994) and adults leave the herbaceous crops moving to trees and bushy plants to look for winter shelters.

In northern Europe, some *Orius* species and/or predatory mites are used by the "multiple release" method, usually by prevention, and non-diapausing strains of *Neoseiulus* (= *Amblyseius*) *cucumeris* (Oudemans) are commercially available. Predatory mites can be introduced also by placing small sachets on plants acting as miniature rearings that continuously release new predators for some weeks (Table 32.1). The use of *Amblyseius degenerans* Berlese is currently under evaluation. Predatory mites are generally not used in the Mediterranean basin, although some species occur naturally on strawberries.

Frankliniella occidentalis is considered rarely harmful in California and treatments are not recommended unless populations exceed ten thrips per blossom. Furthermore, Strand (1994) underlines that, since thrips is a predator of *T. urticae* eggs, any damage it causes usually is outweighed by the benefit rendered in helping keep mites under control. This appears as an evident discrepancy in the relevance of thrips between the areas where the species is endemic and newly-colonized ones.

Lepidopteran Pests
Several species of Lepidoptera (mainly Noctuidae) can damage strawberries, but the intensity of outbreaks can vary in different geographic areas. Strawberry plants can host many polyphagous species, often colonizing the crop after transplanting and overwintering on it. Small plants are especially susceptible to larvae which destroy the crowns; feeding of foliage is generally considered to cause negligible damage, but significant losses can be recorded with the species attacking flowers and fruits. Good cultural practices keep the populations of some species low, such as annual planting and thorough pruning of second-year plantings. The number of species attacking strawberries and the severity of outbreaks increase from northern to southern Europe.

In northern Europe, the noctuid *Phlogophora meticulosa* (L.) has been found on strawberry (Cayrol, 1972; Cross *et al.*, 1994), but is generally considered harmless because the larva feed mainly on foliage. However, certain leaf-rolling species (Tortricidae) can attack fruits, but their density is generally too low to cause damage and *Bacillus thuringiensis* Berliner can eventually provide good control of young larvae.

In northern Mediterranean regions (i.e. northern Italy; unpublished data), larvae of several species have been collected on plants: 17 Noctuidae (including all the most important pests), 2 Tortricidae, 2 Geometridae, 2 Arctidae, 1 Pyralidae and 1 Lasiocampidae. During the first year of the crop cycle, *Agrotis segetum* (Denis & Schiffermüller) and *Agrotis ipsilon* (Hufnagel) can feed on the crowns of plantings, which have to be replaced. *Agrotis ipsilon* is a cosmopolitan species migrating from northern Africa and Asia Minor to Europe in springtime and vice versa in autumn (Tremblay, 1986). Later in the season, *Mamestra suasa* (Denis & Schiffermüller), *Mamestra brassicae* (L.) and *Lacanobia oleracea* (L.) generally feed on foliage and well-developed plants can tolerate some damage. In the second year, *P. meticulosa* feeds mainly on leaves while *Agrochola lychnidis* (Denis & Schiffermüller) often attacks flowers and fruits.

In southern Mediterranean regions, some species of Noctuidae are frequently destructive on several crops (including strawberry), the most important being *Spodoptera littoralis* (Boisduval), *Chrysodeixis chalcites* (Esper), *Autographa gamma* (L.) and *A. segetum* (Sannino *et al.*, 1996; Tropea Garzia, 1997). Some species can reach very high densities due to the migrations of moths from Africa, although it appears reasonable that at least part of the *S. littoralis* population can overwinter in southern Europe, mainly under protection (Tremblay, 1986). A few techniques alternative to chemical control can be used (Tropea Garzia, 1997), such as screens in plastic tunnels to prevent the colonization of moths and soil tillage in the areas surrounding greenhouses, because it seems that weed control destroys eggs and young larvae. Sex-pheromone traps placed outside the tunnels can indicate the need to start checking for eggs and larvae on plants.

Several parasitoids and predators and some viral diseases can contribute to maintaining populations at low levels, but the use of non-selective pesticides can interfere in natural control. *Bacillus thuringiensis* can be used to control some species, but applications must be timed against young larvae (general sprays or baits). When damage is caused by species not susceptible to *B. thuringiensis*, pomace baits or bran baits activated with chemical insecticides can be applied in the evening to prevent drying of baits and favour eating by nocturnal larvae (Table 32.1). General sprays with chemical insecticides are needed to control the noctuid larvae feeding on crowns of the young plants or the heavy outbreaks of foliage-devouring larvae (as *S. littoralis*), but only few selective insecticides are compatible with IPM.

Referring to the coastal areas of southern California, Strand (1994) reports that occasional outbreaks of the noctuid *Helicoverpa* (= *Heliothis*) *zea* (Boddie) can cause damage. Strawberries are not a preferred host plant, but if high populations are present, the first generation larvae can bore into fruits, causing their rejection for processing. Two pheromone traps in each field beginning in early spring are recommended. As soon as eggs are found on leaves, treatments should be commenced, since eggs hatch in 2–3 days and larvae may invade fruits immediately.

Aphids

A number of aphid species have the potential to attack strawberries (Blackman and Eastop, 1984), with the most important ones being: *Chaetosiphon fragaefolii* (Cockerell), *Aphis gossypii* Glover, *Macrosiphum euphorbiae* (Thomas) and *Myzus persicae* (Sulzer). *Myzus ascalonicus* Doncaster is an important pest in the UK (Cross *et al.*, 1994). Aphids are of concern in nurseries and multiple-year plantings as they can transmit viruses and therefore, the exclusion of vectors is critical during propagation of virus-free planting material for nurseries. In fruit production, virus transmission does not cause significant damage: particularly where fields are replanted with certified transplants every one or two years, low populations can be tolerated. Damage results from severe outbreaks producing honeydew (and sooty mould) on fruits. Infestations tend to be concentrated, with groups of plants having much larger populations than surrounding plants. Generally, aphids overwinter as anholocyclic or paracyclic females on strawberry plants and reach peak levels in early spring. Damaging populations usually develop when spring temperatures are moderate and humidity relatively high; later in the season, they decline due also to high temperatures and changes in plant physiology. Although natural control is frequently effective, the release of mass-reared natural enemies or the use of selective insecticides may be necessary.

In the Mediterranean region, the inundative release of second-instar larvae ($c.$ 20/m^2) of the common lacewing $C.$ $carnea$ can provide good control of $C.$ $fragaefolii$ and $M.$ $euphorbiae$ (Celli et $al.$, 1988), as well as the young larvae of the Chinese Coccinellidae $H.$ $axyridis.$ The release threshold for both predators is generally fixed at 20% infestation of young leaves. When outbreaks are largely higher, selective insecticides can be applied mainly to control hot spots: natural pyrethrum is used particularly for synthetic pesticide-free production, and predators are sometimes released a few days after the treatment (Table 32.1).

In northern Europe, mass-reared parasitoids are usually released for prevention, using a mixture of species or choosing the appropriate parasitoid, after the identification of the aphid to control. $Chrysoperla$ $carnea$ larvae are released mainly for controlling hot spots. In California, aphids are not considered key-pests and are eventually controlled by selective insecticides, using a threshold of 30% of infested leaves. If populations reach ten aphids per newly unfolded leaf, treating with an insecticidal soap is advised, instead of conventional selective insecticides (Strand, 1994).

Root Weevils
The root-feeding larvae of weevils (Coleoptera, Curculionidae) may damage plants when adults migrate into strawberry fields from nearby host plants, ornamentals or second-year strawberries. Adults emerge in late spring or summer and feed on the foliage at night; they are about one centimetre long and do not fly, but may crawl into new cultivations from nearby areas. Adult weevils may chew jagged scallops into leaf edges, but they do not affect strawberry significantly. Females lay eggs in the soil, around plant crowns, about a month after emergence. Larvae work their way into the soil and can be destructive, eating root hairs and chewing the bark and cortex of larger roots, whereas some may enter the crown, killing the plant. Damage is usually first seen as wilting with eventual plant death in a localized area of the field, usually in late summer or early autumn, and symptoms often spread down rows as larvae move from dead to healthy plants.

Different species can attack strawberries in various cultivation areas. In Europe, $Otiorhynchus$ $rugosostriatus$ (Goeze) (Servadei et $al.$, 1972) and $Otiorhynchus$ $sulcatus$ (Fabricius) (Malais and Ravensberg, 1992; Cross et $al.$, 1994) are reported as the two potentially more dangerous species. In the UK, the importance of $O.$ $sulcatus$ seems to have recently increased by the widespread use of soil sterilants and methiocarb slug pellets which are harmful to predatory insects, and by the increased use of polythene mulches which provide a protected environment to the pest (Cross et $al.$, 1994). In California, $Nemocestes$ $incomptus$ (Horn) and $Otiorhynchus$ $cribricollis$ Gyll. are the most common, while $O.$ $sulcatus$ and $Pantomorus$ $cervinus$ (Boheman) invade strawberries less frequently (Strand, 1994). In any event, the presence and effect of root-weevils have been generally reduced by rapid turnover of plantings.

In northern Europe, control of weevils has been limited when using entomopathogenic nematodes $Steinernema$ $carpocapsae$ (Weiser) applied through irrigation systems (Cross et $al.$, 1994) or by soil drenching, as they cannot withstand drought, and the soil to which they are applied must be kept as moist as possible. The soil temperature seems to be the key to success. Two windows with soil temperature in excess of 14°C are available for applying nematodes: in late May to control overwintering larvae, and in August to kill larvae before they enter winter hibernation (Kakouli et $al.$, 1994). Also, nematodes of the genus $Heterorhabditis$ can be used to control $O.$ $sulcatus$ (Deseö Kovács and Rovesti, 1992; Malais and Ravensberg, 1992).

In California, sticky barriers are used to prevent infestation of new plantings by adults crawling from their development sites. Preventing invasion is particularly important for autumn harvests or if cultivation is maintained for two years (Strand, 1994).

New opportunities for plant resistance to feeding by adult black wine weevils have been investigated by Doss *et al.* (1987) using a clone of the beach strawberry *Fragaria chiloensis* (L.) Duchesne. Leaf hairs appeared to play an important role, while no evidence of a chemical basis for resistance could be obtained.

32.3.2. FUNGAL AND BACTERIAL DISEASES

Phytophthora Diseases
Phytophthora crown and root rot, caused by *Phytophthora cactorum* (Lebert & Cohn) J. Schröt. may result in serious plant loss world-wide (Wright *et al.*, 1966). Symptoms are manifested by stunted growth and wilting of leaves. The disease progresses until plants eventually collapse and die. *Phytophthora cactorum* attacks other parts of the plant including fruits causing leather rot at all stages of fruit development, from blossoming to maturity (Rose, 1924). Lesions on green fruit are a brown colour, whereas on ripe fruit infected areas appear bleached and are pink to purplish, extending into the flesh. Infected tissue is tough, not soft and fruit are characterized by a bitter taste.

Red stele root rot caused by *Phytophthora fragariae* C.J. Hickman is another major strawberry disease, occurring in areas with cool, moist soil conditions (Maas, 1984; Strand, 1994). The disease is characterized by rotting of young roots from the tips upward with a typical red discoloration of the central cylinder of roots. In certain cultivars, the discoloration may extend into the crown, whereas older roots do not exhibit these symptoms. Affected plants remain stunted and young leaves may turn bluish green, with older leaves having a red, orange or yellow colour.

Both *P. fragariae* and *P. cactorum* form resistant oospores that can survive in the soil for several years (Maas, 1984). In the presence of the host plant, oospores can then germinate and form sporangia, the reproductive structures of Phytophthora diseases. Sporangia release zoospores under moist conditions. Infection takes place when zoospores are splashed with soil particles onto flowers, fruit and root tips. The zoospore germinates and penetrates the appropriate plant tissue. Phytophthora pathogens may be introduced in new growing areas on infected plants and inoculum can be splash dispersed (Yang *et al.*, 1992). The pathogens thrive in poorly drained or over-irrigated soils, or during prolonged wet weather conditions. Both diseases can be managed by appropriate IPM programmes which require soil disinfestation, by timely application of fungicides, by annual planting of healthy propagation material, by good soil tillage for improving drainage, by using plastic mulch to reduce contact of fruit with soil, and by avoiding accumulation of excess irrigation and rain water (Table 32.1).

Anthracnose
Species of the fungal plant pathogen *Colletotrichum* [*Colletotrichum acutatum* J.H. Simmonds, *Colletotrichum fragariae* A.N. Brooks and *Colletotrichum gloeosporioides* (Penz.) Penz. & Sacc. in Penz.] are responsible for strawberry anthracnose (Gullino *et al.*, 1985; Smith and Black, 1990; Howard *et al.*, 1992; Denoyes and Baudry, 1995; Freeman and Katan, 1997). Infection of mother plants with anthracnose may result in collapse of the entire plant due to crown rot. In the nursery, lesions are formed on stolons that eventually girdle the runners, and unrooted daughter plants distal to the lesion wilt and die. Root necrosis caused by *C. acutatum* may also result in plant

stunting (Freeman and Katan, 1997). Infected transplants are capable of spreading the disease from the nursery to the field. Leaves, flowers, green and ripe fruit may be extremely susceptible to anthracnose, which can cause loss of all fruit in some cases (Howard *et al.*, 1992).

Splash dispersal of conidia is the primary source of inoculum (Madden *et al.*, 1992), with optimum conducive temperatures ranging from 15–30°C. Soil disinfestation (solarization, fumigants and steaming) is an important practice for controlling *Colletotrichum* as this pathogen is known to survive in the soil (Table 32.1) (Eastburn and Gubler, 1990). Disease is reduced when drip irrigation is implemented when establishing mother plants in nurseries and transplants in production fields. Heating of bare transplants at 49°C for 5 min is effective for controlling infected plants (Strand, 1994; Freeman *et al.*, 1997). Single-dip prochloraz treatments both at the nursery establishment and transplant stages can be adopted as routine practice, regardless of whether or not plants have visible anthracnose symptoms (Freeman *et al.*, 1997). Accurate identification of species is paramount due to resistance of *C. acutatum* to various fungicides as opposed to *C. gloeosporioides*. This can be determined by methods of molecular biology including the polymerase chain reaction and nuclear DNA probes (Freeman *et al.*, 1993; Freeman and Rodriguez, 1995).

Grey Mould
Grey mould or Botrytis rot caused by *B. cinerea* is a common disease of flowers and fruit causing serious crop damage under moist and rainy conditions, however, quiescent infections which are invisible, occur on leaves and fruit (Powelson, 1960). The fungus spreads mainly by splash dispersal and wind. The pathogen is controlled by carefully timed fungicide applications, drip irrigation, ventilation in plastic tunnels and strict sanitation measures (Table 32.1) (Cooley *et al.*, 1996).

Under Mediterranean conditions, appropriate ventilation of greenhouses, by opening windows early in the morning to reduce nocturnal humidity, is generally very effective in preventing fruit damage. No fungicides are used to control grey mould in the above-mentioned "green-labelled" crop production (Celli, 1987). Biocontrol of grey mould with antagonistic fungi has been evaluated by Sutton and Peng (1993), showing that antagonists can suppress *B. cinerea* primarily when applied to living green leaves. A novel and more efficient approach to apply inoculum of the antagonistic fungi has been tested in Canada, using honeybees (*Apis mellifera* L.). A dispenser with the biocontrol formulation is placed at the exit of the hive and foragers acquire the inoculum on their legs and bodies as they exit the hive, acting as vectors of the conidia and transporting them to flowers (Peng *et al.*, 1992; Sutton, 1994, 1995). A similar technique is under evaluation also in Italy using the antagonistic fungus *Trichoderma harzianum* Rifai (Gullino and Maccagnani, pers. com.).

Powdery Mildew
Powdery mildew caused by *Sphaerotheca macularis* (Wallr.:Fr.) Lind f. sp. *fragariae* Peries is a common foliar and fruit disease of both nursery, greenhouse and field grown strawberry plants (Peries, 1962; McNicol and Gooding, 1979). The fungus proliferates under humid conditions, however, is inhibited by rainy and wet conditions. The pathogen which first appears on the lower surface of leaves progresses to flowers and fruit causing serious crop losses if not treated. Control of powdery mildew can be achieved by timely use of fungicides, such as the protectant sulphur, which should be applied at the first sign of disease or by using resistant cultivars. Disease incidence may

also be reduced by controlling the pathogen in autumn thus avoiding winter and spring fruit infections. The removal of senescent leaves also helps reduce inoculum levels.

Verticillium Wilt
Verticillium wilt, caused by *Verticillium albo-atrum* Reinke & Berthier and *Verticillium dahliae* Kleb., is a common soilborne disease in many plant species besides strawberry (Howard and Albregts, 1982). It is not specific to strawberry but has a wide host range, therefore, prior knowledge of cropping history is an important factor for control. The Verticillium pathogens are found in semi-arid, well irrigated regions. The pathogen invades roots causing eventual wilting of the entire plant. Initially, outer leaves show marginal and interveinal browning and eventually collapse, whereas inner leaves remain green until plant death. *Verticillium* may be spread on infected plant material, and by water and wind dispersal. Soil disinfestation and soil solarization are effective for control of the disease (Table 32.1) (Wilhelm and Paulus, 1980). In addition, inoculum levels of *V. dahliae* may be reduced by planting a cover crop prior to strawberry cultivation (Strand, 1994).

Other Foliar Diseases
Angular leaf spot caused by *Xanthomonas fragariae* Kennedy & King, common leaf spot caused by *Ramularia brunnea* Peck (= *Ramularia tulasnei* Sacc.), Phomopsis leaf blight caused by *Phomopsis obscurans* (Ellis & Everh.) Sutton, and other minor foliar pathogens occur in strawberry cultivation under moist conditions. Most of these diseases can be reduced by avoiding splash dispersal of inoculum via overhead irrigation and by timely fungicide (such as copper hydroxide) regimes.

32.3.3. NEMATODES

Two groups of nematodes can attack strawberries in fruit production and nursery fields: foliar nematodes and root-knot nematodes as listed in Table 32.1 (Tacconi, pers. com.). Strand (1994) provides useful information on IPM of nematodes infesting strawberries.

Foliar nematodes live and reproduce on above-ground parts of host plants, usually in or on crowns and leaves. They infect leaf or stem tissue through stomata and live in the unopened leaves and buds of the crown. Feeding by foliar nematodes causes distorted growth and discoloration of leaves and stems. The number of fruit trusses can be reduced and each inflorescence produces only one or two flowers. Crown buds may be killed. Foliar nematodes can survive on cold stored plants before transplanting, and in this way they can infect fruit production fields. Allowing foliage to dry out between irrigations may help reduce foliar nematodes activity.

Root-knot nematodes must feed inside the roots of host plants to reproduce. Their feeding causes swellings in the roots, called galls or knots. These nematodes reduce plant vigour and fruit production, and plants may wilt and become chlorotic. Root-knot nematodes can cause severe damage in sandy soils where nurseries for plant production are situated. Infestation of fruit production fields may occur via infected transplants.

Annual planting of certified transplants and soil disinfestation usually limits the damage by nematodes. Additional practices that help reduce the threat of nematodes include field selection, sanitation and crop rotation, and they become crucial for fruit production where soil solarization is used, since it is not as effective as fumigation (Strand, 1994). Treating transplants destined for nursery plantings with hot water can prevent the introduction of nematodes, but this procedure is not recommended for fruit-production because this treatment reduces plant vigour.

32.3.4. VIRUS AND MYCOPLASMA DISEASES

Most virus and mycoplasma diseases are spread by insect vectors. Symptoms are usually accompanied by reduced plant vigour and yield. Often, more than a single virus is necessary for substantial yield loss to occur, which is dependent on virus strain and strawberry variety cultivated. Strawberry mottle, crinkle, mild yellow edge and vein banding viruses are all vector transmitted, by either the strawberry aphid, *C. fragaefolii*, and other *Chaetosiphon* species. Mottle and vein banding viruses are transmitted rapidly and remain infective for a few hours or a day, respectively. Crinkle and mild yellow edge viruses are persistent viruses and require latent periods within the vector before acquiring the infective stage. The mycoplasma diseases, lethal decline and green petal, are both persistent being transmitted within leafhoppers. Green petal disease is typified by flowers with green petals which develop into clusters of large, green achenes instead of regular fruits.

Viruses and mycoplasma may be transmitted via infected plants, therefore, using certified transplants is the primary means for preventing these diseases in fruit production fields. Certified disease-free plants are obtained by growing stock at 35 to 37°C (temperatures which prevent virus multiplication) for approximately 40 days (Strand, 1994). Thereafter, meristem tissue is removed from crowns and "clean" plants are regenerated on a nutrient medium. These plants are maintained in screenhouses to protect them from virus and mycoplasma vectors. In the nurseries, virus and mycoplasma infections can be reduced by strict sanitation measures and application of insecticides against their vectors. The use of insecticides to reduce the number of vectors is not needed for production fields on condition that disease-free plants are used in nurseries.

32.3.5. MINOR PESTS

The cyclamen mite, *Steneotarsonemus pallidus* Bks., can also cause damage to strawberries. The mature mites (0.25 mm long) are pinkish orange and not visible without a good hand-lens. Eggs are translucent and about half the size of adults. Adult females overwinter in the crowns and they are moved easily by man and animals, including insects. Feeding by mites causes stunting and distortion of newly emerging leaves; flowers wither and die; fruits are dwarfed and appear "seedy" because the achenes stand out from the surface. Management requires mite-free transplants. Various predatory mites and insects can feed on *S. pallidus*, but their populations often build up too slowly. Sulphur affects the mite and is widely used for control in the Mediterranean region; in northern Europe, good results have been obtained releasing the predatory mite *N. cucumeris*, also as a preventive measure (Table 32.1).

Lygus bugs (Hemiptera, Miridae) are considered major pests in some Californian areas, mainly for summer and autumn outdoor production, causing distortion of fruits. In field production, tractor-mounted vacuum devices are used to remove most adult lygus and fewer nymphs, but also many useful arthropods are removed, sometimes allowing secondary outbreaks of other pests (Strand, 1994). Selective insecticides and insecticidal soaps can be used for reducing populations on strawberries. Similarly, in Europe, with the advent of everbearer cultivars which flower continuously during the summer and the autumn, *Lygus rugulipennis* Poppius and other species can become significant pests (Cross *et al.*, 1994). Destroying weeds and cover crops is under discussion since beneficial species may also be affected and only young bugs are eliminated while adults can move immediately to fresh strawberry plants.

Some whiteflies, leafhoppers and froghoppers can feed on strawberries, sometimes reaching significant population levels. However, winter pruning, natural colonization of plants by wild parasitoids and predators, and the activity of polyphagous predators eventually released, generally keep these species under control.

Damage to ripening fruits by slugs (Mollusca) is widespread in many cultivation areas, despite routine use of methiocarb pellets. In open fields, methiocarb should not be used, as it is toxic to Carabidae, which may prey on weevils, whereas methaldeide is a safer alternative (Cross et al., 1994).

32.4. Perspectives

Strawberry production is one of the agroindustries most dependent on chemicals for pest and disease management. Research needs to be continued on replacement of unsustainable cultivation practices with environmental-friendly integrated pest/pathogen management IPM-based programmes. Of the chemicals most widely used in strawberries, methyl bromide (MBr) as a soil fumigant will be phased out soon, leaving a void to be filled for the control of soilborne pests and diseases. Foliar pesticide sprays are applied in excess in strawberry cultivation and need to be reduced as they have a detrimental affect on beneficial micro-organisms and insects.

Alternative control measures relying on biological control, soil solarization, reduced chemical dosages in combination with selected biocontrol agents, and cultural practices need to be developed urgently and then supported to reach a wide application. Biocontrol of soilborne and foliar pathogens using Trichoderma spp. is under-development in strawberries. These antagonistic fungi have beneficial effects other than pathogen control, principally through increasing soil fertility by the active breakdown of organic matter, a process seriously hindered by MBr and similar fumigants, which destroy natural microbial disease and pest suppression in soil and encourages rapid recolonization of detrimental pathogens and pests. Non-chemical strategies for controlling grey mould appear promising for reducing the use of fungicides prior to fruit harvesting, limiting the risk of residue contamination.

The use of mass-produced bumblebees, Bombus terrestris (L.), is becoming an established practice, mainly in the northern European protected strawberries, providing pollination for the earliest flowers and favouring the production of well-shaped fruits. Honeybees are frequently used in the Mediterranean region by placing the hives near open windows which allows the foragers access to tunnels. Advantages incurred by using pollinators result in enhanced biocontrol and IPM, since no toxic pesticides can be used, at least during flowering. In the future, opportunity for biocontrol of grey mould may be provided by utilizing bees as carriers of antagonistic fungi to flowers.

The release of mass-produced beneficial arthropods has allowed effective control of some key-pests. Nevertheless, problems still exist, particularly in some areas where alternatives to broad-spectrum pesticides are too expensive or do not appear completely reliable. Specific research is necessary in the Mediterranean region, particularly to control F. occidentalis and some Noctuidae, such as S. littoralis. In the future, natural control may provide new opportunities, particularly if new methods will be defined which enhance the multiplication of natural enemies in agroecosystems resulting in improved colonization of crops as soon as the pests' populations increase.

Acknowledgements

Neil Helyer (Fargro Ltd, Littlehampton, GB), Giovanna Tropea Garzia (University of Catania, Italy), Renzo Tacconi (Osservatorio Malattie Piante, Bologna, Italy), Giuseppe Manzaroli and Francesco Bravaccini (Biolab-Centrale Ortofrutticola, Cesena, Italy) are thanked for providing information on the IPM of strawberry pests in Europe. Marcello Sbrighi (CRPV, Cesena, Italy) is thanked for providing data on the world-wide growing surface.

References

Alauzet, C., Dargagnon, D. and Malausa, J.C. (1994) Bionomics of the polyphagous predator: *Orius laevigatus* (Het.: Anthocoridae), *Entomophaga* **39**(1), 33–40.

Benuzzi, M., Manzaroli, G. and Nicoli, G. (1992) Biological control in protected strawberry in northern Italy, *OEPP/EPPO Bulletin* **22**, 445–448.

Benuzzi, M. and Nicoli, G. (1990) *Phytoseiulus persimilis* Athias-Henriot release rate related to the infestation level of *Tetranychus urticae* Koch in protected strawberry, in *Atti Giornate Fitopatologiche 1990*, Clueb Ed., Bologna, pp. 383–391.

Benuzzi, M., Nicoli, G. and Pizzigatti, L. (1989) Effects of some fungicides on the red spider mite (*Tetranychus urticae* Koch) and its predatory mite (*Phytoseiulus persimilis* Athias-Henriot) on strawberry, *Informatore Agrario* **45**(13), 101–105.

Blackman, R.L. and Eastop, V.F. (1984) *Aphids on the World's Crops: An Identification Guide*, John Wiley & Sons, Chichester.

Cayrol, R.A. (1972) Famille des noctuides, in A.S. Balachowsky (ed.), *Traité d'Entomologie Appliquée à l'Agriculture*, Tome II, Vol. 2, Masson et Cie Editeurs, Paris, pp. 1255–1520.

Celli, G. (1987) The so called "organic strawberries", *Informatore Agrario* **43**(49), 66–68.

Celli, G., Benuzzi, M. and Nicoli, G. (1988) Development of an IPM strategy for protected strawberries in the northern Italy's Emilia-Romagna region: 1983–1987, in *Atti Giornate Fitopatologiche 1988*, Clueb Ed., Bologna, pp. 213–222.

Cooley, D.R., Wilcox, W.F., Kovach, J. and Schoelmann, S.G. (1996) Integrated pest management programs for strawberries in the northeastern United States, *Plant Disease* **80**, 228–237.

Cross, J.V., Berrie, A.M., Ryan, M. and Greenfield, A. (1994) Progress towards integrated plant protection in strawberry production in the UK, in *Brighton Crop Protection Conference – Pests and Diseases*, Vol. 2, BCPC, Farnham, pp. 725–730.

Denoyes, B. and Baudry, A. (1995) Species identification and pathogenicity of French *Colletotrichum* strains isolated from strawberry using morphological and cultural characteristics, *Phytopathology* **85**, 53–57.

Deseö Kovács, K.V. and Rovesti, L. (1992) *Microbial Control of Pests: Theory and Practice*, Edagricole, Bologna.

Doss, R.P., Shanks, C.H. Jr., Chamberlain, J.D. and Garth, J.K.L. (1987) Role of leaf hairs in resistance of a clone of beach strawberry, *Fragaria chiloensis*, to feeding by adult black wine weevil, *Otiorhynchus sulcatus* (Coleoptera Curculionidae), *Environmental Entomology* **16**(3), 764–768.

Eastburn, D.M. and Gubler, W.D. (1990) Strawberry anthracnose: Detection and survival of *Colletotrichum acutatum* in soil, *Plant Disease* **74**, 161–163.

Freeman, S. and Katan, T. (1997) Identification of *Colletotrichum* species responsible for anthracnose and root necrosis of strawberry in Israel, *Phytopathology* **87**, 516–521.

Freeman, S., Nizani, Y., Dotan, S., Even, S. and Sando, T. (1997) Control of *Colletotrichum acutatum* in strawberry under laboratory, greenhouse and field conditions, *Plant Disease* **81**, 749–752.

Freeman, S., Pham, P. and Rodriguez, R.J. (1993) Molecular genotyping of *Colletotrichum* species based on arbitrarily primed PCR, A+T-rich DNA, and nuclear DNA analyses, *Experimental Mycology* **17**, 309–322.

Freeman, S. and Rodriguez, R.J. (1995) Differentiation of *Colletotrichum* species responsible for anthracnose of strawberry by arbitrarily primed PCR, *Mycological Research* **99**, 901–905.

Galazzi, D. and Nicoli, G. (1996) Comparative study of strains of *Phytoseiulus persimilis* Athias-Henriot (Acarina Phytoseiidae). I. Development and adult life, *Bolletino dell'Istituto di Entomologia 'Guido Grandi' della Universita degli Studi di Bologna* **50**, 215–231.

Gremo, F., Bogetti, C. and Scarpelli, F. (1997) Thrips harmful to strawberries, *Informatore Agrario* **53**(17), 85–89.

Gullino, M.L., Romano, M.L. and Garibaldi, A. (1985) Identification and response to fungicides of *Colletotrichum gloeosporioides*, incitant of strawberry black rot in Italy, *Plant Disease* **69**, 608–609.

Howard, C.M. and Albregts, E.E. (1982) Outbreaks of Verticillium wilt of strawberries in central Florida, *Plant Disease* **66**, 856–857.

Howard, C.M., Maas, J.L., Chandler, C.K. and Albregts, E.E. (1992) Anthracnose of strawberry caused by the *Colletotrichum* complex in Florida, *Plant Disease* **76**, 976–981.

Kakouli, T., Schirocki, A. and Hague, N.G.M. (1994) Factors affecting the control of *Otiorhynchus sulcatus* with entomopathogenic nematodes in strawberries grown on raised beds, in *Brighton Crop Protection Conference – Pests and Diseases*, Vol. 2, BCPC, Farnham, pp. 945–946.

Maas, J.L. (1984) Anthracnose fruit rots (black spot), in J.L. Maas (ed.), *Compendium of Strawberry Diseases*, APS Press, St Paul, Minn., pp. 57–60.

Madden, L.V., Wilson, L.L., Yang, X. and Ellis, M.A. (1992) Splash dispersal of *Colletotrichum acutatum* and *Phytophthora cactorum* by short-duration simulated rains, *Plant Pathology* **41**, 427–436.

Malais, M. and Ravensberg, W.J. (1992) *Knowing and Recognizing the Biology of Glasshouse Pests and their Natural Enemies*, Koppert B.V. Publ., Berkel en Rodenrijs.

McNicol, R.J. and Gooding, H.J. (1979) Assessment of strawberry clones and seedlings for resistance to *Sphaerotheca macularis* (Wall ex Frier), *Horticultural Research* **19**, 35–41.

Peng, G., Sutton, J.C. and Kevan, P.G. (1992) Effectiveness of honey bees for applying the biocontrol agent *Gliocladium roseum* to strawberry flowers to suppress *Botrytis cinerea*, *Canadian J. of Plant Pathology* **14**, 117–129.

Peries, O.S. (1962) Studies on strawberry mildew caused by *Sphaerotheca macularis* (Wallr. ex Frier) Jaczewski. II. Host-parasite relationships on foliage of strawberry varieties, *Annals of Applied Biology* **50**, 225–233.

Powelson, R.L. (1960) The initiation of strawberry fruit rot caused by *Botrytis cinerea*, *Phytopathology* **50**, 491–494.

Riudavets, J. (1995) Predators of *Frankliniella occidentalis* (Perg.) and *Thrips tabaci* Lind.: A review, *Wageningen Agricultural University Papers* **95-1**, 43–87.

Riudavets, J. and Castañé, C. (1994) Abundance and host plant preferences for oviposition of *Orius* spp. (Heteroptera: Anthocoridae) along the Mediterranean coast of Spain, *IOBC/WPRS Bulletin* **17**(5), 230–236.

Rose, D.H. (1924) Leather rot of strawberries, *J. Agricultural Research* **28**, 357–375.

Sannino, L., Balbiani, A., Lombardi, P. and Avigliano, M. (1996) *Spodoptera littoralis*: A harmful pest of the herbaceous crops, *Informatore Agrario* **53**(18), 76–79.

Scopes, N.E.A. (1985) Implications for biological control on intensively grown outdoor crops. Strawberries, in N.W. Hussey and N.E.A. Scopes (eds.), *Biological Pest Control. The Glasshouse Experience*, Blandford Press, Poole, pp. 190–191.

Servadei, A., Zangheri, S. and Masutti, L. (1972) *General and Applied Entomology*, Cedam Publ., Padova.

Smith, B.J. and Black, L.L. (1990) Morphological, cultural, and pathogenic variation among *Colletotrichum* species isolated from strawberry, *Plant Disease* **74**, 69–76.

Strand, L.L. (1994) *Integrated Pest Management for Strawberries*, University of California, Div. Agric. Nat. Res. Publ. 3351, Oakland, Cal.

Sutton, J.C. (1994) Biological control of strawberry diseases, *Advances in Strawberry Research* **13**, 1–12.

Sutton, J.C. (1995) Evaluation of microorganisms for biocontrol: *Botrytis cinerea* on strawberry, a case study, in J.H. Andrews and I.C. Tommerup (eds.), *Sustainability in Plant Pathology*, Advances in Plant Pathology, Vol. 11, Academic Press, London, pp. 173–190.

Sutton, J.C. and Peng, J. (1993) Biocontrol of *Botrytis cinerea* in strawberry leaves, *Phytopathology* **83**, 615–621.

Tommasini, M.G. and Nicoli, G. (1996) Evaluation of *Orius* spp. as biological control agents of thrips pests. Further experiments on the existence of diapause in *Orius laevigatus*, *IOBC/WPRS Bulletin* **19**(1), 183–186.

Tremblay, E. (1986) *Applied Entomology*, Vol. 2/2, Liguori Ed., Napoli.

Tropea Garzia, G. (1997) Recent advances in the integrated control of noctuids harmful to protected crops, *Notiziario sulla Protezione delle Piante (nuova serie)* **7**, 273–283.

Vacante, V. and Nucifora, A. (1987) Possibilities and perspectives of the biological and integrated control of the two spotted spider mite in the Mediterranean greenhouse crops, *IOBC/WPRS Bulletin* **10**(2), 170–173.

van den Meiracker, R.A.F. (1994) Induction and termination of diapause in *Orius* predatory bugs, *Entomologia Experimentalis et Applicata* **73**, 127–137.

Wilhelm, S. and Paulus, A.O. (1980) How soil fumigation benefits the California strawberry industry, *Plant Disease* **64**, 264–270.

Wright, W.R., Beraha, L. and Smith, M.A. (1966) Leather rot on California strawberries, *Plant Disease Reporter* **50**, 283–287.
Yang, X., Madden, L.V., Reichard, D.L., Wilson, L.L. and Ellis, M.A. (1992) Splash dispersal of *Colletotrichum acutatum* and *Phytophthora cactorum* from strawberry fruit by single drop impactations, *Phytopathology* **82**, 332–340.

CHAPTER 33

SWEET PEPPERS

Aleid J. Dik, Elzbieta Ceglarska and Zoltan Ilovai

33.1. Introduction

Pepper (*Capsicum annuum* L.) is an important crop world-wide. The area planted with this crop is around 1,200,000 ha, about half of which is located in Asia (FAO, 1995). In North America and Europe, pepper is grown on approximately 150,000 ha, only part of which is under protected cultivation. For example, in Israel 30 ha are grown in plastic tunnels, whereas 1200 ha are grown outside. In Hungary, about 2000 ha of greenhouse are used for pepper production, while 8000 ha are planted with pepper outside. In Spain, about 9000 hectares of pepper are grown in protected cultivation compared to around 1000 ha in The Netherlands (García *et al.*, 1997). Protected cultivation of pepper takes place in plastic tunnels, greenhouses and screenhouses. The yields per ha vary enormously between countries. In 1995, yields ranged from 1000 kg/ha in several countries to over 200,000 kg/ha for glasshouse-grown pepper in The Netherlands (FAO, 1995).

33.2. Main Pest and Disease Problems

The main crop protection problems for peppers in greenhouses are caused by arthropods, viruses and fungal diseases. Their occurrence and importance depend on the mode of cultivation. In the case of traditional soil culture, soilborne pathogens and pests are predominant: root-knot nematodes, Fusarium and Verticillium wilt, grey mould and viruses. In soilless cultivation, foliar pathogens and arthropods are more common. Insects pose a major problem for pepper. Spider mites, aphids, thrips and caterpillars can all cause considerable damage. Insects may also transfer viruses, which can be very destructive. Healthy planting material is essential for both growing methods.

The occurrence of pests and diseases may be dependent on the location of the crop. For example, until recently powdery mildew was mainly restricted to warmer countries. However, a shift has occurred in recent years by which powdery mildew has become a major problem in western Europe and Canada. Integrated crop protection, therefore, has to develop continuously, taking into account new problems and developing new solutions.

R. Albajes et al. (eds.), Integrated Pest and Disease Management in Greenhouse Crops, 473-485.
© 1999 *Kluwer Academic Publishers. Printed in the Netherlands.*

33.3. Current Status of Integrated Control

33.3.1. ARTHROPODS

Aphids

Aphids form the most numerous pest group. *Myzus persicae* (Sulzer), *Aphis nasturtii* Kaltenbach, *Aphis gossypii* Glover, *Aulacorthum solani* (Kaltenbach) and *Macrosiphum euphorbiae* (Thomas) cause direct damage to the crop, but their ability to transfer virus diseases (CMV, PVY, AMV, etc.) makes these insects significantly more dangerous. Aphids can spread with seedlings or by flying. Aphids can be controlled by systemic insecticides like oxamyl or dimethoate. Specific aphicides like pirimicarb are no longer effective because of resistance, though imidacloprid and plant extracts may offer temporary respite. Pyrethroids limit the application of biological control agents. Light summer oils/paraffin oil or fatty acid copper salt with oil can be utilized as more selective treatments (Kajati *et al.*, 1989; Ilovai *et al.*, in press).

The parasitoids *Aphidius matricariae* Haliday, *Aphidius colemani* Viereck, *Aphidius ervi* Haliday and *Aphelinus abdominalis* (Dalman) are used for biocontrol of aphids (van Steenis, 1995; Chapter 16). These beneficials are usually combined with the gall-midge *Aphidoletes aphidimyza* (Rondani). In warm summers light oils, soaps and plant extracts help parasitoids and predators to reduce pests (Carnero Hernández *et al.*, 1997b). The use of insect nets prevents or delays an invasion of aphids.

Whiteflies

The whitefly *Trialeurodes vaporariorum* (Westwood), which occurs world-wide, also affects pepper. Recently *Bemisia tabaci* (Gennadius) and *Bemisia argentifolii* Bellows & Perring have also become more common on pepper. In the Mediterranean area, whiteflies can overwinter in open field on a number of wild plants, which poses a permanent threat of infestation. Pepper cultivars possess different susceptibilities to greenhouse whitefly: reproduction of the pest is more successful on Hungarian hybrids than on the "bell"-type hybrids (van Vianen *et al.*, 1987).

For a long time, pesticides were the dominant whitefly control technique. Insect growth regulators (IGR)-type pesticides solve the problem only for a short time. The ability of whiteflies to develop resistance to insecticides rapidly has stimulated research into biological control. *Encarsia formosa* Gahan, a whitefly parasitoid, has been widely applied for decades (see Chapter 14). *Eretmocerus mundus* Mercet and *Eretmocerus eremicus* Rose & Zolnerowich (= *Eretmocerus californicus* Howard) have proved to have good searching ability against *Bemisia* species (Drost *et al.*, 1996).

Thrips

Thrips are also serious pests in pepper. *Thrips tabaci* Lindeman usually causes less problems than *Frankliniella occidentalis* (Pergande), which is also a vector of tomato spotted wilt virus (TSWV). *Thrips palmi* Karny has recently been considered a new pest in Europe (Loomans and Vierbergen, 1997). In thrips control the most important aim is to prevent sucking of flowers and fruits. Prevention of western flower thrips (WFT) damage at the start of the crop can be achieved by direct eradication with dichlorvos

(DDVP) for 5–7 days before planting. Contact insecticides are not very effective as they can only control certain stages; they also harm beneficials. Systemic insecticides like carbofuran, imidacloprid or oxamyl, applied as plant spray or soil treatment, can only be used in case of mass occurrence of pest, when the beneficials have not been able to keep the pest under the economic damage threshold. Mode of application and separation of treatments in space and time will help bring the pest population back to a controllable level.

Biocontrol agents limiting thrips are anthocorids, mirids, predatory mites and entomopathogenic fungi [*Beauveria bassiana* (Balsamo) Vuillemin, *Verticillium lecanii* (A. Zimmerm.) Viégas] (Szabo and Ceglarska-Hodi, 1992; Loomans *et al.*, 1995; Parella and Murphy, 1996).

Several of the predators occur in the natural fauna of the Mediterranean area, while in northern Europe artificial introduction is required. In the case of the predatory mite *Amblyseius degenerans* Berlese, the application of *Ricinus communis* L. as banker plant makes it possible to maintain an open rearing system, from which predators can spread into the crop (Ramakers and Voet, 1996).

Phytophagous Mites

Phytophagous mites are now not as damaging to pepper as they were before. The predatory mite *Phytoseilus persimilis* Athias-Henriot has been widely applied against spider mites (*Tetranychus* spp.) for many years. Its effectiveness made it the most widely used biocontrol agent. In central and eastern Europe, however, mite control is still based on chemicals, which has led to the development of resistant strains (Cs. Budai, pers. com.). The demand for alternative control methods has brought natural pesticides (e.g. sulphur or calcium-polysulphide) and biocontrol agents into the foreground.

The tarsonemid, or broad, mite *Polyphagotarsonemus latus* (Banks) is usually introduced into the crop from nurseries. It significantly limits the success of biocontrol programmes, because its control with sulphur-dusting or spraying decimates the populations of beneficials. Prevention of the introduction of broad mite at the beginning of the vegetative period is crucial.

Lepidopterous Pests

In many pepper-growing regions caterpillars cause considerable damage. Among species reported on pepper, *Spodoptera exigua* (Hübner), *Chrysodeixis chalcites* (Esper), *Autographa gamma* (L.), *Laconobia oleracea* (L.), *Mamestra brassicae* (L.), *Helicoverpa* (= *Heliothis*) *armigera* (Hübner) and *Agrotis segetum* (Denis & Schiffermüller) are predominant. In some areas *Ostrinia nubilalis* (Hübner) causes a special problem. The entrance of Lepidoptera into the greenhouses can be successfully limited by using insect-nets. The fruit-consuming *Helicoverpa* can be controlled by well-timed treatments with *Bacillus thuringiensis* Berliner, based on monitoring with sex-pheromone traps.

Other Pests

In pepper grown under plastic foil, tunnels or in screencages *Arion hortensis* Férussac,

Arion subfuscus (Draparnaud), *Deroceras agreste* (L.) and *Deroceras reticulatum* (Müller) snails damage the crop occasionally. Metaldehyde-baited or environmentally safe beer-baited traps protect the crop effectively. In soil cultivation, especially when the substrate is rich in organic material, the mole cricket *Gryllotalpa gryllotalpa* (L.) may damage the root system. The pest can be controlled by using fenitrotion (Galition) baits.

In order to control root-knot nematodes, soil disinfestation is necessary although pepper can tolerate slight infection. On sandy soils, incorporation of high amounts of organic manure and tolerant varieties of pepper reduce nematodes. Soil solarization in the summer provides a positive effect even in central Europe (Budai, 1994). Application of a biocontrol product based on *Arthrobotrys oligospora* Fresen. fungus for soil treatment in nursery prevents the infection of plants and introduction of nematodes into the greenhouse (Dormanns-Simon and Budai, in press).

33.3.2. VIRUSES

Several viruses cause diseases in pepper (Table 33.1). The main problems are created by viruses with a wide range of host-plants and viruses transmitted in more than one way. Old pepper cultivars are susceptible to a range of viruses, including tobacco mosaic virus (TMV), but intensive breeding programmes have yielded cultivars that are resistant to TMV. In the 70s and 80s, cucumber mosaic virus (CMV) ravaged pepper. Then new pathotypes appeared (P1.2 and P1.2.3 strains). In parallel, tobamovirus problems occurred [dulcamara yellow fleck virus (DYFV), pepper mild mottle virus (PMMV)], due to the susceptibility of varieties possessing the L1 gene. New cultivars armed with the L3 gene are resistant to DYFV, but PMMV strains overwhelm the L3 gene and make the introduction of the L4 gene necessary (Salamon, in press). PMMV is only transmitted mechanically, not by insects. Breeding for resistance is an important approach to integrated control of PMMV. Mechanical transmission can be reduced by the use of skimmed milk by workers (Stijger, 1995). Skimmed milk contains proteins which encapsulate the virus particles and so prevent transmission from plant to plant. This method is common in The Netherlands from the start of the crop to the first harvest, when the danger of PMMV is at its height. The workers dip their hands and tools in milk during all activities in the pepper crop.

CMV is transmitted by aphids and to some extent mechanically. It is mainly a problem in the fall when aphids start forming their wings and transmission takes place more often. TSWV is transmitted only by thrips, not mechanically. The virus caused serious problems in Hungarian pepper. Strains isolated from diseased plants showed a serological relationship to strains in western Europe (Adam and Peters, 1996). Related to TSWV is the impatiens necrotic spot virus, also from the Tospo group and only transmitted by insects.

New, still unidentified viruses occur. Recently, symptoms similar to pepper yellow vein mosaic (suspected to be caused by PYVMV) and a yellow ring and line pattern disease, caused by tomato bushy stunt virus (TBSV) or tobacco rattle virus (TRV) or both of these together, have been described (Salamon, in press).

TABLE 33.1. Main viruses affecting greenhouse pepper (Kiss, in press)

Transmission	Virus name/diseases	Genera/group	Acronym	Disease symptoms[1]	Importance[2]
By seeds/soil/ mechanically	Tobacco mosaic virus	Tobamovirus	TMV	Gm, N, D	+++
	Tomato mosaic virus	Tobamovirus	ToMV	Gm, Ym, N, D	+++
	Tobacco mild green mosaic virus	Tobamovirus	TMGMV	Gm, N	++
	Dulcamara yellow fleck virus	Tobamovirus	DYFV	Ym, N, D	++
	Pepper mild mottle virus	Tobamovirus	PMMV	Gm, N, D	*
	Tomato bushy stunt virus	Tombusvirus	TBSV	D,Yrp	?
By aphids/ mechanically	Cucumber mosaic virus	Cucumovirus	CMV	Gm, R, D, Rs	+++
	Potato virus Y	Potyvirus	PVY	Ym, Vm, N	+++
	Alfalfa mosaic virus	Alfamovirus	AMV	Ym, Yp	++
	Tomato aspermy virus	Cucumovirus	TAV	Ym, Yp, D	+
	Broad bean wilt virus	Fabavirus	BBWV	Gm, Rs	+
By thrips/ mechanically	Tomato spotted wilt virus	Tospovirus	TSWV	Rs, N, D	*
By whiteflies	Pepper yellow/ Golden mosaic diseases	Geminivirus ?	?	Y	?
By soil fungi	Pepper yellow vein mosaic diseases	Varicosavirus ?	PYVMV	Vm	?
By nematodes/ mechanically	Tobacco rattle virus	Tobravirus	TRV	Yrp, N	?
Mechanically	potato virus X	Potexvirus	PVX	Gm, Ym, N	+

[1]Gm = green mosaic, Ym = yellow mosaic, Vm = vein mosaic (vein yellowing/chlorosis/banding), Rs = ring spotting (chlorotic/necrotic), Yp = yellow irregular pattern, Yrp = yellow ring/line pattern, Y = yellowing, D = deformation (leaf/fruit), R = rosetting, N = necrosis (leaf/stem/fruit)
[2]+++: widespread for a long time, still widespread now; ++: widespread for a long time, less so now; +: appeared long ago, under repression; *: appeared recently, still spreading; ?: symptoms of the disease present, identification of the virus in progress

In IPM systems, breeding for virus resistance and strong prophylaxy are crucial. Emphasis should be put on production of healthy planting material and prevention of virus transmission. Growing plants in screencages or in plastic tunnels, with screens on the openings of the tunnel, prevents aphids and other insects from entering. The practice has proved very successful and in certain cases screening has tripled yields. The ability to reduce the transmission of viruses by insects gives protected cultivation a major advantage over outside cropping systems. Integrated control can be used. It is important to remove diseased plants and weeds, so that no source of virus is present. Then, thrips and aphids will not have any great impact on the incidence of virus infection, and biological control of insects can be practised without major problems. In general, viruses do not persist in empty greenhouses except for Tobamo viruses. In the case of Tobamo viruses, therefore, it is very useful to use milk as described above. In order to keep aphid populations at a low level and to inhibit a non-persistent (stylet-born) virus transfer, light summer oils can be used. It has been shown that the oils for plant spraying are fully compatible with beneficials (Ilovai *et al.*, in press).

33.3.3. FUNGAL DISEASES

Plant diseases are mainly caused by the following pathogens: (i) soilborne diseases: damping-off (*Pythium* spp., *Rhizoctonia solani* Kühn, *Fusarium* spp., *Alternaria* spp.), white mould [*Sclerotinia sclerotiorum* (Lib.) de Bary, *Sclerotinia minor* Jagger], Fusarium wilt [*Fusarium solani* (Mart.) Sacc., *Fusarium oxysporum* Schlechtend.:Fr.], Verticillium wilt (*Verticillium dahliae* Kleb., *Verticillium albo-atrum* Reinke & Berthier), basal rots (*Pythium* spp., *Phytophthora capsici* Leonian); and (ii) foliar diseases: powdery mildew [*Leveillula taurica* (Lév.) G. Arnaud], leaf spot [*Phaeoramularia capsicicola* (Vassiljevsky) Deighton (= *Cladosporium capsici* Vassiljevsky)], grey mould (*Botrytis cinerea* Pers.:Fr.).

Powdery Mildew
Powdery mildew, caused by *L. taurica*, has become the most important fungal disease in pepper. Powdery mildew in pepper is controlled by fungicides and sulphur. In western Europe and the North American continent, powdery mildew is a recent problem so not many fungicides are registered for use in this crop. In The Netherlands only two fungicides belonging to the same chemical group can be used, so their control activity is not always adequate. In other countries more fungicides are registered, but only frequent applications, starting very early in the season, give some control over the disease. Sulphur can be applied in different ways. The use of sulphur burners was slightly more effective than sulphur dusting (M.H. Zuijderwijk and A.J. Dik, pers. com.). However, use needs to be optimized. Pepper growers in The Netherlands have switched in recent years to intensive use of sulphur, either applied with burners (hot plates) or with a sulphur cannon. Experiments indicate that sulphur burning is only effective when applied at least three nights per week (Kerssies, unpublished results). However, at that frequency sulphur may have an adverse effect on beneficial insects and so other control methods need to be developed.

   An alternative approach to powdery mildew control is the use of plant extracts. Extracts from *Reynoutria sachalinensis* (F. Schmidt) Nakai (commercial product Milsana, BASF, Germany) and other extracts (commercial product Vital, usually applied in combination with Algan, Europlant, Appelscha, The Netherlands) have given excellent control in a comparative experiment on pepper in The Netherlands when applied weekly in a concentration of 1% and 0.5% (Dik, unpublished results). In Israel extracts from Neem (Neemgard, WR Grace) have also been found to control powdery mildew effectively. Moderate efficacy of potassium bicarbonate was reported (Mor *et al.*, 1997). Biological control of pepper powdery mildew is still a subject of research. In cucumber, *Sporothrix flocculosa* Traquair, Shaw & Jarvis gave much better control than *V. lecanii* Traquair, Shaw & Jarvis and *Ampelomyces quisqualis* Cesati:Schltdl. (Dik *et al.*, 1998), but *S. flocculosa* has not yet been tested on pepper powdery mildew. *Ampelomyces quisqualis* was found to control pepper powdery mildew in Israel, France and Hungary (Diop-Bruckler and Molot, 1987; Sztejnberg and Elad, unpublished results; L. Vajna and F.E. Kiss, pers. com.). Integration of the biocontrol agent *S. flocculosa* and plant extracts that induce resistance has been shown to give very good control in cucumber (Dik, unpublished results), and in the future this approach may also prove useful in pepper.

Fusarium Wilt
Fusarium spp., in particular *F. solani*, can also severely affect pepper. Suppression of symptoms of *F. solani* on fruits is often achieved by harvesting at the green stage instead of the red, but this does not control the fungus itself. The control of this disease is mainly a matter of preventing humid conditions in the greenhouse. The biological agent *Streptomyces griseoviridis* Anderson, Ehrlich, Sun & Burkholder has been found (Lahdenpera *et al.*, 1990) and commercialized by Kemira. It is registered as the product Mycostop in several countries.

Phytophthora Crown and Root Rot
The disease occurs mainly in west and south European countries. The pathogen endangers pepper grown in dense soil with a high level of soil water, so maintenance of good soil condition and drenching are crucial. In recirculating rock-wool culture zoospores of *P. capsici* spread very quickly and are able to destroy the crop in about two weeks. Soil disinfestation or use of specific chemicals are the most commonly adopted methods to control the disease. Of bioagents that could be used against the infection, *Trichoderma harzianum* Rifai, *Bacillus subtilis* (Ehrenberg) Cohn and *S. griseoviridis* may provide good protection in the soil as well as in rock-wool (Park and Kim, 1989; Nemec *et al.*, 1996). As reported by Stanghelini *et al.* (1996), application of non-ionic surfactants in recirculating systems led to the elimination of zoospores and fully controlled the disease. In case of emergency, fungicides traditionally used against phytophthora blight can be used for soil treatment in rhizosphere. Numerous data have been collected on the genetic mechanism of *Capsicum* cultivars for recognizing the pathogen and initiating the defence response by accumulation of phytoalexins (Egea *et al.*, 1996). This feature plays an important role in breeding resistant varieties.

Damping-off
Spreading of the disease largely depends on how seedlings are grown. If they are sown in the soil "carpet"-like, then the mildest infection can cause considerable damage because the disease can spread rapidly out from the source. If they are sown in peat cubes (this is more common), then the spread of the disease is limited. Aeration of the greenhouse, ensuring optimal growth conditions and hygiene, is vital. Application of biocontrol products (Mycostop) can control the disease, but to be effective the biocontrol agent must be in the rhizosphere before the pathogen attacks the plant (Dormanns-Simon *et al.*, in press).

Protected pepper is grown in many countries mainly in soil, not in artificial growing substrates. There is no practical possibility of crop rotation, which means that typically sweet pepper is grown in the same soil year after year. Thus, soilborne pathogens – especially those that can maintain themselves in the soil in the form of sclerotia from which the primary infection originates (*Sclerotinia* spp., *B. cinerea*) – accumulate in the soil and in the end seriously endanger the crop. This problem is especially acute under plastic foil tunnels with very high air humidity and dense plant stand. Soil disinfestation could help, but this method kills the soil microflora indiscriminately and is very harmful to the environment. Prevention and biological control are more desirable methods.

White Mould (Sclerotinia *spp.*)
White mould is one of the most difficult diseases to control chemically because only a few soil disinfectants are able to kill the sclerotia in the soil. The introduction of biological control will hopefully improve the situation in the near future. In Hungary the registration process of a biocontrol product called Micon, based on the hyperparasitic fungus *Coniothyrium minitans* Campbell, has started. This hyperparasite feeds specifically on *Sclerotinia* pathogens, is comparatively easy to produce, adapts well to different soil types, and can maintain and reproduce itself on the sclerotia of the pathogen. Its biological control efficacy can reach 90–100% (L. Vajna, pers. com.).

Botrytis cinerea
Botrytis cinerea is another pathogen for which biological control has been developed. The first commercial product is Trichodex WP (Makhteshim, Israel), based on *T. harzianum* T39. This product has already been registered in many countries and registration is pending in others. *Botrytis cinerea* has become resistant to many chemical fungicides and the danger of resistance is also threatening new fungicides. Therefore, the availability of a biocontrol agent with a completely different mode of action is very useful when employed within an integrated control schedule. Trichodex as a stand-alone treatment controls *B. cinerea* well under a range of climatic conditions in various crops, and is also effective in pepper (Y. Elad, pers. com.). When disease pressure is very high or when conditions are very favourable to *B. cinerea*, alternation with a fungicide improves control (Shtienberg and Elad, 1997).

33.4. Integrated Pest Management – Problems and Perspectives

In Europe, IPM is well-developed in several greenhouse crops. In 1988, van Lenteren and Woets reported the situation of IPM in glasshouses. Several natural enemies of thrips and aphids are recommended for pepper IPM systems. The main protection line for peppers was then still based on chemicals, but the demand for reduction of pesticide use and withdrawal of certain active ingredients had already arisen.

Different countries have prepared regulations concerning pesticide usage on protected pepper. In Hungary at the beginning of the 80s pepper, as the most important protected crop, was considered a candidate for the development of biological control. One continuous problem was the appropriate control of mites. Due to application of acaricides, which limited the efficacy of *Encarsia*, the start of IPM was postponed to the beginning of the 90s. Programmes aimed at decreasing pesticide use stimulated research and increased registration and introduction of biological control agents and alternative means (Ilovai *et al.*, 1996).

So far, IPM programmes, for example the Dutch programme "CAPPA", have mainly been developed for insect control (Ramakers and van der Maas, 1996). The decisions made by the growers are strongly influenced by the predominant pest or disease problems. For example, the major problem posed by powdery mildew has led to the intensive use of sulphur, more or less regardless of the detrimental side-effects it may have on beneficial insects. Also, in the case of thrips, a correctional spray with insecticide is often necessary, but this may affect beneficial insects too. Integrated control of insects would therefore be greatly helped by the development of selective pesticides for thrips.

In Mediterranean countries the natural potential of beneficial insects is effectively used in protected cultivation by conservation methods. In Israel an IPM strategy is based on three compounds: mechanical barriers (screens minimizing pest invasion), biological control and complementary control. Biological control is strengthened by determination of economic thresholds for pests, direct counting or trapping methods for determining pest population densities, use of indicator plants and monitoring of indigenous beneficials. In order to save beneficial insects, complementary control of pests is under development, on the basis of environmentally safe oils, insecticidal soaps, detergents, etc. (Berlinger *et al.*, 1996).

Joint Spanish and Hungarian research has proved that natural pesticides [Vectafid A (a.i.: light summer oil), Bio-sect (a.i.: insecticidal soap), Tiosol (a.i.: calcium-polysulphide)] allow the activity of parasitoids and predators (Carnero Hernández *et al.*, 1997a,b). In Crete the control system of *F. occidentalis* is based on application of colour-traps for monitoring and on introduction of *Orius* predatory bugs combined with oxamyl drenching (Michelakis and Amri, 1997).

Hungarian IPM practice on sweet peppers is based on using beneficial arthropods and microbial biocontrol agents against pests. For controlling soilborne diseases, an antagonistic *Trichoderma* offers a good solution. To prevent virus infection, virus transmission-inhibiting materials (e.g. paraffin oil, light summer oils, soaps) are used. This system has led to a major achievement: the introduction of a Trade Mark Label for pepper as a product of Hungarian origin (Ilovai, 1997).

In Polish pepper-growing, biological control agents are widely used. A range of control agents is available against thrips [*Neoseiulus* (= *Amblyseius*) *cucumeris* (Oudemans), *Orius* spp.], mites (*P. persimilis*, *Therodiplosis persicae* Kieffer), aphids [*A. colemani*, *A. aphidimyza*, *Hippodamia convergens* (Guérin-Méneville)], leaf-miners [*Dacnusa sibirica* Telenga, *Diglyphus isaea* (Walker)] and whiteflies (*E. formosa*). In emergency, selective pesticides are used in the most harmless possible way (Kania, 1996).

Table 33.2 presents the most important pests and diseases of pepper, the biological agents and alternative or supplementary means of control. In integrated systems, viruses are well controlled by sanitation, resistant cultivars and the use of milk as described above. Biological control of insects can be used as long as virus-infected plants are removed. Integrated control of diseases is not yet as well developed as integrated

TABLE 33.2. Integrated Pest Management in sweet peppers

Pest/disease	Control agents/treatments	Supplementary means
Frankliniella occidentalis	*Neoseiulus* (= *Amblyseius*) *cucumeris* *Amblyseius degenerans* *Orius laevigatus* *Orius insidiosus*	Insect screens Tolerant varieties *Beauveria bassiana*
Aphis gossypi *Aulacorthum solani* *Macrosiphum euphorbiae* *Myzus persicae*	*Aphidius colemani* *Aphidius ervi* *Aphelinus abdominalis* *Aphidoletes aphidimyza* *Hippodamia convergens* *Verticillium lecanii*	Insect screens Light summer oils Selective aphicides
Tetranychus urticae	*Phytoseiulus persimilis* *Neoseiulus* (= *Amblyseius*) *californicus*	Selective acaricides Sulphur
Tarsonemideae mites	Ca-polysulphide	Sulphur
Trialeurodes vaporariorum *Bemisia tabaci*	*Encarsia formosa* *Eretmocerus mundus* *Eretomocerus californicus*	Insect screens Insecticidal soaps Paraffin oil Biorational insecticides (IGR)
Lepidopterous caterpillars	*Bacillus thuringiensis*	Insect screens Pheromone traps Baited traps
Viruses	Treatments preventing transmission (oils, skimmed milk)	Resistant varieties and hybrids Insect screens against vectors
Leveillula taurica	*Ampelomyces quisqualis* Paraffin oil	Surfactants Selective, systemic fungicides
Botrytis cinerea	*Trichoderma spp.*	Selective fungicides
Fusarium spp. *Rhizoctonia spp.* *Pythium spp.* *Phytophthora capsici*	*Streptomyces griseoviridis*	Resistant varieties and hybrids Healthy sowing and planting material Soil steaming
Sclerotinia spp.	*Coniothyrium minitans*	Soil solarization Soil steaming
Meloidogyne spp.	*Arthrobotrys oligospora*	Tolerant varieties Soil solarization

control of insects. Prevention of humid conditions controls most diseases, except powdery mildew. Biological control has been developed, but only for a limited number of pathogens. An Israeli system called GREENMAN involves an antagonistic fungus *T. harzianum* T39 for the control of grey mould (Elad and Shtienberg, 1997).

Hungarian experience has also proved the suitability of *Trichoderma* species for control of several soilborne diseases, including damping-off (Ceglarska-Hodi *et al.*, 1998). Hopefully, in the future, a larger number of biocontrol agents will be available for disease control. These biocontrol products need registration and are generally treated in the same way as chemical products in the registration process, which delays their availability. Pepper is too small a crop to justify the commercial development of a biofungicide which needs registration, and only those biocontrol agents that are also effective in other crops will become available to growers. As soon as biocontrol of powdery mildew becomes possible, it will be much easier to grow pepper with a minimum input of chemical pesticides.

Generally, the current status of integrated pest management in pepper depends on the market cultural level and the technological crop level. In west European countries, research is more advanced and concentrates on serving the interests of the consumption triangle: crop-farmer-consumer. In central and eastern Europe, where producers are more interested in short-term economic gains, IPM systems are introduced mostly in farms with good technological facilities. In these countries the mentality of consumers and farmers is now changing whereas in western Europe the demand for healthy food is already well established and, besides, is supported through a variety of mechanisms.

Acknowledgements

The cooperation of Drs Cs. Budai, E. Dormanns-Simon and E. Kiss is gratefully acknowledged.

References

Adam, G. and Peters, D. (1996) An European comparison of Tospovirus isolates and related antibodies, *Report on the action COST 823, Tospovirus ringtest*, Wageningen, 26–31 August 1995.

Berlinger, M.J., Lebiush-Mordechi, S. and Rosenfeld, J. (1996) State of art and the future of IPM in greenhouse vegetables in Israel, *IOBC/WPRS Bullentin* **19**(1), 11–14.

Budai, Cs. (1994) Control of root-knot nematodes (*Meloidogyne* spp.) in Hungary, *OEPP/EPPO Bulletin* **24**, 511–514.

Carnero Hernández, A., Hernández García, M., Torres del Castillo, R., Kajati, I., Ilovai, Z., Kiss, E., Budai, Cs., Hatala-Zseller, I. and Dancshazy, Zs. (1997a) Possibilities of application of preparates on natural basis in the environment saving pest management (IPM) of greenhouse paprika, *IOBC/WPRS Bulletin* **20**(4), 297 (abstract).

Carnero Hernández, A., Hernández García, M., Torres del Castillo, R. and Pérez Padrón, E. (1997b) IPM in protected vegetable crops in Canary Islands, *IOBC/WPRS Bulletin* **20**(4), 293 (abstract).

Ceglarska-Hodi, E., Dormanns-Simon, E. and Aponyi-Garamvoelgyi, I. (1998) Some aspects of biological control of soil-born pathogens in greenhouse paprika: The Hungarian experience and prospects, in *Proceedings of International Workshop on Biological and Integrated Pest Management in Greenhouse Pepper*, Hodmezoevasarhely, Hungary, 10–14 June 1996 (in press).

Dik, A.J., Verhaar, M.A. and Bélanger, R.R. (1998) Comparison of three biological control agents against cucumber powdery mildew (*Sphaerotheca fuliginea*) in semi-commercial-scale glasshouse trials, *European Journal of Plant Pathology* **104**, 413–423.

Diop-Bruckler, M. and Molot, P.M. (1987) Intérêt de quelques hyperparasites dans la lutte contre *Leveillula tauraica* et *Sphaerotheca fuliginea*, *OEPP/EPPO Bulletin* **17**, 593–600.

Dormanns-Simon, E. and Budai, Cs. (in press) Application of *Arthrobotrys oligospora* nematode trapping fungus – New biological control method against root-knot nematodes, in *Proceedings of International Workshop on Biological and Integrated Pest Management in Greenhouse Pepper*, Hodmezoevasarhely, Hungary, 10–14 June 1996.

Dormanns-Simon, E., Ceglarska-Hodi, E., Lahdenpera, M.-L., Mohammadi, O. and Uoti, J. (in press) Biological control of soil-borne plant diseases by the help of MYCOSTOP (*Streptomyces griseoviridis*) – Some results in greenhouse paprika, in *Proceedings of International Workshop on Biological and Integrated Pest Management in Greenhouse Pepper*, Hodmezoevasarhely, Hungary, 10–14 June, 1996.

Drost, Y.C., Fadlelmula, A., Posthuma-Doodeman, C.J.A.M. and van Lenteren, J.C. (1996) Development of criteria evaluation of natural enemies in biological control: Bionomics of different parasitoids of *Bemisia argentifolii*, *IOBC/WPRS Bulletin* **19**(1), 31–34.

Egea, C., Alcazar, M.D. and Candela, M.E. (1996) Capsidiol: Its role in the resistance of *Capsicum annuum* to *Phytophthora capsici*, *Physiologia Plantarum* **98**(4), 737–742.

Elad, Y. and Shtienberg, D. (1997) Integrated manegement of foliar diseases in greenhouse vegetables according to principles of a decision support system GREENMAN, *IOBC/WPRS Bulletin* **20**(4), 71–76.

FAO (1995) *Production Yearbook*, FAO, Rome.

García, F., GreatRex, R.M. and Gómez, J. (1997) Development of integrated crop management systems for sweet peppers in Southern Spain, *IOBC/WPRS Bulletin* **20**(4), 8–15.

Ilovai, Z. (1997) Characteristic of IPM in vegetable crops in Hungary, *IOBC/WPRS Bulletin* **20**(4), 16–26.

Ilovai, Z., Budai, Cs., Dormanns-Simon, E. and Kiss, E. (1996) Implementation of IPM in Hungarian greenhouses, *IOBC/ WPRS Bulletin* **19**(1), 63–66.

Ilovai, Z., Kiss, E., Kajati, I., Budai, Cs., Carnero Hernández, A., Torres, R., Hernández García, M. and Hernández Suarez, E. (in press) Development of IPM for protected paprika with particular attention to beneficial arthropods, in *Proceedings of International Workshop on Biological and Integrated Pest Management in Greenhouse Pepper*, Hodmezoevasarhely, Hungary, 10–14 June 1996.

Kajati, I., Kiss, F., Koelber, M., Basky, Zs. and Nasser, M. (1989) Effect of light summer oils on aphid transmission of viruses in bell and red pepper fields, in *Proceedings of XXXX International Symposium on Crop Protection*, Gent.

Kania, S. (1996) Integrowana ochrona warzyw przed szkodnikami w uprawie pod oslonami, cz. II. Ochrona papryki, *Haslo Ogrodnicze* **6**, 24 (in Polish).

Kiss, F.E. (in press) Virus diseases of pepper grown in glasshouse in Southern Hungary, in *Proceedings of International Workshop on Biological and Integrated Pest Management in Greenhouse Pepper*, Hodmezoevasarhely, Hungary, 10–14 June 1996.

Lahdenpera, M.L., Simon, E. and Uoti, J. (1990) Mycostop – Novel biofungicide based on *Streptomyces* bacteria, in A.B.R. Beemster, G.J. Bollen, M. Gerlagh, M.A. Ruissen, B. Schippers and A. Tempel (eds.), *Proceedings of the First Conference of the European Foundation for Plant Pathology*, Wageningen, 26 February – 2 March 1990, Elsevier Science Publishers, Amsterdam, pp. 258–263.

Lenteren, J.C. and Woets, J. (1988) Biological and integrated pest control in greenhouses, *Annual Review of Entomology* **33**, 239–269.

Loomans, A.J.M., van Lenteren, J.C., Tommasini, M.G., Maini, S. and Riudavets, J. (1995) Biological control of thrips pests, *Wageningen Agricultural University Papers* **95**(1), 1–201.

Loomans, A.J.M. and Vierbergen, G. (1997) *Thrips palmi*: A next thrips pest in line to be introduced into Europe? *IOBC/WPRS Bulletin* **20**(4), 162–168.

Michelakis, S.E. and Amri, A. (1997) Integrated control of *Frankliniella occidentalis* in Crete – Greece, *IOBC/WPRS Bulletin* **20**(4), 169–176.

Mor, N., Mizrahi, S., Zaqes, Y. *et al.* (1997) Test of control agents for pepper powdery mildew, *Gan Sadeh Vameshek* **1**, 68–69 (in Hebrew).

Nemec, S., Datnoff, L.E. and Strandberg, J. (1996) Efficacy of biocontrol agents in planting mixes to colonize roots and control root diseases of vegetables and citrus, *Crop Protection* **15**(8), 735–742.

Parrella, M.P. and Murphy, B. (1996) Western Flower Thrips – Identification, biology and research on the development of control strategies, *IOBC/ WPRS Bulletin* **19**(1), 115–118.

Park, J.H. and Kim, H.K. (1989) Biological control of *Phytophthora* crown and root rot of greenhouse pepper with *Trichoderma harzianum* and *Enterobacter agglomerans* by improved method of application, *Korean Journal of Plant Pathology* **5**(1), 1–12.

Ramakers, P.M.J. and van der Maas, A.A. (1996) Decision support system "CAPPA" for IPM in sweet pepper, *IOBC/WPRS Bulletin* **19**(1), 123–126.

Ramakers, P.M.J. and Voet, S.J.P. (1996) Introduction of *Amblyseius degenerans* for thrips control in sweet peppers with potted castor beans as banker plants, *IOBC/WPRS Bulletin* **19**(1), 127–130.

Salamon, P. (in press) Some little-known and newly-emerged viral diseases in pepper (*Capsicum annuum* L.) produced under cover in Hungary, in *Proceedings of International Workshop on Biological and Integrated Pest Management in Greenhouse Pepper*, Hodmezoevasarhely, Hungary, 10–14 June 1996.

Shtienberg, D. and Elad, Y. (1997) Incorporation of weather fore-casting in integrated, biological, chemical management of *Botrytis cinerea*, *Phytopathology* **87**, 332–340.

Stanghelini, M.E., Kim, D.H., Rasmussen, S.L. and Rorabaugh, P.A. (1996) Control of root rot of peppers caused by *Phytophthora capsici* with a nonionic surfactant, *Plant Disease* **80**(10), 1113–1116.

Stijger, C.C.M.M. (1995) Preventing spread of mechanically transmitted viruses by skimmed milk, in *Proceedings of the 8th Conference on Virus Diseases of Vegetables*, Prague, Czech Republic, 9–15 July 1995, pp. 171–174.

Szabo, P. and Ceglarska-Hodi, E. (1992) Western Flower Thrips (*Frankliniella occidentalis* Pergande) – Occurrence and possibilities for its control in Hungary, *OEPP/EPPO Bulletin* **22**, 377–382.

van Steenis, M. (1995) *Evaluation and application of parasitoids for biological control of Aphis gossypii in glasshouse cucumber crops*, PhD Thesis, Wageningen Agricultural University.

van Vianen, A., Budai, C. and van Lenteren, J.C. (1987) Suitability of two strains of *sweet* pepper, *Capsicum annuum* L. for the greenhouse whitefly *Trialeurodes vaporariorum* (Westwood) in Hungary, *IOBC/WPRS Bulletin* **10**(2), 174–179.

CHAPTER 34

ORNAMENTALS

M. Lodovica Gullino and Leslie R. Wardlow

34.1. Background

The ornamental industry is large and complex, and accurate information on its world-wide value is difficult to obtain. However, it can be estimated at more than five billion US$ world-wide annually. It stands out in the agricultural scenario for the frequency and rapidity of changes in type of product (from cut flowers to flowering and foliage potted plants, landscape and bulb crops) and in production areas and technology.

The unique features of the ornamental industry have been reviewed by Baker and Linderman (1979) and Garibaldi and Gullino (1990). They strongly affect the overall phytopathological picture and the management strategies of pests and diseases.

Diversity of crops and their cultivars and the constant turnover in consumer interests multiply the number of potential pests and diseases and complicate the development of horticulturally useful resistant varieties and plant novelties.

Widespread international and interregional movement of living plants has a tremendous impact on the diffusion of pests and pathogens.

Diverse and sophisticated greenhouse production technologies lead to a variety of environmental and cultural conditions in which different pests and pathogens may thrive in turn. High capital investment and high crop value may justify expensive tactics such as chemical eradication. The use of chemical pesticides from the 60s to the early 80s has enabled growers to produce high yields of good quality, and blemish-free plants. This has set a standard not easy to maintain without severe pesticide pressure. Wide availability of pesticides during this period has been a mixed blessing.

In ornamental crops toxic residues are not so much a problem as in crops for human consumption and a wider range of pesticide has therefore always been available. However, undesirable side effects of chemical control (such as visible pesticide residues, phytotoxicity and development of resistance in pests and pathogens) cannot be neglected. More important, the delicate balance between beneficial and destructive microflora and microfauna may be often upset by inappropriate use of pesticides.

Not only floriculturists must face a great diversity of old pests and diseases that complicates IPM but are also continuously facing new pests and diseases on old and new crops. New pests and diseases can be suddenly brought about by planting material or can arise as a consequence of the introduction of new cultivars. During the past years, a number of diseases spread throughout infected planting material, as happened in the case of *Colletotrichum acutatum* J.H. Simmonds on anemone in the early 80s (Garibaldi and Gullino, 1985). In certain cases, the appearance of new diseases is related to the introduction of new cultivars: for instance, recently developed ranunculus F1 hybrids are particularly susceptible to *Fusarium oxysporum* Schlechtend.:Fr. f. sp.

R. Albajes et al. (eds.), Integrated Pest and Disease Management in Greenhouse Crops, 486-505.
© 1999 *Kluwer Academic Publishers. Printed in the Netherlands.*

ranunculi Garibaldi & Gullino and to a foliar blight, caused by *Ramularia* sp. (Garibaldi *et al.*, 1990). There are similar examples of pests that have been spread around the world, in particular *Liriomyza* spp., *Frankliniella occidentalis* (Pergande) and *Bemisia tabaci* (Gennadius), the later two give the added complication of transmitting plant viruses that can devastate crops.

Artificial substrates present new cultural situations in which new diseases can be expected. Pythium root rot and Penicillium stem rot have become major problems on artificial substrates (Stanghellini and Rasmussen, 1994). On rose, for instance, severe outbreaks of *Phytophthora* spp. and *Gnomonia* sp. have been recently reported (Amsing, 1995). Diseased plants also encourage pests such as fungus gnats (Sciaridae) that are a particular problem on young plants. Also, nematodes are able to colonize roots of roses grown in rockwool.

Due to the great number of cultivated species and varieties, selectivity remains a major problem for many of the chemicals used on ornamental crops. Visible pesticide residues on florist crops diminish their retail value and frequent use of pesticides often hardens, marks or stunts the plants. Because of potential plant growth modifying effects, many of the ergosterol biosynthesis inhibitors (EBIs) must be applied only after evaluation on each new variety. These fungicides are structurally related to some growth regulators, and inappropriate use can lead to shorter stems, a generally negative feature for ornamental crops.

Some insecticides and acaricides are notorious for their phytotoxicity risks particularly on some tender crops such as saintpaulia and poinsettia. Sometimes, even innocuous pesticides can cause damage to crops if they are used too frequently and/or in combination with other products. Mixtures of pesticides can be particularly hazardous. Quite often only certain cultivars of plants may be affected, chrysanthemums are a good example. It is also generally accepted that pesticides which are safe to crops on some nurseries, may not be so on others; this is due to differing cultural conditions which can be even more considerable with the differences in climate between various parts of the world. On the other hand, the cost of generating efficacy and phytotoxicity data for the large numbers of plant species under cultivation discourage chemical companies from expanding the labels on new materials to include ornamental crops (Cline *et al.*, 1988), which are all included into minor crops (Gullino and Kuijpers, 1994; Ragsdale and Sisler, 1994). Moreover, even when chemicals are available, their efficacy is often only partial: for instance, benzimidazoles provide only 50–60% control of *Fusarium oxysporum* Schlechtend.:Fr. f. sp. *dianthi* (Prill. & Delacr.) W.C. Snyder & H.N. Hans. on carnation (Garibaldi and Gullino, 1990).

The use of IPM is therefore likely to expand and only established or new pesticides that integrate well with natural enemies will be used. This probably means a shrinkage in the number of available pesticides in the future.

Currently, particularly in the floriculture sector, growers rely heavily on synthetic chemicals for pest and disease control. However, in spite of the recent introduction of effective chemicals and of improved application techniques, chemical control is often incomplete. Some of the hitherto effective chemicals no longer work due to resistance in the pest and pathogen populations.

Resistance in pests developed seriously during 1960–1980 with many examples in several families of insects and mites, the problem continues to this day. The development of resistance to one group of chemicals made it easier for pests to resist replacement products from different chemical groups by cross-resistance which included multiple resistance spectra: a good example of this occurs in glasshouse whitefly (Wardlow and Ludlam, 1973). In horticulture, it is now a well established fact that spider mites, aphids, thrips and leaf miners, as well as whiteflies, all give serious resistance problems. In some pests, such as *F. occidentalis*, the life history which includes eggs in leaf tissue, larvae on the plant, pupae on the ground and adults free to disperse through the air gives added problems by creating physical resistance factors. Strains of *Botrytis cinerea* Pers.:Fr. with multiple resistance to benzimidazoles and dicarboximides are widespread on several crops (i.e. rose, cyclamen, gerbera, saintpaulia, etc.) (Gullino, 1992). More recently, resistance to phenylcarbamates and anylinopyrimidines also developed. Resistance to benzimidazoles and/or dicarboximides was also reported in the case of *Botrytis tulipae* (Lib.) Lind, causal agent of tulip fire, and of *Botrytis elliptica* (Berk.) Cooke, incitant of blight on lily (Gullino, 1992).

Phytotoxicity, unwanted presence of chemical deposits, lack of safe and sufficiently effective chemicals for certain crops and pests or pathogens are good reasons for expanding non-chemical control strategies. Further reasons to work in this direction is the increasing public support for reduced use of pesticides also in the ornamental industry. A growing number of consumers, particularly in countries such as Germany and Scandinavia, being aware of the need for a safe environment, is asking for organically-grown flowers. Also, their crops of choice are those requiring less external inputs. In the UK and The Netherlands, the supermarkets are increasingly setting up Codes of Practice for growers to observe in crop production; such codes include environmentally-friendly cultivation such as IPM. All these reasons and the chances offered by the high technology of the cultivation under controlled greenhouse conditions stimulate the application of alternative techniques, making IPM very challenging for the ornamental crops.

34.2. Crops and their IPM Programmes

For all the above mentioned reasons, IPM is becoming successful on an increasing number of ornamental plant species. It is evident that, because each plant suffers its own pest complex, each requires its own specially-designed IPM programme. Examples of IPM are given under Table 34.1.

34.2.1. BEDDING PLANTS, FLOWERING POT-PLANTS, CUT-FLOWERS, BULB CROPS, FOLIAGE PLANTS, NURSERY STOCK, AMENITY CROPS: THEIR DISEASE AND PEST COMPLEXES AND POSSIBLE MANAGEMENT

In each cropping area, IPM programmes for insect and mite control vary widely according to the likely pest complex and the method of crop culture. IPM programmes are well-established in northern Europe, but they are often more difficult to implement in hotter climates when intense pest pressure exists outdoors and can threaten greenhouses by massive immigrations. However, the natural enemies for various pests and some standard recommended rates of use are given in Table 34.2. Adjustments to the rates of use will be referred to under the various crop headings.

TABLE 34.1. IPM programmes for ornamental crops

Pest/pathogen (common name)	Method of control	Feasibility
Ivy-leaf geranium (as represenative of bedding plants)		
Bradysia sp. (fungus gnats)	Parasitic nematodes	+++
Frankliniella occidentalis, Thrips tabaci (thrips)	*Amblyseius* spp. broadcast	+++
Aulacorthum solani, Myzus persicae (aphids)	Parasitic wasps	+
Autographa gamma and numerous other moth species (caterpillars)	*Bacillus thuringiensis*	+
Xanthomonas campestris pv. *pelargonii* (bacterial blight)	Healthy material, indexing, sanitation	+++
Pythium spp. (root rot)	Soil/substrate disinfestation; drench with fungicides	++
Verticillium dahliae (Verticillium wilt)	Healthy material, indexing, soil disinfestation, soil drenching with benzimidazoles	++
Puccinia pelargonii-zonalis (rust)	Thermoterapy (stock plants); reduction of RH, ventilation, proper irrigation, fungicides	++
Botrytis cinerea (Botrytis blight)	Reduction of RH, ventilation, heating, balanced N fertilization, proper irrigation, fungicides	++
Cyclamen (as representative of flowering pot planst)		
Otiorhynchus sulcatus (vine weevil)	Imidacloprid compost incorporation or parasitic nematodes	+++
Frankliniella occidentalis, Thrips tabaci (thrips)	*Amblyseius* spp. in sachets	+++
Tetranychus urticae (spider mite)	*Phytoseiulus persimilis*	+++
Numerous species of aphids	Parasitic wasps (usually *Aphidius colemani*)	+
Numerous moth species (caterpillars)	*Bacillus thuringiensis*	++
Erwinia carotovora (soft rot)	Substrate disinfestation; cultural practices (healthy material, proper irrigation, destruction of infected plants)	++
Fusarium oxysporum f. sp. *cyclaminis* (*Fusarium* wilt)	Substrate disinfestation, healthy material, biological control (angonistic *Fusarium* spp.)	++
Thielaviopsis basicola, Cylindrocarpon destructans (= *Cylindrocarpon radicicola*) (root rots)	Substrate disinfestation	++
Botrytis cinerea (Botrytis blight)	Ventilation, heating, proper irrigation, limited application of fungicides	+
Cryptocline cyclaminis (anthracnose)	Ventilation, heating, proper irrigation, limited application of fungicides	+

TABLE 34.1. IPM programmes for ornamental crops (cont.)

Pest/pathogen (common name)	Method of control	Feasibility
Carnation		
Tetranychus cinnabarinus, Tetranychus urticae (spider mites)	*Phytoseiolus persimilis,* higher rates of introduction will probably be necessary in some crops, depending upon the history of this pest on the nursery	+++
Frankliniella occidentalis, Thrips tabaci (thrips)	*Amblyseius* spp. broadcast when plants are small and sachets when a good crop canopy has been attained	+++
Numerous species of aphids	Predatory lacewings and parasitic wasps	++
Numerous moth species, but especially *Cacoecimorpha pronubana* (caterpillars)	*Bacillus thuringiensis*	+++
Burkholderia (= *Pseudomonas*) *caryophylli* (bacterial wilt)	Healthy material, destruction of infected plants, sanitation	++
Erwinia chrysanthemi (bacterial slow wilt)	Healthy material, destruction of infected plants, sanitation	++
Burkholderia andropogonis (= *Pseudomonas woodsii*) (Bacterial leaf soit)	Healthy material, destruction of infected plants, sanitation, RH control	++
Fusarium oxysporum f. sp. *dianthi* (*Fusarium* wilt)	Soil disinfestation, resistant cultivars, biological control (antagonistic *Fusarium* spp.), drench at transplant with benzimidazoles	+++
Phialophora cinerescens (Phialophora wilt)	Soil disinfestation, drench at transplant with benzimidazoles	+++
Rhizoctonia solani (basal rot)	Soil disinfestation, drench with fungicides, biological control (*Trichoderma* spp., *Gliocladium virens*)	+++ +
Phytophthora nicotianae var. *parasitica* (root rot)	Soil disinfestation; drench with phenylamides; suppressive soils	++
Uromyces dianthi (= *Uromyces caryophyllinus*) (rust)	Control of RH, proper irrigation, limited use of fungicides	+++
Cladosporium echinulatum (= *Heterosporium echinulatum*) (Heterosporium leaf spot)	Control of RH, proper irrigation, limited use of fungicides	++
Rose		
Tetranychus urticae (spider mite)	*Phytoseiulus persimilis,* higher numbers will probably be necessary on most crops. The additional use of *Therodiplosis persicae* is recommended until the predator is well-established	+++
Frankliniella occidentalis, Thrips tabaci (thrips)	*Amblyseius* spp. in sachets but extra broadcast introductions if monitoring shows any increase in pest numbers	++
Archips rosanus (caterpillars)	*Bacillus thuringiensis*	+++
Numerous species of aphids	Parasitic wasps and predatory lacewings	++

TABLE 34.1. IPM programmes for ornamental crops (cont.)

Pest/pathogen (common name)	Method of control	Feasibility
Rose (cont.)		
Pratylenchus spp., *Xiphinema* spp., *Longidorus* spp. (nematodes)	Soil disinfestation prior to cropping, chemical nematicides drenches to the growing crop will have only temporary effect on most natural enemies	+++
Agrobacterium tumefaciens (crown gall)	Soil disinfestation, sanitation, biological control (*Agrobacterium radiobacter*)	+++
Rosellinia (root rot)	Soil disinfestation	+++
Verticillium dahliae (Verticillium wilt)	Soil disinfestation, healthy material, soil drench with benzimidazoles, proper plant preparation	++
Diaphorte, Coniothirium (cankers)	Reduction of RH, ventilation, low nitrogen fertilization, local application of fungicides	++
Sphaerotheca pannosa (powdery mildew)	Fungicides, mineral salts, oils, biological control	+++
Botrytis cinerea (Botrytis blight)	Reduction of RH, ventilation, heating, limited application of fungicides	++
Chrysanthemum		
Frankliniella occidentalis, Thrips tabaci (thrips)	*Amblyseius* spp. broadcast for first six weeks plus *Verticillium lecanii* (thrips strain) weekly throughout the life of the crop	++ if humidities high due to blackout screens
Myzus persicae, Aphis gossypii (aphids)	*Verticillium lecanii* (aphid strain) incorporated with thrips treatment weekly	+++ if humidities high
Numerous moth species (caterpillars)	*Bacillus thuringiensis* incorporated with *Verticillium lecanii* treatment weekly	+++
Phytomyza syngenesiae, Liriomyza trifolii, Liriomyza huidobrensis (leaf miners)	Parasitic wasps biased towards *Diglyphus isaea* in a mixture at 3 parasites per 1000 plants three and six weeks after planting	+++
Tetranychus urticae (spider mite)	*Phytoseiulus persimilis* at 1 per 10 plants introduced at three and six weeks after planting	+++
Agrobacterium tumefaciens (crown gall)	Soil disinfestation, sanitation, biological control (*Agrobacterium radiobacter*)	+++
Erwinia chrysanthemi (Erwinia wilt)	Healthy material, soil/substrate disinfestation	+++
Pythium spp., *Rhizoctonia solani, Sclerotinia* spp. (root rots)	Soil/substrate disinfestation	+++
Phoma chrysanthemicola f. sp. *chrysanthemicola*	Soil/substrate disinfestation, resistant cultivars, N and P fertilization	++
Verticillium dahliae (Verticillium wilt)	Soil/substrate disinfestation, healthy material, indexing, soil drenches with benzimidazoles	+++

TABLE 34.1. IPM programmes for ornamental crops (cont.)

Pest/pathogen (common name)	Method of control	Feasibility
Chrysanthemum (cont.)		
Fusarium oxysporum f. sp. *tracheiphilum*, *Fusarium oxysporum* f. sp. *chrysanthemi* (Fusarium wilt)	Soil/substrate disinfestation, resistant cultivars, high pH, soil drenches with benzimidazoles	+++
Puccinia horiana (white rust)	Healthy material, thermoterapy for stock plants, fungicides	+++
Didymella ligulicola (Ascochyta ray blight)	Healthy material, reducing RH, ventilation, proper irrigation, good spacing, destruction of infected plants, limited use of fungicides	++
Gerbera		
Frankliniella occidentalis, Thrips tabaci (thrips) and *Polyphagotarsonemus latus* (tarsonemid mite)	*Amblyseius* spp. broadcast weekly	+++
Tetranychus urticae (spider mite)	*Phytoseiulus persimilis*	+++
Trialeurodes vaporariorum, Bemisia tabaci (whiteflies)	*Encarsia formosa* and/or *Eretmocerus eremicus* (= *Eretmocerus californicus*) depending upon the threat of the whitefly species	+++
	Verticillium lecanii (whitefly strain) is useful if humidities are relevant	+
Numerous moth species (caterpillars)	*Bacillus thuringiensis* every two weeks	++
Numerous species of aphids	Parasitic wasps	++
Phytomyza syngenesiae, Liriomyza trifolii, Liriomyza huidobrensis (leaf miners)	Parasitic wasps biased towards *Diglyphus isaea*	+++
Lygocoris pabulinus, Lygus rugulipennis (capsids)	Heptenophos spray as required	++
Pseudomonas cichorii (bacterial blight)	Healthy material, proper irrigation, limited copper fungicides	++
Phytophthora cryptogea (footl rot)	Soil/substrate disinfestation, suppressive soils, soil drenches with phenylamides	+++
Verticillium spp. (Verticillium wilt)	Healthy material, soil/substrate disinfestation, soil drenches with benzimidazoles	+++
Rhizoctonia solani (basal rot)	Soil/substrate disinfestation, soil drenches with fungicides	+++
Botrytis cinerea (Botrytis blight)	Reduction of RH, balanced N fertilization, ventilation, heating, limited fungicide application	+++
Gladiolus (as representative of bulb crops)		
Thrips simplex (thrips)	*Amblyseius* spp. broadcast on several occasions	+ under research
Other pests	Existing chemical treatments	++
Burkholderia (= *Pseudomonas*) *gladioli* (bacterial leaf spot)	Healthy material, thermoterapy or bulb dipping in formalin	++
Fusarium oxysporum f. sp. *gladioli*, *Penicillium hirsutum* (= *Penicillium corymbiferum*), *Stromatinia gladioli* (bulb rots)	Healthy material, thermoterapy or bulb dipping in fungicides	+++

TABLE 34.1. IPM programmes for ornamental crops (cont.)

Pest/pathogen (common name)	Method of control	Feasibility
Gladiolus (as representative of bulb crops) (cont.)		
Uromyces transversalis (rust)	Reduction of RH, plant spacing, proper irrigation, limited application of fungicides	+++
Azalea and rhododendron (as representatives of nursery stock)		
Otiorhynchus sulcatus (vine weevil)	Compost incorporation of chlopyrifos or imidacloprid. Parasitic nematodes to growing crop if any outbreaks of pest	+++
Numerous species of aphids	Parasitic wasps	++
Caloptilia azaleella (azalea leaf miner)	Heptenophos sprays on sight	+++
Most of the pests that attack a wide range of nursery stock	Use pot plant recommendation for the specific pest	+++
Agrobacterium tumefaciens (crown gall)	Soil/substrate disinfestation, biological control (*Agrobacterium radiobacter*)	+++
Pythium spp., *Phytophthora* spp., *Rhizoctonia solani* (damping-offs)	Healthy material, soil/substrate disinfestation	+++
Phytophthora cinnamomi (basal rot)	Soil/substrate disinfestation, suppressive substrates, drenches with phenylamides	++
Cylindrocarpon destructans (= *Cylindrocarpon radicicola*) (Cylindrocladium blight and root rot)	Sanitation, destruction of infected plants, copper fungicides	++
Ovulinia azaleae (Ovulinia petal blight)	Soil/substrate disinfestation, sanitation, destruction of infected plants, proper irrigation, ventilation, plant spacing, limited fungicide application	++

TABLE 34.2. Parasitoids, predators and pathogens and the rates of their use in ornamental crops

Target pest	Parasitoid, predator or pathogen	Rate of use
Aphids	*Aphidius* sp. and *Aphelinus abdominalis* parasitic wasps; select species according to likely aphid species target	One per two square metres each week for prone crops or token rate of 250 parasites per greenhouse every 14 days for crops at lower risk. Up to 4 per metre square at two-week intervals to cure outbreaks
Aphids	The predatory lacewings, in particular *Chrysoperla carnea*	10–20 per metre square per week until infestation cured
Aphids	The predatory midge *Aphidoletes aphidimyza*	1–2 cocoons per metre square per week as a routine in aphid-prone crops
Aphids, whiteflies and thrips	*Verticillium lecanii* entomopathogenic fungus. Two strains are available, one for aphids and one for the two other pests	Only relevant where relative humidity of the greenhouse is consistently high (90%). Used at maximum rate on frequent occasions as a routine for specific crops
Caterpillars	*Bacillus thuringiensis* bacterial disease of insects	Maximum recommended rate as a high volume spray or fog once per month as a routine whether pests seen or not. A sequence of weekly sprays for three or four occasions if pests discovered

TABLE 34.2. Parasitoids, predators and pathogens and the rates of their use in ornamental crops (cont.)

Target pest	Parasitoid, predator or pathogen	Rate of use
Caterpillars	Parasite of moth eggs, various species of *Trichogramma*	Usually used as a precautionary measure in addition to *Bacillus thuringiensis* at 5 per metre square on several occasions
Citrus mealybug and scales	*Leptomastix dactylopii*	1–2 per metre square every 14 days depending upon the risk from the pests
Fungus gnats and thrips pupae in growing medium	*Hypoaspis* spp. predatory mite	A single introduction of 100 per metre square as a routine preventive measure at planting. Up to 400 per metre square to cure outbreaks of pest
Leaf hopper	The egg parasite, *Anagrus atomus*	Usually purchased as a minimum order of 100 parasitized eggs in leaf tissue and introduced weekly until established
Leaf miners	The parasitic wasps, *Dacnusa sibirica* and/or *Diglyphus isaea*	Weekly introduction of 250 per greenhouse until pest seen, when a rate of one wasp per 10 mines per week is required until pest controlled
Mealybugs and scales	*Cryptolaemus montrouzieri*	1–2 per metre square to establish but up to 10 per metre square to control heavy infestations
Spider mites	Predatory midge, *Therodiplosis persicae*	250 cocoons per week per greenhouse at first signs of pest, continue for 4–6 weeks
Thrips	*Orius* spp., predatory bug	One per two square meters on one or two occasions at intervals of two weeks depending upon the pest situation and degree of flower in the crop. Larger numbers may be justified in some valuable crops
Thrips and tarsonemid mites	*Amblyseius* spp. but particularly *Neoseiulus* (= *Amblyseius*) *cucumeris* predatory mites: see supplier's recommendations for other species	Weekly broadcast over the crop at 50 per metre square and/or sachets 3 per metre square every 6 weeks
Two-spotted spider mite	The predatory mite, *Phytoseiulus persimilis*	In spider mite prone crops a weekly introduction of one predator per 10 plants until the pest is seen, then treatment of affected areas at up to 10 per plant depending upon the level of pest
Vine weevil or fungus gnats	The parasitic nematodes, *Steinernema* sp. or *Heterorhabditis megidis*	One or two treatments of one million nematodes per one or two square metres depending upon the risk from the pest
Whiteflies	The parasitic wasp, *Encarsia formosa*	Weekly introductions of one per 10 plants or 1–2 per square metre whether whitefly seen or not. Numbers boosted to two parasites per adult whitefly for 4 weeks when infestation exceeds one adult whitefly per 10 plants. For tobacco whitefly, rates may need to be five times higher than those for greenhouse whitefly
Whiteflies	The parasitic wasp, *Eretmocerus eremicus* (= *Eretmocerus californicus*)	As for *Encarsia formosa*

Bedding Plants

A high degree of crop uniformity and quality is necessary for bedding plants, and even a low percentage of disease is unacceptable. The major disease problems, affecting a wide range of species of bedding plants, are damping-off incited by *Pythium* spp. and *Rhizoctonia solani* Kühn and Botrytis blight (Jones and Strider, 1985). Where disease is a regular risk, it may be associated with fungus gnats that can exacerbate the situation by damaging plants further and spreading the disease. Routine treatment against *Hypoaspis* spp. or parasitic nematodes at seed sowing can be justified. Growing medium treatment with a chemical such as imidacloprid can integrate well with biological controls that will be used later in the life of the crop. An IPM approach is crucial in order to eliminate the threat of damping-off. Small amounts of chemicals can control it: improved delivery systems are necessary in these production systems. Also, the use of biocontrol agents (BCAs) such as *Gliocladium virens* J.H. Miller, J.E. Giddens & A.A. Foster and *Trichoderma* spp. in the growing mix or as a seed treatment is promising. In the case of geranium, the use of Xanthomonas-free cuttings is important, since there is a zero tolerance for Xanthomonas blight in stock production systems (Cline *et al.*, 1988). The same applies to other crops.

Since many bacterial and fungal pathogens are spread throughout infected seeds, biological, physical and chemical treatments, combined with effective cultural management systems, are important.

Against Botrytis blight, downy mildews and other foliar diseases, heating (when economically sustainable) and ventilation, coupled with a balanced nitrogen fertilization, help reducing the number of chemical sprays (Jarvis, 1992). However, particularly when environmental conditions are very favourable to grey mould, some chemical spray remain necessary.

Once plants are fully emerged they can be protected against thrips by weekly broadcast introductions of *Amblyseius* spp. but rates may be reduced to as low as 10 per square metre if pest levels are low. Bedding plants are often grown in the cool earlier part of the season, but this should not discourage *Amblyseius* spp. use, the predators become active at the same temperature threshold as thrips and so they can take advantage of even short periods of warmth that can occur in winter, this has a marked effect on the spring generation of pest. Regular *Bacillus thuringiensis* Berliner treatment will be required for crops such as primroses and *Phytoseiulus persimilis* Athias-Henriot may be needed for mite control on crops such as impatiens. It is generally a wise precaution to regularly introduce a nominal number of *Aphidius* spp. of several species. Severe outbreaks of aphids can be treated with a light spray of heptenophos or imidacloprid, depending upon the pesticide clearance rules prevailing in the particular country. In the UK, nicotine is still permitted for use against melon-cotton aphid (*Aphis gossypii* Glover) that is resistant to so many products.

Cuttings from stock plants that are also produced under IPM tend to be free of pesticide residues, ensuring more likely success for future IPM.

Flowering Pot Plants
This industry is undergoing continuous changes, with new species becoming popular: poinsettia, chrysanthemum and geranium are the major crops in this sector, followed by cyclamen, fuchsia, cineraria, gloxinia, African violet and *Chrysanthemum frutescens* L. (Cline *et al.*, 1988; Daughtrey *et al.*, 1995).

The control of pests and diseases of flowering pot plants must be based upon a pathogen-free production concept. This means placing high priority on production of propagating material free from root, basal stem, and systemic pathogens. The fungi most frequently involved in root and basal rots are species of *Pythium*, *Phytophthora*, *Fusarium*, *Rhizoctonia*, *Sclerotinia*, *Thielaviopsis* and *Cylindrocladium*. Root and basal rots are important on most crops, particularly during the first growing stage. Species of *Verticillium* and *Fusarium oxysporum* Schlechtend.:Fr. among fungi, *Erwinia* and *Xanthomonas* among bacteria are responsible for wilting. Indexing techniques should be applied whenever possible. The use of pathogen-free growing materials is of primary importance: steam treatment of all growing materials remains the best procedure for producing pathogen-free pot plants. Sanitation too is very important, since it prevents the introduction of root and basal rot as well as vascular wilt pathogens. Matting, benches and generally structures must be disinfected. Soil/substrate disinfestation plays a major role in their management. Also, a well-drained medium is helpful. Soil/substrates suppressive to some of the above-mentioned pathogens have been described (Hoitink *et al.*, 1991): since suppressiveness is generally oriented towards a single pathogen, knowledge of the most important disease on each crop is necessary in order to choose the right substrate. Similarly, the application of BCAs, such as *Trichoderma* spp., *G. virens*, fluorescent pseudomonads and saprophytic *Fusarium* spp., has been attempted and quite a few success stories exist (Jarvis, 1992; Gullino, 1995; Gullino and Garibaldi, 1997).

The quality of water is very important, since through irrigation water several pathogens, such as species of *Erwinia* and *Pseudomonas,* can be easily introduced. In the case of soft rot incited by *Erwinia* spp., management relies on cultural practices. Only plants thought to be free of soft rot *Erwinia* spp. should be vegetatively propagated. Stock plants should not be too soft or have excessive or deficient levels of nitrogen. Cuttings should be removed with a sharp knife. Cutting instruments and bench surfaces should be disinfested.

In the case of foliar diseases, in the most sophisticated greenhouses, by modifying the environmental conditions it is possible to strongly reduce the incidence of downy mildews, rusts, grey mould and several leaf spot agents. Also, proper spacing among plants is important in order to permit good airflow. When necessary, some broad-spectrum fungicides can be sprayed (Jarvis, 1992; Daughtrey *et al.*, 1995).

For pests, each species of plant will fit in to a category where either only a few pests need to be considered or a full complex is likely and a wide range of natural enemies will be necessary. Examples of these programmes are given later in this section. IPM in pot plants has been extremely successful, particularly in the UK, Denmark, Germany and France, where problems with *F. occidentalis* severely threatened the industry in the mid 80s and prompted the development of IPM.

Cut Flowers
The major cut flower crops world-wide are carnation, rose and chrysanthemum. Other important crops are *Eustoma grandiflorum* L., gerbera, *Limonium, Anthirrinum*, alstroemeria, etc.

Vascular wilt pathogens [*formae speciales* of *F. oxysporum*, *Verticillium dahliae* Kleb., *Phialophora cinerescens* (Wollenweb.) van Beyma (= *Verticillium cinerescens*

Wollenweb.), and *Erwinia* and *Xanthomonas* among bacteria] are a very serious threat. They easily spread through contaminated vegetative propagation material: once introduced, they are difficult to eradicate, even when bed or benches are steamed or fumigated. Pathogen-free propagation programmes are routinely carried out, especially in the case of carnation and chrysanthemum (Horst and Nelson, 1997).

On certain crops, such as rose, Verticillium wilt is becoming the cause of serious damage, due to the adoption of cultural techniques which significantly reduce the length of time necessary for plant preparation (from 18–24 months to 4–7 months). This fact strongly increases the risk of dissemination of the infected propagation material. Soil disinfestation, culture-indexed stock when available, and removal and destruction of symptomatic plants are useful practices. Liming soil, using nitrate nitrogen and increasing soil/substrate pH, can help in reducing Fusarium wilt severity. Soils/substrates suppressive to Fusarium wilt have been described: antagonistic, saprophytic Fusaria isolated from such soils provide very satisfactory disease control when applied to conducive soils/substrates, as shown on carnation (Tramier *et al.*, 1983; Garibaldi *et al.*, 1990). Fluorescent *Pseudomonas*, applied alone or in combination with antagonistic *Fusarium* provide good wilt control (van Peer and Schippers, 1991). Moreover, some rhizosphere-competent BCAs are potentially capable of producing growth-stimulating factors, causing significant growth increase response, which could be of interest in the cut flower industry. Benzimidazoles applied to the soil provide some disease suppression in the case of Fusarium and, mostly, Verticillium wilt. On the contrary, no chemicals can assist in the case of bacterial wilts.

When available, resistant cultivars should be planted. In the case of carnation, agronomically acceptable cultivars resistant to the most common races, are available (Garibaldi and Gullino, 1987). *Agrobacterium tumefaciens* (Smith & Townsend) Conn is the causal agent of crown gall on crops such as rose and chrysanthemum. On rose, it has recently become more widespread due to the adoption of new propagation techniques ("mini-greffe"), leading to production of multiple wounds and of modern growing technologies such as soilless cultivation (Aloisi *et al.*, 1994). Management of the disease is based upon exclusion and sanitation. Removing infected plants from the nursery as soon as possible provides the best control. Soil disinfestation is necessary because the pathogen is capable of surviving for years in the soil in the absence of the host. Use of *Agrobacterium radiobacter* (Beijerinck & van Delden) Conn proved very effective on a number of crops, including rose. Such bacterium provides biocontrol of the pathogen through the production of bacteriocin and throughout competition for the infection sites. One single application at transplanting resulted in very good crown gall control (Farrand, 1990).

Certain stem and graft cankers, such as those incited on rose by *Coniothyrium fuckelii* Sacc., have recently regained importance as a consequence of the adoption of cultural practices (such as forcing) which are responsible for the presence of immature tissues that are more susceptible to the pathogen (Gullino and Garibaldi, 1996). Effective management of bacterial and fungal cankers relies on strict sanitation, proper pruning and adoption of cultural techniques which reduce the presence of wounds. Cuttings should be rooted in disinfested media. Wetting the foliage during irrigation, prolonged leaf wetness and poor plant spacing should be avoided. Affected plants should be discarded. In the case of cankers caused by fungi, timely application of fungicides such as chlorothalonil, thiram, zineb and maneb results in disease control.

Foliar diseases, in spite of availability of effective fungicides, remain important. Rose powdery mildew management still relies on repeated application of fungicides, among those EBIs are the most frequently sprayed. However, very interesting and promising results have been obtained by weekly applying natural products such as potassium salts, mineral oils, vinegar, plant extracts (Horst *et al.*, 1992; Pasini *et al.*, 1997) and BCAs, such as *Ampelomyces quisqualis* Cesati:Schltdl. (Pasini *et al.*, 1997) and *Sporothrix flocculosa* Traquair, Shaw & Jarvis (Bélanger *et al.*, 1994). Such management strategies resulted effective on commercially grown roses as well as on other ornamental crops. Also, the possible rotation of biocontrol agents and/or natural products with chemicals looks interesting in crops where very low disease threshold is admitted.

Botrytis blight is a major problem on most crops: its management relies on the careful control of environmental parameters and spray of chemicals limited to when necessary. However, chemical control of Botrytis blight must take into account the easy development of resistance (Gullino, 1992). Humidity control through sensors in the crop canopy and ventilation (and heating) should become the mainstay of Botrytis management (Hausbeck and Moorman, 1996).

Although caused by systematically different pathogens, downy mildews, rusts and fungal leaf spots can be grouped according to the cultural practices that are necessary for their management, mostly based on the control of the environmental parameters through ventilation and heating. Also in the case of such foliar pathogens, the application of salts and horticultural oils could represent a very interesting alternative to continuous use of chemicals. Effective fungicides are generally available and can be applied when necessary on most crops. Special care is taken in many countries in order to avoid the introduction of chrysanthemum plants infected with white rust (*Puccinia horiana* Henn.), which is considered a quarantine organism.

Management of bacterial leaf spots on cut flowers mostly relies on avoiding overhead irrigation, proper plant spacing and discarding infected plants.

Pests are a serious problem for cut flowers, especially since they are usually grown in soil which can harbour life-stages of pests that can be difficult to control biologically. Some crops, such as chrysanthemums, are grown in year-round situations that provide a constant "bridge" for pests to move from older to younger crops; this can put great pressure on natural enemies to perform well. Hence, IPM is not well-established in this area of ornamental production, but there are some successful instances for adoption by those growers who want to depart from the intensive insecticide programmes that are generally preferred. A good example is gerbera; in The Netherlands >40% is under IPM (van Lenteren, 1995). Monitoring of these IPM programmes has to be of a high standard in order to avoid disaster. Most of the cut flower species suffer a wide range of pests necessitating expensive biological control inputs that do not encourage growers to forgo the chemical alternative which is generally cheaper at the present time.

The most serious pests are *F. occidentalis* and the *Liriomyza* spp., these are pests with complicated and severe resistance problems. IPM may be the only solution for their future control.

There is a great international trade in cut flowers and many countries operate a "zero-tolerance" policy for some pests. This also mitigates the development of IPM, since growers feel forced to use pesticides in the mistaken belief that their crops will meet this requirement. It is interesting to note that the distribution of some of the most serious pests of recent times has been on "zero-tolerance" cut flowers.

Bulb Crops

Gladiolus, tulip, narcissus, iris, lily are the most important crops, followed by species of minor importance. Many of the diseases affecting bulb crops have not yet been completely solved and new ones appear since bulb cultivation is moving towards tropical countries.

Soilborne fungi (*formae speciales* of *F. oxysporum*, and species of *Sclerotinia*, *Sclerotium*, *Rhizoctonia* and *Stromatinia*) remain a major problem, although new tissue culture technology and propagation system have changed their relative importance (Chastagner and Byther, 1985; Magie, 1985; Cline *et al.*, 1988).

Fungicide dips for control of soil and bulb-borne disease maintain a great importance: a very limited use of material permits to avoid further soil treatments, which have a much greater economic and environmental impact. However, much care must be taken in order to avoid the application of fungicides to which pathogens became resistant, a phenomenon which often happened in the past, leading to no disease control and spread through the world of fungicide-resistant pathogens (Garibaldi and Gullino, 1990).

New diagnostic tools can help in early detection of bulb-borne pathogens, thus helping in developing prompt control measures (Mes *et al.*, 1994).

Virus diseases can be serious in bulbs but the development of virus-free material has been accomplished through tissue culture (Lawson, 1985).

Little research has been done on IPM for pests in bulbs, although programmes used for pot plants can be practised on the foliage and flowers. In the bulbs, the most important pests are nematodes and mites. The bulb mite *Rhizoglyphus robini* Claparède, an important pest in lilies in The Netherlands, Taiwan and Japan, is now successfully controlled with the predatory mite *Hypoaspis aculeifer* (Canestrini) (Lesna *et al.*, 1995).

Foliage Plants

Hundreds of species are currently grown and sold as foliage plants, and new species or cultivars are continuously being introduced, often from tropical countries. The most economically important species are included within the Araceae, Palmae, Araliaceae, Agavaceae, Moraceae and Polypodiaceae (Chase, 1987; Cline *et al.*, 1988).

Due to the great diversity of foliage plants, complete knowledge of the host-pathogen interactions is very limited, thus representing often a constrain to development of IPM strategies.

For soilborne pathogens, control mostly relies on the use of disinfested substrates and careful avoidance of their reintroduction through infected material.

Important pathogens of foliage include species of *Colletotrichum*, *Xanthomonas* and *Erwinia*: their management mostly relies on correct spacing and reduction of leaf wetness through ventilation and proper irrigation practices.

This sector of the ornamental industry, due to the often uncomplete understanding of its phytopathological problems, still mostly relies on chemicals for disease control. There is a strong need for improved diagnostic methods, better knowledge of aetiology and epidemiology of the most important pathogens, in order to permit the development and implementation of IPM practices for diseases.

Foliage plants suffer similar pest problems to flowering pot plants and IPM programmes are subsequently similar. Pests such as mealybugs and scales are more of a nuisance, especially in amenity situations, and biological control of these pests is more difficult and expensive. Nevertheless, IPM is highly successful, especially against *F. occidentalis* and spider mites.

Recent outbreaks of a new pest, *Thrips palmi* Karny, to Europe threaten the well-established IPM programmes for this section of the industry.

Nursery Stock

This is a very rapidly growing sector, with rhododendron, azalea and juniperus covering a good proportion of the production. In the case of such crops, root diseases are of major importance, with *R. solani* and *Pythium* spp. being more serious on plants at very early stages, and *Phytophthora* spp. affecting plants at all stages. The great importance of root diseases is mostly due to the stresses caused to roots by the small soil volume where they are grown (Cline *et al.*, 1988). In such production system, since in most cases one major soilborne pathogen is present, the use of suppressive container media made from compost of hardwood bark, fir bark, or municipal sewage sludge has become very widespread, particularly in the USA. In order to avoid the problems caused by variability of the raw products used in the compost, since suppressiveness has generally a strong biotic component, controlled recolonization of the compost with microbial inoculants is carried out (Hoitink *et al.,* 1991).

Plants are prone to the wide range of pests suffered by flowering pot plants. However, young plants are often raised in greenhouses (including plastic) and in sheltered screened areas that lend the crop to successful IPM using the full range of natural enemies. In addition, cuttings are usually grown in mist atmospheres that are suitable for the use of *Verticillium lecanii* (A. Zimmerm.) Viégas against whiteflies, aphids and thrips and parasitic nematodes for fungus gnats and vine weevil.

The incorporation of pesticides such as imidacloprid or chlopyrifos in the container compost can control aphids, fungus gnats and vine weevil for a long period of time without affecting biological controls too seriously. IPM programmes are therefore mainly aimed at thrips, caterpillars, whiteflies and mites but also aphids when the effect of the compost pesticides has declined. Pests such as leaf hoppers and capsid bugs become more of a nuisance on these plants and need to be controlled by short-persistence pesticides.

34.2.2. MANAGEMENT OF IPM PROGRAMMES

The integration of chemical and biological treatments, with disease and pest resistant varieties and cultural management systems is a realistic and rational strategy for disease and pest management on a number of ornamental crops. In cases where there is adequate knowledge of the pathogen/pest, growers may have a comprehensive menu of control measures to integrate within a realistic economic environment.

However, while most pesticide programmes are applied routinely and these may control a greater range of pests, natural enemies and diseases than those against which they are targeted, any programme involving reduced pesticide usage, but especially IPM, requires the presence of well trained staff. Although in several countries (i.e. The Netherlands and the UK) a good extension service system (both public and private) exists, in most countries adequate expertise in IPM in this sector is often lacking (Wardlow and O'Neill, 1992).

Comprehensive extension programmes need to be put in place whenever IPM must be applied, in order to help growers gain more confidence in alternative methods (Fransen, 1992). Courses for recognition of pests and diseases and for handling BCAs are necessary, with strong involvement of staff, who are the first to encounter pest or disease problems (Wardlow and O'Neill, 1992).

34.3. Economics of IPM in Ornamentals

A simple IPM programme for pests is used on poinsettia which is mainly affected by only glasshouse whiteflies. Growers therefore introduce *Encarsia formosa* Gahan to the crop as soon as it arrives in the nursery, at a weekly rate of one wasp per ten plants if infestation risk is low; this rate increases to one wasp per three plants if risk is higher. Such a programme for the four months of crop culture may cost US$3300 per ha or just exceeding US$0.016 per plant. This is very acceptable to growers, especially when a pesticide programme is only slightly cheaper. The difference is more evident when the pesticides do not work or are phytotoxic. In the case of a threat of *B. tabaci* rates of parasites need to be multiplied up to fivefold so of course the costs become less acceptable, however this cost may still be economic for most growers.

More complicated IPM programmes such as those used on crops like fuchsia, which suffers from a wide range of pests, are much more expensive. Maybe five or six natural enemies are used and costs of US$0.066 per plant are common. Fuchsias are less valuable than poinsettias and therefore profit margins are affected. However, to produce fuchsias in situations where *F. occidentalis* is a continual threat leaves no alternative to IPM; pesticides do not work or they may damage the plants.

The expensive period for IPM is when growers take the technique up for the first time, rates of natural enemies tend to err on the safe side until growers become more aware of their pest risks and more confident in the performance of the parasitoids and predators. Once pest monitoring and training has attained the necessary level of accuracy and competence, the rates of natural enemy introduction can be manipulated with economy in mind.

Other techniques can economize in the numbers of biological agents used. "Banker plant" systems where colonies of pest and their natural enemies are cultured in the crop can be used successfully to keep a high presence of parasitoids and predators for small initial introduction costs.

It has always been a feature of biological pest control that costs of natural enemies decrease as IPM expands and suppliers compete for the market. This is now occurring in the ornamental sector where growers usually require a wide range of supplies compared with the vegetable industry. Suppliers also benefit by being able to produce hitherto uneconomic BCAs.

34.4. Perspectives

Pressures of pesticide resistance, shortage of new pesticides and environmental factors inevitably mean an expansion of IPM in ornamentals. The day is near when marketing requirements will tip the balance so that it will be difficult to sell pesticide-treated produce.

34.4.1. ADVANTAGES AND SHORTFALLS OF IPM PROGRAMMES

IPM programmes present a number of advantages, which, however, are difficult to quantify:

(i) They permit growers and staff to work in a safer environment.

(ii) Since pesticide application must be done out-of-hours, causing closing down work in crops for a period of time, IPM can avoid this inconvenience.

(iii) They reduce phytotoxicity problems related to use of chemicals.

(iv) From a general point of view, IPM results in a general improvement in crop quality, also due to the fact that regular monitoring makes it easier to follow problems other than those due to pests and pathogens.

(v) The ornamental industry can be seen to be supporting the green trend in modern life. Green products are likely to be favoured in the future (Wardlow et al., 1992). The different needs of ornamentals compared with edible crops are stimulating new ideas and techniques to help IPM work effectively. This benefits the whole industry and can encourage expansion of IPM to more crops outdoors.

There are few limitations to IPM, but these are of great importance:

(i) During the past few years, tomato spotted wilt virus (TSWV) has been a real threat in most ornamentals. This led to the need to apply insecticides against thrips which are their vector. The same problem arises with other virus-transmitted insects.

(ii) Ornamentals are more prone than edible crops to a wider range of pests and these are likely to migrate (usually in large numbers) into glasshouses at certain times during the summer. Using biological control to contain these plagues is difficult since they act too slowly and the numbers required may be uneconomical.

(iii) Growers of ornamentals have a long tradition of using broad-spectrum pesticides and find it difficult to adapt to IPM, particularly in the case of diseases such as the powdery and downy mildew with epidemic potential necessitating rapid action by the grower (Wardlow and O'Neill, 1992). Broad-spectrum fungicides, which are often incompatible with biocontrols, are needed in order to control a number of foliar pathogens. It is significant that a single disease may require fungicide treatments that could be lethal to an IPM programme.

(iv) In spite of good training, especially in how to monitor IPM, any failures of the technique can harden attitudes and dampen enthusiasm more emphatically than occurs with growers of edible crops. It seems more difficult for them to accept a certain amount of crop loss, presumably because damage on ornamentals is (or looks) more dramatic than in edible crops (Wardlow *et al.*, 1992).

34.4.2. AREAS WHERE RESEARCH IS NEEDED

The most urgent area for development is in the control of diseases. The current high dependence upon fungicides often limits the use of parasitoids and predators when harmful products either kill or restrict them. More BCAs are necessary for diseases. Actually, at present only a very few have gone throughout commercial development (see Chapter 26). Even so, only in a few cases they have been registered. This is a strong constraint to their practical application. However, those few registered BCAs are probably the proof that biological control is feasible. Hopefully, genetic engineering will provide a large impact to the development of new resistant cultivars (Loffler and Florack, 1997).

In the case of bacterial and fungal diseases, a better knowledge of host-pathogen interactions will allow better focus of their management. The knowledge of some pathogen requirement will permit, whenever possible, a better manipulation of the greenhouse environment. As already mentioned, this approach is mostly necessary for foliage plants. In order to increase the adoption of BCAs, more effective micro-organisms are necessary. Moreover, understanding the mechanisms of biocontrol of plant diseases is very important in order to develop rational models for exploitation of the antagonists under practical conditions. This knowledge is necessary for the manipulation of parameters affecting BCAs and for their genetic improvement. Manipulation of the environment in order to make it more favourable to BCAs, in principle, should create, on a larger scale, situations of suppressiveness. This approach looks very promising for soilless cultures, where abiotic environmental factors can be controlled and mad favourable to biocontrol.

For pests, there are excellent predatory bugs and beetles that are at present unexploited; there are problems in lack of knowledge for their production but also there is likely to be more expense involved. International funding for research is really necessary; at present these burdens are carried by the commercial suppliers. This state of affairs leads to secrecy that benefits only the successful company in the short-term.

In conclusion, it is possible to imagine a progressive expansion of IPM strategies on most ornamental crops and a very good market acceptance. IPM will steadily expand. Products will have to be labelled with information on the chemicals used on the crop. Consumers will become better informed and will exercise choice for IPM products.

Acknowledgements

This work was supported by grants from Italian Ministry of Agricultural Policies (MIPA, Progetto Finalizzato Floricoltura) and from L.R. Wardlow Ltd, Marsh Road, Ruckinge, Ashford, Kent, United Kingdom. Thanks are due to Angelo Garibaldi and Alberto Matta for critically reading the manuscript.

References

Aloisi, S., Pionnat, S., Jacob, Y., Pelloli, G., Botton, E., Bettachini, A., Hericher, D., Antonini, A., Simonini, L. and Poncet, C. (1994) Crown gall du rosier: Strategies de lutte, *Proc. ANPP Quatrième Conference Internationale sur les Maladies des Plantes*, Vol. 2, Bordeaux, France, 6–8 December, 1994, ANPP, Paris, pp. 791–795.

Amsing, J.J. (1995) *Gnomonia radicicola* and a *Phytophthora* species as causal agents of root rot on roses in artificial substrates, *Acta Horticulturae* **382**, 203–211.

Baker, K.F. and Linderman, R.G. (1979) Unique features of the pathology of ornamental plants, *Annual Review of Phytopathology* **17**, 253–277.

Bélanger, R.R., Labbé, C. and Jarwis, W.R. (1994) Commercial-scale control of rose powdery mildew with a fungal antagonist, *Plant Disease* **78**, 420–424.

Bennison, J.A., Poe, E. and Pickett, J. (1997) When nature knows pest, *Horticulture Week* November 6, 22–25.

Chase, A.R. (1987) *Compendium of Ornamental Foliage Plant Disease*, American Phytopathological Society Press, St Paul, Minn.

Chastagner, G.S. and Byther, R.S. (1985) Bulbs – Narcissus, tulips and iris, in D.L. Strider (ed.), *Diseases of Floral Crops*, Vol. 1, Praeger Scientific, New York, pp. 447–506.

Cline, M.N., Chastagner, G.A., Aragaki, M., Baker, R., Daughtrey, M.L., Lawson, R.H., MacDonald, J.D., Tammen, J.F. and Worf, G.L (1988) Current and future research directions of ornamental pathology, *Plant Disease* **72**, 926–934.

Daughtrey, M.L., Wick, R.L. and Peterson, J.L. (1995) *Compendium of Flowering Potted Plant Diseases*, American Phytopathological Society Press, St Paul, Minn.

Farrand, S.K. (1990) *Agrobacterium radiobacter* strain K 84: A model biocontrol system, in R.R. Baker and P.E. Dunn (eds.), *New Directions in Biological Control*, Alan R. Liss, Inc., New York, pp. 679–691.

Fransen, J.J. (1992) Development of integrated crop protection in glasshouse crops, *Pesticide Science* **36**, 329–333.

Garibaldi, A. and Gullino, M.L. (1985) Wilt of *Ranunculus asiaticus* caused by *Fusarium oxysporum* f. sp. *ranunculi*, *formae speciales nova*, *Phytopathologia Mediterranea* **24**, 213–214.

Garibaldi, A. and Gullino, M.L. (1987) Fusarium wilt of carnation: Present situation, problems and perspectives, *Acta Horticulturae* **216**, 45–54.

Garibaldi, A. and Gullino, M.L. (1990) Disease management of ornamental plants: A never ending challenge, *Mededelingen van de Faculteit Landbouwwetenschappen, Rijksuniversiteit Gent* **55**, 189–201.

Garibaldi, A., Gullino, M.L. and Aloi, C. (1990) Biological control of Fusarium wilt of carnation, in *Brighton Crop Protection Conference – Pests and Diseases*, Vol. 1, British Crop Protection Council, Thornton Heath, pp. 89–95.

Gullino, M.L. (1992) Chemical control of *Botrytis* spp., in K. Verhoeff, N.E. Malathrakis and B. Williamson (eds.), *Recent Advances of Botrytis Research*, Pudoc, Wageningen, pp. 218–222.

Gullino M.L. (1995) Use of biocontrol agents against fungal diseases, in *Proceedings of the International Conference on Microbial Agents in Sustainable Agriculture*, Saint Vincent, Italy, 18–19 October 1995, MAF Servizi, Torino, pp. 50–59.

Gullino, M.L. and Garibaldi, A. (1996) Diseases of roses: Evolution of problems and new approaches for their control, *Acta Horticulturae* **424**, 195–201.

Gullino, M.L. and Garibaldi, A. (1997) Alternatives to methyl bromide for the control of soilborne diseases of ornamental crops in Italy, in A. Bello, J.A. Gonzales, M. Arias, R. Rodriguez Kabana (eds.), *Alternatives to Methyl Bromide for the Southern European Countries*, Graficas Papellone SCV, Valencia, pp. 183–191.

Gullino, M.L. and Kuijpers, L.A.M. (1994) Social and political implications of managing plant diseases with restricted fungicides in Europe, *Annual Review of Phytopathology* **32**, 559–579.

Hausbeck, M.K. and Moorman, G.W. (1996) Managing Botrytis in greenhouse-grown flower crops, *Plant Disease* **80**, 1212–1219.

Hoitink, H.A.J., Inbar, Y. and Bohem, M.J. (1991) Status of composted amended potting mixes naturally suppressive to soil borne diseases of floricultural crops, *Plant Disease* **75**, 869–873.

Horst, R.K., Kawamoto, S.O. and Porter, L.L. (1992) Effect of sodium bicarbonate and oils on the control of powdery mildew and black spot of roses, *Plant Disease* **76**, 347–351.

Horst, R.K. and Nelson, P.E. (1997) *Compendium of Chrysanthemum Diseases*, American Phytopathological Society Press, St Paul, Minn.

Jarvis, W.R. (1992) *Managing Diseases in Greenhouse Crops*, American Phytopathological Society Press, St Paul, Minn.

Jones, R.K. and Strider, D.L. (1985) Bedding plants, in D.L. Strider (ed.), *Diseases of Floral Crops*, Vol. 1, Praeger Scientific, New York, pp. 409–422.

Lawson, R.L. (1985) Viruses and virus diseases, in D.L. Strider (ed.), *Diseases of Floral Crops*, Vol. 1, Praeger Scientific, New York, pp. 253–293.

Lesna, I., Sabelis, M.W., Bolland, H.R. and Conjin, C.G.M. (1995) Candidate natural enemies for control of *Rhizoglyphus robini* Claparède (Acari: Astigmata) in lily bulbs: Exploration in the field and pre-selection in the laboratory, *Experimental and Applied Acarology* **19**, 655–669.

Loffler, H.J.M. and Florack, D.E.A. (1997) Engineering for bacterial and fungal disease resistance, in R.I. Geneve, J.E. Preece and S.A. Merkle (eds.), *Biotechnology of Ornamental Plants*, CAB International, Wallingford, pp. 313–333.

Magie, R.O. (1985) Gladiolus, in D.L. Strider (ed.), *Diseases of Flower Crops*, Vol. 2, Praeger Scientific, New York, pp. 189–226.

Mes, J.J., van Doorn, J., Roebroeck, E.J.A. and Boonekamp, P.M. (1994) Detection and identification of *Fusarium oxysporum* f. sp. *gladioli* by RFLP and RAPD analysis, in A. Schots, F.M. Dewey and R. Oliver (eds.), *Modern Assays for Plant Pathogenic Fungi*, CAB International, Wallingford, pp. 63–68.

Pasini, C., D'Aquila, F., Curir, P. and Gullino, M.L. (1997) Effectiveness of antifungal compounds against rose powdery mildew (*Sphaerotheca pannosa* var. *rosae*) in glasshouses, *Crop Protection* **16**, 251–256.

Ragsdale, N.N. and Sisler, H.D. (1994) Social and political implications of managing plant diseases with decreased availability of fungicides in the United States, *Annual Review of Phytopathology* **32**, 545–547.

Stanghellini, M.E. and Rasmussen, S.L. (1994) Hydroponics: A solution for zoosporic pathogens, *Plant Disease* **78**, 1130–1138.

van Lenteren, J.C. (1995) Integrated Pest Management in protected crops, in D. Dent (ed.), *Integrated Pest Management*, Chapman & Hall, London, pp. 311–343.

van Peer, R., Schippers, B. (1991) Biocontrol of Fusarium wilt by *Pseudomonas* sp. strain WCS417r: Induced resistance and phytoalexin accumulation, in A.B.R. Beemster, G.J. Bollen, M. Gerlagh, M.A. Ruissen, B. Schippers and A. Tempel (eds.), *Biotic Interactions and Soil-Borne Diseases*, Elsevier, Amsterdam, pp. 274–280.

Wardlow, L.R, Brough, W. and Need, C (1992) Integrated pest management in protected ornamentals in England, *EPPO Bulletin* **22**, 493–498.

Wardlow, L.R. and Ludlam, F.A.B. (1973) Insecticide resistance testing and chemical control of glasshouse whitefly, *Proceedings of the 7th British Insecticide and Fungicide Conference*, Vol. 1, Brighton, 19–22 November 1973, BCPC, London, pp. 217–225.

Wardlow, L.R. and O'Neill, T. (1992) Management strategies for controlling pests and diseases in glasshouse crops, *Pesticide Science* **36**, 341–347.

INDEX